Van Nostrand Reinhold Soil Science Series

Editor: Charles W. Finkl, Jnr., Florida Atlantic University

SOIL CLASSIFICATION / *Charles W. Finkl, Jnr.*
CHEMISTRY OF IRRIGATED SOILS / *Rachel Levy*
SOIL SALINITY: Two Decades of Research in Irrigated Agriculture/
    H. Frenkel and A. Meiri
ANDOSOLS / *Kim H. Tan*
PODZOLS / *Peter Buurman*
SOIL NUTRIENT AVAILABILITY: Chemistry and Concepts / *Y. K. Soon*
ADSORPTION PHENOMENA / *Robert D. Harter*
SOIL EROSION AND ITS CONTROL / *R. P. C. Morgan*
SOIL MICROMORPHOLOGY / *Georges Stoops and Hari Eswaran*
SOIL MINERAL WEATHERING / *J. A. Kittrick*
CHEMISTRY OF SOIL SOLUTIONS / *Adel M. Elprince*
LAND EVALUATION / *Donald A. Davidson*
PESTICIDES IN SOIL / *Sarina Saltzman and Bruno Yaron*

Related Titles of Interest

THE ENCYCLOPEDIA OF SOIL SCIENCE, PART I / *Rhodes W. Fairbridge and
    Charles W. Finkl, Jnr.*
SURFICIAL DEPOSITS OF THE UNITED STATES / *Charles B. Hunt*

PESTICIDES IN SOIL

Edited by

SARINA SALTZMAN and BRUNO YARON
Agricultural Research Organization
Bet Dagan, Israel

A Hutchinson Ross Publication

VAN NOSTRAND REINHOLD COMPANY
New York

Copyright © 1986 by **Van Nostrand Reinhold Company Inc.**
Van Nostrand Reinhold Soil Science Series
Library of Congress Catalog Card Number: 86-7839
ISBN: 0-442-28011-4

All rights reserved. No part of this work covered by the copyrights
hereon may be reproduced or used in any form or by any means—
graphic, electronic, or mechanical, including photocopying,
recording, taping, or information storage and retrieval systems—
without written permission of the publisher.

Manufactured in the United States of America.

Van Nostrand Reinhold Company Inc.
115 Fifth Avenue
New York, New York 10003

Van Nostrand Reinhold Company Limited
Molly Millars Lane
Wokingham, Berkshire RG11 2PY, England

Van Nostrand Reinhold
480 La Trobe Street
Melbourne, Victoria 3000, Australia

Macmillan of Canada
Division of Canada Publishing Corporation
164 Commander Boulevard
Agincourt, Ontario M1S 3C7, Canada

15 14 13 12 11 10 9 8 7 6 5 4 3 2

Library of Congress Cataloging in Publication Data
Pesticides in soil.
    (Van Nostrand Reinhold soil science series)
    "A Hutchinson Ross publication."
    Includes indexes.
    1. Soils—Pesticide content.   I. Saltzman, Sarina.   II. Yaron, B.
(Bruno), 1929–   III. Series.
S592.6.P43P47   1986   631.4'1   86-7839
ISBN 0-442-28011-4

CONTENTS

Series Editor's Foreword	xi
Preface	xv
Contents by Author	xvii
Introduction	1

PART I: DISTRIBUTION OF PESTICIDES AMONG THE SOIL PHASES

Editors' Comments on Papers 1 Through 9   8

1. **HADAWAY, A. B., and F. BARLOW:** Sorption of Solid Insecticides by Dried Mud
   Nature **167:**854 (1951)   16

2. **EDWARDS, C. A., S. D. BECK, and E. P. LICHTENSTEIN:** Bioassay of Aldrin and Lindane in Soil
   Jour. Econ. Entomol. **50:**622–626 (1957)   17

3. **SALTZMAN, S., L. KLIGER, and B. YARON:** Adsorption-Desorption of Parathion as Affected by Soil Organic Matter
   Jour. Agric. Food Chem. **20:**1224–1226 (1972)   22

4. **GAILLARDON, P., R. CALVET, and M. TERCE:** Adsorption et désorption de la terbutryne par une montmorillonite-Ca et des acides humiques seuls ou en mélanges
   Weed Res. **17:**41–48 (1977)   25

5. **HUANG, P. M., R. GROVER, and R. B. McKERCHER:** Components and Particle Size Fractions Involved in Atrazine Adsorption by Soils
   Soil Sci. **138:**20–24 (1984)   33

6. **HANCE, R. J.:** The Adsorption of Urea and Some of Its Derivatives by a Variety of Soils
   Weed Res. **5:**98–107 (1965)   38

7. **BRIGGS, G. G.:** Molecular Structure of Herbicides and Their Sorption by Soils
   Nature **223:**1288 (1969)   48

8. **BARLOW, F., and A. B. HADAWAY:** Effect of Changes in Humidity on the Toxicity and Distribution of Insecticides Sorbed by Some Dried Soils
   Nature **178:**1299–1300 (1956)   49

Contents

9 MILLS, A. C., and J. W. BIGGAR: Solubility-Temperature Effect on the Adsorption of Gamma- and Beta-BHC from Aqueous and Hexane Solutions by Soil Materials — 51
Soil Sci. Soc. America Proc. **33**:210-216 (1969)

Editors' Comments on Papers 10 Through 17 — 58

10 BEST, J. A., J. B. WEBER, and S. B. WEED: Competitive Adsorption of Diquat^{2+}, Paraquat^{2+}, and Ca^{2+} on Organic Matter and Exchange Resins — 68
Soil Sci. **114**:444-450 (1972)

11 BAILEY, G. W., J. L. WHITE, and T. ROTHBERG: Adsorption of Organic Herbicides by Montmorillonite: Role of pH and Chemical Character of Adsorbate — 75
Soil Sci. Soc. America Proc. **32**:222-234 (1968)

12 FRISSEL, M. J., AND G. H. BOLT: Interaction between Certain Ionizable Organic Compounds (Herbicides) and Clay Minerals — 88
Soil Sci. **94**:284-291 (1962)

13 SENESI, N., and C. TESTINI: Physico-Chemical Investigations of Interaction Mechanisms between s-Triazine Herbicides and Soil Humic Acids — 96
Geoderma **28**:129-146 (1982)

14 KHAN, S. U.: Equilibrium and Kinetic Studies of the Adsorption of 2,4-D and Picloram on Humic Acid — 114
Can. Jour. Soil Sci. **53**:429-434 (1973)

15 MORTLAND, M. M., and W. F. MEGGITT: Interaction of Ethyl N,N-Di-n-propylthiolcarbamate (EPTC) with Montmorillonite — 120
Jour. Agric. Food Chem. **14**:126-129 (1966)

16 SALTZMAN, S., and S. YARIV: Infrared and X-ray Study of Parathion-Montmorillonite Sorption Complexes — 124
Soil Sci. Soc. America Jour. **40**:34-38 (1976)

17 LEENHEER, J. A., and J. L. AHLRICHS: A Kinetic and Equilibrium Study of the Adsorption of Carbaryl and Parathion upon Soil Organic Matter Surfaces — 129
Soil Sci. Soc. America Proc. **35**:700-705 (1971)

Editors' Comments on Papers 18 Through 21 — 134

18 BRIGGS, G. G.: A Simple Relationship between Soil Adsorption of Organic Chemicals and Their Octanol/Water Partition Coefficients — 138
7th British Insecticide and Fungicide Conference, Brighton, *Proceedings,* The Boots Co., Ltd., Nottingham, 1973, pp. 83-86

19 HANCE, R. J.: Relationship between Partition Data and the Adsorption of Some Herbicides by Soils — 142
Nature **214**:630-631 (1967)

20 LAMBERT, S. M.: Functional Relationship between Sorption in Soil and Chemical Structure — 144
Jour. Agric. Food Chem. **15**:572-576 (1967)

21	MINGELGRIN, U., and Z. GERSTL: Reevaluation of Partitioning as a Mechanism of Nonionic Chemicals Adsorption in Soils *Jour. Environ. Qual.* **12**:1-11 (1983)	149

PART II: DEGRADATION PROCESSES AND PESTICIDE PERSISTENCE

Editors' Comments on Papers 22 Through 31		162
22	AUDUS, L. J.: The Biological Detoxication of 2:4-Dichlorophenoxyacetic Acid in Soil *Plant and Soil* **2**:31-36 (1949)	173
23	AHMED, M. K., and J. E. CASIDA: Metabolism of Some Organophosphorus Insecticides by Microorganisms *Jour. Econ. Entomol.* **51**:59-63 (1958)	179
24	KAUFMAN, D. D., and J. BLAKE: Microbial Degradation of Several Acetamide, Acylanilide, Carbamate, Toluidine and Urea Pesticides *Soil Biol. Biochem.* **5**:297-308 (1973)	184
25	ALEXANDER, M., and B. K. LUSTIGMAN: Effect of Chemical Structure on Microbial Degradation of Substituted Benzenes *Jour. Agric. Food Chem.* **14**:410-413 (1966)	196
26	KEARNEY, P. C.: Influence of Physicochemical Properties on Biodegradability of Phenylcarbamate Herbicides *Jour. Agric. Food Chem.* **15**:568-571 (1967)	200
27	GUENZI, W. D., and W. E. BEARD: The Effects of Temperature and Soil Water on Conversion of DDT to DDE in Soil *Jour. Environ. Qual.* **5**:243-246 (1976)	204
28	WAHID, P. A., C. RAMAKRISHNA, and N. SETHUNATHAN: Instantaneous Degradation of Parathion in Anaerobic Soils *Jour. Environ. Qual.* **9**:127-130 (1980)	208
29	ARMSTRONG, D. E., G. CHESTERS, and R. F. HARRIS: Atrazine Hydrolysis in Soil *Soil Sci. Soc. America Proc.* **31**:61-66 (1967)	211
30	WEBER, J. B., and H. D. COBLE: Microbial Decomposition of Diquat Adsorbed on Montmorillonite and Kaolinite Clays *Jour. Agric. Food Chem.* **16**:475-478 (1968)	217
31	MINGELGRIN, U., and S. SALTZMAN: Surface Reactions of Parathion on Clays *Clays and Clay Minerals* **27**:72-78 (1979)	221
Editors' Comments on Papers 32 Through 35		227
32	VOERMAN, S., and A. F. H. BESEMER: Residues of Dieldrin, Lindane, DDT, and Parathion in a Light Sandy Soil after Repeated Application throughout a Period of 15 Years *Jour. Agric. Food Chem.* **18**:717-719 (1970)	232
33	LICHTENSTEIN, E. P., J. KATAN, and B. N. ANDEREGG: Binding of "Persistent" and "Nonpersistent" ^{14}C-Labeled Insecticides in an Agricultural Soil *Jour. Agric. Food Chem.* **25**:43-47 (1977)	235

Contents

34 KAUFMAN, D. D., and D. E. EDWARDS: Pesticide/Microbe Interaction Effects on Persistence of Pesticides in Soil — 240
Pesticide Chemistry. Human Welfare and the Environment, J. Miyamoto and P. C. Kearney, eds., Proc. 5th Int. Congr. Pestic. Chem., vol. 4, 1982, pp. 177-182

35 WALKER, A.: A Simulation Model for Prediction of Herbicide Persistence — 246
Jour. Environ. Qual. **3**:396-401 (1974)

PART III: TRANSPORT PROCESSES

Editors' Comments on Papers 36 Through 45 — 254

36 LICHTENSTEIN, E. P.: Movement of Insecticides in Soils under Leaching and Non-Leaching Conditions — 265
Jour. Econ. Entomol. **51**:380-383 (1958)

37 GRAHAM-BRYCE, I. J.: Diffusion of Organophosphorus Insecticides in Soils — 269
Jour. Sci. Food Agric. **20**:489-494 (1969)

38 EHLERS, W., W. J. FARMER, W. F. SPENCER, and J. LETEY: Lindane Diffusion in Soils: II. Water Content, Bulk Density, and Temperature Effects — 275
Soil Sci. Soc. America Proc. **33**:505-508 (1969)

39 SCOTT, H. D., R. E. PHILLIPS, and R. F. PAETZOLD: Diffusion of Herbicides in the Adsorbed Phase — 279
Soil Sci. Soc. America Proc. **38**:558-562 (1974)

40 GERSTL, Z., P. H. NYE, and B. YARON: Diffusion of a Biodegradable Pesticide: II. As Affected by Microbial Decomposition — 284
Soil Sci. Soc. America Jour. **43**:843-848 (1979)

41 HELLING, C. S.: Pesticide Mobility in Soils: III. Influence of Soil Properties — 290
Soil Sci. Soc. America Proc. **35**:743-748 (1971)

42 DAVIDSON, J. M., C. E. RIECK, and P. W. SANTELMANN: Influence of Water Flux and Porous Material on the Movement of Selected Herbicides — 295
Soil Sci. Soc. America Proc. **32**:629-633 (1968)

43 VAN GENUCHTEN, M. TH., P. J. WIERENGA, and G. A. O'CONNOR: Mass Transfer Studies in Sorbing Porous Media: III. Experimental Evaluation with 2,4,5-T — 300
Soil Sci. Soc. America Jour. **41**:278-285 (1977)

44 RAO, P. S. C., J. M. DAVIDSON, R. E. JESSUP, and H. M. SELIM: Evaluation of Conceptual Models for Describing Nonequilibrium Adsorption-Desorption of Pesticides During Steady-flow in Soils — 307
Soil Sci. Soc. America Jour. **43**:22-28 (1979)

45 LEISTRA, M.: Computed Redistribution of Pesticides in the Root Zone of an Arable Crop — 314
Plant and Soil **49**:569-580 (1978)

Editors' Comments on Papers 46 Through 48 — 326

46 HALL, J. K., M. PAWLUS, and E. R. HIGGINS: Losses of Atrazine in Runoff Water and Soil Sediment — 329
Jour. Environ. Qual. **1**:172–176 (1972)

47 LORBER, M. N., and L. A. MULKEY: An Evaluation of Three Pesticide Runoff Loading Models — 334
Jour. Environ. Qual. **11**:519–529 (1982)

48 VINTEN, A. J. A., B. YARON, and P. H. NYE: Vertical Transport of Pesticides into Soil when Adsorbed on Suspended Particles — 345
Jour. Agric. Food Chem. **31**:662–664 (1983)

Editors' Comments on Papers 49 Through 51 — 348

49 SPENCER, W. F., M. M. CLIATH, and W. J. FARMER: Vapor Density of Soil-Applied Dieldrin as Related to Soil-Water Content, Temperature, and Dieldrin Concentration — 352
Soil Sci. Soc. America Proc. **33**:509–511 (1969)

50 MAYER, R., J. LETEY, and W. J. FARMER: Models for Predicting Volatilization of Soil-Incorporated Pesticides — 355
Soil Sci. Soc. America Proc. **38**:563–568 (1974)

51 JURY, W. A., W. J. FARMER, and W. F. SPENCER: Behavior Assessment Model for Trace Organics in Soil: II. Chemical Classification and Parameter Sensitivity — 360
Jour. Environ. Qual. **13**:567–572 (1984)

Author Citation Index — 367
Subject Index — 373
About the Editors — 379

SERIES EDITOR'S FOREWORD

The Van Nostrand Reinhold Soil Science Series attempts to provide cogent summaries of the field by reproducing classical and modern papers, ones that provide keys to understanding of critical turning points in the development of the discipline. Scientific literature today is so vast and widely dispersed, especially in a multifaceted discipline like soil science, that much valuable information becomes ignored by default. Many pioneering works are now coveted by libraries, and retrieval from the archives is not easy. In fact, many important papers published in the ephemeral literature are no longer available to serious or committed researchers through interlibrary loan. Other professionals devoted to teaching or burdened with administrative duties must be hard pressed to keep up with comprehensive arrays of technical literature spread through scores of journals. Most of us can, at best, skim only a few select journals to make copies of tables of contents, abstracts and summaries, and reviews in order to remain abreast of specialized and often limited aspects of the robust field of soil science as a whole.

This series in soil science, developed as a practical solution to this problem, reprints key papers and investigative landmarks that relate to a common theme. The papers are reproduced in facsimile, either in their entirety or in significant part, so readers can follow major original events in the field, not peruse paraphrased or abbreviated versions of others. Some foreign works have been especially translated for use in the series. Occasionally short, foreign language articles are reproduced from French or German journals.

Essays by the volume editor provide running commentaries that introduce readers to highlights in the field, provide critical evaluation of the significance of the various papers, and discuss the development of selected topics or subject areas. It is hoped that the volume editor's comments will ease the transition for the seasoned investigator who wishes to step into a new field of research as well as provide students and professors with a compact working library of most important scientific advances in soil science.

Areas of specialization in soil science are divided by the International Society of Soil Science into seven divisions or "commissions." The first six commissions cover soil physics, chemistry, mineralogy, biology, fertility, and technology. Because the scope of the field is so great, we concentrate initially on topics traditionally devoted to the seventh commission: soil

Series Editor's Foreword

morphology, genesis, classification, and geography. The series thus begins with volumes dealing with the major soils of the world: their recognition, characteristics, formation, distribution, and classification. Other volumes concentrate on topics in agronomy, soil-plant relationships, soil engineering topics, or melds of pure science with soil systems. The Van Nostrand Reinhold Soil Science Series plows deeply through the field, picking significant but timely topics on an eclectic basis.

Each volume in the series is edited by a specialist or authority in the area covered by the book. The volume editor's efforts reflect a concerted worldwide search, review, selection, and distillation of the primary literature contained in journals and monographs and in industrial and governmental reports. Individual volumes thus represent an information-selection and repackaging program of value to libraries, students, and professionals.

The books contain a preface, introduction, and highlight commentaries by the volume editor. Many volumes contain rare papers that are hard to locate and obtain, as well as landmark papers published in English for the first time. All volumes contain author citation and subject indexes of the contained papers, usually twenty to fifty key papers in a given subject area.

As the world's population and demand for foodstuffs increases, so does the necessity for maintaining the productive capacity of the soil. Demonstrated many times over throughout history in industrialized and developing countries alike, it is painfully clear that loss of productive capacity of the soil resource base through misuse, abuse, or neglect eventually reduces adequate supplies of food and fiber. Pesticides are now widely used to realize profitable and productive capacities of the soil. Because use of pesticides will likely increase, researchers need to know more about interactions between these chemicals and soil and water. This volume provides a timely review of this complex field of pesticide science. Because of widespread pollution hazards, whether actual or potential, better understanding of this field is needed to focus attention on pathways of pesticides through the soil medium. The persistence of pesticides beyond the critical period for control often leads to residue problems. Threshold levels of hazard in various segments of the environment may be reduced by introducing integrated pest-control programs, rotation of pesticides, use of short-residual pesticides, and application of minimum dosages.

Ideally, pesticides should persist only long enough to effect their intended purpose without harming nontarget organisms or the environment. The objective of getting the proper pesticide, in the proper concentration, on the target organism at the right time is deceptively simple and laced with problems. Achieving this goal requires thorough understanding of pesticide-soil-water interactions. This volume thus considers some of the more important advances in our understanding of complex reactions that involve, for example, aspects of adsorption, volatilization, persistence, plant uptake, curtailment, degradation, and detoxification of pesticides in soil as well as sampling and analytical techniques.

Many of these topics are discussed in the related groups (1) distribution

Series Editor's Foreword

of pesticides among the soil phases, (2) degradation processes and pesticide persistence, and (3) transport processes. This volume contains informative editorial commentaries for benchmark papers that specifically deal with factors affecting phase distributions for aldrin, lindane, parathion, atrazine, urea, and others as well as critical discussions of adsorption mechanisms, diffusion and mass flow, transport in adsorbed phases, and volatilization. These and other aspects of pesticide science are discussed in their historical perspective as they relate, at least in large part, to soils.

CHARLES W. FINKL, JNR.

PREFACE

Pesticides reach the soil either by direct application or indirectly, following their application on plant canopies. Upon reaching the land pesticides are subjected to complex physico-chemical and biological transformations. There are several reasons for trying to understand the fate and behavior of pesticides in soil: the need to improve the efficiency of soil-applied compounds, the need to minimize their potential adverse effects on soil fertility, and the need to minimize the risk of environmental pollution due to their transfer to the groundwater and atmosphere.

The aim of this book is to present the processes governing the fate of synthetic organic pesticides in the soil and their impact on the environment. The editors' approach was to illustrate the scientific development of various topics by representative papers. Because a limited number of research papers had to be selected from the vast quantity published, the choice was very difficult, and we are convinced that there are many other papers of equal worth. We hope, however, that we have succeeded in presenting to the reader a comprehensive (but not exhaustive) illustration of the state-of-the-art in soil-pesticides research. We are grateful to all the authors who graciously allowed us to reproduce their papers in *Pesticides in Soil*.

This book was prepared during sabbatical leaves of the editors: S.S. with the Department of Geoisotopes of the Weizmann Institute of Science, Rehovot, and B.Y. with the department of Agricultural Sciences and St. Cross College of the University of Oxford, Oxford. We extend to the hosting institutions our acknowledgments.

SARINA SALTZMAN
BRUNO YARON

CONTENTS BY AUTHOR

Ahlrichs, J. L., 129
Ahmed, M. K., 179
Alexander, M., 196
Anderegg, B. N., 235
Armstrong, D. E., 211
Audus, L. J., 173
Bailey, G. W., 75
Barlow, F., 16, 49
Beard, W. E., 204
Beck, S. D., 17
Besemer, A. F. H., 232
Best, J. A., 68
Biggar, J. W., 51
Blake, J., 184
Bolt, G. H., 88
Briggs, G. G., 48, 138
Calvet, R., 25
Casida, J. E., 179
Chesters, G., 211
Cliath, M. M., 352
Coble, H. D., 217
Davidson, J. M., 295, 307
Edwards, C. A., 17
Edwards, D. E., 240
Ehlers, W., 275
Farmer, W. J., 275, 352, 355, 360
Frissel, M. J., 88
Gaillardon, P., 25
Gerstl, Z., 149, 284
Graham-Bryce, I. J., 269
Grover, R., 33
Guenzi, W. D., 204
Hadaway, A. B., 16, 49
Hall, J. K., 329
Hance, R. J., 38, 142
Harris, R. F., 211

Helling, C. S., 290
Higgins, E. R., 329
Huang, P. M., 33
Jessup, R. E., 307
Jury, W. A., 360
Katan, J., 235
Kaufman, D. D., 184, 240
Kearney, P. C., 200
Khan, S. U., 114
Kliger, L., 22
Lambert, S. M., 144
Leenheer, J. A., 129
Leistra, M., 314
Letey, J., 275, 355
Lichtenstein, E. P., 17, 235, 265
Lorber, M. N., 334
Lustigman, B. K., 196
McKercher, R. B., 33
Mayer, R., 355
Meggitt, W. F., 120
Mills, A. C., 51
Mingelgrin, U., 149, 221
Mortland, M. M., 120
Mulkey, L. A., 334
Nye, P. H., 284, 345
O'Connor, G. A., 300
Paetzold, R. F., 279
Pawlus, M., 329
Phillips, R. E., 279
Ramakrishna, C., 208
Rao, P. S. C., 307
Rieck, C. E., 295
Rothberg, T., 75
Saltzman, S., 22, 124, 221
Santelman, P. W., 295
Scott, H. D., 279

Contents by Author

Selim, H. M., 307
Senesi, N., 96
Sethunathan, N., 208
Spencer, W. F., 275, 352, 360
Terce, M., 25
Testini, C., 96
Van Genuchten, M. Th., 300
Vinten, A. J. A., 345
Voerman, S., 232

Wahid, P. A., 208
Walker, A., 246
Weber, J. B., 68, 217
Weed, S. B., 68
White, J. L., 75
Wierenga, P. J., 300
Yariv, S., 124
Yaron, B., 22, 284, 345

PESTICIDES IN SOIL

INTRODUCTION

The hazard of environmental pollution, together with consumer safety concerns, may be the most important issues associated with the use of synthetic organic chemicals for crop protection. By definition, such chemicals, generally termed pesticides, are substances used to kill or inhibit the growth and reproduction of species considered as pests. The question of how much harm is done to the nontarget organisms by the use of a continuously increasing assortment of these toxic chemicals is still unanswered. Likewise, the most efficient and safe technologies for their use are not yet defined.

The use of crop protection chemicals is an old practice; however, pesticides were introduced for agricultural use mainly in the second half of the nineteenth century. The first pesticides were preparations of heavy metal salts—copper, mercury, zinc—and natural compounds of botanical origin such as nicotin, pyrethrum, and rotenone. Other naturally occurring compounds such as petroleum oils and sulfur were also used as pesticides.

The first synthetic organic pesticides—substituted phenols—were introduced before 1900; however, the development of these pesticides was initially slow. During World War II, the introduction of the herbicide 2,4-D and the insecticide DDT marked the beginning of a technological revolution. Compared to previous compounds, the new pesticides were extremely efficient at low application rates. Their successful introduction was followed by a series of discoveries of a large number of new classes of biologically active synthetic organic compounds.

The widespread and intensive use of pesticides characterized by high persistence and broad spectrum of activity, such as DDT and related compounds in the first twenty years after their introduction, caused several ecological problems. The main problems were development of resistance in pests, elimination of the natural pest control, and the spreading of persistent residues throughout differing compartments of the environment and their accumulation in living systems. This resulted in the banning of DDT and some related compounds and

Introduction

in a great and successful effort to develop new, nonresidual chemicals effective only against the target organisms.

The increasing use of pesticides in agriculture remains a controversial issue. One of the main concerns is the hazard of soil pollution by pesticides and their impact on soil fertility. Pesticides reach the soil by direct application, to control soil pests and for uptake by plants, and also indirectly, from aerial and ground sprays. The main processes affecting the efficiency and ultimate fate of pesticides in soil are retention by soil materials (adsorption-desorption processes), transformation processes (biological and chemical degradation), and transport into the soil, to the atmosphere, and to surface water.

After reaching the soil, most pesticide formulations are distributed primarily into the soil solution and then onto the surfaces of the solid phase and/or into the soil atomsphere, gravitating toward a dynamic equilibrium. The uptake of pesticides by soils (usually termed sorption or adsorption) and their release (desorption) have been considered from the very beginning of pesticide use as key processes. The availability of pesticides for uptake by the target organisms and for movement in solution or in the gaseous phase, as well as their chemical and biochemical transformation processes, are all affected by adsorption-desorption.

The first reports on soil adsorption of synthetic organic pesticides appeared shortly after World War II as a result of the increasing use of these compounds. It had been observed that the extremely efficient and persistent insecticide DDT and other persistent chlorinated hydrocarbon insecticides used to control malaria vectors and other insects rapidly lost their toxicity when sprayed on the internal surfaces of houses in several African countries. The most rapid inactivation of insecticides was observed when mud blocks made from lateritic soils were used in the construction of the houses. As the insecticides could be quantitatively extracted from the mud blocks, it was assumed that the process responsible for the rapid loss of surface deposits was adsorption (Hadaway and Barlow, 1952). A similar decrease in toxicity of chlorinated hydrocarbon insecticides due to soil adsorption was almost concomitantly reported for soil-applied compounds (Swanson, Thorp, and Friend, 1954). The introduction of herbicides from the phenoxyalcanoic acid group was also accompanied by observations concerning variable behavior in different soils. Weaver (1947) studied the adsorption and elution of these herbicides from several ion-exchange materials and concluded that both toxicity and leaching depend on the adsorptive capacity of these materials.

Since these first observations, and the introduction of new classes of pesticides, adsorption has become a central issue in pesticide studies. The techniques used developed from visual observation to bioassays, to direct adsorption measurements both in the field and in the laboratory, and later to the use of modern sophisticated spectroscopic and microscopic methods and particle-scattering techniques.

Although the subject has been studied intensively, adsorption of pesticides by soils is not yet fully understood. However, the available information provides some understanding of the factors affecting adsorption and its mechanisms and lays the basis for assessing the behavior of pesticides in soil and in the environment.

In contrast to adsorption and transport, which are transfer processes, degradation is the most widespread phenomenon contributing to the disappearance of pesticides from soils. Soil is an ideal medium inducing transformation reactions of pesticides (Graham-Bryce, 1981). The usually moist and aerated upper layer of agricultural soils provides proper conditions for chemical changes (mainly hydrolysis and oxidation reactions) occurring in the soil solution. Under anaerobic conditions, such as those specific to flooded soils, reductive reactions prevail. Possessing a large and active surface area, soil colloids may induce surface-catalyzed degradation reactions of adsorbed pesticides. At the same time, adsorption strongly affects the availability of pesticides for transformation reactions.

However, the most important soil characteristic related to pesticide degradation is probably the rich microbial population, capable of attacking a wide variety of chemical compounds. The first studies of soil persistence of pesticides were carried out with phenoxyacetic acid herbicides in the first decade after their introduction (1945–1955) and indicated microbial degradation. A few years later this disappearance pathway was demonstrated for other groups of pesticides, such as some cyclodiene and organophosphorus insecticides, and the s-triazine herbicides. These accumulating data, and the "principle of microbial infallibility" pervading scientific thought at that time (Alexander, 1965), led to the opinion that microbial degradation is responsible for the detoxification of all the toxic compounds reaching the soil. However, studies of pesticide degradation, carried out in the late 1960s, mainly with organophosphates and s-triazines, showed that, in addition to microbiological processes, nonbiological degradation could play an important role in the transformation of pesticides in soils.

In fact, all the synthetic organic pesticides presently used are transformed in soils by biochemical and chemical reactions but at degradation rates varying from a few days (for some organophosphates) to several months and even years (for some chlorinated hydrocarbon insecticides). During recent years, the concept of persistence was changed by considering the fraction of pesticide residues that is unextractable by conventional procedures because of its binding to soil components.

Pesticide residues (parent compounds and metabolites) remaining in the soil longer than desired may affect the living soil population in several ways: by a direct toxic effect, by producing populations resistant to pesticides, by changing the metabolic or reproductive activity of the living organisms, and by accumulation in these organisms and possible transfer to other compartments of the environment (Edwards, 1973).

The growing concern about the persistence of pesticides in soils is reflected in the continually increasing body of literature on this issue, including efforts to identify the nature and the mechanisms of the transformation processes, to establish the disappearance rate and the factors affecting it, to identify the degradation products, and, finally, to predict persistence.

The soil transport processes determine the redistribution of pesticides within the soil and eventually into the environment—atmosphere, surface waters, and living organisms. Movement of pesticides into the soil can occur in soil solution by diffusion and mass flow, as adsorbed on soil particles or by volatilization.

Diffusion, caused by random thermal motion of molecules, determines pesticide movement from higher to lower concentration sites, in solution and/or in the vapor phase, without the movement of the pesticide-containing phase. In contrast, movement by mass flow occurs by the convective flow of water through the soil, together with the dissolved or suspended pesticides. The relative contribution of these processes to the movement of pesticides into the soil is variable. At low average water fluxes in uniform soil, mass flow is relatively unimportant but becomes dominant over diffusion at high water fluxes or in structured soils, where water velocities vary substantially.

The transport of pesticides in soil is governed by factors such as soil and pesticide properties, rainfall intensity and distribution, irrigation technology, fluctuations in soil moisture content, and temperature. Pesticide movement into the soil is strongly affected by adsorption.

Under field conditions, the movement of pesticides into the soil often does not follow the anticipated general pattern. For example, pesticides may be partially leached into the soil through cracks and large pores. During such transport only a portion of the solid phase comes in contact with the solute, reducing pesticides retention by adsorption.

Pesticides may also be transported when adsorbed on soil particles, often in lateral movement by runoff. The magnitude of transport by runoff is determined by the intensity of the falling water, the soil infiltration capacity, and the slope of the surface.

Volatilization is an additional pattern of transport of pesticides from treated agricultural lands in the vapor phase into the atmosphere. Potential volatility of pesticides is related to their inherent vapor pressure, but actual vaporization rates depend on the environmental conditions and all other factors that control behavior of the chemical at the solid-air-water interface.

The understanding of the behavior and fate of pesticides in soil is especially difficult not only because the processes are generally simultaneous and interrelated but also because they act in an almost infinite number of environmental combinations of almost infinite variability in soil properties and the several hundreds of pesticides in use. However, the research effort invested in this field over the last 25 years, has contributed to a better understanding of the main aspects of soil-pesticide relationship and has enabled the development of models to predict them, and to minimize the adverse effects of the addition of pesticides to soils.

References

Alexander, M., 1965, Persistence and Biological Reactions of Pesticides in Soils, *Soil Sci. Soc. America Proc.* **29:**1-7.

Edwards, C. A., 1973, *Persistent Pesticides in the Environment,* 2nd ed., CRC Press, Cleveland, Ohio 170p.

Graham-Bryce, I. J., 1981, The Behavior of Pesticides in Soil, in *The Chemistry of Soil Processes,* D. J. Greenland and M. H. B. Hayes, eds., Wiley, New York, pp. 621-670.

Hadaway, A. B., and F. Barlow, 1952, Studies on Aqueous Suspensions of Insecticides. III. Factors Affecting the Persistence of Some Synthetic Insecticides, *Bull. Ent. Res.* **43:**281-311.

Swanson, C. L. W., F. C. Thorp, and R. B. Friend, 1954, Adsorption of Lindane by Soils, *Soil Sci.* **78:**379-388.

Weaver, R. J., 1947, Reaction of Certain Plant Growth Regulators with Ion Exchangers, *Bot. Gaz.* **108:**72-84.

Part I

DISTRIBUTION OF PESTICIDES AMONG THE SOIL PHASES

Editors' Comments
on Papers 1 Through 9

1 **HADAWAY and BARLOW**
   Sorption of Solid Insecticides by Dried Mud

2 **EDWARDS, BECK, and LICHTENSTEIN**
   Bioassay of Aldrin and Lindane in Soil

3 **SALTZMAN, KLIGER, and YARON**
   Adsorption-Desorption of Parathion as Affected by Soil Organic Matter

4 **GAILLARDON, CALVET, and TERCE**
   Adsorption et désorption de la terbutryne par une montomorillonite-Ca et des acides humiques seuls ou en mélanges

5 **HUANG, GROVER, and McKERCHER**
   Components and Particle Size Fractions Involved in Atrazine Adsorption by Soils

6 **HANCE**
   The Adsorption of Urea and Some of Its Derivatives by a Variety of Soils

7 **BRIGGS**
   Molecular Structure of Herbicides and Their Sorption by Soils

8 **BARLOW and HADAWAY**
   Effect of Changes in Humidity on the Toxicity and Distribution of Insecticides Sorbed by Some Dried Soils

9 **MILLS and BIGGAR**
   Solubility-Temperature Effect on the Adsorption of Gamma- and Beta-BHC from Aqueous and Hexane Solutions by Soil Materials

FACTORS AFFECTING PHASE DISTRIBUTION

The main factors considered as relevant for the adsorption-desorption of pesticides in soils are the nature and properties of the soil colloids, the chemical and physico-chemical characteristics of pesticides, and the features of the soil environment (Bailey and White, 1964, 1970; Hamaker and Thompson, 1972; Green, 1974; Weed and Weber, 1974; Burchill, Hayes, and Greenland, 1981). Obviously all these factors are interrelated and act simultaneously in the adsorption process, and they are treated so in most adsorption studies. However, the following papers are grouped according to emphasis on specific factors to enable a more systematic discussion.

Nature and Properties of the Soil Components

The soil solids, usually possessing a large and physico-chemically active surface area, provide both a sink for the retention of pesticides and sites for a variety of surface reactions.

One of the earliest works on the adsorption of pesticides on differing types of materials was carried out by a team researching the persistence of chlorinated hydrocarbon insecticides sprayed on mud walls in Uganda (Paper 1). The persistence of the insecticide gamma-benzene-hexachloride was longest on nonadsorbent materials. Although the inseticide was adsorbed by such materials as plaster, fiberboard and wood, adsorption was less than on mud blocks. The variation in adsorption as affected by the type of material to which insecticides are applied is considered of great practical importance because it is related to toxicity.

The soil properties most significant for the adsorptive capacity of soils are usually correlated with each other, so that their individual effect is often difficult to assess. A widespread approach was the use of correlation analysis, which tried to isolate and determine relatively the weight of each soil factor in adsorption. One of the first studies using this approach is that of Edwards, Beck, and Lichtenstein (Paper 2). The results obtained by a bioassay technique for aldrin and lindane adsorption by ten different soils showed that adsorption of both insecticides was correlated with the organic matter content of the soils more than with the water-holding capacity. Linear relationships between the lethal dose of insecticide and organic matter content were obtained with all soils except the muck. In both Papers 1 and 2, adsorption was inferred from the fact that the insecticides could be extracted in a toxic form. However, this fact may be misleading in assessing adsorption, because some compounds, such as dieldrin, are transformed in soil into another toxic form, while other compounds are detoxified by degradation but their degradation products are retained by soil and the chemical analysis could fail to separate and identify them.

Additional soil properties considered in several studies using correlation analysis were the cation exchange capacity, surface area, pH, and moisture content. Because the one most highly correlated with adsorption was the organic matter content, the opinion that this soil fraction provides the most important adsorption sites for pesticides became widely accepted (see Paper 6).

A considerable contribution to the understanding of the relative role of mineral and organic colloids in pesticide adsorption was presented by studies using separated soil fractions and well defined model materials. One approach was the study of adsorption by isolated soil fractions (e.g., humic acid, clay), and another was the comparison of adsorption before and after organic matter removal from soil samples. The second approach was used by Saltzman, Kliger, and Yaron (Paper 3) to assess the relative importance of soil colloids in parathion uptake and release. Although the removal of organic matter from soils by oxidation with hydrogen peroxide could affect the properties of the adsorbent, the results obtained may be considered as qualitative information about the role of the properties of the adsorbent. Similar to other studies, Paper 3 emphasizes that parathion has a greater affinity for organic adsorptive surfaces than for mineral ones. However, the important finding suggested in this work is that adsorption was dependent on the type of association between organic and mineral colloids, which determines the nature and the magnitude of the adsorptive surfaces.

Paper 4 is a representative study, using model materials. Montmorillonite, humic acids, and their mixtures were tested as adsorbents for terbutryn, a herbicide of the s-triazine group. By contrast to the case presented in Paper 3, the picture obtained from this work is complicated by the pH-dependence of adsorption, because terbutryn is a weakly basic compound that could protonate and be adsorbed as a cation. In a slightly acid medium (pH 5.6–6.0), the humic acids were the main adsorbent, but a decrease in pH increased the adsorption by the clay fraction and by the clay-humic acid mixtures. It is suggested that in acid conditions the mixtures have a synergistic effect on adsorption. A hysteresis in desorption was observed for the humic acids and the mixtures but not for the clay. This work also pointed out that the organic matter-mineral colloids relationship, rather than isolated parameters, must be considered in the assessment of pesticide adsorption by soils.

Although the importance of organic matter in pesticides adsorption has been well established, the properties of the organic colloids relevant for adsorption have not yet been thoroughly characterized.

The available information shows that these properties could be the proportion of humic acid, fulvic acid, and humin; the presence of active groups such as carboxyl, hydroxyl, carbonyl, methoxy, amino; and high cation exchange capacity and surface area (Weed and Weber, 1974; Burchill, Hayes and Greenland, 1981).

Contrary to the situation with organic matter, the role of mineral colloids (mainly the clays) as adsorbents for pesticides, has been studied intensively. The main properties affecting the adsorptive capacity of clays are considered to be the available surface area and the cation exchange capacity, as well as the saturating cation, the hydration status, and the surface acidity, which is related to the two preceding properties and to the clay structure (Greenland, 1965). The degree of pesticide adsorption by clays from aqueous solutions is variable, with a general trend of increasing adsorption from non-ionic, nonpolar molecules, to more polar pesticides, to weak bases and cationic compounds.

Although amorphous oxides and hydroxides of iron, aluminum, and silica can adsorb pesticides, very little information concerning this interaction is available. Recently, Huang, Grover, and McKercher (Paper 5) studied the relative importance of organic matter, sesquioxides, and different particle size fractions of soils in the adsorption of atrazine. They found that aluminum, iron, and probably other mineral compounds present in soil fractions ranging from clay to sand provided adsorption sites for atrazine. The removal of aluminum and iron oxides from soils significantly decreased the amount of atrazine adsorbed and changed the adsorption kinetics. However, the amount of extractable Al and Fe in the different particle size fractions was not proportional to the extent of adsorption, suggesting that different forms of sesquioxides could have different reactivities for atrazine. The adsorption capacity of sesquioxide components was attributed to their high specific surface area and proton donor functional groups.

Properties of Pesticides

The important properties affecting the extent and nature of pesticide adsorption by soils are the chemical character of the molecule and its shape, size, conformation, configuration, polarity and polarizability, acidity or basicity, charge distribution, and water solubility (Bailey and White, 1964, and 1970; Greenland, 1965).

Bailey, White, and Rothberg showed (in Paper 11) that the chemical character of a molecule affects adsorption as it determines its acidic or basic nature and strength, affects its water solubility, and determines the importance relative to the formerly mentioned properties of van der Waals-type forces of adsorption. For the basic and acidic compounds, the major factor determining the extent of adsorption was the dissociation constant. The effect of this factor on the adsorption of acidic pesticides by clays was reported by Frissel and Bolt (Paper 12).

Paper 6 is among the first studies of the relationship between the presence of specific functional groups in the pesticide molecule and adsorption by soils. The adsorption of nine substituted urea herbicides by six soils with differing properties was investigated. The results showed that both the aryl and alkyl substituents increased adsorption, compared with the nonsubstituted urea. The increase in the chain length of the alkyl substituents, or chloro and chlorophenoxy substitution in the ring, also increased adsorption. Water solubility of substituted ureas was not closely related to their soil adsorption.

The relationship between water solubility and adsorption has been the subject of much controversy. At present, it is widely accepted that water solubility may be related to the extent of adsorption only within a specific chemical group of compounds. The results reported in Paper 11, for instance, demonstrate that within a chemical group, or an analog series of compounds with a basic character, the magnitude of adsorption by clay is related to water solubility.

Briggs (Paper 7) studied the effect of the chemical structure of two groups of non-ionic, related compounds on their sorption by soils by assuming that the soil organic matter provided most of the adsorption sites. Partition coefficients calculated from the amount adsorbed per unit weight of organic matter for 22 substituted phenylureas and 4 soils varied widely (bewteen 8 and 217). The sorption of substituted phenylureas was well correlated with the Hammet constant, which expresses the acidity of a medium. However, a study of the adsorption of alkyl-N-phenylcarbamates showed a better relationship with a constant derived from partition between octanol and water. In addition, a linear relationship between the adsorption of these compounds and parachor (a measure of molecular size) was observed.

The results presented in Papers 6 and 7 indicate that the extent of adsorption is related to the molecular size of pesticides. The molecular size and shape can affect adsorption in at least three ways: (1) In a homologous series of non-ionic compounds, adsorption by nonpolar adsorbents increases as molecular weight increases, (2) adsorption may be hindered by an unfavorable configuration, and (3) the van der Waals energy of adsorption increases with molecular size (Bailey and White, 1970).

Soil Environment

In addition to the properties of both soil components and pesticides, the distribution of synthetic organic chemicals among the soil phases is influenced by the soil environment. The environment of a specific

Editors' Comments on Papers 1 Through 9

soil is determined not only by its intrinsic properties but also by external factors, mainly climatic conditions and agricultural practices. These factors affect the amount and periodicity of changes in the soil environment.

The soil moisture content affects adsorption in several ways. Pesticides are usually transported to the adsorbing surfaces by water; the moisture content determines the accessibility of the adsorption sites; and water affects the surface properties of the adsorbent. Indirect evidence of the moisture effect on adsorption inferred from bioactivity studies is presented in Paper 8, one of the first studies of this topic. Toxicity and chemical analyses of DDT, gamma-BHC, and dieldrin applied on mud blocks showed that bioactivity followed closely any changes in humidity, that the rate of diffusion of the insecticides into the mud blocks increased with humidity, and that the insecticides sorbed were reactivated by increasing the moisture content. The explanation for this behavior is that competition exists between water and these non-ionic pesticides for adsorption sites. Preferential adsorption of the more polar water molecules by the mud blocks hindered insecticide adsorption at high humidity; the competition was less at low moisture contents, and resulted in increased adsorption. The insecticides sorbed under low humidity conditions were desorbed upon an increase in humidity, thereby increasing both their diffusion to the surface and into the mud blocks. Negative relationships between pesticide adsorption and soil moisture content have been reported often (e.g., Ashton and Sheets, 1959; Yaron and Saltzman, 1972).

Studies of pesticide-clay-water systems have been remarkably useful for achieving an understanding of the effect of water on pesticide adsorption (Green, 1974; Burchill, Hayes, and Greenland, 1981). Although the hydration of clays and the properties of adsorbed water are not yet fully understood, it is generally accepted that water molecules are attracted by the clay surfaces, mainly by the exchangeable cations, forming hydration shells. Adsorbed water provides adsorption sites for pesticide molecules. An important feature of water associated with clay surfaces is its increased dissociation, giving the surfaces a slightly acidic character. Generally, a negative relationship exists between the surface acidity of clays and their water content.

The effect of hydration of the soil organic matter connected to pesticide adsorption has been studied less. It has been suggested that hydration influences the molecular shape of humic substances and thus accessibility for pesticides. Strong, sometimes irreversible, retention of pesticides by hydrated humic substances could be explained by the penetration and trapping of pesticides into the internal structure of the swollen humic substances. The hydrated exchangeable cations and some dissociated functional groups, as well as water held by various polar groups of the humic substances, could also provide adsorption sites. At low moisture contents, the hydrophobic portions of the organic matter structures could bind hydrophobic, non-ionic pesticides (Burchill, Hayes, and Greenland, 1981).

As the adsorption processes are exothermic, changes in soil temperature could have a direct effect on the phase distribution of pesticides. Adsorption usually increases as the temperature decreases, and desorption is favored by increasing temperature. Temperature could indirectly influence adsorption by its effect on pesticide-water interactions. The complex relationship between adsorbent-adsorbate-solvent as affected by temperature is emphasized in Paper 9. The adsorption of lindane (1,2,3,4,5,6-hexachlorocyclohexane) and its beta-isomer by a peaty muck, a clay soil, Ca-bentonite, and silica gel, decreased as the isothermal temperature of the systems increased. The authors suggested that this adsorption-temperature relationship reflects not only the energy contributions in the adsorption process but also the change in solubility of the adsorbate. The activity, and hence the chemical potential of a solute in solution, is more or less dependent on its solubility, as affected by temperature and solvent. They considered that changes in acitivity in solution are important, as the difference between the activity in solution and on the adsorbent is the driving force in the adsorption process. The authors assumed that the change in activity in solution with temperature is related to the change in the reduced concentration, which is the ratio between the actual concentration of the solute at a given temperature and its solubility at the same temperature. Adsorption isotherms obtained by using the reduced concentration showed, contrary to normal adsorption isotherms, an increase in adsorption with an increase in temperature, suggesting that the heat effect involved in the adsorption process was mainly that involved in the solubility of the solute. Similar results, emphasizing the significant influence of temperature on adsorption through its solubility effect, have been obtained by Yaron and Saltzman (1972) for parathion adsorption by soils, and by Yamane and Green (1972) for atrazine and ametryne adsorption by soils and montmorillonite.

References

Ashton, F. M., and T. J. Sheets, 1959, The Relationship of Soil Adsorption of EPTC to Oats Injury in Various Soil Types, *Weeds* **7**:88-90.

Bailey, G. W., and J. L. White, 1964, Review of Adsorption and Desorption of Organic Pesticides by Soil Colloids, with Implications concerning Pesticide Bioactivity, *Jour. Agric. Food Chem.* **12**:324-332.

Bailey, G. W., and J. L. White, 1970, Factors Influencing the Adsorption, Desorption and Movement of Pesticides in Soil, *Residue Rev.* **32**:29-92.

Burchill, S., M. H. B. Hayes, and D. J. Greenland, 1981, Adsorption, in *The Chemistry of Soil Processes,* D. J. Greenland and M. H. B. Hayes, eds., Wiley, New York, pp. 224-400.

Green, R. E., 1974, Pesticide-Clay-Water Interactions, in *Pesticides in Soil and Water,* W. D. Guenzi, ed., Soil Science Society of America, Inc., Madison, Wis. pp. 3-37.

Greenland, D. J., 1965, Interactions between Clays and Organic Compounds in Soils. I. Mechanism of Interaction between Clays and Defined Organic Compounds, *Soils Fertil.* **28:**415-425.

Hamaker, J. W., and J. M. Thompson, 1972, Adsorption, in *Organic Chemicals in the Soil Environment. I.,* C. A. I. Goring and J. W. Hamaker, eds., Marcel Dekker, New York, pp. 49-143.

Weed, S. B., and J. B. Weber, 1974, Pesticide-Organic Matter Interactions, in *Pesticides in Soil and Water,* W. D. Guenzi, ed., Soil Science Society of America Inc., Madison, Wis., pp. 39-66.

Yamane, V. K., and R. E. Green, 1972, Adsorption of Ametryne and Atrazine on an Oxisol, Montmorillonite and Charcoal in Relation to pH and Solubility Effects, *Soil Sci. Soc. America Proc.* **36:**58-64.

Yaron, B., and S. Saltzman, 1972, Influence of Water and Temperature on Adsorption of Parathion by Soils, *Soil Sci. Soc. America Proc.* **36:**583-586.

SORPTION OF SOLID INSECTICIDES BY DRIED MUD

A. B. Hadaway and F. Barlow

In the course of experiments on the persistence of deposits from aqueous suspensions of different particle sizes of volatile insecticides, it has been found that the residual toxicity of particles of any one size is influenced considerably by the type of material to which they are applied.

Most striking results have been obtained on mud blocks made from 'murram', a lateritic ironstone, used in the construction of walls of houses in Uganda. Crystals of all insecticides used rapidly disappear from the surface of these blocks when they are kept at 78° F. (25° C.), and even those of DDT, which is usually regarded as a contact insecticide with a long residual life, are no longer visible after only a few days. As would be expected, the larger the particle the longer it persists; but for any given size-range the insecticides used can be arranged in order of increasing persistence, thus : 'Aldrin', gamma isomer of benzene hexachloride, 'Dieldrin' and DDT.

When DDT and 'Dieldrin' crystals are no longer visible on the surface, the mud blocks lose their toxicity to mosquitoes (*Aedes œgypti*, L.) exposed to them for long contact periods. On the other hand, blocks treated with the gamma isomer of benzene hexachloride or with 'Aldrin' continue to be effective for a considerable time, after the disappearance of the crystals, against mosquitoes resting on the surface.

Chemical tests have shown that with 10–20 micron particles of DDT and gamma-benzene hexachloride, almost the whole of the dosage applied can be recovered from the interior of the block. At a dosage of 25 mgm. of these particles per sq. ft., almost all the DDT is sorbed in the top tenth of an inch, suggesting that considerably higher dosages could be applied before the block becomes saturated. Dosages of 10–20 micron gamma-benzene hexachloride particles of the order of 200 mgm. per sq. ft. do, in fact, disappear from the surface in 24 hr. at 78° F. The rate of disappearance from the surface decreases as the relative humidity of the atmosphere increases. This may be explained by competition between water vapour and the insecticide vapour for the adsorbing surface. An interesting effect observed is that under absolutely dry conditions the sorbed DDT is catalytically decomposed to the ethylene derivative, the soil used having a high iron content. This again suggests that the insecticide is present in the block as a highly active surface layer.

The gamma isomer of benzene hexachloride and 'Aldrin' have, in contrast to DDT and 'Dieldrin', a marked fumigant action against *A. œgypti*, and there is ample evidence to show that the residual toxicity after the disappearance of crystals from the surface of blocks treated with these insecticides is due to the fumigant effect from the material inside the block. Desorption of the volatile insecticides takes several months for completion as compared with a few hours for the sorption process.

Crystals of gamma-benzene hexachloride persist for a much longer time on non-absorbent materials, such as glass and metal plates, than on mud blocks; but toxicity is completely lost as soon as the crystals disappear from the surface. Sorption of gamma-benzene hexachloride occurs on such materials as plaster fibreboard and unpainted wood, but to a much less extent than on mud blocks.

Commercial wettable powders and oil-bound suspensions of these insecticides behave in the same way as the pure materials on Uganda mud. Similar results are obtained with soil from Taveta in Kenya and with Oxford, Weald and Gault clays from Britain. It is intended to test other soils used in the construction of houses in other tropical countries as soon as the materials are available.

These results are of obvious practical importance in the control of adult mosquitoes in houses the walls of which are constructed of dried mud. The difference in effectiveness of DDT and benzene hexachloride wettable powders against *Anopheles gambiæ*, Giles, in houses with mud walls in Africa may be accounted for partly, at least, by the rapid sorption of both insecticides into the wall, resulting in one case in a complete loss of toxicity and in the other a persistent fumigant effect. Residual effects obtained with DDT wettable powders in houses of this type may be due only to the deposit on the roof. Variations in the persistence of insecticides on different types of material may help to explain the conflicting results reported in field-trials against malaria vectors in different parts of the world. Sorption of insecticides may be of significance in treatments against soil insects and in soil fumigation.

It is intended to publish a full report on these experiments.

Colonial Insecticides Research Team,
c/o Ministry of Supply,
Chemical Defence Experimental Station,
Porton, Nr. Salisbury,
Wilts.

Bioassay of Aldrin and Lindane in Soil[1]

C. A. Edwards, S. D. Beck, and E. P. Lichtenstein, *Department of Entomology, University of Wisconsin, Madison*

The rapid development and extensive use of organic insecticides has resulted in a need for suitable analytical techniques to determine insecticidal residues, particularly in foodstuffs. In many cases where chemical analytical techniques have not been adequate, relatively sensitive bioassay methods have been developed. Insecticidal residues have been determined using such insects as *Drosophila melanogaster* Meig. (Morrison 1945), *Musca domestica* L. (Laug 1946), *Tribolium confusum* Duv., *Tenebrio molitor* L. (Busvine & Barnes 1947), and *Aedes aegypti* L. (Hartzell & Storrs 1950).

Similar problems have resulted from the increased emphasis on soil application of insecticides. In addition to direct application soils may become heavily contaminated as a result of foliar applications. In many cases chemical analytical methods indicate the persistence of high concentrations of residues even after several years (Ginsburg & Reed 1954, Ginsburg 1955, Chisholm *et al.* 1950, Lichtenstein 1957). While this information is of considerable value, it is insufficient in itself, since it cannot be assumed that the material remaining in the soil is still insecticidally active. Chemical analytical methods may also be insufficiently specific (*e.g.* organic chloride), to detect chemical degradation that might alter or destroy insecticidal action. Furthermore the insecticide may become bound to some soil fraction in such a way as to become inactivated. Bioassay techniques on the other hand, give no indication as to the nature of the insecticide present, but indicate the residual toxicity. In most past work the residue was assayed after it had been extracted from the soil with a suitable solvent (Fleming *et al.* 1951, 1953; Terriere & Ingalsbe 1953; Lauge & Carlson 1955). Very few results on direct soil bioassay have been reported (Fleming *et al.* 1951, Wylie 1956, Dewey, personal communication 1955). It would seem that to obtain maximum information concerning the persistence of an insecticide in a particular soil, chemical analysis should be used in conjunction with direct soil bioassay as well as bioassay of the extract.

METHODS.—A test insect was required which would combine sensitivity to low concentrations of insecticide in soil or dry deposits with ease of rearing. Adult *Drosophila melanogaster* strain Cinnabar, fulfilled these conditions. Fruit flies were reared in half-gallon preserving jars on a pumpkin medium (Bartlett 1951). Two pounds of pumpkin puree plus 8 millilitres of propionic acid sufficed for three such cultures.

Tests were carried out in small jars, $2\frac{3}{4}$ inches in diameter and 3 inches deep, covered with fine screen inserted into the screw tops. When extracts were assayed, 1 milliliter of the extract and 0.5 milliliter of corn oil solution (1 milliliter corn oil in 200 milliliters hexane) were pipetted into each test jar. The jar was then slowly rotated until the solvent evaporated, so that the extract was deposited evenly over the interior of the jar. An extract of unknown concentration was tested against a range of standard concentrations. If preliminary tests gave too high a mortality, the unknown extract was suitably diluted with solvent. In cases where very exact estimates of the amount of insecticide present in an extract were required, several dilutions were used, and a curve plotted on log-probit paper parallel to the standard curve so that a relative potency determination could be made from LD_{50} values (Litchfield & Wilcoxon 1949).

In most of the work, a direct soil bioassay technique was used. Tests of *Drosophila* over an air dried soil containing insecticide gave erratic results, but the addition of an attractant into the soil counteracted this tendency. Of various fruit products tested as attractants, apple juice proved most satisfactory, giving a suitable smooth mixture with finely screened soil. The method was standardized as follows: Five grams of the soil under assay was weighed into each of two test jars. If preliminary tests indicated that this soil contained sufficient insecticide to bring it outside the range of the assay, less than 5 grams were weighed out, and untreated soil added to give a total of 5 grams. In each assay, at least four levels of standard were used, with two replicates per level. Ten milliliters of solvent were then added to all jars, and the appropriate amounts of insecticide in hexane solution were added to these standards. If the insecticide level in the test soil was expected to be low, then additional insecticide was added to the samples. The contents of each jar was well mixed, and the solvent was removed by standing the jars at the opening of a fume hood in a gentle air flow, care being taken to remove them from the flow as soon as all the solvent had disappeared. Three milliliters of apple juice were added to each jar, and thoroughly mixed into a slurry which evenly coated the bottom of the jar. Surplus moisture was evaporated off to give a firm surface. The high absorptive properties of some muck soils necessitated addition of 4 or even 5 milliliters of apple juice. To each test jar a small shell vial containing apple sauce was added to provide food.

Anaesthesis with either ether or carbon dioxide gave erratic results and a direct method of introducing the flies was employed (fig. 1). The stem of a small glass funnel was drawn out until the outlet was approximately 2 millimeters in diameter, such that the aperture was small enough to allow passage of only one fly at a time. Flies that had been isolated after emergence from pupae 3 days previous to the test were used. Up to one thousand such flies were collected into the reservoir jar and the funnel attached to the end of this by means of a screw cap. The funnel outlet was inserted into a small hole in the screen covering a test jar. With a bright light on the far side of the test jar, flies were counted as they passed down the stem of the funnel. When 50 flies had entered the jar, the funnel was withdrawn and a small cotton plug inserted into the hole. Pilot experiments showed 24 hours to be a suitable time to make mortality counts. By this time, the mortality/time curves had leveled off for both aldrin (12 hours) and lindane (22 hours) under the conditions of the experiments. Consequently, in all

---

[1] Approved for publication by the Director of the Wisconsin Agricultural Experiment Station. This investigation was conducted during the tenure of a Kellogg Foundation Fellowship awarded the senior author. Accepted for publication April, 2, 1957.

FIG. 1.—Introduction of flies into test jar.

experiments, mortality counts were made after a standardized period which was 24 hours of exposure to the insecticide. For each assay run, two controls were prepared and, if necessary, Abbott's correction for mortality in the controls (Healy 1952) was applied.

The effects of using flies of different age, sex, or strain were investigated in initial experiments. More consistent results were obtained with flies of a single sex, but the additional labor involved precluded this as a general practice. Significant differences in mortality between 1-day-old and 10-day-old flies were obtained, and also between flies of Wild and Cinnabar strains. In view of these results, all experiments were standardized on 3-day-old Cinnabar flies of mixed sex. During the course of the experiments it was noted that variations in such factors as light, temperature and humidity during the exposure period caused the results to vary appreciably. To eliminate this as far as possible, all experiments were performed under continuous light and under conditions of temperature and humidity which were nearly constant.

RESULTS AND DISCUSSION.—At an early stage in the work it became apparent that soils differed markedly in their properties as carriers of insecticide. Comparable mortalities of *Drosophila* were obtained in different soils only by the addition of large quantities of insecticide to some soil types. This appeared to be of considerable practical importance and was further investigated. From standard curves plotted for the three soil types; sand loam, loam, and muck, the LD_{50} values recorded in table 1 were obtained, and the ratios of these LD_{50} values to one another are shown in table 2.

Table 1.—Effect of soil type on the insecticidal action of aldrin and lindane.[a]

SOIL TYPE	ALDRIN (4 EXPERIMENTS)	
	Average LD_{50} (p.p.m.)	LD_{50} Range (p.p.m.)
Sandy loam	0.28	0.24–0.38
Loam	0.41	0.34–0.52
Muck	0.90	0.80–1.00
F value	103.70**	
L.S.D. @ 5% level	0.10	
@ 1% level	0.14	

SOIL TYPE	LINDANE (5 EXPERIMENTS)	
	Average LD_{50} (p.p.m.)	LD_{50} Range (p.p.m.)
Sandy loam	1.65	1.44–1.86
Loam	2.25	2.01–2.52
Muck	4.71	3.75–5.55
F value	29.95**	
L.S.D. @ 5% level	0.30	
@ 1% level	0.43	

[a] Test insect *Drosophila* with mortality counts after 24 hours.

A number of important observations can be made from these data. First, aldrin was more toxic to *Drosophila* than lindane. For each soil the LD_{50} for lindane was nearly five times as high as that for aldrin. This was not a soil factor, since a similar relationship was found when similar calculations were made from data obtained by the use of deposits of insecticide on the inner surface of test jars no soil being involved. The dosage/mortality curves (figures 2 & 3) had two interesting properties: curves drawn for aldrin had a greater slope than those obtained for lindane; and standard curves plotted for the same insecticide but in different soils were always parallel with the exception of those plotted for muck which had a lower slope with both aldrin and lindane. The ratios of LD_{50} values in different soils agreed closely for both insecticides (table 2) so that it seems probable that the adsorption of aldrin and lindane by a particular soil is due to the same factors.

Since these results indicated differential adsorption of insecticides by soils, the study was extended to include a wider range of soils, and an attempt was made to correlate the adsorption with physical properties of the soil. The most noticeable properties of soils showing the highest insecticide adsorption were their high organic matter content and high moisture holding capacity. To investigate the relative importance of these factors a range of 10 different soils was used. Estimations were made of the total organic matter content of each of these soils using the colorimetric method of Wilde & Petzer (1940). To obtain a relative value for the moisture holding capacity of each soil, samples were dried in a hot oven at 50° C. and a known volume of water was passed through a weighed amount of soil held in a filter funnel. The water which drained through was collected in a graduated cylinder for 1 hour. From the volume of water which drained through a particular soil, an estimate could be made of its relative

Table 2.—Ratio of LD_{50} values in different soils.

RATIO LD_{50} VALUES	ALDRIN	LINDANE
Sandy loam: muck	0.314	0.352
Loam: muck	0.456	0.478
Sandy loam: loam	0.683	0.734

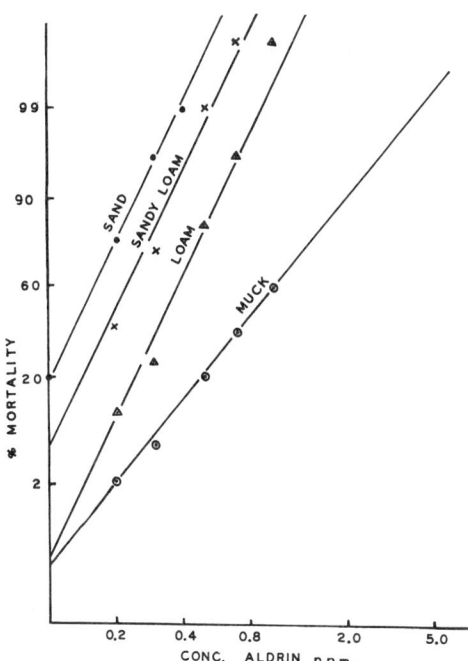

FIG. 2.—Dosage/mortality curves for aldrin in different soils.

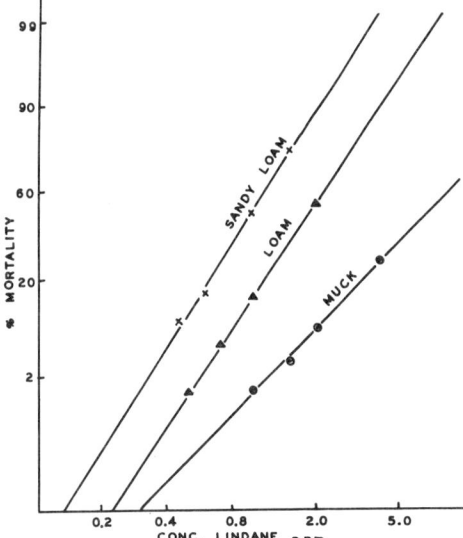

FIG. 3.—Dosage/mortality curves for lindane in different soils.

moisture holding capacity. Standard dosage/mortality curves were obtained for all these soils using both aldrin and lindane. With the exception of muck, the curves for all soils obtained with a particular insecticide were parallel. The LD_{50} for each insecticide in each soil was interpolated from the standard curve and tabulated with data on organic matter content and moisture holding capacity for that soil (table 3).

All these correlations were significant, but the correlation calculations were strongly weighted by the high correlation between organic matter content and moisture holding capacity, which existed in the clay loam and muck soils. When correlation coefficients were calculated omitting the data for these last three soils, significant correlation at 5% level was still obtained between the LD_{50} for both insecticides and the soil organic matter content. In contrast, correlation was no longer found between the LD_{50} values and moisture holding capacity of the soil. Thus it seemed probable that the important factor modifying toxicity of the insecticide in the soil was the organic matter present. Further support to this hypothesis was given by another experiment. Bentonite and Kaolinite are mineral soils containing negligible organic matter, but the former will absorb large quantities of water and the latter only a small amount. Standard dosage/mortality curves obtained in a similar manner for these soils showed no significant difference in LD_{50} value. This indicated that the moisture holding property was not the principal soil factor concerned with adsorption and retention of the insecticide in a non-toxic state.

The relationship between the LD_{50} values of lindane and aldrin and soil organic matter content is demonstrated in figures 4 & 5. It appears to be curvilinear, but could be considered as linear if muck is considered as exceptional. Before this can be finally decided, soils having organic matter contents intermediate between that present in the clay loam and the muck soils studied here, would have to be tested. Not many such soils are found naturally and none were available in the present work. There is some basis for considering muck as exceptional in its behavior to insecticides, in that all standard curves obtained for muck differed in slope from those for other insecticides. It may well be that some additional factor present in muck accounts for the greatly increased adsorption. For practical purposes, e.g. the calculation of amounts of insecticide to be added to field soils, it seems probable that no great error would arise from considering the relationship between soil organic matter content

Table 3.—Relation between lindane and aldrin adsorption and soil factors.

Soil Type	Organic Matter (Per Cent)	Moisture-Holding Capacity (Per Cent)	LD_{50} of (24 Hours)	
			Lindane (p.p.m.)	Aldrin (p.p.m.)
a. Springfield sand	0.5	51	0.25	0.055
b. Silty clay loam	1.0	71	0.38	0.175
c. Light sandy clay loam	1.2	52	0.51	0.065
d. Coarse silt	1.4	58	1.07	0.205
e. Silty clay	1.8	74	0.67	0.22
f. Sandy loam	2.6	49	1.25	0.22
g. Loam	3.8	83	2.65	0.34
h. Clay loam (lower level)	6.4	92	4.10	0.40
i. Clay loam (upper level)	10.0	—	5.9	0.54
j. Muck	40.0	127	8.6	0.85

Correlation coefficients
LD_{50} (aldrin) with organic matter content $r = 0.8935$**
LD_{50} (lindane) with organic matter content $r = 0.9141$**
LD_{50} (aldrin) with moisture holding capacity $r = 0.8468$**
LD_{50} (lindane) with moisture holding capacity $r = 0.8204$**

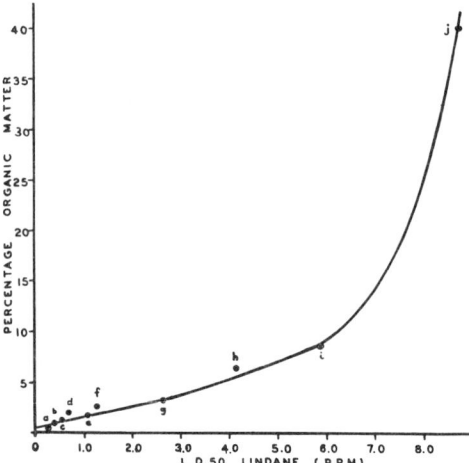

Fig. 4.—Relationship between soil organic matter and LD_{50} of lindane.

and adsorption of insecticide as linear, over the range of soils used in agriculture.

Plots of three typical, but very different, Wisconsin soils were treated with several dosages of aldrin and lindane. The amounts of insecticide applied and worked into these soils in the spring of 1954 were 20 and 200 pounds of aldrin per acre and 10 and 100 pounds of lindane per acre. These soils had been chemically analyzed at intervals to determine the dissipation of insecticide. With these data at hand, careful direct bioassays of the soils were carried out nearly 2 years after treatment. The assays were made on soil samples collected according to a random sampling procedure. Samples were diluted with untreated soil to five levels and measured against five

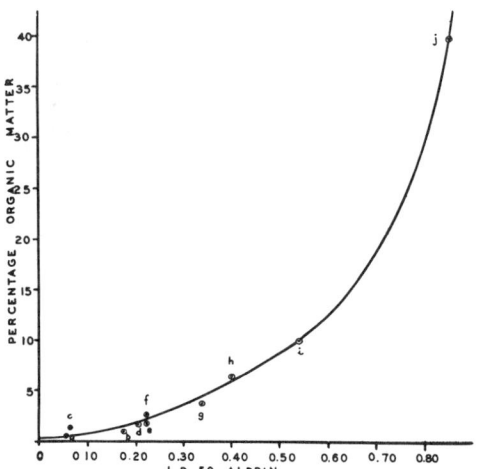

Fig. 5.—Relationship between soil organic matter and LD_{50} of aldrin.

Table 4.—Disappearance of lindane from soil 22 months after treatment.[a]

Soil Type	Treatment (Pounds/Acre)	Chemical Assay (p.p.m.)	Bioassay (p.p.m.)	Bioassay Confidence Limits (p.p.m.)	Bioassay / Chem. Assay ×100
Muck	10	11.6	2.21	1.87– 2.61	19
	100	118.6	28.4	24.1 –33.4	24
Loam	10	1.85	0.5	0.46– 0.56	27
	100	29.8	10.5	9.4 –11.7	35
Sandy loam	10	1.8	1.65	1.47– 1.85	90
	100	29.0	22.2	20.1 –24.0	76

[a] Treated May, 1954; bioassay, March, 1956.

levels of standard by the relative potency method (Litchfield & Wilcoxon). The results are shown in tables 4 and 5 together with data obtained by chemical assay of insecticide residues extracted from corresponding soil samples (Danish & Lidov, 1950) (Lichtenstein et al. 1956).

The considerable discrepancy between chemical estimates of residual lindane and the corresponding results obtained by bioassay is emphasized in the last column of table 4. It has been shown (table 1) that much more insecticide had to be added to a muck soil than to a sandy loam to obtain comparable toxicity to Drosophila. In each bioassay the insecticidal activity present in the assay soil was compared with that of standard concentrations prepared with soil of the same type. Hence, the estimate of the amount of insecticide present in a soil was independent of the immediate capacity of the soil to adsorb it and render it non-toxic. This was borne out by further experiments which traced the reduction in amount of insecticide present in a soil over a period of several weeks after application. Initially there was good agreement between the chemical and bioassay estimates of the amount of insecticide present, which indicated that the discrepancies observed in the later analyses were probably a feature of the time interval between application of the insecticide and the assay. If it is assumed that the extraction of the insecticide from the soil is efficient, the simplest explanation of the results is that degradation products of the gamma isomer lindane may have appeared. The bioassay recorded the actual residual toxic material while the chemical assay probably included degradation products of the insecticide which were no longer toxic (benzene ring structure). This would also account for the greater discrepancy between chemical and bioassay data seen with muck soil. The muck probably adsorbed the degradation materials of the insecticide just as efficiently as the insecticide itself.

Table 5.—Disappearance of aldrin from soil 25 months after treatment.[a]

Soil Type	Treatment (Pounds/Acre)	Chemical Assay (p.p.m.)	Bioassay (p.p.m.)	Bioassay Confidence Limits (p.p.m.)	Bioassay / Chem. Assay ×100
Muck	20	4.95	2.25	2.08– 2.43	45.5
	200	158.0	57.5	53.2 –62.0	36.2
Loam	20	0.69	1.89	1.75– 2.04	274.0
	200	16.5	28.0	26.0 –30.2	169.5
Sandy loam	20	0.24	1.25	1.15– 1.36	522.0
	200	22.5	66.0	60.5 –72.0	293.0

[a] Treated May, 1954; bioassay July, 1956.

Aldrin-treated soils were also assayed for residual insecticide 25 months after soil application. As in the case of lindane, there was considerable discrepancy between bioassay data and that obtained by chemical analysis. However, the estimates of toxic material present in loam and sandy soils made by bioassay were several times larger than the chemical results. This appeared to be the converse of the results with lindane (table 5).

The assay data indicated the possibility that aldrin was being degraded to a more toxic compound. In view of the report that aldrin may be readily oxidized to dieldrin (Bann 1956) samples were submitted to dieldrin analysis. These analyses showed that such a conversion had occurred in the treated soil. Furthermore the percentage conversion was greater for the soils having 20 pounds per acre applied to them than for those having 200 pounds. This showed close correspondence with the discrepancy between chemical assay data and bioassay results which were also greater in the soils having 20 pounds applied per acre. Additional confirmation came from later assays on the same soils. In these assays carried out 6 months after the first ones, there was greater divergence between bioassay and chemical data, which indicated that yet more aldrin had become converted to dieldrin.

CONCLUSIONS AND SUMMARY.—These preliminary studies indicate the value of bioassay as a tool in the study of soil insecticides. Chemical assays of soil insecticides cannot be complete in themselves, but require the additional information about residual toxicity which is provided by bioassay. Toxicity can be demonstrated by bioassay of the same extracts that are chemically analyzed. Direct bioassay of insecticide in soil can provide information on persistence and degradation of the insecticide. The degradation of the insecticide in the soil may result in a lower toxicity (*e.g.*, lindane) or increased toxicity (*e.g.*, aldrin).

Direct bioassay in soil also shows whether the insecticide becomes adsorbed in such a way as to reduce its toxicity. Such adsorption effects occurred with both aldrin and lindane and were correlated with the percentage of organic matter present in the soil. There was no indication of the exact nature of the interaction between the organic matter and the insecticide, but it appeared most likely to be an adsorption effect, since the insecticide could still be extracted in a toxic form. The reduction of toxicity with both aldrin and lindane was of the same order; it is probable that the adsorption is of a similar type for both insecticides. The direct soil method of bioassay provides quantitative data of the relative degree of insecticide adsorption by different soils. This may have extensive practical application since it is obvious that the soil type must be taken into account before the insecticide is applied in the field.

ACKNOWLEDGEMENTS.—Thanks are due to Dr. S. A. Wilde for assistance in estimation of organic matter content of soils, and to Dr. F. D. Hole for help in obtaining a suitable wide range of soils. Dr. J. F. Crow gave valuable assistance in providing a suitable strain of *Drosophila* and advice in culture methods.

REFERENCES CITED

Bann, J. M., T. J. DeCino, N. W. Earle and Y. P. Sun. 1956. The fate of aldrin and dieldrin in the animal body. Jour. Agric. and Food Chem. 4(11): 937.

Busvine, G. R., and S. Barnes. 1947. Observations on mortality among insects exposed to dry insecticidal films. Bull. Ent. Res. 38(1): 81–90.

Chisholm, R. D., L. Koblitsky, J. E. Fahey, W. E. Westlake. 1950. DDT residues in soil. Jour. Econ. Ent. 43(6): 941–2.

Danish, A. A., and R. E. Lidov. 1950. Colorimetric method for estimating small amounts of aldrin. Analyt. Chem. 22: 702–6.

Fleming, W. E., L. W. Coles, and W. W. Maines. 1951. Biological assay of residues of DDT and chlordane in soil using *Macrocentrus ancylivorus* as a test insect. Jour. Econ. Ent. 44: 310–15.

Fleming, W. E., and W. W. Maines. 1953. Persistence of DDT in soils of the area infested by Japanese beetle. Jour. Econ. Ent. 46(3): 445–9.

Ginsburg, M. J. 1955. Accumulation of DDT in soils from spray practices. Agric. and Food Chem. 3(4): 322–5.

Ginsburg, J. M., and J. P. Reed. 1954. A survey of DDT accumulation in soils in relation to different crops. Jour. Econ. Ent. 47(3): 467–474.

Hartzell, A., and E. E. Storrs. 1950. Bioassay of insecticide spray residues in processed food. Contr. Boyce Thompson Inst. 16: 47–53.

Healy, M. J. R. 1952. Abbott's correction. Ann. Appl. Biol. 39(2): 211.

Laug, E. P. 1946. A biological assay method for determining DDT. Jour. Pharmacol. 86: 324–331.

Lauge, W. H., and E. C. Carlson. 1955. Zonal dispersion of chemicals in soil following several tillage methods. Jour. Econ. Ent. 48(1): 61–67.

Lichtenstein, E. P. 1957. DDT accumulation in mid-western orchard and crop soils treated since 1945. Jour. Econ. Ent. 50(5): 545–7.

Lichtenstein, E. P., S. D. Beck and K. R. Schulz. 1956. Colorimetric determination of lindane in soils and crops. Jour. Agric. and Food Chem. 4(11): 936.

Lilly, J. H. 1955. Soil insects and their control. Ann. Rev. Ent. 1: 203–21.

Litchfield, J. R., and F. Wilcoxon. 1949. A simplified method of evaluating dose-effect experiments. Jour. Pharmacol. 96(2): 99–113.

Morrison, F. O. 1945. Comparing the toxicity of synthetic organic compounds. Proc. Ent. Soc. Ontario 76: 18–20.

Terriere, L. C., and D. W. Ingalsbe. 1953. Translocation and residual action of soil insecticides. Jour. Econ. Ent. 46(5): 751–3.

Wilde, S. A., and W. E. Petzer. 1940. Determination of organic matter in soils. Jour. Amer. Soc. Agron. 32(8).

Wylie, W. D. 1956. Determination of insecticide residues in soil by using *Drosophila*. Jour. Econ. Ent. 49(5): 638–40.

3

Copyright © 1972 by the American Chemical Society
Reprinted from *Jour. Agric. Food Chem.* **20**:1224-1226 (1972)

Adsorption–Desorption of Parathion as Affected by Soil Organic Matter

Sarina Saltzman, Lilian Kliger, and Bruno Yaron*

Influence of soil organic matter on the adsorption-desorption of parathion was studied by using tagged insecticide. It was found that the parathion adsorption by soils is dependent on the type of association between the organic and mineral colloids. In aqueous solutions the parathion has a greater affinity for organic than for mineral adsorptive surfaces. Parathion bonding is stronger on organic than on mineral surfaces.

The adsorption–desorption of pesticides by the active soil surfaces is one of the main processes controlling soil–pesticide interactions. Although the organic colloid fraction has been shown to be the most active soil component in affecting pesticide fixation in soils (Wolcott, 1970), it was also noted that it cannot always be used as a single factor to predict the adsorptive capacity of soils (Meggitt, 1970). The character and the interaction between the organic and mineral colloids of the soil are finally defining the nature of the available adsorptive surfaces.

Soil adsorption of the organophosphorus insecticide parathion was found to be related to the organic matter content and to the soil mineralogy (Saltzman and Yaron, 1971). Swoboda and Thomas (1968), studying the adsorption mechanism of parathion by leaching experiments, found that parathion was retained in soils, mainly as a water-insoluble organic constituent of the soil, by partitioning between soil organic matter and the liquid phase. In a recent study, Leenheer and Ahlrichs (1971) stated that parathion affinity for organic surfaces depends on the magnitude of the hydrophobic nature of these surfaces, rather than on the type of organic matter. They assumed parathion adsorption on organic surfaces to be a physical adsorption with formation of weak bonds between the hydrophobic portion of the adsorbent and adsorbed molecule.

Little information on the influence of organic matter on the desorption of pesticides is found in the literature. In the review of Wolcott (1970) concerning the retention of pesticides by organic materials in soils, it is mentioned that there is some evidence that similar treatments may result in complete release of pesticides from clays, but only in a partial desorption from high organic soils.

The aim of this work was to investigate the relative importance of mineral and organic surfaces in the parathion adsorption–desorption process in some semi-arid soils, characterized by low organic matter content and different mineralogy.

EXPERIMENTAL SECTION

Materials. Three mineral soils from various locations in Israel, having a relatively high organic matter content and different mineralogy, and a peat with 95% organic matter content, were selected for this experiment. The analytical characterization of the 20-cm upper layer is given in Table I.

The pesticide studied was the organophosphorus insecticide parathion (O,O-diethyl O-p-nitrophenyl phosphorothioate). Pure parathion (produced by Analabs, Inc.) and ^{14}C-labeled parathion (produced by Amersham Radiochemical Center) with a specific activity of 52 μCi/mg were used. The labeling was done in the alkyl chain.

Apparatus. For counting ^{14}C activity, a Packard 3003 Tricarb liquid scintillation spectrometer was used. The scintillation liquid consisted of 50 g of naphthalene, 7 g of PPO (2,5-diphenyloxazole), and 0.05 g of POPOP [2,2-p-phenylenebis-(5-phenyloxazole)], brought to 1 l. with dioxane. The purity of the material was checked periodically by gas chromatog-

Institute of Soils and Water, Agricultural Research Organization, Volcani Center, P.O. B.6, Bet Dagan, Israel.

Table I. Composition of the Soils Used in the Experiments

Type of soil (and origin)	Predominant clay	Clay, %	OM, % Natural soil	OM, % Oxidized soil	pH Natural soil	pH Oxidized soil
Dark rendzina (Bet Guvrin)	Montmorillonite	72	4.55	2.12	7.1	7.0
Mediterranean soil (Meron)	Mixed mineralogy	63	3.72	1.20	6.6	5.6
Terra rossa (Golan)	Kaolinite	64	4.88	1.95	7.1	5.1

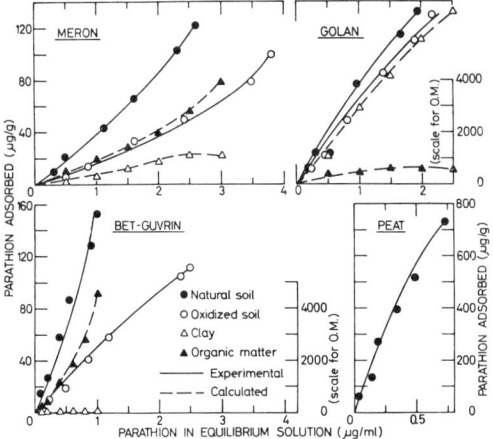

Figure 1. Parathion adsorption from aqueous solutions by soils, oxidized subsamples, organic, and mineral fractions

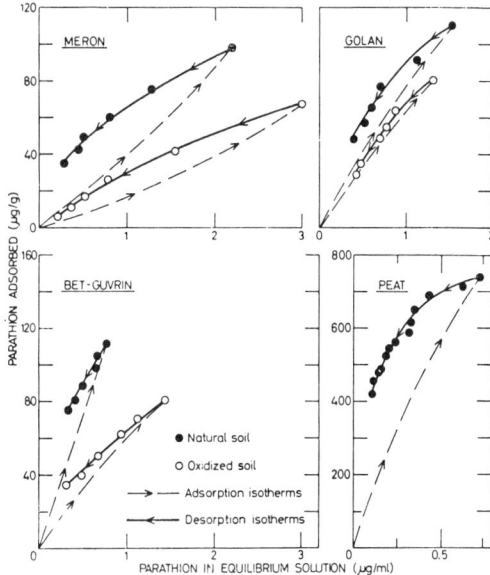

Figure 2. Parathion desorption by water from soils and oxidized subsamples

raphy; no decomposition products were detected during the experiments. An ultracentrifuge was used for sample separation at constant temperature.

Procedure. The destruction of organic matter in soils was obtained by hydrogen peroxide treatment according to the method recommended by Black (1965). For the Bet Guvrin soil, the treatment was interrupted at different phases in order to obtain different organic matter contents. Natural and treated soils were ground to pass a 60-mesh sieve, dried, and stored in a vacuum desiccator on P_2O_5. The organic matter content of the soils was determined by dry combustion, with the CO_2 being adsorbed in potassium hydroxide.

The adsorption equilibria were studied by batch equilibrium experiments using the same technique as described in a previous paper (Yaron and Saltzman, 1972).

Parathion desorption was studied as follows. Soils were equilibrated in weighed centrifugation bottles with an aqueous solution containing 7.2 ppm of parathion, as previously described, and the amounts of parathion adsorbed were determined. After being analyzed, the supernatant was discarded, the bottles were reweighed, and distilled water was added up to the initial weight. The soils were equilibrated for 1 hr, since a previous experiment showed that a longer contact time does not improve parathion extraction. Parathion in the supernatant was determined, and this operation was repeated five times.

RESULTS AND DISCUSSION

The adsorptive affinity for parathion of the experimental soils was initially very different. The distribution coefficient, Kd (μg adsorbed/g of adsorbent at an equilibrium concentration of 1 μg/ml) was 38.5 for Meron, 76.0 for Golan, 164.0 for Bet Guvrin, and 875.0 for the peat soil. The differences between the mineral soils cannot be explained by differences in the organic matter and clay content, which in Golan and Bet Guvrin soils, for instance, are rather similar. Therefore, it is reasonable to suppose that the specific interactions between organic and mineral colloids determine the nature of the adsorptive surfaces for each soil.

Following removal of organic matter by soil treatment with hydrogen peroxide, the adsorptive affinity of the mineral soils decreased (Figure 1). This decrease may be due mainly to the decrease in the organic matter content, and not to other soil modifications which may occur during organic matter oxidation. (The changes in soil reaction were of 2 pH units in Golan, of 1 pH unit in Meron, and of 0.1 pH unit in Bet Guvrin soil, in a pH range between 7.1 and 5.1. In a preliminary experiment, parathion adsorption by soils was not affected by changes in soil reaction in that pH range.)

The decrease in parathion adsorptive capacity of soils following oxidation shows that parathion has a relatively greater affinity for organic adsorptive surfaces than for mineral ones. Although the amounts of organic matter destroyed were similar for all soils, the decrease in parathion adsorption was different: 72% in Bet Guvrin; 60% in Meron; and 22% in Golan soil.

If one assumes that only the changes in organic matter con-

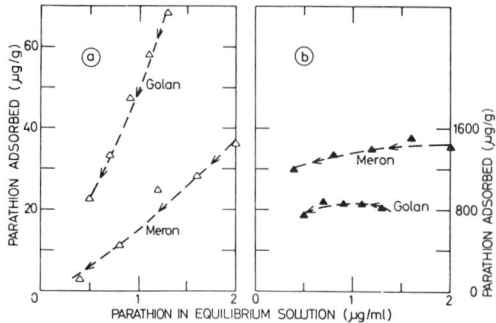

Figure 3. Calculated parathion desorption from the mineral (a) and organic (b) fractions of Golan and Meron soils

tent are responsible for the changes in parathion adsorption, then it is possible to estimate adsorption curves of the organic matter for each soil. The adsorptive capacity for organic matter and clay was estimated as follows. The differences in adsorption between natural and treated soil at several concentrations of the equilibrium solution were calculated. The adsorption per unit weight organic matter was calculated as the ratio between each of these values and the amount of organic matter lost by soil oxidation. Adsorption on the mineral soil fraction was obtained by subtracting the calculated adsorption on organic material from the adsorption on the natural soil.

From the calculated adsorption curves (Figure 1) a quantitative evaluation of the adsorptive capacity of organic and mineral surfaces is also possible. The distribution coefficients for the Meron and Golan mineral fractions are 8 and 60, respectively. The adsorptive capacity of the organic matter expressed as distribution coefficients is 450 for Golan, 950 for Meron, and 4500 for Bet Guvrin soil. The peat soil has a distribution coefficient of 875. The values found by Leenheer and Ahlrichs (1971) for parathion adsorption on organic materials separated from soils are in the same range.

The parathion desorption by water was also dependent on the nature of the soil adsorptive surfaces. After five consecutive desorptions, all the natural soils still contained more parathion than the treated soils. The parathion content of the natural soils was greater (with 8% in Golan, 24% in Meron, and 31% in Bet Guvrin soil) than in the oxidized subsamples. After the same number of desorption, the peat soil retained 2–3 times more parathion than the natural soils. The stronger retention of parathion by natural soils and peat shows that parathion-organic complexes are stronger than parathion-mineral ones.

The desorption isotherms of the natural soils (Figure 2) do not overlap on the adsorption isotherms, showing that it is a "hysteresis" in the parathion adsorption–desorption process. Smaller differences between adsorption and desorption isotherms were found for soils treated with hydrogen peroxide.

With the same assumptions as for parathion adsorption, desorption of parathion from organic and mineral surfaces may be computed in a similar way. From the desorption curves obtained (Figure 3), a net difference between parathion release by organic and mineral colloids may be noted. The slope of the desorption curves of the mineral fraction is rather steep (especially in Golan soil), showing that adsorption is easily and totally reversible. For the organic matter, in the range of concentrations studied, only very small amounts of parathion seem to be released.

In conclusion, it may be emphasized that in aqueous solutions the parathion has a greater affinity for organic than for mineral adsorptive surfaces. The parathion adsorption by soils is dependent on the type of association between organic and mineral colloids, which determines the nature and the magnitude of the adsorptive surfaces. Consequently, oxidation of soil organic matter may have a greater or lesser effect on adsorption. Parathion bonding on organic surfaces is stronger than on mineral surfaces.

LITERATURE CITED

Black, C. A., Methods of Soil Analysis, American Society of Agronomy Monograph 9, (1965).
Leenheer, J. A., Ahlrichs, J. L., *Proc. Soil Sci. Soc. Amer.* **35**, 700 (1971).
Meggitt, W. F., "Pesticides in the Soil: Ecology, Degradation, and Movement," Michigan State University, East Lansing, Mich., 1970, pp 139–141.
Saltzman, S., Yaron, B., "Fate of Pesticides in Environment," London and Breach Science Publishers, New York, N. Y., 1971, pp 88–100.
Swoboda, A. R., Thomas, G. W., J. Agr. Food Chem. **16**, 923 (1968).
Yaron, B., Saltzman, S., *Proc. Soil Sci. Soc. Amer.* **36**, in press (1972).
Wolcott, A. R., "Pesticides in the Soil: Ecology, Degradation, and Movement," Michigan State University, East Lansing, Mich., 1970, pp 128–138.

Received for review May 7, 1972. Accepted July 10, 1972. Contribution from the Agricultural Research Organization, Volcani Center, Bet Dagan, Israel. 1972 Series, No. 1124-E. This research has been financed in part by a grant made by the United States Department of Agriculture, Agricultural Research Service, authorized by Public Law 480.

4

Copyright © 1977 by Blackwell Scientific Publications, Ltd
Reprinted from Weed Res. 17:41-48 (1977)

Adsorption et désorption de la terbutryne par une montmorillonite-Ca et des acides humiques seuls ou en mélanges

P. GAILLARDON,* R. CALVET† ET M. TERCE† *Laboratoire de Malherbologie, INRA—BV 1540, Dijon, France et † Laboratoire de Sciences du Sol, INRA—Route de Saint-Cyr, Versailles, France

Received 3 March 1976

Résumé: Summary: Zusammenfassung

L'effet du pH sur l'adsorption de la terbutryne par des acides humiques, une montmorillonite-Ca ou leurs mélanges montre une certaine similitude de comportement entre ces derniers et l'argile seule. Au voisinage de la neutralité, seuls les acides humiques adsorbent la terbutryne. En milieu acide, les isothermes d'adsorption de la terbutryne par les acides humiques et la montmorillonite sont respectivement de type L et S et traduisent une affinité adsorbat—adsorbant différente; les isothermes d'adsorption correspondant aux mélanges ont une forme différente et font apparaître un effet de synergie, en particulier dans le cas des mélanges pauvres en acides humiques. La désorption s'accompagne d'un phénomène d'hystérésis lié à la présence des acides humiques. Les propriétés particulières des associations argiles—matières organiques sont susceptibles de mieux expliquer l'effet des sols sur l'activité des herbicides.

Adsorption and desorption of terbutryne by a Ca-montmorillonite and humic acids or mixture of both
The effect of pH on the adsorption of terbutryne by humic acids and a Ca-montmorillonite or mixtures of the two shows a certain similarity with that of clay alone. Around the area of neutral pH only humic acids adsorb terbutryne. In an acid environment the isotherms of terbutryne adsorption by humic acids and montmorillonite are types L and S respectively and reflect a different relationship between the adsorbant and the substance adsorbed; in the case of mixtures, isotherms of adsorption are different and reveal a synergistic effect which suggests interaction between the colloids. Desorption is generally accompanied by a hysteresis phenomenon associated with the presence of humic acids. The special relationships of clay with o.m. are likely to provide a better understanding of herbicide activity in soil.

Adsorption und Desorption von Terbutryn durch Ca-Montmorillonit und Huminsäuren—allein, oder in Mischungen
Der Einfluss des pH auf die Adsorption von Terbutryn durch Huminsäuren, Ca-Montmorillonit und Mischungen aus beiden, zeigt eine gewisse Ähnlichkeit zwischen den Mischungen und dem reinen Tonmineral. Im Bereich des Neutralpunktes wurde Terbutryn nur von den Huminsäuren sorbiert. Im sauren Milieu entsprechen die Adsorptionsisothermen für die Huminsäuren und dem Montmorillonit dem L- bzw. dem S-Typ und zeigen damit eine unterschiedliche Adsorbat-Adsorbens-Affinität. Die Adsorptionsisothermen der Mischungen haben eine unterschiedliche Form und zeigen einen synergistischen Effekt an. Das gilt besonders für die Mischungen mit einem geringen Huminsäureanteil. Die Desorption ist mit einem Hystereseiseffekt verbunden, der auf die Huminsäuren zurückzuführen ist. Die speziellen Eigenschaften von Organo-Tonkombinationen könnten zu einem besseren Verständnis der Beziehungen zwischen Böden und der Herbizidaktivität führen.

Introduction

Les colloïdes argileux et humiques peuvent adsorber les triazines. Dans le cas de la montmorillonite l'effet du pH sur l'adsorption permet de penser que la forme protonée des molécules peut se fixer par échange d'ions (Weber, 1966, 1970; Vallet, Calvet & Chaussidon, 1973). Avec des matières organiques ou des acides humiques les observations sont semblables et un phénomène d'hystérésis accompagne souvent la désorption (Weber, Weed & Ward, 1969; Gilmour & Coleman, 1971; Moyer, Kercher & Hance, 1972; Gaillardon, 1975).

Dans les sols les colloïdes minéraux et organiques sont généralement associés; le pouvoir d'adsorption d'un sol peut ainsi se trouver en partie déterminé par les propriétés de ces associations. C'est pourquoi l'objet de cette étude concerne l'adsorption et la désorption d'une triazine herbicide: la terbutryne (2 méthylthio-4 éthylamino-6 t-butylamino-1,3,5 triazine) par divers mélanges d'une montmorillonite-Ca et d'acides humiques ou par ces constituants seuls.

Matériels et méthodes

La préparation des acides humiques et les techniques d'étude de l'adsorption et de la désorption ont été décrites dans une publication antérieure (Gaillardon,

1975). Les mélanges argile—acides humiques sont réalisés à partir d'une solution d'acides humiques bi-ioniques Na-H (7 mg/cm³) à pH 6,5 et d'une suspension de montmorillonite-Ca (7 mg/cm³) d'origine grecque fournie par le Laboratoire de Sciences du Sol de l'INRA à Versailles. Ces mélanges de proportions variables renferment au total 70 mg d'adsorbants pour un volume de 10 cm³; on les prépare 24 h avant l'addition de 20 cm³ d'une solution de terbutryne de concentration connue. Le milieu contient $CaCl_2$ 10^{-2} M apporté avec l'herbicide pour assurer la floculation des colloïdes. Après 24 h de contact le dosage par chromatographie en phase gazeuse de la terbutryne restant en solution permet la détermination des quantités adsorbées. La désorption est effectuée à l'aide d'une solution qui conserve le pH et la salinité du milieu, mais sans herbicide. Le premier équilibre de désorption est obtenu en retirant 27 cm³ de surnageant puis en ajoutant 30 cm³ de la solution de désorption; pour les équilibres suivants le remplacement porte sur 30 cm³. Chaque lavage dure 30 mn, les quantités désorbées pour des temps plus longs restant les mêmes.

L'effet du pH sur l'adsorption de la terbutryne par différents mélanges contenant 0–25–50–75–100% de montmorillonite est mis en évidence en réalisant l'adsorption à différents niveaux d'acidité obtenus par addition d'acide chlorhydrique. La concentration initiale de la terbutryne est 73×10^{-6} M.

Nous nous sommes assurés qu'en condition acide (pH 3–3,2) les propriétés adsorbantes des deux composants des mélanges sont indépendantes de leur concentration dans le milieu. Pour ce faire la solution d'acides humiques est diluée dans les proportions 3/4–1/2–1/4 et pour chacune de ces concentrations on détermine l'isotherme d'adsorption de la terbutryne à partir d'une concentration initiale 71×10^{-6} M. On fait de même avec la suspension de montmorillonite diluée 2 et 4 fois pour une concentration initiale 63×10^{-6} M d'herbicide.

Pour différents mélanges contenant 0–25–50–75–100% de montmorillonite et à partir d'une concentration initiale 69×10^{-6} M de terbutryne on détermine en milieu acide (pH 3–3,2) les isothermes d'adsorption et de désorption. La même expérience est répétée en milieu faiblement acide (pH 5,6–6) à partir d'une concentration initiale 73×10^{-6} M d'herbicide.

Les résultats des expériences précédentes nous ont incités à considérer les propriétés adsorbantes des mélanges à faibles proportions d'acides humiques. Pour des mélanges contenant 0–5–10–20% d'acides

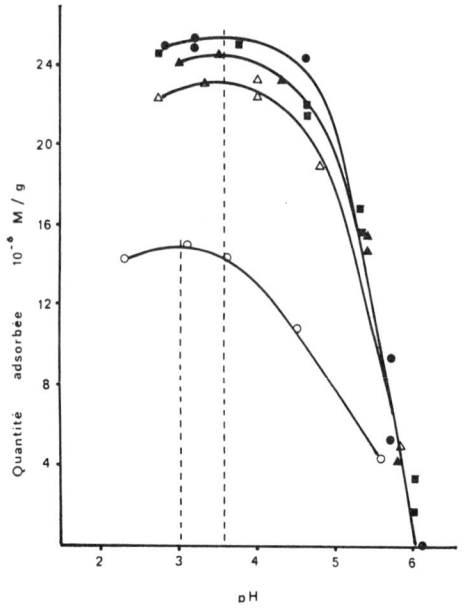

Fig. 1 Effet du pH sur l'adsorption de la terbutryne (concentration initiale élevée) par des acides humiques (○), une montmorillonite-Ca (●) et leurs mélanges contenant 25% (△), 50% (▲), 75% (■) d'argile.

humiques, et à partir d'une concentration initiale 68×10^{-6} M de terbutryne, nous avons déterminé les isothermes d'adsorption et de désorption en milieu acide (pH 3–3,2).

Le rôle éventuel de l'état d'agrégation des particules de montmorillonite est considéré en comparant l'adsorption et la désorption en milieu acide (pH 3,1) de la terbutryne par l'argile en suspension ou préalablement séchée et chauffée à 250° C. Pour chaque condition la distribution granulométrique des particules est établie par sédimentation. Les dosages de la terbutryne sont effectués dans ce cas par spectrométrie UV (225 nm).

Toutes les expériences sont conduites avec deux répétitions et réalisées à la température du laboratoire, à savoir $20 \pm 2°$ C.

Résultats

(1) *Effet de l'acidité*

Conformément aux observations d'autres auteurs, l'effet de l'augmentation de l'acidité est un accroissement des quantités adsorbées qui peuvent ensuite diminuer (Fig. 1) au-delà d'une certaine valeur du pH Lorsque le diminue le maximum d'adsorption est atteint plus vite avec la montmorillonite qu'avec les acides humiques, il est plus bas mais mieux défini pour ces derniers. Les mélanges ont un comportement proche de celui de l'argile seule. Entre pH 6 et pH 5 ils montrent le même accroissement rapide de l'adsorption que la montmorillonite. Leurs maxima d'adsorption, assez bien définis, se situent à des valeurs du pH proches et difficiles à distinguer de celle correspondant à l'argile seule.

(2) *Isothermes d'adsorption*

En milieu acide (pH 3–3,2), les isothermes d'adsorption de la terbutryne par différentes quantités d'acides humiques se superposent parfaitement (Fig. 2). Dans les limites de nos conditions expérimentales on peut admettre que les propriétés adsorbantes des acides humiques sont indépendantes de leur concentration. Pour la montmorillonite les isothermes d'adsorption (Fig. 3) permettent la même conclusion, encore que l'adsorption augmente légèrement avec la dilution de l'adsorbant. Dans certaines conditions physico-chimiques, notamment en milieu très acide, les propriétés adsorbantes de l'argile peuvent en effet dépendre de sa concentration (Vallet *et al.*, 1973).

Selon la nature de l'adsorbant, la forme des isothermes d'adsorption est nettement différente. D'après la classification de Giles *et al.* (1960) on obtient des isothermes de type L avec les acides humiques (Fig. 2) et de type S avec la montmorillonite-Ca (Fig. 3). En conséquence les acides humiques adsorbent plus de terbutryne aux faibles concentrations en herbicide mais beaucoup moins aux fortes concentrations. On remarque que les isothermes d'adsorption d'autres triazines par une montmorillonite-Na du Wyoming (Weber, 1970) sont de type L, donc différentes de celles que nous observons dans le cas de la terbutryne et de la montmorillonite-Ca grecque.

Les Figures 4 et 5 représentent les isothermes d'adsorption de la terbutryne par les acides humiques, la montmorillonite et leurs mélanges en milieu acide (pH 3–3,2) et faiblement acide (pH 5,6–6).

En milieu acide les isothermes d'adsorption correspondant aux mélanges sont différentes de celles relatives aux constituants seuls et l'adsorption n'est pas une fonction simple de la composition des mélanges. La Figure 6 donne les variations de $\log \frac{x}{m}$ en fonction de $\log C_{\hat{e}}$ ($\frac{x}{m}$ étant la quantité adsorbée par unité de masse d'adsorbant et $C_{\hat{e}}$ la concentration de l'herbicide dans la solution à l'équilibre). Les droites obtenues montrent que les isothermes d'adsorption peuvent être décrites numériquement par la relation de Freundlich: $\frac{x}{m} = K C_{\hat{e}}^{1/n}$ dans laquelle K et $1/n$ sont des constantes dont les valeurs sont données dans la légende de la Figure 6. A partir de ces droites on peut représenter les variations de la quantité adsorbée en fonction de la composition des mélanges pour des valeurs fixées de la concentration à l'équilibre (Fig. 7). Ces variations sont obtenues avec une assez mauvaise précision en raison des erreurs expérimentales et de l'approximation qui résulte de l'ajustement aux droites de régression. On peut cependant faire deux observations: 1—les quantités adsorbées par les mélanges ne sont pas égales à la somme des quantités adsorbées par les colloïdes seuls; elles lui sont supérieures; 2—le mélange contenant 25% d'acides humiques tend à se caractériser par une adsorption maximum. Les isothermes d'adsorption non parallèles de la Figure 6 montrent d'ailleurs que cette observation dépend de la concentration de l'herbicide et peut se manifester dans un domaine de concentrations pour lequel les quantités adsorbées par les colloïdes isolés sont voisines; dans le cas de la terbutryne ce domaine se situe plutôt aux basses concentrations.

Les différences observées dans la forme des isothermes d'adsorption soulignent l'importance de la concentration de l'herbicide dans la solution: $C_{\hat{e}}$ détermine la quantité adsorbée par chaque colloïde et par suite sa participation aux propriétés adsorbantes des mélanges. La Figure 7 ne permet pas de situer avec précision la composition du mélange pour lequel l'effet de synergie est maximum mais, dans la limite des valeurs de $C_{\hat{e}}$ considérées, il est évident que ce mélange est plus riche en argile qu'en acides humiques et pourrait se trouver au voisinage du mélange contenant 25% d'acides humiques. La montmorillonite et les acides humiques ne jouent donc pas le même rôle; on note en particulier l'effet de synergie important que ces derniers entraînent lorsqu'ils sont ajoutés en faible quantité à l'argile. L'adsorption de la terbutryne par les mélanges à très faibles proportions d'acides humiques (Fig. 8) ne fournit pas d'informations nouvelles mais confirme ces remarques dans la mesure où le comportement des mélanges, comparable à celui de la montmorillonite seule, s'en distingue par une augmentation des quantités adsorbées aux basses concentrations de terbutryne, dès qu'ils contiennent 10% d'acides humiques.

En milieu faiblement acide les quantités adsorbées

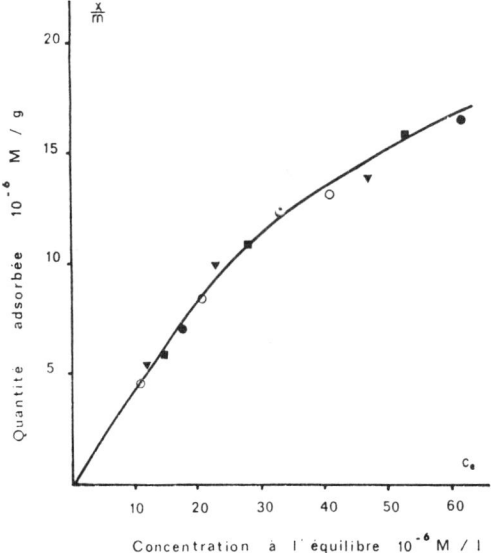

Fig. 2 Isothermes d'adsorption de la terbutryne, en milieu acide (pH 3–3,2), par différentes concentrations d'acides humiques: 70 mg (○), 52,5 mg (▼), 35 mg (■), 17,5 mg (●) pour 10 ml.

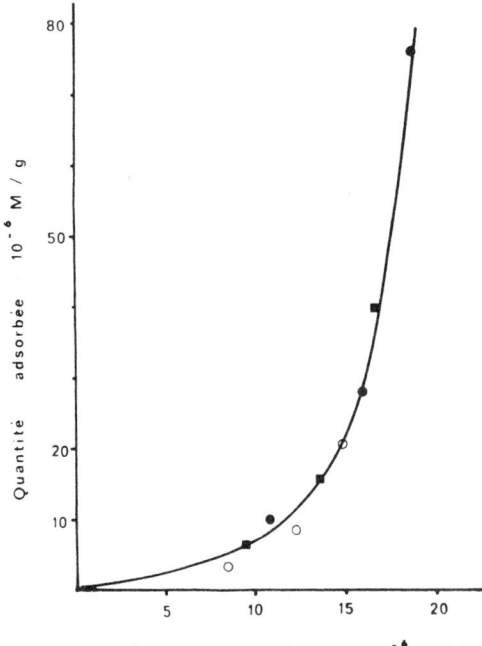

Fig. 3 Isothermes d'adsorption de la terbutryne, en milieu acide (pH 3–3,2), par différentes concentrations de montmorillonite Ca: 70 mg (○), 35 mg (■), 17,5 mg (●) pour 10 ml.

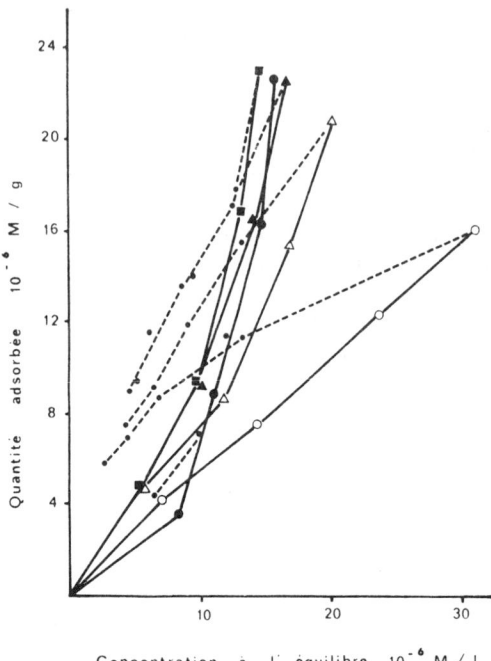

Fig. 4 Isothermes d'adsorption (traits pleins) et de désorption (pointillés) de la terbutryne, en milieu acide (pH 3–3,2), par des acides humiques (○) une montmorillonite Ca (●) et leurs mélanges contenant 25% (△), 50% (▲), 75% (■) d'argile.

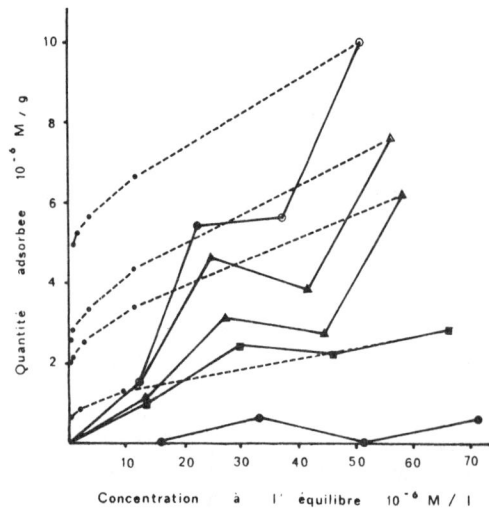

Fig. 5 Isothermes d'adsorption (traits pleins) et de désorption (pointillés) de la terbutryne, en milieu faiblement acide (pH 5,6–6), par des acides humiques (○), une montmorillonite Ca (●) et leurs mélanges contenant 25% (△), 50% (▲), 75% (■) d'argile.

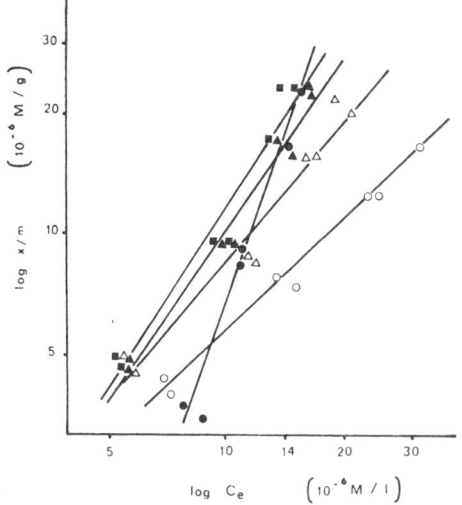

Fig. 6 Représentation linéaire, selon l'expression de Freundlich: $\log \frac{x}{m} = \log K + \frac{1}{n} \log C\hat{e}$, des isothermes d'adsorption de la terbutryne, en milieu acide (pH 3–3,2), par des acides humiques (○){$K = 0,71$; $\frac{1}{n} = 0,90 \pm 0,03$}, une montmorillonite Ca (●){$K = 0,011$; $\frac{1}{n} = 2,74 \pm 0,19$} et leurs mélanges contenant 25% (△){$K = 0,6$; $\frac{1}{n} = 1,14 \pm 0,06$}, 50% (▲) {$K = 0,4$; $\frac{1}{n} = 1,41 \pm 0,08$}, 75% (■){$K = 0,38$; $\frac{1}{n} = 1,51 \pm 0,09$} d'argile.

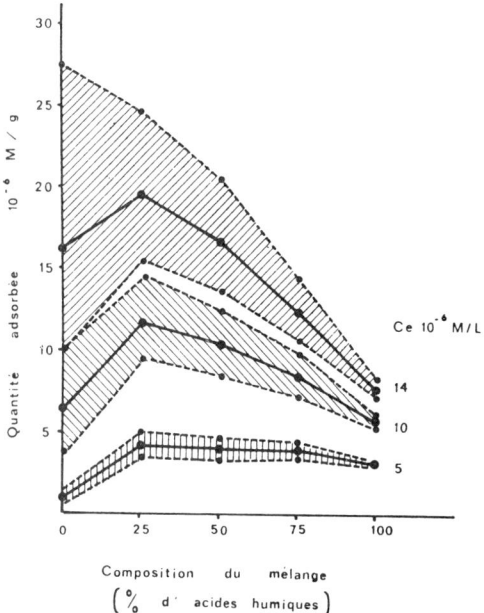

Fig. 7 Adsorption de la terbutryne à pH acide, en fonction de la composition des mélanges, pour différentes valeurs de la concentration à l'équilibre. Les quantités adsorbées sont calculées à partir des équations de régression de la représentation de Freundlich et les zones hachurées matérialisent les intervalles de confiance (t 10%) calculés à partir des mêmes données.

sont petites et leur détermination présente une assez grande variabilité qui peut expliquer la forme irrégulière des isothermes d'adsorption. Ces variations sont imputables à l'erreur relativement grande qui entache la mesure des quantités adsorbées lorsqu'elles sont faibles, mais peuvent également résulter de petites variations du pH qui influencent fortement l'adsorption (Fig. 1). En dépit d'une précision moins bonne il apparaît, vers pH 6, que l'argile n'adsorbe pas la terbutryne et que les mélanges fixent une quantité croissant avec la teneur en acides humiques.

(3) *Isothermes de désorption*

Quelle que soit l'acidité du milieu, les isothermes de désorption (Fig. 4 et 5) indiquent l'existence d'une hystérésis lorsque le mélange contient des acides humiques. Ce phénomène est plus prononcé en milieu faiblement acide, ce qui est conforme à nos observations antérieures (Gaillardon, 1975). L'hystérésis peut être exprimée au moyen du rapport des pentes des isothermes de désorption et d'adsorption en coordonnées logarithmiques, c'est à dire le rapport des paramètres 1/n correspondants. Ainsi, en milieu acide, les faibles valeurs de ce rapport (0,46 pour les acides humiques; 0,58–0,49–0,47 pour les

mélanges contenant 25–50–75% d'argile) traduisent un effet d'hystérésis bien marqué.

(4) *Effet de l'état d'agrégation des particules d'argile*

Les modifications de l'état d'agrégation de la montmorillonite chauffée à 250°C (Tableau 1) se traduisent par une diminution importante des particules inférieures à 2 μm au profit des particules plus grosses des classes 10 à 20 μm et > 20 μm. Les paramètres de Freundlich, calculés à partir des isothermes d'adsorption et de désorption de la terbutryne par la montmorillonite-Ca en suspension ou préalablement chauffée à 250°C (Tableau 1), permettent de penser que l'état d'agrégation des particules de montmorillonite n'a pas d'effet notable sur ses propriétés adsorbantes.

Discussion

Comme toutes les triazines, la terbutryne peut former des cations par protonation (pKa = 4,1).

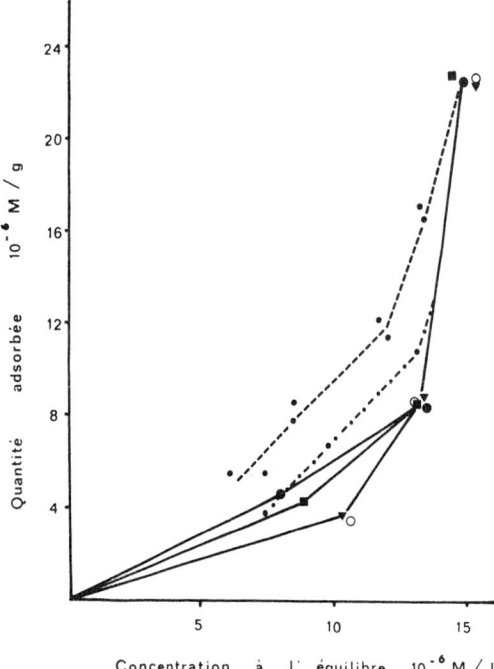

Fig. 8 Isothermes d'adsorption (traits pleins) de la terbutryne, en milieu acide (pH 3-3,2), par une montmorillonite Ca (○) et par des mélanges à faibles proportions d'acides humiques: 5% (▼), 10% (■), 20% (●). Isothermes de désorption de la terbutryne adsorbée sur la montmorillonite ou le mélange à 5% d'acides humiques (– · – · –) et sur les mélanges à 10% et 20% (– – – – –).

Tableau 1 Valeurs des paramètres de Freundlich correspondant aux isothermes d'adsorption et de désorption de la terbutryne par la montmorillonite-Ca en suspension ou préalablement séchée et chauffée à 250° C en milieu acide

	Suspension	Argile chauffée à 250° C
Diamètre des particules		
> 20 μm	0%	35%
10 à 20 μm	0	26,7
2 à 10 μm	4,7	15,6
< 2 μm	95,3	22,7
Adsorption		
K	36×10^{-3}	39×10^{-3}
1/n	2,12	2,15
Désorption		
K		21×10^{-3}
1/n		2,36

L'augmentation de la quantité adsorbée quand le pH diminue peut être attribuée à l'accroissement du nombre de molécules ionisées. Dans le cas de la montmorillonite, le maximum d'adsorption est atteint pour une valeur du pH plus élevée qu'avec les acides humiques. De semblables différences dans le comportement de deux argiles ont été décrites et attribuées à des acidités de surface différentes (Vallet et al., 1973). Celles-ci déterminent l'ionisation des molécules qui ne peut être prévue par la valeur du pH mesuré sur la suspension. Dans ces conditions, les résultats obtenus pourraient être dus à la plus grande acidité de surface de la montmorillonite. Le comportement des mélanges peut être relié à des phénomènes d'acidité de surface et sans préjuger d'autres interactions possibles, on peut penser que les sites d'adsorption des acides humiques bénéficient au moins partiellement de l'acidité de surface des particules d'argile voisines. Entre pH 5 et pH 4, la forme des courbes de la Figure 1 laisse supposer que l'acidité de surface de la montmorillonite pourrait être légèrement plus grande que celle des mélanges.

L'examen des isothermes d'adsorption montre qu'en milieu acide les interactions des molécules de terbutryne avec les acides humiques sont différentes de celles existant avec la montmorillonite. Une isotherme de type L indique une sélectivité d'adsorption des molécules organiques par rapport au solvant. Une isotherme de type S correspond au contraire à une compétition entre l'adsorbat et le solvant. Il en découle deux conclusions: (1—les forces d'adsorption sont plus grandes avec les acides humiques; 2—les mécanismes par lesquels les molécules de terbutryne sont retenues sur l'argile peuvent aussi être responsables de la fixation de certaines molécules d'eau.

Le mécanisme proposé par d'autres auteurs (Bailey, White & Rothberg, 1968; Terce & Calvet, 1975) peut être repris pour expliquer la fixation de la terbutryne par la montmorillonite calcique:

$$T ---H\diagdown \atop H\diagup O --- (cation)\,(argile) \rightleftharpoons {TH^+ \diagdown \atop H \diagup} O^- --- (cation)\,(argile)$$

T = molécule de terbutryne neutre (liaison hydrogène)

TH^+ = molécule de terbutryne protonée (liaison ionique)

Cette hypothèse est conforme aux conceptions sur la protonation des molécules organiques à la surface des argiles (Mortland, 1968; Cruz, White & Russel, 1968; Chaussidon & Calvet, 1974). Elle montre que les molécules d'eau, tout en étant nécessaires à l'adsorption, peuvent entrer en compétition avec les molécules organiques.

Les interactions proposées par Sullivan & Felbeck

(1968) entre l'atrazine et des acides humiques peuvent expliquer l'existence de forces d'adsorption plus fortes. Selon ces auteurs la molécule de triazine serait retenue sur les acides humiques par deux liaisons: une liaison hydrogène et une liaison ionique, au lieu d'une seule dans le cas de la montmorillonite.

Nos observations permettent de penser qu'en milieu faiblement acide (pH 5,6–6) les propriétés des mélanges sont vraisemblablement déterminées par les acides humiques seuls. En milieu acide (pH 3–3,2) les mélanges manifestent un effet de synergie dans l'adsorption de la terbutryne. Des observations analogues ont été faites avec des associations acides fulviques—montmorillonite; les travaux de Khan (1973a; 1974) ont en effet montré que les acides fulviques peuvent favoriser l'adsorption du diquat, du paraquat et du 2,4-D par l'argile. En milieu neutre (pH 7,2–7,9) Hance (1969) observe au contraire un antagonisme entre une montmorillonite—Ca du Wyoming et des acides humiques-Ca vis-à-vis du diuron et de l'atrazine; à ce pH chacun des deux colloïdes est capable d'adsorber ces herbicides, mais leurs mélanges en retiennent une quantité moindre; il conclut que les acides humiques peuvent masquer des sites d'adsorption sur l'argile.

S'il est possible de proposer des mécanismes d'adsorption pour les acides humiques ou l'argile seuls, il est difficile d'en proposer pour les mélanges. Une des principales raisons en est que la nature et la structure des associations acides humiques—argile restent très mal connues. On peut distinguer deux situations qui mettent en jeu des phénomènes différents et éventuellement simultanés. Lorsque le pH est suffisamment bas (inférieur à 3,5) la plupart des molécules de terbutryne sont protonées et la synergie observée à propos de l'adsorption de cet herbicide ne peut résulter que de modifications des propriétés de l'un des colloïdes ou des deux. Pour des valeurs plus élevées du pH, il peut s'y ajouter des phénomènes d'acidité de surface qui modifient l'adsorption par l'intermédiaire du nombre de molécules protonées. Plusieurs hypothèses peuvent être avanéces:

(1) *Action des acides humiques sur l'argile*

Dans le milieu faiblement acide contenant $CaCl_2$ 10^{-2} M les acides humiques fixent du calcium et libèrent des protons qui ne modifient pas sensiblement les propriétés de l'argile puisque la terbutryne n,est pratiquement pas adsorbée par le minéral; le pH est encore probablement trop élevé. Dès que ce dernier diminue, certains protons des acides humiques peuvent s'échanger avec le calcium de l'argile et entraîner un accroissement d'adsorption sur celle-ci, soit directement en augmentant l'acidité de surface soit indirectement en accélérant l'hydrolyse du réseau cristallin. Les acides humiques peuvent également agir en tant qu'agent complexant de l'aluminium, favorisant ainsi son extraction du réseau. Compte-tenu du rôle de l'aluminium dans les phénomènes d'adsorption (Terce & Calvet, 1975) toute action chimique ou physico-chimique modifiant son état et sa répartition sur les surfaces peut influencer la fixation de la terbutryne.

(2) *Action de l'argile sur les acides humiques*

L'échange H-Ca et Na-Ca, qui se produit certainement, modifie la saturation ionique des acides humiques dont la configuration moléculaire peut alors changer (Ling Ong & Bisque, 1968). Ces changements peuvent avoir lieu dans les mélanges en affectant une partie de la matière organique et être ainsi à l'origine de variations des propriétés d'adsorption.

Un autre fait remarquable est l'existence d'un phénomène d'hystérésis apparemment lié à la présence d'acides humiques. Il s'agit probablement d'une conséquence, en partie du moins, de la nature et de la force des liaisons responsables de la fixation de la terbutryne sur les acides humiques (isotherme de type L). Etudiant la cinétique d'adsorption du piclorame et du 2,4-D Khan (1973b) a mis en évidence l'existence d'une diffusion intraparticulaire: des molécules pourraient donc être fixées à l'intérieur de l'arrangement moléculaire des acides humiques. Ceci n'est pas exclu pour la terbutryne et serait un élément d'explication de l'hystérésis, car il se peut que les molécules ainsi "piégées" soient difficiles à désorber. Dans le cas de la montmorillonite seule l'état d'agrégation des particules ne semble pas intervenir (Tableau 1); mais pour les mélanges l'état structural, mal connu, pourrait jouer un rôle dans l'hystérésis observée au cours de la désorption.

Conclusion

En milieu faiblement acide les acides humiques semblent seuls capables d'adsorber la terbutryne; mais de faibles diminutions du pH augmentent rapidement les quantités adsorbées par l'argile et les mélanges dont le comportement est alors semblable.

En milieu acide l'adsorption de la terbutryne par les mélanges fait apparaître une synergie très nette. Les acides humiques et les mélanges présentent au cours de la désorption une hystérésis due vraisemblablement aux acides humiques; la teneur en acides humiques doit toutefois être supérieure à 20% pour que cette propriété apparaisse nettement.

Les isothermes d'adsorption de la terbutryne par les acides humiques et la montmorillonite, en milieu acide, sont respectivement de type L et S; elles traduisent une affinité adsorbat—adsorbant différente. Par suite la contribution de chacun des colloïdes aux propriétés adsorbantes des mélanges varie avec la concentration de l'herbicide dans la solution: le rôle des acides humiques est plus important aux faibles concentrations. Les mélanges contenant moins d'acides humiques que d'argile présentent un intérêt particulier car la synergie qu'ils manifestent est grande. Ces observations suggèrent l'existence d'interactions entre les acides humiques et la montmorillonite. Les phénomènes de synergie et d'hystérésis que nous avons observés se manifestent davantage dans le cas de mélanges bien pourvus en acides humiques. Dans les sols agricoles, le rapport teneur en matières organiques/teneur en argiles est généralement faible; mais, s'il s'accompagne d'une distribution hétérogène des colloïdes, on peut avoir localement des associations argiles—matières organiques assez riches en ces dernières pour présenter des propriétés analogues à celles des mélanges montmorillonite—acides humiques étudiés. Des paramètres isolés tels que le pourcentage de matières organiques, le pourcentage d'argiles ou le pH peuvent donc décrire imparfaitement l'effet du sol sur l'activité d'un herbicide dans la mesure où cet effet est lié aux propriétés particulières des associations argiles—matières organiques différentes de celles des colloïdes seuls. Les interactions observées au laboratoire entre des colloïdes argileux et humiques pourront, dans le sol, dépendre de nombreux facteurs (nature des colloïdes, environnement ionique...) donc présenter une assez grande variabilité. Cela pourrait expliquer les corrélations parfois médiocres obtenues entre l'activité des herbicides et la teneur en argile ou en matières organiques des sols, bien que cette dernière apparaisse généralement déterminante.

Remerciements

Nous remercions Mlle E. Rougetet et Mme A.M. Tabareau pour la collaboration technique qu'elles nous ont apportée. Cette étude a été réalisée dans le cadre du groupe de travail «Pesticides—Sols» de l'INRA.

Références

BAILEY G.W., WHITE J.L. & ROTHBERG T. (1968) Adsorption of organic herbicides by montmorillonite: Rôle of pH and chemical character of the adsorbate. *Soil Sci. Soc. Amer. Proc.* **32**, 222–234.

CHAUSSIDON J. & CALVET R. (1974) Catalytic reactions on clay surfaces. *3rd International Congress of Pesticides Chemistry* (I.U.P.A.C.) Helsinki.

CRUZ M., WHITE J.L. & RUSSEL J.D. (1968) Montmorillonite s-triazine interactions. *Israël J. Chem.*, **6**, 315–323.

GAILLARDON P. (1975) Etude des phénomènes de sorption entre deux triazines herbicides et des acides humiques. *Weed Res.*, **15**, 393–399.

GILES C.H., MACEWAN T.H., NAKHWA S.N. & SMITH D. (1960) Studies in adsorption. Part. XI: A system of classification of solution adsorption isotherms, and its use in diagnosis of adsorption mechanisms and in measurement of specific surface areas of solids. *J. Chem. Soc.*, 3973–3993.

GILMOUR J.T. & COLEMAN M.T. (1971)—S-triazine adsorption studies. *Proc. Soil Sci. Soc. Am.*, **35**, 256–259.

HANCE R.J. (1969) Influence of pH, exchangeable cation and the presence of organic matter on the adsorption of some herbicides by montmorillonite. *Can. J. Soil Sci.*, **49**, 357–364.

KHAN S.U. (1973a) Interaction of bipyridylium herbicides with organoclay complex. *J. Soil Sci.*, **24**, 244–248.

KHAN S.U. (1973b) Equilibrium and kinetic studies of the adsorption of 2,4-D and picloram on humic acid. *Can. J. Soil Sci.*, **53**, 429–434.

KHAN S.U. (1974) Adsorption of 2,4-D from aqueous solution by fulvic acid-clay complex. *Envir. Sci. Technol.*, **8**, 236–238.

LING ONG H. & BISQUE R.E. (1968) Coagulation of humic colloïds by metal ions. *J. Soil Sci.*, **106**, 220–224.

MORTLAND M.M. (1968) Protonation of compounds at clay mineral surfaces. *9th. Int. Cong. Soil Sci.* Adelaïde, **1**, 691–699.

MOYER J.R., KERCHER R.B. & HANCE R.J. (1972) Desorption of some herbicides from montmorillonite and peat. *Can. J. Soil Sci.*, **52**, 439–447.

SULLIVAN J.D. & FELBECK G.T. JR (1968) A study of the interaction of s-triazine herbicides with humic acids from three different soils. *Soil Sci.*, **106**, 42–52.

TERCE M. & CALVET R. (1975) Adsorption de l'atrazine par des montmorillonites-Al. *Ann. Agron.* (à paraître).

VALLET M., CALVET R. & CHAUSSIDON J. (1973) Remarques complémentaires sur quelques aspects physico-chimiques de l'adsorption de l'atrazine par les montmorillonites. *Proc. EWRC Symp. Herbicides and the soil*, 41–50. Versailles.

WEBER J.B. (1966) Molecular structure and pH effects on the adsorption of 13 s-triazine compounds on montmorillonite clay. *The American Mineralogist*, **51**, 1657–1670.

WEBER J.B., WEED S.B. & WARD T.M. (1969) Adsorption of s-triazines by soil organic matter. *Weeds*, **17**, 417–421.

WEBER J.B. (1970) Adsorption of s-triazines by montmorillonite as a function of pH and molecular structure. *Proc. Soil Sci. Soc. Am.*, **34**, 401–404.

COMPONENTS AND PARTICLE SIZE FRACTIONS INVOLVED IN ATRAZINE ADSORPTION BY SOILS[1]

P. M. HUANG, R. GROVER, and R. B. McKERCHER

ABSTRACT

We used two Saskatchewan soils to investigate the relative importance of sesquioxides, organic matter, and a series of particle size fractions in adsorbing atrazine. The DCB (sodium dithionite-citrate-bicarbonate)-extractable Al and Fe are associated with a series of soil particle size fractions ranging from <0.2 to >50 μm. Besides clay particles, the nonclay fractions of the soils have a significant capacity to retain atrazine. After the organic matter has been destroyed, the removal of sesquioxides by the DCB treatment causes further substantial reduction in the degree of adsorptivity and the rate of adsorption of atrazine by the soils. The data indicate that, besides organic matter, the noncrystalline to poorly crystalline Al and Fe components and other inorganic constituents present in a series of particle size fractions of the soils, especially <20 μm fractions, provide adsorption sites for atrazine.

INTRODUCTION

The fate of atrazine in soils and the factors governing its bioavailability and persistence still are not fully understood (Bailey and White 1970; Weber 1970; Adams 1973; Smit and Nels 1977; Rahman and Mathews 1979). The success of atrazine application is being hindered by its erratic performance and the carryover of its residues to affect plants in some soil types and not in others. This is apparently because of the variation in soil properties and climatic conditions (Adams 1968; Terce and Calvet 1975, 1977; Skipper et al. 1978). More recent data indicate that the bioactivity of atrazine seems related not only to soil organic matter, but also to sesquioxides (Grover et al. 1982). The objective of this study was to compare the relative roles of organic matter, noncrystalline to poorly crystalline oxides of Al and Fe, and a series of particle size fractions of soils in the adsorption of atrazine.

MATERIALS AND METHODS

General properties

Two Saskatchewan soils were used (Table 1). The pH values of the soils were determined at the soil-to-water ratio of 1:1. Organic matter was assayed by the Walkley-Black method (Jackson 1958). The calcium carbonate equivalent was determined by the method described by Shaw (1931). The texture of the soils was determined by the pipette method (Day 1965). The extractable Al and Fe of the soils were analyzed by the DCB (sodium dithionite-citrate-bicarbonate) method (Weaver et al. 1968).

Fractionation of soil particles

The soils (<0.5 mm) were dispersed by ultrasonification (Genrich and Bremner 1972). Each of the dispersed soils was separated into fractions—<0.2, 0.2 to 2, 2 to 5, 5 to 20, 20 to 50, and >50 μm (Jackson 1956). Extractable Al and Fe in each size fraction were determined by the

[1] Saskatchewan Institute of Pedology Publication no. R330. Financial support from Agriculture Canada and the Natural Sciences and Engineering Research Council of Canada (Grant A2383-Huang) is acknowledged.

TABLE 1
Selected properties of the soils

Soil	Texture	pH	Organic matter %	DCB-extractable[a] Al (pp m)	Fe
Battrum	Silty clay	7.6	3.31	2031	6868
Swift Current	Loam	6.0	2.85	1794	5361

[a] Dithionite-citrate-bicarbonate.

DCB method, as described by Weaver et al. (1968).

Adsorption isotherms of atrazine by soils

Soil samples (0.5 g) were suspended in 5 ml of solution containing ^{14}C labeled atrazine at initial concentrations of 0.5, 1, 2, and 4 µg/ml. The suspensions were equilibrated for a series of reaction periods (2, 4, 8, 12, 24, and 48 h) in a constant-temperature bath with gentle agitation at $25 \pm 0.2°C$. At the end of each reaction period, the suspensions were centrifuged at 1700 g for 30 min. Atrazine remaining in solution was assayed by liquid scintillation counting techniques using Beckman LS9000. Atrazine adsorbed by the soils were calculated by the difference in atrazine concentrations before and after equilibration with the soils. The increase in the adsorption of atrazine by the soils after a 24-h reaction period was negligible. Therefore, atrazine adsorbed by the soils after 24 h was plotted according to the Freundlich equation

$$\log \frac{x}{m} = \log K + \frac{1}{n} \log C$$

where x/m is the atrazine (µg) adsorbed per g of soil, C is the concentration (µg/ml) of atrazine in equilibrium solution, and K and n are constants.

Adsorption isotherms of atrazine by soils after selective dissolution

The organic matter of the soils was decomposed by the $NaOAc-H_2O_2$ treatment (Jackson 1956). After the $NaOAc-H_2O_2$ treatment, the sesquioxides of a portion of the soils were removed by the DCB treatment, and the residual citrate was oxidized by H_2O_2 (Jackson 1956). Treated soils (0.5 g) were equilibrated with 5 ml of solution containing ^{14}C labeled atrazine at initial concentrations of 0.5, 1, 2, and 4 µg/ml. The suspensions were equilibrated at $25 \pm 0.2°C$ for 24 h as described before. Atrazine adsorbed by the treated soils was plotted using the Freundlich equation.

Comparison of atrazine adsorption at different temperatures

Soil samples (0.5 g) were suspended in 5 ml of solution containing ^{14}C labeled atrazine at an initial concentration of 4 µg/ml. The suspensions were equilibrated for 24 h at 5°C, and atrazine in the supernatant solution was assayed as discussed earlier. The results were compared with the data obtained at 25°C.

Adsorption of atrazine by soil particle size fractions

One hundred milligrams each of the <0.2, 0.2 to 2, 2 to 5, 5 to 20, 20 to 50, and >50 µm particle size fractions of the soils was suspended in 5 ml of solution containing ^{14}C labeled atrazine at the initial concentration of 4 µg/ml. The suspensions were equilibrated for 24 h in a constant temperature bath at $25 \pm 0.2°C$, and the equilibrium concentration of atrazine remaining in solution was assayed as described before.

Time function of atrazine adsorption by soil components

A half-gram of the two soils, before and after the removal of organic matter and organic matter plus sesquioxides (Jackson 1956), was suspended in 5 ml of solution of ^{14}C labeled atrazine at the initial concentration of 4 µg/ml for 2, 4, 8, 12 and 24 h. The suspensions were equilibrated in a constant-temperature bath at $25 \pm 0.2°C$. The concentration of atrazine remaining in solution at each reaction period was determined in the supernatant as described in the previous section.

RESULTS AND DISCUSSION

The adsorption of atrazine by the soils obeys the Freundlich adsorption isotherms. The $1/n$ values of the straight lines of the isotherms are

TABLE 2

The adsorption of atrazine by the soils as influenced by the selected removal of components

Soil	K value, μg/g[a]		
	No treatment	NaOAc—H_2O_2	NaOAc—H_2O_2—DCB—H_2O_2
Battrum	1.7	1.2	0.7
Swift Current	1.9	1.2	0.8

[a] The average standard deviation of quadruplicate analysis of all samples and treatments is 0.1.

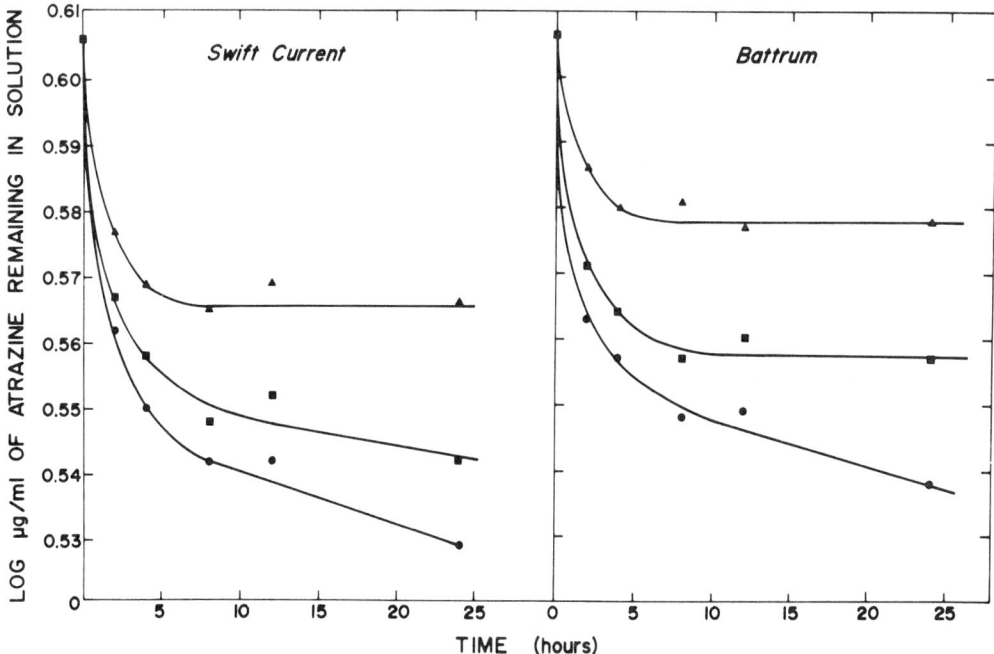

FIG. 1. Dynamics of the adsorption of atrazine by the soils as influenced by selected removal of components.

approximately similar (the slope ranges from 0.96 to 1.00). The K value is thus a measure of the degree of adsorptivity of atrazine by the soil. The K values of the Battrum and Swift Current soils decrease after the NaOAc-H_2O_2 treatment (Table 2). Because calcium carbonates, which are also removed by the NaOAc-H_2O_2 treatment, are not related to the K values (unpublished data), the decrease in the K values upon the treatment (Table 2) is attributed to the removal of organic matter and to the dissolution of certain sesquioxidic components from the soils, it having been shown that the NaOAc—H_2O_2 treatment not only removes organic matter but also extracts Al and Fe (Huang et al. 1977; Huang and Liaw 1979). The further drastic decrease in the K values of the soils after the NaOAc—H_2O_2—DCB—H_2O_2 treatment (Table 2) clearly shows that noncrystalline to poorly crystalline oxides of aluminum and iron play an important role in the adsorption of atrazine by the soils. This study has extended the observations of the role of Al and Fe and their hydroxides in the adsorption of herbicides by pure clay mineral systems (Terce and Calvet 1975, 1977) to natural soils. This finding supports the statistical evidence that noncrystalline to poorly crystalline oxides of Al and Fe are very significantly related to the sorption and desorption of atrazine by soils (unpublished data). Furthermore, the data obtained in this study explain the observation of Adams (1968), who reported that the carryover of atrazine in Minnesota soils appeared to be related to the relatively low amounts of soil extractable Al, which would tie up atrazine and make it less active.

The rate of the adsorption of atrazine by the soils decreases significantly after the removal of the organic matter, especially the sesquioxides (Fig. 1). This indicates that, besides governing the adsorptivity of atrazine, both organic matter and noncrystalline to poorly crystalline oxides of aluminum and iron very significantly enhance the dynamics of the adsorption of atrazine by the soils.

The effect of temperature on the adsorption of atrazine by the soils is not evident (Table 3). This indicates that, contrary to earlier work (Dao and Lavy 1978), the adsorption of atrazine by these soils requires little heat of activation and appears to be insensitive to temperature variations in the temperature range studied. Influence of temperature on the adsorption of atrazine by soils seems to be quite variable with the nature of the soil components (Harris and Warren 1964; McGlamery and Slife 1966).

Atrazine was adsorbed by a series of particle size fractions of the soils ranging from clay to sand (Table 4). The particle size fractions of 5 to 20, 2 to 5, 0.2 to 2, and <0.2 μm are especially noteworthy in the adsorption of atrazine. The adsorption of atrazine molecules by soil components involved van der Waals forces, hydrogen bonding and ion-dipole and coordination types of interaction (Bailey and White 1970). Besides organic matter (Table 2, Fig. 1), the ability of the colloidal and noncolloidal fractions of the soils to adsorb atrazine (Table 4) is attributed to the presence of poorly ordered oxides of aluminum or iron (Table 5) and other mineral components in these particle size fractions of

TABLE 3

Comparison of the adsorption of atrazine by the soils at different temperatures

Soil	μg of atrazine adsorbed/g soil[a]	
	5°C	25°C
Battrum	6.3	6.1
Swift Current	6.4	6.6

[a] The average standard deviation of quadruplicate analysis of all samples and treatments is 0.4.

TABLE 4

The adsorption of atrazine by various particle size fractions of the soils

Particle size fraction, μm	μg of atrazine adsorbed/g of fraction[a]	
	Battrum	Swift Current
<0.2	4.7	7.8
0.2–2	8.6	13.9
2–5	12.3	17.6
5–20	7.9	10.2
20–50	1.1	1.7
>50	2.7	2.5

[a] The average standard deviation of quadruplicate analysis of all samples and treatments is 0.6.

TABLE 5

The dithionite-citrate-bicarbonate-extractable Al and Fe in various particle size fractions of the soils

Particle size fraction, μm	Al		Fe	
	B[a]	SC[b]	B	SC
	ppm			
<0.2	6400	9400	18 600	22 200
0.2–2	4900	4800	19 400	23 300
2–5	2400	2000	10 200	12 200
5–20	400	600	4 800	5 600
20–50	ND[c]	ND	2 200	2 400
>50	700	500	1 300	1 200

[a] B, Battrum.
[b] SC, Swift Current.
[c] ND, not detectable.

the soils. In the >20 μm fractions, the amounts of the DCB-extractable Al and Fe are much less than in the finer fractions. This seems to account for the drastic decrease in the adsorption of atrazine by the >20 μm fractions (Table 4). Although the amounts of the DCB-extractable Al and Fe of the <0.2 and 0.2 to 2 μm fractions are substantially higher than those of the 2 to 5 μm fractions, the adsorption of atrazine by the former was considerably less than by the latter. The DCB-extractable Al and Fe of the <0.2 and 0.2 to 2 μm fractions appear to include substantial amounts of those sesquioxides that are soluble in the dissolution treatment but not very reactive toward atrazine.

Sesquioxidic components have high specific surface and proton donor functional groups (Kwong and Huang 1979; Huang 1980). The uncharged atrazine molecules can be protonated on the heterocyclic ring nitrogen (Russell et al. 1968) and then become adsorbed on the negatively charged surfaces of soil particles. The sources of the protons are attributed to: (1)

partial dissociation of the water molecules surrounding hydroxy-aluminum and iron polymers on the surfaces of minerals and organic matter, and (2) partial dissociation of functional groups of organic components of soils (Huang 1980). Surface acidity has been reported to be about 4 pH units (10 000 times) lower than the pH of the bulk solution (Bailey and White 1970).

In summary, the data indicate that, besides organic matter, noncrystalline to poorly crystalline oxides of Al and Fe present in a series of particle size fractions (<0.2, 0.2 to 2, 2 to 5, 5 to 20, 20 to 50, and >50 µm) of soils deserve close attention in defining soil components governing the degree of adsorptivity of atrazine by soils.

REFERENCES

Adams, R. S., Jr. 1968. Soil factors contributing to atrazine carryover. Minnesota Sci. 25(Fall):9–12.

Adams, R. S., Jr. 1973. Factors influencing soil adsorption and bioavailability of pesticides. Residue Rev. 47:1–54.

Bailey, G. W., and J. L. White. 1970. Factors influencing the adsorption, desorption, and movement of pesticides in soil. Residue Rev. 32:29–92.

Dao, T. H., and T. L. Lavy. 1978. Atrazine adsorption on soil as influenced by temperature, moisture content and electrolyte concentration. Weed Sci. 26:303–308.

Day, P. R. 1965. Particle fractionation and particle-size analysis. *In* Methods of soil analysis, pt. 1. C. A. Black (ed.). Am. Soc. Agron., Madison, Wis., pp. 545–567.

Genrich, D. A., and J. M. Bremer. 1972. A revaluation of the ultrasonic-vibration method of dispersing soils. Soil Sci. Soc. Am. Proc. 36:944–947.

Grover, R., P. M. Morse, and P. M. Huang. 1982. Bioactivity of atrazine in several Saskatchewan soils. Can. J. Plant Sci. 63:489–496.

Harris, C. I., and G. F. Warren. 1964. Adsorption and desorption of herbicides by soil. Weed Sci. 12:120–126.

Huang, P. M. 1980. Adsorption processes in soil. *In* The handbook of environmental chemistry, vol. 2, pt. A. O. Hutzinger (ed.). pp. 47–59.

Huang, P. M., and W. K. Liaw. 1979. Adsorption of arsenite by lake sediments. Int. Rev. Gesamten Hydrobiol. 64:263–271.

Huang, P. M., M. K. Wang, M. H. Wu, C. L. Liu, and T. S. C. Wang. 1977. Sesquioxidic components of selected Taiwan soils. Geoderma 18:251–263.

Jackson, M. L. 1956. Soil chemical analysis—Advanced course. Published by the author. Univ. of Wisconsin, Madison, Wis.

Jackson, M. L. 1958. Soil chemical analysis. Prentice-Hall, Englewood Cliffs, N. J.

Kwong, Ng Kee K. F., and P. M. Huang. 1979. Surface reactivity of aluminum hydroxides precipitated in the presence of low molecular weight organic acids. Soil Sci. Soc. Am. J. 43:1107–1113.

McGlamery, M. D., and F. W. Slife. 1966. The adsorption and desorption of atrazine as affected by pH, temperature, and concentration. Weed Sci. 14:237–239.

Rahman, A., and L. J. Mathews. 1979. Effect of soil organic matter on the phytotoxicity of thirteen *s*-triazine herbicides. Weed Sci. 27:158–161.

Russell, J. D., M. Cruz, J. L. White, G. W. Bailey, W. R. Payne, Jr., J. D. Pope, Jr., and J. I. Teasley. 1968. Mode of chemical degradation of *s*-triazines by montmorillonite. Science 160:1340–1342.

Skipper, H. D., V. V. Volk, M. M. Mortland, and K. V. Raman. 1978. Hydrolysis of atrazine on soil colloids. Weed Sci. 26:46–51.

Shaw, W. M. 1931. Determination of carbon dioxide in soil carbonates—A modification of the official method. J. Assoc. Off. Agric. Chem. 14:283–292.

Smit, N. S. H., and P. C. Nels. 1977. The activity of atrazine on two South African soils. Crop Prod. 6:67–71.

Terce, M., and R. Calvet. 1975. Role of aluminum in the adsorption of atrazine by clay minerals. Proc. Symp. Israel-France, Behavior of Pesticides in Soil, pp. 33–39.

Terce, M., and R. Calvet. 1977. Some observations on the role of Al and Fe and their hydroxides in the adsorption of herbicides by montmorillonite. Z. Pflkrankh. Pfl. Sonderheft 8:237–243.

Weaver, R. M., J. K. Syers, and M. L. Jackson. 1968. Determination of silica in citrate-bicarbonate-dithionite extracts of soils. Soil Sci. Soc. Am. Proc. 32:497–501.

Weber, J. B. 1970. Mechanisms of adsorption of *s*-triazines by clay colloids and factors affecting plant availability. Residue Rev. 32:93–130.

6

Copyright © 1965 by Blackwell Scientific Publications Ltd
Reprinted from *Weed Res.* 5:98–107 (1965)

THE ADSORPTION OF UREA AND SOME OF ITS DERIVATIVES BY A VARIETY OF SOILS

R. J. HANCE

Agricultural Research Council Weed Research Organization,
Begbroke Hill, Kidlington, Oxford, England

Summary. The absorption of urea and a number of its derivatives by different soils was investigated using a slurry-type procedure. The materials could be listed in the following order of increasing tendancy to be adsorbed: urea, fenuron, methylurea, phenylurea, monuron, monolinuron, diuron, linuron, neburon and chloroxuron. Both N-aryl and N-alkyl substituents appeared to play a part in adsorption. Increasing chain length in the alkyl substituents and chloro- and chlorophenoxy substitution in the aryl substituent increased adsorption. There was no relationship between adsorption and water solubility.

Organic matter content was the only soil property that could be related to adsorptive capacity. The evidence of Langmuir isothermal equilibrium plots suggests that only a fraction of the total soil surface is available for the adsorption of substituted ureas.

L'adsorption de l'urée et de ses dérivés sur différents sols

Résumé. L'adsorption de l'urée et d'un certain nombre de ses dérivés sur divers sols, a été étudiée à l'aide d'une méthode du type agitation. Considérant leur tendance à être adsorbées, les substances étudiées peuvent se ranger dans l'ordre croissant suivant: urée, fénuron, méthylurée, phénylurée, monuron, monolinuron, diuron, linuron, néburon et chloroxuron. La substitution N-arylique et N-alkylique paraît affecter l'adsorption. L'adsorption fut accrue par l'allongement de la chaîne dans les substitués alkyliques et par la substitution chloro- et chlorophénoxy- dans le substitué arylique. On n'a pas trouvé de relation entre l'adsorption et la solubilité dans l'eau.

La teneur en matière organique fut la seule caractéristique du sol pouvant être reliée à sa capacité adsorbante. L'examen des courbes d'équilibre isothermique de Langmuir suggère qu'une partie seulement de la surface totale du sol est disponible pour l'adsorption des urées substituées.

Die Adsorption von Harnstoff und einiger seiner Derivate durch eine Anzahl von Böden

Zusammenfassung. Die Adsorption von Harnstoff und eine Reihe seiner Derivate durch verschiedene Böden wurde mit einer Art Schlämmverfahren untersucht. Geordnet nach zunehmender Adsorption ergab sich folgende Reihenfolge: Harnstoff, Fenuron, Methyl-, Phenylharnstoff, Monuron, Monolinuron, Diuron, Linuron, Neburon und Chloroxuron. Sowohl N-Aryl- als auch N-Alkyl-Substition schien einen Einfluss auf die Adsorption zu haben. Zunahme der Kettenlänge in der Alkylsubstitution sowie Chlor- und Chlorphenoxysubstitution der Arylsubstitution erhöhten die Adsorption. Zwischen Adsorption und Wasserlöslichkeit bestand jedoch keine Beziehung.

Der Gehalt an organischer Substanz war die einzige Bodeneigenschaft, die zu dessen adsorptiver Fähigkeit in Beziehung gebracht werden konnte. Die Versuche zeigen weiterhin, dass nur ein Teil der Gesamtbodenoberfläche für die Adsorption von substituierten Harnstoffen frei ist.

INTRODUCTION

Several substituted ureas are soil acting herbicides, useful for the control of a wide range of weeds. Adsorption onto soil particles is one of the processes which influence the effectiveness of such herbicides.

Previous studies indicate that several soil properties may be involved in adsorption. Sheets (1958) and Upchurch (1958), with phytotoxicity experiments and Sherburne & Freed (1954) and Harris & Warren (1964) with direct adsorption measurements made under slurry conditions found that soil organic matter was a major factor. Hilton & Yuen (1963) also showed organic matter to be important, but they considered that only a part of it was actually involved. This fraction was susceptible to a mild oxidation procedure and appeared to consist of 12–60% of the total organic matter. Both Sheets (1958) and Upchurch (1958) found an inverse correlation between phytotoxicity and cation exchange capacity but Hilton & Yuen (1963) found no such relationship with direct adsorption measurements. Sheets also found an inverse correlation between phytotoxicity and clay content, but Upchurch did not. Coggins & Crafts (1959) were able to reduce the toxicity of ureas in culture solution by the addition of bentonite, and this material has also been shown to adsorb chloroxuron and neburon by Geissbuhler, Haselbach & Aebi (1963) and monuron by Harris & Warren (1964). Frissel & Bolt (1962) found that montmorillonite, illite and kaolinite would all adsorb monuron and diuron to some extent.

The relationship between the structure of the adsorbate molecule and its adsorption has received less attention. Sheets & Crafts (1957), working with substituted ureas, considered that the aryl substituent was active in adsorption relationships and that the shift in electron balance caused by the addition of a chlorine atom to the aryl group increased this activity. They suggested that N-alkyl substituents also played a part. Leopold, van Schaik & Neal (1960) were of the opinion that the aryl group was the principal site of the adsorption of phenoxyacetic acids by carbon. Their results indicate that these compounds were similar to the ureas in that chlorine substitution in the aryl group increased adsorption.

The objects of this investigation were to assess which factors in British soil are involved in adsorption and to study the relationship between the structure of substituted urea herbicides and their adsorption.

MATERIALS AND METHODS

The following compounds were studied: urea, methylurea, phenylurea, N,N-dimethyl-N'-phenylurea (fenuron), N'-(4-chlorophenyl)-N,N-dimethylurea (monuron), N'-(3,4-dichlorophenyl)-N,N-dimethylurea (diuron), N'-(4-chlorophenyl)-N-methoxy-N-methylurea (monolinuron), N'-(3,4-dichlorophenyl)-N-methoxy-N-methylurea (linuron), N-butyl-N'-(3,4-dichlorophenyl)-N-methylurea (neburon) and N'-4-(4-chlorophenoxy)phenyl-N,N-dimethylurea (chloroxuron).

Eleven soils were used in the investigation. Eight soils were obtained from N.A.A.S. Experimental Husbandry farms, two from private farms and one from the farm of the A.R.C. Weed Research Organization. Information concerning these soils is given in Table 1.

Sub-samples of each soil were oxidized to remove the bulk of the organic matter. Successive portions of 10% hydrogen peroxide were added to the samples, with gentle heating, until the appearance of the residue and the absence of effervescence indicated that little organic matter remained. Carbon was estimated in the oxidized residues by the dichromate method of Tinsley (1950). The carbon contents of the oxidized soils are given in Table 2.

Experiments with diuron were carried out on all eleven soils, both unoxidized and oxidized, but experiments with the other ureas were confined to unoxidized soils 1–6.

Table 1
Composition of soils

	Soil number										
	1	2	3	4	5	6	7	8	9	10	11
Source	Sunway farm, Woodwalton, Hunts.	Great House E.H.F.,* Helmshore, Lancs.	Toll Farm, Littleport, Cambs.	Trawscoed, E.H.F.,* Cardigan	Weed Res. Orgn., Oxon.	Rosemaunde E.H.F.,* Hereford	Liscombe E.H.F.,* Somerset	Bridget's E.H.F.,* Hants.	Boxworth E.H.F.,* Cambs.	Terrington E.H.F.,* Norfolk	Kirton E.H.F.,* Lincs.
Soil type	Light peat	Dark grey sandy loam	Heavy peat	Silt clay loam	Sandy loam	Sand clay loam	Silt loam	Calcareous silty loam	Chalky boulder clay	Heavy silt	Silt
Mechanical analysis (%)											
Coarse sand	4.0	14.3	3.6	6.9	26.7	1.2	8.3	5.6	8.1	0.6	0.3
Fine sand	3.0	31.2	19.4	27.1	39.3	34.8	35.7	34.4	31.9	41.4	70.7
Silt	10.0	15.9	24.4	33.4	18.4	40.4	33.4	26.4	20.4	23.4	13.4
Clay	10.0	6.6	28.6	32.6	15.6	23.6	22.6	33.6	39.6	34.6	15.6
Organic carbon (%)	36.5	12.0	11.7	3.69	1.93	1.76	3.45	3.09	2.08	1.54	1.50
pH	5.2	6.3	7.4	6.2	7.1	6.7	6.2	8.0	7.9	8.0	7.6
$CaCO_3$ (%)	—	—	—	—	—	—	—	35.0	13.5	3.5	1.5
Cation exchange capacity (mequiv./100 g)	60	18	41	12	11	14	13	24	22	15	13

* E.H.F. = Experimental Husbandry Farm.

Adsorption experiments

Sherburne & Freed (1954) found no relationship between pH and adsorption of monuron by soil and this was confirmed by Hilton & Yuen (1963) with fenuron, monuron, diuron and linuron and Geissbühler *et al.* (1963) with chloroxuron. In addition Frissel & Bolt (1962) showed that above pH 5, pH had very little effect on the adsorption of monuron and diuron by clay minerals. Since the pH values of the soils used in this investigation were in the range 5·2–8·0 no steps were taken to standardize this factor.

A sample of air dry soil (0·1–2 g) ground to pass a 100 mean sieve was shaken with 25 ml of the aqueous adsorbate solution in a 50 ml centrifuge tube at $22\pm2°$ C. Preliminary experiments showed essentially no difference in equilibrium concentration after shaking periods of between 1 hr and 3 days. In most cases the slurry was shaken overnight, but with the soils of high organic content a 3 hr period was generally used in order to keep to a minimum the quantity of

Table 2
Carbon contents of soils before and after peroxidation

	Soil number										
	1	2	3	4	5	6	7	8	9	10	11
C before oxidation (%)	36·5	12·8	11·7	3·69	1·93	1·76	3·45	3·09	2·08	1·54	1·50
C after oxidation (%)	1·77	0·31	0·71	0·42	0·24	0·31	0·21	0·59	0·31	0·75	0·27

extraneous soil substances in the final solution. Five initial concentrations were used in each case, ranging from 10–80 ppm for the more soluble materials to 0·5–3·5 ppm for the least soluble, chloroxuron. A water blank was always included. After shaking, 0·5 g K_2SO_4 was added as a flocculent and the suspension was centrifuged at 2000 *g* for 15 min. Similar experiments in the absence of K_2SO_4, when longer periods of centrifuging were necessary, indicated that the addition of K_2SO_4 had no effect on the results. A suitable aliquot of the supernatant liquid was withdrawn for analysis.

Analytical methods

Urea and methylurea. The α-nitrosopropiophenone colorimetric method described by Snell & Snell (1954) was used.

Phenyl substituted ureas. In most cases the concentration of substituted urea in the supernatant liquid was determined by measuring the u.v. absorbance of a suitably diluted aliquot. In some cases, however, the interference caused by water-soluble soil components was too great for this method to be applicable. Under these circumstances the following clean-up procedure was used. A suitable aliquot of the supernatant liquid was transferred to a 50 ml separating funnel. Water was added to make the total volume 15 ml. One ml ammonia (d. 0·88) was added followed by 10 ml chloroform. The funnel was shaken for 10 min after which the phases were allowed to separate. The chloroform layer was run off into a 15 ml centrifuge tube which was then spun at 2000 *g* for 1

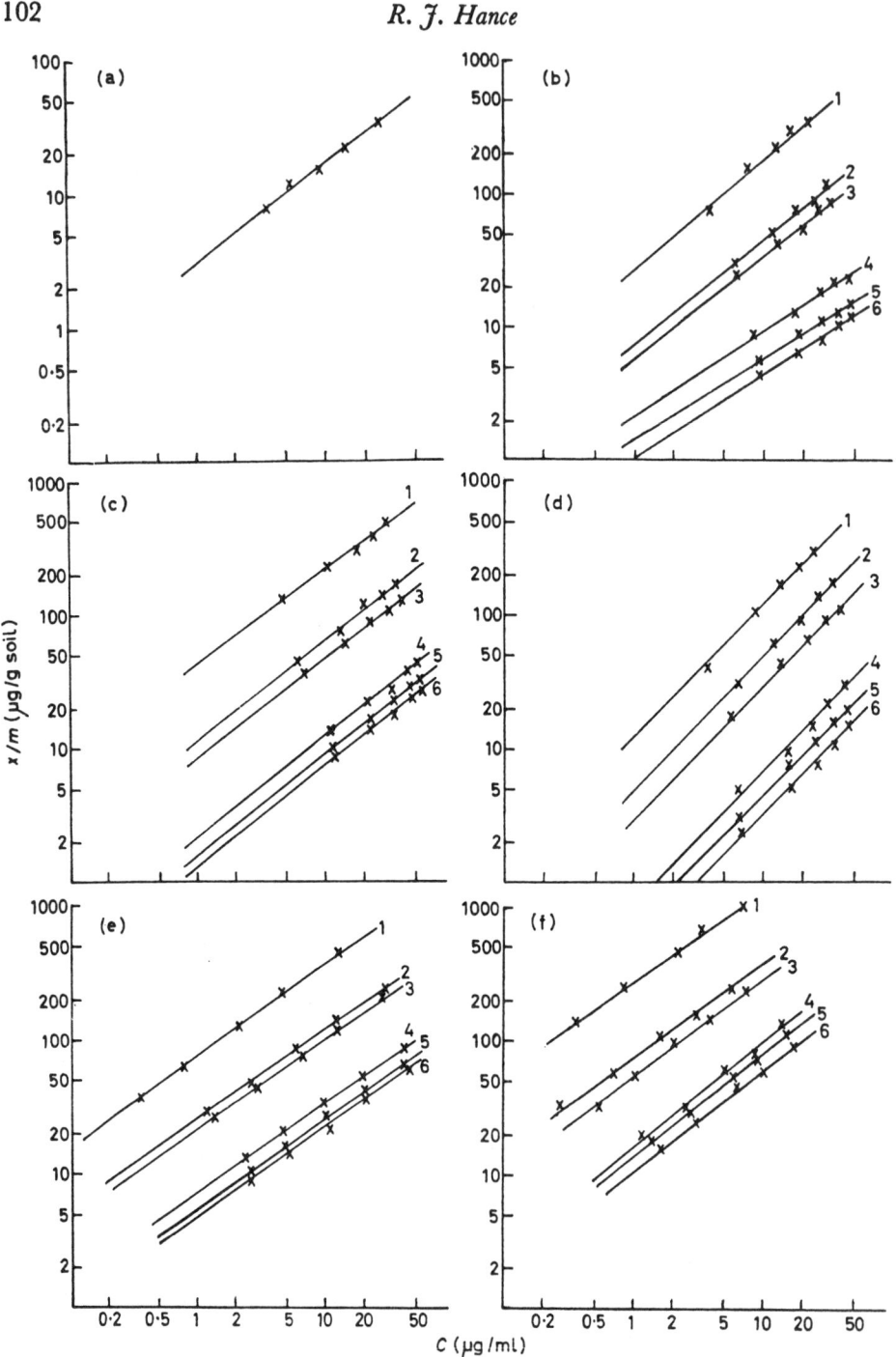

Fig. 1. Isothermal equilibrium adsorption of urea and some of its derivatives: (a) urea by soil No. 1; (b) methyl urea; (c) phenyl urea; (d) fenuron; (e) monuron; (f) diuron.

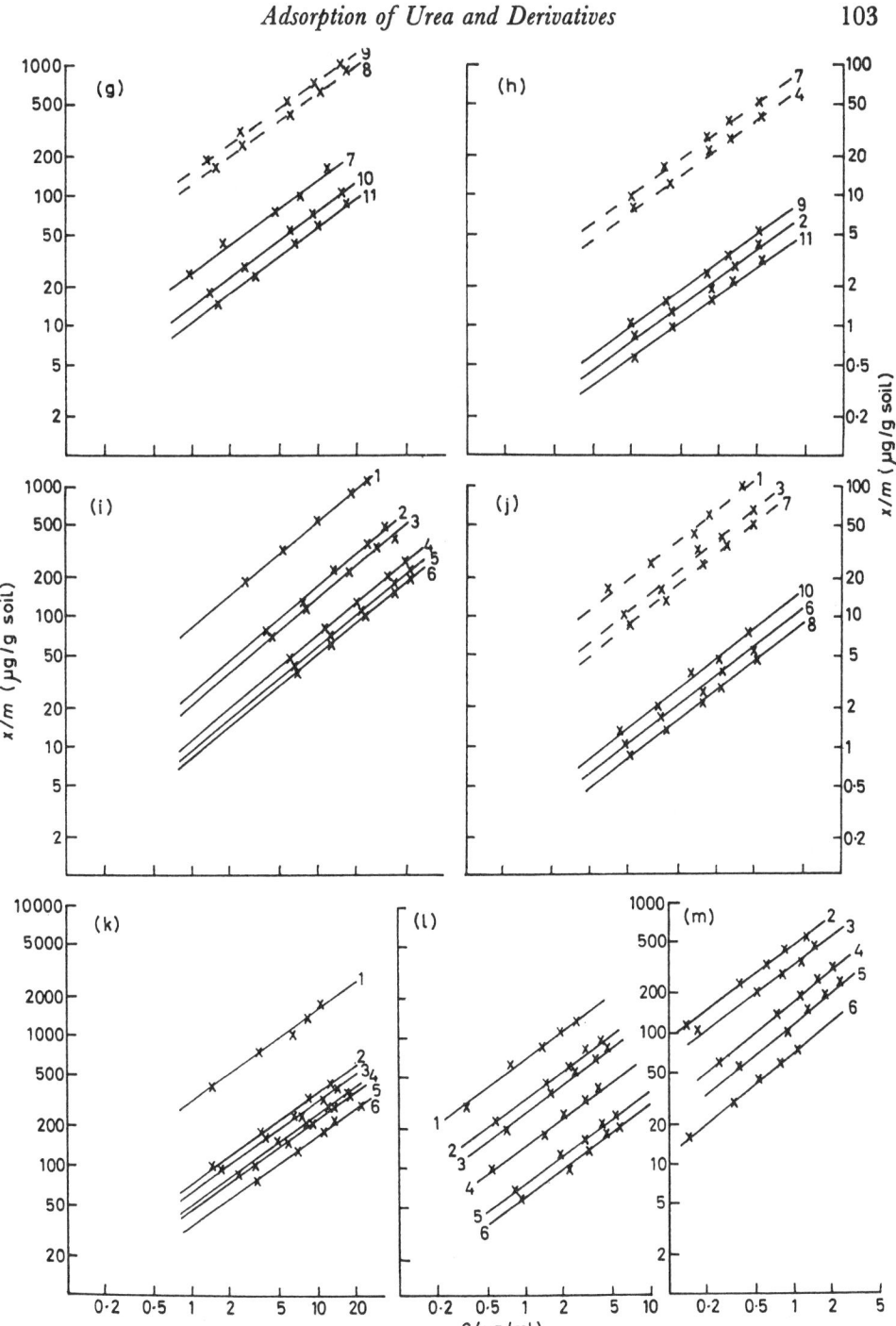

Fig. 1 (*continued*). (g) Diuron (broken curves should be read on right hand scale); (h) diuron by oxidized soils (broken curves should be read on right hand scale); (i) monolinuron; (j) diuron by oxidized soils (broken curves should be read on right hand scale); (k) linuron; (l) neburon; (m) chloroxuron.

min. A 5 ml aliquot of the chloroform solution was transferred to a test tube and the liquid was evaporated on the water bath. Five ml methanol was added to the tube and it was again taken to dryness. After cooling, a further 5 ml methanol was added, the tube stoppered and then shaken for 1 hr. The u.v. absorbance of the solution was then determined. Standard solutions were taken through the same procedure.

The wavelengths used in the determinations were, phenylurea 236 mμ, fenuron 240 mμ, monuron, monolinuron and chloroxuron 245 mμ, diuron, linuron and neburon 250 mμ.

RESULTS AND DISCUSSION

The results were assessed with the aid of the empirical Freundlich adsorption isotherm which may be written

$$\log \frac{x}{m} = \log k + \frac{1}{n} \log C$$

where x/m is the amount adsorbed by unit mass of adsorbent when in equilibrium with a solution of concentration C, k and n being constants.

Fig. 1 (a–m) shows logarithmic plots of x/m against C for each of the compounds. The lines are numbered to correspond with the soil numbers in Table 1.

With the exception of fenuron, the slopes ($1/n$) of the lines are less than 1, with a mean value of 0·75. For these compounds, the slopes are sufficiently similar for comparable values of k to be obtained by reading from the graphs the values of x/m at $C = 1$ (see Table 3). The slope of the fenuron plots ($\sim$1),

Table 3

k values obtained from the Freundlich isotherm graphs

	Soil number										
	1	2	3	4	5	6	7	8	9	10	11
Urea	2·9	0	0	0	0	0					
Methylurea	25·8	7·2	5·6	2·1	1·4	1·0					
Phenylurea	43·0	11·1	8·5	2·2	1·6	1·3					
Fenuron	12·0	4·7	2·9	0·73	0·44	0·34					
Monuron	75·0	26·0	21·8	7·2	5·5	4·8					
Diuron	268	75·0	53·0	16·0	13·6	10·2	25·0	12·0	15·0	14·0	10·0
Diuron (oxidized soils)	10·8	4·7	6·2	4·2	6·2	6·5	4·8	5·0	6·1	7·4	3·5
Monolinuron	81·0	25·0	21·0	11·0	9·6	8·1					
Linuron	315	73·0	63·0	50·0	47·0	35·0					
Neburon	660	325	240	139	72·0	58·0					
Chloroxuron	–	475	330	175	120	70·0					

however, is too disparate for comparisons of this sort. Values of k for fenuron are recorded in Table 3, but comparisons of these figures with those of the other compounds are valid only at $C = 1$, at other concentrations the relationship between the adsorptions of fenuron and the other ureas will be different.

The low values of $1/n$ for compounds other than fenuron are somewhat surprising. Theoretically it might be expected that at the concentrations used in this work, the slopes of all the graphs would approximate to unity, so that

$x/m \propto C$. Phillips (1964) found this was so with phenol and ethylene dibromide, but similar work with herbicides by Geissbühler et al. (1963), Yuen & Hilton (1962) and Hilton & Yuen (1963) has generally produced results in which the slopes of log-log plots have been in the range 0·8–0·98. Similarly, Harris & Warren (1964) obtained graphs of x/m against C which were not linear.

It appears, therefore, that even at low concentrations, solutions of these compounds, with the possible exception of fenuron, do not behave ideally.

On the basis of these results, the ureas may be placed in the following order of increasing tendency to be adsorbed; urea < fenuron < methylurea, < phenylurea < monuron ⩽ monolinuron < diuron ⩽ linuron < neburon < chloroxuron. Both methylurea and phenylurea are adsorbed more than urea which confirms the suggestions of Sheets & Crafts (1957) that both N-aryl and N-alkyl substituents play a part. The contribution of the alkyl substituent appears to increase with increasing chain length as neburon is adsorbed to a greater extent

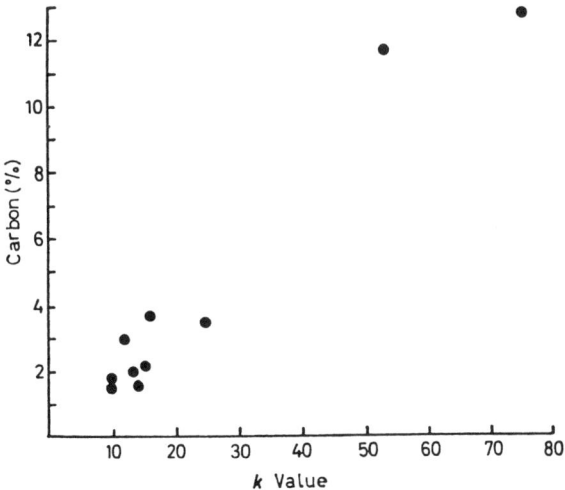

Fig. 2. Variation of k value with carbon content of soil.

than diuron. The anomalous behaviour of fenuron, which, at low concentrations, is adsorbed less than either methylurea or phenylurea, may be explained if it is assumed that substitution on both nitrogen atoms prevents both substituents exerting their full effect, due to the shape of the molecule prohibiting either substituent attaining its optimum orientation at the surface. This suggestion is consistent with the results of Coggins & Crafts (1959) which indicate that N'(3,4-dichlorophenyl)N-methyl urea was more strongly adsorbed than diuron by clay and cellulose. As observed by Sheets & Crafts, chlorine substitution in the aryl group has a considerable effect on adsorption and the introduction of a chlorophenoxy substituent has an even greater effect. Water solubility does not appear to be closely related to adsorption.

There was no correlation between k values and either cation exchange capacity or clay content. There was, however, a general tendency for k values to increase with increasing carbon content (Fig. 2). The importance of organic matter was confirmed by the results obtained with oxidized soils. The reduction in adsorption capacity for diuron produced by oxidation varied between two and twenty times. The size of these reductions must be viewed with caution, however, as the oxidation did not completely remove the organic matter and the side effects of the treatment on the mineral part of the soil have not been evaluated. Nevertheless, the magnitude of the reductions is such as to indicate that organic matter is quantitatively the most important site of adsorption in these soils. There are indications that the adsorptive power of organic matter varies from soil to soil.

The results were also considered with the aid of the Langmuir isotherm which may be expressed in the form

$$\frac{1}{\sigma} = \frac{1}{\sigma_s} + \frac{1}{Ka\sigma_s}$$

where a is the concentration of free adsorbate, σ is the amount adsorbed by unit area at this concentration and σ_s is the value of σ when all sites on the surface are occupied. Thus a plot of $1/\sigma$ against $1/a$ will give a straight line of intercept $1/\sigma_s$ if the relationship is obeyed. Molyneux (1964) has shown that a straight line may be obtained from such a graph even if the underlying assumptions of the equation do not hold, under which circumstances an error of up to 100% may occur in the estimate of σ_s. For this reason the application of the Langmuir equation will not be considered in detail. However, it is of interest to note that the value of σ_s obtained varied from 0·5–24 μM of herbicide per g soil. Assuming the area of these molecules to be of the order of 100 Å², the area covered by 0·5 μM will be approximately 0·3 m² and the area occupied by 24 μM will be approximately 14·4 m². This means that in the most adsorptive soil (No. 1) only 14·4 m² was available to the herbicide. Published figures for soil surface area, such as those of Puri & Murari (1963) and Morin & Jacobs (1964) suggest that 100–200 m²/g would be a reasonable figure to assume, though many soils have a much larger surface area. This indicates that only a fraction of the total soil surface is available to herbicide molecules, even allowing for the errors and approximations inherent in this calculation. One or both of two factors could be responsible for this effect. Either a large part of the soil surface is inherently unable to adsorb substituted urea molecules or this fraction adsorbs water more strongly than the ureas. If water adsorption is largely responsible, slurry-type adsorption experiments will tend to underestimate the adsorptive power of soils in the field where excess water is not always present.

CONCLUSIONS

1. N-aryl and N-aryl substituents play a part in the adsorption of substituted ureas by soils. When both are present the contribution to adsorption of each substituent is reduced, possibly for steric reasons.

2. Increasing chain length in alkyl substituents or chloro and chlorophenoxy substitution in aryl substituents increases adsorption.

3. No close relationship was apparent between adsorption of substituted ureas and their water solubility.

4. Organic matter seems to be the principal site of adsorption.

5. Only a part of the total soil surface appears to be available for adsorption.

ACKNOWLEDGMENTS

The author wishes to express his appreciation to Dr E. K. Woodford, Dr K. Holly and Mr R. G. Powell for their interest and advice. He is also indepted to Miss M. J. House for technical assistance and to Mr S. A. Evans and the Directors of N.A.A.S. Experimental Husbandry Farms who provided the soil samples.

The gifts of pure monuron, diuron and neburon by E. I. du Pont de Nemours & Co. Inc., monolinuron and linuron by Farbwerke Hoechst A.C. and chloroxuron by CIBA Ltd are gratefully acknowledged.

REFERENCES

COGGINS, C. W. & CRAFTS, A. S. (1959) Substituted urea herbicides: their electrophoretic behaviour and the influence of clay colloid in nutrient solution on their phytotoxicity. *Weeds*, **7**, 349–358.

FRISSEL, M. J. & BOLT, G. H. (1962) Interaction between certain ionizable organic compounds (herbicides) and clay minerals. *Soil Sci.*, **94**, 284–291.

GEISSBÜHLER, H., HASELBACH, C. & AEBI, H. (1963) The fate of N'-(4-chlorophenoxy)-phenyl-N,N-dimethylurea (C1983) in soils and plants. I. Adsorption and leaching in different soils. *Weed Res.*, **3**, 140–153.

HARRIS, C. I. & WARREN, G. F. (1964) Adsorption and desorption of herbicides by soil. *Weeds*, **12**, 120–126.

HILTON, H. W. & YUEN, Q. H. (1963) Soil adsorption of herbicides, adsorption of several pre-emergence herbicides by Hawaiian sugar cane soils. *J. agric. Fd Chem.*, **11**, 230–234.

LEOPOLD, A. C., VAN SCHAIK, P. & NEAL, M. (1960) Molecular structure and herbicide adsorption. *Weeds*, **8**, 48–54.

MOLYNEUX, P. (1964) Langmuir isotherm in relation to mobility and molecular size of the adsorbate. *Nature, Lond.*, **202**, 368–370.

MORIN, R. E. & JACOBS, H. S. (1964) Surface area determination of soils by adsorption of ethylene glycol vapour. *Proc. Soil Sci. Soc. Amer.*, **28**, 190–194.

PHILLIPS, F. T. (1964) The aqueous transport of water soluble nematicides through soils. I. The sorption of phenol and ethylene dibromide solutions and the chromatographic leaching of phenol in soils. *J. Sci. Fd Agric.*, **15**, 444–450.

PURI, B. R. & MURARI, K. (1963) Studies in surface area measurement of soils: 1. Comparison of different methods. *Soil Sci.*, **96**, 331–336.

SHEETS, T. J. (1958) The comparative toxicities of four phenylurea herbicides in several soil types. *Weeds*, **6**, 413–424.

SHEETS, T. J. & CRAFTS, A. S. (1957) The phytotoxicity of four phenylurea herbicides in soils. *Weeds*, **5**, 93–101.

SHERBURNE, H. R. & FREED, V. H. (1954) Adsorption of 3-(p-chlorophenyl)-1,1-dimethylurea as a function of soil constituents. *J. agric. Fd Chem.*, **2**, 937–939.

SNELL, F. D. & SNELL, C. T. (1954) *Colorimetric Methods of Analysis*. Van Nostrand, New Jersey.

TINSLEY, J. (1950) The determination of organic carbon in soils by dichromate mixtures. *Trans. 4th int. Congr. Soil Sci.*, **1**, 161–165.

UPCHURCH, R. P. (1958) The influence of soil factors on the phytotoxicity and plant selectivity of diuron. *Weeds*, **6**, 161–171.

YUEN, Q. H. & HILTON, H. W. (1962) Soil adsorption of herbicides: The adsorption of monuron and diuron by Hawaiian sugar cane soils. *J. agric. Fd Chem.*, **10**, 386–392.

(Received 25th September 1964)

MOLECULAR STRUCTURE OF HERBICIDES AND THEIR SORPTION BY SOILS

G. G. Briggs

PARTITION between soil and water of organic compounds that do not ionize in the pH range 4–8 is closely correlated with the soil's content of organic matter. Lambert[1] and Furmidge and Osgerby[2] have expressed sorption in terms of partition coefficients between soil organic matter and water, neglecting mineral constituents. Some investigators have also reported inverse correlations between water solubility and sorption on soil whereas others have found no relationship[2,3]. Little is known about the mechanisms of sorption on soil, but apparently only a small fraction of the surface is responsible for the sorption of certain herbicides[3]. I have attempted to learn more about sorption mechanisms by studying how changes in the chemical structure of related compounds affect sorption.

Sorption isotherms were determined by a wet slurry technique, for twenty-two substituted phenylureas and a homologous series of alkyl-N-phenyl carbamates on four neutral Rothamsted soils containing from 1 to 4 per cent organic matter. The partition coefficient, K, was determined from the linear relationship $x = Kc$, where x is compound sorbed per unit weight of organic matter in the soil and c is the equilibrium concentration. For each compound, K was approximately constant for the four soils and mean values were used.

Table 1. PARTITION COEFFICIENTS OF SUBSTITUTED PHENYLUREAS

X	R_1	R_2	K
4-Cl	Me	Me	29
3,4-Cl	Me	Me	94
3-CF$_3$	Me	Me	22
3-Cl, 4-OMe	Me	Me	32
4-Cl	Me	OMe	40
3,4-Cl	Me	OMe	154
4-Br	Me	OMe	60
3-Cl, 4-Br	Me	OMe	217
3-Cl	Me	H	49
3,4-Cl	Me	H	166
3-Cl, 4-OMe	Me	H	53
3-Cl	H	H	59
3,4-Cl	H	H	178
3-Cl, 4-OMe	H	H	58
3-F	H	H	34
4-F	H	H	19
3-CF$_3$	H	H	53
3-Br	H	H	66
4-Br	H	H	76
3-OH	H	H	21
4-SO$_3^-$	H	H	8
H	H	H	13

Table 2. PARTITION COEFFICIENTS OF ALKYL-N-PHENYLCARBAMATES

R	K
Methyl	31
Ethyl	38
Propyl	66
Iso-propyl	51
Butyl	104
Pentyl	236

The substituted phenylureas (Table 1) gave a linear relationship between log K and σ, the Hammett constant[4] for meta and para substituents in the phenyl ring, and 70·5 per cent of the variation in log K was accounted for by σ. A significant improvement (75·5 per cent of variation) was obtained using a combination of σ and the Taft constants[4] for substituents on the side-chain nitrogen. Of the alternative relationships examined based on free energy contributions of substituent groups, none was better than σ. Hansch's π constant[5], derived from partition between 1-octanol and water, accounted for 50·1 per cent of the variation. Parachor (7·5 per cent of variation), used by Lambert[1] as a measure of the energy required for a molecule to stay in solution, and parachor-$45N$ (24·6 per cent of variation), where N is the number of possible hydrogen bonding sites in the molecule[6], were poor predictors of sorption. Water solubility of the phenylureas used was not a factor in determining their sorption.

In contrast to the results with the substituted phenylureas, those (Table 2) with alkyl-N-phenylcarbamates show a much better linear relationship between log K and π. Each additional methylene group increases parachor, so that the linear relationship between log K and parachor found by Lambert[1] for 2,6-dinitro-4-alkylsulphonyl-NN-dialkylanilines is also observed. A plot of log K against π of Lambert's results is, however, also linear with a slope similar to that given by the phenylcarbamates. Each additional methylene group decreases water solubility so that an inverse relationship with water solubility would also be expected for these compounds.

For the phenylureas without long alkyl side-chains, ring de-activation by substituents in the ring, or by replacement of a methyl group on the side-chain nitrogen by hydrogen or methoxyl, seems the factor controlling sorption, possibly through charge-transfer bonding to activated sites on organic matter. Similar bonding by activated rings to de-activated sites on organic matter could occur. The Hammett and Taft constants are a measure of the de-activating effect of the substituents. The electron distribution in the ring is about the same in the various alkyl-N-phenylcarbamates and changes in sorption are caused by increasing lipophilicity with increasing length of the alkyl chain. Their sorption is probably an accumulation at hydrophobic sites at the organic matter/water interface in a way similar to surface-active agents. π is a measure of lipophilicity or hydrophilic–hydrophobic balance[7] and seems to express these properties better than parachor or water-solubility.

The sorption behaviour of any group of non-ionic organic compounds in soils can probably best be accounted for on the basis of their Hammett, Taft and π constants.

Rothamsted Experimental Station,
Harpenden, Hertfordshire.

Received July 21, 1969.

[1] Lambert, S. M., J. Agric. Food Chem., 15, 572 (1967).
[2] Furmidge, C. G. L., and Osgerby, J. M., J. Sci. Food Agric., 18, 269 (1967).
[3] Hance, R. J., Weed Res., 5, 98 (1965).
[4] Barlin, G. B., and Perrin, D. D., Quart. Rev. Chem. Soc., 20, 75 (1966).
[5] Fujita, T., Iwasa, J., and Hansch, C., J. Amer. Chem. Soc., 86, 5175 (1964).
[6] Hance, R. J., J. Agric. Food Chem., 17, 667 (1969).
[7] Hance, R. J., Nature, 214, 630 (1967).

EFFECT OF CHANGES IN HUMIDITY ON THE TOXICITY AND DISTRIBUTION OF INSECTICIDES SORBED BY SOME DRIED SOILS

F. Barlow and A. B. Hadaway

RECENTLY it has been noticed during field trials in mosquito control that there is an apparent correlation between relative humidity and the insecticidal effectiveness of deposits in houses made of mud walls and thatched roofs. Bordas and Navarro[1] in Mexico found a steady increase in toxicity of dieldrin deposits as the relative humidity increased throughout the season. Burnett[2] in Taveta, Kenya, found that seasonal variations in rainfall, presumably accompanied by variations in humidity, were associated with corresponding changes in the mortalities of mosquitoes exposed to dieldrin deposits.

Relative humidity could presumably exert its influence either upon the insect itself after the insecticide has been picked up or it could modify the availability of the insecticide in the deposit, and these two aspects have now been examined in laboratory experiments.

We have not been able to find any effect of changes in relative humidity upon the kills of mosquitoes which were kept at humidities ranging from 20 to 95 per cent after topical application of DDT, gamma-BHC, dieldrin or 'Diazinon', or after exposure to deposits of wettable powders of DDT, gamma-BHC or dieldrin. Therefore it seems most likely that any influence of relative humidity is upon the availability of the insecticide to the insect.

It has been shown previously[3,4] that the air-dried soils which are frequently used for house construction in the tropics can rapidly adsorb insecticides applied as surface deposits of wettable powders. This process results in a marked decrease in contact action of DDT and dieldrin because there are no longer any particles of insecticide lying on the surface of the house walls. We noticed in this work[3] that relative humidity had an influence upon the rates at which DDT and gamma-BHC thus disappeared; increasing the humidity reduced the rate of loss. It is therefore possible that the behaviour of deposits observed in field trials could be due to variations in the availability of insecticides on and in mud walls and this has been more fully investigated.

It was confirmed that deposits of DDT, gamma-BHC and dieldrin on blocks of Uganda soil are sorbed at rates which decrease as the humidity increases. This was also observed on nine other soils of varying composition and origin. However, dieldrin, which was the insecticide of major interest with field experiments, even at a dosage of 100 mgm. per sq. ft. and at 90 per cent relative humidity, still disappeared from the surface of Uganda mud blocks in about five days, and so for interpretation of the field results it is of more importance to know what happens after sorption is completed. Table 1 shows the results of biological tests with mud blocks sprayed twenty-one days previously with a suspension of dieldrin crystals in the size range 0–10 microns at a dosage of 100 mgm. per sq. ft. and kept for 24 hr. at two different humidities before the exposure of mosquitoes. The positions of the blocks were reversed for a further 24 hr. and re-tested.

It can be seen that the insecticidal activity increases with humidity from 40 to 90 per cent and this difference in activity is almost completely reversible in 24 hr. Similar results were obtained for contact action of DDT and both contact and fumigant action of gamma-BHC. The kills therefore follow closely any changes in humidity and the biological effectiveness of the treated blocks can be rapidly raised or lowered. Because of this close correspondence it seemed unlikely that there could be any marked changes in distribution of insecticide as postulated by Bordas and Navarro[1], who suggested that the insecticide migrated back to the surface under conditions of high humidity. This was borne out by chemical determination of the distributions of the insecticides in the blocks kept at different humidities. These showed that the rate of diffusion of the insecticides into the blocks after sorption was complete actually increased with humidity and this naturally resulted in a decreased concentration of insecticide near the surface of the blocks on which the insects rested. Table 2 gives some results obtained with dieldrin-treated blocks kept at different humidities after spraying.

In the field, therefore, long periods of high humidity would be expected to give a progressive decrease in the concentration of insecticides near the surface of the walls whereas lower humidities would tend to keep the concentration static. At any given time and

Table 1. THE EFFECT OF CHANGES IN RELATIVE HUMIDITY UPON THE TOXICITY OF DIELDRIN-TREATED MUD BLOCKS TO *Aedes aegypti* FEMALES

Relative humidity conditions	Percentage kill 24 hr. after exposure of ... min.			
	4	8	16	32
90 per cent for 24 hr.	55	100	100	
40 per cent for 24 hr.		0	30	68
90 per cent blocks transferred to 40 per cent for 24 hr.			0	5	48
40 per cent blocks transferred to 90 per cent for 24 hr.	38	65	100	

Table 2. AVERAGE PERCENTAGE RECOVERIES OF DIELDRIN FROM SUCCESSIVE LAYERS OF UGANDA MUD BLOCKS

Relative humidity (per cent)	Layer No.	Weight of layer (gm.)	Time after spraying		
			1 hr.	4 days	13 days
10	1	0·25	92	69	52
	2	4·0	8	31	48
	3	4·0	0	0	0
	4	4·0	—	—	0
	5	4·0	—	—	0
50	1	0·25	92	17	10
	2	4·0	8	79	67
	3	4·0	0	4	20
	4	4·0	—	0	3
	5	4·0	—	—	0
90	1	0·25	92	16	6
	2	4·0	8	37	23
	3	4·0	0	27	20
	4	4·0	—	20	15
	5	4·0	—	—	12
	6	4·0	—	—	10
	7	8·0	—	—	14

concentration, however, an increase in humidity would give higher kills, and a decrease lower kills. The enhanced kill at high humidities can more than compensate for decreases in concentration so that kills would remain high even during extended periods of high humidity.

We explain these results on the basis of competition between the water and the insecticides. Water appears to be adsorbed preferentially to the insecticide and the increased amounts which are present in soils at high humidities have two effects. The initial disappearance of the insecticide from solid particles lying on the surface is retarded because water molecules already occupy a large proportion of the 'active' surface of the soil and cannot be displaced. Once the insecticide is adsorbed, however, the same preferential adsorption of water results in an increased mobility of the insecticide molecules and therefore an enhanced rate of diffusion farther into the soil or, alternatively, an enhanced rate of diffusion on to and into an insect resting on the soil.

The laboratory findings therefore provide an explanation for the results obtained in the field by Burnett and by Bordas and Navarro where mosquitoes enter huts and can be killed by contact or fumigant action on treated walls.

Colonial Insecticides Research,
    Porton.
        Sept. 17.

[1] Bordas, E., and Navarro, L., WHO (Malaria), 125 (Geneva, 1955).
[2] Burnett, G., *Nature*, **177**, 663 (1956).
[3] Hadaway, A. B., and Barlow, F., *Bull. Ent. Res.*, **43**, 281 (1952)
[4] Barlow, F., and Hadaway, A. B., *Bull. Ent. Res.*, **46**, 547 (1955).

9

Copyright © 1969 by the Soil Science Society of America
Reprinted by permission from *Soil Sci. Soc. America Proc.* **33**:210-216 (1969)

Solubility-Temperature Effect on the Adsorption of Gamma- and Beta-BHC from Aqueous and Hexane Solutions by Soil Materials[1]

A. C. MILLS AND J. W. BIGGAR[2]

ABSTRACT

Freundlich adsorption isotherms were obtained for gamma- and beta-isomers of 1,2,3,4,5,6-hexachlorocyclohexane adsorbed from aqueous and hexane solvents on Ca-Staten peaty muck, Ca-bentonite Ca-Venado clay, and silica gel in the temperature range of 10 to 40C. Comparison of normal and "corrected" isotherms revealed a significant effect on adsorption due to the influence of temperature on solubility. The $1/n$ constants of the Freundlich equation increased with temperature according to the theory of dilute solutions. When the solubility effect is removed, the gamma isomer has greater adsorbability than the beta isomer, possibly because the former possesses a dipole moment whereas the latter does not. At constant solution fugacities, the gamma isomer on Staten peaty muck competed somewhat more effectively with hexane than with water.

Additional Key Words for Indexing: pesticide, adsorption isotherm, lindane, betaisomer.

---

[1] Investigation supported by Training Grant 5T-1-WP-07 and Research Grant WP-0008 from the Federal Water Pollution Control Administration and Regional Research W-82. Received Oct. 2, 1968. Approved Oct. 21, 1968.
[2] Graduate Student and Associate Professor, respectively, Dept. of Water Science & Engineering, Univ. of California, Davis. The senior author's present address is C-52 N.D. South, c/o AFPRO, Extension II, New Delhi 16, India.

RECENT YEARS have seen great increases in both the types and amounts of pesticides applied to crops. Studies have been numerous on the persistence of herbicides, fungicides, and insecticides applied to crops and soils. Their fate is not limited to the crops and soils, however, particularly if the chemicals are both persistent and somewhat mobile; hence studies on the presence and behavior of these materials involve soils, crop and water supplies (Edwards, 1966; Faust and Suffet, 1966).

Gamma- and beta-isomers of 1,2,3,4,5,6-hexachlorocyclohexane were chosen for study because they have moderate persistence and mobility (Edwards, 1966), and dissipate in water supplies (Faust and Suffet, 1966). Since certain molecular properties are important to their uptake and release by adsorbents, the present study examines the extent to which their differing electrical properties influence the energy of adsorption. The hope is thus to determine the properties which contribute to mobility, persistence and safe use of the chemical. The chemical has eight isomers: two are investigated here.

Grades of the gamma-isomer which have a purity of 99% or higher are known as lindane. This isomer has far greater insecticidal properties than the other isomers. Its lethal action on many insects is typically faster than that of DDT (Shepard, 1957).

The molecule 1,2,3,4,5,6-hexachlorocyclohexane is not planar; instead, the carbon ring assumes a "puckered" chair form. In the gamma-isomer three chlorine atoms are equatorial (extending out roughly from what may be considered the plane of the ring) and three chlorine atoms are axial (projecting above or below the ring). In the beta-isomer all the chlorine atoms are equatorial. Lind et al. (1950) noted that since the beta isomer is centrosymmetric the electric moment is expected to be zero, and they measured it as such. On the other hand, they measured a dipole moment for gamma-BHC in benzene at 30C of 2.84 debye units.

Despite a number of field experiments on lindane, no thorough laboratory experiments have come to light that would provide the information required for predicting their behavior under a wide variety of conditions. Lichtenstein (1958) placed two layers of soil containing 10 ppm lindane over untreated soil layers in cartons. Water treatments revealed that lindane movement was greatest in Plainfield sand and least in muck. Diffusion contributed to movement in unleached soils. Gargantini, Giavotti, and DeTella (1959) applied 1% and 12% BHC to surface acid sandy loam soil in 40-cm-deep lysimeters. After 6 months of gentle application of 1079 mm of water, the BHC was distributed 90% in the top 10 cm and 0.6% in the bottom layer. Swanson et al. (1954), in leaching studies with petroleum ether, found that exchange resins held no lindane, that organic matter seemed to have little effect in increasing adsorption, but that adsorption was closely correlated with clay content. Also, silica gel adsorbed large amounts of BHC.

It was evident from reported work and our own investigations that consideration had to be given to the adsorbent, solvent, and particular properties of the adsorbate since generalizations about these interactions could not be made with the available information.

EXPERIMENTAL METHODS

Preliminary to chosing the adsorbents and solvents used in the adsorption isotherm studies, adsorption studies were performed on a wide range of adsorbents at one temperature. Adsorption levels from benzene and ethanol solution were considered to be too low to permit accurate determination of isotherms over a range of low concentrations. Four adsorbents were chosen as representing particular types of adsorbing surfaces common in soils and having considerable affinity for lindane in hexane and aqueous solutions: Ca-Staten peaty muck, Ca-Venado clay, Ca-bentonite, and silica gel. They respectively represent a high organic soil with mixed mineral fraction, a clay soil high in montmorillonite containing a little organic matter, a montmorillonite clay, and a crystalline silica. Besides being representative of particular soils or soil fractions, these adsorbents showed relatively little or no tendency to catalyse the decomposition of γ-BHC in hexane or water.

Homoionic samples of Wyoming bentonite were prepared by leaching with $CaCl_2$ followed by distilled water until salt free. The 0.2- to 1.0-micron size fraction was selected for study. Venado clay is an alluvial soil found in a coastal mountain valley and formed from the outwash from hills high in serpintine. Surface soil was prepared by saturating with solutions of $2.0N$ Ca acetate, leaching with distilled water until salt free and drying. They were ground to pass through a 1.0-mm sieve. X-ray diffraction analysis showed that Venado clay contained approximately 50% montmorillonite and 6% organic matter. Fisher certified silica gel, 100-to 200-mesh size with a surface area of 890 m^2/g and pH of 4.4 in distilled water was used as received.

Staten peaty muck is a high-organic soil developed from tulereed peat in the delta of the Sacramento river. It was leached with $2.0N$ Ca acetate to constant calcium content, leached free of salt, dried, and sieved to pass a 1.0-mm sieve. It was estimated to contain approximately 22% organic matter after treatment.

All chemicals used in the experiments were reagent grade. The isomers were 99–100% pure, including the ^{14}C labelled isotopes. Solutions were prepared with a constant concentration of radio chemical of either isomer, with the remainder made up with nonradioactive isomer. Because of the limited solubility of the two isomers in water the concentration range for aqueous solutions was kept very low: 0.056–5.0 μg/ml for the initial aqueous γ-BHC solutions and 0.075–0.65 μg/ml for the initial aqueous β-BHC solutions. All aqueous solutions contained 0.5% by volume of ethanol. The range adapted for the hexane-solution isotherms was considerably wider: 0.1–500 μg/liter. Hexane was the solvent used in obtaining other isotherms for the γ-BHC Ca-Venado clay and γ-BHC Ca-Staten peaty muck systems. In all the isotherm studies, unequivocal decomposition occurred only in the γ-BHC Ca-Venado clay-hexane system.

In an experiment, 5 g of oven dried adsorbent was weighed into flasks and placed in an oven at 110C for 36 hours prior to adding solution. After each flask was cooled, 25 ml of the appropriate BHC solution was added to it. Included were duplicates, adsorbent blanks, and standard blanks. Flasks were stoppered, enclosed in polyethylene bags, and immersed in baths controlled to $\pm 0.05C$. After 20 hours of shaking and 50 hours of rest, the solutions were centrifuged for 20 min at 10,000 rpm and constant temperature.

Ten-milliliter portions of aqueous solution were removed and extracted by adding 10 ml of nangrade hexane. This solution was analyzed by both liquid scintillation (LS) and gas-liquid chromatography (GC). When the LS and GC concentrations disagreed by more than 10% the latter was used as a measure of isomer in solution. This provided a means for checking on decomposition even though the LS analyses was more precise. For most experiments, deviation between replicate was less than 0.8% for LS analysis and twice this for GC analysis. In a few cases, particularly where hexane was used or the count rate was very low, the deviations were as high as 4%. The equilibration, centrifugation, and other experiments on methods have been reported elsewhere (A. C. Mills, 1967. The adsorption of the Gamma- and Beta-isomers of 1,2,3,4,5,6-hexachlorocyclohexane on soils. Ph.D. Thesis, University of California, Davis).

The concentrations of γ- and β-isomers in saturated solutions of aqueous and hexane solvents were determined at 20 and 30C for aqueous solutions and at 10 and 20C for hexane solutions. Powdered isomer crystals were added in excess to 400 ml of solvent in glass-stoppered bottles. The bottles were placed in a constant temperature bath for 68 hours with intermittent shaking during this period. Samples of 30 to 40 ml were removed and filtered through a 5μ fritted glass filter into glass-stoppered test tubes. Aqueous solutions were extracted by adding 10 ml of nangrade hexane and inverting the glass-stoppered test tubes back and forth for 1.5 min. Longer mixing times did not improve extraction. The hexane extract was analyzed by GC. The results of these experiments are reported in Table 3. The concentrations of the γ-isomer in water are somewhat higher than those reported by Robeck et al., because of the 0.5% alcohol present in the aqueous solvent.

RESULTS

The normal isotherms (x/m plotted against c) are presented in Fig. 1, 3A, 4A, and 5A for the four adsorbents, two isomers, and two solvents. It is apparent from Fig. 1 that the isomers for γ-BHC in water on all four adsorb-

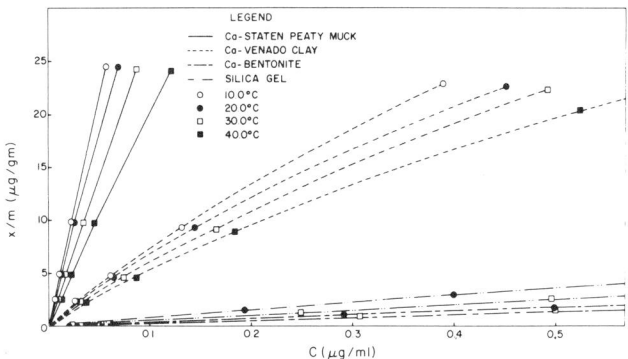

Fig. 1—Normal adsorption isotherms of γ-BHC from water on four adsorbents at 10, 20, 30, and 40C.

ents are nearly linear. The isomers show that an increase in temperature lowers adsorption uniformly for each system. Proportionally, the decrease is least with silica gel and greatest with Ca-bentonite.

The displacement of isotherms downward is commonplace with increases in the temperature at which adsorption is measured. Contributing to this net effect of isotherm displacement are not only the energy contributions in the adsorption reaction itself, but also the change in solubility of the solute as a result of the temperature change. The principle involved is that the fugacity (Lewis and Randall, 1923) of a given solute in solution at a constant concentration will assume different values with different temperatures.

This change in fugacity with temperature may be considered to be closely related to the change in the reduced concentration, c/c_o, with temperature, where c, the solute concentration, may be considered constant, and c_o is the solubility of the solute in the solvent at a specified temperature. Thus the activity, and therefore the chemical potential, of a solute in solution at any specified concentration depends at least in part upon the solubility of the solute in the solvent as affected by the temperature as well as by the particular solvent. Since the difference between the fugacity of the solute in the bulk solution and that on the adsorbent is considered the driving force behind the adsorptive process, it is important to take into account the influence of temperature upon the solute fugacity. The effect of the energy contributions in the adsorption process, apart from solubility-temperature effects, can be seen qualitatively by plotting isotherms as x/m v.s. c/c_o (Kipling, 1965).

Adsorption isotherms, in which x/m values are plotted against the reduced concentration, c/c_o (hereinafter designated as the "corrected" isotherms), are shown in Fig. 2, 2B, 3B, 4B, and 5B.

Figures 2 and 2B, in contrast to Fig. 1, show an increase in adsorption with increase in temperature, the smallest increase occurring in the silica gel system. Accepting the above assumption, the adsorption in this case proceeds with essentially no heat effect, except that involved in the solubility relationship. Thus it appears that the adsorption process, which is indicated to be exothermic by the normal isotherms is at least partially so because of the solubility-temperature interaction. The importance of organic matter in the adsorption of γ-BHC from water is underscored by the higher relative positions of the isotherms for Ca-Staten peaty muck and Ca-Venado clay as seen in Fig. 1. This differs from Swanson et al. (1954) because of the difference in solvent, a point to be discussed later.

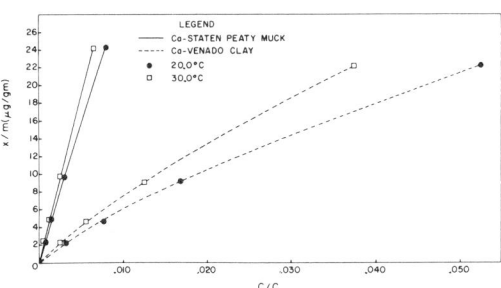

Fig. 2—Corrected adsorption isotherms of γ-BHC from water on Ca-Staten Peaty Muck and Ca-Venado Clay at 20 and 30C.

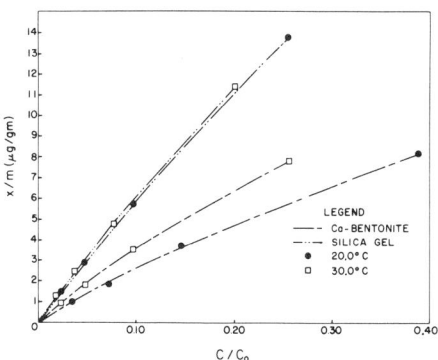

Fig. 2B—Corrected adsorption isotherms of γ-BHC from water on Ca-bentonite and silica gel at two temperatures.

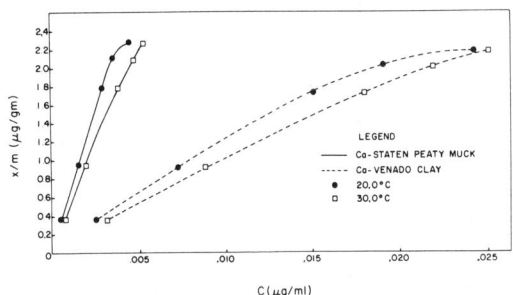

Fig. 3A—Adsorption of β-BHC from water on two adsorbents at 10 and 20C.

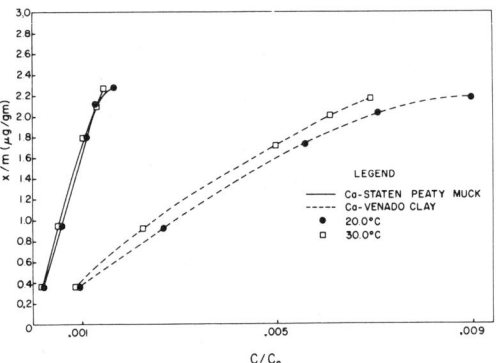

Fig. 3B—Corrected adsorption isotherms of β-BHC from water on the Ca-Staten Peaty Muck and Ca-Venado Clay at two temperatures.

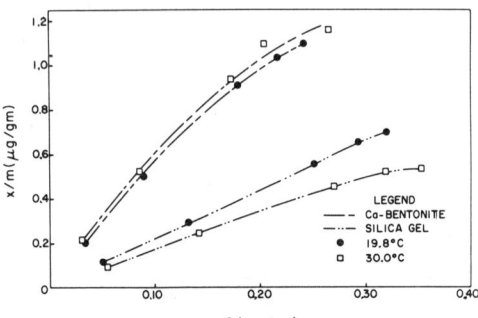

Fig. 4A—Adsorption of β-BHC from water on Ca-bentonite and silica gel at 19.8 and 30C.

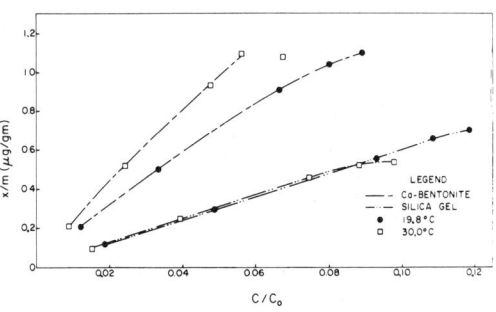

Fig. 4B—Corrected adsorption isotherms of β-BHC on Ca-bentonite and silica gel at 19.8 and 30C.

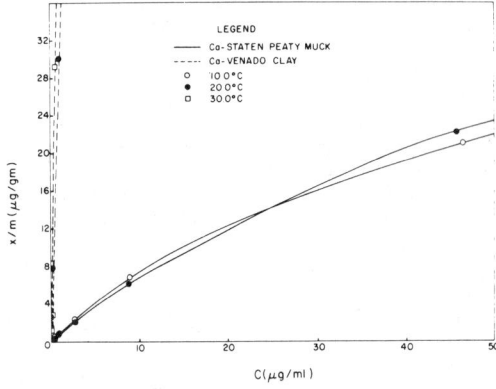

Fig. 5A—Adsorption of γ-BHC from hexane on Ca-Staten Peaty Muck and Ca-Venado Clay at 10, 20, and 30C.

From Fig. 3A the adsorption of β-BHC from aqueous solutions is seen to parallel closely the adsorption of aqueous γ-BHC on Staten peaty muck and Venado clay. The respective slopes of the curves in Fig. 1 and 3A correspond very well. If the highest concentration points are neglected, these β-BHC isotherms also approach a linear form. The corrected isotherms of Fig. 3B show that, as with the aqueous γ-BHC systems, when the solubility-temperature effect is eliminated the adsorption is no longer exothermic. In the corrected isotherms an adsorption on Staten peaty muck, unlike with the aqueous γ-BHC-Staten peaty muck system, showed no change with temperature.

In Fig. 4A the normal isotherms for Ca-bentonite show that β-BHC adsorption increased slightly with temperature. This increase broadened considerably when the solubility effect was removed, as shown in Fig. 4B. A further difference relative to the aqueous γ-BHC systems is in the relative positions of the Ca-bentonite and silica gel isotherms. While silica gel exhibited a greater adsorption affinity for aqueous γ-BHC than did Ca-bentonite (Fig. 1), the reverse is true for the adsorption of β-BHC (Fig. 4A). Similar to the aqueous γ-BHC case, the corrected isotherms for β-BHC on silica gel exhibit little or no temperature effect; i.e., heat effects other than those associated with solution are almost negligible (Fig. 2B, 4B).

The isotherm of γ-BHC adsorption from hexane is shown in Fig. 5. The lesser degree of adsorption on Ca-Staten peaty muck than on Ca-Venado clay is apparent throughout the concentration range. The isotherms for Ca-Staten

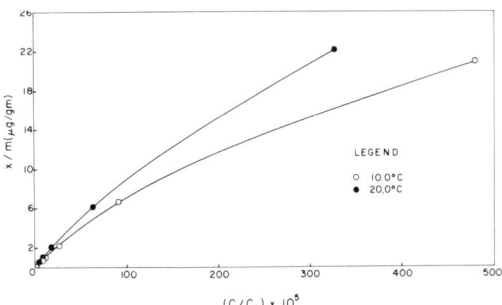

Fig. 5B—Corrected adsorption isotherms of γ-BHC from hexane on Ca-Staten Peaty Muck at two temperatures.

peaty muck approach linearity particularly over the lower part of the concentration range. It is clear from the normal isotherms in Fig. 5A that temperature had no consistent effect on adsorption on Staten peaty muck over the whole range of concentrations studied. It is possible that the overall adsorption process could change from exothermic to endothermic at a certain surface coverage or concentration range, as indicated in Fig. 5A. This fact reflects the heterogeneity of the peat surface. Inspection of the corrected isotherms for the entire concentration range (Fig. 5B), however, reveals that increasing the temperature results uniformly in increased adsorption over the entire range. Over the lower part of the concentration range Fig. 5A and 5B both the normal and corrected isotherms reveal a uniform temperature effect corresponding well with the temperature behavior of aqueous γ-BHC on the same adsorbent.

Freundlich Constants

The Freundlich equation for isotherms is frequently applicable to the adsorption of solutes from solutions. It is written as

$$x/m = kc^{1/n}, \quad [1]$$

where k and n are constants. From the form of the equation it is apparent that unlike with the Langmuir isotherm, no limiting saturation value is obtained. The equation is essentially empirical, based on no theoretical model. Kipling (1965) nevertheless showed that the Freundlich equation can be derived by combining an expression for the free energy of the surface with the Gibbs adsorption equation. But this derivation holds only for dilute solutions where $x/m = \Gamma$, the surface excess. The final result is an equation which is converted into the Freundlich equation if the term $RT(x/m)_m/\sigma_1 - \sigma_2)$ is equated with $1/n$. σ_1 and σ_2 are the respective free surface energies when the surface is in contact with pure solvent and pure solute, $(x/m)_m$ is the maximum monolayer capacity for the solute per gram of adsorbent, T is the absolute temperature, and R is the universal gas constant.

In addition to the constant k obtained when log x/m is plotted against log c, another constant (which we may designate as k') is obtained when log x/m is plotted against log (c/c_o). The k and k' values serve as indices of adsorption. They equal the hypothetical x/m values for a constant c or c/c_o value—c and c/c_o, respectively equaling unity in the case of k and k'. The usefulness of the k' values lies in the assumption that although the curves that relate fugacity and concentration may vary widely with tempera-

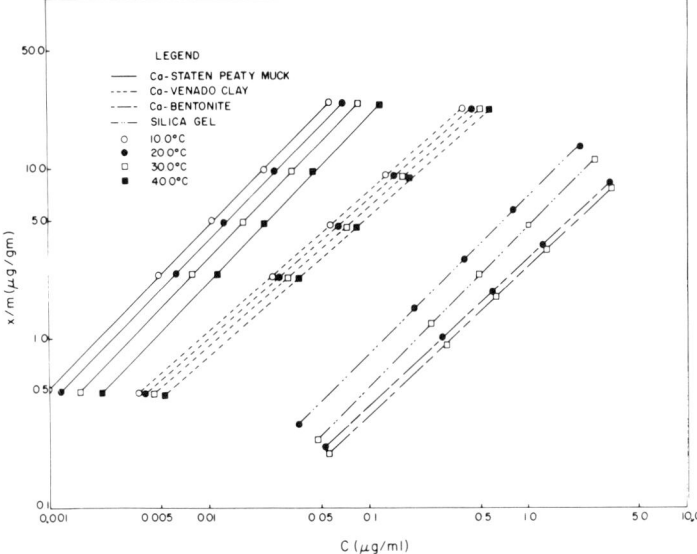

Fig. 6—The adsorption of γ-BHC from water on various adsorbents at four temperatures plotted according to the Freundlich equation.

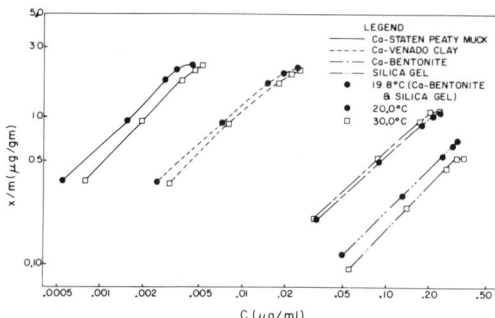

Fig. 7—The adsorption of β-BHC from water on various adsorbents at three temperatures according to the Freundlich equation.

Table 1—Freundlich constants for the isotherms

Isomer and solvent	Adsorbent	°C	k	k'	1/n
Aqueous γ-BHC	Ca-Staten peaty muck	10	377		.956
		20	331	2,630	.969
		30	269	3,350	.981
		40	197		.983
	Ca-Venado clay	10	53.8		.866
		20	45.7	277	.841
		30	41.3	363	.845
		40	36.1		.846
	Ca-bentonite	20	2.92	19.5	.886
		30	2.71	28.2	.911
	Silica gel	20	6.88	61.2	.938
		30	4.62	54.2	.957
γ-BHC in hexane	Ca-Staten peaty muck	10	0.974	3,340	.886
		20	0.823	6,180	.945
Aqueous β-BHC	Ca-Staten peaty muck	20	456	1,170	.950
		30	437	1,550	.990
	Ca-Venado clay	20	62.8	148	.861
		30	60.2	187	.883
	Ca-bentonite	19.8	3.96	9.33	.863
		30	4.46	13.9	.885
	Silica gel	19.8	2.13	5.60	.971
		30	1.64	5.80	.985

Table 2—Relation between Freundlich $1/n$ values and temperature

Isomer and solvent	T_2	T_1	T_2/T_1	$(1/n)_{T_2}/(1/n)_{T_1}$ Staten	Venado	Bentonite	Silica gel	where $T_2 > T_1$
	(°C)	(°C)	(°K/°K)					
Aqueous γ-BHC	20	10	1.035	1.031	.97			
	30	20	1.034	1.012	1.005	1.028	1.020	
	40	30	1.033	1.002	1.002			
γ-BHC in hexane	20	10	1.035	1.005				
Aqueous β-BHC	30	20	1.034	1.042	1.026	1.025	1.014	

Table 3—Concentration of γ- and β-isomers of BHC in saturated aqueous and hexane solutions at 10, 20, and 30C

Isomer	Solvent	Temperature, °C	Concentration of saturated solution
γ	distilled water with 0.5% ETOH	20	8.5 μg/ml
		30	13.1 μg/ml
γ	nanograde hexane	10	9.76 mg/ml
		20	13.97 mg/ml
β	distilled water with 0.5% ETOH	20	2.7 μg/ml
		30	3.6 μg/ml

ture, solvent, or the isomer used as solute, those that relate fugacity and reduced concentration will tend to group closely together or form a single curve. This implies an assumption that the factor determining a solute's fugacity in the solvent is the position of a particular solute concentration relative to the saturated solubility. The k' values can consequently be considered as the hypothetical amount adsorbed from solutions having a nearly constant equilibrium solute fugacity. Since the fugacity of the solute in the bulk phase (and therefore in the adsorbed phase) can thus be considered to be essentially constant, discussion of k' values rules out any consideration of the influence on adsorption of differing solute fugacities in the bulk phase.

Graphs of $c/(x/m)$ vs. c, according to the Langmuir adsorption equation, in no case yielded linear plots. Plots of log x/m vs. log c for all the γ-BHC aqueous isotherms yielded linear plots (Fig. 6). Nearly linear log-log plots were obtained for the lower concentration range of the Staten peaty muck-hexane isotherms. The log-log plots for the β-BHC isotherms (Fig. 7) are essentially linear if the highest concentration point for each isotherm is ignored. Freundlich constants for the aqueous γ-BHC and β-BHC isotherms and for the γ-BHC-Staten peaty muck-hexane isotherms were determined by the least squares method. For the β-BHC isotherms the highest concentration point was neglected in the calculation of these constants. Only the lowest five points were used to calculate the Freundlich constants in the case of the Staten peaty muck-hexane system. These constants are presented in Table 1. In each case k' was determined from the calculated k and $1/n$ values by means of the equality $k' = kc_o^{1/n}$, which follows from the Freundlich adsorption equation if c/c_o is substituted for c. The same slope, $1/n$, for the log-log plots holds regardless whether log c or log (c/c_o) is plotted on the abscissa.

Except with the β-BHC-bentonite-water system, the k values understandably decrease as temperature increases. Similarly, the solubility-corrected intercept values, k', increase in each case as temperature increases.

It is interesting that, except in the γ-BHC-Venado clay-water system at 10C, the $1/n$ values increase uniformly for each adsorbent as temperature increases. This is predicted from Kipling's derivation of the Freundlich equation for dilute solutions. It was observed that by letting $[RT(x/m)_m]/(\sigma_1-\sigma_2)$ equal $1/n$, Kipling's expression reduced to the Freundlich equation. One would therefore expect $1/n$ values for the isotherms of the system to relate to one another as the ratio of the corresponding absolute temperatures T_2/T_1. Table 2 shows that, except with the γ-BHC-Venado clay-water system at 10 and 20C, all the $1/n$ ratios are within 3.2% of the corresponding T_2/T_1 ratios. This validates the nature of the relationship between temperature and $1/n$ given by the above equality.

A consideration of the relationship between $1/n$ values and temperature would lead one to conclude that normal isotherms of the same system for different temperatures will inevitably cross at some equilibrium concentration point since they tend to converge in the lower concentration range. That will be strictly true, however, only if the same $1/n$ values apply to the entire concentration range studied. Thus, the crossing (at about 24 $\mu g/ml$ equilibrium concentration) of the 10 and 20C isotherms for the γ-BHC Staten peaty muck-hexane system (Fig. 5a) can be explained on this basis. In the corrected isotherms, since the relative position of the isotherms for a given system is in general completely reversed from that in ordinary isotherms, corrected isotherms would be expected never to cross, but to diverge throughout the entire concentration range. This divergence is illustrated in Fig. 5B. Corrected isotherms for Venado clay behaved similar to other corrected isotherms.

Except in silica gel-water systems, Table 1 shows k values higher for β-BHC than for γ-BHC. The k values on silica gel that are smaller for β-BHC than for γ-BHC may be discussed in relation to the k values for the bentonite-water systems. It appears that because of the zero dipole moment of β-BHC and the consequent absence of any dipole-dipole type of interaction as an adsorptive force, the net adsorptive forces are less between β-BHC molecules and the adsorbents silica gel and Ca-bentonite than between γ-BHC molecules and the same absorbents. The reason why k values for β-BHC decreased with silica gel but increased somewhat with Ca-bentonite relative to γ-BHC may be related to spatial or geometric factors. It may be that once β-BHC molecules are adsorbed between clay layers and possibly, between the saturating cations on the oxygen surface of Ca-bentonite, their displacement by water molecules is more difficult than such displacement on the relatively "unprotected" oxygen surface of silica gel. Thus even though a higher solution fugacity of β-BHC (than of γ-BHC) provides a greater driving force toward adsorption, the relative energetic ease with which water molecules can displace these β-BHC molecules on most sites on silica gel (because of the dipole moment of the former) results in k values lower than those obtained for γ-BHC. The higher k values of β-BHC than of γ-BHC for the other three adsorbents is probably due primarily to the lower solubility limit of β-BHC and its consequent higher fugacity in solution than that of γ-BHC at equal concentrations.

As seen in Table 1, values of k' increase with temperature for each system. Even when no clear difference between the two corrected isotherms could be observed on the graphs, as in the silica gel-water systems, a definite small increase in k' is seen to result from the temperature increase. Since the solute fugacity is considered to be constant here, temperature may have effected such increases through its influence upon fugacity of the solvent molecules adsorbed on the surface. Comparison of k' values for the γ-BHC-hexane and γ-BHC-water systems containing Ca-Staten peaty muck reveals that on this adsorbent the γ-BHC molecule actually competes more effectively with hexane molecules than with water molecules. This would lead one to conclude that the polar character of γ-BHC may be as important on this adsorbent as we have already assumed to be true on the crystalline adsorbents. It is likely that γ-BHC competes with hexane for sites most effectively on the mineral portion of this soil. Adsorption from the water for corresponding adsorbents gave uniformly lower k' values for β-BHC than for γ-BHC. The reason would appear to lie in the nonpolar character of the β-BHC molecule, and the consequent reduction of its ability to compete with water molecules. The lowest k' values (for β-BHC on silica gel) may be explained as was done above in the discussion of the k values.

LITERATURE CITED

1. Edwards, C. A. 1966. Insecticide residues in soils. Residue Reviews 13:83–132.
2. Faust, S. D., and I. H. Suffet. 1966. Recovery separation and identification of organic pesticides from natural and potable waters. Residue Reviews 15:44–116.
3. Gargantivi, H., O. Gianotti, and R. DeTella. 1959. Leaching of BHC (gamma isomer) from Bauru sandy soil. Bragantia 16:73–79, 1959. Soils Fert. 22(3), 1943.
4. Kipling, J. J. 1965. Adsorption from solutions of non-electrolytes. Academic Press, New York.
5. Lewis, G. N., and M. Randall. 1923. Thermodynamics and the free energy of chemical substances. McGraw-Hill Book Co., New York.
6. Lichtenstein, E. P. 1958. Movement of insecticides in soils under leaching and non-leaching conditions. J. Econ. Ent. 51:380–383.
7. Lind, E. L., M. E. Hobbs, and P. M. Gross. 1950. The electric moments of five of the isomeric hexachlorocyclohexanes. J. Amer. Chem. Soc. 72:4474–77.
8. Shepard, H. 1951. The chemistry and action of insecticides. McGraw-Hill Book Co., New York.
9. Swanson, C. L. W., F. C. Thorp, and R. B. Friend. 1954. Adsorption of lindane by soils. Soil Sci. 78:379–388.
10. Robeck, G. D., K. A. Dostel, J. M. Cohen, and J. F. Kreissal. 1965. Effectiveness of water treatment processes in pesticide removal. J. Amer. Water Works Assoc. 57,181.

Editors' Comments
on Papers 10 Through 17

10 BEST, WEBER, and WEED
Competitive Adsorption of Diquat^{2+}, Paraquat^{2+} and Ca^{2+} on Organic Matter and Exchange Resins

11 BAILEY, WHITE, and ROTHBERG
Adsorption of Organic Herbicides by Montmorillonite: Role of pH and Chemical Character of Adsorbate

12 FRISSEL and BOLT
Interaction between Certain Ionizable Organic Compounds (Herbicides) and Clay Minerals

13 SENESI and TESTINI
Physico-Chemical Investigations of Interaction Mechanisms between s-Triazine Herbicides and Soil Humic Acids

14 KHAN
Equilibrium and Kinetic Studies of the Adsorption of 2,4-D and Picloram on Humic Acid

15 MORTLAND and MEGGITT
Interaction of Ethyl N,N-Di-n-propylthiolcarbamate (EPTC) with Montmorillonite

16 SALTZMAN and YARIV
Infrared and X-ray Study of Parathion-Montmorillonite Sorption Complexes

17 LEENHEER and AHLRICHS
A Kinetic and Equilibrium Study of the Adsorption of Carbaryl and Parathion upon Soil Organic Matter Surfaces

MECHANISMS OF ADSORPTION

The determinant role of the adsorption process for the behavior and fate of the pesticides in soils has resulted in a great effort to understand the mechanisms involved. The most widely used approach was to infer the adsorption mechanism from the equilibrium adsorption data. It has been observed that in most cases the distribution of pesticides between the solid and liquid phases fits an empirical relationship known as the Freundlich equation: $x/m = KC^{1/n}$, where x/m is the amount adsorbed, C is the equilibrium adsorption concentration, and K and n are constants. In some cases (e.g., Paper 6) when a maximum adsorption was attained, the results could be fitted by the Langmuir equation: $x/m = KCX/(1 + KC)$, where X is the maximum adsorption. However, for low solution concentrations, this equation is similar to the Freundlich equation.

A useful relationship between the shape of the adsorption isotherm and the adsorption mechanism is based on the classification of adsorption isotherms of Giles and his co-workers (1960). Green (1974) showed that the adsorption mechanisms of the main chemical classes of pesticides could be explained by the isotherm shape.

Mechanisms of adsorption were generally studied by combined techniques on isolated soil fractions or model materials. As the adsorption mechanisms and the characterization of clay-organic complexes are of vast and general interest, they were treated comprehensively in many books and review articles. By contrast, the mechanisms involved in organic matter-pesticide interactions are little understood. The very complex and variable soil organic materials have not yet been definitively characterized, and moreover some spectroscopic techniques can not easily be applied to study them.

Cationic Pesticides

Best, Weber, and Weed (Paper 10) studied the adsorption of cationic pesticides by organic matter and exchange resins. By comparing the adsorption of diquat^{2+}, paraquat^{2+}, and Ca^{2+} by different soil organic fractions at two pH levels, the authors were able to point out the main features of the adsorption of organic cations by soil organic matter. Both the mineral ion and the organic cations seemed to compete for the same adsorption sites. At low pH levels, at which the adsorption sites must be strong acidic groups, the order of preference was paraquat > diquat > Ca. At neutral pH, calcium was competitive with the organic cations for the weakly acid exchange sites, the order of preference being Ca > paraquat > diquat. An explanation for this effect could be the higher charge density at the higher pH level, which favored the adsorption of the small calcium ion. Khan (1974) demonstrated that the adsorption of theses two organic cations by humic acids is dependent on the saturating cation; divalent-substituted humic acids adsorbed more diquat and paraquat than the trivalent-substituted humic acids did.

Hence, as shown by the results presented in Paper 10, the predominant mechanism for the adsorption of organic cations by soil organic matter is ion exchange. The formation of charge transfer complexes between humic acids ad bipyridylium herbicides has also been demonstrated (Khan, 1974).

A similar main mechanism has been observed for the adsorption of cationic pesticides by clays. Weber, Perry, and Upchurch (1965) showed that these compounds were adsorbed by clays up to their exchange capacities. The charge density of the clay surfaces, expressed as discrete adsorption sites, is a determinant factor in adsorption (Weed and Weber, 1968). In addition to ion exchange, other adsorption mechanisms could be implied: hydrogen bonding, ion-dipole, and physical forces. Adsorption of cationic pesticides by clays is affected by their molecular weight, functional groups, and molecular configuration (Mortland, 1970). Significant differences between the bonding energy of diquat and paraquat and between their molecular configuration in the adsorbed state on clays have been found in thermodynamic and spectroscopic studies (Burchill, Hayes, and Greenland, 1981).

Several differences between adsorption by clays and humic acids were observed and are due to the peculiarities of the humic substances as adsorbents. Unlike clays, which have a rigid structure and fixed charge distribution, the humic substances have flexible exchange sites, and the charge density is pH-dependent. In addition, the conformation and dimensions of the humic polymer in a specific medium determine the availability of the adsorption sites (Burchill, Hayes, and Greenland, 1981).

Basic Pesticides

Because basic pesticides include an important group of herbicides (the s-triazines) that are strongly retained in soils, their mechanism of adsorption has received special attention.

Similar to the organic cations, the basic pesticides could be adsorbed by clays by cationic adsorption. However, in the case of basic pesticides, this adsorption mechanism is conditioned by the acidity of the medium. Bailey, White, and Rothberg (Paper 11) present results for the adsorption of several basic pesticides by montmorillonite and discuss the mechanisms involved. These results, confirmed in many other investigations, unequivocally indicate the predominant role of the acidity of the medium, namely, the surface acidity of the clay, in determining the adsorption mechanism. When the surface acidity was more than about two pH units higher than the dissociation constant of the adsorbate, adsorption of its molecular form occurred mainly by physical van der Waals forces. However, other adsorption mechanisms, such as hydrogen bonding and coordination between adsorbate and the exchangeable cation, could also occur. When the surface acidity of the clay was at least one or two pH units lower than the dissociation constant, chemical adsorption could occur. This reaction is made possible by the protonation of the adsorbate at the acid surface of the clay, due either to the presence of acid exchange sites on the clay or to the dissociation of water associated with the exchangeable cations.

The relative importance of the physical and chemical adsorption mechanisms and their contribution to the magnitude of adsorption of weak bases by clays were emphasized by Frissel and Bolt (Paper 12). Although the adsorption of all the basic compounds investigated increased as the pH of the medium decreased, a steep increase occurred at pH values much higher than the dissociation constants, explainable by the fact that the authors take into consideration the reaction in the bulk solution, and not the surface acidity of the clay. Calculations done according to the Boltzmann distribution law of the concentration of adsorbed ions in the proximity of an adsorbing surface, gave very high values for adsorption by van der Waals forces, as compared with Coulomb forces, perhaps accounting for the strong adsorption of weak bases by Na-montmorillonite at pH values much higher than the dissociation constant.

Although a similar relationship between the adsorption of weak bases and pH was described in both Papers 11 and 12, Bailey, White, and Rothberg suggested that the higher acidity of the montmorillonite surfaces compared with the bulk solution could explain the strong adsorption of weak bases at a pH higher than their dissociation constants.

Infrared techniques were used successfully to obtain information about the nature of the adsorptive bonds formed in the adsorption of basic pesticides by clays. Protonation of 3-aminotriazole at the clay surface was demonstrated by Russel, Cruz, and White (1968).

Many studies of the adsorption of weakly basic pesticides by soil have shown that the most important soil fraction related to adsorption is the organic matter. Several adsorption mechanisms were postulated for the very complex process of interaction between organic colloids and weak bases. Weber, Weed, and Ward (1969) demonstrated that maximum adsorption of seven s-triazines by soil organic matter occurred at pH levels close to the pKa values of the compounds. The molecular structure of the adsorbate and the pH of the adsorbing medium determined the amounts adsorbed. The pH-dependent adsorption and the direct relationship between pH, dissociation constant, and adsorption are strong evidence that protonated basic pesticides could be adsorbed through an ion exchange process.

Senesi and Testini (Paper 13) studied the adsorption of s-triazines by humic acids, by using elemental, thermal, infrared, and spin resonance analyses. The results obtained indicated that s-triazines are adsorbed by humic acids, forming stable complexes. As shown by the infrared analysis, one of the binding mechanisms is formation of ionic bonds (salt linkages), following proton transfer from the humic acids to the s-triazine molecules. An additional mechanism is hydrogen bonding, with the involvement of the carbonyl group of the humic acids and the amino group of s-triazines. The involvement of other functional groups of humic acids (carboxylic, phenolic, alcoholic) in hydrogen bonding with the amino group or the ring nitrogen atoms of the s-triazines is also possible. The authors presented theoretical and experimental evidence for the occurrence of electron donor-acceptor processes involving free radical intermediates and conclude that (a) the adsorptive capacity of humic acids for s-triazines depends on the content of acidic functional groups, capable of forming ionic and hydrogen bonds, and (b) the higher the formation of such bonds, the lower the tendency to form electron-acceptor complexes and covalent bonds. In contrast to the prevailing opinions, they suggested that the basicity of the s-triazines is not the main factor determining their adsorption, probably because, in addition to ionic and hydrogen bonding mechanisms, the electron donor-acceptor processes could play an important role in the adsorption of s-triazines by humic acids.

In a comparative study of the adsorption of the weakly basic herbicide terbutryn by Ca-montmorillonite, humic acids, and their mixtures, Gaillardon, Calvet, and Terce (Paper 4) observed that in acid conditions (pH ~3) the adsorption forces, as shown by the shape of the adsorption isotherm, are stronger in humic acids. This observation was confirmed by the desorption isotherm, which showed a hysteresis phenomenon for the humic acids and the clay-organic mixtures containing more than 20% humic acids. The stronger retention of weak bases by the organic colloids than by clays was explained by the involvement of several simultaneous adsorption mechanisms. H-bonding was suggested in several works as a possible reaction (Hayes, Stacey, and Thompson 1968; Li and Felbeck, 1972).

The role of different fractions of soil organic matter (humic substances, humin, nonoxidizable fraction) in the adsorption of prometryn was studied by Kozak, Weber, and Sheets (1983). They found that humic substances (humic and fulvic acids) showed the highest affinity for prometryn; both the molecular and the protonated form were adsorbed. Humin adsorbed the molecular form preferentially, probably by hydrophobic bonding. The nonoxidizable fraction had the lowest adsorption capacity; in this case, mineral adsorptive surfaces contributed to adsorption.

Acidic Pesticides

Acidic pesticides (such as 2,4-D, 2,4,5-T, picloram and dinoseb) are characterized by their ability to ionize in aqueous solutions, forming anion species. At pH values lower than their dissociation constant, the molecular form will be present, and increasing the pH above pKa will favor dissociation. It has been shown that acidic pesticides are adsorbed by soils and the adsorption is correlated with soil organic matter content (Grover, 1971; Hamaker, Goring, and Youngson, 1966). Although the mineral fraction seems to be of little importance in determining soil adsorption of acidic pesticides, several studies have attempted to clarify the interactions between these compounds and clay minerals.

Acidic pesticides, in their anionic form, are expected to be repelled by the negatively charged clay surfaces. However, a part of the soil clays and amorphous materials have some pH-dependent positive charge at low pH values. Still, under such conditions, most acidic pesticides are in the molecular form, so that significant adsorption by formation of anionic bonds is improbable. Hence, adsorption of acidic pesticides by clays will occur only within a pH range that enables the concomitant presence of positive adsorption sites on clay and dissociation of the adsorbate (Green, 1974).

The effect of pH on the adsorption of acidic pesticides by clays was studied by Frissel and Bolt (Paper 12). They showed that both negative and positive adsorption occurred as a function of pH, which affected at the same time the dissociation of the adsorbate and the charge of the adsorbent. This effect was evident mainly in the case of illite, as montmorillonite had no evident positive charge over the pH range studied.

Bailey, White, and Rothberg (Paper 11) studied the adsorption of several acidic compounds by montmorillonite. With the H-saturated clay, positive adsorption occurred for picloram and 2,4,5-T, both positive and negative adsorption occurred for phenoxyacetic acid, and negative or no adsorption was observed for 2,4-D, benzoic acid, and amiben. With Na-montmorillonite, negative adsorption occurred for all compounds. The authors concluded that positive adsorption of acidic compounds will occur at pH values of the bulk solution about 1 to 1.5 units above the dissociation constant. The postulated adsorption mechanisms are proton association and, for the molecular form, van der Waals adsorption; H-bonding is also a probable mechanism.

As already mentioned, adsorption of acidic compounds by soils is well correlated with the organic matter content. Soil organic colloids are either negatively charged or uncharged, so that significant adsorption by anion bonding is excluded. For some acidic compounds, an inverse relationship between their adsorption and the pH of the bulk solution has been observed, indicating the preferential adsorption of the undissociated species (Grover, 1971). Khan (Paper 14) investigated the mechanism of adsorption of two acidic herbicides by humic acid by studying the equilibrium and kinetic aspects of the reaction. The adsorption of 2,4-D and picloram by humic acid fitted the Freundlich-type isotherm, The adsorption rate was initially fast, but decreased with time, suggesting that diffusion to the surface of the humic acid occurred at the initial times, followed by slow intraparticle diffusions. Calculations of rate constants, activation energies, heats of activation, and entropies of activation indicate that the adsorption mechanism is physical adsorption, probably by van der Waals forces and hydrophobic bonding.

Biggar and Cheung (1973) calculated the thermodynamic parameters associated with the adsorption of picloram by soils. They proposed adsorption of the protonated form by clay surfaces, hydrogen bonding, van der Waals interactions, and electrostatic interactions as possible interaction mechanisms. The dominating mechanisms are variable, depending on pH changes. An additional adsorption mechanism was suggested by Arnold and Farmer (1979). Their data indicated that complex formation between picloram and exchangeable polyvalent cations is likely, especially for Cu^{2+}. However, the practical significance of this mechanism is doubtful at the usually very low Cu^{2+} concentration in soil.

Non-Ionic Pesticides

As most of the pesticides used at present are non-ionic compounds, the understanding of the mechanisms determining their partition between the soil phases had received much consideration. In this class of pesticides are hundreds of compounds belonging to such chemically different groups as chlorinated hydrocarbons, organophosphates, carbamates, ureas, anilines, anilides, amides, uracils, and benzonitriles. The great differences between the properties of these groups, and even between compounds within a group, are reflected in the variability of the adsorption behavior. The only common feature of the adsorption mechanism that has been obviously demonstrated for all the non-ionic pesticides is the predominant role of the organic colloids. However, some additional general characteristics are largely accepted at present.

Neutral organic pesticides are not expected to be considerably attracted by the charged clay surface and must compete with the hydration water for adsorption sites. Consequently, such pesticides are only slightly or practically not adsorbed from water solution by clay minerals. However, at low water contents, significant adsorption could occur. The main adsorption mechanisms postulated are formation of cation-dipole and coordination bonds, hydrogen bonds, and van der Waals attraction (Green, 1974). Mortland and Meggitt (Paper 15) used infrared technique to study the mechanisms of adsorption of EPTC, a carbamate herbicide, by montmorillonite saturated with different metal ions. Adsorption of gaseous EPTC by clays was tested in a vacuum desiccator. The results showed that under these conditions EPTC forms coordination complexes with the metal ions through the carbonyl group. Coordination through the nitrogen atom is also suggested if the molecule is sterically hindered. X-ray and infrared data indicate that the EPTC molecule is oriented with its long axis approximately parallel to the silicate sheets, so that the methylene groups can form H-bonds with the oxygen atoms of the clay surface. The possibility of association of EPTC to the metal ion on clays through a water bridge was not confirmed except for the Al-saturated clay. EPTC-montmorillonite complexes were stable in the atmosphere, but EPTC was quantitatively displaced by water.

Saltzman and Yariv (Paper 16) suggested that several mechanisms could be involved in the adsorption of parathion by montmorillonite. At ambient relative humidity, parathion could be coordinated with the metallic exchangeable cations through a water bridge, the active site in the parathion molecule being the nitro-group. Upon dehydration, direct coordination to the metallic cations occurred with the monovalent cations but not with the divalent ones, which strongly retained water molecules in their hydration shell. With the Al-saturated montmorillonite both the NO_2 and the $P=S$ group, or the $P=S$ group alone could be bound to the cation through a water bridge. This group seems also to be bound to divalent cations by coordination in dehydrated systems. Another suggested mechanism is interaction between the aromatic ring and oxygen atoms of the silicate sheet of the clay mineral.

The infrared and X-ray studies of non-ionic pesticides-clay complexes indicate that the main adsorption sites on clays are the metallic cations. These compounds compete more or less successfully with water for adsorption sites. Usually, more than one adsorption mechanism is involved in the binding of such compounds, and the nature of binding is determined by the hydration status of the clay and the characteristics of the adsorbent. The most important mechanism is likely to be coordination of the polar non-ionic molecules to the exchangeable cation through a water bridge.

Although the predominant role of soil organic matter in the adsorption of neutral organic pesticides is generally accepted, the adsorption mechanisms are not well understood. A kinetic and equilibrium study of the adsorption of two non-ionic insecticides, parathion and carbaryl, by organic matter extracted from soils was carried out by Leenheer and Ahlrichs (Paper 17). The results showed that the different types of organic matter had similar adsorption capacities, but the H-saturated organic matter had a higher adsorptive capacity than the Ca-saturated materials. This difference was explained by the more hydrophobic character of H-organic matter compared with the Ca-saturated adsorbent, which had a higher affinity for water. The authors assumed that functional groups of organic matter are not directly involved in adsorption but could affect the ratio of hydrophilic/hydrophobic sites of the adsorbing surface. As demonstrated by the kinetic study, adsorption proceeded in two stages. The first, fast stage was interpreted to be transference of the solutes from solution to the surface of the adsorbent; the second, slower stage involved diffusion from the surfaces into the adsorbent. The assumption that the mechanism of adsorption of parathion and carbaryl by organic matter is hydrophobic bonding was also confirmed by the low heats of adsorption and the reversibility of the process.

Much evidence shows that a peculiar property of soil organic matter (the presence of hydrophobic structures in addition to other adsorption sites) confers on it the almost ubiquitous character of adsorbent as far as non-ionic pesticides are concerned, because such compounds usually have nonpolar functional groups that could be bound by the nonpolar parts of the humic materials through a hydrophobic interaction. As water competition is minimal in such interactions, the adsorption of non-ionic pesticides from water solution is possible. Numerous studies have suggested that hydrophobic bonding is the most important adsorption mechanism for a large number of pesticides (Carringer, Weber, and Monaco, 1975; Felsot and Dahm, 1979). The often found inverse relationship between the water solubility of pesticides and their adsorption by organic matter is considered to be evidence of hydrophobic interactions.

In addition to hydrophobic bonding, expected to be the main mechanism of adsorption of non-ionic pesticides by soil organic materials, other mechanisms such as H-bonding and coordination to inorganic cations could be involved, as with pesticides possessing functional groups capable of reacting with the polar active sites of the organic matter.

References

Arnold, J. S., and W. J. Farmer, 1979. Exchangeable Cations and Picloram Sorption by Soil and Model Adsorbents, *Weed Sci.* **27**:257-262.

Biggar, J. W., and M. W. Cheung, 1973, Adsorption of Picloram (4-Amino-3,5,6-Trichloropicolinic Acid) on Panoche, Ephrata, and Palouse Soils: A Thermodynamic Approach to the Adsorption Mechanism, *Soil Sci. Soc. America Proc.* **37**:863-868.

Burchill, S., M. H. B. Hayes, and D. J. Greenland, 1981, Adsorption, in *The Chemistry of Soil Processes,* D. J. Greenland and M. H. B. Hayes, eds., Wiley, New York, pp. 224-400.

Carringer, R. D., J. B. Weber, and T. J. Monaco, 1975, Absorption-Desorption of Selected Pesticides by Organic Matter and Montmorillonite, *Jour. Agric. Food Chem.* **23:**568-572.

Felsot, A., and P. A. Dahm, 1979, Sorption of Organophosphorous and Carbamate Insecticides by Soil, *Jour. Agric. Food Chem.* **27:**557-563.

Giles, C. H., T. H. MacEwan, S. N. Nakhwa, and D. Smith, 1960, Studies in Adsorption. Part XI. A System of Classification of Solution Adsorption Isotherms, and its Use in Diagnosis of Adsorption Mechanisms and in Measurement of Specific Surface Areas of Solids, *Chem. Soc. Jour.* 3973-3993.

Green, R. E., 1974, Pesticides-Clay-Water Interactions, in *Pesticides in Soil and Water,* W. D. Guenzi, ed., Soil Science Society of America, Inc., Madison, Wis. pp. 3-37.

Grover, R., 1971, Adsorption of Picloram by Soil Colloids and Various other Adsorbents, *Weed Sci.* **19:**417-418.

Hamaker, J. W., C. A. I. Goring, and C. R. Youngson, 1966, Sorption and Leaching of 4-Amino-3,5,6-Trichloropicolinic Acid in Soils, in *Organic Pesticides in Environment,* Advances in Chemistry Series 60, American Chemical Society, Washington, D. C., pp. 23-37.

Hayes, M. H. B., M. Stacey, and J. M. Thompson, 1968, Adsorption of s-Triazine Herbicides by Soil Organic Matter Preparations, in *Isotopes and Radiation in Soil Organic Matter Studies,* International Atomic Energy Agency, Vienna, pp. 75-90.

Khan, S. U., 1974, Humic Substances Reactions Involving Bipyridylium Herbicides in Soil and Aquatic Environments, *Residues Rev.* **52:**1-26.

Kozak, J., J. B. Weber, and T. J. Sheets, 1983, Adsorption of Prometryn and Metolachlor by Selected Soil Organic Matter Fractions, *Soil Sci.* **136:**94-101.

Li, G., and G. T. Felbeck, Jr., 1972, A Study of the Mechanism of Atrazine Adsorption by Humic Acid from Muck Soil, *Soil Sci.* **113:**145-148.

Mortland, M. M., 1970, Clay-Organic Complexes and Interactions, *Adv. Agron.* **22:**75-117.

Russell, J. D., M. I. Cruz, and J. L. White, 1968, The Adsorption of 3-Aminotriazole by Montmorillonite. *Jour. Agric. Food Chem.* **16:**21-24.

Weber, J. B., P. W. Perry, and R. P. Upchurch, 1965, The Influence of Temperature on the Adsorption of Paraquat, Diquat, 2,4-D and Prometone by Clays, Charcoal and an Anion-Exchange Resin, *Soil Sci. Soc. America Proc.* **29:**678-688.

Weber, J. B., S. B. Weed, and T. M. Ward, 1969, Adsorption of s-Triazines by Soil Organic Matter, *Weed Sci.* **17:**417-421.

Weed, S. B., and J. B. Weber, 1968, The Effect of Adsorbent Charge on the Competitive Adsorption of Divalent Organic Cations by Layer-Silicate Minerals, *Am. Mineralogist* **53:**478-490.

COMPETITIVE ADSORPTION OF DIQUAT^{2+}, PARAQUAT^{2+}, AND CA^{2+} ON ORGANIC MATTER AND EXCHANGE RESINS

J. A. BEST, J. B. WEBER, AND S. B. WEED

North Carolina State University, Raleigh, North Carolina[1]

INTRODUCTION

The organic compounds paraquat (1,1'-dimethyl-4,4'-dipyridilium dichloride) and diquat (1,1'-ethylene-2,2'-dipyridilium dibromide) are soluble and readily ionize in water to form divalent organic cations (Fig. 1).[2] They differ structurally in their charge spacing and degree of flexibility based on atomic scale models. Diquat has a charge spacing of 3-4 A., while paraquat charge spacing is approximately twice that, or 7-8 A. Paraquat is a more flexible molecule than diquat due to a single carbon-carbon bond connecting its two rings. Both compounds are very stable in neutral or acid solutions, but hydrolyze at pH levels above 9.0 to form free radicals.

Diquat and paraquat are used commercially as herbicides. Diquat is used primarily as an aquatic herbicide while paraquat is a nonselective contact herbicide. Both chemicals are adsorbed readily by soil colloids and thus rendered nonphytotoxic.

Adsorption studies with diquat and paraquat on silicate clays (6, 10) and with resins (9) have shown that the competitive adsorption of these compounds is a function of the surface charge density on the adsorbent. Equal adsorption of diquat and paraquat on external surfaces of clay minerals occurred when the surface charge density was about 8.5×10^4 esu/cm^2. Higher or lower values for the surface charge density of an adsorbent resulted in preferential adsorption of diquat and paraquat, respectively. Maximum adsorption of these compounds usually equals or approaches the CEC of the adsorbent, though this may be influenced by the counter-ion initially present (11). With the silicate clays, organic cations are usually more competitive than inorganic cations, apparently due to the supplemental effects of van der Waals forces (3).

Soil organic matter has been shown to have many different types of functional groups contributing to its CEC (5, 7). Estimates of the contribution of various groups to the total CEC placed 54% to carboxyl groups, 36% to phenolic and enolic hydroxyl groups, and approximately 10% or less to imide nitrogen groups (1). As much as 20% of the exchangeable hydrogen in acid peats has been attributed to strong acid groups (4).

Potentiometric titrations of organic materials with bases show the heterogenous character of the acid groups (2). Chatterjee and Bose (2) found the "base-exchange capacity," determined by base titration, to vary in order of Ca(OH)$_2$ > Ba(OH)$_2$ > NaOH calculated at any pH level, thus showing a "cation effect" with inorganic exchange for hydrogen ions with bases.

The purpose of the work reported in this paper was a) to examine the competitive adsorption of diquat versus paraquat and diquat + paraquat versus Ca on various organic matter fractions and materials and b) to use the relative adsorption of the organic cations as calipers for estimating the surface charge densities of the organic fractions.

METHODS AND MATERIALS

Analytical grade diquat, paraquat, CaCl$_2$, and Ca(OH)$_2$ were used. The adsorbents included a Histosol and its humic and humin fractions, Aldrich commercial humic acid, and Amberlite IR-120 (strong sulfonic acid H-form) and IRC-50 (weak carboxylic acid H-form) resins.

The Histosol, obtained near Pantego, North Carolina, was air-dried and sieved to a particle size of 210 μ or less. A portion of this sieved

---

[1] Published with the approval of the Director of the North Carolina Agricultural Experiment Station, Raleigh, N. C. as Paper No. 3606 in the Journal Series. This investigation was supported by grant number EP 00813 from the Environmental Protection Agency.

[2] Supplied by Chevron Chemical Company, Richmond, California.

material was retained as the untreated Histosol adsorbent. The remainder was washed repeatedly with an HF:HCl acid solution for purposes of removing mineral material (1). Afterwards the acid was removed with deionized water washings. A portion of this material was retained as the acid-washed Histosol adsorbent. The remainder was placed in contact with a 0.1 N KOH solution (50 g peat/l) to remove humic materials. After shaking for 24 hr, the solution was centrifuged and decanted. This procedure was repeated several times until the supernatant solution remained clear, indicating that humic materials were essentially completely removed from the peat. The humic material was precipitated with HCl and removed by centrifugation and followed by several deionized water washings. The humin fraction, that which remained after KOH washings, was titrated with HCl to pH 7 and washed repeatedly with deionized water until Cl was removed ($AgNO_3$ test). All four of the Histosol materials were freeze-dried and stored in glass containers. Their ash contents were 13, 8, 3, and 8% for the nontreated Histosol, acid-washed Histosol, and its humic and humin fractions, respectively, as determined by dry combustion at 400° C. The Aldrich humic acid was used as purchased. A low ash content was necessary as mineral matter such as clay minerals is very adsorptive for diquat and paraquat (6, 10).

The IR-120 resin was purchased in the H-form and washed with HCl, followed by deionized water washings until free of excess acid ($AgNO_3$ test). Portions of the resin were K- or Ba-saturated by titration with KOH or $Ba(OH)_2$, respectively. Excess K and Ba were removed with water. The resins, with a moisture content between 30 and 40%, were stored in glass containers. The IRC-50 resin was used as purchased and had similar moisture content.

Diquat versus Paraquat

The organic matter and resin samples were weighed and placed in 50-ml polypropylene screw-cap centrifuge tubes. The amounts were 25 and 10 mg, respectively. Adsorbents used are shown in Table 1.

Forty ml of solution was added to each sample. Each adsorbent received four types of solution; diquat alone, paraquat alone, diquat + paraquat as an equal molar mixture, or deion-

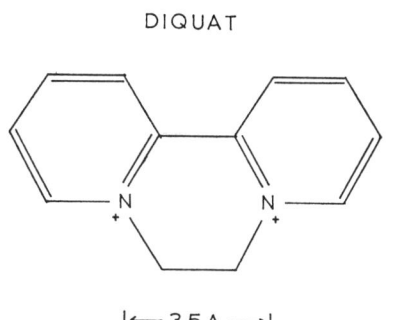

FIG. 1. Physical structures of paraquat and diquat.

ized water. The total adsorbate concentration added was 160 and 800 meq/100 g of adsorbent for the organic matter samples and the resins, respectively. Each sample was shaken for 18 hours on a reciprocating shaker at ambient temperature (25° C). At the end of shaking, the colloidal particles were removed by high speed centriguation and the solution concentrations of diquat and/or paraquat were determined spectrophotometrically according to the methods of Willard et al. (12). The pH of the solutions were measured by employing a glass electrode pH meter and standard reference buffers. The amount of adsorbate adsorbed was considered to be the decrease in solution concentration for each compound. The experiment was run in triplicate.

Diquat + Paraquat versus Ca

The adsorbents used are listed in Tables 2 and 3. The humin fraction differs from the previous study in that it was completely H-saturated. Ten mg of each adsorbent were placed in six 50-ml centrifuge tubes. Two 20-ml aliquots of $CaCl_2$, $Ca(OH)_2$ solutions, or deionized water were placed in the tubes. The Ca solutions were such that the 20 ml was equal to 200 meq/100 g

TABLE 1
The adsorption of diquat and paraquat in competition and alone on organic matter and IR-120 resins

Adsorbent	Chemical Added		Chemical Adsorbed			Ratio[a] $\frac{P}{D+P}$	pH at Equilibrium
	Paraquat	Diquat	Paraquat	Diquat	Total		
			meq/100 g				
(Untreated) Histosol	0	0	0.0	0.0	0.0		5.7
	160	0	56.2	0.0	56.2		4.8
	0	160	0.0	55.2	55.2		4.8
	80	80	28.4	27.3	55.7	0.51	4.8
HF:HCl Washed histosol	0	0	0.0	0.0	0.0		3.8
	160	0	47.2	0.0	47.2		3.4
	0	160	0.0	47.7	47.7		3.4
	80	80	27.3	22.4	49.7	0.55	3.4
Humic fraction	0	0	0.0	0.0	0.0		4.9
	160	0	75.0	0.0	75.0		3.5
	0	160	0.0	76.4	76.4		3.5
	80	80	40.8	35.8	76.6	0.53	3.5
Humin fraction	0	0	0.0	0.0	0.0		7.0
	160	0	78.1	0.0	78.1		6.0
	0	160	0.0	77.3	77.3		6.0
	80	80	42.1	36.1	78.2	0.54	6.0
Aldrich Commercial Humic acid	0	0	0.0	0.0	0.0		5.8
	160	0	88.0	0.0	88.0		5.3
	0	160	0.0	87.6	87.6		5.3
	80	80	44.1	43.1	87.2	0.51	5.3

LSD$_{(.01)}$ = 2.7 meq/100 g (total adsorbed) and 0.5 meq/100 g (within mixture)

IR-120 H-saturated Resin	0	0	0.0	0.0	0.0		5.0
	800	0	354.0	0.0	354.0		3.1
	0	800	0.0	345.0	345.0		3.1
	400	400	127.0	217.0	344.0	0.37	3.1
IR-120 K-saturated Resin	0	0	0.0	0.0	0.0		6.0
	800	0	349.0	0.0	349.0		6.0
	0	800	0.0	343.0	343.0		6.0
	400	400	128.0	218.0	346.0	0.37	6.0
IR-120 Ba-saturated Resin	0	0	0.0	0.0	0.0		5.6
	800	0	203.0	0.0	203.0		6.1
	0	800	0.0	182.0	182.0		6.1
	400	400	70.0	124.0	194.0	0.36	6.1

LSD$_{(.01)}$ = 15.7 meq/100 g (total adsorbed) and 6.3 meq/100 g (within mixture)

[a] Ratio of paraquat (P) and diquat (D) adsorbed.

for the organic materials and 400 meq/100 g for the resins. After shaking for 18 hours, 20 ml of an equal molar mixture of diquat + paraquat, at a level equal to that of Ca previously mentioned, was added to one of each duplicate samples. The other received 20 ml of deionized water. This method of split application was necessary due to the hydrolysis of diquat and paraquat at high pH levels. The tubes were shaken again for 18 hours and analyzed for diquat, paraquat, or Ca, as previously mentioned. Ca concentrations were measured with an atomic absorption spectrophotometer. The amount of Ca or diquat and paraquat adsorbed was determined by differences in the initial and final concentrations. Each sample was run in triplicate.

TABLE 2

The competitive adsorption of diquat, paraquat and Ca as CaCl₂ on organic matter and resins

Adsorbent	Chemical Added			Chemical adsorbed			Ratio[b] $\frac{P}{D+P}$	Ratio[b] $\frac{P+D}{D+P+Ca}$	pH at Adsorption
	CaCl₂	Diquat	Paraquat	Diquat + Paraquat	Ca	Total			
				meq/100 g					
(Untreated)	200	0	0	0.0	74.9	74.9			4.8
Histosol[a]	0	100	100	59.2	8.0	67.2	0.51		4.8
	200	100	100	43.5	25.8	69.3	0.52	0.63	4.9
HF:HCl	200	0	0	0.0	52.3	52.3			4.0
Washed Histosol	0	100	100	49.1	0.0	49.1	0.53		4.0
	200	100	100	43.9	6.6	50.5	0.52	0.87	4.0
Humic	200	0	0	0.0	79.5	79.5			4.0
Fraction	0	100	100	83.2	0.0	83.2	0.53		3.9
	200	100	100	70.4	13.6	84.0	0.54	0.84	3.9
Humin	200	0	0	0.0	48.0	48.0			4.0
Fraction	0	100	100	46.7	0.0	46.7	0.57		3.9
	200	100	100	39.7	7.7	47.4	0.58	0.84	3.9

LSD$_{(.01)}$ = 5.5

IRC-50[c]	400	0	0	0.0	21.3	21.3			4.3
(H⁺)	0	200	200	25.6	0.0	25.6	0.59		4.0
	400	200	200	22.1	5.3	27.4	0.64	0.81	4.1
IR-120[c]	400	0	0	0.0	346.6	346.6			3.1
(H⁺)	0	200	200	344.0	0.0	344.0	0.46		3.1
IR-120[c]	400	200	200	214.1	117.5	331.6	0.42	0.65	3.1
(H⁺)									

LSD$_{(.01)}$ = 13.6

[a] Contained 30 meq/100 g of Ca initially.

[b] Ratio of diquat (D), paraquat (P), and/or Ca adsorbed as mixtures.

[c] The CEC based on manufacturer's data for the IRC-50 and IR-120 resins are 530 and 350 meq/100 g, respectively, on a wet weight basis.

RESULTS AND DISCUSSION

Diquat versus Paraquat

The extent of adsorption of diquat and paraquat as a mixture and separately is shown in Table 1. The same sites on organic matter and on H- and K-saturated resins probably adsorbed both chemicals in a similar fashion. This is evidenced by the equal total adsorption in all three systems. However, Ba was effective in lowering the total adsorption of diquat as compared to paraquat with the resin.

The amount of adsorption of diquat or paraquat when added as an equal molar mixture reveals the relative affinity of the exchange sites for either of the two adsorbates. With the organic matter samples the preference was always slightly in favor of paraquat while the resins strongly preferred diquat. This is illustrated by use of the ratio of paraquat (P) adsorbed to the total of diquat (D) + paraquat adsorbed (Table 1). A value of 0.50 denotes no preference while larger or smaller values indicate the preference is in favor of paraquat or diquat, respectively. The ratio was always larger than 0.50 with the organic matter adsorbents and consistently around 0.37 with the IR-120 resin. Also, the initial counter-ion on the resins did not effect the preference for diquat over paraquat.

The preferential adsorption of diquat over paraquat by the resins is consistent with results of other investigators and is attributed to the closely spaced charges of diquat being more effective than those of paraquat in matching the

TABLE 3

The competitive adsorption of diquat, paraquat, and Ca as Ca(OH)$_2$ on organic matter and resins

Adsorbent	Chemical Added			Chemical Adsorbed			Ratio[b] P	Ratio[b] P + D	pH at Adsorption
	Ca(OH)$_2$	Di-quat	Para-quat	Diquat + Paraquat	Ca	Total	D + P	$\frac{D + P}{D + P + Ca}$	
				meq/100 g					
(Untreated) Histosol[a]	200	0	0	0.0	183.7	183.7			7.2
	0	100	100	59.2	8.0	67.2	0.51		4.8
	200	100	100	67.5	117.1	184.6	0.53	0.37	7.3
HF:HCl	200	0	0	0.0	156.0	156.0			6.7
Washed histosol	0	100	100	49.1	0.0	49.1	0.53		4.0
	200	100	100	67.7	78.7	146.4	0.53	0.46	6.5
Humic fraction	200	0	0	0.0	184.0	184.0			6 3
	0	100	100	83.2	0.0	83.2	0.53		3.9
	200	100	100	103.2	96.3	199.3	0.52	0.52	5.9
Humin fraction	200	0	0	0.0	155.2	155.2			6.8
	0	100	100	46.7	0.0	46.7	0.57		3.9
	200	100	100	68.8	81.1	149.9	0.56	0.46	6.7

LSD$_{(.01)}$ = 5.5

Adsorbent	Ca(OH)$_2$	Di-quat	Para-quat	Diquat + Paraquat	Ca	Total	D + P	$\frac{D + P}{D + P + Ca}$	pH at Adsorption
IRC-50[c]	400	0	0	0.0	354.4	354.4			6.6
(H$^+$)	0	200	200	25.6	0.0	25.6	0.59		4.0
	400	200	200	71.7	258.9	330.6	0.44	0.22	6.8
IR-120[c]	400	0	0	0.0	346.7	346.7			8.0
(H)	0	200	200	344.0	0.0	344.0	0.46		3.1
IR-120[c]	400	200	200	244.1	95.2	339.3	0.42	0.72	7.9

LSD$_{(.01)}$ = 13.6

[a] Contained 30 meq/100 g of Ca initially.
[b] Ratio of diquat (D), paraquat (P), and/or Ca adsorbed as mixtures.
[c] The CEC based on manufacturer's data for the IR-120 and IRC-50 resins are 350 and 530 meq/100 g (wet weight basis), respectively.

charges on the resin surface (9). However, this does not explain the greater amount of paraquat adsorbed by the Ba-saturated resin when diquat and paraquat were adsorbed separately. Since this only occurred with the Ba-saturated resin, then paraquat must be more effective in replacing Ba from the resin even though the charges on the resin surface as an aggregate are better matched to those of diquat.

The similar adsorption of paraquat and diquat when in competition may be a result of the flexibility (coiling and uncoiling) of the polymeric structure of organic matter in contrast to the rigid structure of layer-silicate minerals. The flexibility of the organic matter would allow site mobility to accommodate with equal ease the two charges of either diquat or paraquat. The slight preference favoring paraquat is probably a result of its greater flexibility as compared to diquat and/or the charges being located in positions less likely to create steric hindrance.

Diquat + Paraquat versus Ca

The competition of diquat + paraquat with Ca, as CaCl, is shown in Table 2. The equal total adsorption, i.e., (D + P + Ca), (D + P), and Ca alone, suggests that the same sites are available to all three cations. The competitive adsorption of paraquat over diquat was not affected by Ca or organic matter since the adsorption ratio of diquat and paraquat was essentially unchanged. However, Ca did reduce the extent of adsorption of diquat + paraquat due to site competition. The competitiveness of Ca is depicted by the second ratio. Notice that diquat and paraquat were highly preferred over Ca by each adsorbent. This is probably due to the close proximity that two sites must be in to satisfy the Ca charges as compared to the charges on diquat and paraquat. Similar results occurred with the resins.

The acid groups that adsorbed the cations in this system must be of a strong acid type since the pH at adsorption was relatively low. The much larger extent of adsorption that occurred with the humic fraction reflects its larger amount of such strong acid groups.

In order to increase the pH at adsorption, $Ca(OH)_2$ was used in place of $CaCl_2$, thus the exchanged hydrogen would be removed from the solution (Table 3). As a result, an increase in total adsorption occurred with each cation. However, the increase was two- to threefold with Ca, but only slight with diquat + paraquat (compare Tables 2 and 3). The higher pH levels did not affect the relative adsorption of paraquat to diquat, as illustrated by the similarity of their adsorption ratio at the two pH levels.

Diquat and paraquat were not effective competitors with Ca on the weaker acid groups, as depicted by the decrease in the $(P + D/P + D + Ca)$ ratio for the IRC-50 resin (Tables 2 and 3). These results may reflect the greater relative stability of the Ca bond with carboxyl groups or an increase in charge density of the adsorbents at high pH levels. Steric hindrance would be effective in reducing paraquat and diquat adsorption to these closely spaced sites. Similar results have been noted with silicate clays but the rigid structure of the clays would not allow for changes in the charge distribution as would be the case for the flexible organic polymers of organic matter. Calcium was found to be much more competitive on the higher charged vermiculites than on montmorillonite in terms of diquat and paraquat adsorption (11) and probably reflects on the strength of bonding to the two mineral surfaces. The IRC-50 resin has a relatively high charge density surface. On a dry-weight basis its CEC is 1000 meq/100 g as compared to the IR-120 resin, which is 500 meq/100 g. Calcium is much more competitive on the IRC-50 resin than on the IR-120 resin. However, since no data were available as to the respective surface areas of the two resins at different pH levels inferences cannot be made along this line.

These studies have shown that the order of preference for the adsorption on organic matter is: paraquat $\geqslant$ diquat > Ca at low pH levels (4–5). At higher pH levels (6–7), the order of preference changes to: Ca > paraquat $\geqslant$ diquat. This order may reflect the close proximity of the exchange sites to each other at high pH levels. Other investigators have noted the closeness of carboxyl groups on humic acid as revealed by the formation of cyclic anhydrides (8).

The relatively low adsorption of diquat and paraquat on the IRC-50 resin and on organic matter might be attributed to steric hindrance due to the bulky size of these cations as compared to Ca. Charge spacings have a definite effect on surface adsorption with divalent cations, but whether this is the only factor inhibiting diquat and paraquat adsorption on organic matter is uncertain. Very high concentrations of diquat (4000 meq/100 g) were completely effective in removing Ca adsorbed at high pH levels.[3] This mass action effect implies that the charge spacing on the cations is more important than the bond strength or molecular size in determining the competitiveness of these cations.

No difference was detected in the adsorption behavior of the Histosol and its humic and humin fractions. Such results support other investigators' theories that these materials are composed of the same basic units (5, 7). However, since no attempt was made to characterize the types of functional groups responsible for the competitive adsorption, inferences along this line cannot be made.

SUMMARY

Flexibility in organic matter exchange sites was suggested as the reason for the equal extent of adsorption of paraquat (1,1'-dimethyl-4,4'-dipyridilium dichloride), diquat (1,1'-ethylene-2,2'-dipyridilium dibromide), and Ca as $CaCl_2$. Competitive adsorption revealed the relative affinity of various adsorbents for these cations. A Histosol and its humic and humin fractions showed preference in order: paraquat $\geqslant$ diquat > Ca when adsorption occurred on strong acid sites and Ca > paraquat $\geqslant$ diquat when adsorption occurred with weaker acid groups. The IR-120 resin (strong acid) had an order of preference of: diquat > paraquat > Ca while the IRC-50 resin (weak acid) preferred Ca > diquat > paraquat. Preferential adsorption was

[3] Best, J. A., unpublished data.

attributed to the relationship between surface charge density of the adsorbents and cation charge spacings as well as steric hindrance due to cation size.

REFERENCES

(1) Broadbent, F. E. and Bradford, G. R. 1952. Cation-exchange groupings in the soil organic fraction. Soil Sci. 74: 447–457.

(2) Chatterjee, B. and Bose, S. 1952. The electro-chemical properties of humic acid. J. Colloid Sci. 7: 414–427.

(3) Hendricks, S. B. 1941. Base exchange of the clay mineral montmorillonite for organic cations and its dependence upon adsorption due to van der Waals forces. J. Phys. Chem. 45: 65–81.

(4) Kamprath, E. J. and Welch, C. D. 1962. Retention and cation-exchange properties of organic matter in Coastal Plain soils. Soil. Sci. Soc. Am. Proc. 26: 263–265.

(5) Kononova, M. M. 1966. Soil organic matter. 2nd. ed. Pergamon Press, Inc., New York. 450 p.

(6) Philen, O. C., Jr., Weed, S. B., and Weber, J. B. 1970. Estimation of surface charge density of mica and vermiculite by competitive adsorption of diquat^{2+} vs. paraquat^{2+}. Soil Sci. Soc. Am. Proc. 34: 527–531.

(7) Stevenson, F. J. and Butler, J. H. A. 1969. Chemistry of humic acids and related pigments, In G. Eglinton and M. J. T. Murphy (editors). Organic geochemistry. Springer-Verlag, New York.

(8) Wagner, G. H. and Stevenson, F. J. 1965. Structural arrangement of functional groups in soil humic acid as revealed by infrared analysis. Soil Sci. Soc. Am. Proc. 29: 43–48.

(9) Weber, J. B., Ward, T. M. and Weed, S. B. 1968. Adsorption and desorption of diquat, paraquat, prometone and 2,4-D by charcoal and exchange resins. Soil. Sci. Soc. Am. Proc. 32: 197–200.

(10) Weed, S. B. and Weber, J. B. 1968. The effect of adsorbent charge on the competitive adsorption of divalent organic cations by layer-silicate minerals. Am. Mineral. 53: 478–490.

(11) Weed, S. B. and Weber, J. B. 1969. The effect of cation exchange capacity on the retention of diquat and paraquat by three-layer type clay minerals: I. Adsorption and release. Soil Sci. Soc. Am. Proc. 33: 379–382.

(12) Willard, H. H., Merritt, L. L., Jr., and Dean, J. A. 1965. Instrumental methods of analysis. D. Van Nostrand Co., Inc., New York, 784 p.

11

Copyright © 1968 by the Soil Science Society of America
Reprinted from *Soil Sci. Soc. America Proc.* **32**:222-234 (1968)

ADSORPTION OF ORGANIC HERBICIDES BY MONTMORILLONITE: ROLE OF pH AND CHEMICAL CHARACTER OF ADSORBATE[1]

G. W. Bailey, J. L. White, and T. Rothberg

SINCE World War II, and particularly in the last decade, there has been an increased usage of synthetic organic pesticides in the various segments of our environment. With the advent of chemical pest control in agriculture, there has been increasing awareness of the importance of soil colloids in the adsorption, movement, persistence, degradation, and bioactivity of pesticides. The problems of plant toxicity of residues, residues in food products, and detoxification of pesticides as a result of pesticide-soil colloid interactions are of major importance to the agriculturist, to the consumer, in water pollution control, to research and regulatory agencies, to the Food and Agriculture Organization, and to the World Health Organization.

There is a paucity of information on the basic nature of these interactions between pesticides and soil colloids. The results indicate that the fate and behavior of pesticides in soil systems are dependent upon at least seven factors: (i) chemical decomposition; (ii) photochemical decomposition; (iii) microbial decomposition; (iv) volatilization; (v) movement; (vi) plant uptake; and (vii) adsorption. The phenomenon of adsorption-desorption appears to directly or indirectly influence the magnitude of the effect of the other six factors. Adsorption therefore appears to be one of the major factors affecting pesticide-soil colloid interactions.

The presence of recent review articles on the adsorption of pesticides by soil colloids (Bailey and White, 1964), adsorption of organic matter by soil constituents (Greenland, 1965) and the nature of clay-organic complexes (MacEwan, 1962) obviates the necessity of any detailed literature survey. These reviews clearly point out the importance of pH and the chemical character of the adsorbate on the nature and extent of adsorption.

The initial objective of this research was to determine the role of the chemical character of the adsorbate and pH on adsorption by a well-characterized representative clay mineral with particular reference to the nature of the functional group and the nature and location of substituent groups on the aromatic ring. Attempts were also made to correlate the extent of retention with various physico-chemical properties of the molecule. A second objective was to study the mechanism of adsorption of organic pesticides by soil colloids. Interpretation of the adsorption isotherm data allows certain mechanistic hypotheses to be formulated.

MATERIALS AND METHODS

Preparation and Characterization of Adsorbent

Fractionation and Saturation—The source of montmorillonite was Volclay bentonite from Upton, Wyoming (Supplied by the American Colloid Company, Skokie, Ill.). X-ray diffraction and differential thermal analysis indicated that montmorillonite was the major mineral present with only minor amounts

---

[1] Contribution of Agronomy Department, Purdue Agr. Exp. Sta., Purdue Univ., Lafayette, Ind., as Journal Article no. 3036. Presented before Div. S-3 Soil Science Society of America, Columbus, Ohio, Nov. 1, 1965. The research was supported by Division of Environmental Engineering & Food Protection, US Public Health Service Grant EF-00055. Publication supported by Public Health Service Research Grant CC-00248 from the National Communicable Disease Control Center, Atlanta, Georgia. Received May 22, 1967. Approved Oct. 12, 1967.

[2] Post Doctoral Research Associate, Professor of Agronomy, and Laboratory Technician, respectively. Present address of senior author is Southeast Water Laboratory, Athens, Georgia, Federal Water Pollution Control Admin., US Department of the Interior.

of quartz. The term montmorillonite will be used to designate the bentonite material.

From the original sample the 1–0.2μ fraction was obtained by a sedimentation and centrifugation technique, the original sample being dispersed in distilled water by vigorous stirring.

Saturation of the 1–0.2μ Na-montmorillonite was effected by a batch-type treatment, the clay being added to a 2N NaCl solution until the suspension was approximately 2% by weight; the suspension was then stirred and centrifuged. The supernatant liquid was decanted, 2N NaCl solution again added and the procedure repeated four additional times. The excess salts were removed by repeated washings with water and centrifugation. Washing was continued one additional time following a negative chloride test with silver nitrate. The pH of the suspension was 6.80. The clay was freeze-dried and stored in a plastic bottle. The 1–0.2μ Na-montmorillonite was prepared by Dr. G. R. Dutt. (Dr. G. R. Dutt, Dep. of Water Science, University of California, Davis. Presently, Dr. Dutt is Associate Professor, Dep. of Agricultural & Soil Chemistry, University of Arizona, Tucson.)

H-montmorillonite was prepared by passing a 1% clay suspension through an Amberlite IR-120 cation exchange column saturated with hydrogen. To minimize autolysis the effluent was immediately collected, freeze-dried and stored in the freezing compartment of a refrigerator. The pH of the suspension was 3.35. The hydrogen-saturated montmorillonite will be referred to as an H-montmorillonite throughout the course of this paper although it is realized that the system is actually an H-Al system.

Cation Exchange Capacity (CEC) The neutral 1N-ammonium acetate method as described by Mackenzie (1951) and the operational procedure as outlined by Markham (1942) for the micro-Kjeldahl determination were followed. The cation-exchange capacity (CEC) for the Na- and H-montmorillonite was 87.0/100g and 73.5 meq/100g, respectively.

Surface Area—The procedure proposed by Kinter and Diamond (1958) was followed. The surface area for the Na-montmorillonite was 611 m^2/g.

Surface Acidity—The method used to estimate the acid strength of clay mineral surfaces was the one first suggested by Walling (1950) and extended by Benesi (1956). A 0.1 g sample of oven-dried (two temperatures, 110C and 200C, were used); powdered clay was added to a test tube and 3-5 ml of dry benzene added. Indicator solution (0.1% in dry benzene) was added and the resultant color of the indicator noted. Dry benzene was prepared by reaction with calcium hydride overnight. The Hammett indicators employed and their pK_a's are phenylazonaphthylamine (+4.0), benzene-azodiphenylamine, (+1.5), dicinnamalacetone (−3.0), benzalacetophenone (−5.6) and anthraquinone (−8.2).

Chemical Character of Organic Compounds

Analytical grade chemicals were used as supplied without further purification. The common and chemical name of the compounds, as well as selected physico-chemical properties (Bailey and White, 1965) are given in Table 1. The following manufacturers supplied the respective chemicals: Amchem Products, Inc., phenoxyacetic acid, 2, 4-D, 2, 4, 5-T, benzoic acid, amiben; The Dow Chemical Co., picloram; Pittsburgh Plate Glass Co., Chemical Division, IPC, CIPC; Niagara Chemical Div., dicryl and solan; E. I. duPont de Nemours and Co., 3 phenyl-urea, fenuron, monuron, diuron; Geigy Chemical Corp., simazine, trietazine, atrazine, propazine, simetone, atratone, prometone; Monsanto Chemical Co., propanil.

Adsorption Isotherm Methodology

A stock solution with a concentration in the range of 500 μ moles/liter was prepared for the compounds for which the solubility of the particular compound permitted.

Duplicate sets of herbicide solutions were made by taking suitable aliquots of the stock solution and diluting to volume. Each sample was transferred into a beaker and the necessary amount of acid or base was added in order to adjust the pH to that of the clay. The same volume of acid or base was added to the second set and diluted to volume. Six different initial concentrations of each organic compound were usually prepared.

A 25 ml aliquot of the aqueous herbicide solution was pipetted into a centrifuge tube containing a 0.25 g portion of the adsorbent previously dried overnight under vacuum over phosphorus pentoxide, the tubes stoppered, sealed with electrical insulating tape and immersed in a constant temperature bath (25C ± 0.1C) and shaken end-over-end until equilibrium was attained. Removal from the water bath was followed by centrifugation at 16,000 rpm for 30 min. Following removal from the centrifuge, the temperature of the sample was allowed to re-equilibrate to that of room temperature. The equilibrium solution was decanted into a second 50 ml centrifuge tube, a trace of NaCl crystals added, then the tube was shaken and again centrifuged at 16,000 rpm for another 30 min. Following re-equilibration to room temperature, the equilibrium concentration was determined spectrophotometrically. The solutions were always buffered to pH 9.0 prior to reading. All samples were run in duplicate. In all cases a clay blank was included. Determination of the pH of the equilibrium solution was made at the end of each run with a glass electrode pH meter.

Adsorption Isotherms

The phenomenon of adsorption is usually described by either Langmuir's adsorption equation or by the empirical Freundlich adsorption equation. The data obtained in this study are presented in terms of the empirical isothermal Freundlich relationship in which the concentration of organic compound retained by the adsorbent was plotted against the equilibrium concentration in solution on a log-log scale. The Freundlich adsorption equation is written:

$$x/m = KC^{1/n}$$

where x/m = concentration of constituent (x) adsorbed by a given amount of adsorbent (m), K and n are constants and depend on the nature of the adsorbent, adsorbate, and temperature; C is the equilibrium concentration in solution. Under these conditions, n is represented by the reciprocal of the slope of the line and K is taken to be equal to the concentration of organic compound adsorbed by the adsorbent in equilibrium with a unit concentration of the organic compound (in this case, 1 ppm). The two constants n and K embrace all factors affecting adsorption from solution: properties of the adsorbate, adsorbent, and the solvent, and the equilibria between the adsorbate-adsorbent, adsorbate-solvent, and solvent-adsorbent. Conformity to the Freundlich equation was found for all organic compounds studied on both the H- and Na-montmorillonite with two possible exceptions—the phenoxyacetic acid-H-clay system and the substituted urea–Na-clay systems.

Conformity to the Langmuir type adsorption isotherm was tested using the form of the Langmuir equation as given by Hemwall (1963): none of the compounds showed conformity to the Langmuir adsorption equation. This behavior may be due to the lack of homogeneity of the clay mineral surface.

From a knowledge of the surface area of the adsorbent and adsorbate, and using the value for maximum amount adsorbed, the percent surface coverage of the adsorbent was calculated for both the H- and Na-montmorillonite (Table 2).

The maximum percent of the cation exchange capacity satisfied by the various organic compounds for the H- and Na-montmorillonite was also calculated (Table 2). In making this calculation, the value for the maximum amount adsorbed was used; in the case of the triazines, one mole was equal to 2 equivalents, while for all the other compounds one mole was equal to one equivalent.

Table 1—Physical and chemical properties of organics used in adsorption studies

Family	Common name	Chemical name	Water solubility g/100 ml, (°C)	pK, °C	Analytical wavelength, mμ	Surface area/molecule in Å²
Aniline	aniline	aniline	3, 4 (20)	4.58[23]	230*, 280 H₂O	35.3
Anilide	dicryl	3',4'-dichloro-2-methyl-acrylanilide	8-9 ppm		258 H₂O	58.7
	propanil	3',4'-dichloro-propionanilide	500 ppm		248 H₂O	54.4
Amide	solan	3'-chloro-2-methyl-p-Valerotoluidide	8-9 ppm		247 H₂O	66.4
Phenylcarbamate	CIPC (chloropropham)	m-chloroisopropyl ester, Carbanilic acid	80 ppm 108 ppm (20)		237.5 H₂O	63.5
	IPC (propham)	Isopropyl ester, Carbanilic acid	20-5 ppm 32 ppm		234 H₂O	55.0
Benzoic acid	amiben	3-amino-2,5-dichloro-benzoic acid	700 ppm		297 H₂O, 238 chl.	52.4
	benzoic acid	benzoic acid	0.27 (18)	4.12[20]	225*, 270 H₂O	39.7
Picolinic acid	picloram	4-amino-3,5,6-trichloro-Picolinic acid	430 ppm (25)		223 H₂O	49.5
s-triazine	atratone	2-ethylamino-4-isopropylamino-6-methoxy-s-triazine	1,800 ppm (20-2)		220 H₂O	67.9
	atrazine	2-chloro-4-ethylamino-6-isopropylamino-s-triazine	22 ppm (0) 70 ppm (27) 320 ppm (85)	1.68 22	222, 263 H₂O	64.8
	prometone	2,4-bis (isopropylamino)-6-methoxy-s-triazine	750 ppm		220 H₂O	71.8
	propazine	2-chloro-4,6-bis (isopropylamino)-s-triazine	8.6 ppm (20-2)		221 me. al. 223 H₂O	68.7
	simazine	2-chloro-4,6-bis (ethylamino)-s-triazine	2.0 ppm (0) 5.0 ppm (20) 84.0 ppm (85)	1.65 16	220 H₂O 221 me. al.	55.3
	simetone	2,4-bis (ethylamino)-6-methoxy-s-triazine	3,200 ppm (20-2)	4.17	220 H₂O	58.4
	trietazine	2-chloro-4-diethylamino-6-ethylamino-s-triazine	20 ppm (20-2)	1.88	226 me. al. 228 H₂O	67.6
Substituted ureas	diuron	3-(3,4-dichlorophenyl)-1,1-dimethylurea	42 ppm (25)	-1 to -2	246 H₂O	64.5
	fenuron	3-phenyl-1,1-dimethylurea	2,900 ppm (24)		237 H₂O	55.3
	monuron	3-(p-chlorophenyl)-1,1-dimethylurea	230 ppm (25)	-1 to -2	224 H₂O	60.2
	3-phenylurea	3-phenylurea	sol.		230 H₂O	46.2
Phenylalkanoic acid	2,4-D	2,4-dichlorophenoxyacetic acid	400 ppm 725 ppm (20) 900 ppm	2.64, 2.80 3.22 60 3.31	200*, 230* 283 H₂O	56.1
	phenoxyacetic acid	phenoxyacetic acid	1.2 (10)		228*, 269 H₂O	51.1
	2,4,5-T	2,4,5-trichloro-phenoxyacetic acid	200 ppm 238 ppm (20) 280 ppm	3.14 3.46 60	220*, 289 H₂O 284 n-hex.	60.5

* Analytical wavelength for low concentrations (generally < 10ppm) Chl. = chloroform, me. al. = methylalcohol, n - hex. = normal hexane

Table 2—Percent cation exchange capacity and surface coverage satisfied by the adsorbate at the maximum amount of organic compound adsorbed for Na- and H-montmorillonite

Family	Adsorbate	Maximum amount adsorbed, μmole/g		Percent cation exchange capacity		Percent surface coverage	
		Na-mont.	H-mont.	Na-mont.	H-mont.	Na-mont.	H-mont.
s-triazines	simetone*	113	511	27.2	139	6.51	29.4
	prometone	10.2	40.5	2.45	11.0	0.722	2.86
	atratone	13.2	35.9	3.18	9.77	0.884	2.41
	atrazine*	3.85	22.7	0.927	6.18	0.245	1.45
	trietazine*	2.24	6.20	0.539	1.69	0.149	0.412
	propazine*	0.337	2.70	0.081	0.735	0.023	0.183
Substituted ureas	fenuron	102	381	12.3	51.8	5.56	20.8
	3-phenylurea	1.38	24.6	0.166	3.35	0.063	1.12
	monuron	1.26	17.9	0.152	2.44	0.075	1.06
	diuron*	1.51	6.78	0.182	0.922	0.096	0.430
Aniline, anilides, amides, & phenylcarbamates	aniline	21.4	39.0	2.58	5.31	0.745	1.36
	propanil	7.61	11.1	0.916	1.51	0.407	0.596
	CIPC*	4.14	7.98	0.498	1.09	0.258	0.499
	IPC*	0.173	0.901	0.021	0.123	0.009	0.049
	dicryl*	0.220	0.842	0.027	0.115	0.013	0.049
	solan	0.050	0.842	0.006	0.115	0.003	0.055
Phenoxy acid	phenoxyacetic acid	NA	1.68	NA	0.229	NA	0.085
	2,4-D	NA	NA	NA	NA	NA	NA
	2,4,5-T	NA	2.35	NA	0.320	NA	0.140
Benzoic acid	benzoic acid	NA	NA	NA	NA	NA	NA
	amiben	NA	NA	NA	NA	NA	NA
Picolinic acid	picloram	NA	4.27	NA	0.581	NA	0.208

* Initial concentration approximating maximum solubility. NA - Not adsorbed.

The change in the partial molar free energy of the system, $\overline{F}$, as a result of adsorption, was calculated from the thermodynamic relationship:

$$-\overline{F} = RT \ln(C_e/C_o)$$

where R = molar gas constant, T is absolute temperature in degrees Kelvin, C_e is the equilibrium concentration and C_o is the initial concentration of the solution prior to adsorption.

The free energy changes that may occur when a chemical is adsorbed can be used as a measure of the extent or driving force of the reaction. The greater the absolute magnitude of the $\overline{F}$ value, the greater the extent to which the adsorption reaction will take place. However, it does not follow that because an appreciable amount of compound is adsorbed that it is tightly bound to the surface. It is quite possible to have a substantial amount adsorbed without the compound being bound very tightly. To fully characterize an adsorption reaction, it is nec-

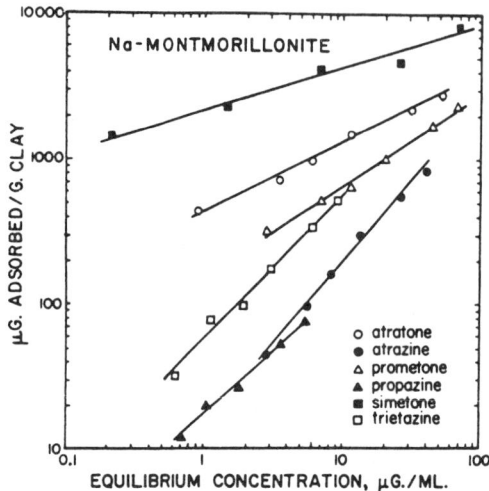

Fig. 1—Freundlich adsorption plots of six s-triazines on Na-montmorillonite.

essary to know: (i) the amount of chemical adsorbed; (ii) strength (enthalpy) with which the molecule is held by the surface, and (iii) degree of order (entropy) attained during the adsorption reaction.

RESULTS

Due to the large number of compounds, it will not be possible to present all the Freundlich and linear isothermal plots. Representative isotherms for the various compounds studied can be seen in Fig. 1–6. Figure 1 indicates the general nature and conformity of the data to the Freundlich plots. The salient points of each isothermal plot will be noted and discussed under the appropriate section.

Adsorption of s-Triazines

The members of the s-triazine family studied and their decreasing order of water solubility (Table 1) were simetone, atratone, prometone, atrazine, propazine, and trietazine. The former three compounds are methoxy-analogues of the latter three s-triazines.

Of particular interest is the relationship between water solubility and molecular structure of the chloro- and methoxy-analogue series. In the methoxy-analogue series, with increasing length of the side chain the water solubility decreases from simetone to atratone, to prometone. An inverse relationship occurs for the chloro series—simazine, propazine, trietazine, atrazine; i.e., the solubility of the series of compounds increases as the side chain length increases, with the exception of atrazine. It should fall between simazine and propazine since it has one ethylamino and one isopropylamino group, while propazine has two isopropylamino groups.

An attempt was made to study the adsorption of simazine but the molar absorbance coefficient was too low over the range of concentrations that could be prepared.

There was complete adsorption of all the triazines studied by the H-montmorillonite (pH 3.35) in the range of concentrations normally applied in the field [an application of 4.5–9 kg/ha (4–8 lb/acre) to the top 6 mm of soil will give a concentration of 50–100 ppm]. It was not possible to determine the concentration at which complete adsorption would cease for any of the triazines except simetone. This was possible with simetone due to the relatively high water solubility exhibited by this compound. For simetone adsorbed on the H-montmorillonite, the "knee" or inflection point of the Brunauer Type I isotherm (Giles et al., 1957) occurred at an adsorption value approximating that of complete saturation of the mineral's CEC, while in the case of the Na-montmorillonite, the "knee" occurred at about 25% of the CEC.

As indicated by Table 2, adsorption of simetone by the acid clay occurred in excess of the CEC (139% of the CEC satisfied). There have been reports in the literature on adsorption in excess of the CEC for adsorbates of widely varying chemical characters. The adsorption of such bases as ethylamine (Farmer and Mortland, 1965), urea (Mortland, 1966), and amides (Tahoun and Mortland, 1966) by montmorillonite in excess of the CEC has been explained by the formation of "hemi" salts where two basic molecules share a single proton. Cowan and White (1958) found in a study of the adsorption of chloro-hydrates of n-primary aliphatic amines by montmorillonite that adsorption in excess of the CEC was related to the chain length. The hydrochloride salt of octylamine was adsorbed in excess of the CEC, but the amount of the hydrochloride salt of heptylamine adsorbed was in proportion to the exchange capacity. The authors believed that the excessive quantity of amine retained was due to the intervention of a complexation reaction:

$$RNH_3^+\text{-clay} + RNH_3Cl \longrightarrow \text{clay-}RNH_3^+(RNH_3Cl),$$

$$\text{clay-}RNH_3^+(RNH_3Cl) \longrightarrow \text{clay-}RNH_3^+(RNH_2) + HCl.$$

For the above mechanism to occur, a decrease in pH of the reaction solution must occur. This has been observed by Grim et al. (1947), who reported that the amount of dodecylamine acetate and ethyl dimethyl octadecenyl ammonium bromide adsorbed was in the excess of the CEC of montmorillonite. The methylene blue cation has been found to be adsorbed by montmorillonite in excess of the CEC (Frissel, 1961; Bergmann and O'Konski, 1963). The nature of the exchange cation present appeared to determine the extent of the adsorption of methylene blue in excess of the CEC (Frissel, 1961). Bergmann and O'Konski (1963) found that the adsorption isotherm of the reaction of methylene blue cation and Na-montmorillonite could be expressed by the equation:

$$x/m = k_1 + k_2\, C^{1/n}$$

where x/m = the amount adsorbed; k_1 = value for the cation exchange capacity of the clay; $k_2\, C^{1/n}$ = amount adsorbed due to physical adsorption (Freundlich adsorption equation). This equation describes two different adsorption processes: (i) ion exchange, and (ii) physical adsorption.

Fripiat et al. (1962) made an infrared study of the adsorption of simple short chain amines and diamines by H-montmorillonite from a benzene medium. They established the existence of alkylammonium and alkyldiammonium cations and found that the adsorption of diamine exceeds the CEC when the diamine $R(HN_3^+)_2$ is transformed into $2[R(NH_3^+)_2(NH_2)]$.

With Na-montmorillonite, complete adsorption occurred only for simetone, the others being adsorbed to a much smaller degree. All the chloro-analogues exhibited a linear isotherm over the concentration range used, while the methoxy-analogues showed a curvilinear relationship (Fig. 2). This would seem to indicate the extent of adsorption is directly related to the initial concentration present in solution in the case of the chloro-analogues but not for the methoxy-analogues.

With regard to Na-montmorillonite, the similarity in the adsorption reaction shown by the members within either the methoxy or the chloro-analogue series can be seen by comparing the Freundlich "n" values in Table 3. It should also be noted that a difference exists between the two separate analogue series.

Using the Freundlich "K" value (Table 3) as a criterion of adsorption, the order of adsorbability by Na-montmorillonite was simetone $\gg$ atratone > prometone > trietazine > propazine > atrazine. A direct relationship was found to exist between water solubility and adsorbability with the exception of atrazine (Table 3). If the assumption is made that the mechanism of adsorption is due strictly to ion exchange, then the criterion of percent saturation of the clay's CEC can be used as a measure of adsorbability. The same order (Table 3) as above was found to apply in the case of the Na-montmorillonite, while in the case of the H-montmorillonite the order is the same except that the order of atratone and prometone are reversed.

From Table 4 it can be seen that the partial molar free energy, $\bar{F}$ data are erratic in nature for the H-montmorillonite system in the case of the methoxy-analogue series, but are very consistent in the case of the chloro-analogue series. No predictive value can be made of $\bar{F}$ for the H-montmorillonite-organic system. The values for atrazine and propazine are essentially the same while the values for trietazine are slightly lower.

Adsorption of Substituted Ureas

The four substituted ureas studied were 3-phenylurea, fenuron, monuron, and diuron. It was not possible to study neburon due to its low molar absorbance coefficient.

Structurally the members differ in the number of substituted chlorines on the phenyl ring and in the presence of two dimethyl groups on the compounds fenuron, monuron, and diuron. The number of chlorine atoms increases from fenuron, which is unsubstituted, to monuron, which has one chlorine in the para-position, to diuron, which has two

Table 3—Effect of water solubility on the magnitude of adsorption of various organic compounds by Na- and H-montmorillonite and the Freundlich "K" and "n" values

Adsorbate	Water solubility, ppm	Na-montmorillonite K	Na-montmorillonite n	H-montmorillonite K	H-montmorillonite n
simetone	3,200	2,200	3.23	*	*
atratone	1,800	440	2.08	*	*
prometone	750	150	1.56	*	*
trietazine	20	58	1.00	*	*
propazine	8	18	0.89	*	*
atrazine	70	15	1.18	*	*
3-phenylurea	Slightly soluble	28‡	2.33	330	1.18
monuron	230	24‡	2.08	100	1.02
diuron	42	23‡	1.08	70	0.95
fenuron	2,900	14‡	1.00	115	1.22
aniline	34,000	130	1.11	1,300	1.24
CIPC	108	27	0.93	30	1.09
propanil	500	16	0.90	65	1.70
IPC	20-32	NA	NA	30	1.56
dicryl	8-9	NA	NA	30	1.61
solan	8-9	NA	NA	58	5.88
2,4,5-T	238	†	†	105	2.38
2,4-D	725	*	*	†	†
phenoxyacetic acid	12,000	†	†	†	†
benzoic acid	2,700	†	†	†	†
amiben	700	†	†	37	1.28
picloram	430	†	†	†	†

NA -- Not adsorbed. * No value due to complete adsorption. † No value, negative adsorption. ‡ Questionable conformity to Freundlich equation.

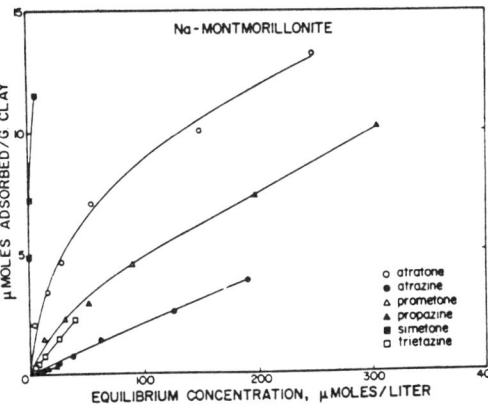

Fig. 2—Adsorption isotherm of six s-triazines on Na-montmorillonite.

Table 4—Averaged $\bar{F}$ Values for each adsorbate and for each chemical family adsorbed by Na- and M-montmorillonite.

Adsorbate	Averaged $\bar{F}$, cal/mole Na-montmorillonite	H-montmorillonite
simetone	2,765	4,336
atratone	2,182	4,372
prometone	2,463	3,763
trietazine	2,170	3,311
propazine	2,636	4,277
atrazine	2,629	4,442
Average*	2,520	4,080
3-phenylurea	2,888	1,925
monuron	2,698	2,345
diuron	2,649	2,337
fenuron	2,666	2,279
Average†	2,680	2,220
aniline	2,680	3,210
CIPC	2,625	2,541
IPC	2,698	2,641
propanil	2,613	2,486
dicryl	2,699	2,552
solan	2,631	2,477
Average‡	2,630	2,650
phenoxyacetic acid	---	2,547
2,4-D	---	---
2,4,5-T	---	2,622
amiben	---	---
benzoic acid	---	---
picloram	---	2,600
Average§	---	2,590

* s-triazine family † substituted urea family ‡ phenylcarbamates, anilides, and amides § acid family (Values for picloram and the phenylalkanoic acid group have been averaged).

chlorines (one-meta and one-para with respect to the side chain).

There was greater adsorption of all four members by the H-montmorillonite than by the Na-montmorillonite (Table 2). However, there was not complete adsorption of any of the substituted ureas by the H-montmorillonite as was the case for the s-triazines.

The isotherm plot was found to be essentially linear for all the four substituted ureas when H-montmorillonite was the adsorbent, but a curvilinear relationship existed when Na-montmorillonite was the adsorbent (Fig. 2).

Using the "K" value as a criterion for extent of adsorption, a direct relationship between water solubility and adsorbability was found to occur in the case of adsorption by the Na-montmorillonite (Table 3).

The above-stated trend is in direct opposition to results reported in the literature from field studies. Wolf et al. (1958) found that the degree of adsorption of four substituted ureas (fenuron, monuron, diuron, and neburon) by soils is inversely related to the order of their water solubilities. These same four substituted ureas were found to have a comparative adsorption ratio inversely related to their solubilities. Studies with Hawaiian sugar cane soils by Yuen and Hilton (1962) and Hilton and Yuen (1963) indicated in all cases that there was greater adsorption of diuron than monuron. Frissel (1961) reported greater adsorption of diuron over monuron from pH 8 to pH 3. However, Ashton (1961) found the order of lateral movement of the substituted ureas and the order of their water solubilities to be the same.

For all four compounds studied, the average-partial molar free energy $\bar{F}$ for adsorption by the Na-montmorillonite was of greater magnitude than the respective $\bar{F}$ for the H-montmorillonite system (Table 4). There was essentially no difference in the average $\bar{F}$ values for the Na-montmorillonite systems for the four different organic compounds. These same trends held for the H-montmorillonite systems except for 3-phenylurea where the average $\bar{F}$ was drastically lower than for the other three compounds.

If the mechanism of adsorption is assumed to be due entirely to ion exchange, then the proportion of the exchange sites occupied is relatively small in the concentration range normally used in field applications. Soil sterilization rates would be required to achieve the occupancy of 52% of the exchange sites as was found in the case of the H-montmorillonite systems (Table 2). As was the case for the s-triazines, a smaller proportion of the exchange sites was occupied in the sodium system than in the hydrogen system.

Adsorption of Aniline, Phenylcarbamates, Anilides, and Amides

Neither IPC, dicryl, nor solan was adsorbed on Na-montmorillonite, while all three were adsorbed on the H-montmorillonite. CIPC was adsorbed by both the Na- and H-montmorillonite. Structurally, the only difference between IPC and CIPC is the addition of a chlorine atom in the latter. With the addition of this chlorine atom, the solubility increased approximately fivefold (Table 1). Dicryl exhibited negative adsorption in the low concentration range on Na-montmorillonite and only a trace was adsorbed at the maximum solubility of the compound. This compound was adsorbed to a slight extent under acid conditions (Fig. 4).

Employing the Freundlich "K" as a criterion of extent of adsorption, no relationship between adsorbability and water solubility, as had been found in the case of the s-triazines and the substituted ureas, was found to exist (Table 3).

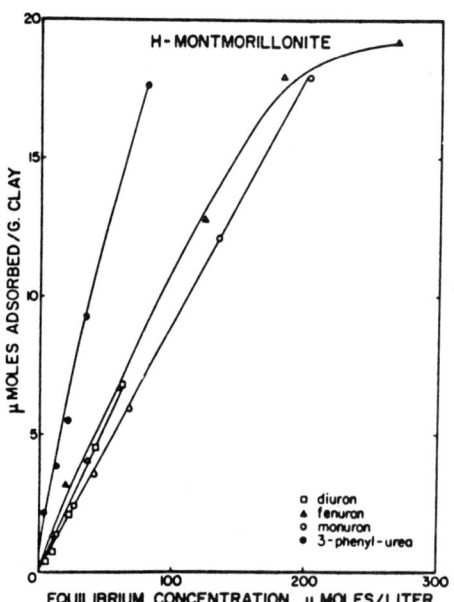

Fig. 3—Adsorption isotherms of four substituted ureas on H-montmorillonite.

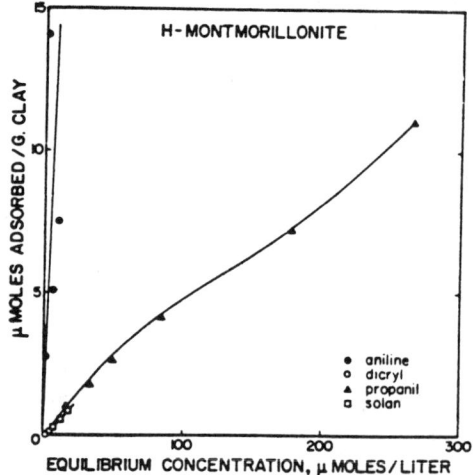

Fig. 4—Adsorption isotherms of aniline, anilides, and amides on H-montmorillonite.

Such a relationship would not be expected to exist since the compounds differ considerably in structure. Note, however, that in the case of the carbamates, CIPC and IPC, such a direct relationship between water solubility and adsorbability exists.

The data for percent saturation of CEC and percent surface coverage are given in Table 2; aniline exhibits maxima for both the Na- and H-montmorillonite.

The average $\bar{F}$ values for adsorption by the Na-montmorillonite are essentially identical (Table 4) except for aniline, while for the H-montmorillonite the average values were not identical and are in the order: aniline > IPC > CIPC = dicryl > propanil > solan. With the exception of aniline, the average $\bar{F}$ was less for the sodium- than for hydrogen-saturated montmorillonite.

Adsorption of Picloram, Phenylalkanoic and Benzoic Acids and Their Derivatives

Negative adsorption occurred for all acidic compounds studied when the adsorbent was Na-montmorillonite. For the H-montmorillonite, the adsorbates picloram (Fig. 5) and 2,4,5-T (Fig. 6) were positively adsorbed, phenoxyacetic acid exhibited both positive and negative adsorption (Fig. 6), while 2,4-D, benzoic acid, and amiben were either not adsorbed or were negatively adsorbed. The work of Frissel (1961) showed that both 2,4-D and 2,4,5-T were negatively adsorbed from pH 10 to pH of approximately 4.0. At pH values below 4.0 positive adsorption occurred:

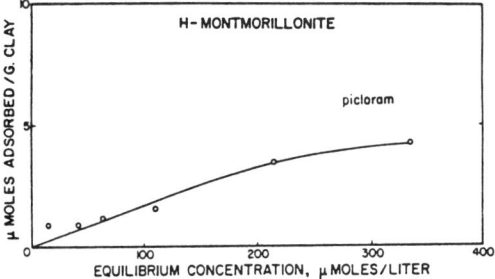

Fig. 5—Adsorption isotherm of picloram on H-montmorillonite.

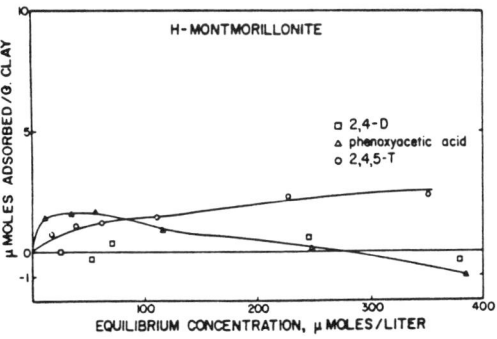

Fig. 6—Adsorption isotherms of phenoxyacetic acid and phenoxyacetic acid derivatives on H-montmorillonite.

the pH value of initial positive adsorption for 2,4-D was about ½ a pH unit lower than that for 2,4,5-T.

Although the difference in average $\bar{F}$ values for the hydrogen system is not too great (Table 4), it appears that there is slightly more of a driving force for the adsorption of picloram than for the phenoxyacetic acid.

DISCUSSION

There was a change in the pH of the supernatant liquid following the adsorption reaction. For all of the chloro s-triazine analogues, the phenylcarbamates, aniline, amides, anilides, phenylalkanoic acids, and benzoic acid the pH of the supernatant liquid following the adsorption reaction with the Na-montmorillonite was lower than that of the initial solution. The pH was higher for the H-montmorillonite systems. The above trend was reversed for atratone, prometone, and fenuron. In the case of simetone and picloram, there was no appreciable increase in the pH of the solution. In general, the magnitude of the pH change was concentration dependent—the higher the initial concentration of the organic compound in solution, the greater the change in pH. The explanation for the change in pH as a result of the adsorption reaction and the difference in the direction of the pH change as a function of chemical character of the adsorbate is not apparent at this time. Further work in this area is being carried out in an attempt to determine the mechanisms responsible for this change in pH.

Data from Table 5 suggest that within a chemical family or within an analog series which is basic in chemical character, the magnitude of adsorption is related to and governed by the degree of water solubility. The major factor governing the magnitude of adsorption by different chemical families basic in character is the dissociation constant of the adsorbate.

The s-triazines, substituted ureas, phenylcarbamates, aniline, anilides, and amides are all chemically basic in nature, the base strength varying between families. Certain similarities in behavior can be noted for these compounds of widely different molecular character.

Using "K" values as a criterion of adsorption magnitude, there was in all cases greater adsorption by the H-montmorillonite than the Na-montmorillonite (Table 3). Generally the differences in magnitude of adsorption were many-fold in nature. In the case of CIPC, IPC, and solan, there was essentially zero adsorption by the Na-montmorillonite.

The magnitude of the $\bar{F}$ values for the four different families of compounds was nearly the same for Na-mont-

Table 5—Role of pK_a and water solubility on the adsorption of weak bases by Na-montmorillonite

Family	Freundlich "K" value	Water solubility, ppm	pK_a*
s-triazine			
Methoxy analog	150-2,200	750-3,200	4.2†
Chloro analog	15-58	9-70	1.5 to 2.0
Substituted urea	14-28	42-2,900	-1 to -2
	70-330‡		
Aniline	130	34,000	4.58
	1,300‡		

* Collated by Bailey and White (1965)   † For simetone only
‡ "K" value for H-montmorillonite

morillonite with the average value for the s-triazines being slightly less than that for the other three families (Table 6). This would indicate that the average driving force for the various adsorption reactions in the case of the Na-montmorillonite is nearly the same. However, a drastic difference is noted in the case of the H-montmorillonite. The average $\overline{F}$ is nearly 60% larger for the adsorption of s-triazines by the H-montmorillonite than for Na-montmorillonite, while for the substituted ureas the average $\overline{F}$ is substantially less than for the Na-montmorillonite. The average $\overline{F}$ value for the phenylcarbamates, anilides, and amides in the two different systems is essentially the same, while the value for aniline adsorbed by the H-montmorillonite is 20% greater than by Na-montmorillonite.

Adsorption Mechanisms for Basic Organic Compounds

Several mechanisms or combinations of mechanisms can be postulated for adsorption of organic compounds by alumino-silicates. Some of these are as follows:

1) Physical adsorption—adsorption due to van der Waals' forces (the summation of dipole-dipole interactions, dipole-induced dipole interactions and induced dipole-induced dipole interactions).
2) Hydrogen bonding.
3) Coordination complexes.
4) Association or bridging complexes.
5) Chemical adsorption.

For the purpose of this discussion, hydrogen bonding, physical adsorption, and chemical adsorption will be taken as three distinct and separate mechanisms. Normally there is a question of whether to classify hydrogen bonding as physical adsorption or chemical adsorption. The purpose of this discussion is to try to ascertain adsorption mechanisms and not their classification.

It appears that chemical adsorption by montmorillonite

Table 6—Partial molar-free energies of adsorption of various chemical families by Na- and H-montmorillonite

Family	Averaged $-\overline{F}$, cal/mole	
	H-montmorillonite	Na-montmorillonite
s-triazine	4,080	2,520
Aniline	3,210	2,680
Carbamates, anilides, & amides	2,540	2,650
Substituted ureas	2,220	2,680
Acids	2,650	*

* Negative adsorption.

can occur in three different manners: (i) protonation at the silicate surface by reaction of a base with the hydronium ion on the exchange site; (ii) protonation in the solution phase with subsequent adsorption of the organic molecule via ion exchange; and (iii) in systems having low water content, protonation by reaction with dissociated protons from the residual water present on the surface or that in coordination with the exchangeable cation.

Possible adsorption mechanisms for the various compounds studied are listed in Table 7. It is realized that not all mechanisms occur simultaneously; however, one or more may occur simultaneously, depending on the nature of the functional group and the acidity of the system.

For those organic compounds possessing a basic chemical character and containing an N-H group, adsorption could occur by formation of a hydrogen bond between the amino group and the oxygen of the clay surface. This would be a prime mechanism for the adsorption of the molecular form of the basic organic compounds. The adsorption of diethylamine (MacEwan, 1948), aliphatic amines (MacEwan, 1948; Jordan et al., 1950; and Cowan and White, 1958) by montmorillonite has been postulated to occur by such a mechanism.

As can be seen from Table 7, the adsorption of substituted ureas could occur by the mechanisms 2, 4, 6, 8, and 9 (Table 7). However, the recent infrared studies of Farmer and Ahlrichs on the adsorption mechanisms of urea and

Table 7—Possible adsorption mechanisms for the various organic compounds adsorbed by Na- and M-montmorillonite

Adsorbate family	Physical adsorption 1*	R-N-H... ..O-Clay 2*	→C-H... ..O-Clay 3*	>C=O.. ..M^{Z+}-Clay 4*	R-C-O.. || O ..HO-Clay 5*	>C=O(H).. ..M^{Z+}-Clay 6*	O || -C-OH... ..O-Clay 7*	B+[H⁺-Clay]→ (HB⁺-Clay) 8*	$M^{Z+}(H_2O)$-Clay+B→ {(M^{Z+}+OH)-Clay] + 9* (HB⁺-Clay)
s-triazine									
chloro-derivatives	A	A_b	$A_a(?), A_b$	--	--	----	--	A_a	A_b
methoxy-derivatives	A	A_b	$A_a(?), A_b$	A_b	--	$A_a(?), A_b$	--	A_a	A_b
Substituted ureas	A	A_b	---	A_b	--	$A_a(?), A_b$	--	A_a	A_b
Aniline	A	A_b	---	--	--	----	--	A_a	A_b
Phenylcarbamates									
IPC†	A_a, N_{Na}	A_b	---	A_b, N_{Na}	A_b	$A_a(?), N_{Na}$	--	A_a, N_{Na}	N_{Na}
CIPC	A	A_b	---	A_b	A_b	$A_a(?), A_b$	--	A_a	A_b
Anilide									
dicryl†	A_b, N_{Na}	A_b	A	A_b, N_{Na}	A_b	$A_a(?), N_{Na}$	--	A_a, N_{Na}	A_b
propanil	A	A_b	A_b	A_b	A_b	$A_a(?), A_b$	--	A_a	A_b
Amide									
solan†	A_a, N_{Na}	A_b	A_b	A_b, N_{Na}	--	$A_a(?), N_{Na}$	--	A_a, N_{Na}	A_b
Phenylalkanoic acids									
Phenoxyacetic acid and 2,4,5-T	A_a, N_{Na}	--	$A_a(?), A_b$	A_a, N_{Na}	--	$A_a(?), N_{Na}$	A_a, N_{Na}	--	--
2,4-D	N	--	N	N	N	N	N	--	--
Benzoic acid									
benzoic acid	N	--	--	--	--	N	A_a, N_{Na}	--	--
amiben	N	N	--	--	--	N	N	N	N
Picolinic acid									
picloram†	A_a, N_{Na}	A_b, N_{Na}	--	A_b, N_{Na}	--	$A_a(?), A_b$	A_a, N_{Na}	A_a, N_{Na}	A_b, N_{Na}

LEGEND:
A — Adsorption mechanism applicable in both acidic and neutral systems
A_a — Adsorption mechanism applicable when acidity of the system is such that pH $\leq$ pK + 2.
A_b — Adsorption mechanism applicable when acidity of system is such that pH > pK + 2.
-- Appropriate group not present.
N — No adsorption in either acidic or neutral systems.
N_{Na} — No adsorption in the Na-montmorillonite system (pH 6.8).
* Adsorption mechanism number.
† pK_a values unavailable, postulated adsorption mechanism(s) based upon: (i) occurrence of adsorption by the H-montmorillonite; (ii) nature of the functional group(s) present.

urea-derivatives by montmorillonite indicates that the predominant adsorption mechanism for urea and possibly phenyl-urea derivatives is the formation of coordination compounds between the adsorbate and the exchangeable cation on the clay, except for acid clay systems (W. J. Farmer and J. L. Ahlrichs. 1966. Infrared studies of the mechanisms of adsorption of urea and its herbicidal derivatives by montmorillonite. *Agron. Abstr.* p. 63.). The coordination of urea to exchangeable metal ions on clay has been similarly reported by Mortland (1966).

The infrared studies of Harter and Ahlrichs[3] of the reaction of urea with acidic montmorillonite systems (H-, Fe-, Al-) give evidence of the protonation of the amino group and subsequent reaction with the silicate surface. The infrared spectra of urea adsorbed on the montmorillonite surface was essentially identical to the hydrochloride salt, thus indicating a reaction with the acid surface by protonation. The conclusion by these authors that urea is protonated on the amino group rather than on the carbonyl group is in direct opposition to the conclusions drawn by Janssen (1961) and Mortland (1966).

Infrared studies of the bonding mechanism of carbamates by Mortland and Meggitt (1966) revealed a decrease in the carbonyl stretching and an increase in the CN stretching frequencies. This was interpreted as indicating coordination of the carbamate, EPTC, to the exchange metal ion through the oxygen of the carbonyl group. This appeared to be the primary mechanism for adsorption except when aluminum was the exchangeable cation. In this case, adsorption of EPTC appeared to occur by association of the carbonyl with the metal ions through a bridging effect of coordinated water. This mechanism has been suggested as responsible for certain pyridine-metal-clay complexes (Farmer and Mortland, 1966).

The results of several investigators show that chemical adsorption can occur under basic conditions due to the unusual dissociative properties of adsorbed water. Pickett and Lemcoe (1959) and Ducros and du Pont (1962) have shown from NMR studies that water adsorbed by montmorillonite has a higher degree of dissociation than that of normal water. By means of infrared spectroscopy, Mortland et al. (1963) were able to show that NH_3, chemisorbed by montmorillonite and vermiculite saturated with various metal cations, existed as NH_4^+. The existence of NH_4^+ was shown to be a result of the interaction of NH_3 with protons dissociated from the residual water in the interlamellar silicate surfaces. Swoboda and Kunze[4] recorded that magnesium-saturated montmorillonites with varying isomorphic substitutions were able to adsorb weakly basic organic amines and pyridine. The reactivity or the acid strengths of the various clays to weak bases of widely differing pK_b's varied. Montmorillonites with tetrahedrally-located substitutions were able to adsorb weaker bases ($pK_b > 11.4$) than those montmorillonites with a low degree of substitution emanating from the tetrahedral layer (reactive with bases of pK_b of 8.8 or less but not with pK_b's > 9.6 or 9.4). Translated in terms of pK_a the montmorillonites were able to react with bases having pK_a values in the range of approximately 2.6 to 5.2. The lower the pK_a the greater the acid strength and the lower the basicity function of the molecule. The importance of this adsorption mechanism and its total contribution to the overall order of magnitude of adsorption for montmorillonite-type clay minerals would depend upon the pK_a of the adsorbate and the origin of the negative charge in the aluminosilicate. If the results of Johnson and Rumon (1965) are applicable to clay-organic complexes, then the degree of acidity of dissociated protons in the residual water is even greater than initially suspected. These workers studied by means of infrared spectroscopy the solid 1:1 pyridine-benzoic acid complex. They found that a critical ΔpK_a (difference between the pK_a of the acid and base) was necessary for proton transfer to occur. When the ΔpK_a between the acid and the base was about 3.75 or greater, proton transfer occurred as evidenced by the presence of the N-H$^+$ band. Therefore, the minimum surface acidity of clays would be at least equal to $(pK_a + 1) - 3.75$. If 100% absorption of the base occurred, then the surface acidity would be at least equal to $(pK_a - 1) - 3.75$.

With regard to the adsorption of aniline, the work of Swoboda and Kunze[4] and Harter and Ahlrichs[3] indicates that aniline can be adsorbed by hydrogen-aluminum clay either by protonation at or near the surface and by base-saturated clays due to dissociation of the proton in residual water on the clay surface and subsequent protonation. Whether protonation occurred directly at the exchange site or in the solution followed by exchange onto the surface could not be ascertained from their data.

In this investigation, no significant amounts of amiben were absorbed by either Na- or H-montmorillonite. However, the work of Harter and Ahlrichs suggests that amiben is adsorbed by an acid clay system by means of protonation of the amine group and subsequent reaction with the silicate surface.

At this time no further statements can be made concerning bonding mechanisms of the s-triazines, the anilides, and the amides.

Adsorption can occur via hydrogen bonding between different functional groups as can be seen in Table 7. With regard to importance of the methylene group, the mechanism of hydrogen bonding between this group and the oxygen on the clay surfaces has been postulated for the complete or partial adsorption of 2,5,8-nonanetrione and 2,5-hexandione (Tensmeyer et al., 1960); glycols, polyglycols, polyglycol-ethers (Bradley, 1945); ethyl and methyl alcohols (MacEwan, 1948); and polyalcohols (Tettenhorst et al., 1960). Laby (R. H. Laby, 1962. Adsorption of amino acids and peptides by montmorillonite. *Ph.D. thesis, University of Adelaide, Adelaide, Australia*) and Brindley and Hoffmann (1962) have pointed out that no lowering of the C-H stretching frequencies, which would be expected if C-H . . . O bonding occurred, was observed.

The hydrogen bonding ability of those functional groups

---

[3] Harter, R. D. and J. L. Ahlrichs. 1966. pH effects on mechanisms of organic acid and amine adsorption by montmorillonitic clay. Agron. Abstr. p. 63.

[4] Swoboda, A. R. and G. W. Kunze. 1965. Measurement of surface acidity of montmorillonite by the adsorption of organic bases. Abstr. 14th North Amer. Clay Mineral Conf., 2nd meeting of the Clay Minerals Soc. p. 20.

which can participate in hydrogen bonding is enhanced or mitigated by several phenomena including resonance, inductive effects, tautomerism, steric effects, and intramolecular bonding.

The presence of phenyl groups, especially those in which substitution of a highly electro-negative group (e.g., chloro) occurs would in general cause a negative inductive effect ($-I$) resulting in a weakening of the N–H bond and therefore increasing hydrogen bond formation. This may explain the adsorption of CIPC (chloro group on the para-position of the phenyl ring) by the Na-saturated clay system and the non-adsorption of IPC (absence of chloro substitution on the phenyl ring).

In the case of CIPC and IPC, the net effect of resonance interaction on the phenyl ring and the COOR group with the nitrogen atom would generally lead to a decrease in the electron density about the nitrogen thereby increasing its acidity. The combined effect of COOR group and the phenyl ring would lead to an enhanced stability of the molecule and a weakening of the N–H bond which would in turn favor the formation of stronger hydrogen bonds with the oxygen of the clay-mineral surface. The same reasoning may apply to dicryl and propanil.

Resonance in CIPC and IPC would also lead to a high electron density on the carbonyl oxygen and enhance its hydrogen bonding potential to the proton-donating portion of the silicate surface.

The ability of a molecule to form hydrogen bonds via both the carbonyl oxygen and the amino hydrogen is enhanced by tautomeric character of the molecule (Hunter, 1945). This effect could be important in the case of the phenylcarbamate (CIPC and IPC), the anilides (dicryl and propanil), substituted ureas, and the amides. It would not be important for anilines since aniline and its derivatives do not have tautomeric structures. Therefore, in the case of IPC, the following tautomeric equilibrium would exist:

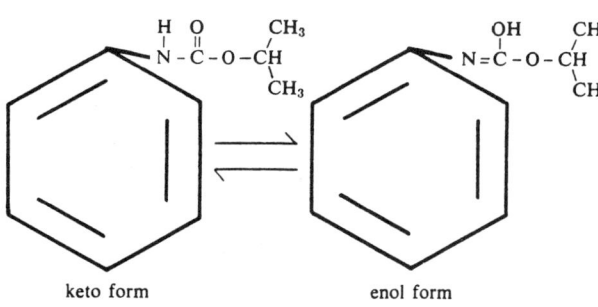

keto form   enol form

The extent of hydrogen bonding by these families would depend on the keto-content of the equilibrium mixture. In turn, the ratio of the keto-form to the enol-form is a function of the nature of the groups attached to each side of the nitrogen atom and the nature of the solvent used.

In the compounds studied, the influence of steric effects on hydrogen bonding would probably be negligible compared to electronic effects. Intramolecular hydrogen bonding would probably not occur with any of the compounds studied.

Adsorption Mechanisms for Acidic Organic Compounds

From Table 7, it can be seen that the possible mechanisms of adsorption of acidic compounds include: (i) physical adsorption; (ii) coordination through the carbonyl; (iii) association of the carbonyl by bridging to coordinated water on the exchange ions; and (iv) hydrogen bonding from the carboxyl group to the clay surface. It is suggested that the primary mechanism of acid adsorption is due to proton association (the anion becomes associated) and adsorption occurring by van der Waals' type adsorption, that is, the compound is adsorbed in the molecular form. Hydrogen bonding between the carbonyl group of the acidic compound and the surface also may occur (Kohl and Taylor, 1961).

Adsorption, pH, and Surface Acidity

The pH of the bulk solution phase and the pH at the clay surface are not the same. This is vividly shown by consideration of adsorption of s-triazines and the substituted ureas by the H-montmorillonite. All the triazines were completely adsorbed by H-montmorillonite at pH 3.35. Atrazine has a pK_2 of 1.68 and pK_1 of approximately 0. A basic compound is 10% associated when the pH is 1 unit higher than the pK and 90% associated when the pH is 1 unit lower than the pK; therefore, a compound will be 100% associated or completely dissociated when the pH is approximately 2 units below or above the pK_a, respectively. If the pH at the surface is actually 3.35 and the dissociation constant is 1.68, then it would be expected that $< 10\%$ of the compounds would be adsorbed. However, experimentally it was found that the material was completely adsorbed. This would indicate that the surface acidity would be approximately 3 units ($1000\times$) lower than the suspension pH.

Bernstein (1959) derived an equation relating internal pH (surface acidity) of a partly neutralized clay and external pH (pH of the bulk solution), this equation being:

$$p\overline{H} = pH - 7 + \tfrac{1}{2}\, pK_a$$

where $p\overline{H}$ is the internal pH or surface acidity, pH is the external pH and pK_a is the dissociation constant of the exchange group on the clay. This dissociation constant of the exchange group for montmorillonite as given by the author is about 10^{-4} and probably an order of magnitude lower for illite. Thus, from this equation it would appear that the pH of the surface would be about four units lower than the external pH for montmorillonite.

For all the substituted ureas except 3-phenylurea it was found that the acid clay adsorbed approximately 50% of the initial concentration (over the entire concentration range studied), indicating that the surface acidity is about four pH units lower than the suspension pH. However, less reliability can be placed upon the surface acidity value calculated from the adsorption of the substituted ureas than for the value calculated from s-triazine adsorption, due to the range in the determined pK_a values (−1 to −2) and therefore the greater degree of uncertainty of these values. A one unit error in the pK_a would result in a one unit error in the calculated surface acidity. Results of previous investigations (McLaren and Esterman, 1957; Harter and Ahlrichs, 1967) and calculations (Hartley and Roe, 1940) indicate the surface acidity to be at least two pH units (100×) more acid than the suspension pH.

In an attempt to assess the degree of surface acidity of montmorillonite, the assumption has been made that the pK_a of the molecule does not change appreciably when the molecule comes into close proximity to the clay mineral surface. This may not be a valid assumption in that the magnitude of the force fields emanating from the surface and/or the presence of high activity protons may induce a substantial change in the pK_a of the molecule. This change in the pK_a would, in turn, be reflected in the degree of association and therefore in the magnitude of adsorption at a given suspension pH.

This three-to-four-fold differential between the surface and the bulk solution is quite compatible with experimental results obtained in this investigation and those presented by Frissel (1961). This investigator studied the reaction of various herbicides (including the s-triazines, substituted ureas, and phenylalkanoic acids) with montmorillonite, illite, and kaolinite as a function of pH. A constant adsorption value was found for basic compounds (probably adsorption due to van der Waals' forces) until the pH of the suspension reached a value such that the surface pH was 1 to 2 pH units above the dissociation constant of basic molecules and then adsorption increased. This increase in adsorption continued as the pH, both in the bulk solution and at the surface, decreased.

Frissel (1961) noted the same phenomenon found in this investigation—that is, adsorption occurred above the dissociation constant of the adsorbate. However, his explanation for the occurrence of this phenomenon was that adsorption occurred on SiOH groups, this mechanism being described earlier by Iler (1952).

With regard to the adsorption of acidic compounds, Frissel (1961) found that negative adsorption occurred until the dissociation constant of the acid was approached and then the acids were positively adsorbed. Similarly, this observation has been made in this study and can be illustrated by considering the adsorption of 2,4-D. This compound has a dissociation constant of 2.64 and, in the case of H-montmorillonite (pH, 3.35), the surface acidity would be approximately −0.65 to +1.4; one would therefore expect 95–100% of the associated form (that is, the molecular form) and positive adsorption should occur. However, the data clearly showed negative adsorption of the compound occurred at pH of 3.35. Therefore, it appears that the pH of the bulk solution rather than the surface acidity determines when positive adsorption of the acid will occur. Theoretically, positive adsorption of the molecules should start to occur when the pH of the bulk solution is approximately 1–1.5 pH units above the dissociation constant (the largest value if the compound has more than one pK_a).

With regard to organic compounds basic in chemical character, it would be expected that if the surface pH is greater than the dissociation constant by a magnitude of 1.5–2 pH units, then adsorption will be principally due to van der Waals' forces and the amount of adsorption will be essentially constant and independent of pH change over the range normally found in soils and sediments. The mechanism of proton association and bond formation to surface predominates when the surface pH is 1–1.5 pH units above the dissociation constant of the adsorbate with maximum retention occurring when the surface acidity is at least 1–2 pH units lower in magnitude than the smallest dissociation constant of the molecules, other factors remaining constant.

Still another measure of surface acidity is gained by the use of Hammett indicators. Hammett indicators have been used by Walling (1950) and Benesi (1956) as a qualitative measure of the relative proton-donating power (surface acidity) of a dry surface in a non-aqueous solvent. In the strictest sense of the Hammett acidity function, H_o is a measure of the ability of a strong acid to donate a proton to a neutral base and convert it to its conjugate acid. This function (Gould, 1959) can be expressed as:

$$H_o = A_{H^+} (\gamma_B / \gamma_{HB^+})$$

where A_{H^+} is the proton activity, γ_B activity coefficient of the neutral base, and γ_{HB^+} activity coefficient of conjugate acid. A_{H^+} is not measurable, but recalling the relationship:

$$A_{H^+} = K_{HB^+} + C_{(HB^+/C_B)} (\gamma_{HB^+}/\gamma_B)$$

where K_{HB^+} is the dissociation constant of the conjugate acid, C_{HB^+} concentration of conjugate acid and C_B concentration of neutral base and combining the two equations and taking the log of both sides, H_o can be defined as:

$$H_o = pK_{HB^+} - \log C_{BH^+}/C_B$$

H_o becomes identical to pH in very dilute solution where the activity coefficients are unity.

The surface acidity of the H-montmorillonite varied between an H_o of −3.0 to −8.2 (−5.6 to −8.2 for the clay dried at 110C and −3.0 to −5.6 for the clay dried at 200C) depending upon the drying temperature. The H_o of the Na-montmorillonite was +4.0 and +1.5.

Based upon the observation that the substituted ureas (pK_a = −1 to −2) were not completely adsorbed even when the initial concentration exceeded the cation exchange capacity, it appears that the acid strength of the clay surface in the clay-water system lies between the suspension pH and the Hammett value found in a dehydrated clay system. One would not expect the H_o value and the surface acidity in the aqueous phase to be identical.

Note also that the H_o value for Na-montmorillonite is similar to the pK_a values of the weak bases used by Swoboda and Kunze[5] (pK_a of 2.6 to 5.2) in adsorption studies with a Mg-saturated montmorillonite where it was found that these weak bases were adsorbed in the protonated state. It is postulated that the H_o values obtained from inorganic base-saturated montmorillonite give a good measure of the acidity due to dissociation of residual water adsorbed onto clay surfaces and in coordination with exchangeable cations.

It would be expected that the H_o values for inorganic base-saturated montmorillonite would vary as the nature of the exchange cation varied.

Although montmorillonite was the only clay mineral studied, the importance of clay minerals such as illite and kaolinite and their adsorptive capacity for pesticides should not be overlooked. Frissel (1961) found in an adsorption study of herbicides of widely different chemical character by the three above mentioned clay minerals that the adsorptive capacity was in the order: montmorillonite > illite > kaolinite. This might be expected since the magnitudes of the exchange capacity of these minerals are in this order and the suggested mechanism of adsorption is dependent upon the number of cation exchange sites.

Also important in the nature and extent of adsorption is the location and distribution of the exchange sites on montmorillonite. Potentiometric titration studies of H-saturated montmorillonite by Pommer (1963) indicated that montmorillonite behaves as a mixture of two acids resulting from the presence of two types of exchange sites, interlayer and edge, both occupied by hydrogen ions. The interlayer sites which hold the hydrogen ions less tightly have the stronger acid function; and edge sites, which form stronger bonds with the proton, are responsible for the weaker acid function. The author indicated that in the case of beidellites, the pattern is reversed and the interlayer position represented the weaker acid function. Swoboda and Kunze[5] found that the acid strength of Mg-saturated montmorillonites with tetrahedral substitution was greater than the ones with predominantly octahedral substitution.

SUMMARY AND CONCLUSIONS

The adsorption studies indicate:

1) That adsorption occurs to the greatest extent on the highly acid H-montmorillonite clay system compared to the nearly neutral Na-montmorillonite system, regardless of the chemical nature of the compounds studied.

2) Within a chemical family or within an analog series basic in chemical character, the magnitude of adsorption is related to and governed by the degree of water solubility.

3) The major factor governing the magnitude of adsorption by different chemical families basic in character is the dissociation constant of the adsorbate.

4) The chemical character of a molecule affects retention by colloidal systems for three reasons: (i) it determines whether or not the molecule is fundamentally acidic or basic in nature and the relative acidic or basic strength, (ii) it affects the water solubility of the molecule, and (iii) it determines the importance relative to (i) and (ii) of van der Waals' type forces.

5) The extent of retention of compounds basic in chemical character will be determined by (i) the total number and distribution (edge or interlayer) of acid sites as well as their acid strength; (ii) the magnitude of the dissociation constant (relative basic strength); and (iii) the water solubility.

6) Maximum retention of basic compounds would be expected to occur when the surface acidity was at least 1-2 pH units lower than the lowest dissociation constant of the molecule.

7) Adsorption of basic compounds will be principally due to van der Waals' forces when the surface acidity is more than 2 pH units larger than the dissociation constant.

8) The adsorption of basic compounds by montmorillonite clay systems is principally dependent upon the surface acidity and not upon the pH of the bulk solution, while the converse is true for the adsorption of acidic type compounds.

9) Positive adsorption of acidic compounds will commence when the pH of the bulk solution is approximately 1–1.5 pH units above the dissociation constant of the acid.

10) For the conditions under which the experiments were conducted, adsorption from aqueous systems appears to obey the Freundlich adsorption equation but not the Langmuir isotherm.

11) The surface acidity of montmorillonite appears to be 3–4 pH units lower than the pH of the bulk solution.

LITERATURE CITED

1. Ashton, F. M., 1961. Movement of herbicides in soil with simulated furrow irrigation. Weeds 9:612–619.
2. Bailey, G. W., and J. L. White. 1964. Review of adsorption and desorption of organic pesticides by soil colloids, with implications concerning pesticide bioactivity. J. Agr. Food Chem. 12:324–332.
3. Bailey, G. W., and J. L. White. 1965. Herbicides: A compilation of their physical, chemical and biological properties. Residue Reviews 10:97–122.
4. Benesi, H. A. 1956. Acidity of catalyst surfaces. I. Acid strength from colors of adsorbed indicator. J. Amer. Chem. Soc. 78:5490–5494.
5. Bergmann, K., and C. T. O'Konski. 1963. A spectroscopic study of methylene blue monomer, dimer and complexes with montmorillonite. J. Phys. Chem. 67:2169–2177.
6. Bernstein, F. 1960. Distribution of water and electrolyte between homionic clays and saturating NaCl solution. Clays and Clay Minerals 8:122–149. Pergamon Press, N. Y.
7. Bradley, W. F. 1945. Molecular associations between mont-montmorillonite and some polyfunctional organic liquids. J. Amer. Chem. Soc. 67:975–981.
8. Brindley, G. W., and R. W. Hoffmann. 1962. Orientation and packing of aliphatic chain molecules on montmorillonite. Clay-Organic Studies. VI. Clays and Clay Minerals 9:546–556. (Pergamon Press, N. Y.)
9. Cowan, C. T., and D. White. 1958. The mechanism of exchange reactions occurring between sodium montmorillonite and various n-primary aliphatic amine salts. Trans. Faraday Soc. 54:691–697.
10. Ducros, P., and M. du Pont. 1962. Etude par résonance magnètique nuclèire des protons dans les argiles. Compt. Rend. 254:1409–1411.
11. Farmer, V. C., and M. M. Mortland. 1965. An infrared study of complexes of ethylamine with ethylammonium and copper ions in montmorillonite. J. Phy. Chem. 69:683–686.
12. Farmer, V. C., and M. M. Mortland. 1966. An infrared study of the coordination of pyridine and water to exchangeable cations in montmorillonite and saponite. J. Chem. Soc. (London) A. 344–351.
13. Fripiat, J. J., A. Servais, and A. Leonard. 1962. Etude de l'adsorption des amines par les montmorillonites. III. La nature de la liaison amine-montmorillonite. Bull. Soc. Chim. France, 635-644.
14. Frissel, M. J. 1961. The adsorption of some organic compounds, especially herbicides, on clay minerals. Verslag Landbouwk. Onderzoek 76:3.

---
[5] Ibid.

15. Giles, G. H., T. H. MacEwan, S. N. Nakhwa, and D. Smith. 1957. Classification of isotherm types for adsorption from solution. Int. Congr. Surface Activity Proc. (2nd Conf.), London, 3:457–461.
16. Gould, E. S. 1959. Mechanism and structure in organic chemistry. Holt-Dryden. p. 103–106.
17. Greenland, D. J. 1965. Interaction between clays and organic compounds in soils. Part I. Mechanisms of interaction between clays and defined organic compounds. Soils Fert. 28:415–425.
18. Grim, R. E., W. H. Allaway, and F. J. Cuthbert. 1947. Reactions of different clay minerals with organic cations. J. Amer. Ceram. Soc. 30:137–142.
19. Harter, R. D., and J. L. Ahlrichs. 1967. Determination of clay surface acidity by infrared spectroscopy. Soil Sci. Soc. Amer. Proc. 31:30–33.
20. Hartley, G. S., and J. W. Roe. 1940. Ionic concentrations at interfaces. Trans. Faraday Soc. 36:101–109.
21. Hemwall, J. B. 1963. The adsorption of 4-tert-butylpyrocatechol by soil clay minerals. Int. Clay Conf. (Stockholm) p. 319–328.
22. Hilton, H. W., and Q. H. Yuen. 1963. Adsorption of several preemergence herbicides by Hawaiian sugar cane soils. J. Agr. Food Chem. 11:230–234.
23. Hunter, J. 1945. Mesohydric tautomerism. J. Chem. Soc. (London), 806–809.
24. Iler, R. K. 1952. Complex of polysilicic acid with N-diethylaniline hydrochloride. J. Amer. Chem. Soc. 74:2929–2931.
25. Janssen, M. J. 1961. The structure of protonated amides and ureas and their thio analogues. Spectrochimica Acta, 17:475–485.
26. Johnson, S. L., and K. A. Rumon. 1965. Infrared spectra of solid 1:1 pyridine-benzoic acid complexes; the nature of the hydrogen bond as a function of the acid-base level in the complex. J. Phys. Chem. 69:74–82.
27. Jordon, J. W., B. J. Hook, and C. M. Finalyson. 1950. Organophillic bentonites. II. Organic liquid gels. J. Phys. Chem. 54:1196–1208.
28. Kinter, E. B., and S. Diamond. 1958. Gravimetric determination of monolayer glycerol complexes of clay minerals. Clays and Clay Minerals. Nat. Acad. Sci., Nat. Res. Council Pub. 566. 5:318–333.
29. Kohl, R. A., and S. A. Taylor. 1961. Hydrogen bonding between the carbonyl group and Wyoming bentonite. Soil Sci. 91: 223–227.
30. MacEwan, D. M. C. 1948. Complexes of clays with organic compounds. I. Complex formation between montmorillonite and halloysite and certain organic liquids. Trans. Faraday Soc. 44:349–367.
31. MacEwan, D. M. C. 1962. Interlammelar reactions of clays and other substances. Clays and Clay Minerals 9:431–443. Pergamon Press, N. Y.
32. Mackenzie, R. C. 1951. A micro-method for determination of cation exchange capacity. J. Coll. Sci. 6:219–222.
33. Markham, R. 1942. A steam distillation apparatus suitable for micro-Kjeldahl analysis. Biochem. J. 36:790.
34. McLaren, A. D., and E. F. Esterman, 1957. Influence of pH on the activity of chymotrypsin at a solid-liquid interface. Archives Biochem. Biophys. 68:157–160.
35. Mortland, M. M. 1966. Urea complexes with montmorillonite: An infrared absorption study. Clay Minerals 6:143–156.
36. Mortland, M. M., J. J. Fripiat, J. Chaussidon, and J. Uytterhoeven. 1963. Interaction between ammonia and the expanding lattices of montmorillonite and vermiculite. J. Phys. Chem. 67:248–258.
37. Mortland, M. M., and W. F. Meggitt. 1966. Interaction of Ethyl-N, N-Di-n-propylthiolcarbamate (EPTC) with montmorillonite. J. Agr. Food Chem. 14:126–129.
38. Pickett, A. G., and M. M. Lemcoe. 1959. An investigation of shear strength of the clay water system by radio-frequency spectroscopy. J. Geophys. Res. 64:1579–1586.
39. Pommer, A. M. 1963. Relationship between dual acidity and structure of H-montmorillonite. Geol. Survey Prof. Paper 386–C.
40. Tahoun, S. A., and M. M. Mortland. 1966. Complexes of montmorillonite with primary, secondary and tertiary amides. I. Protonation of amides on the surface of montmorillonite. Soil Sci. 102: 248–254.
41. Tensmeyer, L. G., R. W. Hoffmann, and G. W. Brindley. 1960. Infrared studies of some complexes between ketones and calcium montmorillonite. Clay-organic studies. Part III. J. Phys. Chem. 64:1655–1662.
42. Tettenhorst, R. C., W. Beck, and G. Brunton. 1960. Montmorillonite-poly-alcohol complexes. Clays and Clay Minerals. 9:500–519. Pergamon Press, N. Y.
43. Walling, C. 1950. The acid strengths of surfaces. J. Amer. Chem. Soc. 72:1164–1168.
44. Wolf, D. E., R. S. Johnson, G. D. Hill, and R. W. Varner. 1958. Herbicidal properties of neburon. Proc. N. Central Weed Control Conf. 15:7–8.
45. Yuen, Q. H., and H. W. Hilton. 1962. The adsorption of monuron and diuron by Hawaiian sugar cane soils. J. Agr. Food Chem. 10:386–392.

INTERACTION BETWEEN CERTAIN IONIZABLE ORGANIC COMPOUNDS (HERBICIDES) AND CLAY MINERALS

M. J. FRISSEL AND G. H. BOLT

State Agricultural University, Wageningen, The Netherlands[1]

The ever-increasing use of organic additives, such as herbicides and pesticides, in agricultural practice warrants an investigation of the fate of these additives when brought into soil. The possibility of a gradual accumulation in the soil should be especially considered with a view to long-range effects on quantity and quality of crop yield. Although it is recognized from the beginning that microbial break-down will be a decisive factor for the overall evaluation of the effects of these compounds in soil, it would seem of interest to study first the physical-chemical aspects of the interaction of organic additives with soil. In the present study only the interaction with clay minerals was investigated.

Although the existing literature (4) seems to indicate that a large spectrum of organic compounds are adsorbed by clay, a closer consideration of the available data shows that only in a limited number of cases was proof actually furnished that the compounds studied were adsorbed in aqueous environment. The information obtained from observed x-ray spacings in the presence of organic compounds must be interpreted with special and great care. In the first place, those observations were limited to montmorillonites, since only with the expanding lattice-type clays can a change in lateral spacing be observed by means of x-rays. In the second place, the samples were studied in the absence of excess water, and any evidence of adsorption under that condition does not necessarily supply any proof for adsorption in aqueous environment. For this reason the present study was entirely based on the determination of adsorption isotherms in the presence of excess electrolyte solution, a condition presumably prevailing in the field.

THEORETICAL CONSIDERATIONS

As a more detailed treatment of the theory concerning the adsorption phenomena studied was presented elsewhere[2], only the main features of the approach used will be discussed here. According to the Boltzmann distribution law the concentration of adsorbed ions in the neighborhood of an adsorbing surface equals:

$$c = gc_e \cdot e^{-E_{ads}/RT} \qquad [1]$$

in which c_e = equilibrium concentration of the adsorbate; E_{ads} = the amount of energy liberated when transferring one mole of adsorbate from the equilibrium solution to a location in the adsorption zone; and g = a statistical weight factor, related to the volume available for occupancy with an adsorbate molecule ("crowding factor"). E_{ads} may be regarded as the sum of the effects of the different types of force fields acting upon the adsorbate molecules. Considering these force fields, it is convenient to distinguish between forces acting over very short distances only (that is, over molecular distances) and those extending over considerable distances. The latter, the "long-range" forces, are mainly those associated with the electrostatic field extending from the charged surface (Coulomb and polarization forces) and the London dispersion

---

[1] Laboratory of Soils and Fertilizers. The senior author is now at the Institute for Atomic Sciences in Agriculture, P. O. B. 48, Wageningen, Netherlands. The authors acknowledge the valuable assistance of Miss D. Wolter and Mr. Wittich, who performed the majority of the analyses.

[2] M. J. Frissel, thesis, Wageningen, 1961.

forces. The "short-range" forces are characterized by a very steep dependency upon distance and comprise the higher-order Van der Waals forces, chemical bonds, and H-bonds. Furthermore, it should be remembered that, in calculating the energy of adsorption in aqueous environment, one has to take into consideration the hydration energy of adsorbate and adsorbent. If water is the solvent, this leads to the conclusion that the energy of adsorption due to short-range forces will only be of importance if its magnitude exceeds at least the hydration energy of the adsorbate particles, because the distances between hydrated adsorbent surface and adsorbate molecules are so large that the short-range forces become vanishingly small.

In order to calculate the distribution of adsorbate molecules, one may thus distinguish the following cases:

a. The short-range energy is less than the combined energy of hydration of adsorbate and adsorbent. The distribution of adsorbate molecules is now fully determined by long-range forces. The adsorption energy due to these forces is fairly well known as a function of distance[3]. For organic molecules of the size of, for example, toluene (which compares reasonably with some herbicides), it can be shown that for surfaces with a surface density of a charge such as that found for clays, the Coulomb forces dominate the long-range forces to such a degree that the existing Gouy theory, in some cases augmented by a small Stern correction, should be applicable as a first approximation. The adsorption of both organic and inorganic ions may then be calculated according to this theory. It should be noted here, that, since many herbicide ions are negatively charged, often a negative adsorption is found in accordance with the existing theory (6, 7).

b. The short-range energy exceeds the energy of hydration. The magnitude of the short-range adsorption energy usually being incalculable, one is thus forced to use the opposite procedure for the analysis of observed adsorption isotherms. Thus calculation of the long-range contribution, according to the above principles, will usually lead to an estimate of the magnitude of the short-range contribution necessary to account for the observed adsorption, which may then be scrutinized to see whether this magnitude is reasonable in the light of the situations to be presented.

Depending on the sign of the charge of the adsorbate ions, several possibilities exist. For cations, the combined effect of short-range attraction and Coulomb forces is usually sufficient to contract the ion distribution to a monolayer. This simplifies the situation considerably, since now it suffices to assign one value to the combined adsorption energy, and one may attempt to describe the adsorption isotherm by means of, for example, the BET theory[4]. If the short-range energy is very large, even a reversal of charge may occur, giving rise again to the formation of a "secondary" diffuse layer of opposite sign. From the measured excess of countercharge due to the organic adsorbate, one may now calculate the distribution of anions in this secondary layer with the help of a Gouy-Stern theory[5].

For anions the situation is quite different. Only if the short-range energy exceeds the sum of hydration energy and Coulomb repulsion, anions will be positively adsorbed. The resulting increase in the electric potential of the surface will limit the adsorption to a fraction of a monolayer, the magnitude of which will depend strongly on the chemical potential of the adsorbate in the equilibrium solution. Beyond this monolayer, short-range forces being negligible, the anions will be negatively adsorbed in accordance with the mentioned theory. Since the magnitude of the negative adsorption is practically independent of the (now un-

[3] *Idem*, p. 27.

[4] *Idem*, p. 34.
[5] *Idem*, p. 30.

known) surface density of charge, one may correct the observed net adsorption for the contribution of the negative adsorption, which then gives a fair estimate of the true positive adsorption.

Outlining the expectations with regard to the adsorption of certain specific organic ions, another facet of the role of Coulomb forces should be considered. Although clay minerals usually carry a net negative charge (arising at least in part from ionic substitution in the clay lattice), the presence of a proton dissociation-association mechanism on the edges of the plates should be taken into account. Thus it seems probable that the alumina groups on the edges of common clay minerals may carry a positive charge at low pH value. In fact it may be expected that the reversal of charge (that is from positive to negative) will occur at pH values between 5 and 7. In addition to this variable sign of charge on the edge of the adsorbent, variation in charge due to dissociation and association of many organic adsorbates must also be taken into account, since the latter are usually weak acids or bases. Accordingly, the adsorption of organic ions by clay should be expected to be strongly dependent upon pH and electrolyte concentration of the system.

EXPERIMENTAL

Materials

The experiments were made with the fraction $<2\mu$ of the following clay minerals: montmorillonite (Wyoming) and illite (Grundite co.) from the U. S. A., illite (Winsum/The Netherlands), and kaolinite (Zettlitz/Czecho-Slovakia). The clays were used as Na-, Ca-, or Al-clays. Unless otherwise stated the inorganic anion present was chloride. The preparation of the homoionic clays and the removal of salt from the clay suspension were done according to the method of Bolt and Frissel (2).

The organic compounds include the acids 2,4-D (pK 2.80), 2,4,5-T (pK 2.65), MCPA (pK 2.90), DNC (pK 4.35), DNBP (pK 4.35), and picric acid (pK 0.40), the base methylene blue and the very weak bases simazine (pK_1 1.65, pK_2 approx. 0), chlorazine (pK_1 1.74), trietazine (pK_1 1.88, pK_2 approx. 0), CMU (pK -1 to -2) and DCMU (pK -1 to -2). The measurements were made with the pure chemical products.

PROCEDURE

The adsorption experiments were executed in aqueous environment under controlled conditions, as pH, salt concentration, temperature, and equilibration time. At low salt concentrations, dialysis was used for the isolation of the equilibrium liquid; at higher salt concentrations, centrifugation was also used. The concentration of the organic compound was determined by means of U.V. spectrophotometry. Blank determinations (that is, clay in the absence of organic compounds, and organic compounds in the absence of clay) were always included.

RESULTS

The experimental data are presented in the form of free-hand curves through observed points (figs. 1–6). As a rule the deviations between experimental points and curves were less than 5 per cent. The number of observations per curve varied for the different experiments, but was never less than 10, and observations were more or less evenly spread over the curve. The amounts adsorbed are given in μmoles/g. of adsorbent, with the exception of methylene blue.

It appeared from the measurements that the adsorption isotherms of chlorazine, trietazine, simazine, DCMU, and CMU are linear up to an equilibrium concentration of 100 μmoles/l. The slope of the isotherms varied with pH, as is shown in figures 1 and 2, where the amount adsorbed at an equilibrium concentration of 30 μmoles/l. ($x_c = 30$) is plotted as a function of the pH in solution.

The adsorption behavior of methylene blue is summarized in figure 3, which shows the isotherms for different clay systems. In

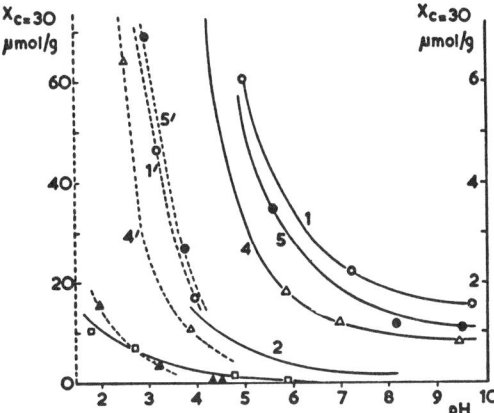

FIG. 1. Adsorption of chlorazine, simazine, and trietazine, at a salt concentration of 0.06 N NaCl, as a function of the pH. Chlorazine on Na-montmorillonite (*curves 1 and 1'*), on Na-illite (Winsum) (*curves 2 and 2'*), and on Na-kaolinite (Zettlitz) (*curve 3*); simazine on Na-montmorillonite (*curves 4 and 4'*); and trietazine on Na-montmorillonite (*curves 5 and 5'*). Curves *1, 2, 3, 4*, and *5* to be read on right-hand scale, and curves *1', 2', 4'*, and *5'* to be read on left-hand scale.

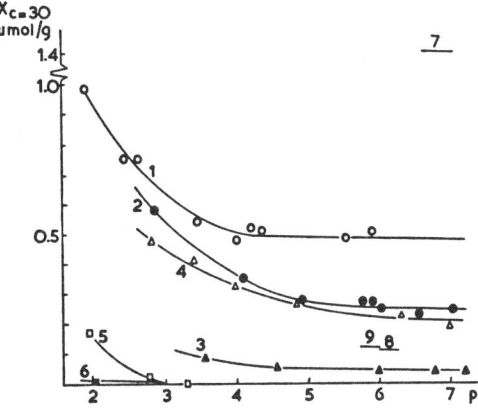

FIG. 2. Adsorption of CMU and DCMU on clay minerals as a function of the pH. DCMU on Na-montmorillonite in 0.06 N NaCl (*curve 1*), on Ca-illite (Grundite, L.) in 0.06 and 0.2 N CaCl$_2$ (*curve 2*), and on Na-illite (Winsum) in 0.06 N NaCl (*curve 3*); and CMU on Na-montmorillonite in 0.06 and 0.02 N NaCl (*curve 4*), on Na-illite (Winsum) in 0.06 N NaCl (*curve 5*), on Na- and Ca-kaolinite (Zettlitz) in 0.06 N NaCl and CaCl$_2$ (*curve 6*), on Na-montmorillonite in 3.6 N NaCl (*curve 7*), on Ca-illite (Grundite, L) in 0.06 N NaCl$_2$ (*curve 8*); and on Na-kaolinite (Zettlitz) in 3.6 N NaCl (*curve 9*). Curves *7, 8*, and *9* are based on single observations (in duplicate).

this case the amount adsorbed is given in units of the cation-exchange capacity.

The adsorption isotherms of 2,4-D, 2,4,5-T, MCPA, picric acid, DNBP, and DNC were again approximately linear at moderate concentrations of salt and organic compounds. In figures 4 and 5 the amounts adsorbed at $x_c = 30$ μmoles/l., are again

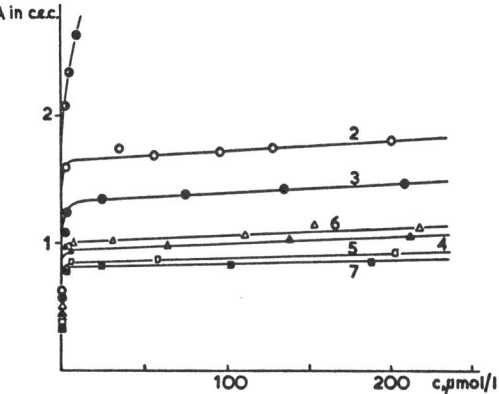

FIG. 3. Adsorption of methylene blue at a pH of 6.0 as a function of the salt concentration and the cationic occupation of montmorillonite. On Na-montmorillonite in 3.6 N NaCl (*curve 1*), in 0.1 N NaCl (*curve 2*), and in 0.005 N NaCl (*curve 3*); on Ca-montmorillonite in 0.1 N CaCl$_2$ (*curve 4*) and in 0.005 N CaCl$_2$ (*curve 5*); and on Al-montmorillonite in 0.1 N AlCl$_3$ (*curve 6*) and in 0.005 N AlCl$_3$ (*curve 7*).

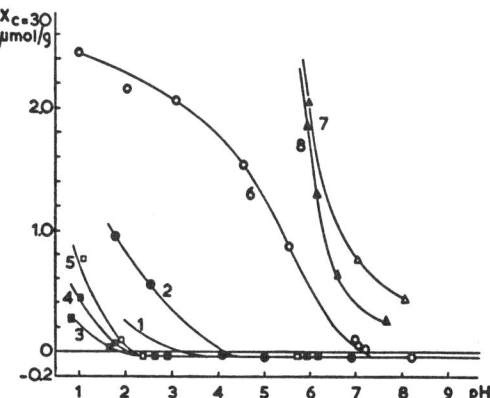

FIG. 4. Adsorption of different acids on montmorillonite as a function of the pH. 2,4-D (*curve 1*) and 2,4,5-T (*curve 2*) in 0.06 N NaCl; picric acid in 0.06 N NaCl (*curve 3*), in 0.006 N NaCl$_2$ (*curve 4*), and in 0.004 N AlCl$_3$ (*curve 5*); DNBP in 0.06 N NaCl (*curve 6*) and in 3.6 N NaCl (*curve 7*); and DNC in 3.6 N NaCl (*curve 8*).

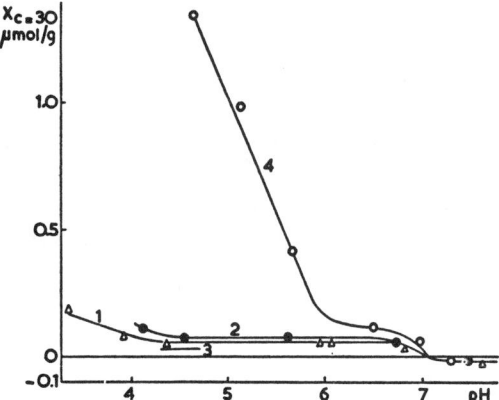

FIG. 5. Adsorption of different acids on illite (Grundite) as a function of the pH. 2,4-D (*curve 1*), 2,4,5-T (*curve 2*), and MCPA (*curve 3*) in 0.06 N NaCl; DNBP in 0.06 N CaCl$_2$ (*curve 4*).

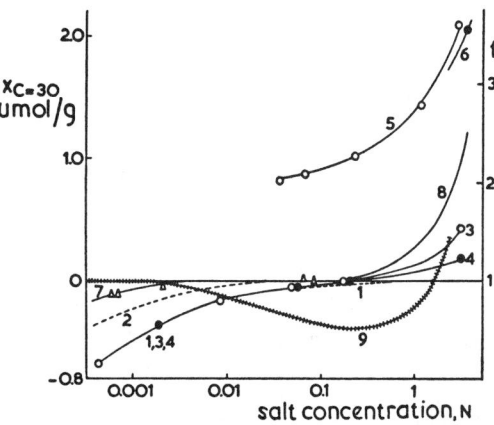

FIG. 6. Adsorption of DNC and DNBP on montmorillonite as a function of the salt concentration, as well as the activity coefficients of α-dinitrophenol. Theoretical curves (*dotted lines, left-hand scale*) for the negative adsorption in NaCl (*curve 1*) and in CaCl$_2$ (*curve 2*). Adsorption in NaCl (*left-hand scale*) of DNBP at a pH of 7.5 (*curve 3*), of DNC at a pH of 7.5 (*curve 4*), of DNBP at a pH of 5.6 (*curve 5*), and of DNC at a pH of 5.6 (*curve 6*), and adsorption of DNC in CaCl$_2$ at a pH of 7.5 (*curve 7*). Activity coefficients f (*right-hand scale*) of undissociated α-dinitrophenol (*curve 8*) and dissociated α-dinitrophenol (*curve 9*) in NaCl. *Curve 6* is based on a single observation (in duplicate).

plotted as a function of the pH, and in figure 6 as a function of the inorganic salt concentration. Apparently both positive and negative adsorption occurs, depending on the pH. There is, however, considerable difference between montmorillonite and illite, the former indicating occurrence of negative adsorption over the entire pH range, the latter exhibiting negative adsorption only above pH 7. In figure 6, some calculated curves are also shown, that is the negative adsorption in NaCl and CaCl$_2$. Furthermore, the activity coefficients of α-dinitrophenol are plotted for comparison.

DISCUSSION

Adsorption of acidic compounds

The negative adsorption of the organic anions by Na-clay, as shown in figures 4, 5, and 6, is in good agreement with the values derived from theory (1). For Ca-montmorillonite the observed values are much smaller than the calculated value. This could be explained by a condensation of elementary montmorillonite platelets into larger "polyplates" in the presence of Ca-ions. Such a condensation was also suggested by van Olphen on the basis of viscosity determinations[6].

With regard to positive adsorption, figures 4 and 5 suggest both an adsorbate-dependent and an adsorbent-dependent pH effect. Thus, for the illite all curves exhibit a change in character at a pH between 6 and 7, whereas the curves for 2,4-D on montmorillonite and illite show a change of slope at a pH of about 4. The latter effect should be caused by the transition of 2,4-D anions to 2,4-D molecules, the anionic form rapidly disappearing below pH 3-4 (pK of 2,4-D is about 2.8). Above pH 4 the adsorption should thus be attributed to an adsorption of 2,4-D anions on positive edges of the illite. The disappearance of the latter at a pH of 6 to 7 is in good agreement with the expectations. Apparently positive edges were not noticeably present in the montmorillonite systems (possibly beyond the pH range of investigation), as was also the case with kaolinite systems (not shown).

As is shown in figure 6, a strong positive

[6] H. van Olphen, thesis, Delft, 1951.

adsorption occurs always at very high salt levels. Probably this is a "salting-out" effect, as is substantiated by a comparison with the known increase in the activity coefficient of a compound closely related to DNC and DNBP, that is α-dinitrophenol. The resemblance between the curves is very striking indeed.

Adsorption of methylene blue

The striking feature of the curves shown in figure 3 is the occurrence of adsorption in excess of the cation-exchange capacity. Emodi (3) suggested that inclusion of smaller (inorganic) cations may occur because of the large surface area occupied by the methylene blue molecule. An estimate of the surface area covered by this molecule in a position parallel to the surface, however, yields only 102 A.^{2}, as compared to the surface area per unit charge in montmorillonite of about 140 A.^{2}. Inclusion, accordingly, does not seem probable.

Based on the above estimate, one finds that complete surface coverage gives an adsorption of about 1470 μmoles/g., which is in agreement with the observed isotherm for 0.1 N NaCl. For this case the net (positive) charge of the covered clay surface would be about two-thirds the original C.E.C. Using the standard double-layer theory[7], one may now estimate the concentration of Cl- counterions in close proximity of the surface, which then yields an estimate of the local value of the electric potential. Using the charge potential difference relationship for the molecular condensor, one finally obtains an estimate for the electric potential in the surface layer of methylene blue. From the known value of the concentration of methylene blue in this layer, one may now attempt to calculate the magnitude of the adsorption energy of methylene blue on clay, provided a reasonable estimate of the crowding factor may be made (equation [1]). Such an estimate was made, assuming that the surface layer is filled at $c_e = 10$ μmoles/l., corresponding to a crowding factor in the surface layer equal to zero. For lower values of c_e, g was then taken proportionally to the area not yet covered. Using the above assumption the value of φ_g, that is the non-Coulombic energy of adsorption of methylene blue on Na-montmorillonite ($\varphi_g = E_{ads} - ze\varphi_r$, with $\varphi_r =$ electric potential in the surface layer) came out to be about 0.013 to 0.014 cal./mole, which seems a very reasonable value. For further details regarding these calculations, see reference.[8]

As is obvious from figure 3, Ca- and Al-montmorillonite behave differently. Saturation is often reached here at values below the C.E.C. It was observed, however, that these systems exhibit the presence of a net positive charge of the clay particles (electrophoretic measurements). Thus it was concluded that in these systems inclusion of inorganic cations did occur, presumably because of the formation of polyplate condensates in these systems, as was suggested before. This was substantiated by the fact that little difference in behavior was observed between Na- and Ca-illites.

Adsorption of weak bases

Here, the striking feature is the large increase in adsorption at pH values about 5 units above the known pH values of the bases. Iler (5) mentioned the possibility of the adsorption of organic compounds via the formation of H-bonds on SiOH groups of polysilicic acid in acidic environment. In order to estimate the adsorption energy of the H-bonds, a calculation was set up for the adsorption of trietazine on Na-montmorillonite (fig. 1, curves 5 and 5'). The ratio of Na- and H-ions in the double layer is taken equal to that in the solution. Thus the concentrations of the monovalent and divalent cations in the neighborhood of the surface can be calculated, making use of $pK_1 = 1.88$ and $pK_2 = 0$. If it is assumed that at a pH of 10 both Coulomb adsorption

[7] M. J. Frissel, thesis, Wageningen, 1961, p. 30.

[8] *Idem*, p. 39.

TABLE 1
Adsorption of trietazine on Na-montmorillonite, in 0.06 N NaCl, as a function of the pH

pH in Equilibrium Dialysate	Calculated Adsorption*						Measured Adsorption
	Molecules (V.W.)	Monovalent ions		Divalent ions		Total	
		(C.)	(V.W.)	(C.)	(V.W.)	(C. and V.W.)	
				μmoles/g.			
10.0	1.00	—	—	—	—	1.00	1.00
9.0	1.00	—	—	—	—	1.00	1.07
8.0	1.00	—	—	—	—	1.00	1.40
7.0	1.00	—	0.02	—	—	1.02	2.05
6.0	1.00	—	0.16	—	—	1.16	2.98
5.0	1.00	0.01	1.62	—	0.07	2.70	5.05
4.0	0.99	0.14	15.9	—	6.69	23.4	19.0
3.0	0.93	1.34	153	0.23	639	795	66.0
2.9	0.91	1.61	185	0.37	989	1180	72.5

* V.W. = according to Van der Waals forces; C. = according to Coulomb forces.

TABLE 2
Predicted values of the adsorption of herbicides on clay minerals present in the soil

Herbicides	Dose kg./ha.	Clay Mineral in Soil	Concentration in Soil Sol'n				Amount Adsorbed			
			pH 4.6	pH 5.5	pH 6.4	pH 7.3	pH 4.6	pH 5.5	pH 6.4	pH 7.3
			mg./l.				%			
DNC	4	illite	0.027	0.065	0.19	6.7	99.6	99.0	97	0
		kaolinite	1.5	2.5	6.7	6.7	77	63	0	0
		montm.	0.031	0.059	0.18	6.7	99.5	99.1	97	0
DNBP	1	illite	0.007	0.016	0.048	1.7	99.6	99.0	97	0
		kaolinite	0.39	0.63	1.7	1.7	77	63	0	0
		montm.	0.008	0.015	0.044	1.7	99.5	99.1	97	0
2,4-D 2,4,5-T MCPA	1	illite	0.048	0.048	0.094	1.7	97	97	95	0
		montm.	1.7	1.7	1.7	1.7	0	0	0	0
CMU DCMU	1	illite	0.054	0.068	0.073	0.079	96	96	96	95
		montm.	0.024	0.028	0.030	0.032	98.6	98	98	98
Trietazine Simazine Chlorazine	1.5	illite	0.007	0.014	0.024	0.037	99.7	99.4	99.0	98.6
		kaolinite	0.037	0.072	0.14	0.14	98.6	97	95	95
		montm.	0.001	0.004	0.007	0.010	100.0	99.8	99.7	99.6

and the mentioned adsorption via H-bonds are absent, the adsorption energy, due to Van der Waals forces, can be calculated. Assuming that the last-mentioned adsorption energy is independent of the pH, the calculation of the energies of various forms of adsorption is now possible.[9] The results are shown in table 1. The calculation explains the increase in adsorption at about 5 units

[9] *Idem*, p. 44.

above the pK. That the measured adsorption at low pH-values is smaller than calculated is understandable, since at these high concentrations of H-ions some of the assumptions made are no longer valid.

GENERAL CONCLUSIONS AND SOME PRACTICAL IMPLICATIONS

As was predicted, the adsorption of organic ions by clays in aqueous environment is strongly dependent on conditions, varying both in sign and in magnitude with pH and electrolyte concentration. Based on the observed pattern of behavior, one may attempt to estimate the adsorption to be expected under field conditions. As an example of such an estimate, table 2 is given, which lists the expected percentage adsorption and concentration in the soil solution for homogeneous distribution of the herbicide in a 20-cm. furrow slice at about 20 per cent moisture content. As was pointed out before, the above considerations refer only to the physical-chemical behavior of the herbicides with respect to clay minerals. For a more general prediction of the fate of herbicides in soil, a thorough study should be made of chemical and microbial breakdown of these compounds under field conditions.

SUMMARY

The adsorption of certain organic ions by clays in aqueous environment was studied by determining adsorption isotherms on clays under various conditions. A comparison with physical-chemical theories concerning the behavior of clay systems often allows a reasonable interpretation of the observed phenomena. The main variables of concern proved to be the pH and electrolyte concentration of the system.

REFERENCES

(1) Bolt, G. H., and Warkentin, B. P. 1958 The negative adsorption of anions by clay suspensions. *Kolloid Z.* 156: 41–46.
(2) Bolt, G. H., and Frissel, M. J. 1960 The preparation of clay suspensions with specified ionic composition by means of exchange resins. *Soil. Sci. Soc. Amer. Proc.* 24: 172–177.
(3) Emodi, B. S. 1949 The adsorption of dye stuffs by montmorillonite. *Clay Minerals Bull.* 1: 76–79.
(4) Grim, R. E. 1953 *Clay Mineralogy*, pp. 250–273. McGraw Hill, New York.
(5) Iler, A. K. 1952 Complex of polysilicic acid with N-diethyl aniline hydrochloride. *J. Am. Chem. Soc.* 74: 2929–2930.
(6) Schofield, R. K. 1947 Calculation of surface areas from measurement of negative adsorption. *Nature* 160: 408–410.
(7) Schofield, R. K., and Wormald Taylor, A. 1954 The hydrolysis of aluminum salt solutions. *J. Chem. Soc.* 4445–4448.

13

Copyright © 1982 by Elsevier Science Publishers
Reprinted from *Geoderma* **28**:129-146 (1982)

PHYSICO-CHEMICAL INVESTIGATIONS OF INTERACTION MECHANISMS BETWEEN S-TRIAZINE HERBICIDES AND SOIL HUMIC ACIDS

N. SENESI and C. TESTINI

Institute of Agricultural Chemistry, University of Bari, Bari (Italy)

(Received July 30, 1981; revised version accepted April 1, 1982)

ABSTRACT

Senesi, N. and Testini, C., 1982. Physico-chemical investigations of interaction mechanisms between s-triazine herbicides and soil humic acids. Geoderma, 28: 129—146.

Multiple binding mechanisms that may occur upon interactions between three soil humic acids of different origins and four s-triazine herbicides, differing in chemical structure and properties, have been investigated.

Results of elemental, thermal, infrared and electron spin resonance analyses of the products of interactions demonstrate that different processes are involved and that they are controlled mainly by the content of the acidic functional groups and hydrogen binding capacity of humic acid and by the basicity of s-triazines.

Experimental and theoretical evidence is given on the occurrence of electron donor—acceptor processes involving free radical intermediates and leading to stable charge-transfer complexes between the adsorbant and the adsorbate. Parallels are suggested between the biological and chemical behaviour of s-triazines towards quinone-like structures in electron donor—acceptor systems.

The feasible formation of covalent bonds, which leads to interaction products of enhanced molecular complexity, capable of stabilizing free radical intermediates, is discussed.

Analysis of our data suggests that the higher the capacity of humic acids to form ionic and hydrogen bonds with s-triazines, the lower their effectiveness in forming electron-transfer complexes, as related to the lower ability to generate free radicals.

Finally, our results show that the basicity of s-triazines, and hence their tendency to form ionic bonds, is not the main factor governing adsorption. Indeed the most basic prometone is not the most adsorbed; nevertheless, it appears to be the most efficient among s-triazines in giving rise to electron donor—acceptor processes with humic acids.

INTRODUCTION

Degradation, transport and the biological activity of herbicides are greatly influenced by the interaction and adsorption phenomena they undergo with soil constituents, in particular with humic substances. The consequence of retention are also very important for application rates, persistence and extraction of the residues of herbicides from soils.

In order to predict and understand the actual effectiveness of herbicides on controlled weeds, a knowledge of retention by soil humic constituents appears necessary. Chemical properties and the electronic and molecular structure of both adsorbate and adsorbant are among the principal factors governing the type and intensity of binding mechanisms involved in the whole interaction process.

Soil humic compounds are rich in acidic (carboxylic and phenolic) functional groups that may be easily involved in acid-base reactions, giving rise to ionically-bound products. The same groups, as well as alcoholic and carbonyl groups, may also give rise to stable hydrogen bonds (Schnitzer, 1978). The presence of quinone units in humic molecules (Schnitzer, 1978) enables these substances to act as electron acceptors in charge-transfer complexes with suitable donor molecules (Ziechmann, 1972, 1977). Furthermore, the different active functional groups on the aromatic nuclei and aliphatic chains of the humic substances are able to form covalent bonds with interacting herbicides (Hayes, 1970).

Symmetric triazines are a widely used class of herbicides, well known electron donors and potent inhibitors of the electron transfer process in photosynthesis (Ebert and Dumford, 1976). Partially substituted amino-groups and nitrogen atoms of heterocyclic ring have a weak basic nature and are also able to form hydrogen bonds and to react with suitable organic molecules thus leading to the formation of covalent bonds (Hayes, 1970).

Preliminary results, obtained by elemental and infrared analysis, have recently been published (Senesi and Testini, 1980) on the formation of ionic and hydrogen bonds between some nitrogenated herbicides and soil humic acids.

The objectives of this work are to elucidate further the multiple interaction mechanisms that may take place between soil humic acids and s-triazines. Studies were made by additional and more advanced analytical and spectroscopic methods. In particular, the possible formation of electron donor—acceptor complexes and covalent bonds between adsorbent and adsorbate molecules are more carefully considered on the basis of single-electron transfer reactions involving humic free radicals. Non biological systems generating free radicals have already been demonstrated as capable of degrading "in vitro" s-triazines (Crosby, 1976) whereas reactions with non-humic free radicals have been suggested as responsible for the chemical degradation of the s-triazines in soils in addition to the activities of soil fungi and higher plants (Plimmer et al., 1968).

MATERIALS AND METHODS

Humic acid (HA) samples were obtained from three types of soils whose origins and characteristics are listed in Table I. Methods of extraction, separation and purification of humic acids have previously been described (Senesi and Testini, 1980).

Four analytical grade s-triazines, which differ in substituent groups on the

TABLE I

Origin and characteristics of soil samples

Sample no.	Geographical origin	Depth of horizon (cm)	Soil texture	pH (Soil:H$_2$O= 1:2.5)	Organic C (%)	Total N (%)	$\frac{C}{N}$	CaCO$_3$ (%)
1	Conversano (Italy)	0—10	Loamy sand	6.9	2.21	0.676	3.3	10.6
2	Monopoli (Italy)	0—20	Sandy loam	7.2	2.52	0.761	3.3	25.0
3	Sassari (Italy)	0—15	Loamy sand	7.3	2.63	0.175	15.0	1.5

heterocyclic ring, nitrogen content, pK$_a$ value and water solubility, were used in this study (Table II).

Interaction products between HA's and s-triazines were obtained according to a tested procedure (Senesi and Testini, 1980). Aliquots (20 mg) of finely ground HA samples were weighed and suspended in distilled water solutions (100 ml) of each herbicide (20 mg). The mixtures were intermittently shaken for 72 h at room temperature and then centrifuged. The residues obtained were submitted twice to the same treatment with fresh herbicide solution (100 ml) and successively washed twice with distilled water (10 ml) to remove non-adsorbed herbicide. Finally, the products were oven-dried under reduced pressure at 40°C. The pH of suspensions ranged from 6.00 to 7.40 and did not vary significantly throughout the treatment. No significant solubilizing effect was observed following both treatments with herbicide solutions and washings. At the end of the experimental procedures, more than 90% of the original product was recovered. Measured losses are attributed to incomplete recovery of products after centrifugation and decantation.

A Hewlett-Packard C,H,N-Analyzer model 185 was used to measure the percentages of these elements in the materials and oxygen was computed by difference. Ash contents in original HA samples were determined by ignition at 750°C for 4 h. Total acidity and carboxyl groups were measured by common analytical methods (Schnitzer and Gupta, 1965); phenolic hydroxyls were calculated by difference.

Infrared (IR) spectra were recorded on KBr pellets prepared by pressing under vacuum suitable mixtures of the sample (1 mg) and KBr, spectrometry grade (400 mg). A Perkin Elmer IR spectrophotometer model 283 was used to determine the spectra.

A Leeds and Northrup apparatus for differential thermal analysis (DTA) was run at a constant rate of heating of 10°C/min under a static air atmosphere to record DTA curves. Pt/Pt-Rh thermocouples were directly immersed in a mixture of sample (10 mg) dispersed in calcined alumina

TABLE II

Structures and analytical data for s-triazines:

$$R_3-HN-\underset{\underset{N}{\parallel}}{\overset{\overset{R_1}{|}}{C}}\underset{N}{=}\underset{}{}-NH-R_2$$

Common name	Chemical name	Substituent group			N (%)	pK$_a$	Water solubility (ppm)
		R$_1$	R$_2$	R$_3$			
Prometone	2-methoxy-4,6-bis(isopropylamino)-s-triazine	CH$_3$O	(CH$_3$)$_2$CH	(CH$_3$)$_2$CH	31.09	4.28	750
Ametryne	2-methylthio-4-ethylamino-6-isopropylamino-s-triazine	CH$_3$S	CH$_3$CH$_2$	(CH$_3$)$_2$CH	30.81	3.12	185
Desmetryne	2-methylthio-4-methylamino-6-isopropylamino-s-triazine	CH$_3$S	CH$_3$	(CH$_3$)$_2$CH	32.83	3.08	580
Methoprotryne	2-methylthio-4-γ-methoxypropylamino-6-isopropylamino-s-triazine	CH$_3$S	CH$_3$O(CH$_2$)$_3$	(CH$_3$)$_2$CH	25.81	3.03	320

(50 mg), sandwiched between layers of calcined alumina against a reference standard (α-Al_2O_3).

Electron spin resonance (ESR) spectra were recorded at room temperature on dried, powdered samples, using a Varian Associates X-band spectrometer model E-109, operating at a 100 KHz modulation and at a nominal frequency of 9.5 GHz. Free radical concentrations (spins/g), linewidths (gauss) and spectroscopic splitting factors (g-values) were computed from the ESR spectra according to Senesi and Testini (1981).

RESULTS

Elemental analysis

Elemental composition, C:N ratios and relative nitrogen percentage increases for the materials are given in Table III. In every case the interaction products have higher nitrogen contents and smaller C:N ratios than the original HA's.

Infrared spectra

Infrared spectra of the original HA's, pure s-triazines and HA—herbicide complexes were recorded. The most significant differences in the IR-absorption bands of the interaction products compared to those of the original HA's lie in the region of 1900—1000 cm^{-1} (Figs. 1, 2 and 3). In particular, the following observations can be made.

(a) A strong reduction in the absorption intensity at 1720—1710 cm^{-1} (attributable to the C=O valence vibration of free carbonyl and carboxyl groups of HA's) and a simultaneous sharp increase and broadening of the superimposed bands in the range of 1660—1600 cm^{-1} (probably due to the antisymmetric stretching of COO^- and also partly to the N—H deformation bending of secondary amino-groups and C=N stretching vibrations of the heterocyclic ring of adsorbed s-triazine).

(b) A slight increase in absorption around 1550 cm^{-1} (mainly due to the skeletal in-plane vibrations of C=C bonds of aromatic rings and C=N stretching of heterocyclic ring).

(c) The appearance of a new band localized between 1370 and 1400 cm^{-1} (preferentially assigned to the symmetric stretching of COO^-).

(d) A marked reduction in the intensity of the broad band centered around 1250 cm^{-1} (attributed to the C—O stretching and O—H deformation vibrations of carboxylic, phenolic and alcoholic groups).

(e) A diffuse enhanced structuration in the range of 1400—1000 cm^{-1}, of difficult assignment to specific groups absorptions.

A peculiar feature of the spectra of HA—s-triazine complexes is represented by the absorption band between 775 and 790 cm^{-1}, attributed to the CH wagging vibration, that appears always to shift towards lower frequencies

TABLE III

Elementary analysis (%, ash free) of original HA's and interaction products of HA's with s-triazines

	HA$_1$ *(ash = 1.5%)					HA$_2$ *(ash = 4.0%)					Ha$_3$ *(ash = 1.4%)				
C	54.41	53.58	53.21	46.11	54.73	55.03	55.50	56.02	53.03	54.51	56.93	56.56	55.07	55.61	51.37
H	4.52	5.28	5.08	4.70	5.05	4.44	5.01	4.83	4.51	4.96	4.38	4.97	4.59	4.89	4.63
N	4.28	6.30	7.22	6.12	4.59	4.72	7.66	9.31	7.55	7.74	3.74	7.28	8.15	8.62	5.11
O	36.34	34.84	34.49	43.07	35.63	35.81	31.83	29.84	34.91	32.79	34.95	31.19	32.19	30.88	38.89
C:N	12.72	8.50	7.37	7.53	11.92	11.66	7.25	6.02	7.02	7.04	15.22	7.77	6.76	6.45	10.05
Relative N% increase	–	47.2	68.7	43.0	7.2	–	62.3	97.2	60.0	64.0	–	94.7	117.9	130.5	36.6

*First column: before interaction;
Second column: afer interaction with prometone;
Third column: after interaction with ametryne;
Fourth column: after interaction with desmetryne;
Fifh column: after interaction with methoprotryne.

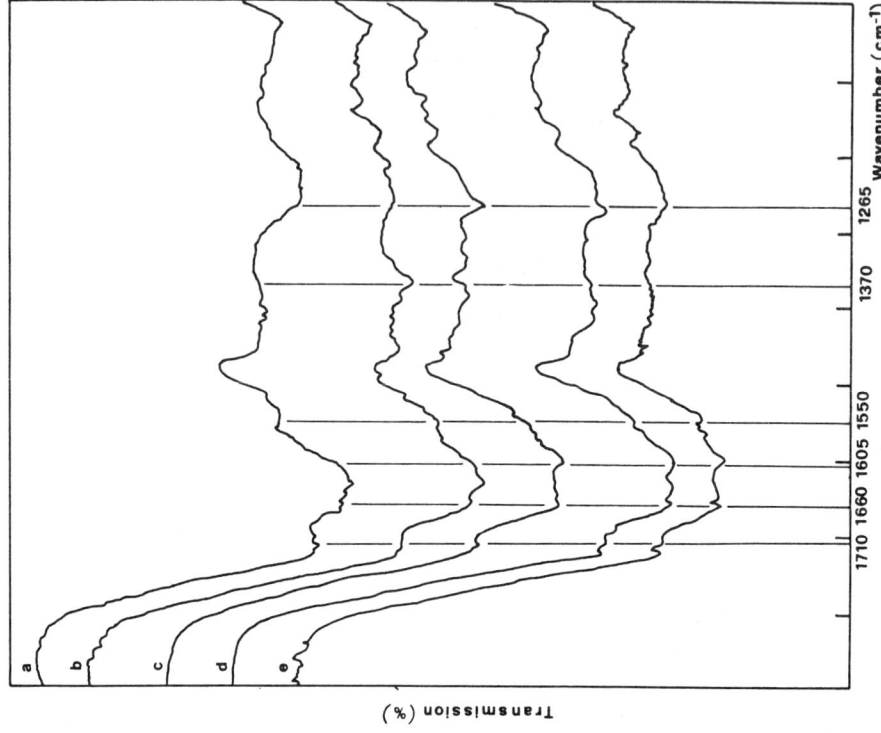

Fig. 2. Infrared spectra in the region between 1800 and 1000 cm^{-1} of: (a) HA$_2$ and its interaction products with (b) prometone, (c) ametryne, (d) desmetryne, (e) methoprotryne.

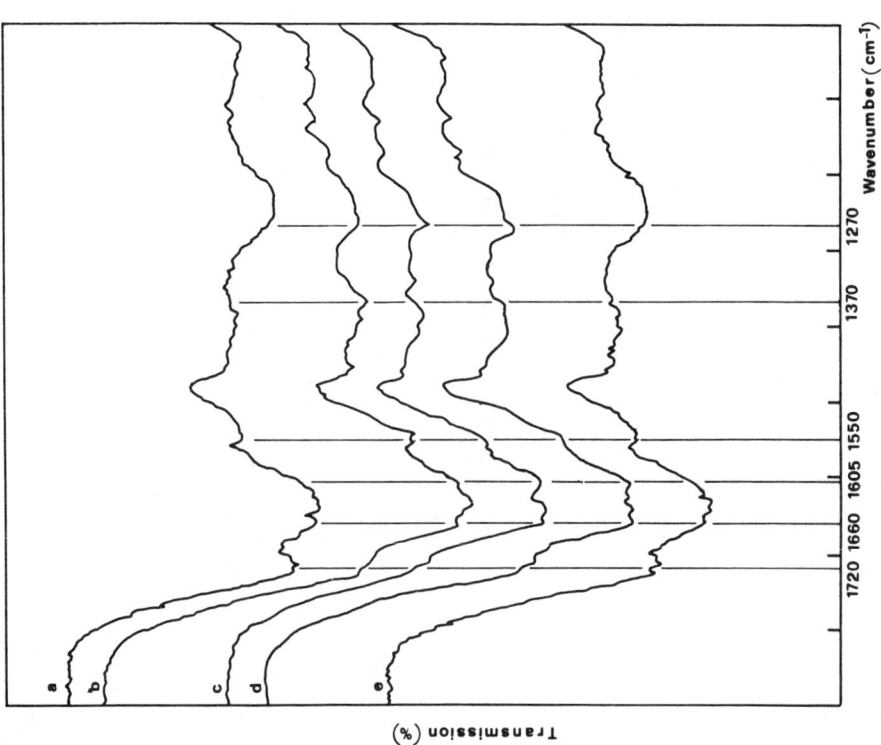

Fig. 1. Infrared spectra in the region between 1800 and 1000 cm^{-1} of: (a) HA$_1$, and of its interaction products with (b) prometone, (c) ametryne, (d) desmetryne, (e) methoprotryne.

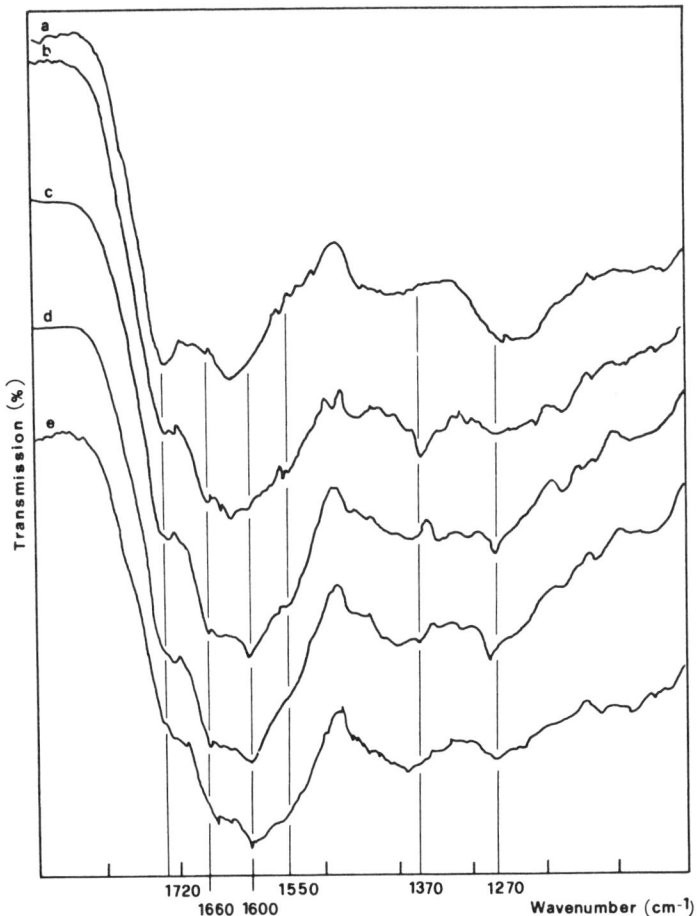

Fig. 3. Infrared spectra in the region between 1800 and 1000 cm^{-1} of: (a) HA$_3$ and its interaction products with (b) prometone, (c) ametryne, (d) desmetryne, (e) methoprotryne.

TABLE IV

Variations of the CH wagging infrared absorption band

Herbicide	Position of infrared band (cm^{-1})			
	pure herbicide	complex with		
		HA$_1$	HA$_2$	HA$_3$
Prometone	815	790	790	790
Ametryne	808	775	783	780
Desmetryne	808	777	780	780
Methoprotryne	809	775	780	780

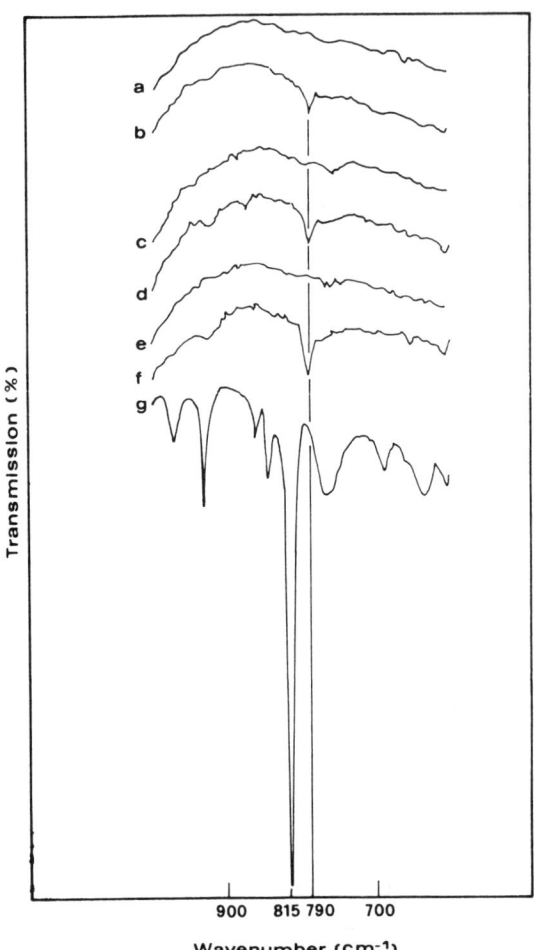

Fig. 4. Infrared spectra in the region 900-700 cm^{-1} of: (a) HA_1, (b) HA_1-prometone complex, (c) HA_2, (d) HA_2-prometone complex, (e) HA_3, (f) HA_3-prometone complex, (g) prometone.

when compared with the absorption value of this group in the spectra of pure s-triazines. Fig. 4 shows the shifting relative to the HA—prometone complexes, whereas the variations observed for all the complexes are listed in Table IV.

Differential thermal analysis curves

Differential thermal analysis curves were recorded both for the original HA samples and for the interaction products with s-triazines. The three HA samples examined show similar behaviour in this aspect so only the representative DTA curves, relative to sample HA_2, and its complexes, are presented in Fig. 5.

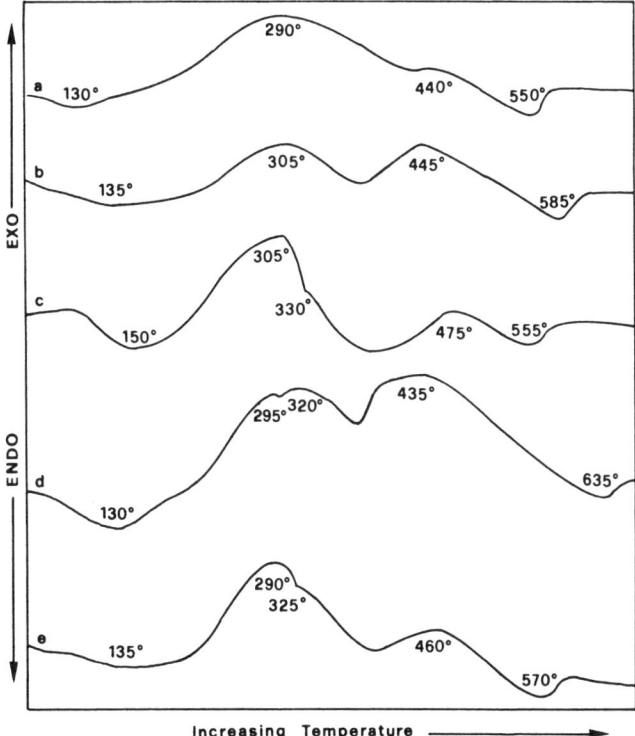

Fig. 5. Differential thermal analysis curves of: (a) HA_2 and of its interaction products with (b) prometone (c) ametryne, (d) desmetryne, (e) methoprotryne.

A common feature is the slight temperature increase shown by the peaks corresponding to the principal thermal decomposition processes (Schnitzer, 1978) that are in this order: water dehydration (endothermic peak at 135—150°C); decarboxylation, phenolic and alcoholic OH dehydration and oxidative thermal decomposition of aliphatic chains (exothermic peaks between 290° and 475°C); and final combustion of the carbon skeleton (endothermic peak ranging from 550° to 635°C).

Electron spin resonance spectra

The ESR spectra of HA—s-triazine complexes are devoid of any hyperfine structure and have single-line shapes, similar to those commonly observed for humic compounds (Schnitzer, 1978). No ESR signal is detectable for the pure herbicides. Spectroscopic splitting factors (g-values) for unreacted HA's and for interaction products appear to have very similar values; whereas a noteworthy broadening of linewidths and a considerable increase of the free radical concentrations are measured for the complexes, with some exceptions (Table V). No significant variation is shown by the ESR parameters, when measured after some weeks, thus indicating that additional, long-lived, free radical species were generated in the HA-herbicide systems.

105

TABLE V

ESR parameters for original HA's and interaction products of HA's with s-triazines

Samples	ESR parameters: Free radical concentration (spin/g $\times 10^{-17}$)	Line widths (gauss)	Spectroscopic splitting factors (g-values)
HA_1	21.75	5.1	2.0035
HA_1-prom.	50.40	5.7	2.0034
HA_1-am.	21.60	7.1	2.0036
HA_1-des.	17.07	6.1	2.0036
HA_1-met.	30.37	.75	2.0035
HA_2	8.62	6.2	2.0035
HA_2-prom.	23.56	8.2	2.0035
HA_2-am.	16.77	8.9	2.0037
HA_2-des.	22.37	8.1	2.0036
HA_2-met.	20.85	8.9	2.0035
HA_3	10.37	5.4	2.0036
HA_3-prom.	12.76	6.7	2.0036
HA_3-am.	9.60	5.4	2.0035
HA_3-des.	13.11	6.9	2.0035
HA_3-met.	15.79	5.9	2.0035

DISCUSSION

Ionic (salt linkage) and hydrogen bindings

The results of elemental analysis, in particular the smaller C:N ratios and the higher nitrogen percentages, found with treated HA's in comparison with values for original HA's indicate that s-triazine herbicides are adsorbed onto soil humic acid molecules, leading to stable final complexes (Senesi and Testini, 1980).

More extended information on different binding mechanisms can be obtained by careful examination of the results from infrared analysis. The conversion of the carboxylic groups to carboxylate ions indicates the formation of ionic bonds (salt linkages) following classical proton transfers (acid-base reaction) from the humic acids to the weak basic s-triazines. The sharp increase of absorption around 1660 cm^{-1} in the IR spectra of the adsorption products may be partially accounted for by the shifting of about 60 cm^{-1} towards the lower frequencies of the absorption of the HA carbonyl, originally at 1720—1710 cm^{-1}, when this group is involved in hydrogen bonds (Williams and Fleming, 1966; Bellamy, 1966) with the amino groups of the s-triazine molecule. Possible involvement of the carboxylic, phenolic and alcoholic OH groups of HA to form hydrogen bonds with the amine and/or ring nitrogen atoms of the s-triazines is also suggested by the observed variations in the infrared region around 1250 cm^{-1}, where the characteristic O—H deformation vibrations of these groups give rise to absorption (Williams and Fleming, 1966; Bellamy, 1966).

The increased thermal stability of the HA—s-triazine complexes in comparison with the unreacted HA's, evident from the DTA curves, supports the suggestion that salt linkages occur in the interaction products. Analogous thermostabilizing effects were observed with HA's polluted by metals (Schnitzer and Kerndorff, 1980) and also when metal ions reacted with fulvic acids (Schnitzer and Kodama, 1972). A possible influence of hydrogen binding on the thermal behavior of humic materials has been suggested as well (Thompson and Chester, 1970).

The lack of clear correlations between the results of elemental and infrared analyses and pK_a values of the s-triazines, on the one hand, and any of the HA's examined, on the other, suggests that (a) molecular properties other than acidity and basicity influence adsorption and (b) mechanisms other than simple acid-base (proton transfer) reactions are involved in the phenomenon as a whole.

Electron donor—acceptor mechanisms

The shifting of the infrared absorption bands relative to the CH wagging of the s-triazines, observed after interactions with HA's (Fig. 4 and Table IV), has been considered an indication of the formation of electron donor—acceptor bonds between the donor, s-triazine, and the acceptor, quinone-like units in HA molecules (Müller Wegener, 1977). This interpretation of ESR results gives further support to the suggestion that charge-transfer systems of this type were formed in the complexes studied.

Free radical species can be generated through three general types of processes: single electron oxidation-reduction reactions, irradiation (UV and visible light) and thermal homolysis. As it appears unlikely that the third type of process could take place under our experimental conditions, the large ESR absorption detected in the majority of our interaction products may be imputed chiefly to the first and possibly to the second process.

The g-values observed for the HA—s-triazine complexes are very similar and do not differ significantly from the g-values of untreated HA samples (Table V). Therefore, the odd electron generated throughout the interaction process should be localized on structural sites very similar to those already known to exist in the original humic substances, i.e. a semiquinone moiety highly conjugated with an extended aromatic network (Senesi and Schnitzer, 1977).

Actually, quinone-like structures in the HA molecule behave as strong oxidizing agents (electron acceptors) and are expected to remove electrons from the donor triazine molecule. Such an electron transfer, one that may take place in the dark whenever an electron acceptor is added to an electron donor, is supposed to produce a charge carrier of the semiquinone type and to increase the charge carrier's lifetime, thus giving rise to the observed modification of the ESR signal. This interpretation is supported by similar effects observed in several different organic electron donor—acceptor systems (Foster, 1969), e.g. o-chloranil and phtalocyanine (Kearns et al., 1960).

On the other hand, mechanisms involving the formation of semiquinone radical species have been shown to occur on a biological scale in chloroplasts. Quinones may act as artificial Hill acceptors in the area of light reaction I by absorbing electrons (Trebst, 1972), antagonizing ferredoxin and/or plastoquinone functions (Black and Myers, 1966; Büchel, 1972); whereas s-triazines, behaving as Hill (electron) donors, may interfere in light reaction II by replacing or inactivating an intermediate electron carrier (Hansch, 1969; Moreland and Hilton, 1976; Ebert and Dumford, 1976).

A further index of the enhanced molecular complexity produced by adsorption might be the commonly observed broadening of the ESR signal linewidths of the interaction products in comparison with the linewidth values of the unreacted HA's (Table V). This behaviour, very marked in the HA_2—s-triazine complexes, may partially account for the nuclear broadening effect exerted by protons. Unresolved electron-nuclear hyperfine interactions have already been suggested as important contributors to the enlargement of ESR linewidths in materials such as peats, lignites and bituminous coals possessing suitable electron acceptor aromatic structures very similar to humic substances and capable of forming charge-transfer complexes, e.g. with electron donor pyridine nucleus (Retcofski and Friedel, 1968; Retcofski et al., 1975).

Donor—acceptor systems, such as HA—s-triazine, may give rise to charge-transfer complexes under the effect of light as well. That may then introduce an unpairing of the involved electrons producing an increase of the ESR signal (photo-induced transfer) (Lagerkrantz and Yhland, 1962, 1963; Foster, 1969). Furthermore, the light-induced free radical oxidation of several organic chemicals, involving humic acids as "photosensitizers", has recently been reported (Mill et al., 1980; Zepp et al., 1981).

Covalent bonds formation (chemisorption)

The increased adsorption of s-triazines on humic acids at higher temperatures (McGlamery and Slife, 1966) was interpreted by Hayes (1970) as an indication of chemisorption or formation of covalent bonds. Later, Hayes et al. (1975) measured a significant increase in the free radical contents, together with an enrichment with N%, when humic acids and fulvic acids were extracted, or treated after extraction, with ethylenediamine (EDA). These effects were tentatively ascribed to the formation of covalent bonds by condensation of the primary amine (EDA) with HA-carbonyl groups and/or by reaction of the amine with HA-quinone units in the presence of oxygen. ESR signals have also been observed in different compounds of diamines and substituted benzoquinones (Bijl et al., 1959).

Chemical reactions leading to the incorporation of aromatic amines into soil organic matter by covalent bonds, which are relatively resistant to hydrolysis, are well known (Cranwell and Haworth, 1971; Hsu and Bartha, 1974, 1976). Parris (1980) has recently proposed a pathway consisting in a Michael (nucleophilic) addition of secondary amines to quinone-like structures in HA, followed by tautomerization and oxidation to yield aminoquinones as the final products.

On the basis of these considerations, some analytical characteristics of our interaction products — i.e. the increased structuration in the "fingerprint" region (1400—1000 cm^{-1}) of the IR spectra, the enrichment of free radical concentrations and the enlargement of linewidths in ESR spectra — may be tentatively interpreted as due, even if but partially, to the more extended and complex aromatic structure attained in consequence of formation of covalent bonds (chemisorption) between HA's and herbicides.

Influence of humic acid and s-triazine structures on adsorption

Contrasting concepts of the influence exerted by the structure of s-triazines and humic acids on the type and intensity of the adsorption mechanisms have been offered.

The prevailing opinion is that humic acids adsorb the s-triazines principally by acid-base reactions with the formation of ionic bonds (Weber et al., 1968; Hayes, 1970; Stevenson, 1972). Studies of the influence of pH and molecular structure on the adsorption of s-triazines by clay and organic matter have suggested that basicity played a very significant role in the adsorption phenomena (Weber et al., 1969; Weber, 1972). The type of substitution in the 2-position strongly influenced the basicity of the s-triazines, whereas the changes of the alkyl groups in the 4-and 6-positions affected this property only secondarily (Weber, 1967, 1970). Therefore the more basic methoxy-substituted triazines should be more intensely adsorbed than the methylthio-compounds. On the other hand, when molecular properties other than electronegativity are considered e.g., the steric hindrance of substituent groups (molecular volume), the water solubility and the capacity to form hydrogen bonds with water — the adsorption of s-triazines by clay colloids followed a reverse order (-SCH$_3$-triazines > -OCH$_3$-triazines) (Weber, 1966, 1970, 1972).

Other authors (Sullivan and Felbeck, 1968; Khan, 1972; Stevenson, 1972) have emphasized the role of hydrogen binding, in addition to ionic bonds, in the retention mechanisms of humic compounds towards s-triazines. Recently, this opinion has been criticized on the basis of a reverse relationship observed between the quantities of adsorbed ametryne and the contents of OH phenolic groups of a series of humic acids (Müller Wegener, 1977).

The same author (Müller Wegener, 1977), by means of an accurate infrared analysis and on the basis of previous experimental and theoretical studies (Ziechmann, 1972), has strongly emphasized the role played by the electron donor—acceptor capacity of the two partners in interaction phenomena.

Possible participation of covalent bonds (chemisorption) in the adsorption process that takes place between s-triazines and HA's has been considered feasible by some authors (McGlamery and Slife, 1966; Hayes, 1970), whereas others have not considered this as possible at all (Stevenson, 1972; Weber, 1972).

The probable involvement of free humic radicals in the phenomenon has been proposed (Dunigan and McIntosh, 1971; Stevenson, 1972; Bartha and Hsu, 1976), but sufficient experimental evidence is lacking up to date. Consequently, the apparently contrasting results and interpretations seem strongly affected by the complexity of the interaction mechanisms that may take place between s-triazines and HA's, both of which possess multiple possibilities for bonds and for activesites.

Our results show that the relative percentage of nitrogen increase, as determined in the products of interaction, is directly related to the carboxylic and phenolic OH contents, as well as to the total acidity of the adsorbent HA's (Fig. 6), whereas an inverse relation is shown between free radical concentrations (spins/g) of interaction complexes and acidic functional groups of HA's (Fig. 7).

If we assume that the relative increase in percent of nitrogen is proportional to the intensity of adsorption of the s-triazine and that the concentrations of free radicals increase with the strength of the donor—acceptor tendency, depending on the electronic structure of the interacting molecules (Biji et al., 1959), it is possible to draw the following conclusions:

(a) The adsorbent strengths of the humic acids toward the basic s-triazines depend primarily on their contents of acidic functional groups, capable also of forming hydrogen bonds.

(b) The higher the capacities of the humic molecules to form ionic (acid-base) and hydrogen bonds, the lower the tendency to give rise to electron

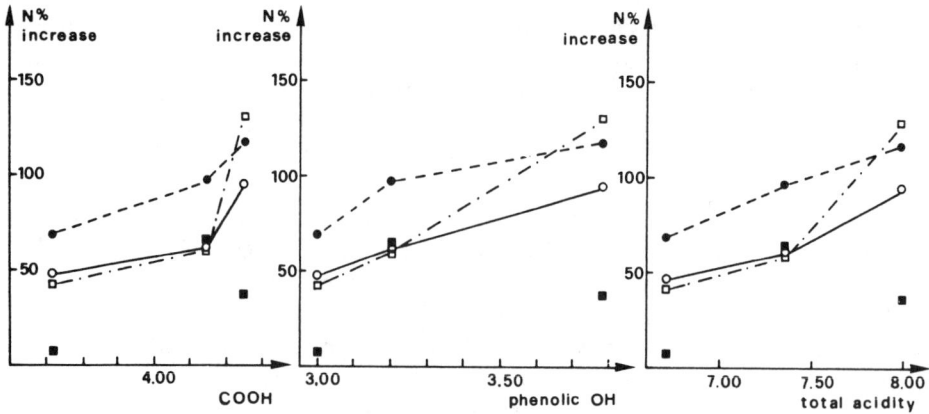

Fig. 6. Relative nitrogen percentage increases in interaction products of HA's with s-triazines as a function of the contents of acidic groups of original HA's.

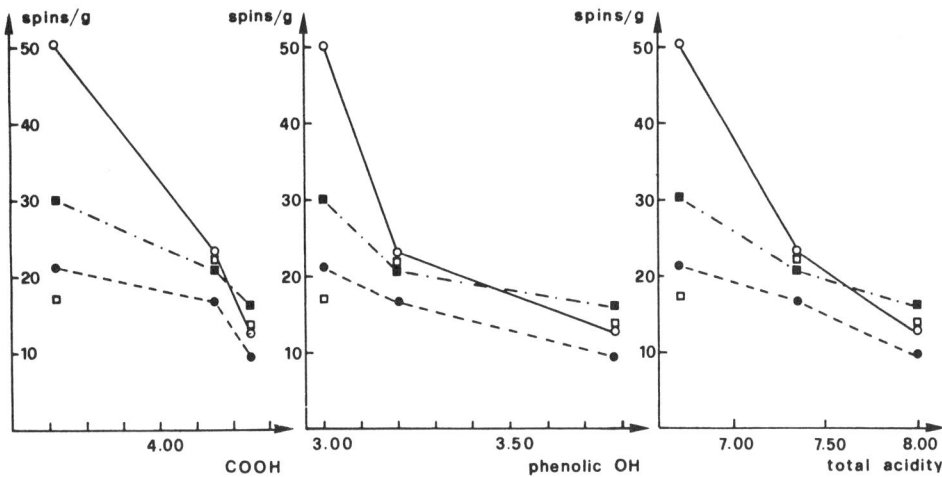

Fig. 7. Free radical contents (spins/g) in interaction products of HA's with s-triazines as a function of acidic function groups contents of original HA's.
(—o—: prometone; —•—: ametryne; —□—: desmetryne; ·—■—·: methoprotryne).

donor—acceptor complexes and covalent bonds, as is shown by the lower production of free radicals.

The influences of chemical structure and properties of the s-triazine specimens that were studied are indicated by various data. The last line in Table III (relative % increases of N found in the interaction products) indicates that the most basic prometone (pK_a = 4.28) does not appear to be the most adsorbed by the 3 HA samples. The less basic ametryne (pK_a = 3.12) seems to be most adsorbed by HA_1 and HA_2 and desmetryne (pK_a = 3.08) by HA_3. At the same time, data for free radical concentrations (Table V) indicate that prometone is the most effective in forming electron donor—acceptor systems with samples HA_1 and HA_2.

These observations, supported also by a semiquantitative analysis of the corresponding infrared spectra, suggest that: (a) the basicity of the s-triazines and hence their tendency to form ionic bonds with humic acid — following proton transfer — is not the main factor governing adsorption; and (b) the electron donor—acceptor processes do play a more important role than previously thought in the adsorption phenomenon of s-triazines on HA's.

ACKNOWLEDGEMENT

Thanks are due to the Italian National Research Council (CNR) for having partially supported this work.

REFERENCES

Bartha, R. and Hsu, T.S., 1976. Spectroscopic characterization of soil organic matter. In: D.D. Kaufman, G.G. Still, G.D. Paulson and S.K. Bandal (Editors), ACS Symposium Series, 29, pp. 258—271.

Bellamy, L.J., 1966. Infrared spectra of complex molecules. Wiley, New York, N.Y., 425 pp.

Bijl, D., Kainer, H. and Rose-Innes, A.C., 1959. Biradical molecular compounds: A study by electron spin resonance. J. Chem. Phys., 30: 765—770.

Black, C.C. Jr. and Myers, L., 1966. Some biochemical aspects of the mechanisms of herbicidal activity. Weeds, 14: 331—338.

Büchel, K.H., 1972. Mechanism of action and structure activity relations of herbicides. Pestic. Sci., 3: 89—110.

Cranwell, P.A. and Haworth, R.D., 1971. Humic acid, IV. The reaction of α-amino acid esters with quinones. Tetrahedron, 27: 1831—1837.

Crosby, D.G., 1976. Non-biological degradation of herbicides in soil. In: L.J. Audus (Editor), Herbicides: Physiology, Biochemistry and Ecology, Vol. 2, 2nd ed. Academic Press, New York, N.Y., pp. 65—97.

Dunigan, E.P. and McIntosh, T.H., 1971. Atrazine—soil organic matter interactions. Weed Sci., 19: 279—281.

Ebert, E. and Dumford, S.W., 1976. Effects of triazine herbicides on the physiology of plants. Res. Rev., 65: 1—103.

Foster, R., 1969. Organic Charge-Transfer Complexes. Academic Press, London, 470 pp.

Hansch, C., 1969. Theoretical considerations of the structure—activity relationship in photosynthesis inhibitors. Progress Photos. Res., III: 1685—1692.

Hayes, M.B.H., 1970. Adsorption of triazine herbicides on soil organic matter, including a short review on soil organic matter chemistry. Res. Rev., 32: 131—174.

Hayes, M.B.H., Swift, R.S., Wardle, R.E. and Brown, J.K., 1975. Humic materials from an organic soil: a comparison of extractants and of properties of extracts. Geoderma, 13: 231—245.

Hsu, T.S. and Bartha, R., 1974. Interaction of pesticide-derived chloroaniline residues with soil organic matter. Soil Sci., 116: 444—452.

Hsu, T.S. and Bartha, R., 1976. Hydrolyzable and nonhydrolyzable 3,4-dichloroaniline—humus complexes and their respective rates of biodegradation. J. Agric. Food Chem., 24: 118—122.

Kearns, D.R., Tollin, G. and Calvin, M., 1960. Electrical properties of organic solids, II. Effects of added electron acceptor on metalfree phtalocyanine. J. Chem. Phys., 32: 1020—1025.

Khan, S.U., 1972. Adsorption of pesticide by humic substances. A review. Environ. Lett., 3: 1—12.

Lagerkrantz, C. and Yhland, M., 1962. Photo-induced electron spin resonance in solutions of some electron-donor—acceptor complexes. Acta Chem. Scand., 16: 1043—1045.

Lagerkrantz, C. and Yhland, M., 1963. Photo-induced free radical reactions in the solution of some tars and humic acids. Acta Chem. Scand., 17: 1299—1306.

McGlamery, M.D. and Slife, F.W., 1966. The adsorption and desorption of atrazine as affected by pH, temperature and concentration. Weeds, 14: 237—239.

Mill, T., Hendry, D. and Richardson, H., 1980. Free-radical oxidants in natural waters. Science, 207: 886—887.

Moreland, D.E. and Hilton, J.L., 1976. Action on photosynthetic systems. In: L.J. Audus (Editor), Herbicides: Physiology, Biochemistry and Ecology, Vol. II, 2nd ed. Acdemic Press, New York, N.Y., pp. 493—521.

Müller Wegener, U., 1977. Über die Bindung von s-Triazinen an Huminsäuren. Geoderma, 19: 227—235.

Parris, G.E., 1980. Covalent binding of aromatic amines to humates, 1. Reactions with carbonyls and quinones. Environ. Sci. Technol., 14: 1099—1106.

Plimmer, J.R., Kearney, P.C. and Rowlands, J.R., 1968. Free-radical oxidation of s-triazines: mechanism of N-dealkylation. Presented Amer. Chem. Soc., Atlantic City, N.J.

Retcofski, H.L. and Friedel, R.A., 1968. Spectral studies of a carbon disulfide extract of bituminous coal. Fuel, 47: 487–498.

Retcofski, H.L., Thompson, G.P., Raymond, R. and Friedel, R.A., 1975. Studies of e.s.r. linewidths in coals and related materials. Fuel, 54: 126–128.

Schnitzer, M., 1978. Humic substances: chemistry and reactions. In: M. Schnitzer and S.U. Khan (Editors), Soil Organic Matter. Elsevier, Amsterdam, pp. 1–64.

Schnitzer, M. and Gupta, U.C., 1965. Determination of acidity in soil organic matter. Soil Sci. Soc. Am. Proc., 29: 274–277.

Schnitzer, M. and Kodama, H., 1972. Differential thermal analysis of metal-fulvic acid salts and complexes. Geoderma, 7: 93–103.

Schnitzer, M. and Kerndorff, H., 1980. Effects of pollution on humic substances. J. Environ. Sci. Health, B15: 431–456.

Senesi, N. and Schnitzer, M., 1977. Effects of pH, reaction time, chemical reduction and irradiation on ESR spectra of fulvic acid. Soil Sci., 123: 224–234.

Senesi, N. and Testini, C., 1980. Adsorption of some nitrogenated herbicides by soil humic acids. Soil Sci., 130: 314–320.

Senesi, N. and Testini, C., 1981. The environmental fate of herbicides: the role of humic substances. Presented 5th Internat. Symp. on "Environmental Biogeochemistry", Stockholm, 1–5 June 1981. In press.

Stevenson, F.J., 1972. Organic matter reactions involving herbicides in soil. J. Environ. Qual., 1: 333–343.

Sullivan, J.D. and Felbeck, G.T. Jr., 1968. The interaction of s-triazine herbicides with humic acids from three different soils. Soil Sci., 106: 42–51.

Thompson, S.O. and Chester, G., 1970. Infra-red spectra and differential thermograms of lignins and soil humic materials saturated with different cations. J. Soil Sci., 21: 265–272.

Trebst, A., 1972. Measurement of Hill reactions and photo-reduction. In: A. San Pietro (Editor), Methods in Enzimology, Vol. 24B. Academic Press, New York, N.Y., pp. 146–165.

Weber, J.B., 1966. Molecular structure and pH effects on the adsorption of 13 s-triazines compounds on montmorillonite clay. Am. Mineralog., 51: 1657–1670.

Weber, J.B., 1967. Spectrophotometrically determined ionization constants of 13 alkylamino-s-triazines and the relationships of molecular structure and basicity. Spectrochim. Acta, 23A: 458–461.

Weber, J.B., 1970. Mechanisms of adsorption of s-triazines by clay colloids and factors affecting plant availability. Res. Rev., 32: 93–130.

Weber, J.B., 1972. Interaction of organic pesticides with particulate matter in aquatic and soil systems. In: Am. Chem. Soc. (Editor), Fate of organic pesticides in the aquatic environment, Advances in Chemistry Series, 111, pp. 55–120.

Weber, J.B., Weed, S.B. and Ward, T.M., 1969. Adsorption of s-triazines by soil organic matter. Weed Sci., 17: 417–421.

Weber, J.B., Perry, P.W. and Ibaraki, K., 1968. Effect of pH on the phytotoxicity of prometryne applied to synthetic soil media. Weed Sci., 16: 134–136.

Williams, D.H. and Fleming, I., 1966. Spectroscopic methods in organic chemistry. McGraw-Hill, London, 236 pp.

Zepp, R.G., Baughman, G.L. and Schlotzhauer, P.F., 1981. Comparison of photochemical behaviour of various humic substances in water: I. Sunlight induced reactions of aquatic pollutants photosensitized by humic substances. Chemosphere, 10: 109–117.

Ziechmann, W., 1972. Über die Elektronen-Donator- und Acceptor-Eigenschaften von Huminstoffen. Geoderma, 8: 111–131.

Ziechmann, W., 1977. Molekülkomplexe bei Huminstoffen durch E-Donator und E-Acceptor Strukturen. Z. Pflanzenernähr. Bodenkd., 140: 133–150.

EQUILIBRIUM AND KINETIC STUDIES OF THE ADSORPTION OF 2,4-D AND PICLORAM ON HUMIC ACID

S. U. KHAN[1]

Research Station, Agriculture Canada, Regina, Saskatchewan. Received 22 Mar. 1973, accepted 19 July 1973.

KHAN, S. U. 1973. Equilibrium and kinetic studies of the adsorption of 2,4-D and picloram on humic acid. Can. J. Soil Sci. 53: 429-434.

Equilibrium and kinetic studies of the adsorption of 2,4-D (2,4-dichlorophenoxy acetic acid) and picloram (4-amino-3,5,6-trichloropicolinic acid) on a humic acid have been made. The equilibrium data followed the Freundlich-type isotherm. Rate constants, activation energies, heats of activation, and entropies of activation were calculated for the adsorption of the two herbicides on humic acid. The rate data indicated a physical type of adsorption. In the overall adsorption process the rate-limiting step for the initial period was shown to be the diffusion of the herbicide molecules to the surface of humic acid. However, the rate-limiting process at longer time intervals was interpreted to be intraparticle diffusion of the herbicide molecules into the interior of the humic acid particles.

On a effectué des études d'équilibre et cinétiques de l'adsorption du 2,4-D (acide 2,4-dichlorophénoxy acétique) et du picloram (acide 4-amino-3,5,6-trichloropicolinique) sur l'acide humique. Les données d'équilibre suivent l'isotherme du type Freundlich. On a calculé les constantes de migration, les énergies, les chaleurs, et les entropies d'activation pour l'adsorption des deux herbicides sur l'acide humique. Les données de migration laissent supposer un type physique d'adsorption. Durant tout le processus d'adsorption, la diffusion des molécules d'herbicide à la surface de l'acide humique s'est avérée être le point limite de migration pour la phase initiale. Cependant, on suppose que le processus limitatif à plus long intervalle résulte de la diffusion intraparticulaire des molécules d'herbicide à l'intérieur des particules d'acide humique.

INTRODUCTION

The importance of humic substances in influencing the activity, behavior, bio-availability and degradability of herbicides in soils has been documented in recent reviews by Hayes (1970), Khan (1972), and Stevenson (1972). Numerous reports have appeared in the literature during the past decade indicating that humic substances, which represent the most active fraction of organic matter in soils, are the most active adsorbent for a wide variety of herbicides (Dunigan and McIntosh 1971; Gilmour and Coleman 1971; Hance 1965, 1969; Hayes et al. 1968; Li and Felbeck 1972; Sullivan and Felbeck 1968; Weber et al. 1969). Several mechanisms or combination of mechanisms have been postulated for the adsorption of herbicides by humic substances (Hayes 1970; Khan 1972; Stevenson 1972). These include ion exchange, hydrogen bonding, protonation, charge transfer, ligand exchange, coordination through a metal ion, van der Waals forces, and hydrophobic bonding.

Despite the considerable amount of research undertaken, very little attention has been paid to the equilibrium and kinetic aspect of the problem. These fundamental studies are necessary for a better understanding of the mechanism(s) of adsorption. The work described here attempts to obtain information on the equilibrium and kinetics of adsorption of 2,4-D and picloram on a humic acid (HA). The two herbicides have been used extensively for the control of broad-leaved weeds in a variety of crops.

MATERIALS AND METHODS

Materials

The HA originated from a Black Chernozemic soil of Western Canada. Methods of extraction,

---

[1]Present address: Chemistry and Biology Research Institute, Agriculture Canada, Research Branch, Ottawa, Ontario K1A 0C6.

separation, and purification of the HA were identical to those described previously (Khan 1971). The purified HA contained, on a dry ash free basis, 56.4% C, 5.5% H, 4.1% N, 1.1% S, and 33.0% O. Functional group analysis showed 4.5 meq COOH: 2.1 meq phenolic OH, 2.8 meq alcoholic OH; 4.5 meq C = O; and 0.3 meq OCH_3 per g of HA. The extracted and purified HA accounted for about 28% of the organic matter in the original soil.

Analytically pure 2,4-dichlorophenoxyacetic acid (2,4-D) and 4-amino-3,5,6-trichloropicolinic acid (picloram) were used in this study.

Equilibrium Study

A known amount of finely ground HA (20 mg) was weighed into several glass-stoppered centrifuge tubes and shaken for 48 h with a known volume of aqueous herbicide solution (20 ml) of varied concentration. The experiments were conducted at 5 and 25 C. The pH of the suspensions was in the range of 3.3–3.6. HA was removed from the suspension by centrifugation at the appropriate temperature. The supernatant solution was analyzed for the herbicide concentration. The amount of the herbicide adsorbed by HA was determined by subtracting the concentration remaining in solution after equilibrium from the initial concentration.

Kinetic Study

Finely ground HA was weighed into glass-stoppered centrifuge tubes and a known quantity of the herbicide solution was added to each tube. The samples were shaken continuously at 5 and 25 C. At appropriate time intervals a tube was removed, centrifuged immediately for a very short period of time, and the supernatant liquid analyzed for herbicide concentration. During the short centrifugation period the temperature variation was minimal.

Analysis

The concentration of 2,4-D or picloram in the above experiments was determined as follows. An aliquot of the solution was acidified to pH 1 with 6 N H_2SO_4 and extracted several times with ether. After drying the ether extract over anhydrous Na_2SO_4, the volume of the solvent in the flask was reduced to about 0.5 ml using a rotary evaporator. The residue was taken up in a small volume of methanol and methylated with an ether solution of diazomethane, generated from Diazald. A few drops of hexane were added in the flask and the excess of diazomethane removed by allowing it to evaporate just to dryness in the fumehood. The resulting material was taken up in hexane and an appropriate aliquot of the solution injected into the gas chromatograph for quantitative analysis of the herbicide. Esters of 2,4-D and picloram present in the samples were calculated by comparing the sample peak heights (in the linear response region) with those of the appropriate standards. Analysis of untreated blanks confirmed the absence of interfering substances.

A Hewlett-Packard model 7610 A gas chromatograph equipped with a Ni^{63} electron capture detector was used in this study. The glass column (1,200 × 4 mm) was packed with 5% ethyl acetate fractionated Dow Corning high vacuum grease on Chromosorb W DCMS, 80–100 mesh. The carrier gas was 95% argon and 5% methane mixture, which was also used for purging the detector.

RESULTS AND DISCUSSION

Equilibrium Study

An analysis of the data established the fact that they can be best represented in terms of the empirical Freundlich adsorption isotherm. The Freundlich equation can be expressed as (Glasstone and Lewis 1960):

$$X = KC^n \quad (1)$$

where X is the amount of adsorbate taken up per unit mass of the adsorbent; C is the equilibrium concentration in solution; n and K are constants representing the slope and the intercept of the isotherm, respectively. In the present study, the data obtained gave reasonably good straight lines by plotting log X against log C. A typical Freundlich plot for adsorption of 2,4-D on humic acid at 5 C is shown in Fig. 1. The values of n and log K ($C = 1$ ppm) were estimated from the Freundlich plots by using the method of least-square fit (Table 1).

The constants K and n may provide rough estimates of the adsorbent capacity and the intensity of adsorption, respectively (Adamson 1967). For both 2,4-D and picloram, the value of slope n decreased with decrease in temperature (Table 1). This is in accordance with the Freundlich-type isotherm

Table 1. Freundlich isotherm constants for the adsorption of 2,4-D and picloram on humic acid

Herbicide	Slope n		Intercept log K	
	5 C	25 C	5 C	25 C
2,4-D	0.748	0.789	1.034	0.869
Picloram	0.877	0.912	1.132	1.046

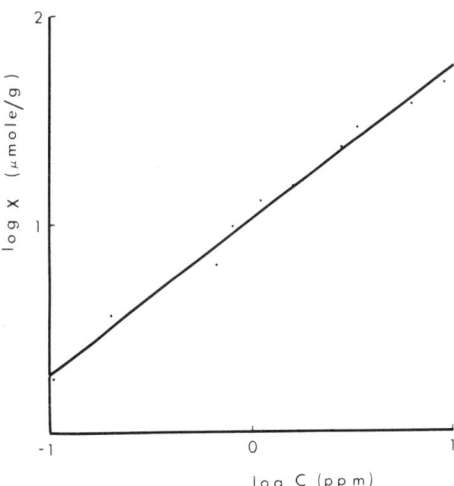

Fig. 1. Freundlich plot for the adsorption of 2,4-D on humic acid at 5 C.

(Hayward and Trapnell 1964) and indicates that the intensity of the herbicide adsorption was lower at 5 C than that at 25 C. At both temperatures the value of n was less than unity, indicating a convex or L-type isotherm. This kind of isotherm may arise due to minimum competition of solvent for sites on the adsorbing surface (Hayward and Trapnell 1964). However, it is possible that in the adsorption process a few layers of water molecules may be always present between the herbicide and humic acid surface. Examination of K values in Table 1 shows that the adsorptive capacity of humic acid for picloram was slightly greater than for 2,4-D at both temperatures. An increase in temperature resulted in a decrease of K values for both herbicides, thereby indicating the exothermic nature of adsorption (Kipling 1965). Yamane and Green (1972) obtained similar results in a study of s-triazine herbicides adsorption on soil material. However, they attributed the apparent exothermic nature of adsorption of triazines to the temperature dependence of the herbicide–water interaction.

Kinetic Study

An examination of the data revealed that there was an initial rapid rate of adsorption of 2,4-D and picloram on humic acid at both temperatures. This was followed by slower rates at longer times. Weber and Gould (1966) studied adsorption of 2,4-D and several other organic pesticides from dilute aqueous solution by porous activated charcoal and have suggested a mechanism involving intraparticle transport of the solute in the pores and capillaries of the adsorbent. For such systems, the amount of solute adsorbed from solution is directly proportional to the square root of the time elapsed (Crank 1965). In accordance with this the amount of herbicide adsorbed X was plotted as a function of the square root of time t. A typical curve for 2,4-D at 5 C is shown in Fig. 2. In each instance the linearity in the plots was usually attained after about 1 or 1.5 h. Thus it appears that, at longer times, intraparticle transport was the dominant rate-limiting process in the adsorption of 2,4-D and picloram on humic acid. These findings fit in well with the surface geometry and structural concepts of humic substances. In the solid state the humic acid is considered to have a laminated, textured makeup (Orlov and Glebova 1972). It has been postulated that the structure of humic substances is loose or open (Kodama and Schnitzer 1967) and contains voids or holes of different molecular dimensions (Schnitzer and Khan 1972). It follows, therefore, that at longer times, adsorption is governed by the diffusion of the herbicide molecules from the exterior surface to the interior of the pores of humic acid structure.

As linearity of the data was not observed for times less than about 1.5 h, the rate-limiting adsorption mechanism, during this initial period, was considered to be different than predicted by intraparticle transport. Therefore, the kinetic data for this initial period was examined in the light of the generalized equilibrium theory proposed by Fava and Eyring (1956). Accordingly, the following equation takes into account both adsorption and desorption in obtaining a rate equation for adsorption of a solute from solution (Fava and Eyring 1956; Haque and Sexton 1968).

$$\frac{d\phi}{dt} = 2K' (1 - \phi) \text{Sinh} \{b (1 - \phi)\} = y, \quad (2)$$

where ϕ is the fraction adsorbed (amount adsorbed at time t divided by the amount

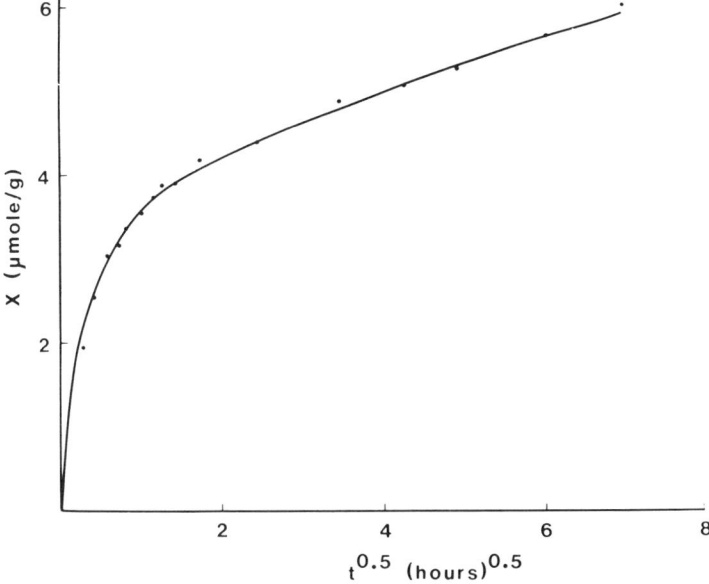

Fig. 2. Adsorption rate curve of 2,4-D on humic acid at 5 C as a function of the square root of time.

adsorbed at equilibrium); K' is the rate constant for adsorption; and b is a constant that yields a measure of the surface stressing energy due to loading with molecules. Plots of ϕ vs. t were obtained. An example for the adsorption of 2,4-D on humic acid at 5 C is shown in Fig. 3. From these plots the differential $d\phi/dt$ for various values of ϕ were obtained. A computer program was used to estimate the rate constant K' for a series of substituted b values so that $f(K',b)$ is minimized, where

$$f(K',b) = \sum_{i=1}^{n} [Y_i - 2K'(1-\phi_i) \sinh\{b(1-\phi_i)\}]^2. \quad (3)$$

The rate constants K' thus estimated for the adsorption of 2,4-D and picloram on humic acid at two different temperatures are shown in Table 2.

The activation energy ΔE was calculated from the Arrhenius equation (Castellan 1964):

$$K' = Ae^{-\frac{E}{RT}}, \quad (4)$$

where K' is the rate constant, T the absolute temperature, R the gas constant and E the activation energy for the process. A is the constant related to the frequency of collision. Converting equation (3) to logarithmic form, we have

$$\ln K' = \ln A - \frac{E}{RT}. \quad (5)$$

Table 2. Kinetic parameters for the adsorption of 2,4-D and picloram on humic acid

Herbicide	Temperature, C	Rate constant K, sec^{-1}	Energy of activation ΔE K cal mole^{-1}	Heat of activation $\Delta H^{\neq}$ K cal mole^{-1}	Entropy activation $\Delta S^{\neq}$ e.u.
2,4-D	5	5.15×10^{-5}			
	25	6.53×10^{-5}	1.95	1.36	−118.9
Picloram	5	8.13×10^{-5}			
	25	10.13×10^{-5}	1.81	1.22	−118.5

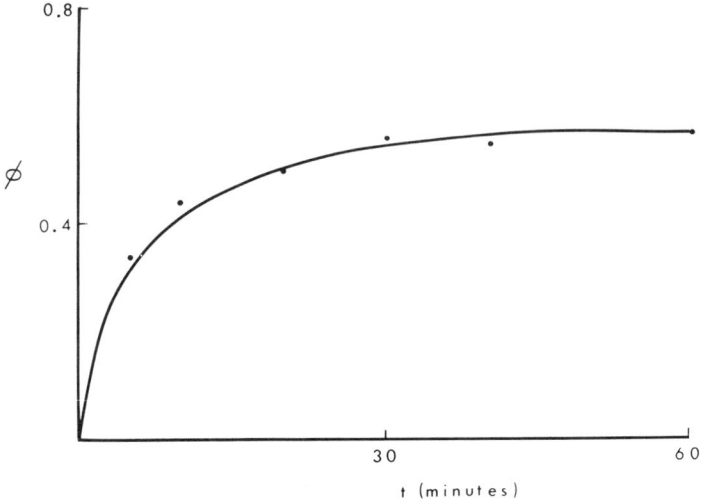

Fig. 3. Rate of approach to equilibrium of 2,4-D on humic acid at 5 C.

For two rates, K'_1 and K'_2, corresponding to absolute temperatures T_1 and T_2, the equation for the activation energy based on the rate constants may be written as:

$$\ln \frac{K'_2}{K'_1} = -\frac{\Delta E}{R}\left(\frac{1}{T_2} - \frac{1}{T_1}\right). \quad (6)$$

Equation (6) was used to estimate the activation energy ΔE for 2,4-D and picloram adsorption on humic acid.

The heat of activation $\Delta H^{\neq}$ is related to the activation energy ΔE as:

$$\Delta E = \Delta H^{\neq} + RT. \quad (7)$$

Applying the absolute reaction rate theory we can calculate the entropy of activation $\Delta S^{\neq}$ from the following relationship (Castellan 1964):

$$K' = \left(\frac{kT}{h}\right)\exp\left(-\frac{\Delta H^{\neq}}{RT}\right)\exp\left(\frac{\Delta S^{\neq}}{R}\right). \quad (8)$$

where k is the Boltzman constant and h the Plank's constant.

A summary of the rate parameters calculated for the adsorption of 2,4-D and picloram on humic acid is given in Table 2. The values of the rate constant K' are of the order that indicates that the initial rate was controlled by the herbicide movement to the humic acid surface involving a physical type of adsorption. The transference rate would be dependent upon diffusion of the herbicide molecules across the water film surrounding the humic acid particles and on the shaking rate of the suspension. The attractive forces in physical adsorption process involve activation energies that are less than a few Kcal/mol (Adamson 1967). The values of the activation energy ΔE reported in Table 2 are small, which further suggests physical adsorption, and characterize diffusion-controlled processes. The low magnitude of heat of activation $\Delta H^{\neq}$ ruled out the possibility of chemisorption (Hayward and Trapnell 1964) and suggests that the most probable nature of the adsorption is physical. The large negative values for entropy of activation $\Delta S^{\neq}$ (Table 2) indicate a lesser degree of freedom for 2,4-D and picloram in the transition state.

CONCLUSIONS

The adsorption of 2,4-D and picloram on humic acid followed the Freundlich-type isotherm. The adsorption at the initial times appears to be a diffusion of the herbicide molecules to the surface of humic acid. However, at longer times the adsorption rate becomes slow as it is controlled by the intraparticle diffusion of the herbicide molecules into the interior of the humic acid particles.

The relatively fast rate of adsorption, low values of activation energy, and heat of activation suggest the physical type of adsorption possibly involving van der Waals forces and hydrophobic bonding between the herbicide molecules and humic acid surface in an aqueous system.

ACKNOWLEDGMENTS

The technical assistance of R. Mazurkewich is much appreciated. I express my thanks to L. P. Lefkovitch of Statistical Research Service, Ottawa, for his valuable help in computing work.

LITERATURE CITED

ADAMSON, A. W. 1967. Physical chemistry of surfaces. Academic Press, Inc., New York, N.Y. 402 pp.

CASTELLAN, G. W. 1964. Physical chemistry. Addison-Wesley Publishing Co. Inc., London. pp. 607–643.

CRANK, J. 1965. The mathematics of diffusion. Clarendon Press, London. 147 pp.

DUNIGAN, E. P. and McINTOSH, T. H. 1971. Atrazine soil organic matter interaction. Weed Sci. **19**: 279–282.

FAVA, A. and EYRING, H. 1956. Equilibrium and kinetic of detergent adsorption — a generalized equilibrium theory. J. Phys. Chem. **60**: 890–898.

GILMOUR, J. T. and COLEMAN, N. T. 1971. s-Triazine adsorption studies: CA-H humic acid. Soil Sci. Soc. Amer. Proc. **35**: 256–259.

GLASSTONE, S. and LEWIS, D. 1960. Elements of physical chemistry. D. Van Nostrand Company, Inc., New York, N.Y. pp. 567.

HANCE, R. J. 1965. Observation on the relationship between the adsorption of diuron and the nature of the adsorbent. Weed Res. **5**: 108–114.

HANCE, R. J. 1969. The adsorption of linuron, atrazine and EPTC by model aliphatic adsorbents and soil organic preparations. Weed Res. **9**: 108–113.

HAQUE, R. and SEXTON, R. 1968. Kinetic and equilibrium study of the adsorption of 2,4-dichlorophenoxyacetic acid on some surfaces. J. Colloid Interface Sci. **27**: 818–827.

HAYES, M. H. B. 1970. Adsorption of triazine herbicides on soil organic matter, including a short review on soil organic matter chemistry. Residue Rev. **32**: 131–174.

HAYES, M. H. B., STACEY, M. and THOMPSON, J. M. 1968. Adsorption of atrazine herbicides by soil organic matter preparations. Pages 75–90 in Isotopes and radiation in soil organic matter studies. Int. Atomic Energy Agency, Vienna, Austria.

HAYWARD, D. O. and TRAPNELL, B. M. W. 1964. Chemisorption. Butterworths, London. 91 pp.

KHAN, S. U. 1971. Distribution and characteristics of organic matter extracted from the black solonetzic and black chernozemic soils of Alberta. The humic acid fraction. Soil Sci. **112**: 401–409.

KHAN, S. U. 1972. Adsorption of pesticide by humic substances: a review. Environ. Lett. **3**: 1–12.

KIPLING, J. J. 1965. Adsorption from solutions of non-electrolytes. Academic Press, Inc., New York, N.Y. 129 pp.

KODAMA, H. and SCHNITZER, M. 1967. X-ray studies of fulvic acid, a soil humic compound. Fuel **47**: 87–94.

LI, G. and FELBECK, G. T. Jr. 1972. A study of the mechanism of atrazine adsorption by humic acid from muck soil. Soil Sci. **113**: 140–148.

ORLOV, D. S. and GLEBOVA, G. I. 1972. Electron-microscopic investigation of humic acids. Sov. Soil Sci. (Engl. Transl. Pochvovedenie) **4**: 445–452.

SCHNITZER, M. and KHAN, S. U. 1972. Humic substances in the environment. Marcel Dekker, Inc., New York. pp. 137–201.

STEVENSON, F. J. 1972. Organic matter reactions involving herbicides in soil. J. Environ. Qual. **1**: 333–343.

SULLIVAN, J. D. and FELBECK, G. T. Jr. 1968. A study of the interaction of atrazine herbicides with humic acids from three different soils. Soil Sci. **106**: 42–52.

WEBER, W. J. Jr. and GOULD, J. P. 1966. Sorption of organic pesticides from aqueous solutions. In R. F. Gould, ed. Organic pesticides in the environment. Advan. Chem. Ser. **60**: 280–304.

WEBER, J. B., WEED, S. B. and WARD, T. M. 1969. Adsorption of atrazine by soil organic matter. Weed Sci. **17**: 417–421.

YAMANE, V. K. and GREEN, R. E. 1972. Adsorption of ametryne and atrazine on an oxisol, montmorillonite, and charcoal in relation to pH and solubility effects. Soil Sci. Soc. Amer. Proc. **36**: 58–64.

Interaction of Ethyl N,N-Di-n-propylthiolcarbamate (EPTC) with Montmorillonite

M. M. MORTLAND and
W. F. MEGGITT
Departments of Soil Science and Crop Science, Michigan State University, East Lansing, Mich.

Results of infrared studies of EPTC-montmorillonite complexes show a decrease in the CO stretching and an increase in the CN stretching frequencies indicating coordination of the EPTC to the exchangeable metal ion through the oxygen of the carbonyl group. The amount of shift was determined by the kind of metal ion on the exchange complex. The EPTC-montmorillonite complexes were stable against atmospheric humidity but, when immersed in water, the EPTC could be completely displaced. Bioassay showed that EPTC-montmorillonite complexes exerted herbicidal activity against germination and growth of rye grass.

The literature concerned with the interaction of organic pesticides and soil colloids has been reviewed recently by Bailey and White (1). Considerable work has been done in investigating the various factors affecting adsorption of various pesticides such as pH, moisture content, kind of exchangeable ion, nature of clay mineral, and effect of organic matter. However, relatively little has been reported on the status of the adsorbed molecule concerning exactly what sort of interaction exists between the adsorbent and the adsorbate. Infrared absorption provides a tool which can both establish such interactions as hydrogen bonding, coordination, salt formation, and actual breakdown or conversion of the compound to other forms. The work reported here is concerned with the mechanism of adsorption of ethyl N,N-di-n-propylthiolcarbamate (EPTC) by montmorillonite, a clay mineral of high specific surface (800 sq. meters per gram), as revealed by infrared absorption.

The EPTC compound is a selective herbicide especially active against annual grasses and certain broadleaf weed species. The structural formula is:

$$CH_3CH_2-S-\underset{\underset{O}{\|}}{C}-N\underset{CH_2CH_2CH_3}{\overset{CH_2CH_2CH_3}{\diagup}}$$

Its boiling point is 232° C. at 760 mm., and the vapor pressure is 0.15 mm. at 25° C. The solubility in water is 375 p.p.m., and it is miscible in such organic solvents as benzene, toluene, xylene, acetone, methanol, and 2-propanol.

Methods

The montmorillonite used in this study was H-25 from Upton, Wyo., supplied by Wards Natural Science Establishment. Homoionic clays were prepared by treating the <0.5-micron fraction with chloride salts of the various cations in excess of the exchange capacity. After flocculation had occurred, the supernatant liquid was siphoned off, distilled water was added to bring it to the original volume, and the chloride salts were added again. This process was repeated three times at the end of which no more salts were added, but the clays were allowed to settle and were redispersed with distilled water until the clays showed signs of not reflocculating, at which time they were placed in dialysis bags and dialyzed against distilled water until the conductivity of the dialyzate approached that of distilled water.

Thin films (2 to 5 mg. per sq. cm.) of montmorillonite were prepared by evaporating in dishes of aluminum foil. The films, which could be readily stripped from these surfaces, were placed in a vacuum desiccator containing a dish of liquid EPTC. The desiccator was evacuated with a rotary pump for an hour, the stopcock to the vacuum line closed off, and the samples were allowed to adsorb EPTC from the gaseous state provided by the vapor pressure of the liquid EPTC. Five days were apparently sufficient to give maximum adsorption since liquid as well as complexed EPTC was noted in the infrared spectra after this period. Infrared spectra were obtained by mounting the films at right angles to the beam in a Beckman IR-7 spectrophotometer.

Infrared Spectra of EPTC-Montmorillonite Complexes

Figure 1 shows the infrared spectra of EPTC complexed with montmorillonite of varying cation saturation in the region 1150 to 1750 cm.$^{-1}$ The top spectrum is that of pure liquid EPTC. Great changes in the spectra of the compound are indicated. The intense peak at 1655 cm.$^{-1}$ in the liquid is the carbonyl stretching band which shifts to much lower frequencies upon complexation with montmorillonite. The peak at about 1630 cm.$^{-1}$ is the deformation band for water. This figure shows that the amount of shift is a function of the kind of metal ion on the clay exchange sites. There is a relationship between the amount of shift of the carbonyl stretching band and the known complexing abilities of the various ions; copper, aluminum, and cobalt cause a greater shift than the alkali metal and alkaline earth cations represented. The band at 1222 cm.$^{-1}$ is allocated mostly to C—N stretching in accordance with the observations of Nyquist and Potts (6) in their work on thiocarbamates where they reported this strong band in the 1152 to 1275 cm.$^{-1}$ region. The

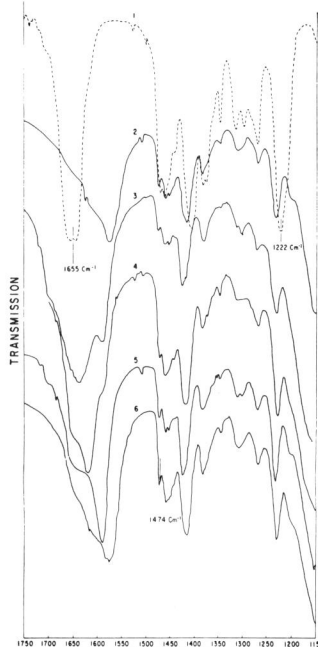

Figure 1. Infrared spectra of EPTC and its complexes with montmorillonite in the region 1150–1750 cm.$^{-1}$

1. Liquid EPTC
2. Al-clay
3. Mg-clay
4. Li-clay
5. Ca-clay
6. Cu-clay

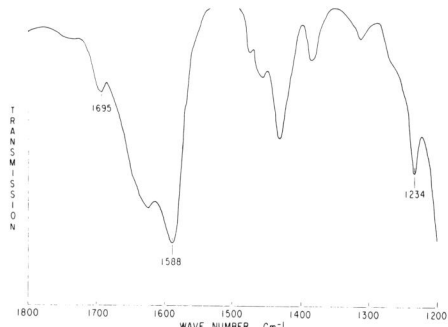

Figure 2. Infrared spectrum of an EPTC-Mg montmorillonite complex subjected to desiccation by evacuation and heating at 70° C. for one hour

Table I. Infrared Absorption Bands of EPTC Complexes with Montmorillonite

Exchangeable Cation on Clay	C—O Stretching, Cm.$^{-1}$	C—N Stretching, Cm.$^{-1}$
Li$^+$	1593	1227
Na$^+$	1590	1222
Ca^{+2}	1594	1232
Mg^{+2}	1587	1232
Al^{+3}	1570	1234
Cu^{+2}	1566	1232
Co^{+2}	1583	1231
EPTC (liquid)	1655	1222

spectra and, more precisely, in Table I show that this band is shifted to a higher wave number when the EPTC is complexed with the montmorillonite. These results strongly suggest that the coordination of the EPTC with the metal ion is through the carbonyl group. Nakamoto (5) points out that in the case of urea, thiourea, and related compounds, the CO stretching frequency is reduced when coordination occurs through this group and concurrently the CN stretching frequency increases. Both of these changes were observed in the EPTC-clay complexes as shown in Table I. These results agree with work done in the senior author's laboratory on complexes of tertiary amides with montmorillonite where it was observed that coordination to the metal was through the oxygen. Apparently the two unshared electrons on the nitrogen atom are not very accessible for coordination probably because of steric factors brought about by relatively large groups on the nitrogen atom–propyl groups, in the case of EPTC. There may be a resonance hybrid of the form:

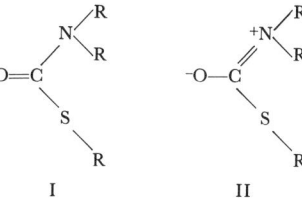

Coordination through the oxygen would enhance the resonance and structure II, thus lowering the CO and increasing the CN stretch frequencies. The CS stretching frequency might also be affected but this band could not be identified in the EPTC-clay complex.

An attempt was made to obtain some information on the orientation of the EPTC molecule in the clay complex. X-ray diffraction results showed that the stable EPTC-clay systems had 001 reflections at 14 A. (10 A. of this is the silicate structure) regardless of the kind of metal cation on the exchange complex. Since there was relatively little water in these systems, as indicated by the low intensity of the H$_2$O deformation band at 1620 cm.$^{-1}$ and the OH stretching in the 3000 to 3500 cm.$^{-1}$ region, the above spacing can reasonably be assumed to result from the bulk of the EPTC rather than water of hydration. Noting the relative length of the molecule compared with the small distance between successive silicate sheets (about 4 A.), it must be concluded that the molecules are lying approximately flat between the sheets and not in a perpendicular position.

The method of preparation of the EPTC-clay films leads to a well-oriented sample with the 001 planes of the clay mineral predominantly in the plane of the film. Absorption bands whose component is perpendicular to the clay sheets could be enhanced if the film is positioned at an angle to the infrared beam. The only band enhanced in intensity was at 1474 cm.$^{-1}$ This was true on all the clay-EPTC complexes. The absorbance of the 1474 cm.$^{-1}$ band was increased about 38% when the film was positioned at 45° with respect to the beam. The assignment of this particular band is not certain but in this region of the spectra is likely to be a CH$_2$ or CH$_3$ deformation of some kind. It may be a CH$_2$ scissoring vibration which Farmer and Mortland (2) found (at 1472 cm.$^{-1}$) to be enhanced in the case of ethylamine complexes of montmorillonite. Considering their reasoning in the case of the ethylamine clay complexes, and the observed x-ray and infrared data here, the EPTC molecule is oriented with its long axis approximately parallel to the silicate sheets, and one or more methylene groups are oriented with the bisector of the H—C—H angle pointed to some extent in the C direction. Since the dipole change is along this bisector, enhancement should be observed with the film at an angle to the beam. It may be that the hydrogens of the methylene groups are attracted to some extent by the oxygen atoms of the silicate sheet (hydrogen bonding), thus accounting for the observed orientation.

The band at 1460 cm.$^{-1}$ (unaffected by film angle) may be the antisymmetric vibration of CH_3.

It is impossible to account for all the changes in the spectrum of EPTC when it is adsorbed on montmorillonite. Some of the changes have been described above, but major shifts in methyl and methylene bands and skeletal vibrations may result from changes in molecular symmetry and force field effects of the silicate sheets on the electron distribution in the molecular structure.

A relatively weak but discrete band appeared at 1695 cm.$^{-1}$ in fresh preparations of Mg and Al-clay-EPTC complexes that were highly dehydrated is shown in Figure 2. This is attributed to CO stretching of some EPTC which is coordinated through the nitrogen rather than the oxygen. This would tend to enhance structure I and to increase the frequency of the CO bond. The authors believe that, while the EPTC tends to coordinate through the carbonyl group and most of its does, some of the EPTC molecules are sterically hindered by crowding of adjacent molecules and limitations imposed by the clay structure so that a small proportion may coordinate through the nitrogen to the metal ion.

There is the possibility that some of the EPTC is associated with the metal ions through a bridging effect of coordinated water as has been suggested by Farmer and Mortland (2) for certain pyridine-metal-clay complexes:

$$\begin{array}{c} R \quad R \\ \diagdown \diagup \\ N \\ | \\ C=O \cdots H \\ \diagup \quad \diagdown \\ S \quad\quad O\ M^+\ -clay \\ | \quad\quad | \\ R \quad\quad H \end{array}$$

It is felt, however, with the possible exception of the Al-clay system, that relatively little of this takes place because the OH stretching bands of H_2O have not been displaced to lower frequencies nor has the deformation band increased much in frequency as is the case in hydrogen bonding. In the EPTC Al-clay system, OH stretching bands were somewhat displaced downward (a broad band with its peak near 3250 cm.$^{-1}$) compared with the other EPTC-metal-clay systems but not to the extent observed by Farmer and Mortland (2) in some pyridine-metal-clay complexes.

Stability of EPTC-Montmorillonite Complexes

Table II shows the EPTC content of clay films exposed to the atmosphere and warmed, as a function of time after formation of the complexes. These complexes were stable during the 35-day period, as there was no significant

Table II. EPTC Content[a] of Montmorillonite after Initial Treatment[b]

Time after Treatment, Days	Exchangeable Cation			
	Cu^{+2}	Mg^{+2}	Ca^{+2}	Al^{+3}
10	72	62	...	93
18	62	...	79	82
35	79	62	76	100

[a] Millimoles of EPTC per 100 grams of clay.
[b] Analyses of very small samples for nitrogen by microkjeldahl. Samples were exposed to atmosphere and heated on a steam radiator during this period.

change in the EPTC content. Under these conditions, the water vapor was not able to displace the organic compound. Farmer and Mortland (3) and James and Harward (4) have noted that ethylamine and ammonia, respectively, may be displaced from certain clay complexes by water in that they all compete for coordinating sites with the metal ion. In some copper-montmorillonite systems, coordinated ethylamine was stable and could not be displaced by water vapor.

However, when the EPTC-clay complexes were immersed in distilled water, the EPTC was readily displaced as shown in Figure 3. The clay films retained their form during immersion after which they were dried in the air and scanned with the spectrophotometer. In both Ca^{+2} and Al^{+3} systems, the EPTC could be completely displaced as indicated by the disappearance of the EPTC bands. These results suggest that, as far as the clay mineral fraction o soil is concerned, residues of EPTC will not remain on the clay surfaces in the presence of liquid water, yet under dried conditions, a stable EPTC-clay complex may exist. In the latter condition, the compound would not be expected to be very active biologically if it were in the interlamellar regions of swelling clay minerals since it would be removed from the proximity of plant roots. An influx of water would be expected, then, to displace the material again into the soil solution where it would exert its influence on growing plants. Apparently, water and EPTC compete for coordination position with the metal ions. Water is successful in displacing coordinated EPTC, however, mainly in an aqueous medium where the clay will expand to its limit, allowing easier entrance of water and egress of EPTC, and where the concentration of water is very great compared with that of EPTC.

Quantitative Estimation of EPTC Complexed with Montmorillonite by Infrared Absorption

An attempt was made to utilize infrared absorption to determine quantita-

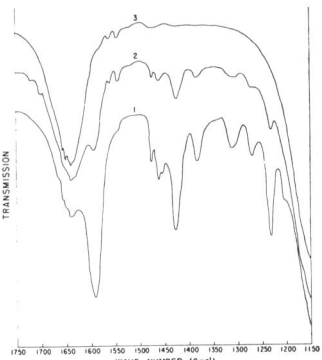

Figure 3. Infrared spectra of EPTC-Ca-montmorillonite complex in 1150 to 1750 cm.$^{-1}$ region

1. Stable complex in air
2. After immersion in water for 16 hours
3. After immersion in water for 32 hours

tively the amount of EPTC complexed with montmorillonite. Freeze-dried Ca-montmorillonite was placed in thin layers in Petri dishes and then positioned in a vacuum desiccator containing a dish of liquid EPTC. The system was evacuated for 30 minutes, then maintained at about 40° C. for 3 days. The Petri dishes were then removed from the desiccator and exposed to the atmosphere at a temperature of 40° C. for 40 days to remove all free EPTC as indicated by the infrared spectrum. The EPTC-clay complex was then mixed with KBr in a Wig-L-Bug dental amalgam mixer at a concentration of 4 mg. per 600 mg. KBr. Then, 400 mg. of this mixture was made into standard circular KBr pellets. This pellet represented the highest EPTC concentration which was found by microkjeldahl analysis of the complex to be 0.76 mmole of EPTC per gram of clay. Lower concentrations were made by using less than 4 mg. of the EPTC-clay complex and making up the differences with the original freeze-dried Ca-montmorillonite containing no EPTC. All mixtures then contained a total of 4 mg. of clay.

Figure 4 shows the plot of the absorbance of the band at 1420 cm.$^{-1}$ vs. concentration of complexed EPTC in the clay. Apparently this technique can be used to obtain quantitative information on amount of EPTC complexed in clay minerals. A plot of the absorbance of the carbonyl band near 1590 cm.$^{-1}$ vs. concentration also gave a relatively good linear plot; however, the deformation band of H_2O at 1620 cm.$^{-1}$ interfered to some degree with the determination of the absorbance.

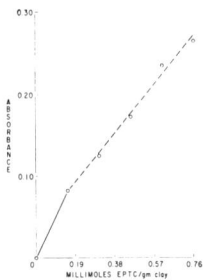

Figure 4. Plot of absorbance of the 1420 cm.$^{-1}$ band of EPTC-Ca-montmorillonite vs. m-mole of EPTC in the clay

Table III. Biological Activity of EPTC on Ryegrass Seedlings in a Bioassay

Treatment	Concentration of EPTC, P.P.M.	Germination, per 20 Seeds		Mean Shoot Length per Plant, Mm.	
		Sand	Filter paper	Sand	Filter paper
		TEST I			
EPTC (technical)	0	12	10	10.6	12.7
	1	15	10	10.3	10.4
	2.5	10	8	8.5	6.4
	5	9	1	6.2	1.0
EPTC-clay complex	0	14	11	11.1	9.6
	10	15	6	12.5	7.4
	25	6	9	4.3	5.5
	50	6	7	5.1	4.5
		TEST II			
EPTC (technical)	0	16	16	6.6	6.7
	5	0	0	0	0
	10	0	0	0	0
	25	0	0	0	0
EPTC-clay complex	0	17	14	7.2	8.2
	50	14	13	2.1	3.0
	100	15	12	2.0	2.4
	250	8	11	1.5	1.8

Biological Activity of Clay-Adsorbed EPTC

A bioassay of EPTC-montmorillonite complex that was stable in the atmosphere was conducted and compared directly with a standard using technical EPTC. The bioassay was carried out in Petri dishes using annual ryegrass (*Lolium multiflorum*) as a test plant. Twenty seeds of ryegrass were put in the Petri dish on both filter paper and sand media. Clay that contained EPTC at 0.76 mmole per gram was added to provide 10, 25, and 50 p.p.m. of EPTC in 10 cc. of water. In another series, technical EPTC was included at 1, 2.5, and 5 p.p.m. in 10 cc. of water. In a second experiment, the EPTC-clay complex was increased to 50, 100, 250 p.p.m., and the pure EPTC series was 0, 5, 10 and 25 p.p.m. The seeds were germinated in controlled environment incubators for 7 days at 80° F. in the first experiment and for 5 days in the second. Readings were also taken in the second experiment at the end of 11 days. The number germinated and the shoot length in millimeters were recorded.

The data shown in Table III show little biological activity from the EPTC-clay complex at concentrations below 25 p.p.m. The range of activity from this complex is approximately 10% of the technical EPTC used as a standard. Table III shows the data at higher concentrations of both the EPTC-clay complex and the technical EPTC. At 5 p.p.m., the technical EPTC inhibited germination of ryegrass seeds. At 50 p.p.m. of the complex, some seeds germinated; however, a high degree of biological activity was noted indicating a substantial release of active EPTC from the EPTC-clay complex in the presence of water used in the bioassay. These results are in accord with those obtained above in the study of the stability of the EPTC-clay complex where it was shown that liquid water could displace EPTC from the clay system.

Conclusions

The results presented above suggest the mode of adsorption of EPTC on the clay mineral montmorillonite may involve at least three mechanisms: coordination of the EPTC to exchangeable metal cations through the carbonyl group; coordination of the EPTC to the metal cations through the nitrogen if the molecule is sterically hindered; and hydrogen bonding of hydrogen on methylene groups to surface oxygen atoms on the clay mineral surface. Complexes of EPTC with montmorillonite were stable in the atmosphere, yet when immersed in water, the EPTC could be quantitatively displaced from the clay. Bioassay results indicated that the EPTC-clay complex, which was stable in the air, exerted bioactivity when placed in the assay medium. Quantitative estimation of EPTC complexed with clay was accomplished utilizing infrared absorption techniques.

Literature Cited

(1) Bailey, G. W., White, J. L., J. AGR. FOOD CHEM. **12**, 324 (1964).
(2) Farmer, V. C., Mortland, M. M., J. Chem. Soc. to be published.
(3) Farmer, V. C., Mortland, M. M., J. Phys. Chem. **69**, 683 (1965).
(4) James, D. W., Harward, M. E., Soil Sci. Soc. Am. Proc. **28**, 636 (1964).
(5) Nakamoto, K., "Infrared Spectra of Inorganic and Coordination Compounds," p. 184, Wiley, New York, 1963.
(6) Nyquist, R. A., Potts, W. J., Spectrochim. Acta **17**, 679 (1961).

Received for review July 6, 1965. Accepted October 15, 1965. Authorized for publication by the Director as Journal Article No. 3676 of the Michigan Agricultural Experiment Station, East Lansing, Mich.

Infrared and X-ray Study of Parathion-Montmorillonite Sorption Complexes[1]

Sarina Saltzman[2] and S. Yariv[3]

ABSTRACT

Infrared spectra indicate that parathion sorbed by montmorillonite coordinates through water molecules to the metallic cations in the interlayer space of the clay. By dehydrating the clay-parathion complexes, parathion becomes directly coordinated with the monovalent cations. The main interaction is through the oxygen atoms of the nitro-group, although interactions through the P=S group were also observed, especially for complexes saturated with polyvalent cations.

Additional Index Words: pesticides, organophosphate insecticides, infrared spectroscopy, adsorption mechanisms by clay.

Since pesticides may reach the soil either directly (soil pesticides) or indirectly (foliar application), the understanding of their behavior in soil is very important in controlling both their biological activity and pollution hazards.

There are several processes responsible for the fate of a pesticide in soil but adsorption on soil colloids is considered the most important. This is because adsorption strongly affects the magnitude of all the other processes.

Sorption of organic compounds by clays has been studied thoroughly, so that the behavior of relatively simple, organic molecules at a well-defined surface may be anticipated. This is not yet possible for larger, complicated molecules such as those of pesticides. In a review on clay-organic complexes and interactions, Mortland (1970) showed that most or probably all of the bonding mechanisms known in clay-organic complexes may be applied to organic pesticides-clay interactions. Consequently, much work is needed in order to understand pesticide behavior in the adsorbed state.

Parathion, a plant and soil insecticide, was selected for our study as it is representative of a group of phenolic phosphorus esters used as insecticides. A previous work of our research group, concerning parathion behavior in soils, showed that soils and soil constituents adsorb parathion from water and hexane solutions (Saltzman and Yaron, 1971; Yaron and Saltzman, 1972). The chemical degradation of parathion in soils was found to be mainly a surface catalyzed process (Saltzman et al., 1973; Yaron 1975). Consequently, the study of the adsorption mechanism of parathion by model soil materials may be helpful for the understanding of the complex soil-parathion interactions.

Since parathion has a relatively complicated structure, previous studies carried out with simpler, related compounds such as nitrobenzene (Yariv et al., 1966), phenol and *p*-nitrophenol (Saltzman and Yariv, 1975), were used as references. It was shown that phenols may form different associations with the exchangeable ions and the hydration water in the interlayer space of the clay. Phenols act either as proton donors or acceptors, depending on the acidity of the interlayer water, and on the substituent group. The nitro-group was found to enhance the acidic character of the phenol molecule. This group acts as an electron donor. Following dehydration the nitro-group may become directly coordinated with the metallic cations, and a direct interaction between the phenol group and the oxygen sheets of the clay may also occur.

Parathion is the diethylthionophosphoric acid ester of *p*-nitrophenol.

$$\begin{array}{c} C_2H_5O \\ \diagdown \\ P-O-\!\!\!\bigcirc\!\!\!-NO_2 \\ \diagup\!\!\!\!\parallel \\ C_2H_5O S \end{array}$$

Compared with *p*-nitrophenol, the parathion molecule has a less polar character, so that a weaker interaction with the clay surfaces is to be expected. Both functional groups of parathion, NO_2 and P=S, may interact with the clay. For the nitro-group, although a less basic character than in the *p*-nitrophenol molecules is to be expected, a rather similar behavior can be assumed.

EXPERIMENTAL

Materials—Samples of almost monoionic montmorillonite saturated with Li, Na, K, Ca, Mg, and Al were prepared as described previously (Saltzman and Yariv, 1975). Pure parathion (Analabs, Inc.) was used for the IR and X-ray studies, and ^{14}C-labeled parathion (Amersham Radiochemical Centre) was used as a tracer in the quantitative study of adsorption.

Instruments—The infrared and X-ray spectrometers used in this study have been described elsewhere (Saltzman and Yariv, 1975.) Radioactivity was counted with a Packard 3003 Tri-carb liquid scintillation spectrometer.

Procedure—Clay-parathion complexes were prepared by immersing monoionic air-dried, self-supporting clay films in solutions containing about 10 mg/ml parathion. The solvents used were carbon tetrachloride, benzene, and hexane. For the IR and X-ray studies the samples were prepared using a hexane solution and a contact time of 7 days.

The quantitative study of adsorption was carried out by bringing in contact about 40 mg of clay film with 5 cm^3 of a hexane solution containing 2,000 ppm ^{14}C-labeled parathion, for 7 days, at room temperature (about 22C). Since during this period some evaporation of the hexane occured (up to 20%), the solution was made up to the initial volume and the ^{14}C remaining was counted. After removing the films from the hexane solution, desorption was carried out by shaking the parathion-saturated films twice, for 2 hours with 5 cm^3 benzene. (Benzene was more effective than hexane in extracting the sorbed parathion.)

---

[1] Contribution from the Agric. Res. Organization, The Volcani Center, Bet Dagan (1975 Series, No. 120-E), and the Dep. of Geology, The Hebrew Univ. of Jerusalem, Israel. This research was financed in part by a grant made by the USDA under Public Law 480. Received 2 April 1975. Approved 4 Sept. 1975.

[2] Scientist, Division of Soil Residues Chemistry, Inst. of Soils and Water, Agric. Res. Organ., The Volcani Center, Bet Dagan, Israel.

[3] Senior Lecturer in Geochemistry, Dep. of Geology, The Hebrew Univ. of Jerusalem, Israel.

Table 1 The effect of the saturating cation on adsorption and desorption of parathion by montmorillonite.

Cation	Adsorption from hexane solution	Desorption by benzene (remaining)
	μmoles/g	
Li	96.5	51.4
Na	36.5	18.8
K	89.7	0
Mg	243.2	27.0
Ca	313.6	15.1
Al	335.8	31.6

RESULTS AND INTERPRETATIONS

Sorption and Desorption of Parathion

Parathion sorption by montmorillonite is greatly dependent on the nature of the solvent. The amounts adsorbed decrease in the order: hexane >> CCl_4 > benzene >>> H_2O. When considering the dielectric constants of these solvents (hexane, 1.89; CCl_4, 2.24; benzene, 2.28; water, 80.37), it is evident that with the increasing polar character of the solvent, parathion molecules are less adsorbed. Although the dielectric constants of CCl_4 and benzene are rather similar, the difference in parathion adsorption is probably due also to the chemical properties of the solvent. The infrared spectra show that benzene is preferentially sorbed by montmorillonite (the ring vibration band at 1478 cm^{-1} is characteristic for the presence of sorbed benzene). Sorption of hexane was detected only with Na-montmorillonite.

Compared with p-nitrophenol, parathion sorption by clay films is a slower process. With polyvalent cations it takes 7 days for parathion to be sorbed in quantities suitable for the IR-study, while similar band intensities were obtained for p-nitrophenol in only 3 days. The sorption is highly dependent on the cation and on its hydration state. With air-dried Na-montmorillonite, for instance, less adsorption was noted even after 8 days. This clay, after losing most of its hydration water by overnight drying at 80C, did not sorb parathion. For Ca-montmorillonite, which has a greater affinity for water, the equilibrium was reached within 2-5 days and overnight drying at 80C had no effect on the sorption rate.

The dependence of the sorption on the saturating cation was also shown by the amounts of parathion taken out by the various monoionic montmorillonites from a hexane solution containing 2000 ppm parathion (Table 1). Generally, the adsorption was greater for the clay saturated with polyvalent cations than for the clay saturated with monovalent ones. These results are typical for sorption of very weak bases (Sofer et al., 1969).

Parathion may be desorbed from the clay by benzene (Table 1). Except for the K-clay, with which the sorption is totally reversible, all the other clays retained some of the sorbed parathion.

Infrared Study

The interpretation of the IR spectra is based on previous works (Saltzman and Yariv 1975; Yariv et al., 1966). When heated under vacuum, up to 200C, clay–parathion complexes were less stable than clay–p-nitrophenol ones (Saltzman and Yariv, 1975). It was observed that the stability is

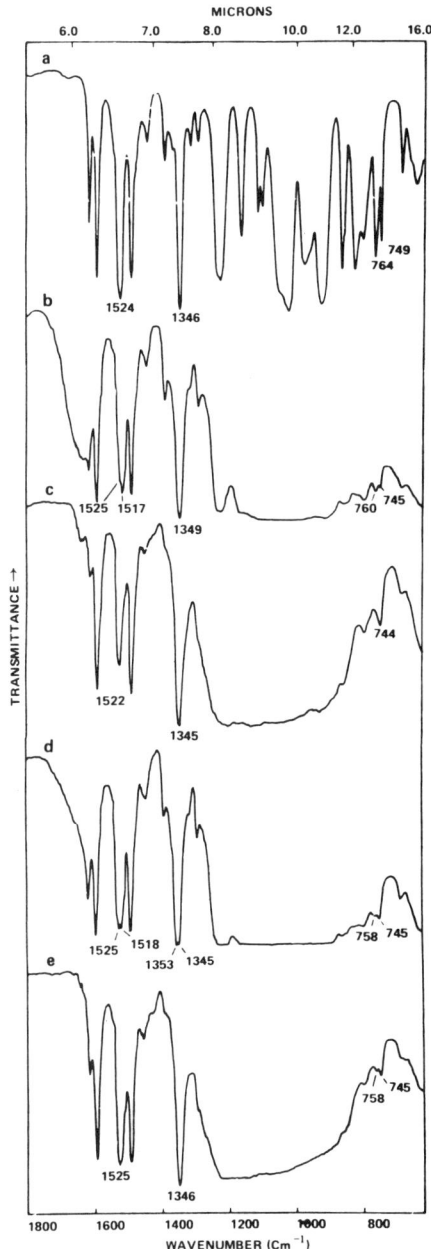

Fig. 1 Infrared spectra of (a) liquid parathion pressed between two KBr windows; (b) Mg-montmorillonite-parathion air-dried; (c) Mg-montmorillonite-parathion heated at 180C under vacuum; (d) Al-montmorillonite-parathion air-dried; (e) Al-montmorillonite-parathion heated at 180C under vacuum.

dependent on the saturating cations. Almost no losses were observed with polyvalent cations. Losses were very great

Table 2 Vibrational frequencies of parathion in the free state and when complexed with montmorillonite.

	Parathion		Montmorillonite-parathion complexes										
			Li		Na		K		Mg		Ca		Al	
Vibration type	Liquid	Hexane solution	air-dried	heated under vacuum	air-dried	heated under vacuum	air-dried	heated under vacuum	air-dried	heated under vacuum	air-dried	heated under vacuum	air-dried	heated under vacuum
							cm^{-1}							
NO_2 antisymmetric stretching	1524	1529	1518	1512	1518	1514	1517	1515	1517	1522	1518	1523	1518 1525	1525
NO_2 symmetric stretching	1346	1344	1349	1338	1350	1342	1350	1346	1349	1345	1350	1345	1353 1345	1346
Ring vibrations	1614 1592 1492	1614 1595 1491	1614 1590 1490	1614 1592 1487	1614 1592 1492	1614 1590 1492	1614 1589 1487	1614 1584 1484	1614 1589 1491	1614 1587 1487	1614 1590 1492	1614 1590 1492	1614 1590 1492 1487	1614 1591 1492
P=S stretching	764	762	nd*	nd	764†	nd	763	760‡	760	744	764	746	758	758 745

* Not detected.   † Film left 50 days in hexane solution.   ‡ Disappears on heating above 70C.

Table 3 O–H stretching frequency of H_2O in clay-parathion complexes.

Saturating cation	Untreated clay, air-dried	Parathion-clay complexes*			
		Air-dried	Heated under vacuum at:		
			70C	120C	200C
		cm^{-1}			
Li	3460-3440	3410 br. sh.‡	d. i.	d. i.	d. i.
Na	3460-3440	3450 v. br.	3450 v. br.	3450 sh.	d. i.
K	3460-3440	3445	d. i.	d. i.	d. i.
Mg	~3360	v. br. d. i.	3475 sh.	3475 sh.	d. i.
Ca	~3360	v. br. d. i.	3535 sh.	d. i.	d. i.
Al	~3360	v. br. d. i.	v. b. d. i.	3465 sh.	3465 sh.

* v. br. = very broad; br. sh. = broad shoulder; d. i. = difficult to identify.
† Samples heated 30 min in a vacuum cell before recording the spectra. Spectra were recorded under vacuum.
‡ Disappears under vacuum at room temperature.

with monovalent cations and the decreasing order of stability was Li > Na > K.

Parathion loss under these conditions is due to its desorption together with water molecules during dehydration. A similar process was recently suggested for another organophosphorus molecule (dasanit) adsorbed on montmorillonite (Bowman, 1973). In Na- and K-montmorillonites, since the electrostatic attraction forces between silicate sheets and interlayer cations are weak, parathion molecules may move freely during dehydration.

Heating clay–parathion complexes under vacuum caused most of the water to be lost, while the same treatment failed to remove similar amounts of water from clay-p-nitrophenol complexes.

Infrared absorption bands of parathion in hexane solution of liquid parathion and of parathion–clay complexes are given in Table 2. Some of the spectra are given in Fig. 1. Sorption of parathion by montmorillonite results in perturbations of several characteristic vibrations.

In all of the air-dried complexes the NO_2 antisymmetric stretching frequencies were shifted to lower values (11-12 cm^{-1} shift), while the symmetric were generally shifted 5-9 cm^{-1} to higher frequencies. As suggested by Nakamoto (1963), this fact indicates that only one of the O-atoms of the nitro-group is involved in coordination. In air-dried Al-complexes, additional bands were observed at frequencies similar to those of liquid parathion, indicating a different sorption mechanism. The other complexes showed only weak shoulders at 1525 cm^{-1}.

Following dehydration by heating under vacuum, clear differences appeared between monovalent and polyvalent cations. With monovalent cations, the antisymmetric stretching frequencies were shifted 2-6 cm^{-1} to lower values. The same trend was observed for the symmetric NO_2 vibrations. Both values are lower than those obtained for either a benzene or hexane solution of parathion. With polyvalent cations, the stretching frequencies of both the symmetric and antisymmetric nitro-groups were similar to those obtained for liquid parathion.

Nakamoto (1963) suggested that the decrease in the frequencies of both nitro-group absorption bands indicates the involvement of both O-atoms of this group, in coordination. Since this coordination was obtained only after dehydration, and only with cations of low hydration energy, we suggest that it is due to a direct interaction between the cation and the nitro-group.

Since the C–O and C–N stretching frequencies of parathion sorbed on montmorillonite appeared as broad bands or as shoulders, the determination of these frequencies was in most cases difficult.

The P=S band of liquid parathion appears at 764 cm^{-1}. McCaulley and Cook (1960) found this peak rather inconvenient for measuring absorbance. In the air-dried complexes, a perturbation of this band was observed with the Al-clay and less with the Mg-clay. During the dehydration process this band shifts to lower frequencies and, (except for Al), in the dehydrated complexes it disappears. This is because it probably overlaps neighboring CH out-of-plane bands. With Li-clay the amount of sorbed parathion was too small to enable its identification. These results indicate that with Al and Mg there is an interaction between the S-atom and the water of hydration. The strong perturbation observed after dehydration indicates a direct coordination of the S-atom with the metal. From Fig. 1 it follows that this interaction occurs with Al only to a limited extent.

The ring vibration at 1592 cm^{-1} was observed in all the air-dried clay complexes, except for Na and Al, at 1590-1589 cm^{-1}. Slightly lower frequencies of this vibration were observed in the dehydrated complexes, mainly with K- and Mg-clays.

Following Hair (1967), the vibration at 1483 cm^{-1} of benzene sorbed on silica is due to ring stretching vibration and is much more intense in the sorbed state than in the liquid state. This was observed clearly with p-nitrophenol, with which the intensity of this band increases in the order: benzene solution < KBr disc < clay complexes (Saltzman and Yariv, 1975). With parathion, the ratio between the intensity of this band and the neighboring band at 1524 cm^{-1} (the antisymmetric NO$_2$ vibration) was 0.83 for free parathion and about 1.00 for all the clay-parathion complexes, evidencing the same phenomena of increase in the intensity of this ring vibration in the adsorbed state. In the air-dried K- and Al- clay complexes this band was shifted to lower frequencies, and the shift was increased by dehydration. In the dehydrated Mg- and Ca- clay complexes, a shift of this band was also noted. This indicates interaction between clay O sheet and the aromatic ring.

The band at 1292 cm^{-1} is attributed to the ethoxyl radical (Gore, 1950). This band, the position of which is almost independent of the exchangeable cation, is very broad with air-dried Na- and K-clays and very sharp with the other samples. Drying under vacuum causes this band to broaden with all the samples. This is probably associated with "keying" of the ethyl group into the hexagonal holes of the oxygen sheet.

Although poorly defined, absorption bands of water in the air-dried complexes could be observed at 3370-3450 cm^{-1}. The effect of partial dehydration on the absorption band of water is shown in Table 3. With untreated hydrated clays, the H$_2$O absorption band was observed at 3460-3440 cm^{-1} for the monovalent cations, and at about 3360 cm^{-1} for the polyvalent ones. Partial dehydration of clay-parathion complexes had little or no effect on the position of the water band in monovalent saturated complexes but resulted in significant shifts of this band in Ca-, Mg-, and Al-clays. As the shift is to higher frequencies, it may be assumed that in these complexes water is more weakly H-bonded than normal water in clay. The same trend was observed in the case of nitrobenzene sorption by clays (Yariv, et al., 1966). It may be concluded that water-water hydrogen bonding in clays is replaced by water H-bonded to nitro- or thio-groups and coordinated with the cation in the clay complexes. This is not the case with the monovalent cations, since the amount of parathion sorbed is very small.

X-ray Study

The c-spacing of monoionic clays and of the clay-parathion complexes is shown in Table 4. As previously shown, clay-parathion complexes are rather unstable, especially when the saturating cations are monovalent. Therefore, integral series of reflection were not obtained for monovalent clays or for any of the heated clay complexes. A comparison between untreated clays and clay-parathion complexes shows that due to sorption of parathion the basal spacings may increase (Li- and Na-clays), decrease (Ca- and Al-clays), or remain unchanged (K- and Mg-clays). Comparing c-spacings with the amounts of parathion sorbed (Table 1) indicates that the c-spacings of clay-parathion complexes is largely dependent on the water content of the clays. The low spacings obtained for Na- and K-clays indicate some kind of "keying" of the molecule. This is in agreement with infrared observations on broadening of bands of ethoxyl groups. Bodenheimer et al. (1962) reached a similar conclusion for some substituted diamines. Similar to previous observations on indoles (Sofer et al., 1969), depending on its charge, the aromatic ring in the interlayer space can be either tilted or parallel to the oxygen sheet of the silicate layer. This will be further discussed in connection with the various configurations. The hydration of clay-parathion complexes at 40% relative humidity resulted in a decrease of the c-spacing of the clay complexes saturated with monovalent cations. Although no quantitative determinations were undertaken the loss of parathion from these complexes was inferred from the infrared study.

After heating the basal spacing of all the complexes decreased, due to loss of water as well as of organic material. The c-spacings of the dehydrated clay complexes indicate a single-layer water structure for the clays saturated with monovalent cations and a double-layer water structure for the polyvalent saturating cations.

Table 4 Basal spacing of parathion-montmorillonite complexes (a—before heating; b—hydrated at 40% relative humidity; and c—after heating at 200C under vacuum)

Saturating cation	Parathion-clay complex			Untreated clay	
	a	b	c	a	c
			Å		
Li	15.2	14.2	12.3	14.5	12.8
Na	13.6	12.4	11.9	12.8	9.8
K	12.3	12.1	11.9	12.3	10.4
Mg	15.2*	15.2†	14.7	15.2	10.0
Ca	15.2*	15.2†	14.7	15.5	9.7
Al	15.5*	15.8†	15.2	15.8	10.3

* Almost integral order of reflection.
† Integral order of reflection.

DISCUSSION

Parathion sorption by montmorillonite is affected by several factors such as the polarity of the solvent, the competition with water present on the clay, and the nature of the saturating cations. The magnitude of adsorption was not in perfect agreement with any of the ionic series obtained, when considering different ionic properties such as size, hydration energy, and electronegativity. When plotting, for example, adsorption vs. the electronegativity of ions, Na, Li, Mg, and Al fall on a straight line but K and Ca could not be fitted in this correlation. This shows that in most cases the adsorptive capacity of clays for parathion is dependent on the nature of bonds between the saturating cations and the hydration water. For Ca and K it seems that other effects interfere with this effect.

The IR and X-ray data presented show that several cation-water-parathion assemblages may be formed in the clay interlayer space. The following configurations may be inferred from the results obtained.

Configuration A

This configuration was observed for all the air-dried complexes. The infrared absorption bands of the nitro-group show that, similar to p-nitrophenol, in hydrated complexes parathion is involved in H-bonding through the nitro-group with water associated with the cations. However, unlike p-nitrophenol, with which both oxygens of the NO$_2$-group were involved in this interaction, with parathion only one oxygen is H-bonded to the clay. Therefore, the following assemblage may be assumed:

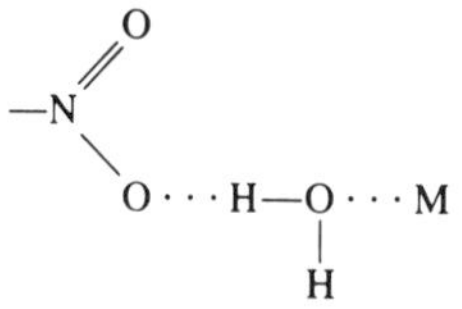

In the case of K-montmorillonite the polarization of the water is low, so that parathion is easily desorbed by benzene (Table 1).

Configuration B

In dehydrated clay-parathion complexes saturated with monovalent cations, a direct interaction between the metallic cation and the organic molecules could be assumed:

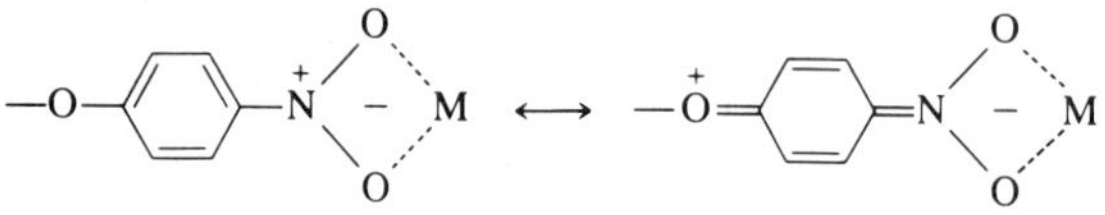

Scheme 1

Evidence for this configuration lies in the shift of both nitro absorption bands to lower frequencies.

In Configuration B the benzene ring does not carry a negative charge, so that it can be parallel to the silicate sheets. This results in relatively low values of the c-spacing obtained for these complexes.

Configurations AC and C

The shift of the band at 764 cm^{-1} in the air-dried Mg- and Al-clay–parathion complex shows that the following configuration may exist:

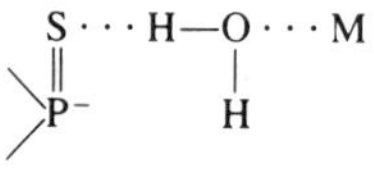

The same parathion molecule could be simultaneously bound through both functional groups (Config. AC). In the Al complex each of the two absorption bands of the nitro-group vibrations appears with two maxima, one of them at frequencies similar to those in free parathion. Therefore, it seems that parathion is partly bound to the Al-clay only through the P=S group (Config. C). Very weak shoulders at 1525 cm^{-1} were also observed for the other cations, except Na and Li.

Configuration D

When clay-parathion complexes saturated with divalent cations are heated under vacuum, the values obtained for the stretching frequencies of the NO$_2$-group were similar to those of free parathion. This suggests that no interaction occurred between the cations and the nitro-group. At the same time, there is a tendency for the P=S band to shift to lower frequencies, so that a direct interaction between the P=S group and the divalent cations may be assumed:

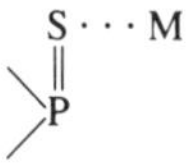

It may be expected that in this configuration the benzene may carry a negative charge (Scheme 2), so that it would be repelled by the clay surfaces. The c-spacing obtained shows that possibly, parathion molecules form an angle with the clay layers. With Al this interaction occurs to a small extent.

Besides the bonding mechanisms previously mentioned, the perturbations in the vibrations of the aromatic ring in the adsorbed state indicate that interaction between the aromatic ring and the silicate oxygen sheet could also occur. This was observed with K-clay complexes in an air-dried as well as in a dehydrated state, and with dehydrated complexes saturated with Mg, Ca, and Al.

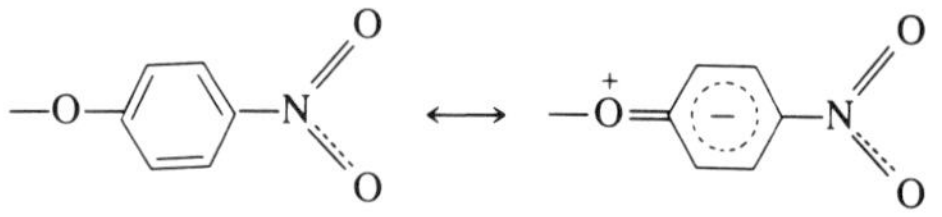

Scheme 2

Since no parathion degradation was observed in the present study it is likely that contrary to the behavior of parathion-kaolinite complexes (Saltzman et al., 1974), these configurations contribute to the stabilization of parathion in the montmorillonite interlayer space.

LITERATURE CITED

1. Bodenheimer, W., L. Heller, B. Kirson and S. Yariv. 1962. Organometallic clay complexes. Part II. Clay Miner Bull. 5:145–154.
2. Bowman, B. T. 1973. The effect of saturating cations on the adsorption of dasanit (O, O-diethyl O-p-(methylsulfinyl) phenyl phosphorothioate), by montmorillonite suspensions. Soil Sci. Soc. Am. Proc. 37: 200–207.
3. Gore, R. C. 1950. Infrared spectra of organic thiophosphates. Faraday Disc. Chem. Soc. 9:710–717.
4. Hair, M. L. 1967. Infrared spectroscopy in surface chemistry. M. Dekker Inc., New York, 315 p.
5. McCaulley, D. F., and H. W. Cook, 1960. The infrared spectra of organic phosphate pesticides and their application to some problems in phosphate pesticides analysis. J. Assoc. Off. Agric. Chem. 45: 710–717.
6. Mortland, M. M. 1970. Clay-organic complexes and interactions. Adv. Agron. 22:75–117.
7. Nakamoto, K. 1963. Infrared spectra of inorganic and coordination compounds. John Wiley & Sons, Inc., New York. p. 197–201.
8. Saltzman, S., and S. Yariv. 1975. Infrared study of the sorption of phenol and p-nitrophenol by montmorillonite. Soil. Sci. Soc. Am. Proc. 39:474–479.
9. Saltzman, S., and B. Yaron, 1972. Parathion adsorption from aqueous solutions as influenced by soil components. In Fate of pesticides in environment. Int. Congr. IUPAC, Proc. 2nd Pesticide Chemistry 6: 87–100. Gordon and Breach Science Publishers, New York.
10. Saltzman, S., B. Yaron and U. Mingelgrin. 1974. The surface catalyzed hydrolysis of parathion on kaolinite. Soil Sci. Soc. Am. Proc. 38:231–234.
11. Sofer, Z., L. Heller and S. Yariv. 1969. Sorption of indoles by montmorillonite. Israel J. Chem. 7:697–712.
12. Yariv, S., J. D. Russell and V. C. Farmer. 1966. Infrared study of the adsorption of benzoic acid and nitrobenzene in montmorillonite. Israel J. Chem. 4:201–213.
13. Yaron, B. 1975. Chemical conversion of parathion on soil surfaces. Soil Sci. Soc. Am. Proc. 39:639–643.
14. Yaron, B., and S. Saltzman, 1972. Influence of water and temperature on adsorption of parathion by soils. Soil Sci. Soc. Am. Proc. 36:583–586.

17

Copyright © 1971 by the Soil Science Society of America
Reprinted from *Soil Sci. Soc. America Proc.* **35**:700-705 (1971)

A KINETIC AND EQUILIBRIUM STUDY OF THE ADSORPTION OF CARBARYL AND PARATHION UPON SOIL ORGANIC MATTER SURFACES[1]

J. A. Leenheer and J. L. Ahlrichs[2]

ABSTRACT

Insight into the mechanisms of carbaryl (1-Napthyl-N-methyl carbamate) and parathion (O, O-Diethyl-o-p-nitrophenylphosphorothioate) adsorption upon organic matter derived from the Romney silty clay loam, Zanesville silt loam, and Carlisle muck soils was obtained by twofold kinetic and equilibrium study of adsorption in nonflow aqueous systems. The differences in adsorptive characteristics of the various types of organic matter were small in both the kinetic and equilibrium studies, but changing the saturating cation from calcium to hydrogen greatly increased the adsorptive capacities for both insecticides. The magnitude of the adsorptive capacities was explained in terms of the magnitude of the hydrophobic natures of the insecticide adsorbates and the organic matter adsorbents.

Kinetic adsorption studies conducted at 5, 25, and 40C showed the rate to increase as the temperature increased with the magnitude of the initial rate constant being 10^{-4} sec^{-1}. The rate-limiting step was interpreted to be diffusion of the insecticide solute molecules to the surface of the adsorbent for the first 10 min of adsorption. At longer times, intraparticle diffusion of the adsorbate into the interior of the adsorbent particles was rate limiting.

HUMIFIED organic matter is one of the most active adsorbents for many pesticide molecules added to soils. In general, the adsorption of insecticides of low water solubility from soils into crops is smallest in those soils that contain high percentages of organic matter (10). A knowledge of the mechanisms of adsorption on organic matter is needed for evaluating the potential for water pollution by certain pesticides applied to a watershed and for estimating the extent to which pesticides are deactivated by the soil through adsorption.

Adsorption mechanisms involving ion exchange for

---

[1] Contribution from the Department of Agronomy, Purdue Agricultural Exp. Sta. (Journal Paper no. 4282, Lafayette, Ind. 47907. Received Jan. 4, 1971. Approved April 30, 1971.
[2] Former Graduate Assistant (now Research Hydrologist, US Geological Survey, Denver, Colo.), and Professor of Agronomy, Purdue University, respectively.

organic bases (1), and H bonding for organic acids (11) have been well elucidated for these classes of pesticides. However, for pesticides with less reactive functional groups, such as the carbamates and the organic phosphates, adsorption mechanisms are more complex and are less clearly defined.

For the adsorption of nonionic organic solutes in aqueous systems, the importance of the competitive nature of the water solvent in adsorption processes is emphasized by Kipling (7). Hance (4) claims that in aqueous systems the functional groups of organic-matter adsorbents are involved in the adsorption of diuron only indirectly by affecting the hydrophilic/hydrophobic balance of the surface. Barlow and Hadaway (2) state that the principal effect of increase in soil moisture upon many insecticides adsorbed upon dry soils is the displacement of the insecticide from the soil particles by adsorbed water. From the adsorbate standpoint, Leopold (9) found a good inverse correlation between water solubility and charcoal adsorption within a class of substituted phenoxyacetic acid compounds. Therefore, maximum adsorption of organic pesticides of low solubility appears to occur on hydrophobic adsorbents where the competitive aspect of water adsorption is minimized.

Both equilibrium and kinetic studies give insight into adsorption mechanisms and bond strengths that determine whether physical or chemical adsorption has occurred. The purpose of this study is to determine the strength and mechanisms of carbaryl and parathion adsorption upon various soil organic-matter substrates in aqueous systems. This will give a predictive capability in assessing the extent to which similar pesticides will be adsorbed and deactivated by various soils.

METHODS AND MATERIALS

Carbaryl (1-Naphthyl-N-methyl carbamate) is a contact insecticide of low mammalian toxicity. It is used against a broad spectrum of insects on a large number of fruit, vegetable, and forage crops. Application rates range from 0.6 to 2.2 kg/ha (0.5 to 2.0 lb/acre). The formulations used are

129

dusts and wettable powders since the crystalline solid is only slightly soluble in solvents commonly used in pesticide formulations.

Parathion (O, O-Diethyl-o-p-nitrophenylphosphorothioate) is a strong systemic organic phosphate insecticide used against a broad spectrum of insects. The compound exists at room temperature as a viscous, pale yellow liquid that should be handled with care due to its high toxicity. It is applied to crops greatly diluted in formulations of wettable powders, dusts, and flowable emulsions.

Soil organic-matter adsorbents were obtained from the Zanesville silt loam, a Typic Fragiudalf with an organic matter content of 1.4%, the Romney silty clay loam, a Typic Argiaquoll with organic matter content of 7.0%, and the Carlisle muck, which is a Typic Medisaprist with an organic matter content of 58%. Organic matter was obtained as an organic residue after the sand and silt fractions had been removed by sedimentation from a soil-water suspension, and the clay removed by HCl-HF acid dissolution treatments (8). The mineral soils were pretreated with the H_2SiF_6 treatments of Chapman et al. (3) to remove the biotite and vermiculite resistant to HF-HCl dissolution. The ash contents of the organic-matter preparations averaged 35% for the Zanesville soil, 10% for the Romney soil, and 1.5% for the Carlisle muck. X-ray diffraction of the inorganic constituents indicated that most of the material was composed of the titanium oxides, rutile and anatase, which were found to have little effect in the adsorption studies.

Characterization of the organic-matter adsorbents by functional group analysis of the carboxyl, aliphatic hydroxyl, phenolic hydroxyl, and carbonyl functional groups indicated a higher state of humification (oxidation) for the Romney organic matter than for the Zanesville organic matter (8). The muck soil was included in the equilibrium studies to check whether the adsorptive properties of organic matter derived from mineral soils were similar to those derived from an organic soil. After the separation procedures, the organic-matter adsorbents were either washed with $1N$ HCl or titrated with a saturated $Ca(OH)_2$ solution to obtain H- and Ca- saturated adsorbents. The adsorbents were then filtered from suspension, freeze-dried, and passed through a 140-mesh sieve.

Synthetic resins of known composition also were used as adsorbents for comparison with soil organic-matter adsorbents. A H- and Ca- saturated ion-exchange resin, Amberlite CG-50, was used for comparison with the exchange characteristics of soil organic matter. It is a weakly acidic, carboxylic (polymethylacrylic), 100-200 mesh, cation exchange resin with an exchange capacity of 10 meq/g. Amberlite XAD-2, a hydrophobic resin, was used to test the effect of a hydrophobic surface upon adsorption. This resin is an insoluble crosslinked polystyrene polymer supplied as 20-50 mesh beads that are porous agglomerations of a larger number of very small microspheres.

Adsorption studies were conducted in 90-ml glass centrifuge tubes immersed in a constant-temperature water bath. Adsorption isotherms were run at 25C, and kinetic adsorption studies were run at 5, 25, and 40C. Agitation of the adsorbent suspensions was maintained by a magnetic stirring bar in each. The centrifuge tubes were suspended around the periphery of small, cylindrical air-driven magnetic stirrers immersed in the water bath that provided the impetus for the stirring bar in each centifuge tube.

Samples of adsorbent were weighed into the centrifuge tubes, and 25 ml of water were added and agitated for 24 hours to allow complete adsorption of water on the adsorbent. At the beginning of an adsorption isotherm or kinetic adsorption study, 25 ml of 10-12 ppm aqueous solution of the insecticide was added to give a 5-6 ppm solution for the initial concentration of the insecticide. Duplicate blank samples with no adsorbent were simultaneously run to exactly determine initial pesticide concentrations. For the equilibrium studies, adsorption was considered complete after 24 hours, and the adsorbents were removed from the suspensions by centrifugation. The samples were centrifuged at 3,000 rpm in an International Model UV centrifuge for 1 min. The greatest temperature change in the samples during centrifugation was 3C for those samples run at 5C. Temperature variation for the 25-C and 40-C samples was minimal during the short period of centrifugation. For kinetic studies, adsorption was terminated at specified times by removal of the adsorbent from suspension by centrifugation.

Due to the presence of small amounts of dissolved organic matter that interfered with direct spectroscopic analysis of aqueous solutions of the insecticides, it was necessary to extract the insecticides with methylene chloride. After centrifugation two 20-ml portions of the supernatant solution were immediately pipetted from a centrifuge tube into two 125-ml separatory funnels, and 20 ml of doubly distilled methylene chloride were added to each funnel. Vigorous shaking of the funnels for 2 min extracted better than 97% of the carbaryl or parathion into the methylene chloride. The phases were allowed to separate for 2 min, and the lower methylene choride phase was drawn off into 5-cm quartz UV adsorption cells. Slight heating of the methylene chloride solutions in the adsorption cells was sometimes necessary to clear up the cloudiness of the water not in solution.

The UV spectra of carbaryl and parathion were taken in a Beckman DK-2 spectrophotometer scanning the 340 nm to 240 nm range, which is the cut-off point for the methylene chloride solvent. The analytical wave-length for carbaryl was 281 nm; $\epsilon = 5{,}890$ 1/mol-cm; the analytical wave-length for parathion was 276 nm; $\epsilon = 10{,}837$ 1/mol-cm.

A small amount of organic matter was extracted by the methylene chloride and caused significant absorbance in the UV spectrum. This absorbance had to be subtracted from the spectrum before a quantitative determination of the insecticide concentration could be made from the standard curve. Since the ratio of the spectrum maximum to the minimum was a constant for each insecticide regardless of the concentration, and since the organic matter spectrum increased roughly linearly as the wavelength decreased, a sloping baseline approximating the organic matter spectral contribution was computed so that the characteristic maximum to minimum ratio was obtained.

Since large quantities of the organic matter adsorbents were difficult to prepare, only six to eight samples with no replicates were used in each run. However, insecticide analysis of each sample was determined as the average result of two duplicate determinations.

With each set of adsorption samples, two blank samples were run without organic matter in suspension. The blank samples were buffered with phosphate buffers to the same pH that existed in the organic matter suspensions, and duplicate determinations were made on each blank. Adsorption studies were run at pH 6.5 for the Ca-saturated organic matter adsorbents, and 2.8 for the H-saturated organic matter adsorbents. At these pH values, both pesticides were stable during the adsorption study, since the hydrolysis products (α naphthol & p-nitrophenol) were not detected in the UV absorbance spectrum. The amount of insecticide adsorbed was determined as the difference in insecticide concentrations between the blank samples and the organic matter suspensions.

The hydrophobic and hydrophilic characteristics of the adsorbents were determined by water vapor adsorption isotherms as suggested by Kipling (6). After the adsorbent had been dried over P_2O_5, duplicate 250-mg samples were weighed on a microbalance into previously tared weighing bottles, and placed in desiccators whose relative humidities were controlled at 22, 53, 75, and 93% by the respective saturated salt solutions of KAc, $Na_2Cr_2O_7$, NaCl and $NH_4H_2PO_4$ (13). Adsorption was allowed to proceed for 1 week, after which the samples were again weighed to determine the amount of water adsorbed. The weighing bottles were immediately capped during removal from the desiccators to prevent adsorption or desorption of water from the adsorbent. Microbial growth on the adsorbents was prevented by a few crystals of thymol

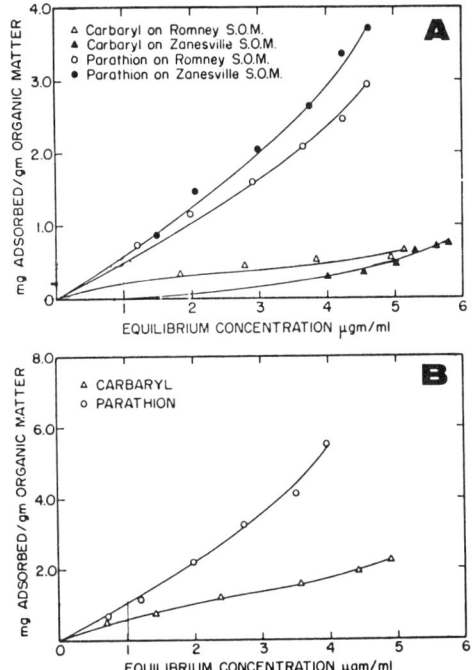

Fig. 1—Insecticide adsorption isotherms (25C) on: (a) Ca-saturated soil organic matter; (b) H-saturated Romney soil organic matter.

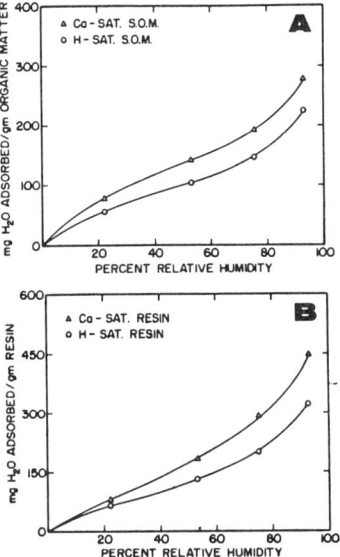

Fig. 2—Water vapor adsorption isotherms (25C) for Ca- and H-saturated organic matter and cation exchange resins.

placed in each desiccator. Adsorption of thymol vapor by the adsorbents was insignificant since the weight loss of thymol due to volatilization was slight compared to the weight of adsorbed water.

RESULTS AND DISCUSSION

The adsorptive capacities of the various organic matter adsorbents for carbaryl and parathion were compared by determining adsorption isotherms. For every adsorbent tested in the equilibrium studies, the adsorptive capacities on a molecular basis for parathion were better than double the adsorptive capacities for carbaryl. This difference is probably related to the lesser water solubility of parathion (24 ppm, 25C) as compared to carbaryl (40 ppm, 30C).

Soil organic matter with H on the exchange sites adsorbed significantly larger amounts of carbaryl and parathion than with Ca on the exchange sites. The adsorption isotherms for Ca-saturated Romney and Zanesville organic matter, and for H-saturated Romney organic matter are given in Fig. 1. A similar difference in adsorptive capacities between Ca and H saturation was observed for the cation exchange resin, but the data is not included here.

The competitive role of the water solvent in adsorption processes is shown by the water vapor adsorption isotherms of Fig. 2. For both the organic matter and cation exchange resin adsorbents the H-saturated adsorbents have greater hydrophobic characteristics and higher insecticide adsorptive capacities than do the more hydrophilic Ca-saturated adsorbents. The cation exchange resin adsorbent with its higher exchange capacity had a greater hydrophilic nature and lower adsorptive capacity for the insecticides than did the organic-matter adsorbents. The hydrophobic resin that adsorbed only 3 mg H_2O at 93% relative humidity had very high insecticide adsorptive capacities of greater than 40 times the amount adsorbed by H-saturated organic matter. This study indicates carbaryl and parathion are adsorbed in much greater amounts in acid than in neutral or alkaline soils due to the decreasing competition of the solvent for adsorption sites as the pH drops.

The differential heat of adsorption was calculated by the van't Hoff equation (12) from the maximum levels of adsorption at equilibrium for the temperatures of 5, 25, and 40C. Heats of adsorption ranged from around 400 cal/g-mole for parathion to about 2,000 cal/g-mole for carbaryl. The greater heat of adsorption for carbaryl may be due to its greater solubility change with temperature change. The low heats of adsorption are on the order one would expect for physical adsorption rather than chemisorption (6).

Differences in the adsorptive capacities between the organic-matter adsorbents derived from different soils were relatively slight. Although functional group analysis showed that the Romney organic matter existed in a higher degree of oxidation than did the Zanesville organic matter, the total number of O-containing functional groups was about the same on both types of organic matter. Therefore, both types of organic matter should have surfaces exhibiting similar adsorption of the water solvent,

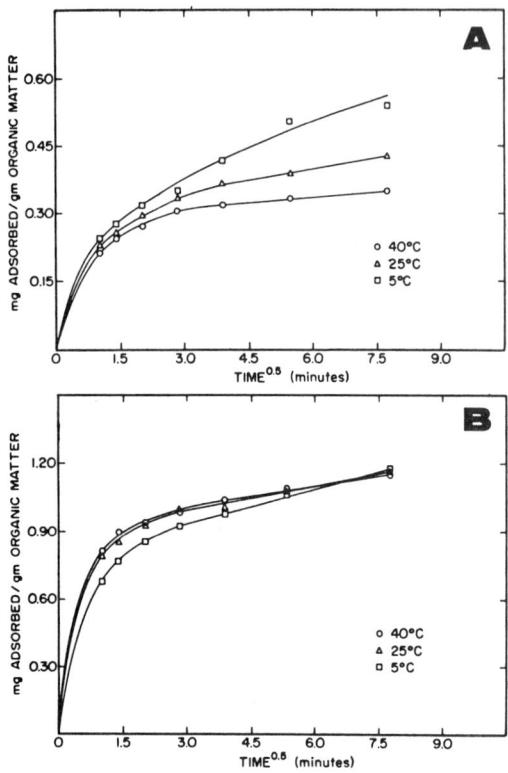

Fig. 3—Adsorption rate curves of (a) carbaryl and (b) parathion on Ca-saturated Romney organic matter as a funtion of the square root of time.

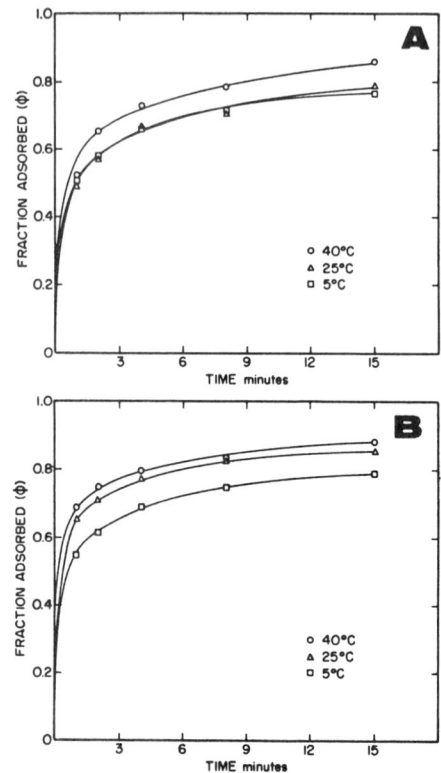

Fig. 4—Rate of approach to equilibrium of (a) carbaryl and (b) parathion on Ca-saturated Romney organic matter.

and consequently, similar adsorption of the insecticide solutes. The functional groups are probably involved in adsorption only indirectly by affecting the hydrophilic/hydrophobic balance of the surface as claimed by Hance (4). The organic matter derived from the muck soil gave adsorptive capacities similar to the organic matter derived from the two mineral soils. Therefore, it is unlikely that the extended treatments removing mineral matter from the organic-matter preparations had much effect on the adsorptive nature of the organic matter other than making more surface area available for adsorption.

Kinetic Studies

Kinetic studies of carbaryl and parathion adsorption on organic matter adsorbents were conducted at 5, 25, and 40C. The shortest time at which adsorption measurements could be obtained was around 1 min, at which time significant amounts of adsorption had already occurred. Equilibrium was attained in 2 hours or less, and the amount of insecticide adsorbed at equilibrium decreased as temperature increased. Desorption studies indicated that adsorption was completely reversible, and the amount of insecticide desorbed at equilibrium in a nonflow system increased as temperature increased.

Weber and Gould (12) found that the rate-limiting step for the removal of a series of organic pesticides from dilute aqueous solution by porous-activated charcoal was one of intraparticle transport of the solute from exterior adsorption sites into the pores and capillaries of the adsorbent. Systems with this adsorption mechanism give a linear plot for adsorption as a function of the square root of time. When the adsorption of carbaryl and parathion on soil organic matter adsorbents was plotted as a function of the square root of time, as shown in Fig. 3, linearity in the plots was attained only for times greater than about 10 min for both insecticides with all the adsorbents. For times of less than 10 min, where the adsorption rates were faster than predicted by intraparticle transport, insight into the rate-limiting adsorption mechanisms was obtained by the following treatment of the kinetic data.

Since adsorption and desorption take place simultaneously in the nonflow system used in this experiment, the rate of adsorption was proportional to the distance of the system from equilibrium. The following equation given by Haque and Sexton (5) takes into account both ad-

Table 1—Adsorption rate constants and activation energies

Soil organic matter adsorbents	Adsorption rate constants, k (sec^{-1}) × 10^{-4}			Activation energy, E (kcal/mole)
	5C	25C	40C	
Carbaryl Adsorption				
Ca-sat. Romney	1.91	1.99	7.09	5.9
H-sat. Romney	0.96	5.04	8.78	10.9
Ca-sat. Zanesville	2.78	1.60	4.13	1.5
Parathion Adsorption				
Ca-sat. Romney	1.07	1.86	4.06	6.3
H-sat. Romney	2.55	8.32	8.27	6.0
Ca-sat. Zanesville	1.44	5.57	2.46	3.2

sorption and desorption in obtaining a rate equation for adsorption:

$$d\Phi/dt = 2k(1-\Phi)\sinh b(1-\Phi)$$

where Φ is the distance from equilibrium as a fraction of initial distance from equilibrium, b is a constant, and k is the rate constant for adsorption when $\Phi = 0$. A plot of Φ vs. time is given in Fig. 4 for carbaryl and parathion adsorption on calcium-saturated Romney organic matter. The differential $d\Phi/dt$ for various values of Φ from 1 to 15 min was measured as the slope of the tangent to the plot at certain values of Φ and t. A computer program was prepared that solved the rate equation for the rate constant k for a series of substituted b values raised in increments of 0.1 from 1 to 10. The variance of k for six known values of $d\Phi/dt$ and $1-\Phi$ was determined for each substituted b value. The solution of the rate equation was obtained when the ratio of the variance of k over the mean of k reaches a minimum for a certain value to b. The rate of change of k with temperature was used to determine the activation energy of adsorption as determined by the Arrhenius equation (12). The rate constants and adsorption activation energies are given in Table 1.

The fast adsorption rates are on the order one would expect for physical adsorption; however, the activation energies are of sufficient magnitude where chemisorption appears to be a possibility. This possibility is discounted by Hayward and Trapnell (6), who state that distinctions between chemical and physical adsorption on the basis of the magnitude of the activation energy cannot be made with porous adsorbents since diffusion of the solute within the adsorbent is temperature-dependent and gives an appreciable activation energy.

The initial pesticide adsorption rates appear to be controlled by solute movement to the adsorbent surface. This transference rate is dependent upon solute diffusion through the water film surrounding the adsorbent particles and on the stirring and mixing rate of the suspension. The suspensions were stirred as vigorously as possible, but the rate of mixing may still be the limiting factor in the initial adsorption rates. As adsorption proceeds, the rate slows as adsorption becomes governed by the solute diffusing within the adsorbent particle, and after 10 min, intraparticle transport is the dominant rate-limiting process.

Precise determinations of initial adsorption rates are complicated by the difficulty of obtaining adsorption measurements at very short times, the change in rate-limiting mechanisms with time, and the heterogeneous nature of the adsorbent surface providing sites with different activation energies of adsorption. These factors obscure the effects of the different functional group compositions of the organic matter adsorbents upon the pesticide adsorption rates. Physical characteristics such as surface area and porosity are of great importance in kinetic studies; and since all the organic matter adsorbents were obtained in a similar manner, there are probably only small physical differences between the adsorbents. Therefore, the rates of adsorption should apply to soil organic matter as an entity without regard to different functional group composition. The complexities involved in kinetic studies have provided useful estimates of the rate constants and activation energies of adsorption and, in addition, have pointed out the complexities with which one must deal in kinetic adsorption studies involving soil organic matter.

CONCLUSIONS

The adsorption mechanisms for carbaryl and parathion adsorption on soil organic matter adsorbents were found to involve pesticide transference rates of the solutes from solution to the surface of the adsorbents at initial times followed by diffusion from exterior adsorption sites to the interior of the porous organic matter adsorbents at longer times. The fast adsorption rates, the low heats of adsorption, the reversibility of adsorption, and the high adsorptive capacities on hydrophobic surfaces tend to rule out chemisorption and point rather to physical adsorption with formation of Van der Waals bonds between hydrophobic portions of the adsorbate molecules and the adsorbent surface in aqueous systems.

LITERATURE CITED

1. Bailey, G. W., and J. L. White. 1964. Review of adsorption and desorption of organic pesticides by soil colloids, with implications concerning pesticide bioactivity. J. Agr. Food Chem. 12:324–332.
2. Barlow, F., and A. B. Hadaway. 1956. Effect of changes in humidity on the toxicity and distribution of insecticides sorbed by some dried soils. Nature (London) 178:1299–1300.
3. Chapman, S. L., J. K. Syers, and M. L. Jackson. 1969. Quantitative determination of quartz in soils, sediments, and rocks by pyrosulfate fusion, and hydrofluosilicic acid treatment. Soil Sci. 107:348–355.
4. Hance, R. J. 1965. Observations on the relationship between the adsorption of diuron and the nature of the adsorbent. Weed Res. 5:108–114.
5. Haque, F., and R. Sexton. 1968. Kinetic and equilibrium study of the adsorption of 2,4-dichlorophenoxy acetic acid on some surfaces. J. Colloid Interface Sci. 27:818–827.
6. Hayward, D. O., and B. M. W. Trapnell. 1964. Chemisorption. Butterworths, London.
7. Kipling, J. J. 1965. Adsorption from solutions of non-electrolytes. Academic Press, New York.
8. Leenheer, J. A., and P. G. Moe. 1969. Separation and functional group analysis of soil organic matter. Soil Sci. Soc. Amer. Proc. 33:267–269.
9. Leopold, A. S., P. van Schaik, and M. Neal. 1960. Molecular structure and herbicide adsorption. Weeds 8:48–54.
10. Lichtenstein, E. P. 1969. Pesticide residues in soils, water, and crops. Ann. New York Acad. Sci. 160:155–161.
11. Ward, T. M., and F. W. Getzen. 1970. Influence of pH on the adsorption of aromatic acids on activated carbon. J. Environ. Sci. Tech. 4:64–67.
12. Weber, W. J., and J. P. Gould. 1966. Sorption of organic pesticides from aqueous solution. In R. F. Gould (ed.) Organic pesticides in the environment. Advan. Chem. Ser. 60:280–304.
13. Winston, P. W., and D. H. Bates. 1960. Saturated solutions for control of humidity in biological research. Ecology 41:232–237.

Editors' Comments on Papers 18 Through 21

18   BRIGGS
A Simple Relationship between Soil Adsorption of Organic Chemicals and Their Octanol/Water Partition Coefficients

19   HANCE
Relationship between Partition Data and the Adsorption of Some Herbicides by Soils

20   LAMBERT
Functional Relationship between Sorption in Soil and Chemical Structure

21   MINGELGRIN and GERSTL
Reevaluation of Partitioning as a Mechanism of Nonionic Chemicals Adsorption in Soils

PREDICTING PHASE DISTRIBUTION OF PESTICIDES

Because uptake of pesticides by soils affects their performance and fate, adsorption is considered a benchmark property needed to predict their environmental behavior (Jury, Spencer, and Farmer, 1983). Adsorption data are usually obtained from equilibrium adsorption measurements and are expressed as distribution coefficients (the ratio between the concentration in adsorbed state and in solution), derived from the adsorption isotherms, which at low concentrations are usually linear. As adsorption measurements are generally tedious, and sometimes difficult due to the low water solubility of some compounds, much effort has been directed toward finding simple parameters and suitable models that can predict adsorption accurately. However, because of the variability of both adsorbents and adsorbates and the complexity of the process, this goal seems hardly attainable.

The models developed are concerned mainly with non-ionic pesticides and are based on the common assumption that, for these compounds, soil organic matter is the dominant adsorbing fraction. Soil adsorption has been described by two basic, contrasting concepts: that of adsorption as a surface reaction and that of adsorption as a partition process.

One of the first models, developed by Lambert, Porter, and Schieferstein (1965), suggested that the adsorption and movement of pesticides in soil could be described by the theory of partition chromatography. In this model soil was considered as a chromatographic column with organic matter as the stationary phase; an important conclusion was that the fraction of organic matter involved in adsorption seemed to be of the same nature in soils of various compositions and sources. Many other works supported the idea of the uniform adsorbing properties of soil organic matter (e.g., Paper 17). This conclusion, and the successful verification of the model for a few non-ionic compounds, contributed to the development of other models based on the partition hypothesis. Most of these models postulated that adsorption could be predicted either by measuring partition parameters correlated with adsorption, or directly from structural parameters.

The basic assumption in the first approach is that partition of a non-ionic compound between organic matter and water could be described by its partition between a polar and a nonpolar phase. Briggs (Paper 18) tested the relationship between the octanol-water partition coefficients and the soil behavior of 30 chemicals. He assumed that if soil organic matter behaves like an "organic solvent," adsorption could be described by Collander's equation, relating the partition between one pair of solvents to that of any other pair. The octanol-water partition coefficients were well correlated with the organic matter-water partition coefficients and with the mobility of the compounds on thin-layer plates. He suggested that this simple parameter could be used to predict the distribution of pesticides in soil.

Based on the same assumption that adsorption could be described by liquid-liquid partition, solubility was suggested as another parameter predicting adsorption (water solubility may be considered as the partition of a compound between water and itself). Chiou and his co-workers (1977) established an empirical equation relating water solubility to octanol-water partition coefficients for various chemicals. These two parameters, as related to soil adsorption and bioaccumulation, were studied intensively both experimentally and theoretically (Chiou, Peters, and Freed, 1979; Karickhoff, Brown, and Scott, 1979; Briggs, 1981; Chiou, Schmedding, and Manes, 1981).

The approach relating adsorption to structural parameters was initially based on the relationship between the partition coefficient and the R_F value, a chromatographic parameter describing the migration of a substance in a partition system. Hance (Paper 19) studied the correlation between soil adsorption of several pesticides belonging to various chemical groups and their hydrophilic-hydrophobic balance.

This property was assessed by a thin-layer partition chromatographic technique. As good correlations were obtained with all the chemicals and for two different soils, the author suggested that the hydrophobic-hydrophilic balance could be used as an adsorption predictor.

Lambert (Paper 20) considered the effect of the chemical structure of the adsorbate on adsorption by utilizing extrathermodynamic relationships. This theory is based on the assumption that the free energy of a compound in solution is the sum of additive contributions from each of its molecular groups. The molecular property parachor, which is proportional to the molecular volume, was proposed as a suitable parameter to predict adsorption. The choice of this parameter is based on the volume energy concept, assuming that the distribution of a solute in a solvent depends on the energy necessary to make a hole in the solvent. Tests of this parameter showed that adsorption by soil organic matter could be predicted accurately for certain classes of compounds that are adsorbed without H-bond formation. Later, Hance (1969) proposed an empirical relationship between adsorption and chemical structure, based on a modification of the parachor. This relationship takes into consideration the number of sites in the adsorbate molecule that could participate in hydrogen-bonding. Briggs (1982) suggested that parachor could be used to calculate partition parameters and developed an empirical relationship between the water-octanol partition coefficient and parachor that takes into consideration the presence of H-bonding groups and halogen atoms in the molecule.

Recently Sabljic (1984) proposed another structural parameter, molecular connectivity indexes, to predict adsorption. In the calculation of such indexes, molecular properties such as size, cyclization, branching, unsaturation, and heteroatom content are included. The author reported results showing a very good correlation between the organic matter-water partition coefficients and the first-order molecular connectivity index (the branching index) for 37 polycyclic aromatic hydrocarbons and halogenated hydrocarbons. Because this index is a two-dimensional representation of the molecular structure, soil sorption could be considered as an interaction between two planes, its magnitude being proportional to the surface area of the solute. This concept supports the view of adsorption as a surface area dependent process. A poorer correlation between the connectivity index and octanol-water partition coefficients indicated that soil adsorption and partitioning are not identical mechanisms.

Mingelgrin and Gerstl (Paper 21) evaluated comparatively the basic concepts of adsorption as a surface reaction and as a partition process. In a comprehensive discussion the authors point out the oversimplification on which the partition concept is based, and the fact that it could lead to erroneous conclusions. Equilibrium adsorption data and thermodynamic calculations show that, the complex soil adsorption is not generally fitted by the partitioning model. Even the basic assumption that organic matter as the principal adsorbent is similar in all soils is discussed and is shown to be sometimes incorrect. Although the authors agree that partition parameters could be used to estimate the uptake of non-ionic pesticides by soils, they show the limitations and argue against generalization. They suggest that correlations between soil adsorption and the octanol-water partition coefficients and/or water solubility should be used carefully, avoiding extrapolation of data for a homologous series of compounds to the general behavior of pesticides. The best prediction of soil uptake of non-ionic pesticides should probably be based upon the structural parameters of the adsorbate.

References

Briggs, G. G., 1981, Theoretical and Experimental Relationships between Soil Adsorption, Octanol-Water Partition Coefficients, Water Solubilities, Bioconcentration Factors and the Parachor, *Jour. Agric. Food Chem.* **29**:1050-1059.

Chiou, C. T., V. F. Freed, D. W. Schmedding, and R. L. Kohnert, 1977, Partition Coefficient and Bioaccumulation of Selected Organic Chemicals, *Environ. Sci. Technol.* **11**:475-478.

Chiou, C. T., L. J. Peters, and V. H. Freed, 1979, A Physical Concept of Soil-Water Equilibria for Nonionic Compounds, *Science* **206**:831-832.

Chiou, C. T., D. W. Schmedding, and M. Manes, 1981, Partitioning of Organic Compounds in Octanol-Water Systems, *Environ. Sci. Technol.* **16**:4-10.

Hance, R. J., 1969, An Empirical Relationship between Chemical Structure and the Sorption of Some Herbicides by Soils, *Jour. Agric. Food Chem.* **17**:667-668.

Jury, W. A., W. F. Spencer, and W. J. Farmer, 1983, Behavior Assessment Model for Trace Organics in Soil: I. Model Description. *Jour. Environ. Qual.* **12**:558-564.

Karickhoff, S. W., D. S. Brown, and T. A. Scott, 1979, Sorption of Hydrophobic Pollutants on Natural Sediments, *Water Research* **13**:241-248.

Lambert, S. M., P. E. Porter, and R. H. Schieferstein, 1965, Movement and Sorption of Chemicals Applied to the Soil, *Weeds* **13**:185-190.

Sabljic, A., 1984, Predictions of the Nature and Strength of Soil Sorption of Organic Pollutants by Molecular Topology, *Jour. Agric. Food Chem.* **32**:243-246.

18

Copyright © 1973 by Blackwell Scientific Publications
Reprinted from *7th British Insecticide and Fungicide Conference Proc.*, The Boots Co., Nottingham, 1973, pp. 83-86

A SIMPLE RELATIONSHIP BETWEEN SOIL ADSORPTION OF ORGANIC CHEMICALS AND THEIR OCTANOL/WATER PARTITION COEFFICIENTS

G. G. Briggs
Rothamsted Experimental Station, Harpenden, Herts.

Summary The adsorption of un-ionised organic chemical by four Rothamsted soils, expressed as soil organic matter/water partition coefficients (Q), is related to the octanol/water partition coefficients (P) by log Q = 0.524 log P + 0.618. A similar relationship exists between P and Rf values on soil thin layers. Good prediction of soil behaviour of organic chemicals can be obtained using mobility classes based on octanol/water partition coefficients.

INTRODUCTION

The correlation between adsorption of un-ionised organic compounds by soil and the soil's organic matter content is well established. Several attempts have been made to predict adsorption by soils by relating the organic matter/water partition coefficient. (Q), to the parachor (Lambert 1967, Hance 1969) or free energy substituent constants within related series of compounds (Briggs 1969). In practice, these relationships need to be determined for each series of compounds and they cannot be simply applied to a wide range of chemical structures. This paper reports a relationship between Q and octanol/water partition coefficients that can be easily and widely used.

Previous work has shown that adsorption by soils can be related systematically to chemical structure and supports the hypothesis that organic matter behaves like an organic 'solvent' (Lambert 1967). If it does then equation 1, found by Collander (1950), which relates partition between one organic solvent and water and a second organic solvent and water should hold, as Lambert (1968) has suggested, when organic matter/water is one of the solvent pairs.

$$\log P_1 = a \log P_2 + b \qquad (1)$$

RESULTS

This hypothesis was tested using average values of Q determined as previously described (Briggs 1969) on four Rothamsted soils for 30 chemicals with a wide range of polarities and values of P, the octanol/water partition coefficient, either taken from the literature (Leo et al 1971) or determined spectrophotometrically (Fujita et al 1964) Octanol/water was chosen because of its wide use as a model for biophases in correlations between biological activity and chemical structure. Equation 2 gave a good fit to the data ($r^2 = 0.84$).

$$\log Q = 0.524\ (\pm .048) \log P + 0.618\ (\pm .113) \qquad (2)$$

A similar test of the hypothesis can be applied to results obtained from soil thin-layer chromatography if it is assumed that relative mobility on the plates is governed only by adsorption on soil organic matter as the stationary phase. The Martin and Synge (1941) equation may then be written:

$$\log (1/R_f - 1) = \log Q + \log A_s / A_m \qquad (3)$$

For a given soil A_s and A_m, the cross-sectional areas of the organic phase and water phase, are constants and substituting for log Q from equation 1 gives equation 4 where P is again the octanol/water partition coefficient.

$$\log (1/R_f - 1) = a \log P + \text{constant} \qquad (4)$$

Frontal Rf values for 25 un-ionised pesticides on Hagerstown soil from Maryland U.S.A. were taken from Helling and Turner (1968), Helling (1971 a,b,c,) and Helling et al (1971). The regression line is:

$$\log (1/R_f - 1) = 0.517\ (\pm .022) \log P - .951\ (\pm .075) \qquad (5)$$

The slope is almost the same as that in equation 2 as it should be if organic matter from different soils behaves relatively uniformly as an adsorbing surface. Where comparable values of Q are available they are similar for soils from England, Europe and North America despite obviously different origins and probable differences in the detailed chemical structure of the soil organic matter. The close similarity of the slopes of equations 2 and 5 indicates that log P and equation 2 can be used to give an estimate of Q that is widely applicable. It is interesting that the slopes are similar to the average value (0.55) found by Leo et al (1971) in equations relating log P to the partition of organic compounds onto a number of biological macromolecules.

DISCUSSION

Helling and Turner (1968) divided pesticides into five mobility classes based on Rf values on Hagerstown soil (2.5 per cent organic matter) and Helling (1971c) concluded that this classification adequately described behaviour in 14 soils containing 1-8 per cent organic matter, a range common to most agricultural soils. Rf values for a given compound decreased as organic matter content increased; however the range of Rf used to define the mobility class on Hagerstown soil was wide enough to cover most of the soils examined. Using equation 5 these mobility classes can be defined in terms of log P and Q.

Class	Rf	log P	Q
Immobile	0 – 0.09	>3.78	>398
Low	0.10 – 0.34	3.78 – 2.39	398 – 74
Intermediate	0.35 – 0.64	2.39 – 1.36	74 – 29
Mobile	0.65 – 0.89	1.36 – 0.08	29 – 4.5
Very mobile	0.90 – 1.00	<0.08	<4.5

The mobility class from the above table or the approximate values of Q from equation 2 can readily be obtained for new compounds or possible soil metabolites either using log P values from the review by Leo et al (1971), calculated using Hansch's π constant, (Fujita et al 1964) or a simple experimental determination. Many π constants in the literature are derived from the phenoxyacetic acid series where electronic effects of substituents are small; in a series such as the phenylureas where P is very sensitive to electronic effects, experimental values of P, which include these effects, predict Q much better than calculated values.

Three examples illustrate the use of the octanol/water partition coefficient to predict behaviour of three different types of compound in soil.

The herbicide 2,6-dichlorothiobenzamide is converted in soils to 2,6-dichlorobenzonitrile which in turn is hydrolysed to 2,6-dichlorobenzamide (Beynon and Wright, 1972). Using the value for an aliphatic amide (because of steric inhibition of resonance), calculated values of log P are 2.3, 3.0 and 1.3 and indicate mobility classes of 'intermediate', 'low' and 'mobile' for the thioamide, nitrile and amide. This agrees well with field observations. Calculated values of Q are 66, 155 and 20 and experimental values are 57, 135-165 and 5-15 (Beynon and Wright, 1972; Briggs 1968).

The oxime carbamate insecticide aldicarb is rapidly oxidised in soil to the sulphoxide. Experimental values of log P of 0.8 and −0.7 classify the parent compound as 'mobile' and the metabolite as 'very mobile' which is the leaching behaviour observed in practice (Goring, 1972).

The polychlorinated biphenyls, important environmental contaminants are a third example of a different structural type, whose behaviour in soil has not been closely examined. The experimental value (Leo et al, 1971) of log P for biphenyl itself is 4.0 so that it would be classed as immobile in soil. The bulk of the material in commercial PCB samples contains more than 3 chlorine atoms per molecule and has a calculated log P > 6. Extensive metabolism would have to occur before any leaching in soil would be expected.

Octanol/water partition coefficients provide a good prediction of soil behaviour using equation 2 and the mobility classes derived from it. Organic matter in soils has similar properties to the humic materials in lakes, streams and rivers so that a good indication of the likely redistribution in the environment by water transport of un-ionised compounds from any source can also be obtained from a single measurement or calculation.

Acknowledgements

I thank Drs. N. F. Janes, K. A. Lord and I. J. Graham-Bryce for helpful discussions and criticism.

References

BEYNON, K. I. and WRIGHT, A. N. (1972) The fates of the herbicides chlorthiamid and dichlobenil in relation to residues in crops, soils and animals. *Residue Reviews*, 43, 23-53.

BRIGGS, G. G. (1968) Unpublished results.

BRIGGS, G. G. (1969) Molecular structure of herbicides and their sorption by soils. *Nature*, 223, 1288

COLLANDER, R. (1950) The distribution of organic compounds between isobutanol and water. *Acta Chemica Scandinavica*, 4, 1085-1098.

FUJITA, T., IWASA, J. and HANSCH, C. (1964) A new substituent constant, pi, derived from partition coefficients, Journal of the *American Chemical Society*, 86, 5175-5180.

GORING, C. A. I. (1972) in 'Organic Chemicals in the soil environment' edited by GORING, C. A. I. and HAMAKER, J. W. (Marcel Dekker, New York), 604.

HANCE, R. J. (1969) An empirical relationship between chemical structure and the sorption of some herbicides by soils. *Journal of Agricultural and Food Chemistry*, 17 667-668.

HELLING, C. S. and TURNER, B. C. (1968) Pesticide Mobility: Determination by soil thin-layer chromatography. *Science*, 162, 562-563.

HELLING, C. S. (1971a) Pesticide mobility in soils I. Parameters of Thin layer chromatography. *Soil Science Society of America Proceedings*, 35, 732-737.

HELLING, C. S. (1971b) Pesticide mobility in soils II. Applications of soil thin-layer chromatography. *Soil Science Society of America Proceedings*, 35, 737-743.

HELLING, C. S. (1971c) Pesticide Mobility in soils III. Influence of soil properties. *Soil Science Society of America Proceedings*, 35, 743-748.

HELLING, C. S., KAUFMAN, D. D. and DIETER, C. T. (1971) Algae bioassay detection of pesticide mobility in soils. *Weed Science*, 19 685-690.

LAMBERT, S. M. (1967) Functional relationship between sorption in soil and chemical structure. *Journal of Agricultural and Food Chemistry*, 15, 572-576.

LAMBERT, S. M. (1968) Omega, a useful index of soil sorption equilibria. *Journal of Agricultural and Food Chemistry*, 16, 340-343.

LEO, A., HANSCH, C. and ELKINS, D. (1971) Partition coefficients and their uses. *Chemical Reviews*, 71, 525-616.

MARTIN, A. J. P. and SYNGE, R. L. M. (1941) A new form of chromatogram employing two liquid phases. *Biochemical Journal*, 35, 1358-1368.

RELATIONSHIP BETWEEN PARTITION DATA AND THE ADSORPTION OF SOME HERBICIDES BY SOILS

R. J. Hance

THE widespread use of organic pesticides has led to an increasing interest in the adsorption of such solutes from aqueous solution by soil colloids because of the influence of this process on pesticidal performance, mobility in the soil and residue problems. The mechanisms involved are not clearly understood because of the complicated balance between soil–solute, solute–water and water–soil interactions. It seems possible, however, that in systems in which ionic forces are unimportant, adsorption may be related to the hydrophilic–hydrophobic balance of the solute. An inverse relationship has been observed[1] between the hydrophilic–hydrophobic balance of some proteins and their interaction with a paraffin surface as measured by paraffin–water interfacial pressures. Ward and Holly[2] have recently found partition between *cyclo*-hexane and water to be related to the adsorption of a number of *s*-triazine herbicides by nylon and cellulose triacetate, although they considered that with these adsorbents the process involved the formation of a solid solution. The object of the present investigation was to determine whether a relationship exists between hydrophilic–hydrophobic balance and the adsorption of a number of non-ionic solutes by soil.

Twenty-nine compounds were investigated; seven phenylureas, nine chloro-*s*-triazines, three methoxy-*s*-triazines, five methylthio-*s*-triazines, one ethylthio-*s*-triazine, one cyano-*s*-triazine, one phenyl carbamate and two uracil derivatives. The triazines all contained alkyl amino substituents in positions 4 and 6. All compounds are herbicides or potential herbicides. Two soils were used, a dark grey sandy loam containing 12·0 per cent carbon and 6·6 per cent clay, and a chalky boulder clay containing 2·1 per cent carbon and 39·6 per cent clay.

Assessments of hydrophilic–hydrophobic balance were made by the thin-layer partition chromatographic method of Boyce and Milborrow[3], modified in that the developing solvent was 40 per cent aqueous ethanol. It was found that although $\bar{R}_F$ values were inconsistent from plate to plate, mobility with respect to a reference compound was reproducible. The reference compound used was fenuron (NN-dimethyl-N'-phenylurea). After development the chromatograms were sprayed with 0·05 normal silver nitrate in 50 per cent aqueous acetone and exposed to the ultra-violet light from a Hanovia 'Chromatolite' for 15 min. Adsorption was determined by the slurry method previously described[4] in which a series of soil samples are shaken overnight with aqueous solutions containing a range of solute concentration. After centrifugation the solute concentration in the equilibrium solution is determined and the amount adsorbed is estimated by difference. In each case the analysis was carried out by measurement of the absorbance at a suitable ultra-violet wavelength and the application of an appropriate blank correction made. In the case of the *s*-triazines measurement was preceded by hydrolysis and the resultant hydroxy-triazine was determined using the background correction procedure described by Gysin and Knüsli[5].

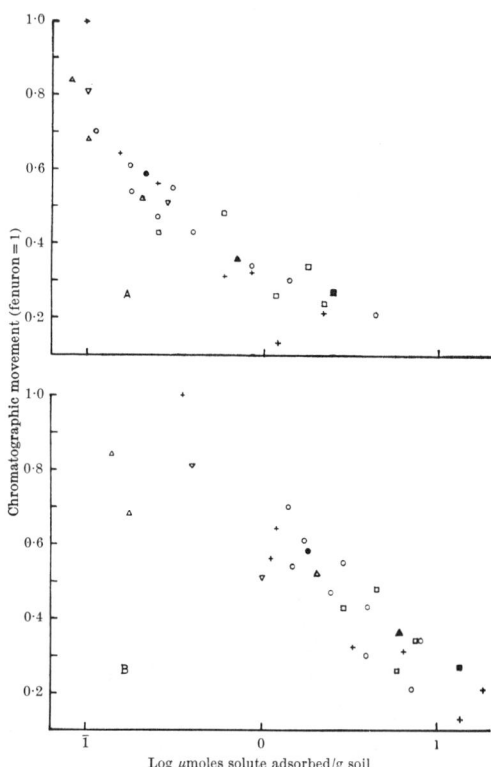

Fig. 1. The relationship between chromatographic movement and the logarithm of the amount of solute adsorbed in μmoles/g at an equilibrium concentration of 1×10^{-4} molar; (A) for the clay soil, and (B) for the loam soil. +, Ureas; ○, chloro-triazines; △, methoxy-triazines; □, methylthio-triazines; ●, cyano-triazines; ▲, ethylthio-triazines; ■, carbamates; ▽, uracils.

The results were evaluated using the Freundlich relationship, $x/m = kc^{1/n}$, where x is the amount of solute adsorbed by a weight m of adsorbent when the equilibrium solution concentration is c. Straight lines were obtained when log x/m was plotted against log c. For comparative purposes the amount of solute adsorbed by 1 g of soil in equilibrium with an arbitrarily chosen solution concentration was used. The slopes of the lines obtained with all compounds but the ureas were one, and thus comparisons of their adsorption behaviour made at one concentration will be valid at other concentrations. As noted previously[4], the slopes obtained with the ureas, with the exception of fenuron, were less than one, so that any comparison made between these compounds and the others will only apply to the one particular equilibrium concentration chosen. Comparisons, therefore, were made using the amounts of solute adsorbed at three equilibrium concentrations, 1×10^{-4} molar, 1×10^{-5} molar, and 1×10^{-6} molar.

Using the figures obtained at $c = 1 \times 10^{-4}$ molar, the coefficients of correlation between log μmolar solute adsorbed/g of soil and chromatographic movement relative to fenuron were 0·851 with the clay soil and 0·903 with the loam soil. The relationships are illustrated in Fig. 1. The corresponding figures at $c = 1 \times 10^{-5}$ were 0·911 and 0·845 and at $c = 1 \times 10^{-6}$ were 0·901 and 0·886. The existence of these correlations for such a variety of compounds suggests that the basic mechanism of their adsorption involves forces of a non-specific nature.

It is proposed, therefore, that an estimate of hydrophobic–hydrophilic balance may give a useful indication of the soil adsorption behaviour of non-ionic organic compounds and may commend itself as a test to be included in screening or evaluation programmes for pesticides.

Agricultural Research Council,
Weed Research Organization,
Begbroke Hill,
Kidlington, Oxford.

[1] Ghosh, S., Breese, K., and Bull, H. B., *J. Colloid Sci.*, **19**, 457 (1964).
[2] Ward, T. M., and Holly, K., *J. Colloid Sci.*, **22**, 221 (1966).
[3] Boyce, C. B. C., and Milborrow, B. V., *Nature*, **208**, 537 (1965).
[4] Hance, R. J., *Weed Res.*, **5**, 98 (1965).
[5] Gysin, H., and Knüsli, E., *Adv. Pest Control Res.*, **3**, 301 (1960).

20

Copyright © 1967 by the American Chemical Society
Reprinted from *Jour. Agric. Food Chem.* **15**:572-576 (1967)

Functional Relationship between Sorption in Soil and Chemical Structure

SHELDON M. LAMBERT

By utilizing an equilibrium constant which considers soil–organic matter as the sorbing medium, a functional relationship between sorption in soil and chemical structure has been developed for certain classes of chemicals. This relationship is based upon extrathermodynamic linear free energy approximations and uses the parachor as an approximate measure of the molar volume of the chemical under consideration. Distribution equilibria between soil and water for a number of chemical homologs of two different chemical classes were used to establish the relationship. The result is a natural consequence of the informative chromatographic model for movement and sorption of chemicals applied to the soil.

Sorption of chemicals in soil is an extremely complex phenomenon and any explanation of what is, in fact, occurring should reflect this complexity. Numerous articles concerning sorption characteristics of herbicides have recently appeared in the literature; these have been for the most part descriptive in nature. The authors have expressed an almost unanimous concern for the lack of an existing relationship between soil sorption and chemical structure (Bailey and White, 1964; Hance, 1965; Harris, 1966; Ward and Upchurch, 1965).

In attempting to describe a functional relationship between soil sorption and chemical structure, the use of suitable models for explanatory purposes may decrease the magnitude of the complexity involved. A simple but useful model describing the movement and sorption of certain types of chemicals in the soil has been described by Lambert *et al.* (1965). Their model is based upon the theory of chromatography and considers the soil as a chromatographic column carrying a stationary phase of organic matter. One natural consequence of this informative chromatographic model is that it predicts a relationship between chemical structure and soil sorption phenomena. The applicability of chromatographic principles to soils is not new and its utility for herbicidal behavior has been discussed by Hartley (1964).

Extension of Conceptual Model

If one considers the chromatographic model to be an adequate representation of the phenomenon involved, then certain consequences, in terms of the model, will arise naturally. These consequences can be considered from the point of view of the effect of chemical structure on soil sorption equilibria. One method by which this may be accomplished is the utilization of extrathermodynamic relationships; specifically, those classified as linear free energy relationships.

A theory of the effects of substituents or structural changes on the rates or equilibria of organic chemical reactions was developed by Hammett in 1940. The general form of the correlations he obtained is a linear relationship between the logarithms of the rate or equilibrium constants for a large number of aromatic side-chain reactions and the variation in reactant structure. More recently, Martin (1949), during the development of the theory of partition chromatography, proposed a relationship between the logarithm of the partition coefficient and a set of parameters, which are characteristic of those functional groupings constituting the molecule under consideration. Certainly there are many assumptions made in deriving the above relationships and, as a consequence, specific restrictions are imposed for their use. However, if the limitations are borne in mind, the relationships are valid and of great practical importance.

Functional Relationship

The general theory of extrathermodynamic linear free energy relationships will not be reviewed here. Mathematically this theory is based upon the premise that quantities such as standard free energies are additive functions of molecular structure.

Let us represent a molecule as $A\text{-}B\text{-}H$, where A, B, and H refer to atoms or functional groupings. The standard free energy, $\bar{F}^\circ$, may be expressed as

$$\bar{F}^\circ = F_A + F_B + F_H + F_{A,B} + F_{A,H} + F_{B,H} \quad (1)$$

The quantities F_A, F_B, and F_H are independent functions of A, B, and H, respectively. $F_{A,B}$, $F_{A,H}$, and $F_{B,H}$ are terms resulting from the interaction of A with B, A with H, and B with H. These F_i terms are both constitutive and additive functions of molecular structure.

Now let us consider a process: the transfer of molecule $A\text{-}B\text{-}H$ from liquid phase 1 to liquid phase 2. The distribution equilibria may be described as

$$A\text{-}B\text{-}H^1 \underset{K_1}{\rightleftharpoons} A\text{-}B\text{-}H^2 \quad (2)$$

where the superscripts refer to phases 1 and 2. The equilibrium constant, K_1, for the reaction is

$$K_1 = \frac{(A\text{-}B\text{-}H^2)}{(A\text{-}B\text{-}H^1)} \quad (3)$$

The change in standard free energy for the reaction

$$\Delta \bar{F}^\circ = \bar{F}^\circ_{A\text{-}B\text{-}H^2} - \bar{F}^\circ_{A\text{-}B\text{-}H^1} \quad (4)$$

Shell Development Co., Agricultural Research Division, Modesto, Calif. 95353 is related to the equilibrium constant, K_1, by the equation

$$\Delta \bar{F}^\circ = -RT \ln K_1 \qquad (5)$$

If next we consider the same distribution equilibria for another molecule, $A\text{-}B\text{-}X$, differing from $A\text{-}B\text{-}H$ only by the introduction of a substituent X for H, we write

$$A\text{-}B\text{-}X^1 \underset{K_2}{\rightleftharpoons} A\text{-}B\text{-}X^2 \qquad (6)$$

where

$$K_2 = \frac{(A\text{-}B\text{-}X^2)}{(A\text{-}B\text{-}X^1)} \qquad (7)$$

The ratio of K_2 to K_1, K_2/K_1, may be expressed as

$$K_2/K_1 = \frac{(A\text{-}B\text{-}X^2)\,(A\text{-}B\text{-}H^1)}{(A\text{-}B\text{-}X^1)\,(A\text{-}B\text{-}H^2)} \qquad (8)$$

Using the standard free energy notation, Equation 8 may be expressed as

$$\Delta\Delta\bar{F}^\circ = \bar{F}^\circ{}_{A\text{-}B\text{-}X^2} + \bar{F}^\circ{}_{A\text{-}B\text{-}H^1} - \bar{F}^\circ{}_{A\text{-}B\text{-}X^1} - \bar{F}^\circ{}_{A\text{-}B\text{-}H^2} \qquad (9)$$

or using the notations introduced in Equation 1,

$$\Delta\Delta\bar{F}^\circ = F_{A^2} + F_{B^2} + F_{X^2} + F_{(A,B)^2} + F_{(A,X)^2} + F_{(B,X)^2} +$$
$$F_{A^1} + F_{B^1} + F_{H^1} + F_{(A,B)^1} + F_{(A,H)^1} + F_{(B,H)^1} -$$
$$F_{A^1} - F_{B^1} - F_{X^1} - F_{(A,B)^1} - F_{(A,X)^1} - F_{(B,X)^1} -$$
$$F_{A^2} - F_{B^2} - F_{H^2} - F_{(A,B)^2} - F_{(A,H)^2} - F_{(B,H)^2} \qquad (10)$$

Interaction terms involving the solvent have not been included. These terms vanish in the first approximation for reasons discussed in a later section. Combining terms one obtains

$$\Delta\Delta\bar{F}^\circ = F_{X^2} + F_{(A,X)^2} + F_{(B,X)^2} + F_{H^1} + F_{(A,H)^1} +$$
$$F_{(B,H)^1} - F_{X^1} - F_{(A,X)^1} - F_{(B,X)^1} - F_{H^2} - F_{(A,H)^2} -$$
$$F_{(B,H)^2} \qquad (11)$$

If one assumes that the interaction terms, $F_{(i,j)q}$, are factorable (This need not be argued here. Suffice it to say that this arises from the assumption that the structural changes being considered are small. Other theoretical considerations when dealing with solute-solvent interactions show this postulate to be of general validity.) so that $F_{(A,X)^2} = F_{A^2} + F_{X^2}$ and $F_{(B,X)^2} = F_{B^2} + F_{X^2}$, etc., then by combining terms one obtains

$$\Delta\Delta\bar{F}^\circ = aF_{X^2} + aF_{H^1} - aF_{X^1} - aF_{H^2} \qquad (12)$$

where a is a constant. This may be written as

$$\Delta\Delta\bar{F}^\circ = a[\Delta F_{(X\text{-}H)^2} - \Delta F_{(X\text{-}H)^1}] \qquad (13)$$

where $\Delta F_{(X\text{-}H)^2} = F_{X^2} - F_{H^2}$ and $\Delta F_{(X\text{-}H)^1} = F_{X^1} - F_{H^1}$. Expressing Equation 13 in terms of the equilibrium constant one obtains

$$\ln K_2/K_1 \propto \Delta F_{(X\text{-}H)^2} - \Delta F_{(X\text{-}H)^1} \qquad (14)$$

A qualitative interpretation of Equation 14 would be that the logarithm of the ratio of the distribution coefficients, $\ln K_2/K_1$, for molecules $A\text{-}B\text{-}H$ and $A\text{-}B\text{-}X$ between phases 1 and 2, is proportional to the differences in the free energies required to transport atoms (or groups) H and X between the two phases. At this point it might be wise to emphasize that the original postulate—i.e., that the free energy of a substance in solution is the sum of additive and constitutive term contributions from each of its molecular groups—is only an approximation.

The approximate function expressed by Equation 14 will be helpful in the following development in that it allows one to focus attention on individual atoms or functional groups constituting the molecule. In 1925, Langmuir voiced the opinion that a major factor contributing to the solubility of a solute in a solvent was the energy necessary to make a hole in the solvent. This idea was pursued by several men and used successfully in the development of a theory of physical toxicity by McGowan in 1952. Recent experimental verification of McGowan's hypothesis was made by Deno and Berkheimer (1960) in an investigation of activity coefficients of hydrocarbons.

This volume energy concept may be applied to the processes expressed by Equation 9 and the corresponding free energy change given by Equation 13. If the transfer of molecules $A\text{-}B\text{-}H$ and $A\text{-}B\text{-}X$ between phases 1 and 2 is made under conditions where the concentration of solute in each phase remains unchanged, and if only London dispersion forces are operative, entropy terms vanish and the free energy of transfer may be expressed as

$$\Delta\Delta\bar{F} = \Delta E_{X^2\text{-}H^2} - \Delta E_{X^1\text{-}H^1} \qquad (15)$$

where $\Delta E_{X^2\text{-}H^2}$ is the difference in energy required to form a hole in solvent 2 to accommodate X or H. Similarly $\Delta E_{X^1\text{-}H^1}$ is the energy difference in solvent 1. Interaction terms for a given group, X or H, with the solvent may be thought of as a measure of London dispersion forces. For a given solute (or group) in a given two-phase system these values are nearly the same and therefore tend to cancel. The experimental verification of McGowan's (1952) work by Deno and Berkheimer in 1960 is added substantiation that this is generally valid for solute-solvent interactions. One restriction should be noted—i.e., when there is appreciable hydrogen bonding of the solute to one of the solvents. In this case a constant term is required in Equation 15. If one is interested in the magnitude of the solute-solvent interaction terms, this constant must be evaluated separately. For the development which follows, this aspect may be disregarded, as it would result only in changing the graphical intercept of the final equations.

Consider again Equation 15, as an approximation, the energy required to form a hole in the solvent is proportional to the volume of the hole, V_s, multiplied by the internal pressure of the liquid phase, U^i. The product, $V_s U^i$, is termed the volume expansion energy as opposed to the cavity formation energy, a more exact energy expression. Substituting this proportionality into Equation 15 and converting $\Delta\Delta\bar{F}^\circ$ to the appropriate equilibrium expression, one obtains

$$\ln K_2/K_1 = a'(V_X - V_H)U^2 - a'(V_X - V_H)U' \quad (16)$$

and by combining terms

$$\ln K_2/K_1 = a'[(V_X - V_H)\Delta U] \quad (17)$$

where ΔU is the difference in internal pressures of the two solvent phases.

One convenient measure of V_s is the parachor, P, a constitutive and additive function of molecular structure, which is defined for a liquid as

$$P = \frac{M\gamma^{1/4}}{\rho - \rho°} \quad (18)$$

where M is the molecular weight; γ the surface tension; ρ the density of the liquid; $\rho°$ is density of the vapor. The vapor density $\rho°$ is usually neglected in comparison with ρ so that

$$P = \frac{M\gamma^{1/4}}{\rho} \text{ or } P = V_m \gamma^{1/4} \quad (19)$$

where V_m is the molar volume. Comparison of parachors of various substances is then essentially a comparison of molecular volumes modified to eliminate the influence of internal pressures. A table of recommended group and bond contributions useful in calculating P was tabulated by Quayle in 1953. In addition, he cataloged the parachors of a wide variety of organic compounds. The substitution of P_i for V_i in Equation 17 yields

$$\ln K_2/K_1 = a[(P_X - P_H) \Delta U] \text{ or } \Delta \ln K = a \Delta P \Delta U$$
$$(20)$$

For operational convenience Equation 20 may be divided into two separate energy functions,

$$\ln K_2 = aP_X \Delta U \quad (21)$$

and

$$\ln K_1 = aP_H \Delta U$$

These equations are defined as referring to a process in which the group X or H is transferred between the two appropriate phases. They are similar in concept to the use in electrochemistry of half reactions and corresponding half cell, $E°$, potentials. In the same sense they may be used operationally in determining group contributions to complete equilibria expressions.

Application of Derived Equations

The fact that Equation 21 expresses a relationship between equilibrium concentrations of a solute between two phases and the molar volume of that solute is in itself sufficient grounds to attempt an application to soil sorption equilibria. The choice of K_e, an equilibrium constant, to use in these equations, requires some explanation. By way of example, let us restrict our consideration to soil-applied herbicides, and in particular, the "uncharged" organic chemicals. It is this class of compounds which has proved so amenable to the chromatographic model developed by Lambert et al. in 1965. In this model description of the interaction of chemicals with soil, the character of the equilibrium constant, K_e, has been altered from the classical constant, in that the sorbing medium is now considered to be the soil–organic matter rather than the total mass of soil. This aspect was incorporated explicitly into the equilibrium coefficient which becomes, for all practical purposes, a true constant independent of soil type, within the framework of the model.

The use of K_f, the constant from the Freundlich equation, to correlate with chemical structure is not justified from theoretical considerations. If one assumes that Henry's law is obeyed in sufficiently dilute solution (infinite dilution in the limit), it would be desirable to use the limiting slope of the isotherm. However, at these lower concentrations, if $n \neq 1$, the Freundlich equation breaks down, since the limiting slope is infinite or zero. Certainly when $n = 1$ the use of K_f in Equation 21 requires no explanation. A much better value for K_e might be obtained from thermodynamic considerations by use of a power series to express the equilibrium between the concentration of solute sorbed, x/m, and the concentration of solute in solution, C_e. An expression of the type

$$\frac{x}{m} = \alpha C_e + \beta C_e^2 + \gamma C_e^3 + .. \quad (22)$$

should be highly applicable. α, β, and γ are the adjustable coefficients used to fit the data, x is the quantity of chemical sorbed, and m is the mass of sorbing medium.

If an additional restriction is imposed—i.e., that C_e be no greater than 1 μmole per liter—the third term of the series will in all probability vanish. In fact the series will for most practical purposes converge under these conditions without the second term, β probably being a small number. Equation 22 is useful in that under the boundary conditions specified (concentration limit), the proportionately constant α may be taken as a measure of K_e. In any event, the data obtained will provide the justification for using or rejecting the equation.

The application of Equation 22 will of course require more effort than the use of the Freundlich equation, and under certain conditions, K_f is a sufficiently good approximation of the required K_e for use in Equation 21. How good an approximation depends upon both K_f and the value of n, since K_e is a function of both K_f and $\frac{1}{n}$—i.e., the slope for the Freundlich case is

$$\frac{d(x/m)}{dC_e} = \frac{K_f}{n} C_e^{1/n-1} \quad (23)$$

The Freundlich isotherm has been discussed here because of its current widespread use with herbicide-soil sorption equilibria. The point to be made is, however, that its use should be discouraged in favor of an equation as represented by Equation 22. Once computational methods have been established for use of Equation 22, it becomes a matter of routine to evaluate the required K_e.

Functional Relationship between Sorption in Soil and Chemical Structure

Results

Only two independent studies are accessible which will allow a test of Equation 21 for soil sorption equilibria. One such study has been conducted at Modesto specifically for this purpose. The second study conducted for other purposes contains enough data which are amenable to the aforementioned treatment to provide an independent test of the proposed relationship. The data are in a paper by Hance (1965) in which a number of Freundlich isotherms are plotted for a series of urea derivatives on a variety of soils. Hance found no quantitative correlation between soil sorption and water solubility or soil sorption and chemical structure.

Soil sorption isotherms of seven analogs of SD 11831 [aniline, 4-(methylsulfonyl)-2,6-dinitro-N,N-dipropyl-] have been investigated at Modesto.

$$CH_3\overset{O}{\underset{O}{\overset{\uparrow}{\underset{\downarrow}{S}}}}\!\!-\!\!\underset{NO_2}{\overset{NO_2}{\diamondsuit}}\!\!-\!\!N(CH_2CH_2CH_3)_2$$

SD 11831

Plots of x/m vs. C (concentrations expressed on a molar basis) were linear on our standard Ripperdan (1% organic matter) and Sacramento (5% o.m.) soils. A plot of K_e vs. P for these compounds is depicted in Figure 1. The observed linearity confirms the validity of the proposed relationship. The agreement between K_e values obtained from Ripperdan (1% o.m.) and Sacramento (5% o.m.) soils is given in Table I. This is added substantiation for the relationship between soil organic matter content and the extent of soil sorption developed by Lambert *et al.* in 1965.

Data from the paper by Hance were also subjected to the above treatment by converting the isotherm data to the corresponding K_e values. The K_e values for each compound were averaged for the six soils used. These average values for the seven substituted ureas are plotted in Figure 2. Only the tertiary amides were compared because of hydrogen bonding restrictions previously discussed. It is surprising how well the data fit, since the range of K_e values spanned for any single compound on all six soils was in general a factor of about 2 times. This variation might be attributed to the difficulty in obtaining a representative soil sample since very small samples, 0.1 to 2 grams, were used to obtain the isotherms. In any event, the plot reaffirms the results obtained from our own study.

Table I. Comparison of K_e Values for SD 11831 Analogs Obtained from Different Soils

$$R_2\overset{O}{\underset{O}{\overset{\uparrow}{\underset{\downarrow}{S}}}}\!\!-\!\!\underset{NO_2}{\overset{NO_2}{\diamondsuit}}\!\!-\!\!N(R_1)_2$$

			K_e Values	
SD	R_1	R_2	Ripperdan,[a] 1% o.m.	Sacramento,[b] 5% o.m.
11830	CH_3	CH_3	125	145
12639	C_2H_5	CH_3	230	193
11831	C_3H_7	CH_3	500	520
13207	C_2H_5	C_2H_5	320	269
12030	C_3H_7	C_2H_5	750	702
12346	C_3H_7	C_3H_7	1170	
12400	CH_3	iso-C_3H_7	222	

[a] From multipoint isotherms (K_e values plotted in Figure 1).
[b] By replicated single point determinations.

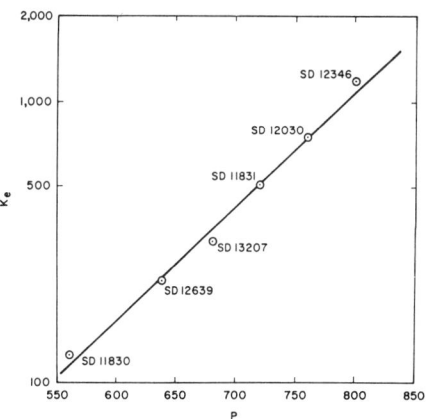

Figure 1. Plot of K_e vs. parachor for analogs of SD 11831

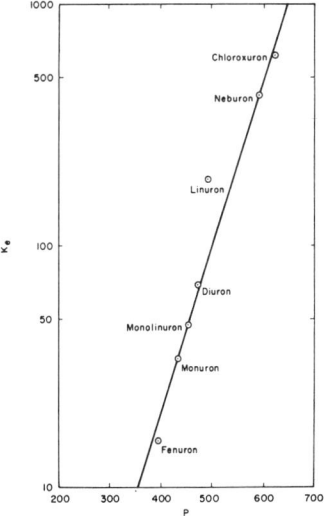

Figure 2. Plot of K_e vs. parachor for substituted phenylurea herbicides

The differences in the slope between Figures 1 and 2 are construed at this time as evidence of the two families of compounds sorbing with different energies on the soil organic matter. This is only speculation, and more work must be done before the cause of the differences is established.

A certain degree of caution must be exercised in interpreting the results obtained, especially because of complications arising from the fact that both adsorption and absorption phenomena are occurring, and the probability in this type of study of the formation of mixed solute–water–organic matter complexes.

Determinations of herbicide-soil sorption isotherms are normally carried out using high ratios of solvent (water) to sorbent (soil) in the presence of very low concentrations of solute. Under these conditions where the mole fraction of water is extremely high, the water may be an effective competitor for sites on the organic matter. If this situation exists, then the sorption phenomena should be more suitably described as a replacement reaction. Mechanistically this may be formulated as

$$nS_x(H_2O)_y + m(H_2O)_p(o.m.)_q \rightleftharpoons S_{nx}(H_2O)_{mp-nx}(o.m.)_{mq} + n^2xy(H_2O) \quad (24)$$

S represents the solute, H_2O the water, and o.m. the soil–organic matter.

There may be secondary effects which when operative might produce some curvature in the plots of Figures 1 and 2. Within experimental error and for all practical purposes these effects may be neglected.

Significance and Conclusions

The rationale behind the use of parachor in Equation 21 stems from its being an approximate measure of the molar volume. This was based upon Langmuir's (1925) energy considerations for solubility. The functional relationship developed here in conjunction with the equations relating biological activity to K_e (Lambert *et al.*, 1965) offers an explanation for some of the observations made by other investigators in the past. More precisely, it accounts for the inverse correlation between water solubility and dosage of herbicides observed by Freed and Burschel (1957) and the strong inverse correlation between adsorption and solubility found by Leopold *et al.* (1960).

The significance of this correlation is that for certain classes of compounds the distribution coefficients, which describe sorption equilibria, are predictable functions of molecular structure. It means that for these compounds we can predict where in the soil the chemical will reside under the influence of certain environmental conditions. The importance of this type of prediction is predicated upon the fact that the sorption of pesticide in the soil mediates its biological activity and that the loci of chemical in the soil will determine what factors are capable of operating upon it to effect its disappearance.

In practice these concepts may be utilized with supplementary information of a physicochemical and/or biological nature to account for and in some cases predict the fate and biological activity of pesticides added to the soil (Lambert *et al.*, 1965; Lambert, 1966).

This type of treatment represents a step towards establishing a quantitative relationship between soil sorption equilibria and chemical structure. In so doing, it emphasizes the importance of utilizing the partition or distribution coefficient, defined with respect to organic matter, as the most representative index of soil sorption equilibria. As is always the case when a new relationship is established, more questions may be posed which require additional study to answer. The relationship established here does, however, provide a theoretical justification for our conceptual model, and perhaps a deeper insight into the fundamental processes contributing to soil sorption equilibria.

Literature Cited

Bailey, G. W., White, J. A., J. AGR. FOOD CHEM. **12**, 324 (1964).
Deno, N. C., Berkheimer, H. E., *J. Chem. Eng. Data* **5**, 1 (1960).
Freed, V. H., Burschel, P., *Z. Pflanzenkrankh. Pflanzenschutz* **64**, 477 (1957).
Hammett, L. P., "Physical Organic Chemistry," p. 184, McGraw-Hill, New York, 1940.
Hance, R. J., *Weed Res.* **5**, 98 (1965).
Harris, C. I., *Weeds* **14**, 6 (1966).
Hartley, G. S., "Physiology and Biochemistry of Herbicides," L. J. Audus, Ed., p. 111, Academic Press, London, 1964.
Lambert, S. M., *Weeds* **14**, 273 (1966).
Lambert, S. M., Porter, P. E., Schieferstein, R. H., *Weeds* **13**, 185 (1965).
Langmuir, I., *Colloid Symp. Monograph* **3**, 48 (1925).
Leopold, A. C., van Schaik, P., Neal, M., *Weeds* **8**, 48 (1960).
Martin, A. J. P., *Biochem. Soc. Symp.* **3**, 4 (1949).
McGowan, J. C., *J. Appl. Chem.* **2**, 323 (1952).
Quayle, O. R., *Chem. Rev.* **53**, 439 (1953).
Ward, T. M., Upchurch, R. P., J. AGR. FOOD CHEM. **13**, 334 (1965).

Received for review November 7, 1966. Acceptde April 24, 1967. Division of Agricultural and Food Chemistry, 152nd Meeting, ACS, New York, N.Y., September 1966.

21

Copyright © 1983 by ASA-CSSA-SSSA
Reprinted from *Jour. Environ. Qual.* **12**:1-11 (1983)

Reevaluation of Partitioning as a Mechanism of Nonionic Chemicals Adsorption in Soils[1]

U. MINGELGRIN AND Z. GERSTL[2]

ABSTRACT

A critical evaluation of recent suggestions that the adsorption of nonionic organic compounds by soils can be well-correlated with the partition of the compounds between an aqueous and a nonpolar phase (usually 1-octanol), or with water solubility is presented. The physical basis proposed for these assumptions is that adsorption of nonionic organic compounds on soil organic matter is a "partition" process between the aqueous phase and a hydrophobic surface phase. The evidence presented in the literature for the dominance of a partition mechanism is insufficient to prove its general applicability. For example, systems in which a distinct hydrophobic phase at the solid surface does not exist are shown to exhibit the uptake behavior presumed to characterize partition. Theoretically, as well as practically, surface uptake cannot be simply defined as "adsorption" or "partition", but rather there is a continuum of possible interactions starting with fixed site adsorption and ending with true partition between three-dimensional phases.

Additional Index Words: organic matter, solubility, pesticides, pollutants.

Mingelgrin, U., and Z. Gerstl. 1983. Reevaluation of partitioning as a mechanism of nonionic chemicals adsorption in soils. J. Environ. Qual. 12:1-11.

Numerous investigators (10, 11, 16, 45) have stated that the uptake of nonionic organic compounds by soil from water can be well-correlated with their partition between octanol and water. In a recent paper (14), the uptake data of nonionic organic compounds from aqueous solution by soils were interpreted in terms of a partitioning of the solute between the solution and a hydrophobic phase found in the soil organic matter. This concept of partitioning was claimed by some authors (11, 14) to be universally applicable to nonionic compounds, thus allowing accurate estimation of soil-water distribution coefficients from solvent-water partition coefficients or from aqueous solubilities. If not evaluated critically, such statements may lead to serious misconceptions regarding the nature of the interaction of nonionic compounds in soils and to erroneous predictions regarding the fate of such compounds there, thus creating possible economic and ecological damage.

In many cases where it is not feasible to obtain an accurate value for sorption in a given soil system, it is necessary to settle for a reasonable estimate. Such estimation techniques should not be overextended or used without regard to the limitations of the techniques. There is a need for a better understanding of the limitations of presently recommended estimation procedures, and for better defined user guidelines. It is not the purpose of this paper to negate the use of estimation procedures, but rather to point out some limitations of such techniques. It will be demonstrated that the available data do not prove a "partition" mechanism at all. The resulting implications to the use of the correlation of the partition into a nonpolar solvent, or of the solubility in water with the uptake of nonionic compounds in soil for estimating the uptake, will be discussed.

MATERIALS AND METHODS

The chemical names of the substances referred to in this work are given in the appendix. The adsorption data for EDB were obtained on the same soils and by methods identical to those used by Saltzman and

---

[1] Contribution no. 320-E, 1982 series from the Agricultural Research Organization, The Volcani Center, Bet Dagan, Israel. Received 12 June 1981.
[2] Senior Scientist and Scientist, Division of Soil Residues Chemistry, Institute of Soils and Water, Bet Dagan, Israel.

Kliger (59) for DBCP. Adsorption and partition studies of napropamide and bromacil were carried out on six Israeli soils varying in clay and organic matter content by the batch method.[3] The literature was searched for adsorption data of nonionic organic compounds. Published adsorption data where linear isotherms were presented, or from which K_{om} (distribution coefficient on an organic matter basis) could be calculated were collected. The K_{om} values were then correlated with aqueous solubility and octanol-water partition coefficients (K_{ow}). The sources for the solubility and partition data are given in the text and are often the same as for the adsorption data.

RESULTS AND DISCUSSION

Adsorption and Partition—A Comparison

It would be useful to define partition and adsorption and their appropriate distribution coefficients before proceeding to interpret the relevant experimental data. Adsorption is generally defined as the excess of solute concentration at the solid-liquid interface over the concentration in the bulk solution (accordingly, negative adsorption is defined as the deficit in solute concentration at the interface), regardless of the nature of the interface region or of the interaction between the solute and the solid surface causing the excess. Adsorption can be viewed (23) as a two-dimensional process, in which the adsorbed molecules are assumed to be in the plane of the surface of the solid. These molecules are either attached to fixed sites on the surface, or are free to move about its plane. A second model for adsorption (31) is that of a three-dimensional interfacial region bordering on the solid surface, in which the solute has (due to the effect of the solid surface) thermodynamic properties different from those of the bulk phase. We then have a distinct phase of a finite volume, in which the excess concentration is defined as adsorption. A model unifying both concepts is that of a decaying solution-solid interaction potential. A very fast decay of the potential with distance approaches the two-dimensional model; a slowly decaying interaction is well-represented by the three-dimensional model. Partition in the present context is defined simply as the ratio between the activities of a solute in two bulk phases in equilibrium.

It is easy to see the parallels between a two-bulk-phase partition process and adsorption when the decay is sufficiently slow (a large volume of the interface region). However, both the mathematical treatment and experimental behavior of "adsorption" and "partition" bear great similarity, even for adsorption according to the two-dimensional interface model.

The partition coefficient is defined as:

$$K_p = \frac{a_1}{a_2} \cong \frac{c_1}{c_2} = K_p', \quad [1]$$

where a is the activity and c is the concentration of a solute in solvents 1 and 2. Due to the problematic nature of defining activities of the wide variety of nonionic compounds, and even more so due to the fact that most studies are made at relatively low concentrations, the experimental partition coefficients are defined as the ratios between the concentrations. As a result, all discussions in the relevant literature as well as in the following will deal with K_p'.

Generally, K_p' is not a constant throughout the whole concentration range of the solute in both solvents (e.g., 25, 63). As a result,

$$K_p' \neq s_1/s_2, \quad [2]$$

where s is the solubility of the solute in solvents 1 and 2. Yet, often there is a good correlation between the solubilities ratio and K_p' at lower concentrations. In different systems, therefore, there may or may not be a linear relation between c_1 and c_2 at equilibrium. The likelihood of deviation from linearity will increase as the concentration approaches saturation.

It is well-known that adsorption isotherms are not linear throughout the whole concentration range. Adsorption isotherms may have various shapes (e.g., 24) depending on the interactions in the solvent-adsorbent-solute system. The most common isotherm, obeyed up to some limiting concentration, can be described by the Langmuir equation:

$$S = ac/(1 + bc)$$
or
$$\theta = bc/(1 + bc) \quad [3]$$

where S is the quantity of adsorbate adsorbed per unit weight of adsorbent, c is the concentration in the bulk solution, a and b are constants, and θ is the fraction of adsorption capacity occupied by the adsorbate. The original derivation of Eq. [3] is based on the two-dimensional model (50), and thus represents a situation supposedly different from "partition." Yet, when only a small fraction of the adsorption capacity is occupied by the solute (i.e., $\theta \ll 1$), then

$$1 \gg bc \quad \text{and} \quad S \cong ac. \quad [4]$$

In this case, the distribution coefficient (K_d) between the adsorbed and bulk concentrations is well-defined and is equal to the constant a. Such a linear relation between S and c is frequently observed in adsorption with a wide variety of adsorbents including clay minerals, which are major soil components. Thus, at low surface concentration (low θ), the adsorption isotherm is very often linear. Similarly, the partition of a solute between two solvents is, in general, constant only at sufficiently low bulk concentrations. Hence, it is not possible to distinguish functionally between the two processes. Note that Eq. [4] holds for low θ, and does not depend directly on the proximity of the bulk concentration to saturation.

The aforementioned linear relation between the uptake by soil and the equilibrium concentration in water of nonionic compounds (e.g., 14, 38) was taken by Chiou et al. (14) as proof of a "partition" process. The fact that the isotherms presented by Chiou et al. (14) were extended to concentrations 0.3-0.95 of the solubility of the various solutes, and were still linear, was used as strong evidence for "partition." Those isotherms that extended very close to the solubility limit

---
[3] The soils' properties, as well as detailed results, are presently being prepared for publication in a separate manuscript: "Behavior of Bromacil and Napropamide in Soils I. Adsorption and Degradation" by Z. Gerstl and B. Yaron.

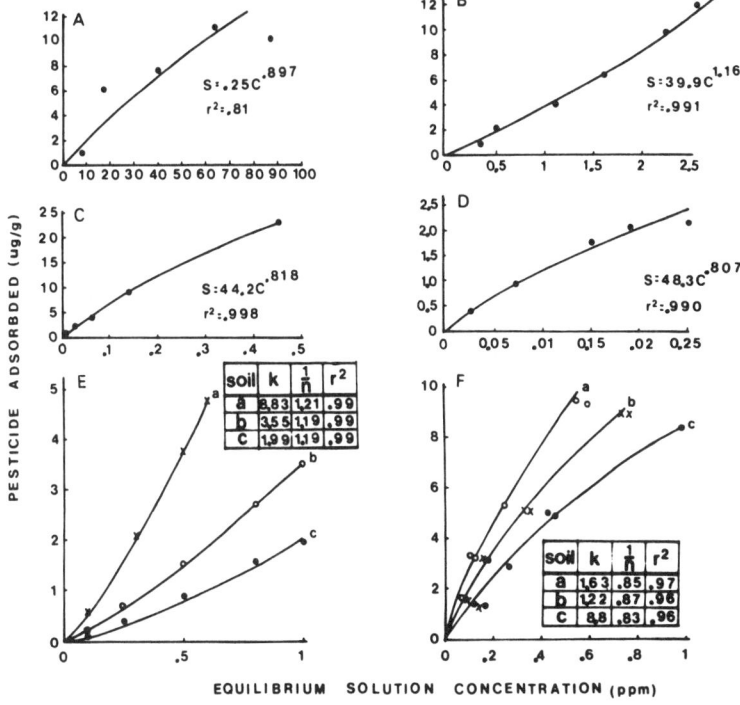

Fig. 1—Nonlinear adsorption isotherms for the uptake of nonionic pesticides by soils. (*A*) EDB (present study); (*B*) Parathion (60); (*C*) γ-BHC (55); (*D*) β-BHC (55) (*E*) Napropamide (68); (*F*) Terbufos (42). *K* and 1/*n* are the Freundlich constants.

were of the solutes with the lowest solubility ($\cong 150$ ppm). As stated above, this linearity is not evidence for partition, if, throughout the concentration range investigated, θ was rather low. Assuming a monolayer adsorption at surface saturation, the adsorption data presented in that paper, as well as in other papers where similar isotherms were studied (e.g., 38), would likely be in the linear range. This can be shown from the surface areas of the different soils and sediments used in the above studies, even if this area was estimated from the organic matter content of the samples, neglecting clay and other adsorbing surfaces. Monolayer adsorption does not necessarily take place, and solute-solvent and solvent-adsorbent interactions will also affect the adsorption at equilibrium. Therefore, the linearity of the adsorption isotherms at the range of concentrations studied should not always exist. Figure 1 presents adsorption isotherms of a few nonionic organic compounds in various soils which are not linear. Figure 2, on the other hand, demonstrates the linear behavior of such compounds on adsorbents (clays), which are very different from soil organic matter and do not possess the presumed solvent action of the lipid fraction of the soil organic matter. This lipid fraction is invoked as the phase into which the partition from the aqueous solution supposedly occurs in soil (37). Linear adsorption on clays was also reported by Greenland et al. (30).

Attempts have also been made to differentiate between partition and adsorption by comparing the relevant thermodynamic data (14). It is assumed that an adsorption process implies a decreased entropy ($\Delta S < 0$), and hence requires a high (exothermic) enthalpy (ΔH) to make adsorption significant. Partitioning by the same reasoning should not be necessarily as exothermic as adsorption, even if a transfer of a low solubility solute into a solvent in which it is very soluble is involved. This is so since ΔS may have any value.

In reality, the above assumptions are not valid. Adsorption from solutions is frequently a competitive process, in which solvent molecules or other adsorbed molecules compete with solute molecules for sites at the interface (e.g., 2). The entropy can therefore be positive or negative, depending on the balance between the ΔS of the solvent and that of the solute. Other contributions to the entropy changes upon adsorption (e.g., 48) may also be positive. Although, often $\Delta S < 0$, examples of positive ΔS of adsorption (i.e., endothermic adsorption) exist in systems where a solvent-like phase is not available at the solid surface, so that partition is not feasible. This is often the case with adsorption in clays (e.g., 54, 64). Accordingly, ΔH of adsorption may have any magnitude or sign. This is exemplified by the uptake of parathion from hexane on various soils (69). The distribution coefficient from hexane is much larger than that from water. The large uptake from the nonpolar hexane in which parathion is very soluble, as compared with uptake from water in which parathion is slightly soluble, contradicts the hypothesis that uptake by soil is a partition process between the aqueous phase and a hydrophobic phase present in the soil organic matter. The adsorption from hexane is endothermic (as opposed to adsorption from water), implying a large positive entropy of adsorption from hexane.

The accurate calculation of ΔS is difficult. Most data for ΔS are based on the relationships $\Delta G = \Delta H - T\Delta S$ and $\Delta G = RT \cdot \ln K$. Since K, the equilibrium constant for the distribution between the bulk and surface phases is not well-defined due to the uncertainty in the thickness (volume) of the adsorption layer, the values of ΔG are only approximate and the values of ΔS can therefore be in considerable error (7, 19, 56). Hence, great care must be taken in the interpretation of experimental ΔS values.

The thermodynamic argument used to prove partition is too simplified and may lead to erroneous conclusions. Since in both partition and adsorption, ΔH and ΔS may vary greatly in magnitude and sign, the value of these parameters does not prove or disprove the adsorption or partition models.

Correlation Between Uptake by Soil and Aqueous Solubility, Water-Octanol Partition Coefficient or Organic Matter Content

For many nonionic organic compounds applied to the soil, organic matter is the principal adsorbent (26, 32). In these cases, uptake in soil can be discussed in terms of interaction with the organic matter. There are many nonionic compounds for which the clay inorganic fraction participates, and may even dominate, the adsorption in soil (5, 56, 60, 70). Saltzman et al.[4] have shown that removal of the organic matter from soils reduces the adsorption capacity for parathion in some soils; but in the case of a high organic-matter soil, nearly no reduction was observed. Other studies have shown (5, 70) that removal of organic matter from soils and sediments had relatively little effect on the adsorption of some nonionic pesticides, whereas in some cases, adsorption actually increased after the removal. Furthermore, selective removal of the lipid fraction of the soil organic matter (ether extraction) actually increased DDT adsorption on two soils (62). This does not imply that the contributions of the soil organic and mineral fractions to the uptake are additive, only that soil minerals by themselves, or by interacting with the soil organic matter, can significantly affect the uptake of nonionic molecules. Hence, the picture of a hydrophobic lipid fraction at the adsorbing surfaces in soils into which the nonionic molecules are partitioned is not supported by these findings. In most soils (except the special and limited group of organic soils), organic matter content is only a few percent, while clay content is an order of magnitude higher. Clay may therefore strongly influence the adsorption, even if the affinity of the nonionic compound to the clay is significantly lower than to the organic matter.

The often-encountered correlation between organic matter content and adsorption is cited in support of the

[4] S. Saltzman. 1977. Sorption and nonbiological degradation of parathion in soils and clays. Ph.D. Thesis, Univ. Catholique De Louvain, Belgium.

Table 1—K_{om} values of nonionic compounds in various soils.

Compound	K_{om}	Organic matter	Reference
EDB	mL/g OM	%	
EDB	21– 93	0.5–21.7	33, present study
Piperophos	72– 7,627	1.8–10.2	39
Napropamide	110– 1,223	0.1– 2.4	68, present study
Parathion	182– 9,200	0.2– 6.1	8, 47, 69[a]
Phorate	211– 3,980	0.2–31.7	16, 47
Lindane	427– 1,502	1.2–20.5	46, 52, 64
Disulfoton	470– 5,060	0.2– 4.6	27, 28, 47
Chlorpyrifos	1,255– 20,400	1.2– 6.6	16, 52
Dieldrin	2,302– 7,122	0.7–28.9	17
DDT	76,300– 257,040	1.6– 3.9	14, 52, 62
3-Methycholanthrene	211,000–3,710,000	0.8– 4.1†	38
Dibenzanthracene	328,500–1,779,000	0.8– 4.1†	38

† Lowest organic matter samples not included.

dominant contribution of the soil organic matter to adsorption. However, this correlation is far from perfect (49) and, in addition, there is often a good correlation between the organic matter content and the clay content (1, 4). The assumption that soil organic matter is similar in all soils is not always correct. The chemical and physical nature of the organic matter varies from soil to soil (49, 67). Furthermore, the organic matter is usually present in mineral soils in various complexes with the inorganic soil components, a fact which strongly affects adsorption (29, 61). Since soil organic matter is not uniform in all soils, it cannot be treated as a well-defined organophilic phase. Reported K_{om} values (the ratio between the uptake per unit weight of soil organic matter and the equilibrium concentration in solution) vary appreciably between soils for many nonionic compounds. This holds true, even after the differences between measurement procedures used by various investigators are taken into consideration (e.g., Table 1). The above implies that the uptake by the soil organic matter can often be estimated to no better than an order of magnitude from water solubility or water-octanol partition. The total uptake by soil may be even less well-correlated with the above partition and solubility.

It has been shown (14, 38) that the K_{om} of many adsorbates correlates significantly with the aqueous solubility. Previous work (13) demonstrated the correlation that exists between the aqueous solubility of a significant number of compounds and their octanol-water partition coefficient. This correlation permitted calculation, within an order of magnitude, of a compound's partition coefficient from its aqueous solubility. Other workers (10, 16, 45) have presented cases in which the octanol-water partition and uptake by soils and sediments are also well-correlated. These correlations were presented as additional proof for a partition mechanism for the uptake of nonionic organic compounds by soils. On the other hand, a number of investigators (e.g., 13, 55) have pointed out the correlations between adsorption of nonionic substances on various surfaces on which a partition-like interaction is not possible, and their solubility or octanol-water partition coefficients. As stated above, adsorption of organic molecules on a solid surface from a solution is frequently a competitive process in the sense that an adsorbed solute molecule replaces a solvent molecule or another

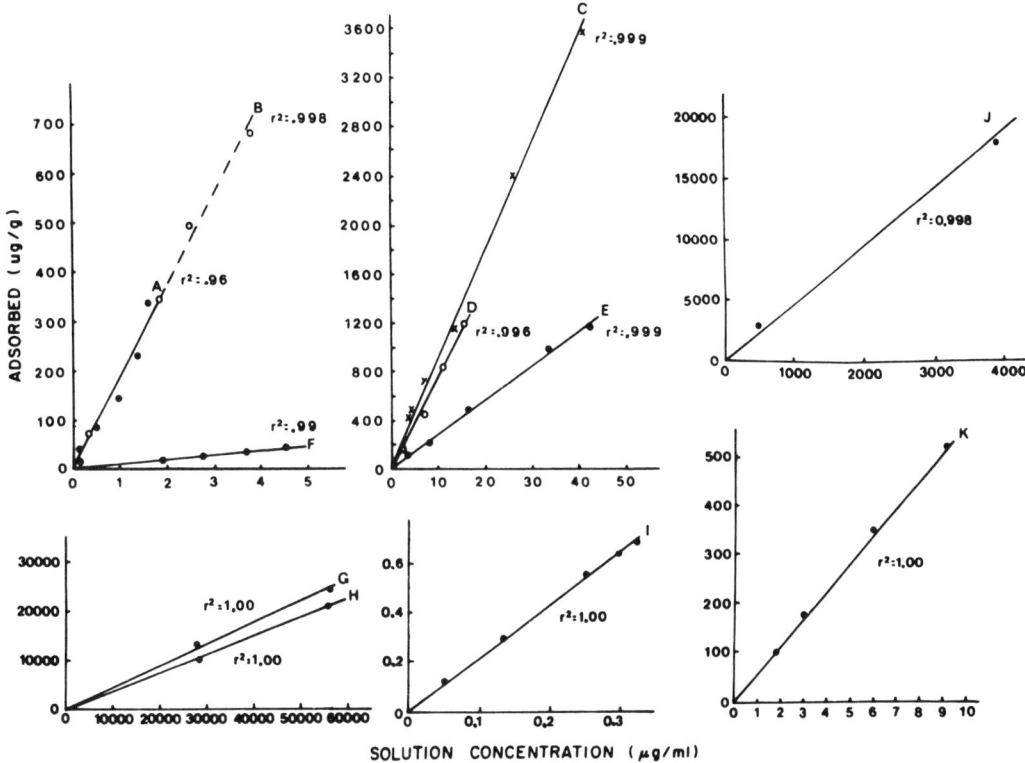

Fig. 2—Linear adsorption isotherm for the uptake of nonionic pesticides by inorganic adsorbents. (*A*) Parathion-Ca-montmorillonite[4]; (*B*) Parathion-Ca-attapulgite (hydrated) (22); (*C*) Monuron-Na-montmorillonite (3); (*D*) Methyl-Parathion-Na-montmorillonite[4]; (*E*) Monuron-H(Al)-montmorillonite (64); (*F*) Parathion-Ca-attapulgite (850°) (22); (*G*) Oxydipropionitrile-Kaolinite (12); (*H*) Oxydipropionitrile-gibbsite (12); (*I*) Trietazine-Na-montmorillonite (3); (*J*) Hexanol-Na-montmorillonite (63); (*K*) β-BHC-silica gel (55).

adsorbed molecule in the interface region. The less polar the molecule, the more it will tend to adsorb on a "hydrophobic" surface from a polar solvent, while removing solvent molecules from that surface. Such a situation may exist in the case of adsorption from aqueous solutions on soil organic matter. The more nonpolar the molecules, the larger will also tend to be its nonpolar solvent-water partition coefficient and the lower its water solubility. Correlation between soil uptake and low water solubility is thus expected, but the complexity of surface interactions both on clays and on organic matter in soil can result in large deviations from any proposed fit. The correlation of K_{om} with the aqueous solubility for a large number of nonionic pesticides is presented in Fig. 3. The weak correlation ($r^2 = 0.635$) can be readily seen, and for several compounds (e.g., MIT and carbophenothion), deviations from the best fit are more than an order of magnitude. Similarly, the large fluctuations observed in the ratio between soil uptake and octanol-water partitions as presented in Fig. 4, are also to be expected. Additional recent data (e.g., 17, 38) further demonstrate that deviations of an order of magnitude or more, from the calculated fit between the K_{om} and solubility or octanol partition exist.

The use of aqueous solubility alone, for predicting K_{om} should be done carefully. Ellgehausen et al. (15) have reported several instances of compounds with approximately the same aqueous solubility, but which differ greatly in their K_{ow}. Thus, for example, fluorodifen and fenchlorphos have aqueous solubilities of 6.1 and 3.1 μmole/L, respectively, but have solubilities in octanol of 30,500 and 685,000 μmol/L, respectively. These data yield K_{ow} values of 5,012 and 218,800 for fluorodifen and fenchlorphos, respectively. When compared to the values of 25,545 and 40,612, respectively, calculated from the equation of Chiou et al. (13), these values demonstrate that predictions concerning the K_{ow}, based on the aqueous solubility of compounds can be in great error. Additional cases in point are bromacil (sol = 4,080 μmol/L) and molinate (sol = 4,706 μmol/L) (43). The K_{ow} values for bromacil and molinate were, respectively, 70 and 1,628, neither of which are similar to the value of 280, which is predicted for such compounds by the equation of Chiou et al. (13).

The correlation between K_{om} and solubility is sometimes modified by correcting for the crystal energy of solids (11, 44). This correction succeeds, because it correlates the K_{om} with the interaction between the solute molecules and water, eliminating the effect of the interactions in the pure solid on the solubility. If the

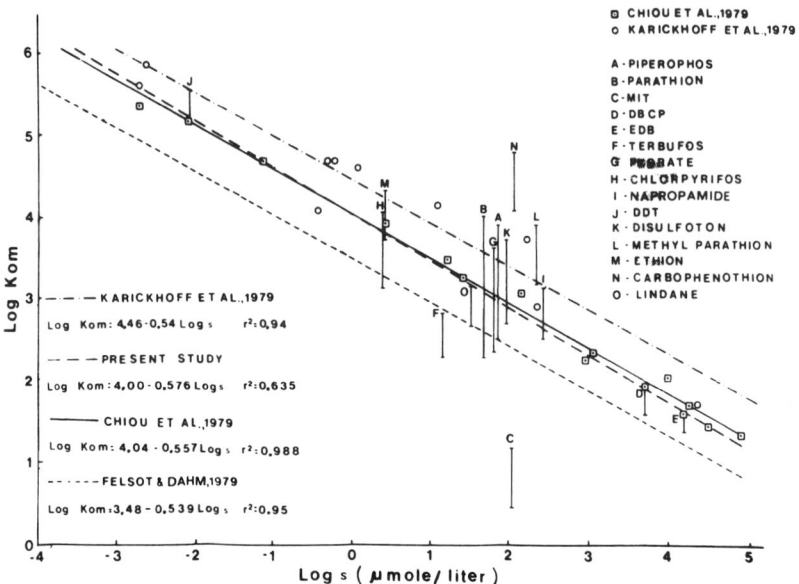

Fig. 3—Soil organic matter distribution coefficient (K_{om}) as a function of the aqueous solubilities (s) of selected nonionic compounds. (A) (39); (B) (8, 47, 69).ᵃ (C) (21); (D) (59); (E) (32 and present study); (F) (16); (G) (16, 47); (H) (16, 52); (I) (68 and present study); (J) (52, 62); (K) (27, 28, 52); (L) (47); (M) (47); (N) (47); (O) (46, 52, 64).

fugacity coefficient in the adsorbed phase is approximately constant (44), this correction is a good one. If, on the other hand, this coefficient is not constant, it may be correlated to some extent with the crystal energy of the hydrophobic solid. In this case, the crystal energy correction will fail to improve the K_{om}-solubility correlation and may greatly overshoot the experimentally measured value (e.g., 44).

When $K_p' \gg 1$ or $K_p' \ll 1$, the experimental accuracy in determining the partition coefficient is very poor, and data in the literature for the same system may vary by several orders of magnitude (4, 5, 11, 13, 45). Therefore, partition coefficients may be inconvenient parameters for predicting adsorption, even when they correlate well with K_{om}, and when K_{om} is approximately constant within a group of soils. The adsorption in one soil of a group of related soils, coupled with the organic matter content of all other soils may be an experimentally more reliable parameter for predicting adsorption, assuming adsorption correlates well with organic matter content (K_{om} similar for all members of the group). Other estimation procedures (see below) may, in fact, be preferable to K_{om}-K_{ow}, or K_{om}-solubility relations.

Although partition and adsorption may be expressed functionally in a similar manner at a sufficiently low θ, the dependence of each process on molecular properties may be different. For example, the solvent medium may often be viewed as being isotropic, while the adsorbing medium is not. The adsorbing surface is relatively rigid, implying that the conformation of the molecule will greatly affect its adsorption, and less so its partition into an organic liquid phase. Molecules with similar functional groups, but with different spatial arrangement, may have similar partition coefficients since the interaction with the easily redistributed solvent molecules may not be greatly affected. At the rigid surface, the spatial arrangement of the molecule may affect the interaction more strongly. Thus, the molecular structure is an example of a property which may affect adsorption and solution differently, and hinder the expected correlation between K_{om} and K_{ow} or solubility. This also demonstrates the possible pitfalls in defining soil uptake as a partition process when a three-dimensional, non-rigid interface phase is not available at the surface into which uptake is taking place. A case in point is the adsorption of γ- and β-BHC on several adsorbents as reported by Mills and Biggar (55). The γ and β isomers of BHC differ only in that, in the β isomer all the chlorine atoms are equatorial, while in the γ isomer three are equatorial and three are axial. In agreement with the above, relative differences in adsorption between the two isomers of 60–250% on various organic and inorganic adsorbents were observed, while their K_{ow} (44) differed by only 20%.

Conditions for the Applicability of a Partition Process

In order to invoke a "partition" mechanism for soil uptake, it is necessary to demonstrate the presence of a phase into which partition could take place. One possibility is to consider the volume of the solution under the influence of the surface as the "non-bulk phase". This is acceptable as a conceptual aid, since the experimental distinction between partition and three-dimensional adsorption is not simple. Yet, even this adsorption is not an identical process to partition between two bulk phases. The differences arise from the continuous, rather than abrupt, transfer from one phase to another,

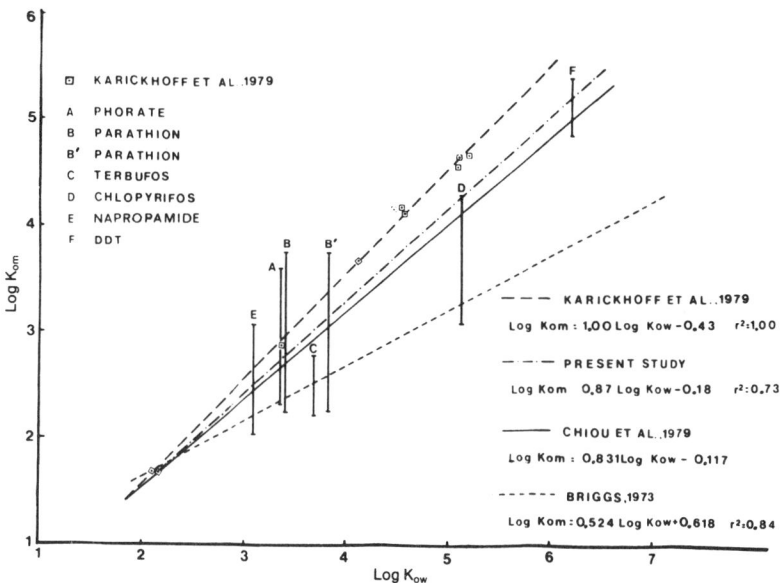

Fig. 4—Soil organic matter distribution coefficient (K_{om}) as a function of octanol/water distribution coefficients (K_{ow}). (A) (16, 47); (B) (8, 16, 47, 69)[4]; (B') (8, 13, 47, 69)[4]; (C) (16); (D) (16, 52); (E) (68 and present study); (F) (52, 62).

the free exchange of solvent molecules between both phases, the surface-induced restrictions on the change in the "non bulk phase" volume, and the often small volume of that interfacial phase. Another possibility is the actual presence of a nonaqueous solvent phase at the surface. This is true, for example, when adsorption occurs on cation exchangers with sufficiently long-chain alkyl-ammonium exchangeable cations (20, 63). Hartley (37) suggests that "oily" constituents of the soil organic matter are responsible for soil uptake of nonionic compounds. The existence of such a lipid phase is in agreement with Schnitzer and Khan (61), who described the presence of fatty acids and alkanes at the surface of the soil organic matter with resulting long alkyl chains protruding from that surface. Some interactions described as hydrophobic bonding may actually be uptake by this lipid portion (61).

A general, or dominant, lipid-phase partitioning mechanism for soil uptake of many nonionics is unlikely, however. This claim is supported, for example, by the strong adsorption of many nonionic substances, both on soils after organic matter removal and on pure inorganic soil components such as clays (e.g., 3, 22, 64)[4], by the cases of poor correlations between organic matter content and adsorption (e.g., 39, 40, 53), and by other data presented above.

GENERAL DISCUSSION

It is well-known that the dissolution of a solute in a solvent of a "similar" chemical nature is greater than in a "dissimilar" solvent. More generally, apolar solutes dissolve well in apolar solvents, whereas they do not dissolve well in polar solvents. The inverse correlation (13) between solubility in water and octanol-water partition of many organic compounds is in agreement with the above. The more similar a solute molecule is to the solvent molecule, the more similar are the potential surfaces describing their interactions with any molecule. It is likely, therefore, that when two solvents do not attract each other (are immiscible), a solute similar to one of them will partition into that solvent. The term "similar" is very qualitative. The polarity of molecules is often taken as a measure of similarity. Indeed, often (but with notable exceptions), the more polar the solvent, the more likely it is to dissolve polar molecules and to repel nonpolar molecules. Similarly, a surface will often display attraction to adsorbates of a chemical nature similar to its own. Since solid state implies strong interaction between the units occupying the lattice points, adsorption of species of a similar nature to that of the solid's basic units may be very strong indeed. A case in point is the coating of metals by similar metals. On the other hand, many surfaces (including soil organic matter and clays) are rather heterogeneous and many types of interactions may control adsorption on them. As a result, general measures of similarity between the adsorbent and adsorbate, such as their polarity, hydrophilicity, or any other single measure of similarity may not always predict the extent of adsorption well.

It was previously shown that there is a well-defined sequence of adsorption, as well as a good correlation between adsorption and solubility, within homologous series. This was stated by Freundlich (18) as a form of Traube's rule and by Holmes and McKelvey (41) and demonstrated by Bartell and Fu (6) and Hansen and Craig (35). A similar observation was made within a group of substituted molecules with an ordered series of substituents (e.g., 34, 49). It is not safe, however, to extrapolate from correlations within such well-defined groups to the general behavior of heterogeneous groups of the sort of the group of nonionic organic compounds.

Conclusions derived from a well-defined subgroup when applied to the wider group are likely to fail or at least to weaken as demonstrated in Fig. 1, 3 and 4, and in the other examples cited throughout this text. The closer the substances compared, the greater the correlation. Much of the data presented regarding the correlations between adsorption, water solubility and octanol-water partition (10, 13, 14, 16, 45), suggest that the regressions obtained are too specific to be applied to the large group of nonionic organic substances. Even when a good correlation between octanol-water partition, solubility, and soil uptake was found, the regression equations obtained from the various sets of data differed significantly (Fig. 3, 4). This was expected, due to the difference in the nature of the sets of adsorbates studied by the different investigators. The discrepancies, often by more than an order of magnitude throughout much of the adsorption range of interest (e.g., 11), became less obvious due to the log-log scale used for the above regressions.

If within a group of substances, adsorption (K_d) is inversely related to the solubility, the surface concentration of all members of the group at a given reduced concentration should be similar. The reduced concentration (c/c_o) is the ratio between the actual concentration and the solubility of the substance. This was shown to be true for the longer-chain members of a homologous series of fatty acids on certain adsorbents (36). Addition of a CH_2 group to a long chain may affect the solubility in an apolar phase or the activity near an apolar surface much less than it affects the solubility in a phase in which the solubility of the long-chain molecule is low. Both adsorption and partition will then depend directly on the solubility of the members of the homologous series in the low solubility phase. Neglect of the effect of an additional methylene group on the interactions in the other phase is not always possible, however. Also, for highly soluble members, solute-solute interactions may become important as c/c_o approaches one, and cause deviation from the adsorption predicted for the group. Indeed, the shorter chain (and more soluble) members of the homologous series of the fatty acids mentioned above do not exhibit the same adsorption at a given c/c_o as the longer-chain members. At any rate, the presence of a constant uptake at a given c/c_o within a group of substances (i.e., the presence of an inverse dependence of uptake on solubility), or its absence cannot be used as an indication of an adsorption or partition mechanism.

Recently it was suggested that retention time in reverse phase HPLC can be used for the estimation of chemical mobilities in soils, or of the octanol/water partition coefficients of organic chemicals (49, 66). All the reservations which were presented for correlations between related but not analogous processes discussed in this paper apply also to the use of HPLC for predicting interactions in soil. The more-than-an-order-of-magnitude deviations (and in some cases, almost two orders of magnitude) from the suggested relations between retention time in HPLC and the properties of interest (66) emphasize this point. Again, the frequent use of log-log scales should be cautioned against, since activity and mobility in soil are approximately inversely, and not logarithmically, related to adsorption (e.g., 52, 58, 65). The inverse relationship often noted between the uptake of nonionic compounds and their solubility (44) suggests that an inverse, rather than a log, fit may be more suitable in those cases where a good correlation is expected between uptake and solubility. Indeed, the data in Fig. 3 (K_{om} vs. solubility), which gave a log-log fit with $r^2 = 0.635$, fit the relationship $K_{om} = a + b/s$, with $r^2 = 0.75$. Although the two coefficients of determination derived from statistical least-square fitting are not directly comparable, they do indicate that functional relationships other than the commonly employed log-log fit may sometimes be preferable. At any rate, a proper choice of the solid (a likely first try is the reverse phase C18 column) and liquid phases (e.g., water or water-polar solvent mixture) in HPLC determinations of the mobility of nonionic compounds may make HPLC a useful tool for predicting adsorption of groups of chemically related compounds on soils with similar properties.

It is often sufficient, or even necessary, to estimate the uptake of nonionic molecules in soils or sediments rather than perform the actual adsorption measurements. Empirical or semi-empirical relations between adsorption and other properties are the tools used for such estimates. The K_{ow}-K_{om} or the K_{om}-s relationships are often used for this purpose. We have attempted to demonstrate that these relationships have their limitations and to caution against the too general applicability assigned to them. It is not certain, even when "real world estimates" are required, that the above correlations are the best choice for deriving such estimates. Even if the use of K_{om}-s correlation is justified, the log-log plots presently used are not necessarily the best functional relationship. As pointed out before, an inverse relationship between K_{om} and s may be more (or at least equally) accurate and has theoretical justification (11, 44).

Karickhoff (44) has presented a comprehensive discussion of the semi-empirical estimation of sorption of hydrophobic pollutants. That paper includes the theoretical basis for the above correlations, as well as an example of direct K_{om} estimation from structural properties of the solutes. Karickhoff (44) formulated, in thermodynamic terms, our contention that the success of these correlations hinges on the assumption that either the fugacity coefficients of the nonionic compounds in the adsorbed phase (i.e., the soil organic matter) or the ratio between them and the fugacity coefficients in octanol are approximately constant. The data and discussion presented there demonstrate that this assumption may hold well for groups of closely related compounds (e.g., the polycyclic hydrocarbons studied by Karickhoff (44) and the other groups of solutes tabulated there). But, as the chemical (or structural) similarity diminishes, so does the applicability of this assumption. The above assumption is also more likely to be applicable to the less polar (more hydrophobic) solutes, since in this case the hydrophobic sites of the soil organic matter are more likely to dominate adsorption. Yet, as seen from the data in Table 1 and Fig. 3 and 4 in the present paper, even for highly apolar substances, large deviations from the proposed regressions are observed. Even if hydrophobic interactions (or van der Waals forces) do dominate the adsorption on soil organic matter, their magnitude relative to that of the interactions in octanol and their absolute values may be rather different for the different solutes. The more polar the solute, the more likely are other interactions (e.g., with the soil mineral fractions) to contribute significantly to adsorption and the above correlations are more likely to fail.

In the past, a number of attempts were made to correlate adsorption directly with structural parameters of the adsorbates. Examples for these approaches are Lambert's (49) use of the parachor, the LFET approach relating molecular fragments to adsorption (35, 44, 49) and the inorganic/organic ratio proposed by Matsuo (51). Reasonable correlations were obtained in all these cases. Such correlations with structural parameters have the advantage of a reduced need for laboratory measurements with their associated experimental errors. Only relatively concise tabulations of the required data, and the functional relations between uptake and the tabulated properties are needed. This approach is also more flexible, as the required precision may be obtained by including finer structural and chemical details. Therefore, if "real world estimates" are required, approaches that depend directly on the molecular structure may sometimes be more accurate and easier to use by environmental engineers. Correlations of this nature should, therefore, be further developed.

Since estimates are often made of the mobility of toxic materials, the possibility of "irregular behavior" by even a few compounds could be critically important. The use of estimates should, therefore, be made with great care and only after their range of applicability is well-understood.

CONCLUSIONS

Correlations between solubilities, liquid-liquid partitions and soil uptake are, contrary to claims made in the past (14), insufficient proof for a partition process. Furthermore, these correlations are not of a universal nature. They are better the more closely related are the adsorbates used to formulate the regressions. The wider the group of compounds to which these correlations are applied, the weaker the correlation. Deviations of up to a few orders of magnitude from the proposed functional dependence of the soil uptake on the solubility or partition were observed. Yet, differences in uptake smaller than an order of magnitude may strongly affect the solute's mobility and effectiveness in soil. Due to the inverse relation between soil uptake and mobility, small differences in K_d may greatly affect the mobility for low K_d (<1).

The correlations between K_{om}, K_{ow} and s, which do exist, arise from the expected similarity in the behavior of substances of similar chemical properties. The less polar the solute, the more likely it is that K_{om} will be a good parameter for soil uptake, and that K_{ow} will correlate well with the soil uptake, since the hydrophobic sites of the organic matter will then be more likely to dominate the uptake process. The complexity of adsorption processes on soil surfaces, however, does not allow predictions based on these correlations to apply to all members of the very diverse group of nonionic organic substances. It is sometimes necessary to study separately a particular group of substances or even an individual substance (57) before drawing conclusions regarding their interactions in various soils. The best empirical or semi-empirical relationship for predicting uptake may often be that which is directly based upon structural parameters of the adsorbate, and not necessarily on the correlations between K_{ow} and aqueous solubility with K_{om}. Estimates based on K_{om} correlations with K_{ow} or s (with or without crystal energy corrections) are acceptable as long as the limitations of these correlations are taken into consideration. The argument that the partition model is the physical basis for the uptake of nonionic organic molecules is not justified by the available data. On the other hand, the acceptance of this model may obscure the limitations in the use of the above correlations for soil uptake estimates.

APPENDIX

Nomenclature of Pesticides Mentioned in Text, Table, and Figures

Common name	Chemical name
BHC	1,2,3,4,5,6-hexachlorocyclohexane
Bromacil	5-bromo-3-*sec*-butyl-6-methyluracil
Carbophenothion	*S*-{[(*p*-chlorophenyl)thio]methyl} o,o-diethyl phosphorodithioate
Chlorpyrifos	*O*,*O*-diethyl *O*-(3,5,6-trichloro-2-pyridyl) phosphorothioate
DBCP	1,2-dibromo-3-chloropropane
DDT	1,1,1-trichloro-2,2-bis(4-chlorophenyl)ethane
Dieldrin	1,2,3,4,10,10-hexachloro-*exo*-6,7-epoxy-1,4,4a,5,6,7,8,8a-octahydro-1,4-*endo*-*exo*-5,8-dimethanonaphthalene
Disulfoton	*O*,*O*-diethyl *S*-2-(ethylthio)-ethyl phosphorodithioate
EDB	1,2-dibromoethane
Ethion	*O*,*O*,*O'*,*O'*-tetraethyl *S*,*S'*-methylene bis(phosphorodithioate)
Fluorodifen	*p*-nitrophenyl α,α,α-trifluoro-2-nitro-*p*-tolyl ether
Lindane	γ-1,2,3,4,5,6-hexachlorocyclohexane
Methyl parathion	*O*,*O*-dimethyl *O*-*p*-nitrophenyl phosphorothioate
MIT	methyl isothiocyanate
Molinate	*S*-ethyl hexahydro-1H-azepine-1-carbothioate
Monuron	3-(4-chlorophenyl)-1,1-dimethylurea
Napropamide	2-(α-naphthoxy)-*N*,*N*-diethylpropionamide

Parathion	O,O-diethyl O-p-nitrophenyl phosphorothioate
Phorate	O,O-diethyl S-(ethylthio)methyl phosphorodithioate
Piperophos	O,O-di-n-propyl S-2-methylpiperidinocarbonyl-methyl phosphorodithioate
Ronnel	O,O-dimethyl O-(2,4,5-trichlorophenyl) phosphorothioate
Terbufos	S-{[(1,1-dimethylethyl)thio]methyl} O,O-diethyl phosphorodithioate
Trietazine	2-chloro-4-diethylamino-6-ethylamino-s-triazine

LITERATURE CITED

1. Adams, R. S., Jr. 1972. Effect of soil organic matter on the movement and activity of pesticides in the environment. p. 81-93. In D. D. Hemphill (ed.) Trace substances in environmental health. Univ. of Missouri, Columbia.
2. Adamson, A. W. 1976. Physical chemistry of surfaces. John Wiley and Sons. 3rd ed. p. 698. New york.
3. Bailey, G. W., J. L. White, and T. Rothberg. 1968. Adsorption of organic herbicides by montmorillonite: role of pH and chemical character of adsorbate. Soil Sci. Soc. Am. Proc. 32:222-234.
4. Banin, A., and A. Amiel. 1969. A correlative study of the chemical and physical properties of a group of natural soils of Israel. Geoderma 3:185-198.
5. Banwart, W. L., A. Khan, and J. J. Hassett. 1980. Effect of sample pretreatment on sorption of acetophenone by soils and sediments. J. Environ. Sci. Health B15:165-179.
6. Bartell, F. E., and Y. Fu. 1929. Adsorption from aqueous solutions by silica. J. Phys. Chem. 33:676-687.
7. Biggar, J. W., and M. W. Cheung. 1973. Adsorption of picloram on Panoche, Ephrata and Palouse soils: a thermodynamic approach to the adsorption mechanism. Soil Sci. Soc. Am. Proc. 37:863-868.
8. Biggar, J. W., U. Mingelgrin, and M. W. Cheung. 1978. Equilibrium and kinetics of adsorption of picloram and parathion with soils. J. Agric. Food Chem. 26:1306-1312.
9. Bowman, B. T., and W. W. Sans. 1977. Adsorption of parathion, fenitrothion, methyl parathion, aminoparathion and paraoxon by Na^+, Ca^{2+}, and Fe^{3+} montmorillonite suspensions. soil Sci. Soc. Am. J. 41:514-519.
10. Briggs, G. G. 1974. A simple relationship between soil adsorption of organic chemicals and their octanol/water partition coefficients. Proc. 7th Br. Insectic. Fungic. Conf. 1973. Brighton p. 83-86. The Boots Co. Ltd., Nottingham.
11. Briggs, G. G. 1981. Theoretical and experimental relationships between soil adsorption, octanol-water partition coefficients, water solubilities, bioconcentration factors, and the parachor. J. Agric. Food Chem. 29:1050-1059.
12. Brindley, G. W., R. Bender, and S. Ray. 1963. Sorption of nonionic aliphatic molecules from aqueous solutions on clay minerals. Clay-organic studies VII. Geochim. Cosmochim. Acta 27:1129-1137.
13. Chiou, C. T., V. H. Freed, D. W. Schmedding, and R. L. Kohnert. 1977. Partition coefficients and bioaccumulation of selected organic chemicals. Environ. Sci. Technol. 11:475-478.
14. Chiou, C. T., L. J. Peters, and V. H. Freed. 1979. A physical concept of soil-water equilibria for nonionic organic compounds. Science 206:831-832.
15. Ellgehausen, H., C. D'Hondt, and R. Fuerer. 1981. Reversed-phase chromatography as a general method for determining octan-1-ol water partition coefficients. Pestic. Sci. 12:219-227.
16. Felsot, A., and P. A. Dahm. 1979. Sorption of organophosphorus and carbamate insecticides by soil. J. Agric. Food Chem. 27:557-559.
17. Felsot, A., and J. Wilson. 1980. Adsorption of carbofuran and movement on soil thin layers. Bull. Environ. Contam. Toxicol. 24:778-782.
18. Freundlich, H. 1926. Colloid and capillary chemistry. Methuen, London. p. 147.
19. Fu, Y., R. S. Hansen, and F. E. Bartell. 1948. Thermodynamics of adsorption from solutions. J. Phys. Chem. 52:374-386.
20. Gerstl, Z., and U. Mingelgrin. 1979. A note on the adsorption of organic molecules on clays. Clays Clay Miner. 27:285-290.
21. Gerstl, Z., U. Mingelgrin, and B. Yaron. 1977. Behavior of Vapam and methylisothiocyanate in soils. Soil Sci. Soc. Am. J. 41:545-548.
22. Gerstl, Z., and B. Yaron. 1977. Adsorption and desorption of parathion by attapulgite as affected by the mineral structure. J. Agric. Food Chem. 26:569-573.
23. Gibbs, J. W. 1928. The collected works of J. W. Gibbs, W. R. Longley and R. G. Van Name (ed.) Longmans and Green, New York. Vol. I. p. 219.
24. Giles, C. H., D. Smith, and A. Huitson. 1974. A general treatment and classification of the solute adsorption isotherm. I. Theoretical. J. Colloid Interface Sci. 47:755-765.
25. Classtone, S. 1946. Textbook of physical chemistry. D. van Nostrand (ed.) New York. p. 1320.
26. Goring, C. A. I. 1967. Physical aspects of soil in relation to the action of soil fungicides. Annu. Rev. Phytopathol. 5:285-318.
27. Graham-Bryce, I. J. 1968. Movement of systemic insecticides through soil to plant roots. p. 251-267. In Physico-chemical and biophysical factors affecting the activity of pesticides. Soc. Chem. Ind. Monogr. no. 29.
28. Graham-Bryce, I. J. 1969. Diffusion of organophosphorus insecticides in soils. J. Sci. Food Agric. 20:489-494.
29. Greenland, D. J. 1971. Interactions between humic and fulvic acids and clays. Soil Sci. 111:34-41.
30. Greenland, D. J., R. H. Laby, and J. P. Quirk. 1962. Adsorption of glycine and its di-, tri-, and tetra-peptides by montmorillonite. Trans. Faraday Soc. 58:829-841.
31. Guggenheim, E. A. 1940. The thermodynamics of interfaces in systems of several components. Trans. Faraday Soc. 36:397-412.
32. Hamaker, J. W. 1975. The interpretation of soil leaching experiments. p. 115-134. In V. H. Freed (ed.) Environmental dynamics of pesticides. Plenum Press, New York.
33. Hamaker, J. W., and J. M. Thompson. 1972. Adsorption. p. 49-144. In C. A. I. Goring, and J. W. Hamaker (ed.) Organic chemicals in the soil environment. Marcel Dekker Inc., New York.
34. Hance, R. J. 1965. The adsorption of urea and some of its derivatives by a variety of soils. Weed Res. 5:98-107.
35. Hance, R. J. 1969. An empirical relationship between chemical structure and the sorption of some herbicides by soils. J. Agric. Food Chem. 17:667-668.
36. Hansen, R. S., and R. P. Craig. 1954. The adsorption of aliphatic alcohols and acids from aqueous solutions by non-porous carbons. J. Phys. Chem. 58:211-215.
37. Hartley, G. S. 1960. Physico-chemical aspects of the availability of herbicides in soils. p. 63-78. In E. K. Woodford and G. R. Sagar (ed.) Herbicides and the soil. Blackwell Scientific Publishing Co., Oxford.
38. Hassett, J. J., J. C. Means, W. L. Banwart, and S. G. Wood. 1980. Sorption properties of sediments and energy-related pollutants. EPA Report 600/3-80-041.
39. Hata, U., and Y. Isozaki. 1980a. The influence of soil properties on the adsorption and the phytotoxicity of piperophos. J. Pestic. Sci. 5:23-27.
40. Hata, Y., and K. Akashi. 1980b. Effect of properties of soil colloids on adsorption of piperophos by soils. J. Pestic. Sci. 5:473-479.
41. Holmes, H. N., and J. B. McKelvey. 1928. The reversal of Traube's rule of adsorption. J. Phys. Chem. 32:1522-1523.
42. Jamet, P., and M. A. Piedaller. 1978. Comportment du C^{14}-terbufos dans le sols. I. Adsorption et migration. Phytiatri. Phytopharm. 27:111-122.
43. Kanazawa, J. 1981. Measurement of the bioconcentration factors of pesticides by freshwater fish and their correlation with physicochemical properties or acute toxicities. Pestic. Sci. 12:417-424.
44. Karickhoff, S. W. 1981. Semi-empirical estimation of sorption of hydrophobic pollutants on natural sediments and soils. Chemosphere 10:833-846.
45. Karickhoff, S. W., D. S. Brown, and T. A. Scott. 1979. Sorption of hydrophobic pollutants on natural sediments. Water Res. 13:241-248.

46. Kay, B. D., and D. E. Elrick. 1967. Adsorption and movement of lindane in soils. Soil Sci. 104:314-322.
47. King, P. H., and P. L. McCarty. 1968. A chromatographic model for predicting pesticide migration in soils. Soil Sci. 106: 248-261.
48. Kipling, J. J. 1965. Adsorption from solutions of nonelectrolytes. Academic Press, London. p. 328.
49. Lambert, S. M. 1967. Functional relationshpi between sorption in soil and chemical structure. J. Agric. Food Chem. 15:572-576.
50. Langmuir, I. 1918. The adsorption of gases on plane surfaces on glass, mica and platinum. J. Am. Chem. Soc. 40:1361-1403.
51. Matsuo, M. 1980. The i/o—characters to correlate bio-accumulation of some chlorobenzenes in guppies with their chemical structures. Chemosphere 9:409-414.
52. McCall, P. J., R. L. Swann, D. A. Laskowski, S. M. Unger, S. A. Vrona, and H. J. Dishburger. 1980. Estimation of chemical mobility in soil from liquid chromatographic retention times. Bull. Environ. Contam. Toxicol. 24:190-195.
53. Mercado, A., and Y. Kahanovitch. 1977. Groundwater pollution from pesticides in the coastal plain: hazard evaluation. In M. Horowitz (ed.) Behavior of pesticides in soil. Israel-France Symposium, Bet Dagen, Israel. 1975. Publ. no. 82, Agric. Res. Organ.
54. Mills, A. C., and J. W. Biggar. 1969a. Adsorption of 1,2,3,4,5,6-hexachlorohexane from solution: the differential heat of adsorption applied to adsorption from dilute solutions on organic and inorganic surfaces. J. Colloid Interface Sci. 29:720-731.
55. Mills, A. C., and J. W. Biggar. 1969b. Solubility-temperature effect on the adsorption of gamma- and beta-BHC from aqueous and hexane solutions by soil materials. Soil Sci. Soc. Am. Proc. 33:210-216.
56. Moreale, A., and R. van Bladel. 1979. Soil interactions of herbicide-derived aniline residues: a thermodynamic approach. Soil Sci. 127:1-9.
57. Mortland, M. 1980. Surface reactions of low-molecular-weight organics with soil components. p. 67-72. In A. Banin and U. Kafkafi (ed.) Agrochemicals in soil. International Irrigation Information Center, Israel, and Pergamon Press, New York.
58. Rhodes, R. C., I. J. Belasco, and H. L. Pease. 1970. Determination of mobility and adsorption of agrochemicals on soils. J. Agric. Food Chem. 8:524-528.
59. Saltzman, S., and L. Kliger. (1979). Volatilization of 1,2-dibromo-3-chloropropane (DBCP) from soils. J. Environ. Sci. Health B14:353-366.
60. Saltzman, S., L. Kliger, and B. Yaron. 1972. Adsorption-desorption of parathion as affected by soil organic matter. J. Agric. Food Chem. 20:1224-1226.
61. Schnitzer, M., and S. U. Khan. 1972. Humic substances in the environment. p. 240. Marcel Dekker Inc., New York.
62. Shin, Y., J. J. Chodan, and A. R. Walcott. 1970. Adsorption of DDT by soils, soil fractions and biological materials. J. Agric. Food Chem. 18:1129-1133.
63. Stuhl, M. S., J. B. Uytterhoeven, J. de Bock, and P. L. Huyskens. 1979. The adsorption of N-aliphatic alcohols from dilute aqueous solutions on RNH_3-montmorillonites. II. Interlamellar association of the adsorbate. Clays Clay Miner. 27:377-386.
64. van Bladel, R., and A. Moreale. 1974. Adsorption of fenuron and monuron (substituted ureas) by two montmorillonitic clays. Soil Sci. Soc. Am. Proc. 38:244-249.
65. van Genuchten, M. T., and P. J. Wierenga. 1976. Mass transfer studies in sorbing porous media. I. Analytical solutions. Soil Sci. Soc. Am. J. 40:473-480.
66. Veith, G. D., N. M. Austin, and R. T. Morris. 1979. A rapid method for estimating log P for organic chemicals. Water Res. 13:43-47.
67. Weed, S. B., and J. B. Weber. 1974. Pesticide-organic matter interactions. p. 39-66. In W. D. Guenzi (ed.) Pesticides in soil and water. Soil Sci. Soc. Am., Madison, Wis.
68. Wu, C. H., N. Buehring, J. M. Davidson, and P. W. Santelmann. 1975. Napropamide adsorption, desorption and movement in soils. Weed Sci. 23:454-457.
69. Yaron, B., and S. Saltzman. 1972. Influences of water and temperature on adsorption of parathion by soils. Soil Sci. Soc. Am. Proc. 36:583-586.
70. Yaron, B., A. R. Swoboda, and G. W. Thomas. 1967. Aldrin adsorption by soils and clays. J. Agric. Food Chem. 15:671-675.

Part II

DEGRADATION PROCESSES AND PESTICIDE PERSISTENCE

Editors' Comments
on Papers 22 Through 31

22   AUDUS
The Biological Detoxification of 2:4-Dichlorophenoxyacetic Acid in Soil

23   AHMED and CASIDA
Metabolism of Some Organophosphorus Insecticides by Microorganisms

24   KAUFMAN and BLAKE
Microbial Degradation of Several Acetamide, Acylanilide, Carbamate, Toluidine and Urea Pesticides

25   ALEXANDER and LUSTIGMAN
Effect of Chemical Structure on Microbial Degradation of Substituted Benzenes

26   KEARNEY
Influence of Physicochemical Properties on Biodegradability of Phenylcarbamate Herbicides

27   GUENZI and BEARD
The Effects of Temperature and Soil Water on Conversion of DDT to DDE in Soil

28   WAHID, RAMAKRISHNA, and SETHUNATHAN
Instantaneous Degradation of Parathion in Anaerobic Soils

29   ARMSTRONG, CHESTERS, and HARRIS
Atrazine Hydrolysis in Soil

30   WEBER and COBLE
Microbial Decomposition of Diquat Adsorbed on Montmorillonite and Kaolinite Clays

31   MINGELGRIN and SALTZMAN
Surface Reactions of Parathion on Clays

DEGRADATION PROCESSES

Biological Versus Abiotic Transformations

Biochemical and chemical transformations of pesticides are usually simultaneous processes, and their relative contribution, which is usually difficult to define, has been a controversial subject.

The first, beautifully demonstrated evidence for microbial degradation of 2,4-D has been provided by Audus (1951; Paper 22). Using a simple technique in which soil samples were continuously percolated with 2,4-D solutions and changes in the percolating solution concentrations with time were monitored, the author observed three steps in the disappearance curve. The initial small decrease in concentration was assumed to be due to adsorption. This step was followed by a relatively long period with no change in concentration (the lag period) and then by a fast decrease in concentration. Additions of fresh solutions to the same soil resulted in the rapid disappearance of 2,4-D without the lag phase. Successive application of 2,4-D strongly increased the degradative capacity of the soil; in the soil previously treated with this herbicide, a 100 ppm solution was completely detoxified in 3 days, and a lag period of 10 days was observed in the detoxification of the same solution in the untreated soil. It has been assumed that this disappearance pattern, a lag phase followed by rapid degradation specific to microbial degradation, reflects the growth and adaptation of the microbial population and is therefore a general phenomenon.

Hamaker (1972) summarized the data showing the lag period in the degradation rate of several pesticides and suggested that, although undeniable, this phenomenon is not as prominent and widespread under field conditions as could be expected from the laboratory studies.

In Paper 29, which will be discussed further under other aspects, Armstrong, Chesters, and Harris evaluated the chemical hydrolysis of atrazine as a possible degradation pathway. Chemical degradation was demonstrated by the fact that addition of sterilized soil to atrazine solutions caused an almost tenfold increase in the hydrolysis rate. The authors concluded that chemical degradation was catalyzed by contact with soil and that chemical hydrolysis was an important pathway of atrazine degradation in soil. Evidence for microbial degradation in nonsterilized soils was not obtained. However, microbial degradation is not excluded; probably the analytical techniques used at that time failed to identify metabolites produced by microbial attack.

The unequivocal evidence for microbiological degradation of a specific pesticide is provided by the isolation of the microorganisms capable of attacking it. The distinction between chemical and biochemical processes is often difficult to make. The sterilization procedures used for this purpose are either too mild and probably do not destroy the extracellular enzymes (irradiation and chemical sterilants) or too severe and affect other soil properties in addition to the microbial population (autoclaving). Extracellular enzymes and possibly other products of biological origin, free or adsorbed, are present in soils and could be involved in pesticide degradation (Burns and Edwards, 1980). In fact, biochemical reactions could be considered as simple chemical reactions determined by enzymes (Hamaker, 1972). The precise distinction between chemically and biologically induced degradation is not only difficult but not very meaningful (Graham-Bryce, 1981).

Degradation Pathways

Because of the great diversity in chemical structures, the possible combinations of the transformation pathways of pesticides in soil are multitudinous. The most common reactions involved in the degradation of pesticides are hydrolysis, oxidation, reduction, alkylation, dealkylation, dehalogenation, dehydrogenation, hydroxylation, ring cleavage, ether cleavage, condensation, conjugation, and isomerization. Usually, several of these reactions may be detected in the degradation pathway of a given pesticide. Because transformation by chemical or biological reactions is not always synonymous with detoxification, and toxic and persistent metabolites can be formed, knowledge of the degradation pathway and identification of metabolites are of special practical interest. The degradation pathways of the various classes of pesticides have been widely and thoroughly investigated, and for some of them (e.g., the phenoxyacetic acid herbicides) the degradation reactions and resulting metabolites have been well established (Kaufman, 1974).

Paper 23 is an example of a pioneering work in the investigation of the degradation reactions involved in the microbial degradation of some organophosphate insecticides. In this study, pure cultures of microorganisms, yeast, bacteria, a green alga—usually present in all normal soils—were used to investigate metabolic action on several phosphate and phosphorothioate insecticides. The main degradation reactions identified were hydrolysis and oxidation, and the authors established the relative degradation rate as a function of the chemical structure of the compound and the organism involved. The yeast *Torulopsis utilis* and the alga *Chlorella pyrenoidosa*, for example, hydrolyzed phorate, phorate sulfoxide, and sulfone, and oxidized phorate to its sulfoxide. However, only hydrolysis and no oxidation reactions were detected in the degradation of phorate by both bacteria tested.

Later studies showed that the possible degradation pathways of organophosphate pesticides include reduction reactions and isomerization. Both abiotic and biologically induced reactions were shown to be involved in the degradation of organophosphates in soils (e.g., Walker and Stojanovic, 1973). Microbial hydrolytic reactions are very common and have been demonstrated for several classes of pesticides, such as carbamates, amides, and benzoic acid esters. Hydrolytic enzymes are excreted by many microorganisms and metabolize large molecules to smaller fragments that can enter the cell membrane. Oxidative reactions observed by Ahmed and Casida are also known to be induced by microorganisms and were observed in the metabolism of various groups of pesticides (Matsumara, 1982). In 1958, reductive reactions in the degradation of organophosphates in soil had not yet been observed. A few years later, Lichtenstein and Schulz (1964) demonstrated that parathion degradation could proceed by reduction to its amino form and that yeasts were the main organisms involved in this reaction.

A large number of pesticide compounds from such differing chemical groups as phenylurea, acetamide, acylanilide, carbanilate, and toluidine have an aniline moiety (C_6H_n NH-R) as a common feature of their structure. This moiety is usually liberated following hydrolytic reactions and is further metabolized with possible formation of azobenzene compounds that are structurally related to some carcinogenic substances, prompting interest in further study. Kaufman and Blake (Paper 24) studied the rate of disappearance of the parent compound and of aniline formation for two carbanilates (propham and swep), an acylanilide (propanil), and a toluidine compound (solan) in two differing soils. All these compounds disappeared at a fast rate (5–16 days), being biologically degraded. The degradation of both propanil and swep resulted in the liberation of 3,4-dichloroaniline, which was not further metabolized during the experiment. However, traces of azobenzene were detected at the end of the incubation period. Various degradation patterns were observed for propham, resulting either in aniline accumulation, with or without further degradation, or in nondetectable aniline formation. The authors suggested that various microorganisms developing in soil during incubation could have differing capacities to use the aniline moiety as a carbon source. The presence of populations that hydrolyze propham but do not use the aniline moiety could result in aniline accumulation. Several soil fungi and two bacteria species were isolated from the soils treated with the compounds mentioned and were tested for their capacity to degrade pesticides from the phenylamide and carbamate groups. All the isolated microorganisms were capable of decomposing a variety of pesticides. Besides the hydrolytic reaction with aniline liberation, dehalogenation with or without hydrolysis could occur. The microbial metabolism of chloroanilines has been widely investigated and several degradation pathways have been proposed. It has been demonstrated that chloroanilines can be bound by soil components (Kaufman, 1974).

Factors Affecting the Degradation Rate

In several studies it had been observed that the degradation rate is proportional to the concentration of pesticides in soil. Thus, degradation fitted the simple first-order kinetics: $-dc/dt = kc$, where c is the concentration, t the time, and k the rate constant. However, in some cases the degradation rate was better described by a hyperbolic expression. Hamaker (1972) suggested that generally the degradation reactions in soil, which at low concentrations seem to be of the first order, are probably hyperbolic. Apparent orders of reaction greater than one were often obtained and were explained either by the complexity of the systems or by experimental limitations (Kempson-Jones and Hance, 1979).

The degradation rate of pesticides in soils is determined by complex physicochemical and biological processes that are affected by two main groups of factors: the inherent stability of the compounds (which is related to the molecular structure) and the properties of the soil environment.

Because of the enormous variety of pesticides and the complexity of soil processes, it is extremely difficult to find general relationships between soil persistence and molecular structure. One of the first systematic studies of the effect of chemical structure on the microbial degradation rate of some persistent compounds was presented by Alexander and Lustigman (Paper 25). Despite the shortcomings of the experimental conditions pointed out by the authors, their data show the strong influence of the ring substituent on the degradation rate of simple substituted benzenes by a mixed population of soil microorganisms. For the monosubstituted benzenes, the hydroxyl- and carboxyl-substituted compounds were more readily degradable than those with sulfonate and nitro substituent groups. For the disubstituted benzenes, both the type and the position of the substituent affected the degradation rate. The more persistent were again the sulfonates and the nitro, as well as the chlorosubstituted benzenes. Substitution in the para position enhanced the degradation rate, compared with the same orto- and meta-substitution.

Carbamate pesticides are an important class of compounds, with a broad spectrum of biological activity and a relatively short residual life in soil. Chemical hydrolysis of carbamates under alkaline conditions is a well known reaction. The effect of the alkyl group and ring substitution on the biologically induced degradation rate of phenylcarbamate herbicides was studied by Kearney (Paper 26), using an isolated soil microorganism known for its capability to degrade phenylcarbamates. Steric effects, imposed by the size of the alcohol group, had a strong effect on degradation; the rate decreased with increasing size of this group. The nature of the substituent in the meta-position also affected degradation, which decreased in the order $NO_2 > CH_3\text{-}CO > Cl > CH_3\text{-}O > H$, the same as the relative acidity of the compounds. Positional effects of the ring substituents showed, as in Paper 25, that the degradation of para-substituents was faster than the degradation rate of compounds substituted in the meta-position. The results show that the hydrolysis rate of phenylcarbamates is affected by steric, inductive, and resonance effects of the functional groups in the molecule, suggesting that the principal factor determining the hydrolysis rate is the enzyme-substrate fit, reflected in the size of the molecule.

These two studies demonstrate how the consideration of the molecular steric and electronic properties could enable the assessment of the degradability of pesticides. However, such studies could indicate only the relative rate of degradation of compounds from a specific chemical class under controlled conditions. Their extrapolation to other groups of pesticides and to field conditions must be cautiously considered.

The major soil properties known for their effect on pesticide degradation rate are moisture content, aeration, temperature, pH, and the amount and nature of the colloidal organic and mineral fractions. These factors affect both the soil microbial population and the conditions determining chemical transformation processes. For example, it has been observed often that the presence of soil organic matter enhances pesticide degradation, explainable by the increased microbial activity associated with this soil component. The metabolic products of soil microorganisms contribute to the aggregation of microorganisms among themselves and on soil aggregates. It is considered that 80% of the soil microorganisms are adsorbed by the organic matter and clay particles (Haider, 1982). In addition, soil organic matter can induce surface-catalyzed reactions of adsorbed pesticides but also hinder degradation of some pesticides by decreasing both their availability to microbial attack and their solution concentration.

The moisture content and temperature influence pesticide transformations because they have a direct effect on the microbial activity and affect adsorption (and hence pesticide availability for enzymatic and chemical attack) and the nature and kinetics of the chemical reactions. In addition, the moisture content determines the aerobic or anaerobic conditions of the soil environment. The strong effect of these factors on pesticide degradation rate was initially observed for the dichlorophenoxyacetic acid herbicides. Brown and Mitchell (1948) showed that the rate of inactivation of 2,4-D in soil increased as the moisture content and temperature increased. A similar trend of the moisture effect was later observed for the degradation of organophosphates (Lichtenstein and Schulz, 1964) and other pesticides (e.g., Lichtenstein and Schulz, 1960).

Guenzi and Beard (Paper 27) studied the effect of soil temperature and moisture on the transformation of DDT to the dechlorinated product DDE. The degradation of DDT and other related chlorinated hydrocarbon insecticides known for their relatively high persistence in the environment has been studied intensively. In addition to their resistance to microbiological degradation, the availability of these compounds is reduced by strong adsorption in soil. The conversion of DDT to the dehalogenated compound DDE was demonstrated in numerous studies and was shown to be both biologically and chemically induced (Kaufman, 1974). The results of Guenzi and Beard show that, under their experimental conditions, the chemical process was predominant. DDT conversion was much faster in moist than in air-dried soil; however, no significant difference was observed between the degradation rates in the flooded soil and in the soil kept at 1/3 bar humidity. The authors suggest that the conversion of DDT to DDE is a surface-catalyzed reaction, the active sites being Fe and Al oxides. Water is necessary for the diffusion of DDT molecules to the active sites. Soil temperature strongly influenced the degradation rate of DDT. Equations describing DDT disappearance and DDE formation as functions of temperature and time were calculated.

Several studies carried out with DDT and other chlorinated hydrocarbon pesticides demonstrated that both aerobic and anaerobic degradation could contribute to the transformation of these compounds in soil and that generally the anaerobic processes are faster (e.g., Guenzi and Beard, 1968). Similarly, pesticides possessing nitro groups are more rapidly transformed under anaerobic than aerobic conditions. Alternating aerobic and anaerobic conditions were also shown to enhance pesticide degradation.

An interesting case of very rapid degradation of a pesticide in anaerobic conditions is described by Wahid, Ramakrishna, and Sethunathan et al. (Paper 28). It has already been demonstrated that parathion degradation is faster in flooded than in upland soils

(Sethunathan and Yoshida, 1973), and microbial populations degrading parathion under these conditions have been isolated and identified. Wahid, Ramakrishna, and Sethunathan report the results of parathion degradation in a soil prereduced by flooding, simulating real conditions in a rice field. Of the three soils investigated, the fastest, almost instantaneous degradation, was obtained in the soil with the highest organic matter content. In this soil the drop in the redox potential upon flooding was much faster than in the other soils. Because the major degradation pathway was reduction of the nitro group, it was assumed that this pathway was favored by lower redox potentials. Plimmer, Kearney, and Von Endt (1967) also observed that the presence of organic matter accelerates the decrease of the redox potential of soils, enhancing the microbiological reductive reactions. In an attempt to clarify the nature of the degradation process, the authors compared various sterilization procedures and found that degradation was faster in soils sterilized by irradiation than by autoclaving. They suggested that the fast parathion degradation in prereduced soils was due to enzymes and/or other heat-labile soil substances that are destroyed by autoclaving but not by irradiation. Glass (1972) showed that the iron redox system in reduced soil was capable of degrading DDT and proposed a mechanism whereby electrons furnished by the reduced organic substrate are transferred to DDT via the Fe^{+2} ions. The lack of free electrons in aerobic soils could explain the high persistence of DDT under these conditions.

The practical importance of pesticide degradation under anaerobic conditions could probably be extended beyond the flooded soils. Haider (1982) showed that even when the soil surface is exposed to a normal oxygen-containing atmosphere, centers of soil aggregates can be oxygen-depleted. Consequently, the composition of the microflora will vary. These anaerobic microhabitats could increase the degradation rate of pesticides in aerobic soils. Indeed, prolonged aerobic incubation of some pesticides yielded, in addition to the known degradation products, metabolites specific for an anaerobic degradation pathway.

Soil reaction has a very complex influence on pesticide transformation processes. Besides its direct effect on the proliferation and activity of the soil microflora, soil reaction strongly affects the chemical stability of some important groups of pesticides (organophosphates, carbamates, s-triazines). By their effect on the adsorption-desorption of numerous pesticides by soil, the reaction in the bulk solution and the surface acidity influence the availability of pesticides for degradation reactions as well as for surface-induced transformations.

Armstrong, Chesters, and Harris (Paper 29) present an investigation of the effect of pH and other soil properties on the degradation rate of atrazine. This compound is a herbicide from the widely used s-triazine pesticide group. The microbiological degradation of s-triazines has been widely studied and is well established. Chemical hydrolysis has also been recognized as an important degradative mechanism (Kaufman, 1974). The authors found that the degradation rate of atrazine was pH-dependent and that fast hydrolysis occurred under both highly acid and alkaline conditions. However, the degradation mechanisms differed; alkaline hydrolysis resulted in the displacement of chlorine by OH, while acid hydrolysis resulted in the cleavage of the C-Cl bond by H_2O, following protonation of a ring or chain N-atom.

Surface-Induced Reactions

An interesting observation in Paper 29 is related to the combined effect of acidity and adsorption on the degradation rate of atrazine: at similar pH, atrazine hydrolysis was faster in soil than in soil-free systems, indicating that degradation was adsorption-catalyzed. For the three soils investigated, atrazine adsorption could be correlated to the organic matter content, and degradation was faster in the two soils with high adsorption capacities. However, a comparison of these two soils showed that organic matter is not the determinant factor but that the combination between organic matter and a sufficiently acidic pH determined the fastest degradation rate. Adsorption of atrazine by soil organic matter, by H-bonding to the weak acid groups, could catalyze hydrolysis by a mechanism similar to H-ion catalysis.

The important role of organic matter in the biological and chemical decomposition of pesticides in soil, both by contributing to faster degradation rates and by decreasing the availability of pesticides through adsorption, is generally accepted. However, the mechanisms involved in the degradation of pesticides adsorbed by organic matter are not yet understood. Much more information is available on the surface-induced degradation of pesticides adsorbed by clay minerals.

An important effect of the adsorption-degradation relationship is the decrease in the availability of pesticides upon adsorption. This aspect is emphasized by Weber and Coble (Paper 30) in their study of the microbial degradation of diquat adsorbed on clay minerals. This herbicide, an organic base, is rapidly deactivated by contact with soil, and both adsorption and microbial degradation are possible pathways. Additions of montmorillonite clay to diquat solutions containing a mixed soil microflora completely inhibited diquat mineralization if a sufficient amount of clay for total removal of diquat from solution was provided. However, the microbial degradation of diquat was not influenced by adsorption by the kaolinite clay. Because diquat is readily desorbed from this clay, it was not possible to distinguish between microbial degradation in solution and in the adsorbed state. This investigation demonstrates that pesticides adsorbed in the crystalline lattice of swelling clays could be unavailable for microbial attack.

Surface-catalyzed reactions of pesticides adsorbed on clays were observed initially by formulation chemists using clays as carriers and diluents (e.g., Fawkes et al., 1960). Adsorption-catalyzed degradation on clay surfaces was later demonstrated for several organophosphate and s-triazine pesticides (e.g., Brown and White, 1969; Saltzman, Yaron, and Mingelgrin, 1974; Mingelgrin, Saltzman, and Yaron, 1977).

Mingelgrin and Saltzman (Paper 31) showed that parathion adsorbed on clays was decomposed by hydrolysis, either directly or through an intermediary rearrangement step. The factors affecting this process were the nature of the clay, the saturating cation, and the hydration status of the clay. The nature of the clay determined the degradation mechanism; the predominant degradation pathway was hydrolysis on kaolinite and isomerization on bentonite surfaces, indicating the importance of steric factors in such reactions.

The surface-catalyzed degradation of several pesticides has been related to the surface acidity of clay minerals (e.g., Brown and White, 1969). According to Theng (1974), the catalytic activity of clays is generally correlated to their ability to act as proton donors (Bronsted acids) or electron acceptors (Lewis acids). The first type of acidity derives from the dissociation of water at the clay surface and is affected by the saturating cation and the water content. The water polarization effect of the cation, and the effect of the dissociated water molecules on the catalytic activity of clays, become more important at low moisture contents. Mingelgrin and Saltzman demonstrated that both the exchangeable cation and the hydration status of the clay determined the degradation rate and mechanism in the case of parthion. They suggested that the hydration water of the cations participates in both the adsorption and degradation at adsorption sites. The degradation pathway—direct hydrolysis or rearrangement—was also influenced by steric and inductive effects induced by parathion-clay surface interactions.

References

Audus, L. J., 1951, The Biological Detoxification of Hormone Herbicides in Soil, *Plant and Soil* **3**:170-192.

Brown, J. W., and J. W. Mitchell, 1948, Inactivation of 2,4-Dichlorophenoxyacetic Acid in Soil as Affected by Moisture, Temperature, the Addition of Manure, and Autoclaving, *Bot. Gaz.* **109**:314-323.

Brown, C. B., and J. L. White, 1969, Reactions of 12 s-Triazines with Soil Clays, *Soil Sci. Soc. America Proc.* **33**:863-867.

Burns, R. G., and J. A. Edwards, 1980, Pesticide Breakdown by Soil Enzymes, *Pestic. Sci.* **11:**506-512.

Fowkes, F. M., H. A. Benesi, L. B. Ryland, W. M. Sawyer, K. D. Detling, E. S. Loeffler, F. B. Flockemer, M. R. Johnson, and Y. P. Sun, 1960, Clay-Catalyzed Decomposition of Insecticides, *Jour. Agric. Food Chem.* **8:**203-210.

Glass, B., 1972, Relation between the Degradation of DDT and the Iron Redox System in Soils, *Jour. Agric. Food Chem.* **20:**324-327.

Graham-Bryce, I. J., 1981, The Behavior of Pesticides in Soil, in *The Chemistry of Soil Processes,* D. J. Greenland and M. H. B. Hayes, eds., Wiley, New York, pp. 621-670.

Guenzi, W. D., and W. E. Beard, 1968, Anaerobic Conversion of DDT to DDD and Aerobic Stability of DDT in Soil, *Soil Sci. Soc. America Proc.* **32:**522-524.

Haider, K., 1982, *Anaerobic Microsites in Soils and Their Possible Effect on Pesticide Degradation,* in Proc. of the 5th Int. Congr. Pesticide Chemistry, Kyoto, Japan, J. Myiamoto and P. C. Kearney eds., Pergamon Press, Oxford, vol. 3, pp. 351-356.

Hamaker, J. W., 1972, Decomposition: Quantitative Aspects, in *Organic Chemicals in the Soil Environment,* C. A. I. Goring and J. W. Hamaker, eds., vol. I, Marcel Dekker, Inc., New York, pp. 253-340.

Kaufman, D. D., 1974, Degradation of Pesticides by Soil Microorganisms, in *Pesticides in Soil and Water,* W. D. Guenzi, ed., Soil Science Society America Inc., Madison, Wis., pp. 133-202.

Kempson-Jones, G. F., and R. J. Hance, 1979, Kinetics of Linuron and Metribuzin Degradation in Soil, *Pestic. Sci.* **10:**449-454.

Lichtenstein, E. P., and K. R. Schulz, 1960, Epoxidation of Aldrin and Heptachlor in Soils as Influenced by Autoclaving, Moisture and Soil Types, *Jour. Econ. Entomol.* **53:**192-197.

Lichtenstein, E. P., and K. R. Schulz, 1964, The Effects of Moisture and Microorganisms on the Persistence and Metabolism of Some Organophosphorus Insecticides in Soils, with Special Emphasis on Parathion, *Jour. Econ. Entomol.* **57:**618-627.

Matsumara, F., 1982, Degradation of Pesticides in the Environment by Microorganisms and Sunlight, in *Biodegradation of Pesticides,* F. Matsumara and C. R. Krishna Murti, eds., Plenum Press, New York, pp. 67-87.

Mingelgrin, U., S. Saltzman, and B. Yaron, 1977, A Possible Model for the Surface-Induced Hydrolysis of Organo-Phosphorus Pesticides on Kaolinite Clays, *Soil Sci. Soc. America Jour.* **41:**519-523.

Plimmer, J. R., P. C. Kearney, and D. W. Von Endt, 1967, Mechanism of Conversion of DDT to DDE by Aerobacter Aerogenes, *Jour. Agric. Food Chem.* **16:**594-597.

Saltzman, S., B. Yaron, and U. Mingelgrin, 1974. The Surface-Catalyzed Hydrolysis of Parathion on Kaolinite, *Soil Sci. Soc. America Proc.* **38:**231-234.

Sethunathan, N., and T. Yoshida, 1973, Parathion Degradation in Submerged Rice Soils in the Philippines, *Jour. Agric. Food Chem.* **21:**504-506.

Theng, B. K. G., 1974, *The Chemistry of Clay-Organic Reactions,* A. Holger, London, pp. 261-291.

Walker, W. W., and B. J. Stojanovic, 1973, Microbial versus Chemical Degradation of Malathion in Soil, *Jour. Environ. Qual.* **2:**229-232.

22

Copyright © 1949 by Martinus Nijhoff Publishers
Reprinted from *Plant and Soil* **2**:31-36 (1949)

THE BIOLOGICAL DETOXICATION OF 2:4-DICHLOROPHENOXYACETIC ACID IN SOIL

by L. J. AUDUS

Botany Department, Bedford College University of London, England

Introduction. The persistence in the soil of the new selective herbicide 2:4-dichlorophenoxyacetic acid (2:4-*D*) is one of its most important properties (Nutman, Thornton and Quastel)[9], but the ultimate recovery of the soil from its toxic action is of equal importance in practice. The velocity of this detoxication varies considerably with soil conditions, being accelerated by high soil moisture content, high temperatures and high organic content (Brown and Mitchell[3], Kries[6], Mitchell and Marth)[8]. There are also indications that high soil pH (excess lime) (Hanks[5], Kries)[6] greatly retards the detoxication. It has been suggested (Brown and Mitchell[3], de Rose[4], Hanks)[5] that leaching, chemical or physical combination with soil constituents and biological decomposition may all be effective in normal soil detoxication. It would seem that leaching might play an important role under conditions of heavy rainfall and good drainage (Hanks)[5], but cannot account for 2:4-*D* diaappearance in soil under storage (Brown and Mitchell)[3]. The part played by adsorption seems doubtful since here an immediate, not a slow delayed, action would be expected. This leaves chemical and biological decomposition as the two most probable alternatives. Studies on the persistence of 2:4-*D* in autoclaved soil (Brown and Mitchell)[3] tend to support the latter.

Methods. In an attempt to obtain more direct evidence on this point, the soil perfusion technique of Lees and Quastel[7] has been employed to study the kinetics of 2:4-*D* disappearance from soil solution. In this technique, in which a solution of a known metabolite can be circulated in solution through normal soil for any required period of time, the fate of the metabolite can be followed with an accuracy dependent only on the analytical methods employed. It has been shown (Lees and Quastel)[7]) that in biological processes in soils the time progress curve of the reaction has a characteristic form, showing a considerable initial lag followed by an exponential rise, both of which are characteristic of bacterial proliferation in pure culture. Such curves are regarded as strong evidence of a similar bacterial proliferation in the soil in response to the presence of the specific metabolite concerned. After these initial phases a steady maximum rate is reached at which point the soil is regarded as "saturated" with the particular bacteria responsible for the process concerned. In many cases this bacterially enriched soil retains its activity after drying and can be subsequently used to study the kinetics of the reaction by the Warburg manometric technique (Quastel)[11]).

In the present preliminary investigations 200 ml of dilute solutions of 2:4-*D* (10–1,000 p.p.m.) have been perfused through garden loam (50 g of 1–4 mm fractions of the air-dried sieved soil) using a modified form of perfusion apparatus (Audus)[1]) at room temperatures (20°C). Daily samples of the perfusate were taken during the perfusion. By using a suitable range of dilutions, reasonably accurate estimates of the 2:4-*D* content could be obtained by a biological assay depending on the growth inhibition of Cress (*Lepidium sativum*) roots studied by a technique already described (Audus and Quastel)[2]).

Experimental Results. The results obtained appear in Fig. 1, where the growth inhibition produced by relevant dilutions of perfusate are plotted against the duration of the perfusion. With an initial concentration of 10 p.p.m. (Fig. 1a) there is a slight but significant decrease in toxicity over the first 3 to 4 days but from there onwards there is no significant change until after 10 days perfusion. Then a rapid disappearance of 2:4-*D* sets in, all toxicity being lost after another 4 days. In a subsequent perfusion through the same soil with the same amount and concentration of 2:4-*D* all toxicity was destroyed within two days. In a third perfusion with 200 ml of ten times the concentration (100 p.p.m.) all toxicity was lost in

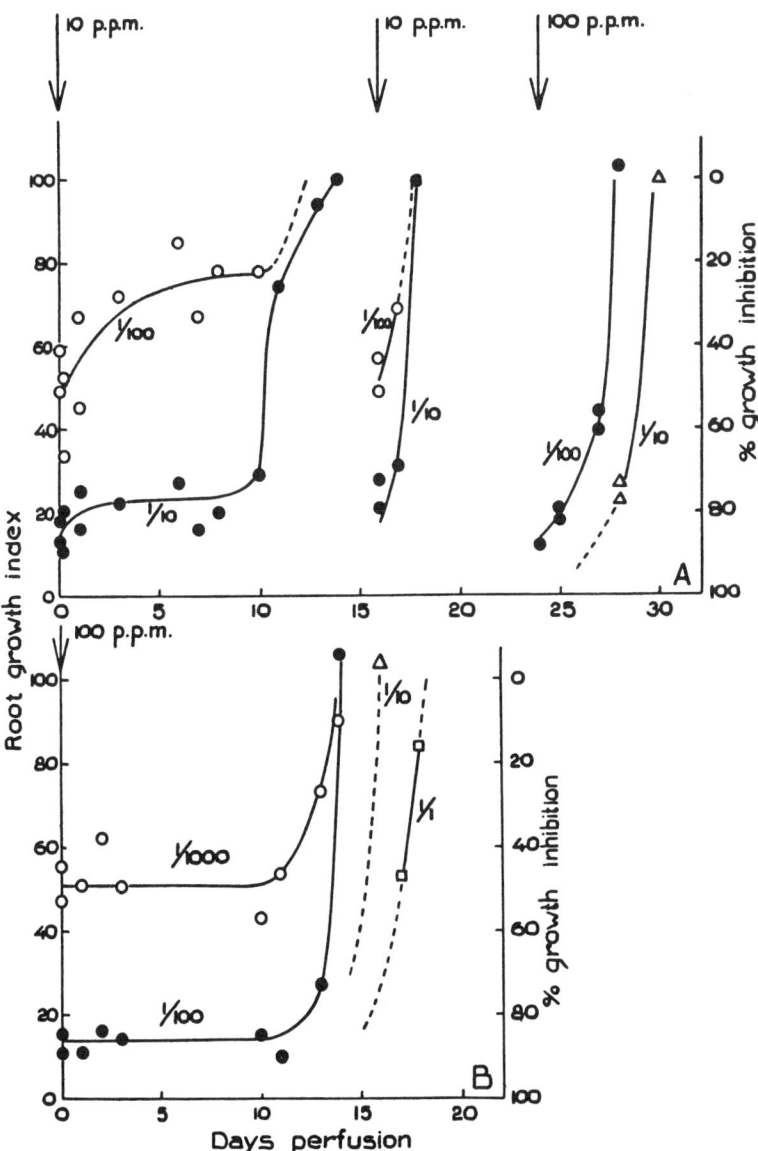

Fig. 1 (a) Curves showing the time course of toxicity of 2: 4-D solution on perfusion through garden loam. The fractions against the curves represent degrees of dilution of the assay solutions. The vertical arrows denote the time of introduction of fresh perfusate, the concentration of which is marked at the side. Initial perfusion with 10 p.p.m.

(b) As in (a) but with initial perfusion with 100 p.p.m.

5 days. Similar results were observed with an initial perfusion of 100 p.p.m. (Fig. 1*b*) where a lag of 11 to 12 days was obtained and complete detoxication within a further 6 days. Fig. 2 shows these results expressed as concentration-time curves. This long lag, followed by a rapid disappearance of the metabolite, is typical of the results of Q u a s t e l [10] for biological processes in the soil and suggest that the detoxication of 2:4-*D* in soil is primarily biological. Proof of this should be obtained when the above experiments are repeated in the presence of a bacterial poison which should inhibit the detoxication process. This must await the adoption of a new

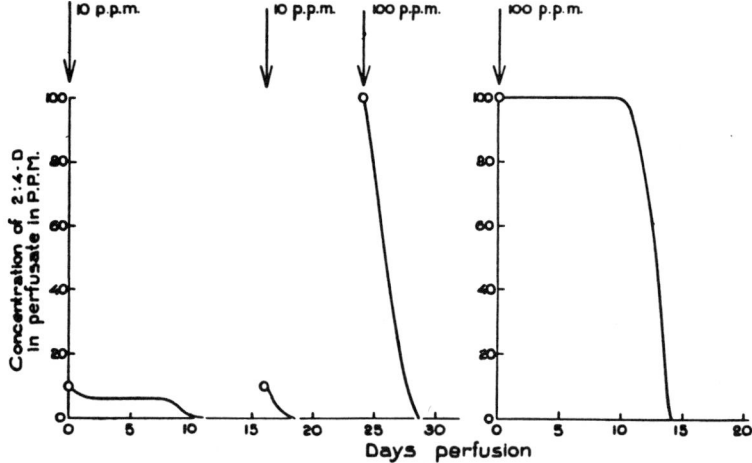

Fig. 2. The results of Fig. 1 smoothed and expressed as 2: 4-*D* concentration curves.

technique for the estimation of 2:4-*D*, since these poisons preclude, or at least greatly complicate, the use of biological assay methods.

The initial small decrease in toxicity of the dilute perfusate (10 p.p.m.) in the first few days of perfusion is probably due to adsorption onto soil colloids. A similar small adsorption followed by a typical biological decomposition has been shown to occur when coumarin is perfused through soil (A u d u s and Q u a s t e l) [2]).

After enriching the soil by perfusion with a concentration of 100 p.p.m., a solution of 1,000 p.p.m. was employed. The toxicity of this disappeared completely in 13 days. Assuming therefore a linear rate of decomposition by the organism concerned (Bacterial saturation of the soil) this corresponds to a maximum rate of approximately 0.01 mg 2:4-*D* per g dry soil per hour. If this decomposition is due to an oxidative breakdown then it should be possible to detect or even measure the oxygen utilisation by the Warburg technique and thus determine the stoichiometric relationships between oxygen and 2:4-*D* in this breakdown. Using therefore a method devised by Q u a s t e l [11]), this enriched soil was rapidly dried in a current of warm air and then placed, after sieving, in 5 gm aliquots in Warburg vessels. The oxygen uptakes of these samples were then followed at 35°C after moistening with 3 ml of 50, 100, 150 and 1,000 p.p.m. 2:4-*D*, using a similar soil sample moistened with 3 ml of distilled water as a control. All samples gave the same initial oxygen uptake of approximately 20 mm^3/g/hour, falling to half that value after 5 hours. These results suggest, although they are not offered as proof, that the detoxication of 2:4-*D* is not brought about by a process involving the absorption of oxygen but by one which may be hydrolytic in nature, It is of course possible that the activity of the organisms concerned is lost on drying and not restored on subsequent remoistening, but this is a subject for further research.

It is interesting that the detoxicated perfusate, when tested in the undiluted state, showed, in some cases, distinct evidence of a stimulation of root growth in Cress. With three perfusates (Original concentration of 2:4-*D* 10 p.p.m.) stimulations of 11, 26 and 57 per cent. were observed on a fresh weight basis, the roots being thinner but on the average twice the length of the controls. Similar observations have been made in soil (K r i e s) [6], T a y l o r) [12]) and indicate that the products of 2:4-*D* breakdown may function as root growth stimulants. It is obvious that detailed microchemical studies on 2:4-*D* breakdown in these perfusion experiments is the next step to be undertaken.

Summary

By making use of a soil perfusion technique it has been shown that the detoxication of 2: 4-dichlorophenoxyacetic acid in garden loam is due almost entirely to the activity of microorganisms. Preliminary experiments suggest that the process does not involve oxidation. There are also indications that the decomposition products may include a root growth stimulant.

Received February 18, 1949.

LITERATURE CITED

1) Audus, L. J., A new soil perfusion apparatus. Nature, **158**, 419 (1946).
2) Audus, L. J. and Quastel, J. H., Coumarin als a selective phytocidal agent. Nature, **159**, 320–324 (1947).
3) Brown, J. W. and Mitchell, J. W., Inactivation of 2,4-dichlorophenoxyacetic acid in soil as affected by soil moisture, temperature, the addition of manure and autoclaving. Botan. Gaz. **109**, 314–323 (1948).
4) De Rose, H. R., Persistance of some plant growth-regulators when applied to the soil in herbicidal treatments. Botan. Gaz. 107, 583–589 (1946).
5) Hanks, R. W., Removal of 2,4-dichlorophenoxyacetic acid and its calcium salt from six different soils by leaching. Botan. Gaz. **108**, 186–191 (1946).
6) Kries, O. H., Persistance of 2,4-dichlorophenoxyacetic acid in soil in relation to content of water, organic matter and lime. Botan. Gaz. **108**, 510–525 (1947).
7) Lees, H. and Quastel, J. H., Biochemistry of nitrification in soil. I. Kinetics of, and the effects of poisons on, soil nitrification as studied by a soil perfusion technique. Biochem. J. **40**, 803–828 (1946).
8) Mitchell, J. W. and Marth, P. C., Effects of 2,4-dichlorophenoxyacetic acid on the growth of grass plants. Botan. Gaz. **107**, 276–284 (1945).
9) Nutman, P. S., Thornton, H. G. and Quastel, J. H., Plant growth-substances as selective weed-killers. Inhibition of plant growth by 2,4-dichlorophenoxyacetic acid and other plant growth substances. Nature, **155**, 497 (1945).
10) Quastel, J. H., Soil Metabolism, Lecture to The Royal Institute of Chemistry (1946).
11) Quastel, J. H. (In the press).
12) Taylor, D. L., Growth of Field crops in soil treated with chemical growth regulators. Botan. Gaz. **108**, 432–455 (1947).

23

Copyright © 1958 by the Entomological Society of America
Reprinted from *Jour. Econ. Entomol.* 51:59–63 (1958)

Metabolism of Some Organophosphorus Insecticides by Microorganisms[1]

MOSTAFA KAMAL AHMED[2] and J. E. CASIDA, *Department of Entomology, University of Wisconsin, Madison*

ABSTRACT

The metabolism by some soil microorganisms of Thimet, Am. Cyanamid 12008, their oxidation products and certain other organophosphates was investigated. The yeast, *Torulopsis utilis*, the bacteria, *Pseudomonas fluorescens* and *Thiobacillus thiooxidans*, and the green alga, *Chlorella pyrenoidosa* were grown in pure cultures under aseptic conditions. The first two organisms were grown in a glucose-yeast extract medium and the latter two in mineral solutions. Rate studies were made on the hydrolysis and oxidation of organophosphates introduced into the cultures as emulsions at 1,000 p.p.m.

Torulopsis and *Chlorella* reacted similarly with the thiophosphates. It appeared as if the organophosphates were rapidly absorbed by the organisms, and then slowly released from the living and dead cells in the cultures. With both organisms the S-(alkylthio)methyl derivatives were much more rapidly hydrolyzed than the S-(alkylsulfonyl)methyl, and the latter only slightly faster than the S-(alkylsulfinyl)methyl compounds. In all cases the phosphorodithioate sulfoxides were the most stable and the phosphorothiolate sulfides were the least stable derivatives in the presence of the organisms. With *Chlorella*, Thimet was oxidized to the phosphorodithioate sulfoxide, which was very stable to hydrolysis but slowly converted to the phosphorothiolate sulfoxide with little if any phosphorothiolate sulfide or sulfones being formed. Both *Torulopsis* and *Chlorella* oxidized the sulfides to sulfoxides, but *Chlorella* was more effective in oxidizing the phosphorodithioates to phosphorothiolates. Little if any oxidation was found with parathion, Dow ET-57, dimefox and schradan in *Chlorella* cultures.

The two bacteria failed to oxidize Thimet but were effective at hydrolyzing this compound. *Thiobacillus* could not utilize sulfur from the Thimet molecule.

Metabolism of organophosphorus insecticides has been extensively studied in insects, mammals and plants. In their recommended usages, these insecticides frequently come in contact with the soil. Little information is available on their breakdown by microorganisms, particularly those of importance in the soil.

Mounter *et al.* (1955) demonstrated the presence in lyophilized bacterial cells of a phosphatase which hydrolyzed dialkylfluorophosphates with the liberation of fluoride ions. He further showed the properties of these phosphatases to vary with different bacteria, since they differed in their response to metal ion activators and inhibitors. Wäckers (1955) showed that the green algae, *Ankistrodesmus braunii* (Näg.) Collins, could store demeton inside their cells and liberate this organophosphate into demeton-free nutrient solutions. Cells rich in fat and deficient in nitrogen stored about 30% more demeton than the normal cells. Wäckers further demonstrated that the phosphorothioate demeton isomer doubled the respiration rate of both the algae *Ankistrodesmus* and *Chlorella* at a concentration at which the phosphorothiolate demeton isomer did not influence the oxygen consumption.

This study concerns the metabolism of certain organophosphorus insecticides by several soil microorganisms.

METHODS AND MATERIALS.—The organisms used in this study were two bacteria, *Pseudomonas fluorescens* Migula and *Thiobacillus thiooxidans* Waksman and Joffe, the yeast, *Torulopsis utilis* (Henneberg) Lodder, and the green alga *Chlorella pyrenoidosa* (Emerson's strain). All organisms were used in pure cultures only.

P. fluorescens is an aerobic bacterium which is quite abundant in soils and plays an important role in the decomposition of carbohydrates and proteins (Waksman, 1932). *Thiobacillus* is present in all normal soils, but is most active where sulfur-containing fertilizers are used (Waksman 1932). *Torulopsis utilis* has been isolated from many different types of soils and is present in Wisconsin soils (based on records of the Bacteriology Department of the University of Wisconsin). *T. utilis* is used as a supplement in cattle and human diets because of its ability to grow rapidly on the hydrolysate of waste wood (Underkofler & Hickey 1954) and the high protein content of the dried yeast, averaging 45 to 50% (Thaysen 1943). The *Chlorella* group has been frequently isolated from soils. Bristol Roach (1927) recognized the presence of two forms of *Chlorella* in English soils and found that this genus made up 35% of the common 10 species of algae in the upper 6 to 12 inches of soil. Moore & Carter (1926) observed *Chlorella* as far down as 5 feet in Missouri Botanical Garden soils. It has also been reported from south German and Danish soils (Waksman 1932).

For bacteria and yeast cultures, 50-ml. quantities of medium were placed in 125-ml. Erlenmeyer flasks which were plugged with cotton and sterilized by autoclaving for 1 hour at 15 lbs. After cooling, each flask was inoculated under aseptic conditions with 1 ml. of the pure culture suspension. The flasks were then shaken for 18 hours with the *Torulopsis* and *Pseudomonas* and for 2 to 3 days with the *Thiobacillus* to provide aeration as the original inoculum grew into a heavy population before introducing the organophosphates. For the larger culture volumes needed to yield sufficient metabolites for isolation and characterization, 200 ml. of medium were used in 500-ml. Erlenmeyers.

Torulopsis was grown in a medium of 2.0% glucose and 0.5% yeast extract in distilled water. This medium kept the cultures at about pH 6.0. At the time the organophosphates were added there was about 5.2 mg. of dry *Torulopsis* cells/ml. The growth progressed until the fourth day when 7.2 mg./ml. were present and then declined so that after 8 days only 4.7 mg. dry cells/ml. remained. In such determinations the cells were separated from the medium and washed by centrifugation. *Thiobacillus* was grown in the inorganic medium described by

---

[1] Approved for publication by the Director of the Wisconsin Agricultural Experiment Station. This investigation was supported in part by a research grant from the American Cyanamid Company. Accepted for publication July 29, 1957.

[2] Present address: Department of Plant Protection, University of Cairo, Giza, Egypt. The authors wish to acknowledge the generous help of Drs. Elizabeth McCoy and G. C. Gerloff of the University of Wisconsin in organizing and carrying out these studies, and of Dr. R. L. Starkey of Rutgers University for furnishing the *Thiobacillus* cultures.

Vogler & Umbreit (1941) consisting of 300 p.p.m. NH_4Cl, 3,000 p.p.m. KH_2PO_4, 250 p.p.m. $CaCl_2$, 500 p.p.m. $MgSO_4$, and 10 p.p.m. $FeSO_4$. Colloidal sulfur was sterilized separately for 3 hours with flowing steam and 0.5 gm. was added to each flask. The *Thiobacillus* cultures were held at pH 5.0 by adding sterilized Na_2CO_3 solution because of reduced *Thiobacillus* activity at higher pH values. *Pseudomonas* was grown in a culture medium of 0.5% yeast extract, 0.5% glucose and 0.7% KH_2PO_4, a medium which kept the pH between 5.5 and 6.2 during the experiment. In the 8-day experimental period, the *Pseudomonas* population dropped from about 2.3 mg. dry cells/ml. to 0.82 mg./ml.

Chlorella was grown by a method (Gerloff 1956) for which the apparatus is illustrated in figure 1. The cultures were aerated with a mixture of air filtered through cotton (A) and CO_2, the composition of the mixture being adjusted to 5% CO_2 as measured by a manometer (B). The CO_2 was washed with water (C) before mixing with the air (D) and passing through another filter (E) packed half with cotton to absorb the excess moisture and half with glass wool to pick up oil droplets which might pass through the system from the air compressor. The 5% mixture of CO_2 in air then passed through "T" and capillary tubes (H and J) adjusted in capillary size and length to allow the same rate of flow for the gas mixture into each of the 1.5×12 in. culture tubes (F). Rubber tubing thoroughly washed with distilled water was used for all non-glass connections. The culture tubes (F) were mounted in a stainless steel rack (G) and illumination provided by two 40-watt tubular fluorescent bulbs on either side of the rack. *Chlorella* was grown in Knop's solution as modified by Myer & Clark (1944) consisting 3792 p.p.m. KNO_3, 2456 p.p.m. $MgSO_4 \cdot 7H_2O$, 1,255 p.p.m. KH_2PO_4, 74 p.p.m. $CaCl_2 \cdot 2H_2O$, 30 p.p.m. ferric citrate and 30 p.p.m. of citric acid. Pyrex distilled water was used throughout and Hoagland's A_5 solution (Hoagland & Arnon 1950) was added to provide 9.5 p.p.m. B, 9.5 p.p.m. Mn, 0.02 p.p.m. Zn, 0.02 p.p.m. Cu and 0.01 p.p.m. Mo. The culture medium was adjusted to pH 6.8 and 150 ml. added to each tube (F), which was then plugged with cotton and the series of tubes connected as indicated in figure 1. A clamp closed the rubber tubing at position "I" to prevent the nutrient solutions from sucking from one tube to another from pressure changes on autoclaving. The system of tubes in the rack along with the cotton-glass wool filter (E) was autoclaved for 20 minutes at 15 lbs., cooled, the clamps at position "I" re-

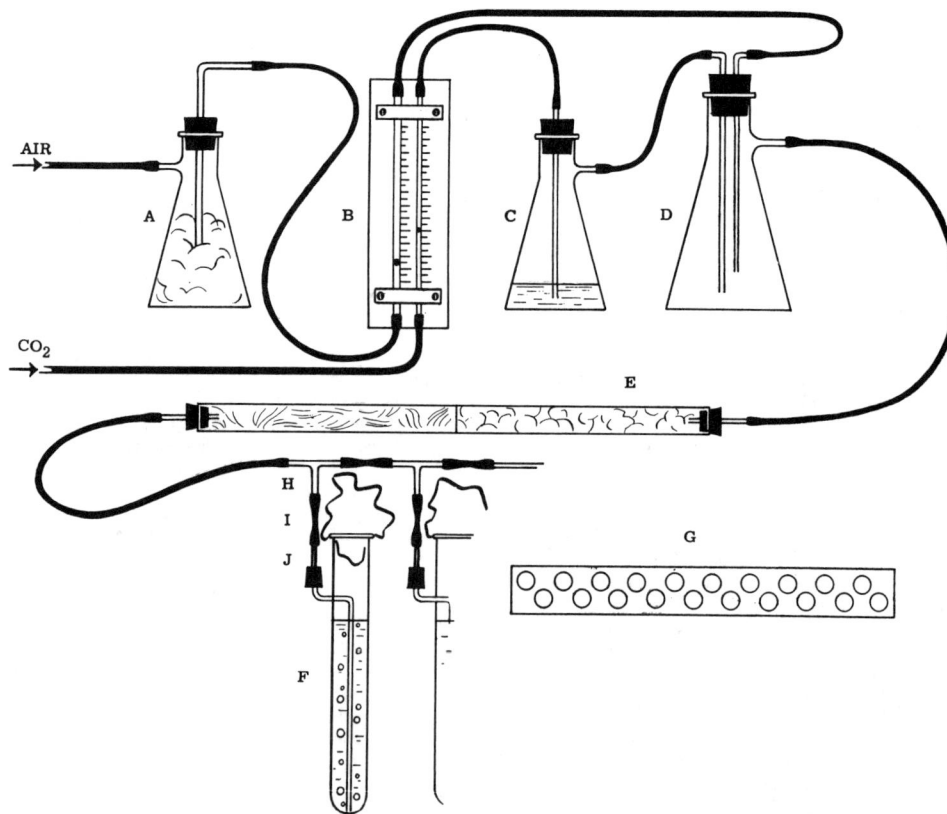

FIG 1.—Aeration apparatus used for *Chlorella* cultures.

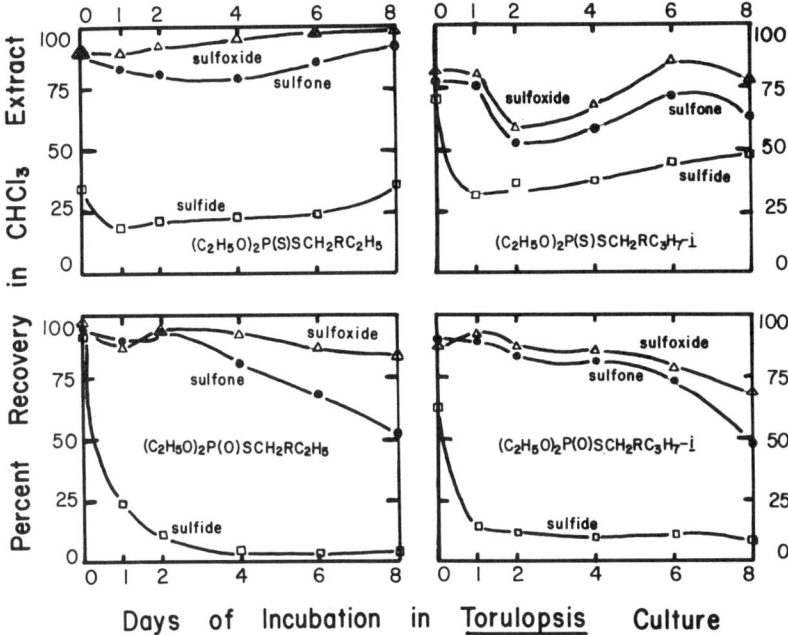

Fig. 2.—Metabolism of Thimet and Am. Cyanamid 12008 derivatives in *Torulopsis* cultures.

moved and each tube inoculated under aseptic conditions with a 1.0 ml. suspension of *Chlorella* cells. The remaining parts of the aeration equipment were then connected and the gas mixture passed through the cultures at the rate of 1.5 liters of gas/min./liter of medium. After 4 to 6 days growth when the cell population was about 1.2 to 1.5 mg. of dry cells per ml. of culture, a small drop of Antifoam KCM (Hodag Chemical Co., Chicago, Illinois) was introduced into each tube by a sterilized bacteriological transfer needle to minimize foaming in the cultures. After 8 days of metabolism time for the organophosphates, when the experiment was terminated, there were about 4.0 mgm. of dry *Chlorella* cells per ml. of medium. Similar determinations in the presence of the compounds showed that none of the materials at the 1,000 p.p.m. level used affected the growth rate of the algal cultures.

Emulsifiable concentrates of the organophosphates were prepared by mixing equal weights of the phosphates with Triton X-155 and emulsifying in sterilized distilled water. The emulsifiable concentrates were introduced under aseptic conditions into the cultures to give a final concentration of 1,000 p.p.m. for each the emulsifier and the compound. Samples of the cultures from 3.0 to 5.0 ml. each were harvested with sterile volumetric pipettes at 0, 1, 2, 4, 6 and 8 days after introducing the compounds and the samples were extracted with equal volumes of chloroform in 15 ml. tubes with centrifugation to aid the separation of the solvent phases. Aliquots of the chloroform extracts were then analyzed for total phosphorus by the method of Allen (1940). Direct analyses were made on the emulsifiable concentrates for total phosphorus to calculate the original amount of organophosphate introduced into the culture, so that the per cent of compound recovered and that lost due to hydrolysis or binding in the cells could be calculated. At the end of the 8-day experimental periods, the remainder of the cultures were extracted with chloroform and the partitioning of the chloroform-soluble phosphorus compounds determined between hexane, acetone and water in a volume ratio of 5:4:7. The chloroform-soluble materials from larger cultures were chromatographed on celite columns using the technique described by Bowman & Casida (1957). Fractions from these columns were located by total phosphorus and anticholinesterase analyses and characterized by their infra-red spectra (10% chloroform solutions of the organophosphates in a sodium chloride prism with a Baird spectrophotometer). The techniques used in studying the phosphoramides were those described by Arthur & Casida (1957).

The compounds used were several dialkyl phenyl-phosphates and phosphorothioates, dimefox, schradan, Thimet, Am. Cyanamid 12008 (common names based on Haller 1957) and the sulfinyl and sulfonyl derivatives of Thimet and Am. Cyanamid 12008 and their phosphorothiolate analogs.

RESULTS AND DISCUSSION.—The yeast, *Torulopsis utilis*, and the alga, *Chlorella pyrenoidosa*, were quite similar in their action on Thimet, Am. Cyanamid 12008, their phosphorothiolate, sulfinyl and sulfonyl derivatives. Considerable amounts of the compounds were unextractable soon after introducing them into the cultures (figures 2 and 3). This inability to recover the added organophos-

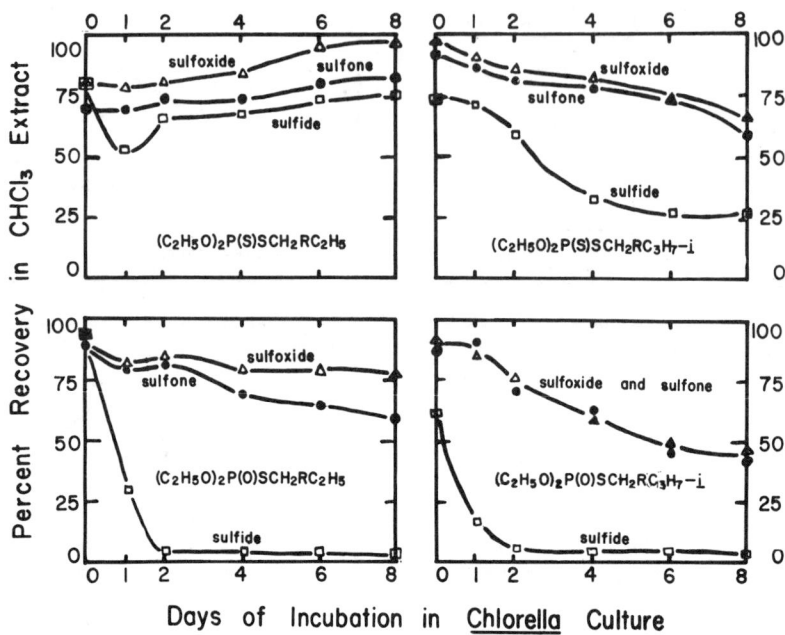

FIG. 3.—Metabolism of Thimet and Am. Cyanamid 12008 derivatives in *Chlorella* cultures.

phates was more evident with the sulfides than with the sulfoxides and sulfones, and was most pronounced with Thimet added to *Torulopsis* cultures where only 34% of the Thimet could be recovered by chloroform extraction immediately after addition. In certain cases the absorbed materials were later released from the living and dead cells in the culture. Wäkers (1955) found that demeton was stored by *A. braunii* and then liberated into demeton-free nutrient solutions.

Both *Torulopsis* and *Chlorella* hydrolyzed many of the Thimet and Am. Cyanamid 12008 derivatives, with the *Chlorella* being somewhat more effective (figures 2 and 3). The sulfides were attacked much more readily than the sulfones, and the latter only slightly faster than the sulfoxides. The phosphorothiolates were always less stable than the phosphorodithioates, the greatest difference occurring with their respective sulfides. In all cases the phosphorodithioate sulfoxides were the most stable and the phosphorothiolate sulfides the least stable derivatives in the presence of the organisms.

Certain Thimet and Am. Cyanamid 12008 derivatives were subject to oxidation as well as hydrolysis. The solubility properties of the residual organophosphates after metabolism by the organisms indicated the formation of more polar oxidation products (table 1). *O,O*-diethyl *S*-(alkylthio)methyl phosphorodithioates and phosphorothiolates were recovered from *Chlorella* and *Torulopsis* cultures mixed with their sulfoxides. Twenty-two per cent of the Thimet incubated 8 days in the yeast cultures was recovered as *O,O*-diethyl *S*-(ethylsulfinyl)-methyl phosphorodithioate based on column chromatography of the metabolites and infra-red and partition coefficient determinations on the fractions obtained. No phosphorothiolate derivatives and sulfones were detected from the Thimet metabolized by the yeast. *O,O*-diethyl *S*-(ethylsulfinyl)methyl phosphorothiolate incubated with *Torulopsis* was recovered as the same compound in the chloroform-soluble fraction (within the sensitivity of the column chromatography and infra-red absorption techniques). Seventy-eight per cent of Thimet and its chloroform-soluble metabolites were recovered from a *Chlorella* culture, and column chromatography, partition coefficients and infra-red spectra on the metabolites indicated that 12.3% *O,O*-diethyl *S*-(ethylsulfinyl)methyl phosphorodithioate and 2.9% *O,O*-diethyl *S*-(ethylsulfinyl)-methyl phosphorothiolate were formed with less than 1% total of the other three possible oxidation products being formed. *O,O*-Diethyl *S*-(ethylsulfinyl)methyl phosphorodithioate was recovered without *in vivo* oxidation after 8 days in *Chlorella* culture based on the sensitivity of the column chromatography and infra-red techniques. Similarly incubation of the phosphorothiolate sulfoxide with *Chlorella* formed little if any phosphorothiolate sulfone based on infra-red spectra. These studies considered in relation to the information in table 1 indicate that with *Chlorella* the phosphorodithioate sulfide (Thimet) was oxidized to the phosphorodithioate sulfoxide, which was very stable to hydrolysis but slowly converted to the phosphorothiolate sulfoxide with little if any phosphorothiolate sulfide or sulfones being recovered. The phosphorothiolate sulfide was rapidly hydrolyzed and slowly oxidized to the phosphorothiolate sulfoxide. Torulopsis reacted nearly the same as *Chlorella* (table 1) but was less effective in oxidizing the phosphorodithioate to the phos-

Table 1.—Partitioning properties of metabolites from Thimet, Am. Cyanamid 12008 and their oxidation products after 8 days' incubation in cultures of *Torulopsis* and *Chlorella*.[a]

	Thimet Derivatives (R=ethyl)			Am. Cyanamid 12008 Derivatives (R=isopropyl)		
		Metabolites from			Metabolites from	
Compound	Known	*Torulopsis*	*Chlorella*	Known	*Torulopsis*	*Chorella*
$(C_2H_5O)_2P(S)SCH_2SR$	100	55	48	100	83	89
$(C_2H_5O)_2P(S)SCH_2S(O)R$	30	34	48	55	55	37
$(C_2H_5O)_2P(S)SCH_2S(O)_2R$	57	60	55	68	68	58
$(C_2H_5O)_2P(O)SCH_2SR$	62	21	38	83	60	14
$(C_2H_5O)_2P(O)SCH_2S(O)R$	2	—	—	9	—	—
$(C_2H_5O)_2P(O)SCH_2S(O)_2R$	4	—	—	8	—	—

[a] All figures represent the per cent of total phosphorus partitioning in the hexane layer with a hexane, acetone, water mixture in a volume ratio of 7:5:4.

phorothiolate. Thimet and Am. Cyanamid 12008 derivatives responded similarly in respect to oxidative metabolism by *Chlorella* and *Torulopsis*.

It is interesting to note that the metabolism of Thimet and Am. Cyanamid 12008 by the *Chlorella* and *Torulopsis* was similar except for the extent of oxidation to that reported for Thimet, Bayer 19639, and Systox in higher plants (Fukuto et al. 1956, Bowman & Casida 1957, Metcalf et al. 1957).

Several dialkyl phenylphosphates were incubated with *Torulopsis* cultures for 8 days. From 45 to 65% of the following materials were recovered after this metabolism time: methyl parathion, Am. Cyanamid 4124 and its oxygen analog, Chlorthion, and Dow ET-57 and its oxygen analog.

After 8 days incubation in a *Chlorella* culture, parathion was recovered 37% as chloroform-soluble materials. Chromatography and infra-red determination of the recovered organophosphate failed to indicate the formation of para-oxon. Dow ET-57 was 72% recovered from *Chlorella* cultures after 8 days, but under the test conditions the Dow ET-57 was in excess of the saturation point for the culture and crystallization occurred within 24 hours. Of the chloroform-solubles recovered greater than 98% of the organophosphate was ET-57 per se based on chromatography and infra-red spectra. Under similar conditions with *Chlorella*, schradan was recovered to the extent of 71% and dimefox 84% after 8 days. The *Chlorella* formed no chloroform-soluble metabolites that yielded formaldehyde on acid degradation from either dimefox or schradan nor formed any metabolites differing from the original phosphoramides in solvent partitioning, chromatographic characteristics or infra-red spectra (techniques after Arthur & Casida 1957).

Pseudomonas cells hydrolyzed 58% of the Thimet incubated with them for 8 days but no oxidation occurred at any site on the organophosphate molecule since the residual organophosphate recovered partitioned completely into hexane from an acetone-water mixture.

Thiobacillus hydrolyzed 75% of the Thimet within 8 days and the residual organophasphate recovered was Thimet per se. Although *Thiobacillus* requires sulfur for growth, it failed to utilize the sulfur from the Thimet molecule. When Thimet was added to the *Thiobacillus* medium as the only energy source, the introduced inoculum failed to grow, but when elemental sulfur was added to these cultures, a heavy *Thiobacillus* population was rapidly formed.

References Cited

Allen, R. J. L. 1940. The estimation of phosphorus. Biochem. Jour. 34: 858–65.

Arthur, B. W., and J. E. Casida. 1957. Biological and chemical oxidation of tetramethyl phosphorodiamidic fluoride (dimefox). Jour. Econ. Ent. (In press)

Bowman, J. S., and J. E. Casida. 1957. Metabolism of the systemic insecticide 0,0-diethyl S-ethylthiomethyl phosphorodithioate (Thimet) in plants. Jour. Agric. Food Chem. 5: 192–7.

Bristol Roach, B. M. 1927. On the algae of some normal English soils. Jour. Agric. Sci. 17: 563–88.

Fukuto, T. R., J. P. Wolf, III, R. L. Metcalf and R. B. March. 1956. Identification of the sulfoxide and sulfone plant metabolites of the thiol isomer of Systox. Jour. Econ. Ent. 49: 147–51.

Gerloff, G. C. 1956. Personal communication.

Haller, H. L. 1957. Common names of insecticides. Jour. Econ. Ent. 50: 226–8.

Hoagland, D. R., and D. I. Arnon. 1950. The water-culture method for growing plants without soil. California Expt. Sta. Circ. 347: 31 pp.

Metcalf, R. L., T. R. Fukuto and R. B. March. 1957. Plant metabolism of dithio-Systox and Thimet. Jour. Econ. Ent. 50: 338–45.

Moore, G. T., and N. Carter. 1926. Further studies on the subterranean algal flora of the Missouri Botanical Gardens. Ann. Missouri Bot. Garden 13: 101–40.

Mounter, L. A., R. F. Baxter and A. Chanutin. 1955. Dialkylfluorophosphatases of microorganisms. Jour. Biol. Chem. 215: 699–704.

Myer, J., and L. B. Clark. 1944. Culture conditions and the development of the photosynthetic mechanism. II. An apparatus for the continuous culture of *Chlorella*. Jour. Gen. Physiol. 28: 103–12.

Thaysen, A. C. 1943. Value of micro-organisms in nutrition (food yeast). Nature 151: 406–8.

Underkofler, L. A., and R. J. Hickey. 1954. Industrial fermentations. Vol. I. Chem. Publ. Co. Inc. N. Y.

Vogler, K. G., and W. W. Umbreit. 1941. The necessity for direct contact in sulfur oxidation by *Thiobacillus thiooxidans*. Soil Sci. 51: 331–7.

Wäckers, R. W. 1955. Phytophysiological effects of the systemic insecticide Systox (diethyl thionophosphoric ester of B-oxyethyl thioethyl ether) Höfchen-Briefe (English edition) Bayer Leverkusen, Germany. 8: 265–324.

Waksman, S. A. 1932. Principles of soil microbiology. The Williams and Wilkins Co., Baltimore, Md.

MICROBIAL DEGRADATION OF SEVERAL ACETAMIDE, ACYLANILIDE, CARBAMATE, TOLUIDINE AND UREA PESTICIDES

D. D. KAUFMAN and J. BLAKE

Agricultural Environmental Quality Institute, Agricultural Research Center, ARS, USDA, Beltsville, Md 20705

(*Accepted* 12 *October* 1972)

Summary—Soil enrichment was used to isolate soil microorganisms capable of degrading isopropyl carbanilate (propham), 3′,4′-dichloropropionanilide (propanil), 3′-chloro-2-methyl-*p*-valerotoluidide (solan), and methyl 3,4-dichlorocarbanilate (swep) in a muck and a silty clay loam. Degradation of the pesticides in enrichment solutions, and by pure cultures of effective microbial isolates was demonstrated by the production of the corresponding aniline, chloride ion liberation and disappearance of the parent compound. Degradation products were identified by gas–liquid and thin-layer chromatography.

Organisms isolated include *Pseudomonas striata* Chester, *Achromobacter* sp., *Aspergillus ustus* (Bain) Thom and Church, *A. versicolor* (Vuill. Tirabaschi), *Fusarium oxysporum* Schlecht, *F. solani* (Martius) Appel and Wollenweber, *Penicillium chrysogenum* Thom, *P. janthinellum* Biourge, *P. rugulosum* Thom and *Trichoderma viride* Pers. Each organism demonstrated a unique substrate specificity and was capable of degrading other aniline-based pesticides of the acetamide, acylanilide, carbamate, toluidine and urea classes.

INTRODUCTION

CARBANILATES, acylanilide and phenylurea herbicides contain aniline moieties as a basic part of their chemical structure. Numerous other pesticides also contain either anilines or chemical moieties which may be degraded to aniline type residues. Considerable interest has developed toward understanding the fate and behavior of these aniline products in the environment. Carbanilate herbicides are relatively nonpersistent (2–5 weeks) in soil, whereas phenylureas are generally more persistent (2–10 months), some persisting as long as a year or more in soil. Most acylanilide herbicides appear to remain intact only briefly in soil.

Microbial degradation is an important factor affecting the persistence of these pesticides in soil. The phenylcarbamate herbicides chlorpropham (Kaufman and Kearney, 1965; Kearney and Kaufman, 1965; Clark and Wright, 1970a, b), propham (Clark and Wright, 1970a, b), and swep (Bartha and Pramer, 1969), are hydrolyzed to their corresponding alcohol and aniline. Further decomposition of both the alcohol and aniline moieties also occurs. The acylanilide herbicides propanil and solan are hydrolyzed to their corresponding anilines and acids (Bartha, 1969; Chisaka and Kearney, 1970). Dealkylation followed by hydrolysis of the urea linkage is necessary before aniline is liberated from the dialkyl-phenylurea herbicides (Geissbuhler, 1969). More recent investigations, however, indicate that methoxy-substituted phenylurea herbicides may also be degraded directly by enzymic hydrolysis to CO_2 and the corresponding aniline and alkylalkoxyamino residue (Engelhardt *et al.*, 1971).

The ultimate fate of the chloroaniline moiety in soil is only partially understood at present. Adsorption to soil particles is known to account for some of the chloroaniline (Linke and Bartha, 1970). Bartha and Pramer (1967) reported the isolation of large quantities of

3,3',4,4'-tetrachloroazobenzene (TCAB) from soil treated with excessively high rates of 3'4'-dichloropropionanilide (propanil). This product was formed by the condensation of two 3,4-dichloroaniline metabolites. Kearney *et al.* (1970) subsequently detected TCAB residues in rice fields treated at recommended propanil application rates 2 and 3 years before sampling.

Several additional condensation products including 1,3-bis(3,4-dichlorophenyl)triazene (Plimmer *et al.*, 1970), 4-(3,4-dichloroanilino)-3,3',4'-tetrachloroazobenzene (Rosen and Siewerski, 1971) and 3,3',4,4'-tetrachloroazoxybenzene (Kaufman *et al.*, 1972) have also been identified as products of 3,4-dichloroaniline metabolism. Mixed azobenzenes, or azobenzenes resulting from the condensation of two unlike chloroaniline molecules have also been reported (Bartha, 1969; Kearney *et al.*, 1969). An understanding of the formation mechanism of such products and their fate in soil is essential if one is to prevent or reduce the production of environmental pollutants. The purpose of this investigation was to observe aniline-based herbicide degradation in several soil systems and to isolate soil microorganisms active in the degradation of these compounds.

METHODS

Pesticides

Common and chemical names, and chemical purity of all pesticides used in this investigation are listed in Table 1. All of the pesticides listed are used as herbicides with the exception of chlorphenamidine which is a miticide.

TABLE 1. COMMON AND CHEMICAL NAMES, AND PURITY OF PESTICIDES USED

Common name	Chemical name	Purity %
Alachlor	2-chloro-2',6'-diethyl-N-(methoxymethyl)acetanilide	98
Barban	4-Chloro-2-butynyl m-chlorocarbanilate	92·2+
Benefin	N-butyl-N-ethyl-a,a,a,trifluoro-2,6-dinitro-p-toliudine	98+
CDAA	N,N-diallyl-2-chloroacetamide	95+
CDEC	2-Chloroallyl diethyldithiocarbamate	95+
Chlorbromuron	3-[4-Bromo-3-chlorophenyl]-1-methoxy-1-methylurea	98
Chloroxuron	3-[p-(p-Chlorophenoxy)phenyl]-1,1-dimethylurea	98
Chlorphenamidine	N-(2-methyl-4-chlorophenyl)-N',N'-dimethylformamidine	95+
Chlorpropham	Isopropyl m-chlorocarbanilate	100·0
Cycloate	S-ethyl N-ethylthiocyclohexanecarbamate	95·2
Diallate	S-(2,3-dichloroallyl)diisopropylthiocarbamate	95+
Dicryl	3',4'-Dichloro-2-methylacrylanilide	98·2
Diphenamid	N,N-dimethyl-2,2-diphenylacetamide	96·5
Diuron	3-(3,4-Dichlorophenyl)-1,1-dimethylurea	100
DMU	3-(3,4-Dichlorophenyl)1-methylurea	100
Fenuron	1,1-Dimethyl-3-phenylurea	100
Fluometuron	1,1-Dimethyl-3-(a,a,a-trifluoro-m-tolyl) urea	98
Metobromuron	3-(p-Bromophenyl)-1-methoxy-1-methylurea	98·2
Neburon	1-Butyl-3-(3,4-dichlorophenyl)-1-methylurea	100
NIA-11092	3-[3-(N-tert-buthylcarbamyloxy)phenyl]urea	87
Nitralin	4-(Methylsulfonyl)-2,6-dinitro-N,N-dipropylaniline	95
Norea	3-(Hexahydro-4,7-methanoindan-5-yl)-1,1-dimethylurea	95+
Propachlor	2-Chloro-N-isopropylacetanilide	95·6
Propanil	3',4'-Dichloropropionanilide	99+
Propham	Isopropyl carbanilate	100·0
Solan	3'-Chloro-2-methyl-p-valerotoluidide	96·4
Swep	Methyl 3,4-dichlorocarbanilate	99

Enrichment cultures

A soil-solution enrichment technique was used to examine the microbial degradation of propham, propanil, solan and swep. Five grams of soil were added to an aqueous suspension (100 ml) of the pesticide (100 parts/10^6) in a 250 ml flask on a reciprocating shaker. Two soils were used: a Celeryville muck and a Hagerstown silty clay loam (Table 2). Duplicate flasks were used for each soil and pesticide. Flasks with sterile soil and water were used for controls.

TABLE 2. CHEMICAL AND PHYSICAL PROPERTIES OF THE SOILS USED IN ENRICHMENT STUDIES

Soil	pH	Cation exchange capacity (m-equiv./100 g)	Moisture field capacity (%)	Organic matter (%)	Clay (%)
Celeryville muck	5·0	165·3	51·2	74·9	ND*
Hagerstown silty clay loam	6·8	14·7	25·8	2·5	39·4

* ND—not determined.

Pesticide analysis

Pesticide degradation in the enrichment solutions was demonstrated by removing samples from each flask at 2-day intervals and measuring the disappearance of the parent material, the production of the corresponding aniline and the release of chloride ion. Aniline residues were measured colorimetrically using a modification of the Bratton–Marshall reaction (Pease, 1962). Residual concentrations of the parent material were determined by first submitting a sample to alkaline hydrolysis and subsequent measurement of total aniline content (microbially evolved aniline plus hydrolyzed aniline). Chloride ion determinations were made by the procedure of Iwasaki, Utsumi and Ozawa (1952).

At the conclusion of the incubation period the enrichment solutions (approx. 70 ml) were extracted twice with 50 ml of petroleum ether. The extracts were combined, dried with powdered $MgSO_4$, filtered, and concentrated to a volume suitable for gas–liquid and thin-layer chromatographic analysis. Thin-layer chromatograms were developed on silica gel HF_{254} in hexane:benzene:acetone (7:3:1). Gas chromatographic analyses were performed on an F and M Model 700 gas chromatograph with a flame ionization detector and a 180 cm stainless steel column packed with 10 per cent methylvinyl silicone gum rubber on diatoport S 80–100 mesh. The carrier gas (N_2) flow rate was 40 ml/min. Injection port and detector temperatures were 270 and 310°C, respectively. A column temperature of 180°C was used for detecting propham, propanil, solan and swep, and their corresponding anilines. A column temperature of 250°C was used for detecting azobenzene type compounds.

Pure cultures

Pure cultures of effective microorganisms were isolated from the enriched soils by a soil dilution plate method. Serial dilutions were prepared from the enriched soils and a set of 5 plates was prepared from each dilution. The plating medium contained: K_2HPO_4, 0·8 g: KH_2PO_4, 0·2 g; $MgSO_4.7H_2O$, 0·2 g; $CaSO_4$, 0·1 g; $(NH_4)_6Mo_7O_{24}.4H_2O$, 1 mg $(NH_4)_2SO_4$, 5·0 g; bacto-agar, 20 g; and distilled water 1000 ml. The corresponding herbicide was supplied (100 parts/10^6) as essentially the sole source of carbon. The chemicals were introduced in 0·1 ml acetone to the sterilized and cooled (50°C) medium. One ml of the serial dilution was added to each plate. The agar medium containing the pesticide was added to each plate, and the plates swirled to assure even distribution of the organisms throughout the plate. Plates prepared in this manner were incubated for 1–3 weeks at 24°C. Microbial cultures appearing during the incubation period were isolated, purified and maintained on the basal medium described above plus 100 parts/10^6 of the herbicide and 0·1 g/l. of yeast extract as carbon sources. All effective cultures were identified at least to genus.

Pesticide degradation by pure cultures of microorganisms

The organisms isolated were examined for their ability to degrade chemicals analogous to the one used for their isolation. The medium used for these experiments contained the basal salts listed above in addition to 0·1 g of yeast extract and a 100 parts/10^6 concentration of the pesticide. The pesticides were aseptically introduced in 0·1 ml acetone to each 100 ml of sterile media. One ml of actively growing cell suspensions was used as the inoculum. Incubation was at 24°C on a rotary shaker. Samples were removed daily and analyzed as described previously. Duplicate flasks were used for each experimental parameter. Controls, with and without pesticides or microorganisms were included. At the conclusion of the incubation period, the contents of one flask were extracted twice with petroleum ether and the extract analyzed as described previously. The second flask was used in an oat bioassay method to determine whether the isolated microorganisms reduced the pesticides to non-phytotoxic or less phytotoxic compounds. The addition of 100 ml of the incubated culture solution to 300 g per pot of the Hagerstown soil was equivalent to an initial application rate of 71·7 kg/ha. Inoculated pesticide solutions were compared with sterile pesticide solutions. The treated soils were planted with oat (*Avena sativa* L. var. Markton) seeds. The seedlings were harvested after a 3-week growing period, and the fresh weight was expressed as percentage of the sterile control with no pesticide.

RESULTS AND DISCUSSION

Soil enrichment

Microbial degradation of propham, propanil, solan, and swep was observed in soil enrichment cultures (Figs. 1–3). Lag periods of only 2–5 days were observed for the various chemicals. No degradation was observed in similar sterilized culture systems. The rate of microbial degradation of the pesticides varied with the pesticides, soil type and in the case of propham, with the replication. Considerable variation was observed between replications and in experiments in the extent of propham biodegradation (Fig. 1). Several patterns of propham biodegradation were observed. Propham degradation occurred more readily and more completely in enrichment cultures containing muck soil. In some instances propham degradation occurred readily with no detectable accumulation of aniline, whereas in others aniline accumulated before it was degraded. Carbanilate pesticides and their corresponding aniline and alcohol are hydrolyzed to CO_2 by soil microorganisms (Kaufman and Kearney, 1965; Kearney and Kaufman, 1965; Clark and Wright, 1970a, b). The variation in experimental results observed with propham could be characteristic of the population developing during the enrichment period. Populations capable of rapidly metabolizing aniline may have developed in some enrichment solutions but not in others. Alternatively, microbial populations capable of degrading propham by mechanisms not involving aniline liberation could have been present in some enrichment cultures. Other experiments we have conducted which will be reported subsequently, indicate some soil microorganisms may hydrolyze propham to aniline, isopropyl alcohol and CO_2, and subsequently use the aliphatic moiety as a carbon source and ignore the aromatic moiety. A predominance of this type of organism in the enrichment flora would result in the accumulation of aniline which may then be degraded by a subsequent enrichment flora of its own. No attempt was made to characterize such microbial population differences in these enrichment cultures.

Propanil and swep degradation resulted in the liberation of 3,4-dichloroaniline. Propanil was readily degraded in both soil types, whereas swep degradation occurred only in enrichment cultures containing muck. Further degradation of the 3,4-dichloroaniline moiety was not observed. Trace amounts (1 per cent) of the 3,3′,4,4′-tetrachloroazobenzene

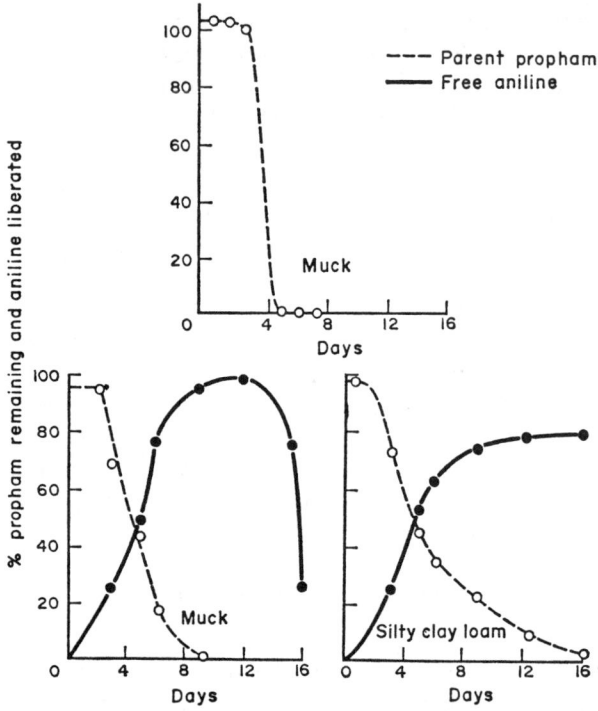

FIG. 1. Patterns of propham biodegradation observed in soil enrichment cultures.

(TCAB) were observed in ether extracts of the enrichment solution at the conclusion of the incubation period. Identity of TCAB was established through gas–liquid and thin-layer chromatographic comparisons with a known standard.

Solan degradation occurred in both muck and silty clay loam enrichment solutions. 3-Chloro-p-toluidine was a product which accumulated during the enrichment period, but was subsequently degraded in enrichment cultures containing muck. Degradation of the toluidine moiety was accompanied by a nearly quantitative release of the chloride ion. No azobenzene formation was observed in solan enrichment cultures.

Isolation of pesticide degrading microorganisms

Numerous soil microorganisms were isolated from the various enrichment cultures. In several instances similar organisms were isolated on different compounds. Morphological comparisons and metabolic characterizations of these organisms with several analagous pesticides indicated that differences between these similar species were relatively minor.

Therefore, only single cultures were maintained for further study. Soil fungi isolated from the enrichment cultures included *Aspergillus ustus* (Bain) Thom and Church, *A. versicolor* (Vuill. Tirabaschi), *Fusaruim oxysporum* Schlecht, *F. solani* (Martius) Appel and Wollenweber, *Penicillium chrysogenum* Thom, *P. janthinellum* Biourge, *P. nigulosum* Thom and *Trichoderma viride* Pers. Only two distinctly different bacterial strains were isolated: *Pseudomonas striata* Chester and *Achromobacter* sp. The source of these cultures is shown in Table 3.

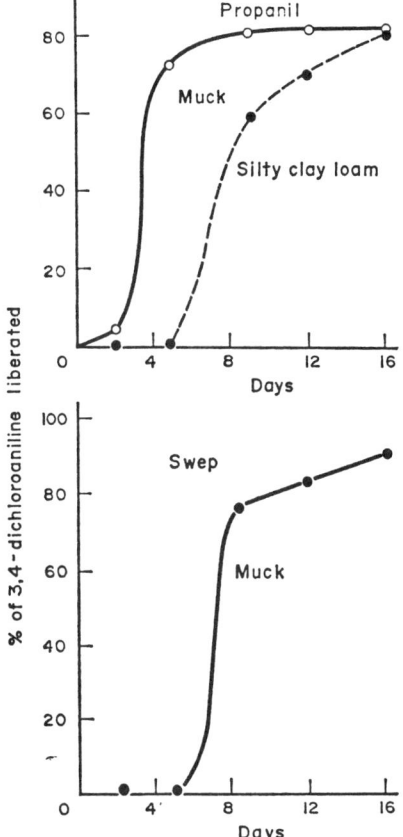

FIG. 2. Liberation of 3,4-dichloroaniline from propanil and swep in soil enrichment cultures.

Soil fungi were predominately isolated on solan and propanil whereas bacteria were more common in propham and swep enrichment cultures. *Fusarium oxysporum* Schlecht was the most prevalent soil fungus isolated, whereas *Pseudomonas striata* Chester was the most prevalent bacterium. The ability of these organisms to metabolize other aniline-based, acetamide, or thio- and dithiocarbamate herbicides was determined under pure culture conditions.

Substrate specificity of microbial isolates

All of the microbial isolates obtained from the enrichment cultures were examined for their ability to degrade other phenylamide and carbamate pesticides. Differences were observed in the abilities of all soil fungus isolates to metabolize chlorphenamidine, chlorpro-

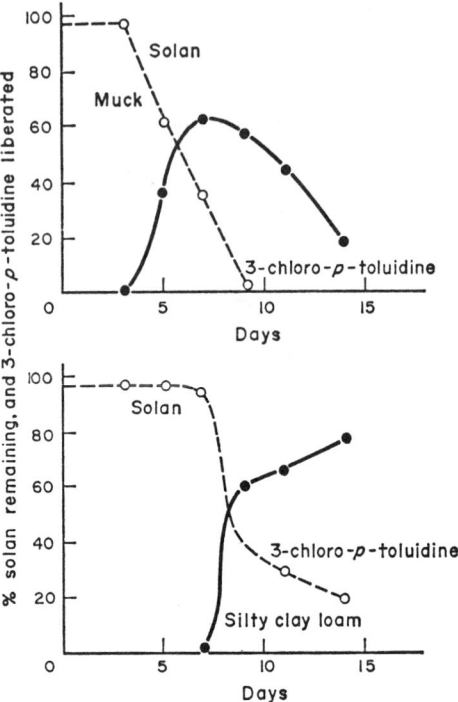

FIG. 3. Microbial degradation of solan in soil enrichment cultures.

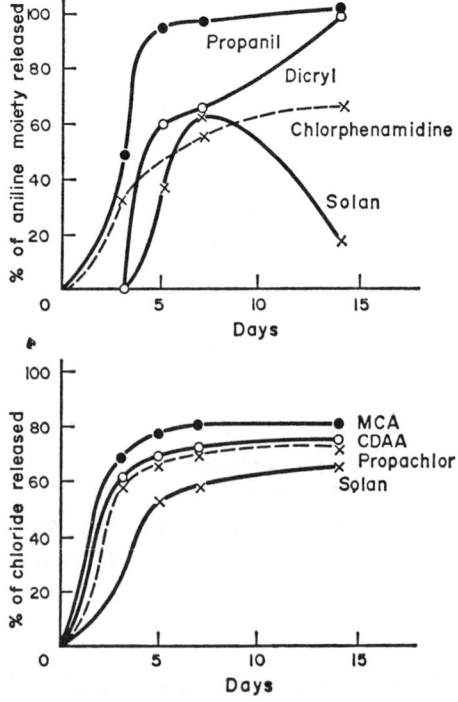

FIG. 4. Liberation of chloroanilines and chloride ion from several pesticides by *F. oxysporum*. (MCA = monochloroacetate.)

pham, dicryl, diuron, propanil, propachlor, propham and solan (Table 4). Each of the fungus isolates appeared unique in their substrate specificity. All of the fungi metabolized several aniline-based herbicides and liberated the corresponding aniline, or Cl^-. Each appeared unable to metabolize one or more of the pesticides. Several organisms appeared to liberate Cl^- without the liberation of detectable amounts of the corresponding aniline. Whether this was a result of rapid utilization of the liberated aniline or a dehalogenation without hydrolysis of the phenylamide linkage was not determined. Subsequent investigations (Kaufman et al., 1971) with propachlor indicate that dehalogenation to hydroxypropachlor is a mechanism of microbial degradation for this molecule. Thus the possibility for dehalogenation without hydrolysis of the phenylamide linkage cannot be discounted.

Fusarium oxysporum was selected for further investigation to determine its ability to metabolize a more inclusive group of aniline-based or carbamate pesticides (Table 5). This organism possessed the ability to degrade a large number of carbanilate, acylanilide,

TABLE 3. SOIL MICROORGANISMS ISOLATED FROM SOIL ENRICHMENT CULTURES CONTAINING PESTICIDES

Organism	Enrichment cultures from which organisms were obtained			
	Propham	Propanil	Solan	Swep
Aspergillus ustus		×		
A. versicolor		×	×	
Fusarium oxysporum	×	×	×	×
F. solani		×	×	
Penicillium chrysogenum	×	×	×	
P. janthinellum	×	×	×	
P. rugulosum		×	×	
Trichoderma viride	×	×	×	
Pseudomonas striata	×	×	×	×
Achromobacter sp.	×		×	

phenylurea, acetamide and dithiocarbamate pesticides. This metabolism led to some detoxication of the pesticides when oat seedlings were used as a bioassay plant for culture solutions added to soil. No attempt was made in this investigation to establish identity of all metabolic products detected on t.l.c. plates. The corresponding anilines were detected from most aniline-based pesticides. The monoalkylated derivative (DMU) of diuron was detected in culture solutions initially containing diuron.

Extensive release of Cl^- occurred during the metabolism of solan, CDAA and propachlor by *F. oxysporum* (Tables 4–5, Fig. 4). 3-Chloro-*p*-toluidine was positively identified as a product of solan metabolism through co-chromatography in both gas–liquid and t.l.c. systems. This moiety was also rapidly degraded with the liberation of Cl^- (Fig. 4). The degradation of propachlor by *F. oxysporum* has been examined in detail and reported elsewhere (Kaufman et al., 1971). The major route of degradation involved dehalogenation and the formation of hydroxypropachlor, thus hydrolysis of the amide linkage was not necessary for degradation to occur. Hydrolysis of the amide linkage of CDAA and propachlor would result in the formation of monochloroacetate (MCA) and diallylamine or isopropylaniline, respectively. *F. oxysporum* was capable of dehalogenating MCA (Fig. 4). Whether dehalogenation of CDAA occurred with or without hydrolysis of the amide

TABLE 4. EFFECT OF SEVERAL SOIL FUNGI ON SELECTED ANILINE-BASED PESTICIDES

Organism	Substrate	Aniline moiety (%) detected on day			Chloride ion (%) detected on day 7
		3	5	7	
Aspergillus ustus	Chlorphenamidine	0	2·8	19·5	4·7
	Chlorpropham	0	0	0	5·7
	Dicryl	0	5·0	13·5	13·0
	Diuron	0	0	0	3·0
	Propanil	1·3	6·1	33·0	12·9
	Propachlor	0	0	0	22·2
	Propham	0	0	0	0
	Solan	0	0	0	0
Aspergillus versicolor	Chlorphenamidine	0	12·5	29·9	36·9
	Chlorpropham	0	0	0	0
	Dicryl	1·5	2·1	4·3	13·4
	Diuron	0	0	0	0
	Propanil	5·4	14·8	16·2	4·9
	Propachlor	0	0	0	0
	Propham	0	0	1·0	0
	Solan	1·7	5·9	11·9	72·6
Fusarium oxysporum	Chlorphenamidine	0	5·6	29·2	3·1
	Chlorpropham	0	0	0	39·2
	Dicryl	0	9·2	7·1	5·7
	Diuron	0	0	0	0
	Propanil	45·1	56·5	49·8	4·5
	Propachlor	0	0	0	29·9
	Propham	0	0	1·0	0
	Solan	0	18·6	25·4	5·7
Fusarium solan	Chlorphenamidine	0·7	1·4	1·7	13·9
	Chlorpropham	0	0	0	0
	Dicryl	9·9	30·5	32·7	6·1
	Diuron	0	0	0	0
	Propanil	8·8	36·0	38·7	0
	Propachlor	0	0	0	100·0
	Propham	0	0	2·4	0
	Solan	2·1	22·0	10·2	25·3
Penicillium chrysogenum	Chlorphenamidine	0	0	0	0
	Chlorpropham	0	0	0	0
	Dicryl	3·6	48·3	87·4	8·1
	Diuron	0	0	0	0
	Propanil	10·1	29·6	71·3	4·0
	Propachlor	0	0	0	0
	Propham	0	19·3	14·5	0
	Solan	49·2	57·6	60·2	11·8
Pencillium janthinellum	Chlorphenamidine	0	0	4·9	0
	Chlorpropham	1·7	0	0·8	0
	Dicryl	10·7	26·9	60·4	1·1
	Diuron	0	0	0	0
	Propanil	14·1	19·5	30·3	0
	Propachlor	0	0	0	5·8
	Propham	36·6	38·5	14·5	0
	Solan	0	25·4	29·7	1·7

TABLE 4—continued

Organism	Substrate	Aniline moiety (%) detected on day			Chloride ion (%) detected on day 7
		3	5	7	
Pencillium rugulosum	Chlorphenamidine	0	0	6·0	3·3
	Chlorpropham	0	0	0	0
	Dicryl	0	2·1	9·2	0·2
	Diuron	0	0	0	0
	Propanil	1·4	0	7·4	0
	Propachlor	0	0	0	0
	Propham	4·8	0	0	0
	Solan	2·6	25·4	41·5	0·3
Trichoderma viride	Chlorphenamidine	0	0	2·1	0
	Chlorpropham	0	0	0	0
	Dicryl	1·8	21·0	54·0	19·7
	Diuron	0	0	0	0
	Propanil	1·4	19·5	77·4	4·7
	Propachlor	0	0	0	34·5
	Propham	0	0	1·0	0
	Solan	0	6·4	19·9	8·8

TABLE 5. DEGRADATION OF PESTICIDES BY *F. oxysporum* IN 20-DAY INCUBATION PERIOD

Pesticide	Aniline moiety % detected at 20 days	Halide % detected at 20 days	Oat seedling bioassay, fresh weight: control %	Products detected on t.l.c. plates
Alachlor	0	22·0	ND*	+
Barban	0·5	31·8	37·0	+
Benefin	0	ND	100·0	−
CDAA	ND	100·0	96·3	+
CDEC	ND	23·6	14·8	−
Chlorobromuron	1·8	9·5	7·4	+
Chloroxuron	0	30·7	63·0	+
Chlorphenamidine	11·8	13·9	88·9	+
Chlorpropham	0	11·0	0	+
Cycloate	0	ND	0	−
Diallate	ND	90·3	0	−
Dicryl	19·2	8·1	92·6	+
Diphenamid	ND	ND	0	−
Diuron	0	0	7·4	+
DMU	0·3	69·4	25·9	+
Fenuron	0	ND	11·1	+
Fluometuron	0	ND	7·4	−
Metobromuron	3·0	12·2	7·4	+
Neburon	0	26·7	96·3	+
NIA-11092	1·3	ND	7·4	+
Nitralin	0	ND	59·3	+
Norea	9·7	ND	14·8	+
Propachlor	0	100·0	88·9	+
Propanil	34·3	23·0	77·8	+
Propham	7·2	ND	0	+
Solan	0·9	100·0	100·0	+
Swep	0·7	5·8	22·2	+

* ND—Not determined.

linkage was not determined in the present investigation. Based on the results of this investigation and others (Kaufman et al., 1971) it seems probable that dehalogenation of CDAA or its products could occur either before or after hydrolysis of the amide linkage.

Both bacterial isolates were examined for their ability to degrade other phenylamide, acetamide and carbamate herbicides. The two organisms were similar in their substrate specificity. Both were most active on chlorpropham and propham and only slightly active on phenylureas (diuron, DMU, fenuron). Oat seedling bioassay of culture solutions of *P. striata* originally containing chlorpropham, dicryl, propham and solan indicated not only

TABLE 6. DEGRADATION OF HERBICIDES BY TWO BACTERIAL ISOLATES IN 6-DAY INCUBATION PERIOD

Herbicide	Organism	Aniline moiety % detected at 6 days	Halide ion % detected at 6 days	Oat seedling bioassay, fresh weight: control %
CDAA	1*	ND†	36·8	0
	2	ND	24·5	0
CDEC	1	ND	7·9	42·9
	2	ND	31·5	56·3
Chlorpropham	1	0	100·0	114·3
	2	0	100·1	93·8
Cycloate	1	0	ND	0
	2	0	ND	0
Dicryl	1	67·5	8·1	121·4
	2	56·8	12·1	93·8
Diuron	1	0	0	7·1
	2	0	4·1	3·1
DMU	1	0	0	6·3
	2	0	3·9	0
Fenuron	1	0	ND	7·1
	2	0	ND	6·3
Propachlor	1	0	0	67·9
	2	0	15·3	75·0
Propanil	1	55·2	0	89·3
	2	58·6	0	81·3
Propham	1	0	ND	107·1
	2	0	ND	100·0
Solan	1	0	67·6	110·7
	2	0	84·5	93·8
Swep	1	6·1	7·7	96·4
	2	8·8	31·0	93·8

* Organism 1 = *P. striata*; 2 = *Achromobacter* sp.
† ND—Not determined.

that detoxication had occurred, but that some growth promoting substances may have been produced. Although the exact nature of this growth stimulation was not characterized, the influence of various herbicides on production of growth-regulator type substances has been reported (Sobieszczanski, 1970).

The results of this investigation indicate that numerous acetamide, acylanilide, carbamate, toluidine and urea pesticides are biodegraded by a variety of soil microorganisms. Each of the soil microorganisms isolated demonstrated a unique range of substrate specificity, but all were capable of degrading and dehalogenating a variety of pesticides. Such results illustrate the adaptability and omnivorous nature of soil microbial populations with respect to certain chemical classes of pesticides.

REFERENCES

Bartha R. (1969) Transformation of solan in soil. *Weed Sci.* **17**, 471–472.
Bartha R. (1969) Pesticide interaction creates hybrid residue. *Science, N.Y.* **166**, 1299–1300.
Bartha R. and Pramer D. (1967) Pesticide transformation to aniline and azo compounds in soil. *Science, N.Y.* **156**, 1617–1618.
Bartha R. and Pramer D. (1969) Transformation of the herbicide methyl-N-(3,4-dichlorophenyl)–carbamate (swep) in soil. *Bull. Environ. Contam. Toxicol.* **4**, 240–245.
Chisaka H. and Kearney P. C. (1970) Metabolism of propanil in soils. *J. agric. Fd Chem.* **18**, 854–858.
Clark C. G. and Wright S. J. L. (1970a) Detoxication of isopropyl-N-phenylcarbamate (IPC) and isopropyl N-3-chlorophenylcarbamate (CIPC) in soil, and isolation of IPC-metabolizing bacteria. *Soil Biol. Biochem.* **2**, 19–26.
Clark C. G. and Wright S. J. K. (1970b) Degradation of the herbicide isopropyl N-phenylcarbamate by *Arthrobacter* and *Achromobacter* sp. from soil. *Soil Biol. Biochem.* **2**, 217–226.
Engelhardt G., Wallnofer P. R. and Plapp R. (1971) Degradation of linuron and some other herbicides and fungicides by a linuron-inducible enzyme obtained from *Bacillus sphaericus*. *Appl. Microbiol.* **22**, 284–288.
Geissbuhler H. (1969) The substituted ureas. In *Degradation of Herbicides* (P. C. Kearney and D. D. Kaufman, Eds.) pp. 79–111. Marcel Dekker, New York.
Iwasaki I., Utsumi S. and Ozawa T. (1952) New colorimetric determination of chloride using mercuric thiocyanate and ferric ion. *Bull. Chem. Soc. Japan* **25**, 226.
Kaufman D. D. and Kearney P. C. (1965) Microbial degradation of isopropyl N-3-chlorophenylcarbamate and 2-chloroethyl N-3-chlorophenylcarbamate. *Appl. Microbiol.* **13**, 443–446.
Kaufman D. D., Plimmer J. R. and Iwan J. (1971) Microbial degradation of propachlor. *American Chemical Society Abstracts 162nd Meeting*, Washington, D.C. A-21.
Kaufman D. D., Plimmer J. R., Iwan J. and Klingebiel U. I. (1972) 3,3',4,4'-Tetrachloroazoxybenzene from 3,4-dichloroaniline in microbial culture. *J. agric. Fd Chem.* **20**, 916–919.
Kearney P. C. and Kaufman D. D. (1965) Enzyme from soil bacterium hydrolyzes phenylcarbamate herbicides. *Science, N.Y.* **147**, 740–741.
Kearney P. C., Plimmer J. R. and Guardia F. B. (1969) Mixed chloroazobenzene formation in soil. *J. agric. Fd Chem.* **17**, 1418–1419.
Kearney P. C., Smith R. J., Plimmer J. R. and Guardia F. B. (1970) Propanil and TCAB residues in rice soils. *Weed Sci.* **18**, 464–466.
Linke H. A. B. and Bartha R. (1970) Transformation products of the herbicide propanil in soil: a balance study. *Bacteriol. Proc.* **70**, 9.
Pease H. L. (1962) Separation and colorimetric determination of monuron and diuron residues. *J. agric. Fd Chem.* **10**, 279–281.
Plimmer J. R., Kearney P. C., Chisaka H., Yount J. B. and Klingebiel U. I. (1970) 1,3-bis(3,4-dichlorophenyl)triazene from propanil in soils. *J. agric Fd Chem.* **18**, 859–861.
Rosen J. D. and Siewerski M. (1971) Synthesis and properties of 4-(3,4-dichloroanilino)3,3',4'-trichloroazobenzene. *J. agric. Fd Chem.* **19**, 50–51.
Sobieszczanski J. (1970) Influence of herbicides on the production of gibberellin-like substances by *Pseudomonas* sp. and *Arthrobacter* sp. In *International colloquim: Action des pesticides et herbicides sur la microflore et la faunule du sol. Biodegradation tellurique de leurs molecules.* (J. Pochon and J. P. Voets, Eds) *Mededelingen Faculteit Landbouw Wetenschappen*, Gent. Vol. 35, pp. 681–688.

Effect of Chemical Structure on Microbial Degradation of Substituted Benzenes

MARTIN ALEXANDER and B. K. LUSTIGMAN

Department of Agronomy, Cornell University, Ithaca, N. Y.

> The rate of degradation of mono- and disubstituted benzenes by soil microorganisms was determined by a spectrophotometric technique. Chloro, sulfonate, and nitro groups retarded the rate of biodegradation whereas carboxyl and phenolic hydroxyl groups favored decomposition of the substituted benzenes. The meta isomer was commonly the most resistant to attack by soil microorganisms, but the ortho isomer was the most resistant for certain classes of compounds.

The persistence of synthetic chemicals in natural environments, particularly in soil and water, is a problem of considerable concern. Pesticides, detergents, packaging materials, and industrial wastes may reside in a particular ecosystem or move through a number of environments because of the inability of microorganisms to degrade the unnatural compound at significant rates, if at all. Despite the concern about the lack of rapid biodegradation of many compounds that are potential or actual soil or water pollutants, there is surprisingly little information about the influence of chemical structure upon the microbial degradation of classes of synthetic compounds which are appearing in significant amounts as environmental pollutants (1).

Considerable attention has been directed to determining the influence of the structure of unsubstituted hydrocarbons on their residence times in natural environments, but few studies have been concerned with the effect of the type, number, or position of substituents on the rate of decomposition of organic compounds by a mixed microflora. An influence of meta substitution on the rate of degradation of chlorophenols and chlorophenoxyalkanoic acids has been noted (2, 3), and benzoic acid metabolism by a soil population was observed to be delayed by the introduction on the ring of a halogen substituent (11). Biological attack on the triazine ring appears to be retarded by the addition of amino groups to the molecule (6), phenylurea destruction in soil is slowed down by insertion of chlorine on the aromatic moiety (12), and the position of the sulfonate group alters significantly the disappearance rate of alkylbenzene sulfonates (13, 14).

The present investigation was designed as a systematic study of the effect of type, position, and number of substituents on the rate of decomposition of the benzene ring. To date, few such systematic studies have been made, except for the work of Kameda, Toyoura, and Kimura (9) with pure cultures of Pseudomonas strains. To study the biological potential for the degradation of synthetic chemicals and to avoid idiosyncrasies and physiological limitations of individual microbial strains, a mixed population of soil microorganisms was selected as the appropriate assay system.

Methods

The procedure employed, essentially the same as that used earlier with the chlorophenols and chlorophenoxyalkanoic acids (2), relies upon the loss of ultraviolet absorbancy when the benzene ring is cleaved by microorganisms derived from a soil inoculum. The solution contained the test compound as the sole carbon source to support microbial proliferation and had, in addition to the aromatic compound, 1.6 grams of K_2HPO_4, 0.40 gram of KH_2PO_4, 0.50 gram of NH_4NO_3, 0.20 gram of $MgSO_4 \cdot 7H_2O$, 25 mg. of $CaCl_2 \cdot 2H_2O$, 2.3 mg. of $FeCl_3 \cdot 6H_2O$ and 1000 ml. of distilled water.

The absorption spectrum of each compound was determined by dissolving it in a solution containing 0.16% K_2HPO_4 and 0.04% KH_2PO_4 and recording the ultraviolet absorbancy obtained with the Beckman spectrophotometer, Model DB. The wavelength selected for measuring the rate of degradation was at or near the absorption maximum for each of the test substances, except where no distinct peak was noted; under such circumstances, the wavelength chosen was one at which the light absorption was sufficiently high for convenient use.

Forty-milliliter aliquots of the medium were placed in 4-oz. screw-cap bottles, 45 mm. diameter × 80 mm. high, and these were inoculated with 1.0 ml. of a 1% suspension of Niagara silt loam. A parallel series of reaction vessels was set up identical to the first except that each bottle also contained 8 mg. of $HgCl_2$ and $5 \times 10^{-7}M$ Tween 80. Readings were made on these flasks at the same time intervals. Another series identical to the first was set up to determine if the chemicals at the concentrations employed were toxic to the microflora; these vessels received glucose to a final concentration of 1%, and growth in the tubes was recorded visually. The bottles were incubated in the dark at 25° C.

At intervals of 3 to 6 hours and at 1, 2, 4, 8, 16, 32, and 64 days after inoculation, the solutions were mixed, an aliquot was removed, and the suspension was centrifuged 10 minutes at 825 × G. The absorbancy of the supernatant was read at the selected wavelength against the supernatant from the reaction vessel containing a soil-medium mixture free of the chemical but incubated in an identical fashion. The liquid and the soil residue were returned to the bottles and incubated further. Readings were made on the Beckman spectrophotometer, Model DU. All reaction vessels were set up in duplicate, and absorbancies measured on the replicates at each sampling period.

After selecting the appropriate wavelength for each of the test solutions, the concentration of chemical added to the soil-medium mixture was adjusted to give an absorbancy of 0.2 to 0.6 when read against the soil-medium mixture containing no aromatic compound. The wavelengths and chemical concentrations employed are presented in Table I.

Results

For the sake of brevity, the data on degradation rates are presented as the

Effect of Chemical Structure   411

Table I. Wavelength and Concentration of the Substrates Employed in Decomposition Studies

Second Substituent		First Substituent[a]							
Type	Position	COOH	OH	NO_2	NH_2	OCH_3	SO_3H	Cl	CH_3
None		250 (25)	269 (25)	266 (5)	283 (10)	270 (25)	264 (100)	...	...
COOH	o	250 (25)	300 (25)	267 (10)	311 (10)	280 (50)	269 (100)	250 (25)	245 (25)
	m	245 (20)	250 (25)	267 (10)	304 (10)	289 (25)	273 (200)	250 (25)	250 (25)
	p	260 (20)	270 (25)	274 (8)	240 (5)	253 (5)	275 (100)	260 (25)	260 (25)
OH	o		276 (25)	275 (15)	280 (25)	276 (15)	...	274 (25)	272 (10)
	m		275 (25)	272 (10)	285 (25)	274 (15)	...	274 (25)	273 (10)
	p		...	400 (5)	...	288 (15)	260 (75)	279 (25)	278 (10)
NO_2	o			263 (10)	285 (10)	282 (10)	250 (25)	260 (10)	265 (10)
	m			250 (5)	280 (10)	274 (10)	263 (10)	265 (10)	...
	p			268 (5)	380 (5)	318 (10)	267 (10)	283 (10)	285 (8)
NH_2	o				294 (10)	295 (25)	295 (25)	287 (15)	265 (25)
	m				293 (25)	285 (20)	294 (30)	287 (20)	275 (25)
	p				...	295 (25)	270 (25)	285 (25)	285 (30)
OCH_3	o					273 (15)	...	...	...
	m					273 (15)	...	...	...
	p					289 (15)	...	...	...
SO_3H	m						269 (200)	...	...
	p						...	266 (20)	263 (200)

[a] The first number represents the wavelength. The number in parenthesis indicates the substrate concentration in μg./ml.

Table II. Decomposition of Monosubstituted Benzenes by a Soil Microflora

Compound	Substituent	Decomposition Period, Days
Benzoate	COOH	1
Phenol	OH	1
Nitrobenzene	NO_2	>64
Aniline	NH_2	4
Anisole	OCH_3	8
Benzenesulfonate	SO_3H	16

time interval, in days, necessary for the absorbancy in the supernatant of the soil-medium mixture containing the chemical to fall essentially to the level found in the reaction vessels having none of the test substance. In most instances, a small quantity of ultraviolet-absorbing materials remained for some time after this period, possibly a result of the microbial formation of other ultraviolet-absorbing metabolites. The designation >64 indicates that significant ring cleavage was not detected even on the 64th day.

A marked influence of chemical structure on biodegradation rates was noted with the monosubstituted benzenes, all of which were soluble in the medium at the concentration used (Table II). Phenol and benzoate were degraded rapidly, aniline and anisole were attacked less readily, and benzenesulfonate and nitrobenzene appeared to be the most resistant to decomposition by the mixed population. The observed loss of the compounds was considered to be a result of biological activity because the absorbancy was still high in the vessels containing $HgCl_2$ when aromatic ring cleavage was complete in flasks containing no inhibitor. Moreover, the long persistence of nitrobenzene or benzenesulfonate, or aromatic products formed from them, could not be attributed to any significant suppression of microbial activity, as suggested by the lack of inhibition of glucose breakdown by these compounds. Thus, it appears that the hydroxyl and carboxyl groups are the most favorable and the sulfonate and nitro substituents are least favorable to microbial degradation of benzene rings containing only a single substituent.

An effect of chemical structure on biodegradability is also evident among the disubstituted benzenes (Table III). An effect of type and position of the substituent is readily apparent. Not one of the 13 sulfonates and none of the 13 chloro-substituted compounds were destroyed in less than 2 weeks, and most persisted for periods in excess of 64 days; the only two of this group which had been metabolized in 16 days were para - substituted—namely, p - chlorophenol and p-chlorobenzenesulfonic acid. Nitro compounds were also quite difficult to degrade, and the ultraviolet absorbancy of all solutions was retained for periods in excess of 2 months, except for the compounds also containing a carboxyl or phenolic hydroxyl.

Most of the substituted anilines were not suitable substrates under the test conditions. The resistance of the anilines was rather surprising because amino compounds are universal cellular constituents, by contrast with nitro, chloro, and sulfonate compounds. Seven of the 15 anisoles were also largely inert when exposed to the mixed soil population, but six of the seven contained either a nitro or an amino substituent. Similarly, although four of the 13 toluenes had not been destroyed within a 64-day period, the four resistant molecules possessed either a nitro or a sulfonate group. Thus, as a first approximation, it seems that sulfonate, chloro, and nitro substitution tends to increase the resistance of these chemicals to biodegradation, as previously observed in two instances with the monosubstituted benzenes.

On the other hand, the amino and methoxy groups either exert a retarding influence or, alternatively, they do not possess a marked enhancing effect on the rate of degradation.

In marked contrast are the phenols and benzoic acids. In each class of compounds except the chlorobenzenes, ring cleavage was most rapid when the molecule contained a carboxyl or phenolic hydroxyl. This generalization holds regardless of the second substituent, and is in agreement with the findings with the monosubstituted benzenes.

Not only is the type of substituent of importance in conferring resistance or susceptibility to the aromatic ring but also its position. Most commonly, it is the meta isomer which is associated with the greatest resistance—e.g., in the phthalates, nitrobenzoates, aminobenzoates, methoxybenzoates, methoxyphenols, dihydroxybenzenes, aminophenols, and dimethoxybenzenes. Such an effect was noted earlier for several chloro compounds (2), although the meta effect with chlorophenols is not evident here. Greater resistance is associated with the ortho isomer only in the toluic acids, nitrophenols, and toluidines.

Some changes in addition to the loss in ultraviolet absorbancy were noted during the incubation. For example, the o-phenylenediamine-containing solution was yellow at the eighth day, and a color change was noted with p- and m-toluidine on the fourth and sixteenth days, respectively.

Biodegradation of a particular material will be evident by the method herein described when the test compound is metabolized by a prototroph which is capable not only of degrading the compound but also of using it as a carbon and energy source to sustain growth. However, evidence exists that microorganisms may destroy readily compounds which they cannot apparently

Table III. Decomposition of Disubstituted Benzenes by a Soil Microflora

Second Substituent		First Substituent[a]							
Type	Position	COOH	OH	NO_2	NH_2	OCH_3	SO_3H	Cl	CH_3
COOH	o	2	2	8	2	4	>64	>64	16
	m	8	2	>64	>64	16	>64	32	2
	p	2	1	4	8	2	>64	64	8
OH	o		1	>64	4	4	...	>64	1
	m		8	4	>64	16	...	>64	1
	p		...	16	...	8	32	16	1
NO_2	o			>64	>64	>64	>64	>64	>64
	m			>64	>64	>64	>64	>64	>64
	p			>64	>64	>64	>64	>64	>64
NH_2	o				>64	>64	>64	>64	64
	m				>64	>64	>64	>64	8
	p				...	64	>64	>64	4
OCH_3	o					8	...	...	...
	m					>32	...	...	...
	p					8	...	...	...
SO_3H	m						>64	...	...
	p						...	16	>64

[a] Values reflect days for total loss of ultraviolet absorbancy at the designated wavelength.

utilize as carbon and energy sources for growth—e.g., certain phenoxy compounds (10), aliphatic hydrocarbons (5), and benzoates (8). Supplemental available carbon will be required to permit proliferation of such organisms. To determine whether available carbon would enhance the decomposition of an apparently resistant disubstituted benzene, o-anisidine was selected because either it or an aromatic compound derived from it persists for more than 2 months. When glucose was added to the o-anisidine-containing solution, the ultraviolet absorption resulting from the benzene ring disappeared within 4 days. Presumably, the sugar served as an energy source for the population which was responsible for ring cleavage.

When $HgCl_2$ was included in most of the test solutions, there was little or no loss of ultraviolet light absorbancy at the designated wavelengths in the time interval required for biological destruction. With a few compounds, however, a slow loss of absorbancy was noted in the presence of the microbial inhibitor. Disappearance of these compounds may have resulted from their volatilization, chemical degradation, or from the development of organisms resistant to the mercury salt. For all compounds except one, however, either the rate of loss in the mercury-containing reaction mixtures was appreciably slower than in samples free of the inhibitor or no loss was detectable until some time after the ultraviolet absorbancy had disappeared from the mercury-free solutions. The sole exception was o-aminophenol, soil-inoculated solutions containing this compound losing their absorbancy at similar rates in the presence or absence of $HgCl_2$. However, if a second increment of o-aminophenol was added to solutions previously incubated 8 days with the same compound in the absence of the germicide, the absorbancy disappeared in about 4 hours, suggesting that microorganisms are capable of metabolizing the chemical and cleaving the benzene ring.

At the levels used, the disubstituted benzenes did not appear to alter the rate of microbial development in solutions containing glucose as the major carbon source. Although it is possible that the less readily degraded chemicals were selectively toxic to those species capable of destroying them, it is more likely that the failure to observe degradation is attributable to the inherent resistance to biodegradation of either the compounds themselves or aromatic products formed from them.

Discussion

The present report constitutes a systematic examination of the effect of structure of certain simple aromatic compounds on their susceptibility to degradation by a mixed population of soil microorganisms. The limitations of the techniques employed should be cited, however. A chief shortcoming is the possible unsuitability of the test conditions. For example, the active organisms may require growth factors, anaerobic conditions, solutions of different pH, or the presence of an organic compound suitable as an energy and/or carbon source. Moreover, the present assay technique makes use of a small soil inoculum, so selected to minimize interference by soluble aromatic substances derived from the soil, but inocula of larger size or those obtained from other soils or other ecosystems might contain biochemically active species degrading substrates herein found to be refractory. The method also neither reveals conversions of the parent compound to aromatic products nor indicates the specific aromatic structures which are recalcitrant, and the use of different substrate concentrations might have biased the results.

Despite these shortcomings, however, the data reveal a marked favorable effect of certain substituents, notably the carboxyl and phenolic hydroxyl groups, on microbial decomposition. The significance of position of substituent is also pronounced. Likewise, although larger soil inocula might have revealed the presence of microorganisms degrading some of the compounds, such prototrophs must be relatively rare by comparison with the strains degrading the short-lived chemicals such as the carboxylic acids and phenols.

The results also are in agreement with the few studies of pure cultures and mixed or natural populations. For example, Cartwright and Cain (4) experienced great difficulty in isolating organisms metabolizing m-nitrobenzoate, although strains active on o- and p-nitrobenzoate were found with ease. Investigations of 34 soil pseudomonads revealed that some were capable of utilizing o- and p-methoxybenzoate, o- and p-aminobenzoate and p-nitrobenzoate, but not one of the 34 was active on m-methoxy-, m-amino-, or m-nitrobenzoates (9). Likewise, aniline but not nitrobenzene is readily oxidized by a sewage microflora (7), and a phenol-adapted culture was found by Tabak, Chambers, and Kabler (15) to be capable of oxidizing dihydric phenols and cresols, but mono- or disubstituted benzenes containing nitro, chloro, or sulfonate groups were degraded slowly if at all by the bacteria.

The results have considerable bearing on the persistence of synthetic chemicals in natural environments. Pesticides, household and industrial wastes, and other materials are entering terrestrial and aquatic environments at ever increasing rates, and many are slowly degraded biologically in the environments into which they are introduced or through which they pass. The data presented herein demonstrate the significance of chemical structure of a variety of simple aromatic compounds on the susceptibility of these molecules to attack by a diverse microbial flora.

Literature Cited

(1) Alexander, M., *Advan. Appl. Microbiol.* **7,** 35 (1965).
(2) Alexander, M., Aleem, M. I. H., J. Agr. Food Chem. **9,** 44 (1961).
(3) Burger, K., MacRae, I. C., Alexander, M., *Soil Sci. Soc. Am. Proc.* **26,** 243 (1962).
(4) Cartwright, N. J., Cain, R. B., *Biochem. J.* **71,** 248 (1959).
(5) Foster, J. W., *Antonie van Leeuwenhoek J. Microbiol. Serol.* **28,** 241 (1962).
(6) Hauck, R. D., Stephenson, H. F., J. Agr. Food Chem. **12,** 147 (1964).
(7) Heukelekian, H., Rand, M. C., *Sewage Ind. Wastes* **27,** 1040 (1955).
(8) Hughes, D. E., *Biochem. J.* **96,** 181 (1965).
(9) Kameda, Y., Toyoura, E., Kimura, Y., *Kanazawa Daigaku Yakugakubu Kenkyu Nempo* **7,** 37 (1957); *C.A.* **52,** 4081 (1958).
(10) Loos, M. A., Alexander, M., Cornell University, Ithaca, N. Y., unpublished data, 1965.
(11) MacRae, I. C., Alexander, M., J. Agr. Food Chem. **13,** 72 (1965).
(12) Sheets, T. J., *Weeds* **6,** 413 (1958).
(13) Swisher, R. D., *Develop. Ind. Microbiol.* **4,** 39 (1963).
(14) Swisher, R. D., *J. Water Pollution Control Federation* **35,** 877 (1963).
(15) Tabak, H. H., Chambers, C. W., Kabler, P. W., *J. Bacteriol.* **87,** 910 (1964).

Received for review December 6, 1965. Accepted February 10, 1966. Work supported in part by a grant from the Division of Environmental Engineering and Food Protection of the U. S. Public Health Service (EF00547). Agronomy Paper No. 698.

Influence of Physicochemical Properties on Biodegradability of Phenylcarbamate Herbicides

PHILIP C. KEARNEY

Enzymic cleavage of the phenylcarbamate herbicides is influenced by physicochemical properties of the substrates. Inductive effects exerted by meta substitution of electron-withdrawing groups and steric effects imposed by increasing the size of the alcohol group significantly altered the reaction rate. A positive relationship between the relative acidities of the phenylcarbamates, as influenced by the inductive effects of meta substituents, and hydrolytic rates was demonstrated. Resonance interactions imposed by para substitution of strong inductive groups increased the velocity of the reaction. With the nitrophenylcarbamates, the rate of reaction was inversely proportional to the basicity of the parent anilines. Increasing the over-all size of the ring portion of the molecule decreased the velocity. Molecular parameters were studied in an attempt to determine those properties of a compound that influence decomposition by microbial enzymes.

Biologists and chemists are constantly seeking empirical relationships by which to predict chemical, physical, or biochemical properties of molecules on the basis of their structure. From a pesticide residue standpoint, it would be desirable to have some information on certain physicochemical properties of a herbicide as an indicator of its general soil persistence or its biodegradability by soil microorganisms. There have been a number of problems, however, that have made structure *vs.* persistence studies difficult to perform and interpret in soils. First, the herbicide's structural and electronic properties must be well understood before any reasonable basis for predicting its susceptibility to microbial attack can be established. Second, soils are complex systems in which more than one process may affect the persistence or degradation of a herbicide; and finally, the mechanism of the degradation reaction under study is often not well understood.

Owing to these complications, it has been impossible to correlate rates of disappearance with some physical property of the herbicide. Modern physical organic chemistry, however, provides a wealth of information for studying the molecular reactivity of many herbicides. For example, detailed information on physical properties of the substituted benzenes has been obtained in an attempt to evaluate the separate contributions of induction and resonance to the electronic effects of ring substituents. A detailed consideration of steric and electronic properties of certain herbicides might yield valuable information as to their rates of breakdown under laboratory conditions and their persistence under field conditions.

Recently, the isolation and identification of several soil microorganisms capable of metabolizing the widely used herbicide CIPC [isopropyl *N*-(3-chlorophenyl)-carbamate] was reported (Kaufman and Kearney, 1965). Subsequently, the isolation and purification of the enzyme within the organism responsible for catalyzing the hydrolysis of CIPC was described (Kearney, 1965; Kearney and Kaufman, 1965). The partially purified enzyme has the capacity to hydrolyze phenylcarbamates other than CIPC. The ability of the enzyme to cleave a large number of structurally related phenylcarbamates suggested an ideal situation for studying the effect of various molecular parameters on the enzymatic rate of hydrolysis under carefully controlled conditions. Most chemical decomposition studies of the carbamates have been done by alkaline hydrolysis. Rate constants for the alkaline hydrolysis of several carbamate insecticides, including both the *N*-alkyl and *N,N*-dialkylcarbamates, have been determined (Casida *et al.*, 1960). Detailed mechanistic studies have recently been carried out on ethyl, phenoxy, and *p*-nitrophenoxy esters of carbamic acid, *N*-methylcarbamic acid, and *N,N*-dimethylcarbamic acid (Dittert, 1961). Alkaline hydrolysis of several aromatic *N*-substituted carbamates has been investigated in strongly basic or in buffered solutions and rate constants determined at three or four different temperatures (Christenson, 1964). The present paper deals with the enzymatic rate of hydrolysis of several structurally related *N*-phenylcarbamates to determine effect of various ring substituents and alcohol groups on the velocity of reaction:

Methods

The enzymatic conversion of several phenylcarbamates to their respective anilines was measured as previously described (Kearney, 1965). The enzyme catalyzes the following reaction:

$$\text{Cl-C}_6\text{H}_4\text{-NH-CO-O-CH(CH}_3)_2 \longrightarrow \text{Cl-C}_6\text{H}_4\text{-NH}_2 + \text{CO}_2 + (\text{CH}_3)_2\text{CHOH} \quad (1)$$

The assay system used in the previous studies was modified to include some of the less soluble phenylcarbamates in the survey. The volume of the phenylcarbamate substrate solution was increased from 2.5 ml. (containing 1 μmole of substrate) to 4.0 ml. The increased volume facilitated the solubilization of many phenylcarbamates less soluble than CIPC. The phenylcarbamate substrates (Table I) were obtained from commercial sources and purified by recrystallization with appropriate solvent solutions. The infrared spectra of several phenylcarbamates were examined to determine the presence of the desired functional groups.

The enzyme used was the precipitate collected from a 30 to 60% $(NH_4)_2SO_4$ fraction of a crude soluble portion of harvested, lysed cells of *Pseudomonas striata*. The rate of the reaction refers to the number of millimicromoles of aniline produced per 20 minutes (at pH 8.0 in $0.1 M$ Tris buffer) with a constant number of enzyme units.

Standard curves were established for each of the respective substituted anilines by a colorimetric procedure (Pease, 1962). The identity of several of these anilines produced by the enzymatic reaction was verified as in the previous experiment (Kearney, 1965). Unfortunately, several of the corresponding anilines could not be obtained and thus precluded an examination of hydrolysis rates of the parent phenylcarbamate. For example, the isopropyl ester of N-(3-cyanophenyl)-carbamic acid yielded a positive aniline reaction, but 3-cyanoaniline was unavailable.

Several of the dichlorophenylcarbamates were extremely insoluble in water and could not be compared on an equal molar basis. Unfortunately, surfactants had an inhibitory effect on the enzyme, and consequently could not be used to solubilize several of the substrates.

Results and Discussion

The substrates were grouped into several classes to study the effect of certain molecular parameters on the rates of enzymatic hydrolysis. The steric effects, imposed by increasing the size of alcohol group, are shown in Table II. As the size of the alcohol group increases from ethyl to n-propyl to the isopropyl, the rate of the reaction steadily decreases. The bulky benzyl ester of the N-phenylcarbamic acid decreased the velocity by a factor of 2 when compared with the isopropyl ester of the same compound. The steric effects, imposed by the geometry of the alcohol groups, have a profound effect on the ease with which the carbonyl carbon in the molecule can be attacked by some reactive group on the enzyme surface. The unique spatial relationships, imposed by alcohol groups of varying size, suggest that the enzyme-substrate fit might be an important factor in determining the enzymatic reaction rate. Previously (Kearney, 1965), the rate of reaction was shown to decrease by introduction of chlorines into the isopropyl alcohol moiety. The reaction velocity was shown to decrease in the following order: isopropyl > 1-chloroisopropyl > 1,3-dichloroisopropyl.

The effect of various meta substituents on hydrolysis rates is shown in Table III. Since meta and para substituents are probably too far removed from the reaction site to have a noticeable steric effect on the

Table I. Phenylcarbamate Substrates Examined in Enzymatic Rate Studies

Ethyl N-(3-chlorophenyl)carbamate
Propyl N-(3-chlorophenyl)carbamate
Isopropyl N-(3-chlorophenyl)carbamate
Benzyl N-phenylcarbamate
Isopropyl N-phenylcarbamate
Isopropyl N-(2-chlorophenyl)carbamate
Isopropyl N-(3-nitrophenyl)carbamate
Isopropyl N-(3-acetylphenyl)carbamate
Isopropyl N-(3-methoxyphenyl)carbamate
Isopropyl N-(4-nitrophenyl)carbamate
Isopropyl N-naphthylcarbamate
Isopropyl N-(3-carboxyphenyl)carbamate

Table II. The Effect of Increasing the Size of the Alcohol Group on the Rate of Enzyme Hydrolysis of the Various Esters of N-(3-Chlorophenyl)carbamic Acid

Group	Rate
C_2H_5	80
$n\text{-}C_3H_7$	68
$iso\text{-}C_3H_7$	55
⬡—CH_2 [a]	13

[a] Hydrolysis rate for the benzyl alcohol group should be compared to the rate for the isopropyl N-phenylcarbamate (at 26 mμmoles) since both compounds are not chloro-substituted on the ring.

Table III. The Effect of Different Meta Substituents on the Rates of Enzyme Hydrolysis of the Isopropyl Esters of N-Phenylcarbamic Acid

Substituent	Rate
NO_2	65
$CH_3\text{—}CO$	60
Cl	56
$CH_3\text{—}O$	39
H	26

rate, the values presented in Table III apparently must reflect the inductive or electron-withdrawing effects. The reaction rate decreases with the following meta substituents on the ring: $NO_2 > CH_3\text{—}CO > Cl > CH_3\text{—}O\text{—} > H$.

There are several methods by which these inductive effects can be quantitatively compared to the reaction rates. One approach is to use the well known Hammett values. Another more direct approach is the use of relative acidities of the various substrates. Cluett (1959, 1962) reported that substituted phenylamides could be titrated as acids in n-butylamine. The method measures the ease with which the proton is removed from the amide nitrogen (Equation 2). Camper and Moreland (1965) measured the relative acidities of many of the herbicidal phenylcarbamates used in the present enzyme studies. Relative acidities of the phenylcarbamates are obtained by a potentiometric titration and the results expressed as half-neutralization potentials or HNP values.

$$\langle\!\!\!\!\bigcirc\!\!\!\!\rangle\!-\!\overset{H}{\underset{}{N}}\!-\!\overset{O}{\underset{}{C}}\!-\!R\ +\ (n\text{-butyl})_4N^+,\ OH^- \longrightarrow$$

$$\langle\!\!\!\!\bigcirc\!\!\!\!\rangle\!-\!\overset{-}{\underset{}{N}}\!-\!\overset{O}{\underset{}{C}}\!-\!R,\ (n\text{-butyl})_4N^+\ +\ HOH \quad (2)$$

Benzoic acid serves as a reference standard (at 500 mv.). As the HNP decreases, the phenylcarbamate becomes more acid. The technique is of interest here, since it expresses the contribution of various ring substitutions to the intrinsic acidity of many of the phenylcarbamates used in the enzyme studies.

A previous publication shows that as the HNP decreases, the enzymatic rate of hydrolysis increases (Kearney, 1965). The HNP effect on the hydrolysis of several phenylcarbamates not previously reported is shown in Figure 1. The inductive effect of the meta substituents correlates positively with HNP or relative acidity. The m-nitro carbamate is the strongest acid in the meta series, followed by the acetyl, chloro, methoxy, and hydrogen. The HNP value of the methoxy carbamate was an estimate based on the ethoxy-analog value previously reported (Camper and Moreland, 1965).

There was a general tendency for the hydrolysis rate to increase with increasing relative acidity. Two other phenylcarbamates that are not meta substituted, but which show a similar relationship between HNP and hydrolysis, are found in Figure 1. The isopropyl ester of 2-chlorophenylcarbamate is a stronger acid than CIPC; however, it does not follow the expected increase in hydrolysis rate. Only one ortho-substituted carbamate was available for examination in this survey, but the single observation suggests that steric hindrance may decrease enzymic cleavage because the ortho chlorine is in close proximity to the carbon undergoing attack. In the thiono carbamate, the sulfur exerts a strong influence on the amide hydrogen, making it relatively easy for the proton to leave. Again the general tendency for the hydrolysis rate to increase with increasing acidity was noted. The thionocarbamate was the most readily cleaved compound studied. Some abnormalities arise, however, when attempts are made to interpret a strict relationship between hydrolysis rate and HNP. Still unexplained, for example, was the inability of the enzyme to attack the CIPC analog bearing a meta-carboxyl substituent.

Positional effects of certain ring substituents are also important in governing the hydrolysis rate. The hydrolysis rates for the isopropyl esters of the m- and p-nitrophenyl carbamic acids are shown in Table IV. Hydrolysis of the p-nitrophenyl carbamate is considerably faster than for the meta compound. This was not particularly surprising, because a resonance interaction occurs when a strong electron-withdrawing group is substituted in the para position. The ability of the nitro group to pull electrons away from the reactive site of the molecule no doubt has an influence on the relative activity of the compound, and consequently on the ease with which it is hydrolyzed. Unfortunately,

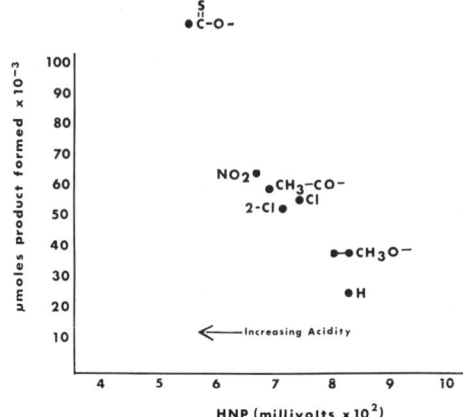

Figure 1. Rates of enzyme hydrolysis as a function of HNP for o-chloro and various meta-substituted isopropyl esters of the N-phenylcarbamic acids

Table IV. Rates of Enzyme Hydrolysis for the Isopropyl Esters of the m- and p-Nitrophenylcarbamic Acids

Structure	Rate
NO_2-C$_6$H$_4$-NH-CO-O-CH(CH$_3$)$_2$ (m-nitro)	65
NO_2-C$_6$H$_4$-NH-CO-O-CH(CH$_3$)$_2$ (p-nitro)	85

HNP values were not available for the p-nitro compound.

In p-nitrophenyl carbamates, the charge density is reduced in the vicinity of the carbonyl group by the strong electron-withdrawing power of the nitro group translated through the resonating structure.

The electron flow away from the active site is shown by a series of arrows denoting the shift in electrons away from the amide nitrogen through the aromatic system to the nitro group. In the case of the nitrophenyl carbamates, the velocity of hydrolysis is inversely proportional to the basicity of the parent aniline.

Thus far the inductive, steric, and resonance effects on the rates of hydrolysis have been discussed. The size of the molecule also has an effect on hydrolysis rate. The hydrolysis rates of the isopropyl esters of phenyl and 2-naphthyl carbamic acids were compared (Table V). The bulky naphthyl group has a profound decreasing effect on the hydrolysis rate, since only 7 mμmoles of 2-naphthylamine were produced. The

Table V. Enzymic Rates of Hydrolysis of Phenyl and 2-Naphthyl Isopropyl Esters of Carbamic Acid

Structure	Rate
Phenyl isopropyl carbamate	26
2-Naphthyl isopropyl carbamate	7

results strongly suggest that the actual size of the molecule may be the principal factor retarding the rate of hydrolysis.

Soil microorganisms are responsible for detoxifying many organic herbicides by reactions that, in a few cases, have been studied in detail (Kearney, 1966; Kearney et al., 1967). Ester and/or amide hydrolysis is the reaction responsible for cleavage of CIPC and its related analogs. The influence of steric hindrance on this type of reaction is fairly well understood (Newman, 1956), while the electronic effects caused by ring substitutes have been discussed previously for this same enzyme (Kearney, 1965). Although studies on the enzymic rates of hydrolysis of selected phenyl carbamates may be far removed from the actual conditions in soils, an examination of relative rates of hydrolysis of model compounds may serve as an indicator of the longevity of herbicides that are largely detoxified by soil microorganisms. Conclusions drawn from this type of study, however, may have broader implications outside of a consideration of the phenylcarbamates in terms of biodegradability. An understanding of the effects of various substituents in close proximity to the site of reaction of the pesticide molecule undergoing reaction may give a keener insight into the rapidity with which other soil-applied herbicides are decomposed.

Literature Cited

Camper, N. D., Moreland, D. E., *Biochim. Biophys. Acta* **94**, 383 (1965).
Casida, J. E., Augustinsson, K.-B., Jansson, G., *J. Econ. Entomol.* **53**, 205 (1960).
Christenson, I., *Acta Chem. Scand.* **18**, 904 (1964).
Cluett, M. L., *Anal. Chem.* **31**, 610 (1959).
Cluett, M. L., *Anal. Chem.* **34**, 1491 (1962).
Dittert, L. W., "The Kinetics and Mechanisms of the Base Catalyzed Hydrolysis of Organic Carbamates and Carbonates," dissertation, University of Wisconsin, Madison, Wis., 1961, *Dissertation Abst.* **22**, 1837-8 (1961).
Kaufman, D. D., Kearney, P. C., *J. Appl. Microbiol.* **13**, 443 (1965).
Kearney, P. C., *Advan. Chem. Ser.* **60**, 250–62 (1966).
Kearney, P. C., *J. Agr. Food Chem.* **13**, 561 (1965).
Kearney, P. C., Kaufman, D. D., *Science* **147**, 740 (1965).
Kearney, P. C., Kaufman, D. D., Alexander, M., Biochemistry of Herbicide Decomposition in Soils, pp. 318–42, in "Soil Biochemistry," A. D. McLaren, G. H. Peterson, Eds., Marcel Dekker, Inc., New York, 1967.
Newman, M. S., Ed., "Steric Effects in Organic Chemistry," p. 221, Wiley, New York, 1956.
Pease, H. L., *J. Agr. Food Chem.* **10**, 279 (1962).

Received for review February 1, 1967. Accepted June 1, 1967. Division of Agricultural and Food Chemistry, 152nd Meeting, ACS, New York, N.Y., September 1966.

The Effects of Temperature and Soil Water on Conversion of DDT to DDE in Soil[1]

W. D. Guenzi and W. E. Beard[2]

ABSTRACT

A laboratory study was conducted to determine the rates of DDT (1,1,1-trichloro-2,2-bis[p-chlorophenyl]ethane) degradation and DDE (1,1-dichloro-2,2-bis[p-chlorophenyl]ethylene) formation in soil. Degradation rates increased with higher temperatures and in the presence of water. Of the DDT mixed with Raber silty clay loam, 82.1, 74.5, 53.2, and 38.3% was recovered as DDT and 6.7, 12.5, 21.6, and 34.8% as DDE after 140 days incubation at 30, 40, 50, and 60C, respectively. A comparison of DDE formation in sterile and nonsterile soil showed that 84% of the conversion was due to a chemical process at 30C and 91% at 60C. In sterile systems at 30C, rates of DDE formation were similar in submerged soil and soil at 1/3 bar suction, and both were much higher than in air-dried soil.

Additional Index Words: pesticide, insecticide, chemical degradation, persistence, decomposition, microbial degradation.

The main degradation product of DDT (1,1,1-trichloro-2,2-bis[p-chlorophenyl]ethane) in soils under aerobic conditions is DDE (1,1-dichloro-2,2-bis[p-chlorophenyl]ethylene). DDT loses hydrogen chloride in strong basic solutions to form DDE, and qualitative and quantitative procedures have been developed utilizing this reaction. Research has shown that the conversion of DDT to DDE is catalyzed by heavy metals and certain minerals, especially at high temperatures. Fowkes et al. (1960) found that DDT rapidly decomposed at high temperatures in the presence of clay diluents with high surface acidity and that adding urea effectively reduced the decomposition rate. Downs et al. (1951) studied the degradation of DDT at 130C, and found that soils containing a high concentration of Fe were more effective in catalyzing the reaction than soils low in Fe. They concluded that the readily available iron oxide fraction was primarily responsible for the catalytic activity. The conversion of DDT to DDE is also catalyzed by anhydrous ferric and aluminum chlorides, iron, iron oxides, and certain minerals at 110 to 120C (Fleck and Haller, 1944). Lord (1948) increased DDT degradation by adding ferrous sulfate, ferric ammonium alum, copper sulfate, and manganese sulfate to basic acetone solutions at 30C. Birrell (1963) found that thermal decomposition of DDT in soil and clays at 112C in a nitrogen atmosphere resulted in the formation of DDE. In the presence of certain forms of iron oxides, a minor breakdown product, 4,4'-dichlorobenzil, was sometimes found.

Lopez-Gonzalez and Valenzuela-Calahorro (1970) found that as DDT diffused through clay minerals, a considerable amount degraded to DDE. They suggested that the degradation resulted from the interaction of DDT with active zones on the surface of homoionic clay minerals during diffusion through the pesticide-free clay. More DDT decomposed in a homoionic sodium clay than in a hydrogen clay, which they attributed to the higher pH in the sodium system. In a study using sterile glass microbeads, Smith and Parr (1972) showed that DDT was stable up to pH 12, with very little converted to DDE at pH 12.5, but considerable converted at pH 13. Conversion of DDT to DDE was very slow in nonamended, moist soil in the pH range 4 to 6.8, but increased rapidly in the 6.8 to 7.5 pH range (Nash et al., 1973). However, when MgO was added to increase pH, DDT was slowly converted to DDE as pH approached 9 and conversion increased rapidly above pH 9. They concluded that the main mechanism for DDT conversion was microbial at lower pH, and probably chemical at high (above 9) pH values when MgO was added and microbial activity was suppressed.

Previous studies have shown that DDT is converted to DDE in soil; but information is lacking on the mechanism, and the influence of temperature and soil water on the conversion process. Therefore, the objectives of this study were to determine the influence of temperature and soil water on the chemical and microbial conversion of DDT to DDE in soil.

MATERIALS AND METHODS

This study was designed to determine the rates of DDT degradation at 30, 40, 50, and 60C in a soil under aerobic conditions with water held at 1/3 bar suction during a 140-day period. In addition, a set of sterile samples was incubated at 1/3 bar water content at 30C, at air dry water content at 30 and 60C, and flooded at 30 and 60C. A comparable set of sterile samples at 1/3 bar water content at 60C was not included because the necessity of frequently replacing evaporated water during the 140-day incubation period prevented the maintenance of sample sterility. Soil was sterilized by autoclaving at 120C for 1 hour, left at room temperature for 2 days, and again autoclaved for 1 hour. Soils were checked for sterility before extraction by routine plating techniques. Each treatment was replicated three times.

The soil, Raber silty clay loam collected from native grass pasture, was air dried and ground to pass through a 2-mm sieve. The Raber soil is a Typic Argiustoll of the fine, mixed, mesic family. Table 1 gives some of the soil's physical and chemical properties.

Uniformly ring-labeled ^{14}C-p,p'-DDT and carrier p,p'-DDT were dissolved in n-hexane so that 2 ml of standard solution, which was added to 10 g of soil, contained 0.1 mg of DDT with a ^{14}C activity of 0.5 µCi. Solvent was evaporated from the samples by exposing them in a fume hood for 24 hours. One set of samples was wet with water to 1/3 bar suction, and others were submerged with water. Four glass vials, each containing a 10-g soil sample, were placed in a larger glass bottle for incubation. Each of the aerobic soil vials was weighed so that water evaporated during the experiment could be replaced. Water was added as required to keep the anaerobic samples submerged.

DDT was normally applied to cotton (*Gossypium hirsutum* L.) at a rate of approximately 1.1 kg/ha per application, and the number of applications during a growing season varied from 2 to 12, depending on the infestation problem. The concentration of DDT used in this study was equivalent to about 22.4 kg/ha, which represents about twice the concentration that would be applied to cotton, assuming 10 applications during the season.

Incubation bottles were continuously flushed with moist, CO_2-free air at a rate of 5 ml/min. Exhaust gases were first passed through a n-hexane trap and then a 1N KOH trap. Only very small

---

[1] Contribution from U. S. Dep. of Agric., Agric. Res. Serv., Western Region. P. O. Box E, Fort Collins, CO 80522, in cooperation with Colorado State Univ. Exp. Stn., Sci. J. Ser. No. 2049. Received 2 June 1975.

[2] Soil Scientist and Chemist, USDA.

Table 1—Selected physical and chemical properties of the Raber soil

Property	Value
pH (1:1, H_2O)	6.4
Clay, %	30
Organic matter, %	3.1
Surface area, m^2/g	175
CEC, meq/100 g	28
Extractable iron, % Fe†	0.74
Extractable aluminum, meq/100 g‡	0.70
Water content at:	
1/3 bar suction, %	32.9
air dry, %	2.4

† Dithionite-citrate extraction (Holmgren, 1967).
‡ Ammonium acetate extraction (McLean, 1965).

amounts of DDT or degradation products were detected by radioactivity measurements in the n-hexane traps or as $^{14}CO_2$ in the KOH traps, and these amounts were insignificant in relation to the objectives of this study.

Soils were extracted with n-hexane-acetone (41:59 vol/vol) solvent in a Soxhlet extractor at near 1/3 bar suction water content for 4 hours. Aerobic samples incubated at that water content were transferred directly to Soxhlet thimbles, water was added to the dry samples, and submerged samples were spread on aluminum foil, dried, and rewet to 1/3 bar before extraction. Quantitative data for specific compounds were obtained by gas-liquid chromatography (GLC) using an electron capture detector, and ^{14}C activity was determined by liquid scintillation. More detailed information on extraction and GLC analysis can be found in Guenzi and Beard (1970). Confirmation of degradation products was obtained from thin-layer chromatography (TLC). Individual compounds (spots) were removed from the two directional TLC plates by suction directly into scintillation vials and counted in a scintillation counter (Guenzi and Beard, 1968). Nonextractable ^{14}C activity was determined by dry combustion of the extracted soil (Beard and Guenzi, 1975).

Using both GLC and TLC, the following seven possible DDT degradation products could be detected: DDD (1,1-dichloro-2,2-bis[p-chlorophenyl]ethane), DDE, DDMU (1-chloro-2,2-bis[p-chlorophenyl]ethylene), DDMS (1-chloro-2,2-bis[p-chlorophenyl]ethane), DDNU (unsymbis[p-chlorophenyl]ethylene), DDC=O (4,4'-dichlorobenzophenone), and DDOH (1,1-bis[p-chlorophenyl]-2,2,2-trichloroethanol) Since DDE was the only degradation product found, a complete analysis by TLC for the other possible degradation products was conducted on only one replicate. DDT and DDE values given in the text are the means of three replicated samples that were determined by GLC analysis. Throughout the manuscript, DDE concentrations are expressed as a mole percentage of the DDT added.

RESULTS AND DISCUSSION

Mechanism

The conversion of DDT to DDE in a Raber soil was predominantly a chemical process with some microbial contribution. Sterile and nonsterile treatments were used to evaluate the magnitude of microbial and chemical processes. In addition, the influence of water in sterile systems was studied by using air dried, 1/3 bar, and flooded soils. The amount of DDE accumulation after 140 days was relatively small at 30C and represented only 6.7 (1/3 bar nonsterile), 5.7 (flooded sterile), 5.6 (1/3 bar sterile), and 0.4% (air dried sterile) of the DDT added (Fig. 1). An overall significantly greater amount ($P < 0.1$) was converted in the nonsterile 1/3 bar treatment than in the sterile 1/3 bar or the sterile flooded soils. There was no overall significant difference ($P > 0.01$) between the conversion rates in sterile 1/3 bar and sterile flooded treatments, but both were significantly greater ($P < 0.1$) than those in sterile air dried soil. At 60C, the same types of relationships existed but considerably more conversion resulted (Fig. 2). After 140 days, the amount of DDE formed was 34.8 (1/3 bar nonsterile), 31.6 (flooded sterile), and 17.0% (air dried sterile) of the applied DDT. Overall differences were significant between the 1/3 bar nonsterile and flooded sterile treatments ($P < 0.1$), and between the flooded sterile and air dried sterile treatments ($P < 0.01$) (Fig. 2).

Assuming the DDE recovery values in the 1/3 bar nonsterile soil after 140 days represented the total of both chemical and microbial contributions, then the difference between 1/3 bar nonsterile and 1/3 bar sterile at 30C should reflect the microbial component. At 60C, the difference between 1/3 bar nonsterile and flooded sterile was used to evaluate the microbial component. This is true assuming that sterilization produced no chemical changes in the soil that influenced the results. Based on these assumptions, 16.4% of the conversion at 30C was due to microbial degradation and 83.6% due to a chemical mechanism. The microbial portion decreased to 9.2% and the chemical increased to 90.8% at 60C, which would be expected at the higher temperature.

Water had an influence on the rate of conversion in sterile soil systems. The accumulation of DDE in the air dried soil after 140 days at 30C amounted to only 0.4% of the applied DDT. However, in sterile 1/3 bar soils or flooded soils, DDE accumulations were 5.6 and 5.7% under the same conditions. Trends were similar at 60C, but a considerably larger amount of DDE was formed at 140 days in the air dried sterile soil (17.0%) and in the flooded sterile soil (31.6%).

The chemical conversion of DDT to DDE was influenced by water, temperature, and the physical and chemical soil properties. Differences among soils have been shown by Guenzi and Beard (1970). A logical assumption is that the conversion reaction in soils occurs at active sites (Lopez-Gonzalez and Valenzuela-Calahorro,

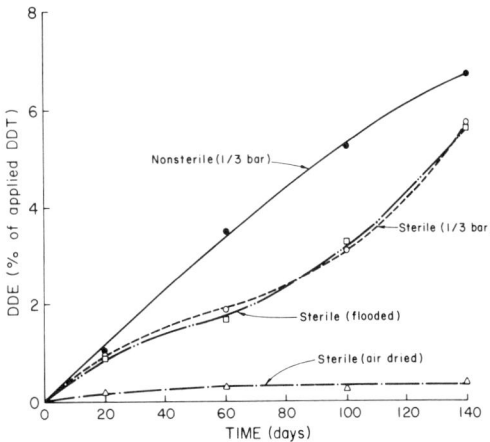

Fig. 1—Conversion of DDT to DDE in sterile and nonsterile soil at 30C.

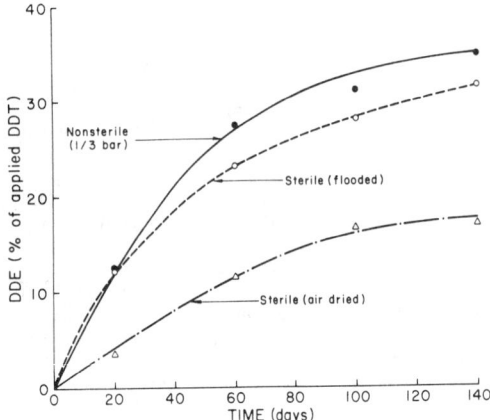

Fig. 2—Conversion of DDT to DDE in sterile and nonsterile soil at 60C.

1970) or more specifically, on the iron oxides (Downs et al., 1951; Birrell, 1963). Since Fe and Al oxides may exist as positively charged sites and the DDT molecule possesses some electronegative character due to the three chlorine atoms on the β carbon, this could be the site for an adsorbed catalytic reaction. The role of water may be to provide the medium for diffusion of molecules to active sites. Our data showed that conversion depended more on the presence of water at 30C than at 60C. Although this study was not designed to elucidate the mechanism of catalytic degradation of DDT, we feel that it provided insights which could be used in future studies with pure systems to provide better information on the mechanism.

Degradation-Formation Rates

The degradation of DDT and the formation of DDE in a Raber soil at field capacity were related to length of incubation and soil temperature (Table 2). After 140 days, 82.1, 74.5, 53.2, and 38.3% of DDT was recovered at 30, 40, 50, and 60C, respectively. Although we analyzed for several known DDT degradation products, DDE was the only detectable product found in the acetone-hexane extract. DDE concentrations increased as duration of incubation and temperature increased. After 140 days, 6.7, 12.5, 21.6, and 34.8% DDE was formed at 30, 40, 50, and 60C, respectively.

Curve fitting by computer analysis yielded the following equations describing DDT degradation and DDE formation at 1/3 bar suction as functions of temperature and time for the Raber soil:

DDT degradation ($r^2 = 0.97$)

$$y = 96.17 - 2.0391 T^2 D \times 10^{-4} + 7.0183 T^2 D^2 \times 10^{-7} - 2.0141 T^3 \times 10^{-5}$$

DDE formation ($r^2 = 0.97$)

$$y = -0.65 + 1.5924 T^2 D \times 10^{-4} - 6.2134 T D^2 \times 10^{-5} + 8.8411 D^3 \times 10^{-6}$$

where T = temperature in °C, D = time in days, and y = % degradation (DDT) or formation (DDE).

An average of the ratios of DDT degradation to DDE formation for the 30, 40, and 50C treatments indicated that for every two molecules of DDT degraded, approximately one molecule of DDE was recovered. This ratio was relatively constant near two for both the short and long incubation periods, indicating that the amount of DDE degradation was very small. If DDE degradation was very fast, the ratio should have increased with length of incubation. The consistency of this ratio over a range of three temperatures and four sampling periods suggested that two molecules of DDT may be involved in an intermediate which yields only one molecule of DDE. Obviously, this is an over-simplified approach to a mechanism which probably also involves interactions with the soil inorganic fraction. At 60C, the DDT to DDE ratio averaged 1.4, which would require a higher yield of DDE from the proposed intermediate.

Recovery

Total recovery of applied DDT was not obtained from the sum of residual DDT and DDE by GLC analysis of the soil extract (Table 2). The percent recovered decreased as temperature and length of incubation increased. At 30C, when little degradation occurred, the recovery based on both compounds (DDT and DDE) was relatively high and

Table 2—Recovery of C-14 activity, DDT, and DDE from soil incubated at 1/3 bar suction

Time of incubation	Temperature	Recovery of			
		Specific compounds†		^{14}C Activity	
		DDT	DDE	Extract	NE‡
Days	°C	% of applied DDT			
0	30	93.5 a§	0	94.9 a	4.8 b
	40	95.2 a	0	94.2 a	1.3 a
	50	95.2 a	0	95.6 a	2.4 a
	60	92.3 a	0	95.4 a	1.7 a
20	30	91.0 a	1.0 a	94.6 a	3.7 a
	40	90.3 a	4.3 b	92.6 a	4.3 a
	50	84.2 a	5.8 b	92.8 a	5.8 b
	60	77.3 b	12.4 c	91.9 a	5.8 b
60	30	85.6 a	3.5 a	93.8 b	5.4 a
	40	77.0 b	7.7 b	96.8 a	5.7 a
	50	69.6 c	11.6 c	93.5 b	6.6 ab
	60	55.5 d	27.5 d	85.8 c	7.4 b
100	30	84.6 a	5.2 a	96.4 a	5.7 a
	40	74.9 b	8.4 a	92.7 b	5.5 a
	50	62.7 c	17.6 b	92.6 b	7.0 b
	60	44.4 d	31.2 c	88.5 c	6.6 ab
140	30	82.1 a	6.7 a	98.4 a	5.6 a
	40	74.5 b	12.5 b	92.9 b	6.1 a
	50	53.2 c	21.6 c	94.7 b	8.2 b
	60	38.3 d	34.8 d	83.3 c	5.9 a

† Compounds detected by gas liquid chromatography.
‡ Nonextractable ^{14}C activity determined by dry combustion of the extracted soil sample.
§ Values in the same column for each sampling date which have a letter in common do not differ in character measured at the 5% probability level (Tukey's test).

ranged from 92.0% after 20 days to 88.8% after 140 days. However, at 60C, the percent recovered ranged from 89.7% after 20 days to 73.1% after 140 days, indicating that other degradation products were created even though none were identified or observed as unknown peaks on the GLC chromatogram.

Essentially all of the applied radioactivity could be accounted for in the soil extract and in the combusted soil after extraction (Table 2). Evidently, decomposition products other than DDE were relatively nonpolar, since generally more than 90% of the activity was found in the acetone-hexane extract. If polar or acidic compounds (insoluble in the acetone-hexane solvent) were formed during the degradation process, activity in the nonextractable fraction would have been expected to increase during incubation. However, the amount of nonextractable activity remained somewhat constant (4 to 8%) after successive incubation periods

CONCLUSIONS

Since the results from this study indicated that the conversion of DDT to DDE in a Raber soil was predominantly a chemical process, then future studies should be directed toward identifying the physical and chemical properties in relation to the reaction mechanism. The influence of temperature and soil water content was shown in this study and additional information on the active site or sites would be needed to develop a predictive model for describing the conversion reaction of DDT to DDE in a variety of soils.

LITERATURE CITED

1. Beard, W. E., and W. D. Guenzi. 1975. Microcombustion method for the determination of carbon-14-labeled pesticide residues in soil. Soil Sci. Soc. Am. Proc. 39:63-65.
2. Birrell, K. S. 1963. Thermal decomposition of DDT by some soil constituents. N. Z. J. Sci. 6:169-178.
3. Downs, W. G., E. Bordas, and L. Navarro. 1951. Duration of action of residual DDT deposits on adobe surfaces. Science 114:259-262.
4. Fleck, E. E., and H. L. Haller. 1944. Catalytic removal of hydrogen chloride from some substituted α-trichloroethanes. J. Am. Chem. Soc. 66:2095.
5. Fowkes, F. M., H. A. Benesi, L. B. Ryland, W. M. Sawyer, K. D. Detling, E. S. Loeffler, F. B. Folckemer, M. R. Johnson, and Y. P. Sun. 1960. Clay-catalyzed decomposition of insecticides. J. Agric. Food Chem. 8:203-210.
6. Guenzi, W. D., and W. E. Beard. 1968. Anaerobic conversion of DDT to DDD and aerobic stability of DDT in soil. Soil Sci. Soc. Am. Proc. 32:522-524.
7. Guenzi, W. D., and W. E. Beard. 1970. Volatilization of lindane and DDT from soils. Soil Sci. Soc. Am. Proc. 34:443-447.
8. Holmgren, G. G. S. 1967. A rapid citrate-dithionite extractable iron procedure. Soil Sci. Soc. Am. Proc. 31:210-211.
9. Lopez-Gonzalez, J. D., and C. Valenzuela-Calahorro. 1970. Associated decomposition of DDT to DDE in the diffusion of DDT on homoionic clays. J. Agric. Food Chem. 18:520-523.
10. Lord, K. A. 1948. Decomposition of DDT by basic substances. J. Chem. Soc. (London), Oct.-Dec., Pap. No. 335. p. 1657-1661.
11. McLean, E. O. 1965. Aluminum. In C. A. Black (ed.) Methods of soil analysis. Part 2. Agronomy 9:994-997. Am. Soc. of Agron., Madison, Wis.
12. Nash, R. G., W. G. Harris, and C. C. Lewis. 1973. Soil pH and metallic amendment effects on DDT conversion to DDE. J. Environ. Qual. 2:390-394.
13. Smith, S., and J. F. Parr. 1972. Chemical stability of DDT and related compounds in selected environments. J. Agric. Food Chem. 20:839-841.

Instantaneous Degradation of Parathion in Anaerobic Soils[1]

P. A. WAHID, C. RAMAKRISHNA, AND N. SETHUNATHAN[2]

ABSTRACT

In flooded rice (*Oryza sativa* L.) culture, the pesticides are applied to rice fields after several days of submergence when the soil is already in a reduced state. Parathion (*O,O*-diethyl *O,p*-nitrophenyl phosphorothioate) was, therefore, equilibrated with soils previously reduced by flooding with water and then analyzed by isotope technique or by gas-liquid chromatography. Instantaneous surface-catalyzed degradation of parathion occurred when the insecticide was shaken for as little as 5 sec with soils prereduced by flooding. Aminoparathion was the major product of this reaction. The interaction of parathion with prereduced soil appeared to be mediated by soil enzymes and/or other heat-labile substances produced by soil anaerobiosis.

In flooded rice (*Oryza sativa* L.) culture, pesticides reach the soil when broadcast as granules or incorporated in the root zone in capsules, and when sprayed on the foliage or the lower part of the leaf sheath. Since the first convincing demonstration of the instability of γ-hexachlorocyclohexane in predominantly anaerobic flooded soil (5), there has been considerable research devoted to the fate and behavior of pesticides in anaerobic systems (6, 7). Parathion (*O,O*-diethyl *O,p*-nitrophenyl phosphorothioate), an organophosphate of generally low persistence in soil and water systems, decomposes even more rapidly in flooded soils (8). In almost all reported studies on the persistence of pesticides in flooded soils, pesticides were applied at the time of flooding and then monitored for their disappearance. The major disadvantage of this procedure is that the pesticides are exposed to aerobic conditions for 1 week or more before anaerobiosis sets in, although in transplanted rice culture the soil is already in a reduced state when pesticides are applied to the standing crop. This study is concerned with the instantaneous degradation of parathion by soils prereduced by flooding.

MATERIALS AND METHODS

Soils

Three soils, an acid sulfate soil known as *pokkali* (no. 13), an alluvial soil (no. 10) from our Institute's experimental farm, and a laterite soil (no. 11), were used in the study. Some properties of these soils, shown in Table, were measured as described earlier (9).

Parathion

Technical grade (99.1%) nonlabeled parathion (Bayer Farbenfabriken, Leverkusen, Germany) and ^{14}C-ethoxy-1-labeled parathion (Radiochemical Centre, Amersham, England) with specific activity of 15.9 mCi/mmole were used in the study.

Prereduced Soil-Parathion Interaction

Air-dried soils, ground to pass through a 2-mm sieve, were prereduced by flooding as follows before their exposure to parathion: 10-g portions of the soils were placed in round-bottomed, screw-capped glass tubes (30 by 120 mm, 30 ml capacity) and then flooded with 10 ml of distilled water to provide a water column of about 20 mm. The soils were incubated under undisturbed flooded conditions at room temperature (28 ± 4°C) for 60 days to allow reduction. The reduction of the soil was confirmed by the redox potentials before equilibration with parathion.

The soils + water, prereduced by flooding for 60 days, were transferred from the test tubes to 250-ml Erlenmeyer flasks and 0.3 ml of ^{14}C-ethoxy-1-labeled parathion in acetone containing 5 μg of the insecticide was added. The flask was swirled by hand for 5 sec or equilibrated on a shaker for 30 min to allow reaction between the insecticide and the prereduced soils. Parathion was reacted with aerobic soils by flooding 10-g samples of the soils with 10 ml of distilled water just before equilibration with parathion.

To determine the relative role of soil and water in the instantaneous degradation of parathion by prereduced soils, 100 g of soil no. 13 was incubated with 100 ml of distilled water in glass bottles. After 60 days of flooding, the reduced soil and water were mixed and the resulting soil suspension was filtered through Whatman no. 42 filter paper. Fifty ml of the filtrate were equilibrated with ^{14}C-parathion in a 50-ml glass bottle for 30 min on a shaker.

To determine whether the instantaneous degradation of parathion by prereduced soil was chemical and/or biological, soils, prereduced by flooding for 60 days, were sterilized by autoclaving (121°C for 1 hour), or by irradiation with ^{60}Co at 2 or 4 Mrad or by treating with 100 or 500 ppm of sodium azide. Redox potentials of the soil suspension were measured before or after sterilization. Nonlabeled technical parathion in acetone (0.2 ml of 800 ppm, vol/vol) was added aseptically to nonsterilized and sterilized prereduced soils contained in glass tubes and the contents were shaken for 30 min. Soils treated with parathion just at the time of flooding served as aerobic controls.

Extraction and Analysis

In isotope studies, the residues of parathion and its degradation products in each of the two replicates were extracted by shaking once with 40 ml of 1:1 chloroform-diethyl ether for 45 min and two times with 30-ml portions of the same solvent mixture for 30 min each on an orbital shaker. After centrifugation of the soil suspension, the organic solvent fractions from three successive extractions were pooled. The organic solvent was evaporated at room temperature and the residues were dissolved in 2 ml of methanol. Thin-layer chromatographic separation of the residues was performed on 300 μm thick silica gel-G plates using hexane-chloroform-methanol (7:2:1, vol/vol) as the solvent system. Parathion from the silica gel areas of the samples opposite to the authentic standard was quantified by liquid scintillation after scraping the entire spot into liquid scintillator [5 g of PPO (2,4-diphenyl oxazole) and 0.3 g of POPOP [1,4-bis-(2,5-phenyl oxazolyl)-benzene] in 1 liter of toluene]. Radioactive zones on thin-layer plates were detected by autoradiography using Kodak X-ray no-screen film.

Table 1—Some physicochemical characteristics of the soils used.

Soil	pH†	Organic matter‡	Clay‡	Free Fe‡
		——————— % ———————		
Alluvial (no. 10)	6.2	0.75	15.6	0.83
Laterite (no. 11)	6.3	2.88	23.6	2.40
Pokkali (no. 13)	5.2	5.52	45.6	1.38

† 1:2.5 soil/water ratio.
‡ Measured as described earlier (9).

---

[1] Contribution from the Central Rice Res. Inst., Cuttack-753006, India. This work was supported, in part, by funds from the Dep. of Sci. and Technol., Government of India, and the Div. of Tech. Assistance Program of the Int. Atomic Energy Agency, Vienna, Austria. Received 2 Mar. 1979.

[2] Research Fellows (supported by a fellowship from the Dep. of Atomic Energy, Government of India) and Soil Microbiologist, respectively.

In studies (sterile vs. nonsterile soils, soils flooded for different lengths of time) using nonlabeled technical parathion, the residues of parathion and its degradation products in the soil samples were analyzed by gas-liquid chromatography. After equilibration with parathion, the soil suspensions contained in glass tubes were transferred to 250-ml Erlenmeyer flasks with the help of 5-ml aliquots of distilled water. The residues were extracted by shaking the soil suspension first with 30 ml of acetone + 20 g of anhydrous sodium sulfate for 15 min and then with 50 ml of hexane for 45 min on an orbital shaker. The residues of parathion and its degradation products partitioned into the hexane layer were analyzed in a Perkin Elmer gas chromatograph model 3920 equipped with flame photometric detector specific to phosphorus. The spiral glass column (6.25 mm O.D., 2 m length) was packed with 2% SE-30 on Gaschrom Q (60 to 80 mesh). The flow rates for argon (carrier gas), hydrogen, and air were 40, 70, and 180 ml per min, respectively. The injector, column, and detector were maintained at 220, 210, and 220°C, respectively.

Redox Potential

The soil, prereduced by flooding in a glass tube, was transferred to a 100-ml beaker. The redox potential of the soil suspension in duplicate samples was measured by a compound calomel-platinum electrode attached to a portable redox meter (TOA Electronics Ltd., Tokyo).

RESULTS AND DISCUSSION

Equilibration of parathion with prereduced pokkali soil, but not with aerobic soil, led to its extensive degradation (Fig. 1). The reaction appeared to be instantaneous since degradation occurred within 5 sec of equilibration with prereduced soil leading to the formation of three products carrying the ^{14}C label. Parathion decreased to 44.4 and 11.6% of the original level after 5 sec and 30 min of equilibration, respectively. More than 90% of the added parathion was recovered after equilibration with aerobic soil. The redox potential of the prereduced soil was not significantly affected by shaking for 30 min since the potential increased from -200 mV to only -180 mV at the end of 30-min equilibration.

The extent of parathion degradation in soils as influenced by period of flooding and soil characteristics is shown in Fig. 2. Three soils (no. 10, 11, 13) were first flooded for different periods ranging from 0 to 190 days and then shaken with nonlabeled technical parathion for 30 min. Pokkali soil (no. 13) with the highest organic matter content was the most effective in catalyzing the degradation of parathion in terms of the rate of the reaction. Within 6 to 11 days of flooding, pokkali soil developed a remarkable capacity to degrade parathion as compared to the 20 to 40 days required for similar effects in the lateritic and alluvial soils. Following flooding, the redox potentials dropped at a faster rate in organic matter-rich pokkali soil than in the lateritic and alluvial soils with lower organic matter contents. The redox potential of pokkali soil at the end of 6 days of flooding was -130 mV as compared to a potential of -60 mV in the lateritic soil and $+30$ mV in the alluvial soil. This would probably explain the shorter flooding period required to produce detectable rates of parathion degradation with pokkali soil since the major pathway of its degradation, nitro group reduction, was presumably favored by low potentials.

Striking transformation of parathion within 5 sec of its equilibration with prereduced pokkali soil raises the possibility of its chemical instability during the extraction and analytical procedures employed. To examine this, organic extractant (chloroform-diethyl ether, 1:1, 40 ml) was mixed with prereduced soil (no. 13). ^{14}C-Parathion was added to this suspension and the residues were extracted immediately. Autoradiography revealed that added parathion remained unaltered during extraction and analysis. Clearly, this precludes the chemical decomposition of parathion during extraction and analysis.

In another study, parathion was equilibrated with the filtrate of a suspension of prereduced pokkali soil. The soil suspension before filtration and the clear filtrate showed potentials of -200 and -180 mV, respectively. The potential of the filtrate after equilibration with parathion was -120 mV. No appreciable degradation of parathion occurred in the filtrate and >97% of added insecticide was recovered at the end of 30-min equilibration with the filtrate. Certainly, reduced soil, but not water, was responsible for the degradation of parathion and the reaction was surface-catalyzed.

To ascertain whether instantaneous degradation of parathion in prereduced soils was chemical or biological, nonlabeled parathion was equilibrated with

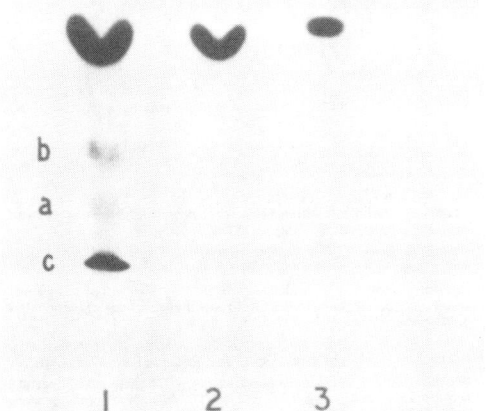

Fig. 1—Autoradiograph of parathion and its degradation products (a: aminoparathion; b: desethyl aminoparathion; c: unidentified) formed within 5 sec of its equilibration with prereduced pokkali soil. (1) prereduced soil; (2) aerobic soil; (3) standard parathion

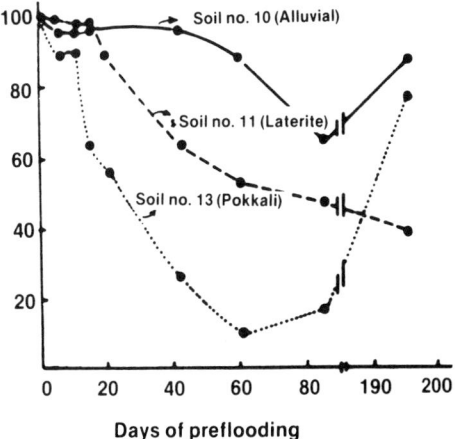

Fig. 2—Parathion recovered after 30-min equilibration with soils preflooded for different periods.

Table 2—Degradation of parathion after 30-min equilibration with sterilized and nonsterilized reduced soils.

Treatment	Parathion recovered		Redox potential‡	
	Soil no. 11	Soil no. 13	Soil no. 11	Soil no. 13
	%		mV	
Aerobic soil	95.0	92.4	+260	+200
Reduced† soil				
Nonsterile	52.1	13.8	−140	−200
Irradiated (2 Mrad)	68.1	46.9	−100	−150
Irradiated (4 Mrad)	63.1	43.6	−120	−160
Autoclaved	92.3	86.9	−120	−180
Sodium azide (100 ppm)	55.6	30.5	−160	−180
Sodium azide (500 ppm)	54.2	27.5	−140	−180

† Soils, prereduced by flooding for 60 days, were equilibrated with ^{14}C-parathion for 30 min.
‡ Soil potentials were measured after equilibration with parathion.

nonsterile and sterile samples of soils prereduced by flooding. Prereduced soils were sterilized by gamma irradiation, autoclaving, or sodium azide. Soil treated with parathion just at the time of flooding served as the aerobic control. In aerobic soils, parathion remained undegraded with recoveries as high as 92 to 95% from both the pokkali and lateritic soils after 30-min equilibration (Table 2). In contrast, rapid degradation of parathion occurred in nonsterile samples of prereduced soils. Its concentration declined to 13.8 and 52.1% of the original levels in pokkali and laterite soils, respectively. Among the sterilizing agents, autoclaving effected almost complete inhibition of the degradation of parathion by prereduced soils. Gamma irradiation was less effective while sodium azide was the least effective. The increase in the potentials of the prereduced soils upon sterilization by the agents used was negligible.

Gas chromatograms (Fig. 3) of parathion residues from sterile and nonsterile samples of prereduced soils yielded interesting results. Rapid destruction of parathion by prereduced soils led to the formation of two products, A and B in pokkali soil, and only A in the lateritic soil; in aerobic soils, no degradation product was detected. Product A was identified as aminoparathion and product B as desethyl aminoparathion (P. A. Wahid. 1978. Behavior of pesticides in soils. Ph.D. Thesis. Utkal Univ., Bhubanewar, India) by cochromatography with authentic compounds. Among the sterilizing agents, autoclaving completely inhibited the formation of both products, irradiation inhibited the formation of product B, and sodium azide was not inhibitory to the formation of either. Thus, the degradation of parathion by prereduced pokkali soil was more pronounced and extensive in irradiated soil than in autoclaved soil. Product C, detected in the autoradiograph of residues from nonsterile samples of prereduced pokkali soil (Fig. 1), but not in the gas chromatograms (Fig. 3), was not identified.

Autoclaving is a drastic means of sterilization and destroys not only the microbial activity but also changes several chemical and physical characteristics of the soil (1). Most of the enzymatic and biochemical activities would continue in irradiated, but not in autoclaved, soil (3, 4). Thus fairly rapid destruction of parathion by irradiated prereduced soil could be ascribed to the involvement of enzymes. Other heat-labile soil substances may also be implicated as reported in the nonenzymatic degradation of malathion under aerobic soil conditions (2). The role of soil enzymes and/or other heat-labile soil substances in the degradation of soil-applied pesticides merits more intensive study.

A — Aminoparathion (R_t = 2.9)
P — Parathion (R_t = 3.5)
B — Desethyl aminoparathion (R_t = 4.0)

Treatment:
Aerobic (soil nos. 11 and 13)
Anaerobic (soil no. 13)
Anaerobic (soil no. 11)
Autoclaved anaerobic (soil nos. 11 and 13)
Irradiated anaerobic (soil nos. 11 and 13)
Sodium azide treated anaerobic (soil no. 13)
Sodium azide treated anaerobic (soil no. 11)
Aminoparathion standard
Desethyl aminoparathion standard

Retention time (min)

Fig. 3—Gas chromatograms of parathion and its products formed in nonsterile aerobic soil and in nonsterile and sterile samples of prereduced (anaerobic) soil after 30-min equilibration.

ACKNOWLEDGMENTS

The authors are grateful to Dr. H. K. Pande, Director for facilities, and to Dr. K. Raghu, Biology and Agriculture Centre, Bombay, for irradiating the soil samples.

LITERATURE CITED

1. Eno, C. F., and H. Popenoe. 1964. Gamma irradiation compared with steam and methyl bromide as a sterilizing agent. Soil Sci. Soc. Am. Proc. 28:533-535.
2. Getzin, L. W., and I. Rosefield. 1968. Organophosphorus insecticides degradation by heat-labile substances in soil. J. Agric. Food Chem. 16:598-601.
3. McLaren, A. D., L. Rastetko, and W. Huber. 1957. Sterilization of soil by irradiation with an electron beam, and some observations on soil enzyme activity. Soil Sci. 83:497-501.
4. Peterson, G. H. 1962. Respiration of soil sterilized by ionizing radiations. Soil Sci. 94:71-74.
5. Raghu, K., and I. C. MacRae. 1966. Biodegradation of the gamma isomer of benzenehexachloride in submerged soils. Science 154:263-264.
6. Sethunathan, N. 1973. Microbial degradation of insecticides in flooded soil and in anaerobic cultures. Residue Rev. 47:143-165.
7. Sethunathan, N., and R. Siddaramappa. 1978. Microbial degradation of pesticides in rice soils. p. 479-497. In F. N. Ponnamperuma (ed.) Soils and rice. Int. Rice Res. Inst., Los Banos, Philippines.
8. Sethunathan, N., R. Siddaramappa, K. P. Rajaram, Sudhakar-Barik, and P. A. Wahid. 1977. Parathion: Residues in soil and water. Residue Rev. 68:91-122.
9. Wahid, P. A., and N. Sethunathan. 1978. Sorption-desorption of parathion in soils. J. Agric. Food Chem. 26:101-105.

ATRAZINE HYDROLYSIS IN SOIL

D. E. Armstrong, G. Chesters, and R. F. Harris

ABSTRACT

An important pathway of atrazine degradation in perfusion systems of three soils was chemical hydrolysis to hydroxyatrazine. Ultraviolet spectrophotometric analyses of the perfusates showed the presence and accumulation of hydroxyatrazine. Atrazine degradation followed first-order kinetics in soil-free, sterilized soil and perfusion systems. An increased rate of atrazine hydrolysis in an acid soil was consistent with the effect of pH on hydrolysis. No microbial degradation of atrazine was detected following inoculation of a soil-free atrazine medium with perfusates. An increased rate of hydrolysis in the presence of sterilized soil was postulated to result from soil adsorption of atrazine. Soil pH and organic matter content largely controlled the rate of atrazine hydrolysis; for soils of similar pH, atrazine degradation rates increased with increased atrazine adsorption.

MICROBIAL METABOLISM has been regarded generally as the major mechanism of triazine herbicide degradation in soil. Direct relationships between the rate of triazine degradation and soil temperature and organic matter content (1, 2, 17) have been explained largely on the basis of triazine degradation by microorganisms.

Recent experiments (3, 10) have shown that certain microbial isolates are capable of metabolizing the side chains but not the triazine ring of simazine. Investigations employing C^{14}-labeled traizines to detect microbial degradation in soil are somewhat contradictory. The evolution of $C^{14}O_2$ from soil amended with C^{14}-labeled simazine indicated that microbial metabolism of simazine had occurred (15, 16). However, appreciable evolution of $C^{14}O_2$ was not detected from soils treated with C^{14}-labeled simazine, atrazine, or propazine (12). MacRae and Alexander (12) suggested that the $C^{14}O_2$ evolved from ring-labeled triazines (15) may have originated from labeled contaminants rather than from the triazine ring.

Because of the persistence and accumulation of the triazines in soils, the possible importance of slow chemical degradation should not be overlooked (6, 13). The effect of soil properties on triazine degradation can be interpreted on the basis of chemical as well as microbial mechanisms. Chemical hydrolysis of the chlorotriazines is pH and temperature dependent (5) and may be affected by the presence of soil colloids (6, 13).

This paper describes an investigation of the relationship between the rate of atrazine degradation in soil and pH, organic matter, and clay content of the soil. Factors affecting the hydrolysis of atrazine in soil-free systems were also investigated to evaluate chemical hydrolysis as a possible mechanism of atrazine degradation in soils.

MATERIALS AND METHODS

The properties and classification of the three soils used in perfusion experiments are described in Table 1. Soil pH was determined on a thin aqueous paste by the glass electrode method. Organic matter contents were obtained by chromic acid oxidation (18) and clay contents by the centrifugation method (8).

The perfusion technique consisted of continuously recycling or perfusing a soil column with a basal microbial medium containing atrazine. Perfusion was conducted at room temperature under laboratory lighting conditions. The apparatus consisted of an Erlenmeyer flask (1 liter) over which a 3 by 25 cm glass column was mounted (11). Samples (100 g) of each of the three soils

---

[1] Contribution from the Department of Soils, Univ. of Wisconsin, Madison. Published with the approval of the Director, Wisconsin Agr Exp. Sta. This investigation was supported in part by PHS research grant No. WP-00751 from The Water Polution Division, Public Health Service. Presented before Div, S-3, Soil Science Society of America, Nov. 1, 1965, at Columbus Ohio. Received June 13, 1966. Approved Oct. 18, 1966.
[2] Project Associate, Associate Professor, and Assistant Professor of Soils, respectively.

were perfused with medium supplemented with 0, 5.3, 15.4, and 47.7 ppm atrazine. The basal medium contained K_2PHO_4, $CaSO_4$, $MgSO_4$ (each 0.2 g), NH_4NO_3 (0.3 g), and sucrose (0.1 g) per liter of solution (9). The perfusate from each soil column was sampled at intervals (usually weekly) and analyzed for atrazine.

Atrazine was determined by the pyridine-alkali-ethyl cyanoacetate method (14); standard curves were prepared from purified atrazine supplied through the courtesy of Geigy Chemical Corporation. Complete hydrolysis of atrazine to hydroxyatrazine was affected by treatment with 0.01N HCl at 90C for 5 days, as verified by total disappearance of the 223 mμ absorption maximum of atrazine and the appearance of the 240 mμ absorption maximum of hydroxyatrazine.

Attempts to obtain microbial enrichment cultures were made by inoculating the basal medium containing atrazine with samples removed from the perfusion systems and following the atrazine concentration in solution over a period of 1 to 2 months.

Hydroxyatrazine in perfusates was determined by ultraviolet spectrophotometry after treatment to remove atrazine and other interfering components as follows: The perfusate was centrifuged to remove any suspended material and an aliquot (5 ml) was transferred to a separatory funnel (30 ml) and acidified to pH 3 with HCl to protonate hydroxyatrazine to the cationic form. The solution was extracted three times with chloroform (3 ml each time) to remove atrazine. To concentrate the sample and remove the last traces of chloroform, the aqueous fraction was evaporated to approximately 3 ml and transferred to a chromatographic column (10 by 1 cm) containing a bed of the anion-exchange resin Dowex 21 K (5 cm) in the Cl form. The sample was leached with water, and the hold-up volume (1.5 ml) was discarded. The subsequent fraction (5 ml) was collected, and the absorbance was measured against a water blank at 240 mμ. The hydroxyatrazine solutions obeyed Beer's Law; absorbance was equivalent to 0.115/ppm hydroxyatrazine. Recovery of hydroxyatrazine from standard solutions averaged 95% with a range of 89 to 99%.

The rate of atrazine hydrolysis in buffered aqueous systems was followed at pH 1.3, 2.2 (buffered with HCl); 3.1 (0.01M citrate); 11.1 (0.05M carbonate); 11.9 and 12.9 (NaOH). The solutions were maintained at 25±2C for 4 weeks and their atrazine contents determined at weekly intervals.

The effect of soil particles on atrazine hydrolysis was investigated following the addition of atrazine to 0.01M phosphate buffer (100 ml) containing Poygan sil (50 g) to provide a pH of 3.9 and a concentration of atrazine of 10 ppm after equilibration. Sterilization was accomplished by boiling the buffered soil for 15 minutes prior to the addition of atrazine, and aseptic conditions were maintained throughout the experiment. The system was maintained on a temperature controlled (25±2C) reciprocating shaker; samples were withdrawn at 10-day intervals and analyzed for solution atrazine.

RESULTS

Atrazine Degradation in Soil Perfusion Systems

Atrazine concentration in solution decreased rapidly during the first few days as a result of adsorption by the soil. Adsorption of atrazine by the three soils increased in the order of their organic matter contents, namely, Kewaunee < Ella < Poygan (Table 2). Following adsorption, a slow but continued decrease in atrazine concentration was observed and attributed to degradation. Adsorption is generally rapid and was expected to reach equilibrium within a few days. Volatilization losses are believed to be negligible in aqueous systems (17). After 225 days of perfusion, degradation had occurred to the greatest extent in the 5.3 and 47.7 ppm atrazine systems of the Ella soil; degradation in the Poygan soil was somewhat less and degradation in the Kewaunee was extremely slow (Table 2).

Atrazine degradation in the Ella and Poygan perfusion systems followed first-order kinetics with respect to concentration of atrazine in solution, i.e., the log of the atrazine concentration decreased linearly with time (Fig. 1, 2). The degradation rate in the Ella perfusion unit at 15.4 ppm atrazine deviated sharply from first-order kinetics after 125 days because faulty operation of the perfusion apparatus prevented circulation of the medium (Fig. 2).

The half-life values of atrazine ($t_{1/2}$), the time required to degrade half the atrazine in solution after initial adsorption was complete, were obtained from Fig. 1 and 2 by extrapolation. For the same atrazine concentration the $t_{1/2}$ values for

Table 1—Description of soils used in perfusion experiments

Parent material	Drainage	pH	Organic matter	Clay	Classification
			%	%	
			Poygan silt loam		
Lake sediments	Very poor	6.9	13	26	Humic Gley (Mollic normaquept)
			Ella loamy sand		
Outwash sand	Moderately well	4.9	4	5	Brunizem (Paraquentic hapludoll)
			Kewaunee clay		
Dolomitic till	Well	7.3	2	44	Gray-Brown Podzol (Typic normudalf)

Table 2—Atrazine adsorbed and degraded in perfusion systems and atrazine half-life

Soil type	pH of perfusate	Initial amount of atrazine		Amount of atrazine after 225 days			$t_{1/2}$
		Added	Adsorbed	In solution*	Adsorbed†	Degraded‡	
		ppm	ppm	%	ppm	ppm	%	days
Ella ls	4	5.3	1.7	32	0.7	0.7	74	95
	4	15.4	4.4	20	2.6	1.4	74	115
	4	47.7	12.7	27	13.6	5.3	60	165
Poygan sil	7	5.3	2.2	42	1.0	0.8	66	145
	7	15.4	6.4	42	4.5	3.2	50	220
	7	47.7	17.7	37	19.5	12.1	34	350
Kewaunee c	8	5.3	0.8	15	4.0	0.7	11	3–5 years
	8	15.4	2.4	16	11.0	2.0	16	3–5 years
	8	47.7	3.7	8	40.0	3.5	9	3–5 years

* Obtained from first-order rate plots with extrapolation where necessary
† Estimated from atrazine in solution and Fig. 3.
‡ % degraded = $\frac{\text{added} - (\text{solution} + \text{adsorbed})}{\text{added}} \times 100$

the Ella soil were shorter than those for the Poygan. However, for a given soil, $t_{1/2}$ values increased with increasing initial atrazine concentration. Because atrazine adsorption varied almost linearly with solution concentration (Fig. 3), the slopes of the lines showing the adherence of atrazine degradation to first-order kinetics (Fig. 1 and 2) would be changed little by correcting for adsorbed atrazine.

To determine the extent of atrazine degradation in perfusion systems, it was necessary to estimate changes in adsorption during the course of the experiment. Adsorption is expected to decrease with decreasing solution concentration of atrazine (R. O. Radke, G. Chesters, and L. E. Engelbert. 1965. Adsorption and desorption of atrazine and atratone by Wyoming bentonite, humic acid and lignin. Agronomy Abstracts. p. 75). The amount of atrazine adsorbed at any given time can be estimated from Fig. 3, recognizing that irreversible adsorption of atrazine or replacement of atrazine on adsorption sites by degradation products would affect the estimate. Because irreversibility would lead to underestimation and replacement would lead to overestimation of adsorption, the two effects are counteracting.

The most rapid rate of atrazine degradation occurred in the most acid soil investigated, namely the Ella loamy sand (Table 2). For the two soils of similar pH, the Poygan silt loam and the Kewaunee clay, degradation was more rapid in the more organic Poygan soil. Although the rate of degradation was related inversely to the clay contents of the three soils studied, the hypothesized mechanisms of atrazine degradation described in the Discussion section suggested that soil pH and organic matter content were the principal factors controlling the rate of degradation.

The accumulation of hydroxyatrazine in the perfusate of the Ella soil initially treated with the 47.7 ppm atrazine medium was considerably higher than that found in the comparable Poygan system (Table 3); only trace amounts of hydroxyatrazine were found in the Kewaunee perfusate. Of the atrazine degraded after 244 days, 42% and 11% appeared as hydroxyatrazine in the Ella and Poygan perfusates, respectively. The equivalent of 0.5 ppm hydroxyatrazine was found in control systems (not amended with atrazine); the data in Table 3 were corrected accordingly.

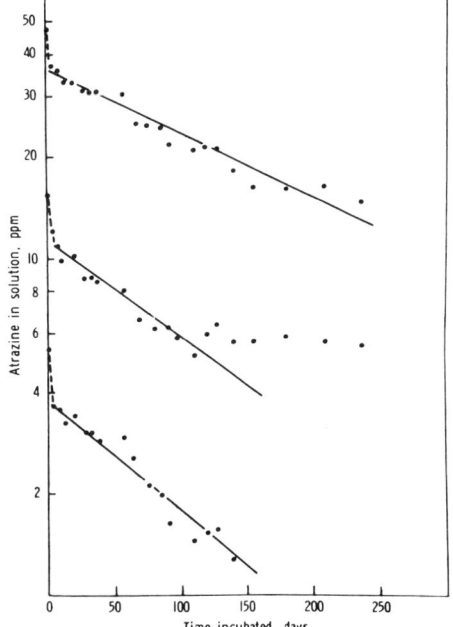

Fig. 2—Atrazine degradation in three Ella ls perfusion systems.

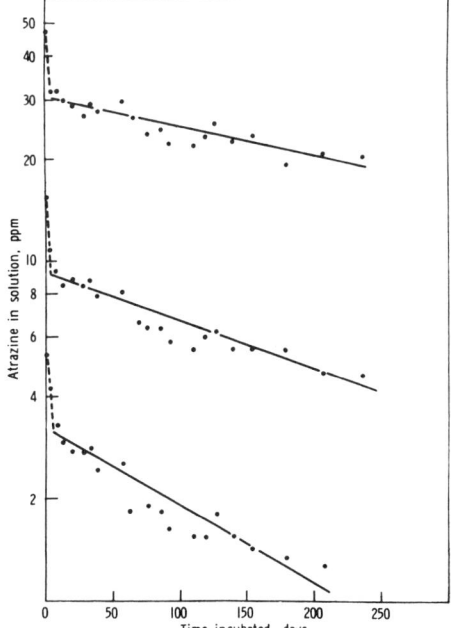

Fig. 1—Atrazine degradation in three Poygan sil perfusion systems.

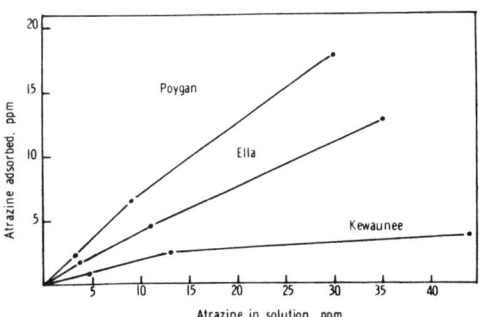

Fig 3—Initial atrazine adsorption as related to the equilibrium solution concentration of atrazine.

Table 3—Atrazine and hydroxyatrazine in solution in Ella and Poygan soil perfusion systems initially containing 47.7 ppm atrazine

Perfusion time	In solution*	Atrazine		Hydroxyatrazine in solution	Degraded atrazine unaccounted for§
		Adsorbed†	Degraded‡		
days	ppm				
			Ella loamy sand		
121	21.2	8.0	18.5	5.0	13.5
150	18.8	7.0	21.9	8.5	13.5
222	13.8	5.4	28.5	9.0	19.5
244	12.8	5.0	29.9	12.5	17.5
317	9.4	3.8	34.5	15.0	19.5
			Poygan silt loam		
121	23.8	14.3	9.6	0.5	9.0
152	22.3	13.6	11.8	1.5	10.5
244	18.5	11.5	17.7	2.0	15.5

* Obtained from Fig. 1 and 2.
† Estimated from atrazine in solution and Fig. 3.
‡ Degraded atrazine = added atrazine − (solution + adsorbed) atrazine.
§ Degraded atrazine unaccounted for = degraded atrazine − hydroxyatrazine in solution.

Atrazine Hydrolysis

Atrazine hydrolysis in aqueous buffered systems followed first-order kinetics, i.e., the log of the atrazine concentration decreased linearly with time, showing that rate of hydrolysis was dependent on atrazine concentration (Fig. 4). However, the rate of hydrolysis was also influenced by pH as shown by a change in the slope of the first-order plots with pH. The half-life ($t_{1/2}$) of atrazine was calculated at a given pH-value from the slope of first-order plots as described by Daniels and Alberty (4). The $t_{1/2}$ values were least at pH-values 1 and 13 and increased as pH approached neutrality (Fig. 5). Half-life values for atrazine can be predicted at any pH using equations derived from Fig. 5, providing the relationship between log $t_{1/2}$ and pH remains linear:

For acid systems, $\log t_{1/2} = 0.62 \, \text{pH} - 0.1$ [1]

For alkaline systems, $\log t_{1/2} = 5.64 - 0.91 (\text{pH} - 7)$ [2]

The rate of atrazine hydrolysis in a soil-free system followed first-order kinetics and was slower than in a comparable system system to which sterilized soil had been added (Fig. 4). Using equation [1], the $t_{1/2}$ value for the aqueous atrazine system at pH 3.9 is 209 days. In the system of the same pH to which sterilized soil (50 g) was added to atrazine solution (100 ml) the $t_{1/2}$ value was 22 days. Thus, the presence of the sterilized soil caused an almost tenfold increase in the rate of atrazine hydrolysis.

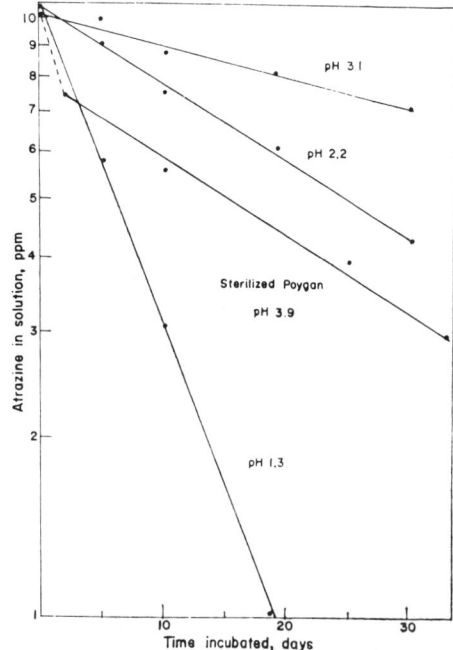

Fig. 4—Atrazine hydrolysis in aqueous and sterilized soil systems.

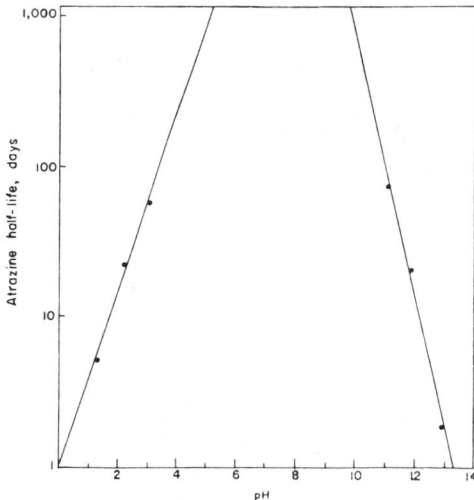

Fig. 5—Relationship between atrazine half-life and pH in aqueous systems.

DISCUSSION

Atrazine hydrolysis followed first-order kinetics with respect to atrazine concentration at constant pH, but the rate of hydrolyis was also pH dependent. Although atrazine was stable at neutral pH-values, rapid hydrolysis occurred under highly acid or alkaline conditions. Atrazine hydrolysis mechanisms are probably similar to those proposed by Horrobin (7) for hydrolysis of related chlorotriazines. Alkaline hydrolysis likely involves direct nucleophilic displacement of Cl by OH. Acid hydrolysis may result from protonation of a ring or chain N atom followed by cleavage of the C-Cl bond by H_2O. Protonation of N would increase the electron deficiency of the C bonded to Cl and increase the tendency for nucleophilic displacement of Cl by H_2O.

Considerable evidence was obtained to show that chemical hydrolysis is an important mechanism of atrazine degradation in soil.

Although rates of atrazine hydrolysis in soils predicted from rates in acid and alkaline systems showed that hydrolysis would not be expected to be appreciable in soils on the basis of pH only (Fig. 5), the studies in a sterilized soil system indicated that hydrolysis of atrazine was catalyzed by contact with soil and that the occurrence of chemical hydrolysis in soils is feasible (Fig. 4). Several explanations are available for the more rapid rate of atrazine hydrolysis in sterilized soil systems than in soil-free systems. It is postulated that nucleophilic compounds and/or Fe and Al dissolved from the soil are capable of catalyzing atrazine hydrolysis because the nucleophilic compounds will behave in a manner analagous to alkalies and the Fe and Al are capable of forming complexes which behave as Lewis-type acids. Also, the concentration of counter ions near the surfaces of soil particles in the double layer (19) may influence the rate of hydrolysis. Concentration of H ions around the soil particles would create a zone of lower pH than that measured in the bulk system, resulting in more rapid acid hydrolysis than predicted from the pH of the system. Finally, adsorption of atrazine through H bonding between the amino groups of atrazine and acidic protons on the soil colloidal surfaces could catalyze hydrolysis by a mechanism similar to that suggested for acid hydrolysis.

The formation of hydroxyatrazine, the hydrolysis product of atrazine, in atrazine soil perfusion systems is further evidence for the occurrence of atrazine hydrolysis in soil (Table 3). The accumulation of hydroxyatrazine indicates that it, like atrazine, is quite resistant to microbial degradation. All material absorbing at 240 mμ and not removed by the clean-up procedure was considered to be hydroxyatrazine. Although other atrazine derivatives having ultraviolet spectra similar to hydroxyatrazine could have been included in the hydroxyatrazine determination, chlorodiamino-s-triazine and hydroxydiamino-s-triazine, two likely atrazine degradation products, do not interfere appreciably with the method of hydroxyatrazine determination employed. Because hydroxyatrazine is adsorbed to a greater degree than atrazine, a large portion of the atrazine described as unaccounted for in Table 3 is expected to be hydroxyatrazine adsorbed on the soils. Substitution of the OH for the Cl group causes the pKa value of the amino group to be higher for hydroxyatrazine than for atrazine, and thus increases the tendency for hydroxyatrazine to bond to soil adsorptive sites through H bonding.

Both atrazine hydrolysis in buffered soil-free systems and atrazine degradation in soil perfusion systems followed first-order kinetics at constant pH, although the $t_{1/2}$ values in perfusion systems decreased with large increases in initial atrazine concentration (Fig. 2, 3, 4). Although the similarity in kinetics alone does not necessarily mean that similar degradative mechanisms were operative, consistency with the hypothesis of chemical hydrolysis is maintained.

Evidence supporting the occurrence of microbial degradation in soil perfusion systems was not obtained. Because degradation of atrazine was not detected in atrazine media inoculated from perfusion systems, it did not appear that microorganisms were involved in atrazine hydrolysis. Evidence in the literature for the ability of soil microorganisms to metabolize atrazine is sparse. Investigations with ring C^{14}-labeled atrazine and simazine indicate that the triazine ring is quite resistant to microbial attack (10, 12). However, investigations with chain C^{14}-labeled simazine have shown that certain organisms are able to metabolize the side chains of simazine (3, 10). Studies involving chain C^{14}-labeled atrazine are not reported, and the relative resistance of atrazine and simazine to microbial attack is uncertain. Although the occurrence of hydrolysis in the sterilized soil system and the lack of hydrolysis in enrichments indicated that atrazine hydrolysis was not due to microbial activity, it is possible that microbial alteration of the amino side chains of atrazine occurred but could not be detected by the colorimetric analysis for atrazine (14). Because the method involves reaction with the C-Cl bond of the triazine, only molecular alterations involving dechlorination or destruction of the aromatic ring would prevent color formation. However, hydroxyatrazine was the only dechlorinated product detected by ultraviolet analysis, and other dechlorinated products would be expected if atrazine side-chain degradation had occurred. Such compounds would have been detected, providing their ultraviolet spectra were different from the spectrum of hydroxyatrazine.

Atrazine hydrolysis in soil was apparently catalyzed by adsorption. Adsorption of atrazine by the Poygan and Ella soils was considerably greater than in the Kewaunee, and degradation was much slower in the Kewaunee than in the other two soils. Furthermore, the decrease in the half-life of atrazine with decreasing initial atrazine concentration in the Ella and Poygan systems appeared to be associated with the increased relative (%) adsorption of atrazine in the more dilute systems; adsorption was about 5% greater at 5.3 than at 47.7 ppm atrazine (Table 2).

Adsorption of atrazine could catalyze hydrolysis by a mechanism similar to H ion catalysis. Adsorption of atrazine probably resulted largely from H bonding between the ring or side-chain N atoms of atrazine and weak acid groups in the soil organic fraction. The H bonding would further increase the electron deficiency of the C atom in the two positions of atrazine already surrounded by electronegative N and Cl atoms. The increased electron deficiency would increase the tendency for attack by weakly nucleophilic H_2O, thereby increasing the rate of hydrolysis.

Organic matter content and pH appeared to be the soil properties influencing the rate of hydrolysis of atrazine.

Organic matter evidently catalyzed hydrolysis by adsorption of atrazine, accounting for the more rapid rate of hydrolysis in the Ella and Poygan than the Kewaunee systems. Because hydrolysis was catalyzed by H ion, the rate of hydrolysis increased with increased acidity. The combined effects of adsorption and acidity apparently caused hydrolysis to proceed more rapidly in Ella than Poygan systems, even though adsorption by the Poygan was higher. Evidently the slow rate of degradation in the Kewaunee soil was a result of the almost neutral pH and low atrazine adsorption, resulting in unfavorable conditions for hydrolysis.

LITERATURE CITED

1. Burnside, O. C., E. L. Schmidt, and R. Behrens. 1961. Dissipation of simazine from the soil. Weeds 9:477–484.
2. Burschel, P. 1961. Untersuchungen über das Verhalten von Simazin im Boden. Weed Res. 1:131–141.
3. Couch, R. W., J. V. Gramlich, D. E. Davis, and H. H. Funderburk. 1965. The metabolism of atrazine and simazine by soil fungi. Southern Weed Control Conf., Proc. 18:623–631.
4. Daniels, F., and R. A. Alberty. 1961. Chemical kinetics, p. 294–349. In F. Daniels and R. A. Alberty. Physical chemistry 2nd ed. John Wiley and Sons, Inc., New York.
5. Gysin, H., and E. Knüsli. 1960. Chemistry and herbicidal properties of triazine derivatives. Advance. Pest Control Res. 3:289–358.
6. Hartley, G. S. 1960. Physico-chemical aspects of the availability of herbicides in soil, p. 63–78. In E. K. Woodford and G. R. Sagar [ed.] Herbicides and the soil. Blackwell Scientific Publications Ltd., Oxford, England.
7. Horrobin, S. 1963. The hydrolysis of some 1,3,5-triazines: mechanism, structure and reactivity. J. Chem. Soc. 4130–4135.
8. Jackson, M. L. 1956. Mineral fractionation for soils, p. 101–164. In M. L. Jackson. Soil chemical analysis-advanced course. Publ. by the author, Dept. of Soils, University of Wisconsin, Madison.
9. Kaufman, D. D., P. C. Kearney, and T. J. Sheets. 1963. Simazine: degradation by soil microorganisms. Science 142:405–406.
10. Kaufman, D. D., P. C. Kearney, and T. J. Sheets. 1965. Microbial degradation of simazine. J. Agr. Food Chem. 13:238–242.
11. Lees, H. 1947. A simple automatic percolator. J. Agr. Sci. 37:27–28.
12. MacRae, I. C., and M. Alexander. 1965. Microbial degradation of selected herbicides in soil. J. Agr. Food Chem. 13:72–76.
13. Nearpass, D. C. 1965. Effects of soil acidity on the adsorption, penetration and persistence of simazine. Weeds 13:341–346.
14. Radke, R. O., D. E. Armstrong, and G. Chesters. 1966. Evaluation of the pyridine-alkali colorimetric method for determination of atrazine. J. Agr. Food Chem. 14:70–73.
15. Ragab, M. T. H., and J. P. McCollum. 1961. Degradation of C^{14}-labeled simazine by plants and soil microorganisms. Weeds 9:72–84.
16. Stroube, E. W. 1962. The movement and persistence of simazine and atrazine in soil and some related studies. Diss. Abstr. 22:3339–3340.
17. Talbert, R. E., and O. H. Fletchall. 1964. Inactivation of simazine and atrazine in the field. Weeds 12:33–37.
18. Walkley, A. 1947. A critical examination of a rapid method for determining organic carbon in soils ... effect of variations in digestion conditions and of inorganic soil constituents. Soil Sci. 63:251–264.
19. Wicklander, L. 1964. Cation and anion exchange phenomena. p. 163–205. In F. E. Bear [ed.] Chemistry of the soil. Reinhold Publishing Corp., New York.

30

Copyright © 1968 by the American Chemical Society
Reprinted from *Jour. Agric. Food Chem.* **16**:475–478 (1968)

Microbial Decomposition of Diquat Adsorbed on Montmorillonite and Kaolinite Clays

J. B. Weber and H. D. Coble

Nutrient solutions treated with tagged diquat [6,7-dihydrodipyrido(1,2-a:2′,1′-c)-pyrazidiinium dibromide] and soil microorganisms readily released $^{14}CO_2$. Sterile controls and standard solutions showed no nonbiological decomposition of diquat. Additions of montmorillonite clay in an amount calculated to adsorb one half of the ^{14}C-diquat reduced $^{14}CO_2$ evolution to approximately one half. When enough montmorillonite clay was added to adsorb all of the diquat, no $^{14}CO_2$ was detected. Additions of kaolinite clay to the nutrient solutions had no significant effect on the total diquat decomposed in these systems.

Knowledge concerning the microbial decomposition of pesticides which are applied to soils is of great importance. It is desirable that these compounds be decomposed by some agent, biological or nonbiological, and not accumulate and cause problems in the future. Several requirements must be fulfilled if a compound is to be decomposed by soil microbes (Alexander, 1965). Briefly, to be decomposed a pesticide must be metabolizable and available. The present study is most strongly concerned with availability. Several workers have shown that the dipyridylium herbicides can be microbiologically degraded (Baldwin *et al.*, 1966; Funderburk and Bozarth, 1967; Tu, 1966). These compounds ionize completely in aqueous solutions to yield organic cations which react very strongly with soil particulates, especially the clay minerals. The dipyridylium compounds, and all organic cationic pesticides, are readily inactivated once they reach the soil. However, inactivation and decomposition are not synonymous, and it is important to know if the compounds are being decomposed in the soil.

Considerable research has been performed on the interaction of pure organic compounds with clay minerals, especially montmorillonite, because of its expanding lattice structure and high cation exchange capacity. The adsorption of organic molecules by montmorillonite clay involves the entry of the organic molecules between the silicate sheets of clay, causing an expansion of the crystalline lattice structure. Observations have shown the presence of mono-, di-, and trimolecular layers of organic molecules in the expanded crystal lattice (MacEwan, 1948). Other investigators (Gieseking, 1939; Hendricks, 1941) showed that large aromatic organic compounds are adsorbed in the interlayer spaces of montmorillonite in various positions, and they are held by several types of adsorption forces. Several workers (Bower, 1949; Ensminger and Gieseking, 1942; Estermann *et al.*, 1959; Goring and Bartholomew, 1949; Pinck and Allison, 1951) showed that the adsorption of organic compounds by clay minerals influences their availability to soil microbes. The adsorption of organic compounds by clay minerals is dependent upon the chemical properties of the compounds and the types of clay minerals involved (Pinck *et al.*, 1961a, 1961b). The dipyridylium herbicides are strongly adsorbed in the interlayer spaces of montmorillonite clay and are not readily extractable using $1M$ salt solutions (Weber *et al.*, 1965). The adsorption of the dipyridylium herbicides by several clay minerals reduces their availability to plants (Coats *et al.*, 1966; Weber and Scott, 1966) and their photochemical decomposition by ultraviolet light (Funderburk *et al.*, 1966).

The present experiments were performed to determine the effects of two clay minerals on the microbial mineralization of ^{14}C-tagged diquat [6,7-dihydrodipyrido(1,2-a:-2′,1′-c)-pyrazidiinium dibromide].

PROCEDURE

To determine whether soil microorganisms are active in decomposing diquat, 10-ml. aliquots of filtrate from a mixture of 10 grams of a Norfolk sandy loam soil and 250 ml. of distilled water were added to 500-ml. Erlenmeyer flasks containing 100 ml. of nutrient solution (0.25 gram of polypeptone and 0.15 gram of beef extract) and varying levels of ring-labeled ^{14}C-diquat. The polypeptone consisted of a mixture of casein (Trypticase) and animal tissue (Thiotone) peptones and was obtained from the Baltimore Biological Laboratory. The flasks were incubated at 25° C. Air was slowly bubbled through the nutrient solutions and then passed through a $CaCl_2$ dryer and finally into vials containing 10 ml. of a CO_2-trapping solution of $0.01M$ hyamine [*p*-(diisobutylcresoxyethoxyethyl)-dimethylbenzylammonium chloride] in methanol. The scintillators PPO (2,5-diphenyloxazole) and POPOP [2,2-*p*-phenylenebis(5-phenyloxazole)] in toluene were added to the vials containing the hyamine-carbonate-methanol mixture, and the vials were then counted in a Tri-Carb scintillation spectrometer for 10 minutes at 0° C. The $^{14}CO_2$ was collected for 8 hours each day; the counts were then multiplied by three to obtain c.p.m. per day. $^{14}CO_2$ was collected over a 24-hour period for random samples, and the counts obtained were in agreement with the standard 8-hour $^{14}CO_2$ collections. Rates of diquat employed were 10, 1, and 0.1 p.p.m. At the end of 10 days, the cumulative release of $^{14}CO_2$ for the high rate of diquat amounted to 6100 c.p.m., and the authors concluded that diquat was being decomposed. Experiments were then established to determine the effects of the clay minerals on the diquat mineralization. Flasks with actively growing microorganisms were used for

Department of Crop Science, North Carolina State University, Raleigh, N.C.

inoculation in the experiments involving the clay minerals.

Previous studies (Weber et al., 1965) showed that diquat was strongly adsorbed on the clay minerals montmorillonite and kaolinite to approximately their cation exchange capacity (C.E.C.). To determine the effects of microbial decomposition of diquat adsorbed on these two clay minerals, experiment 1 was established. Ten milliliters of deionized water containing 10 μc. (3.7 μmoles) of ^{14}C-diquat were added to varying amounts of each of the clay minerals and shaken for 1 hour. The samples were then centrifuged, and the amount of diquat present in the supernatant was determined spectrophotometrically employing a recording spectrophotometer equipped with a deuterium lamp. An analytical wavelength of 307 mμ was used. The samples were then resuspended and added to the 500-ml. flasks containing similar nutrient solutions as employed above. Microbes were introduced by adding 2 ml. of solution from flasks with actively growing microbes from the preliminary experiment discussed above. Sterile (autoclaved) controls containing 10 μc. of ^{14}C-diquat were included to detect possible nonbiological mineralization of the compound. Spectrophotometric analysis of autoclaved diquat solutions showed that the sterilization process had no effect on the herbicide. Samples were also included for background purposes. Duplicate, and in some cases triplicate, treatments were employed. Standard solutions of diquat which were analyzed weekly were included to determine if diquat was photochemically decomposed under these conditions. ^{14}CO$_2$ was collected in the manner described above. The amounts of each clay mineral employed were based on their cation exchange capacities as determined by the ammonium acetate method of Jackson (1958) using Na as the saturating cation and NH$_4$ as the replacing cation. Two levels of each clay mineral were employed. The low level was calculated to adsorb one half of the diquat from solution and to reduce the free diquat available to the microorganisms by one half. The high rates of each clay mineral were calculated to adsorb all of the diquat from solution and make it necessary for the microbes to decompose the herbicide on the clay surfaces. The following calculations show how the amounts of each clay mineral employed were determined: For montmorillonite clay with a C.E.C. of 0.847 meq. per gram:

0.847 meq. per gram ÷ 2 (divalent cations) =
0.423 mmole per gram
3.7 μmoles ^{14}C-diquat ÷ 423 μmoles per gram =
8.7×10^{-3} gram

The high rate of montmorillonite employed for experiment 1 was 9.0 mg. For kaolinite clay with a C.E.C. of 0.051 meq. per gram:

0.051 meq. per gram ÷ 2 (divalent cations) =
0.025 mmole per gram
3.7 μmoles ^{14}C-diquat ÷ 25 μmoles per gram =
0.148 gram

The high rate of kaolinite employed for experiment 1 was 160 mg.

Experiment 2 was performed in a manner similar to experiment 1, except that the amounts of ^{14}C-diquat were doubled, and the amounts of each clay mineral were slightly more than doubled. Microbial inoculation was made by adding 2-ml. aliquots from solutions with actively growing microorganisms from experiment 1.

Experiment 3 was similar to experiment 2 except that the amounts of ^{14}C-diquat and clay were those employed in experiment 1. Inoculation was made employing solutions with actively growing microorganisms from experiment 2. At the end of experiment 3, the solutions were centrifuged, and isotopic determinations were made on the ^{14}C remaining in the supernatant and in the clay pellets. The clay pellets were suspended in a thixotropic gel, and the ^{14}C in the two fractions was determined using a naphthalene-dioxane mixture and employing PPO and POPOP as the scintillators according to the method of Bray (1960).

^{14}C-diquat was extracted from the clay pellets by use of solutions of $10^{-4}M$ paraquat (1,1'-dimethyl-4,4'-bipyridinium dichloride) (Weber et al., 1968).

RESULTS AND DISCUSSION

The sterile control [10 μc. (3.7 μmoles) of ^{14}C-diquat] of experiment 1 yielded counts of 186 c.p.m. per 8-hour ^{14}CO$_2$ collection, and the background treatments (nutrient solution only) gave counts of 100 c.p.m. for the same period. Both treatments gave similar results for all experiments. Chemical determinations of standard diquat solutions for the same period showed no loss of diquat resulting from nonbiological decomposition. All of the treatments in Table I are corrected according to the sterile control for the respective experiment. The acidity of the nutrient solutions ranged from pH 7 to 8.5.

Mineralization of diquat, as measured by ^{14}CO$_2$ evolution, occurred in all treatments in experiment 1 (Table I). Additions of montmorillonite clay to the nutrient solutions effectively reduced ^{14}CO$_2$ evolution as it was calculated to do. The 9-mg. rate of montmorillonite decreased ^{14}CO$_2$ to a very low rate, but not to zero. Chemical determinations showed that 9 mg. of montmorillonite was not enough to adsorb all of the diquat in the solutions in experiment 1, and explain why the small amount of diquat mineralization resulted from this treatment. Results in experiments 2 and 3 show that when enough montmorillonite clay was added to adsorb all of the diquat from solution, ^{14}CO$_2$ evolution did not occur and mineralization of diquat was inhibited. Additions of kaolinite clay to the nutrient solutions appeared to stimulate ^{14}CO$_2$ evolution initially in experiment 1, but by the end of the 6-week period ^{14}CO$_2$ evolution was similar for the no-clay and kaolinite clay treatments.

Microbial decomposition of diquat on the no-clay treatment of experiment 2 was approximately twice as high as for experiment 1. This resulted from the higher rate of diquat employed (7.4 vs. 3.7 μmoles for experiments 2 and 1, respectively). Additions of 10 mg. of montmorillonite to the flasks resulted in ^{14}CO$_2$ evolution of one half to one third that of the no-clay treatment. The high rate of montmorillonite, which adsorbed all of the diquat from solutions, completely inhibited ^{14}CO$_2$ evolution. Mineralization of diquat was very similar

Table I. Cumulative Release of $^{14}CO_2$ Resulting from Application of Tagged Diquat to Nutrient Solutions Containing Montmorillonite (M) and Kaolinite (K) Clays

Diquat Added, μmoles	Clay Added, Mg.	Radioactivity in Evolved $^{14}CO_2$, C.P.M.[a]				
		1 week	2 weeks	4 weeks	6 weeks	8 weeks
Experiment 1						
3.7	0	1,450	15,800	89,700	178,000	
3.7	4 M	1,660	28,000	81,100	115,000	
3.7	9 M	90	1,510	4,020	5,310	
3.7	60 K	13,100	69,600	183,000	262,000	
3.7	160 K	12,300	72,900	139,000	180,000	
Experiment 2						
7.4	0	2,090	8,710	191,000	424,000	
7.4	10 M	658	8,760	87,600	134,000	
7.4	20 M	0	0	0	0	
7.4	160 K	1,880	58,200	380,000	589,000	
7.4	340 K	775	65,000	426,000	628,000	
Experiment 3						
3.7	0	37,000	80,800	159,000	217,000	254,000
3.7	5 M	16,300	27,300	44,300	60,200	72,800
3.7	10 M	0	0	0	0	0
3.7	80 K	38,800	58,700	106,000	150,000	187,000
3.7	170 K	24,400	46,000	94,500	152,000	204,000

[a] Corrected for $^{14}CO_2$ derived from sterile control (<200 c.p.m.).

on both kaolinite clay treatments and was from 50 to 100% higher than the no-clay treatment for experiment 2.

Experiment 3 yielded results very similar to experiments 1 and 2 (Table I). Additions of montmorillonite clay inhibited diquat mineralization by soil microorganisms. $^{14}CO_2$ evolution on the kaolinite clay treatments was similar to the no-clay treatment and indicated that diquat adsorbed on this clay mineral was completely available for microbial decomposition.

At the conclusion of experiment 3, isotopic ^{14}C determinations were made on the supernatant from centrifuged samples of the nutrient solutions. No diquat was present in the solutions containing the high level of montmorillonite clay, confirming that no free diquat was present in the solutions for the microbes to decompose (Table II). ^{14}C in the solutions containing the low rate of montmorillonite was approximately one half of that initially added, and confirms the calculations that 5 mg. of montmorillonite adsorbed one half of the diquat added and left one half free in solution to be decomposed. Isotopic determinations of ^{14}C-diquat in the clay pellets showed that virtually all (97%) of the diquat adsorbed by the high rate of montmorillonite was still present on the clay particles. Adsorption-desorption studies (Weber et al., 1968) showed that diquat adsorbed on montmorillonite clay could not be displaced by $1M$ $BaCl_2$ solutions, but could be displaced by $10^{-1}M$ paraquat solutions. Diquat thus displaced with paraquat from the clay pellets in experiment 3 (Table II) yielded ultraviolet spectra identical with the pure compound in the range from 290 to 360 mμ. The inhibition of diquat mineralization from adsorption by montmorillonite clay is analogous to results obtained when cucumber plants were employed to determine whether paraquat adsorbed on the same clay mineral was available to plant roots (Weber and Scott, 1966).

Adsorption of ^{14}C-diquat by the kaolinite clay did not reduce the microbial decomposition of the herbicide in this system (Table I). In experiments 1 and 2, diquat mineralization in the solutions containing kaolinite clay was considerably higher in the first 3 weeks than in the solutions containing no clay, but not in experiment 3. Chemical determinations showed that the nutrient solutions displaced some of the diquat from the kaolinite clay particles, so it is not possible to distinguish whether the microbes decomposed only free diquat in solution or diquat adsorbed on the kaolinite clay surfaces. An analogous situation resulted in studies where paraquat was adsorbed on kaolinite clay and was found to be readily available to cucumber plants (Weber and Scott, 1966). One half of the added ^{14}C-diquat was present in the solutions where 170 mg. of kaolinite clay was calculated to adsorb all of the added diquat (Table II). Adsorption-desorption studies (Weber et al., 1968) showed that deionized water did not displace diquat from kaolinite, but that $1M$ $BaCl_2$ solutions displaced 80 to 85% of the adsorbed herbicide.

These experiments indicate that diquat free in solution

Table II. Tagged Diquat Remaining in Nutrient Solutions and in Montmorillonite (M) and Kaolinite (K) Clay Pellets at Termination of Experiment 3

Diquat Added, μc.	Clay Added, Mg.	C^{14}-Diquat Present,[a] μc.	
		In solution	In pellet[b]
10	0	8.4	0.4
10	5 M	5.1	4.5
10	10 M	0	9.7
10	80 K	8.1	1.0
10	170 K	5.4	3.6

[a] Corrected for background.
[b] Pellet contained microbial cells and clay.

was microbiologically decomposed, but diquat adsorbed in the crystalline lattice of montmorillonite clay particles was not available for mineralization. Only small amounts of the dipyridylium herbicides are applied to agricultural soils, so this phenomenon is at present only academic, but because the use of cationic pesticides is increasing, we need to be aware of this problem. To demonstrate this effect conclusively, it will be necessary to employ diquat adsorbed on montmorillonite clay particles in the soil system.

ACKNOWLEDGMENT

The authors thank F. T. Corbin for his assistance and the Chevron Chemical Co., San Francisco, Calif., for the gift of ^{14}C-tagged diquat.

LITERATURE CITED

Alexander, M., *Adv. Appl. Microbiol.* **7**, 35 (1965).
Baldwin, B. C., Bray, M. F., Geoghegan, M. J., *Biochem. J.* **101**, 15 (1966).
Bower, C. A., *Iowa Agr. Expt. Sta. Bull.* **362**, 961 (1949).
Bray, G. A., *Anal. Biochem.* **1**, 279 (1960).
Coats, G. E., Funderburk, H. H., Jr., Lawrence, J. M., Davis, D. E., *Weed Res.* **6**, 58 (1966).
Ensminger, L. E., Gieseking, J. E., *Soil Sci.* **53**, 205 (1942).
Estermann, E. F., Peterson, G. H., McLaren, A. D., *Soil Sci. Soc. Am. Proc.* **23**, 31 (1959).
Funderburk, H. H., Jr., Bozarth, G. A., J. AGR. FOOD CHEM. **15**, 563 (1967).
Funderburk, H. H., Jr., Negi, N. S., Lawrence, J. M., *Weeds* **14**, 240 (1966).
Gieseking, J. E., *Soil Sci.* **47**, 1 (1939).
Goring, C. A. I., Bartholomew, W. V., *Soil Sci. Soc. Am. Proc.* **13**, 152 (1949).
Hendricks, S. B., *J. Phys. Chem.* **45**, 65 (1941).
Jackson, M. L., "Soil Chemical Analysis," Prentice-Hall, Englewood Cliffs, N. J., 1958.
MacEwan, D. M. C., *Trans. Faraday Soc.* **44**, 349 (1948).
Pinck, L. A., Allison, F. E., *Science* **114**, 130 (1951).
Pinck, L. A., Holton, W. F., Allison, F. E., *Soil Sci.* **91**, 22 (1961a).
Pinck, L. A., Soulides, D. A., Allison, F. E., *Soil Sci.* **91**, 94 (1961b).
Tu, Chin-Ming, Ph.D. dissertation, Oregon State University, Corallis, Ore., 1966.
Weber, J. B., Perry, P. W., Upchurch, R. P., *Soil Sci. Soc. Am. Proc.* **29**, 678 (1965).
Weber, J. B., Scott, D. C., *Science* **152**, 1400 (1966).
Weber, J. B., Weed, S. B., Ward, T. M., *Soil Sci. Soc. Am. Proc.*, in press (1968).

Received for review January 18, 1968. Accepted March 14, 1968. Investigation supported by Public Health Service Research Grant CC 00282-01 from the National Communicable Disease Center, Atlanta, Ga. Paper No. 2545 of the Journal Series of the North Carolina State Agricultural Experiment Station.

31

Copyright © 1979 by The Clay Minerals Society
Reprinted from *Clays and Clay Minerals* **27**:72–78 (1979)

SURFACE REACTIONS OF PARATHION ON CLAYS[1]

U. MINGELGRIN AND SARINA SALTZMAN

Institute of Soils and Water, Agricultural Research Organization,
The Volcani Center, Bet Dagan, Israel

(Received 6 February 1978)

Abstract—The adsorption-catalyzed degradation of parathion on clay surfaces is a hydrolysis process, proceeding either directly or through a rearrangement step. The rate and mechanism of degradation are dependent on the nature of the clay, its hydration status, and saturating cation. A mechanism for parathion degradation at adsorption sites on clay surfaces, in the absence of a liquid phase, is proposed.

Key Words—Adsorption, Catalysis, Hydrolysis, Insecticide, Parathion.

INTRODUCTION

Parathion (O,O-diethyl O-p-nitrophenyl phosphorothioate) is one of the most widely used plant and soil insecticides. Its metabolism in both biotic and abiotic media proceeds through one or more of the following reactions: isomerization, hydrolysis, oxidation, reduction (Melnikov, 1971). Clays are well known as potential catalyzers of various kinds of reactions of the adsorbed molecules (Mortland, 1970; Theng, 1974). As clay-parathion complexes are often formed, either in soils or in formulations using clays as carriers, their effect on parathion conversion may play an important role in parathion alteration in the environment.

The catalytic effect of clays on the metabolism of some organophosphate pesticides, such as malathion, dursban, diazinon, ronnel, zytron, and pyrimiphos ethyl, was reported by Polon and Sawyer (1962), Mortland and Raman (1967), and Mingelgrin et al. (1975). The only degradation mechanism observed in all these cases was the hydrolysis of the phosphate ester bond of the adsorbed molecule. Such factors as the nature of the clay, the moisture content, the saturating cations, and the incubation temperature, were found to affect the rate of the process. In recent years, some results on kaolinite-parathion interactions were reported. Kaolinite was found to enhance parathion degradation; this process also proceeds by the hydrolysis of the phosphate ester bond (Saltzman et al., 1974; Mingelgrin et al., 1977).

Rosenfield and Van Valkenburg (1965) observed that various homoionic bentonites, dried at elevated temperatures (300–950°C), induced the degradation of a thiophosphate (ronnel), the process occurring through a molecular rearrangement. The possible rearrangement products of parathion are known to be much more toxic to mammals than is the parent compound (Joiner et al., 1973). Therefore, it is important to check if such a process is possible in the case of clay-parathion complexes and, if so, to understand its mechanism and study the specific conditions favoring it.

MATERIALS AND METHODS

The clays used in this study were Wyoming bentonite (B-235, Fisher Scientific Co., Fair Lawn, N.J., U.S.A.) and kaolinite (Peerless No. 2, R. T. Vanderbilt, Export Corp., Norwalk, Conn., U.S.A.). The clays investigated were the natural commercial bentonite (which is a Na-saturated clay), and homoionic Ca-bentonite and Ca-kaolinite, prepared by a method described by Shainberg and Otoh (1968). Other materials used as adsorbents were silica gel and anionotrop and cationotrop aluminum oxide (chromatography grade, activity grade I, M. Woelm, Eschwege, W. Germany).

Parathion-^{14}C, labeled in the alkyl chain (Amersham Radiochemical Center, Arlington Heights, Illinois, U.S.A.), and high grade parathion (Analabs, Inc., North Haven, Conn., U.S.A.) were used; paraoxon (Koch Light Labs., England), diethyl thiophosphoric acid, ammonium salt (Ciba-Geigy, AG, Basle, Switzerland), and p-nitrophenol (BDH, Poole, England), were used as standards.

Procedure

The persistence of parathion when adsorbed on various adsorbents was investigated by two procedures:

a) Clay-parathion-^{14}C complexes were prepared by shaking for 30 min 0.3 g air-dried clay with 5 ml hexane solution containing 10,000 ppm parathion-^{14}C. The supernatant was checked for parathion, and discarded; the clay was washed with 5 ml hexane, which was also checked for parathion and discarded. The clay-parathion complex obtained was dried in an air stream, and divided into subsamples of 0.05 g, which were incubated in an oven, at different temperatures, for various periods of time. After incubation the samples were extracted twice, each time for 1 hr, with 4 ml hexane, together with 2 ml deionized water.

b) The clays were dried for 24 hr at 110° and 200°C and clay-parathion complexes were prepared by adding

---

[1] Contribution from the Agricultural Research Organization, The Volcani Center, Bet Dagan, Israel. 1977 Series, No. 216-E.

Table 1. Composition of the hexane extracts of parathion complexes (TLC data).

Adsorbent	Pretreatment	Incubation conditions		Components[1]	
		Temp (°C)	Time	Parathion	SEP
Bentonite (crude)	none	180	10 min	—	m
Ca-bentonite	none	180	10 min	—	m
Ca-kaolinite	none	110	10 min	s	w
Aluminum oxide cationotrop	none	110	2 min	m	—
Aluminum oxide anionotrop	none	110	2 min	m	—
Silica gel	none	140	10 min	s	w
Bentonite (crude)	none			s	m
	110°C			s	m
	200°C	22	60 days	m	w
Ca-bentonite	none			s	m
	110°C			m	m
	200°C			m	s
Ca-kaolinite	none			m	w
Aluminum oxide cationotrop	none		50 days	w	—
Aluminum oxide anionotrop	none			m	—

[1] The letters s (strong), m (medium), and w (weak) denote the relative intensities of the spots.

parathion to the predried, as well as to air-dried clays. The samples were kept in a desiccator and checked at various intervals. The preparation and the extraction procedures were similar to those described previously. All incubations were in the absence of a liquid phase.

Analytical methods

Several analytical methods were used to identify and quantitate parathion and its metabolites in both the water and hexane used for extraction.

Thin-layer chromatography. Silica gel, fluorescein-containing precoated plates (Riedel-De-Haën AG, Seelze-Hannover, W. Germany), were activated for 1 hr at 105°C. The extracts were spotted and the spots were either air-dried or dried with the aid of an air stream. The solvent systems used were petroleum ether:chloroform:methanol, 7:3:0.3 for the hexane extracts, and 6:2:2 for the water extracts. The spots were viewed under UV light and after spraying with a palladium chloride solution, and an additional spray with NaOH (Lichtenstein and Schulz, 1964). Parathion, paraoxon, diethylthiophosphate, and p-nitrophenol solutions were used as standards.

Infrared spectroscopy. Hexane extracts were applied as continuous bands on TLC plates and developed as described previously. The separated bands were scraped off and eluted with water:hexane, 1:2. The concentrated hexane extracts were evaporated on AgBr windows and the IR spectra were recorded. Parathion and paraoxon standards were run in a similar way. These two analytical methods were used for the separation and identification of parathion metabolites.

Gas-liquid chromatography. Five μl aliquots of the hexane extracts and standard solutions were injected into a gas chromatograph equipped with a glass column 90 cm long, filled with 3.4% QF-1 + 6.2% DC-200 on Gaschrom Q. The operating conditions were 225°, 205°, and 225°C for the inlet, column, and detector, respectively, and the flow rate of the gas carrier (N_2) was 70 ml/min. Under these conditions, a good separation of parathion, paraoxon, and O,S-diethyl O-p-nitrophenyl phosphate was obtained.

^{14}C-*counting*. Both water and hexane extracts obtained from the experiment in which labeled parathion was used, were checked in the presence of a scintillation fluid containing 60 g naphthalene, 4 g PPO, 0.2 g POPOP, 100 cc methyl alcohol, and 20 cc ethylene glycol to 1 liter dioxan. The last two methods were used to quantitate parathion and its metabolites, which were identified by TLC and IR spectroscopy.

Apparatus. The IR spectra were recorded on a Perkin-Elmer 257 infrared spectrophotometer; for the radioactivity measurements a Packard model 3003 Tricarb scintillation spectrometer was used. Gas-liquid chromatography was performed by a Packard gas chromatograph equipped with an alkali flame ionization detector.

RESULTS AND INTERPRETATION

The behavior of parathion as affected by temperature was checked by IR spectroscopy by heating pure parathion at 180°C for up to 2 hr in a heating cell. No alterations in the spectra were noted, other than a decrease in the intensity of all the peaks, due to volatilization. However, incubation of clay-parathion complexes at similar or lower temperatures resulted in marked and rather fast modifications of the adsorbed parathion.

Thin-layer chromatography. All the hexane extracts obtained from parathion incubated with different adsorbents and in different conditions were checked by

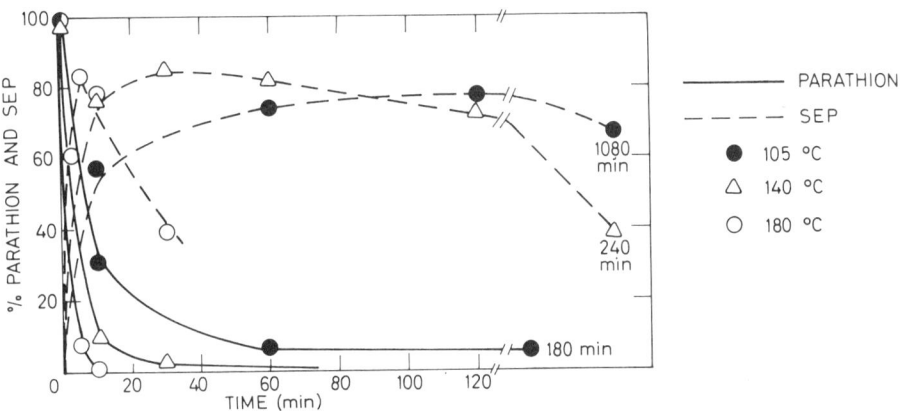

Fig. 1. Parathion loss and SEP evolution at various temperatures on Ca-bentonite.

TLC. Some representative data are given in Table 1. The qualitative nature of the TLC data enables the comparison between the intensities of the spots obtained from one sample, but less so between different samples. Only one or two spots could be detected in all the samples investigated. One of them (Rf 0.87) was identified as parathion (O,O-diethyl O-p-nitrophenyl phosphorothioate), by using a standard parathion solution, while the second (Rf 0.56) was identified as the parathion isomer, O,S-diethyl O-p-nitrophenyl phosphate (SEP). This identification was carried out by isolating the metabolite by TLC, and comparing its IR spectra with the IR spectra of standards (Joiner and Baetcke, 1974).

The qualitative analysis of the hexane extracts shows that SEP is always found in the presence of bentonites in which under some specific conditions it could be the predominant or the only compound present. Relatively small amounts of SEP were also found in the hexane extracts obtained from kaolinite and silica gel, and no such compound was detected in the case of the aluminum oxides investigated.

The TLC examination of the water extracts indicated that 2–3 spots are separated by the solvent system used. Two of the spots, with Rf 0.56 and Rf 0.16, were identified as p-nitrophenol (PNP) and diethyl thiophosphate (DETP), respectively. A more polar, unknown, compound (Rf 0.12) could be seen by UV-light and gave a brownish-yellow color upon spraying with the detection reagents, which indicates the presence of a S atom. This spot was tentatively identified as the phosphate hydrolysis product of SEP. This identification is supported by the fact that it appeared after SEP was formed and degraded. For example, in the water extract of ben-

Table 2. Composition of the water extracts of parathion complexes (TLC data).

Adsorbent	Pretreatment	Incubation conditions		Components[1]		
		Temp (°C)	Time	PNP	DETP	Unknown
Bentonite (crude)	none	180	10 min	m	—	s
Ca-bentonite	none	180	10 min	m	—	s
Ca-kaolinite	none	110	10 min	m	m	m
Aluminum oxide cationotrop	none	110	2 min	m	m	w
Bentonite (crude)	none			w	vw	—
	110°C			w	vw	—
	200°C			vw	vw	—
Ca-bentonite	none	22	60 days	w	m	vw
	110°C			vw	w	vw
	200°C			vw	vw	vw
Ca-kaolinite	none			m	w	m
Aluminum oxide cationotrop	none	22	50 days	s	s	m
Aluminum oxide anionotrop	none			m	m	m

[1] The letters s (strong), m (medium), w (weak) and vw (very weak) denote the relative intensities of the spots.

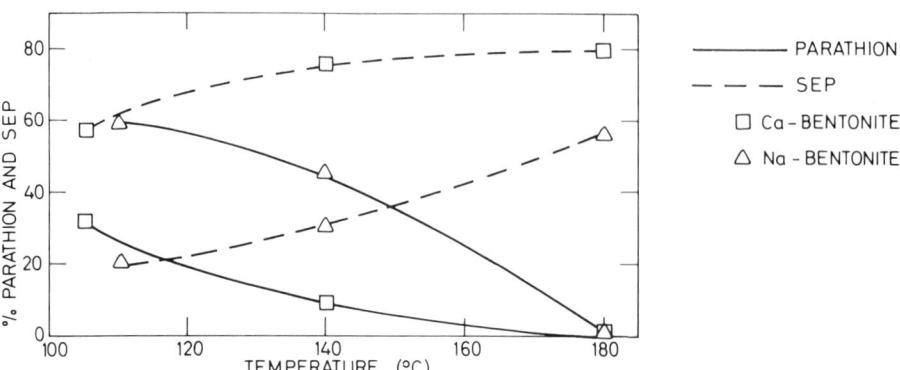

Fig. 2. Comparison between parathion loss and SEP evolution on Na- and Ca-bentonite.

tonites heated at 180°C for 10 min this spot is rather strong (Table 2 and Figure 1).

Table 2 gives some of the representative results obtained by the TLC analysis of the water extracts. The presence of either DETP or the unknown polar compound, or both, is an indication of the degradation mechanism. DETP is the phosphate-containing, hydrolysis product of parathion, while the polar, unknown, water-soluble compound seems to be the hydrolysis product of SEP, which was identified in the hexane extracts. Both SEP and parathion hydrolysis produces p-nitrophenol. The results obtained suggest that the degradation pathway is dependent on both the nature of the adsorbents and the incubation conditions. In bentonite-parathion complexes incubated at temperatures above 100°C no DETP, but only the unknown more polar compound and PNP were detected. Under similar conditions, Ca-kaolinite and aluminum oxide yielded both DETP and the unknown compound. This indicates that under these conditions, formation of SEP and its subsequent hydrolysis is the predominant mechanism for bentonites, while for kaolinite and aluminum oxide this mechanism and direct parathion hydrolysis seem to proceed simultaneously, with direct hydrolysis of parathion dominating in the case of the aluminum oxide. The rate of hydrolysis was about 5 times larger on the cationotrop as compared with the anionotrop aluminum oxide.

With all the adsorbents except the crude bentonite, long term incubation for 50–60 days at room temperature resulted in a degradation pattern similar to that mentioned previously in the case of kaolinite and aluminum oxides (the aluminum oxide cationotrop is one of the most active adsorbents in promoting parathion hydrolysis). On the other hand, the water extracts of the Na-bentonite complexes, irrespective of the predrying temperature, displayed only weak spots of DETP, and no spots of the unknown polar compound could be detected. With the homoionic Ca-bentonite direct parathion hydrolysis seems to be the dominant process in the air-dried samples, whereas in samples preheated at higher temperatures the relative importance of this degradation pathway declined (Table 2).

Gas-liquid chromatography—^{14}C-counting. The information obtained by TLC was used further to quantitate the rate of these processes by a combined GLC-^{14}C counting technique. In the hexane extracts in which parathion only was detected by TLC, this compound was measured by ^{14}C-counting. When both parathion and SEP were present in the same extract, the total amount of both compounds was accounted for by ^{14}C counting. In these extracts two peaks were obtained by GLC. One of the peaks (Rt 4.2 min) was identified as parathion and quantitatively measured by using standard parathion solutions. The retention time of the second peak was 8.2 min. The amount of SEP was calculated as the difference between the total amount of ^{14}C labeled compound and the amount of parathion found by GLC. The results obtained in such a way for Ca-bentonite-parathion complexes heated at different temperatures are presented in Figure 1.

The increase in temperature leads, as expected, to an increase in the rate of the degradation process, which in the case of bentonite proceeds through the formation of the rearrangement product SEP. The stability of this product is highly temperature-dependent. After 5 min at 180°C SEP concentration declined sharply, whereas at 105°C it was stable even after 18 hr. This shows that, under some conditions, the formation of SEP could be the rate-limiting step in parathion degradation. As the decrease in parathion concentration, the formation and disappearance of SEP, and the formation of the water-soluble hydrolysis product are consecutive reactions, each one with its own kinetics, the order of the overall reaction is complicated. Even the rate of the rearrangement reaction as extracted from the slope of parathion disappearance at time = 0 (Figure 1) does not obey simple zero or first-order reaction kinetics.

The effect of the saturating cation of the bentonites may be inferred from Figure 2, which presents a plot of the amounts of parathion disappearing, and of SEP recovered from bentonites after 10 min of incubation, versus the incubation temperature. Although the effect of temperature is rather similar—by increasing the temperature the amount of parathion decreases and that of SEP increases—there is an obvious difference between the two clays. The fraction of parathion remaining is higher with the crude bentonite, which is Na-saturated, than with the homoionic Ca-bentonite; also the fraction of SEP present is smaller with the Na-clay.

The extent of the hydrolysis of both parathion and SEP was determined from the ^{14}C-counting in the water extracts. However, total ^{14}C recovery (water plus hexane) at temperatures above 110°C decreased with time of heating. The loss in ^{14}C-counting is due to three factors: a) volatilization of parathion and SEP; b) fixation of the phosphate degradation product (Saltzman et al., 1976); and c) formation of volatile products of further degradation. At 180°C the presence of the ethyl mercaptan gas was detected by the lead acetate test.

DISCUSSION

The results obtained show that the adsorption catalyzed degradation of parathion on clays is a hydrolysis process, which proceeds directly and/or through an intermediary step, which was identified as a molecular rearrangement:

(1)

The rearrangement product may be rather long-lived.

The degradation mechanism and the velocity of the process are dependent on the nature of the clay, the saturating cation, and the hydration status of the clay, which is determined by the incubation conditions. The effect of the hydration status is emphasized by the results obtained with predried Ca-bentonite. By reducing the moisture content, the relative rate of the rearrangement reaction increases (Table 1), while the direct hydrolysis rate decreases (Table 2). It should be noted that an IR study of the dehydration of homoionic bentonites demonstrated that Ca-bentonite loses most of its cation hydration water at about 150°C, while some of this water is retained even above 200°C (Mingelgrin and Saltzman, unpublished data).

The results obtained with silica gel and aluminum oxides as adsorbents shed some light on the degradation mechanism, as related to the surface structure. Although the process was very slow in the presence of silica gel, the degradation proceeded through the molecular rearrangement exclusively, and no traces of the water-soluble degradation products were observed. With aluminum oxide, relatively very fast, direct hydrolysis was the predominant mechanism. This provides the explanation for the difference in behavior between bentonite and kaolinite clays. Bentonites, the adsorbing surfaces of which are almost exclusively silica sheets, behaved in a way similar to silica gel, although the specific features of the bentonite surfaces (the presence of counter-cations, the characteristics of the interlayer spaces, etc.) greatly enhanced the velocity of the process, as compared with silica gel. The presence of both alumina and silica surfaces in kaolinite may explain the simultaneous occurrence of the two degradation mechanisms, although direct hydrolysis is the predominant pathway at lower parathion concentrations (Mingelgrin et al., 1977). This is probably due to the stronger adsorbing capacity of alumina surfaces, as compared with silica ones, due to the surface hydroxyls-benzene ring interactions (Sahay and Low, 1974).

The importance of the cation in the degradation process is exemplified by the difference between Na- and Ca-bentonites. Calcium, as the saturating cation, enhances the degradation process, as compared with sodium. This cation effect was also evident with kaolinites (Saltzman et al., 1974). Another indication of the importance of the cation is provided by the much larger rate of hydrolysis on the cationotrop, as compared with the anionotrop aluminum oxide.

Based on the above observations that both the exchangeable cation and hydration status determine the mode of degradation, it may be suggested that ligand water of the exchangeable cation participates in both direct hydrolysis and rearrangement processes through the following mechanism:

(2)

where M is an exchangeable cation.

The above sequence (2) explains both hydrolysis and rearrangement at the same site in the absence of a liquid phase and is in agreement with the previously proposed mode of adsorption on clays (Mingelgrin et al., 1977). The occurrence of direct hydrolysis or rearrangement is a function of the precise conformation of the ester on the surface. Minor energy changes may affect the conformation of the phosphate, which in turn may affect considerably the chemical reactivity of the species (Fest and Schmidt, 1973). If the oxygen of the ligand water of the cation distorts the phosphate conformation towards a trigonal bipyramid structure, putting the three P–O bonds of the parathion molecule closer to one plane, the S–C bond will tend to form more easily, as the proper C-atom will get closer to the S-atom. The extent of distortion will depend on the total surface interactions. On the alumina surfaces the tendency towards a trigonal bipyramid structure may be smaller than on the silica surfaces, thus making the direct hydrolysis of parathion the dominant process on these surfaces, and the rearrangement dominant on silica and bentonites. An inductive effect of the conjugated system interaction with the hydroxyl surface may further induce hydrolysis on the alumina surfaces (Mingelgrin et al., 1977). The significantly larger rate of rearrangement on bentonites as compared with silica may result from the increased tendency of all the P–O bonds to get into one plane in the interlayer space, due to steric constraints imposed by the interlayer space.

Finally, it is interesting to note that the peak at 1265 cm^{-1} prominent in the IR spectrum of free SEP and assigned to P=O (Gore, 1950), shifts in the adsorbed SEP to 1282 cm^{-1} (Mingelgrin and Saltzman, 1978). This suggests that the newly formed P=O bond interacts strongly with the surface, in agreement with equation (2). In the region assigned to the P=S bond (Gore, 1950), no such large shift was observed, which is additional support for the mechanism proposed.

REFERENCES

Fest, C. and Schmidt, K. J. (1973) *The Chemistry of Organophosphorus Pesticides:* Springer-Verlag, Berlin.

Gore, R. C. (1950) Infrared spectra of organic thiophosphates: *Discuss. Faraday Soc.* **9**, 138–143.

Joiner, R. L. and Baetcke, K. P. (1974) Identification of the photoalteration products formed from parathion by ultraviolet light: *J. Assoc. Off. Anal. Chem.* **57**, 408–415.

Joiner, R. L., Chambers, H. W., and Baetcke, K. P. (1973) Comparative inhibition of boll weevil, golden shiner and white rat cholinesterases by selected photoalteration products of parathion: *Pestic. Biochem. and Physiol.* **2**, 371–376.

Lichtenstein, E. P. and Schulz, K. R. (1964) The effects of moisture and microorganisms on the persistence and metabolism of some organophosphorus insecticides in soils, with special emphasis on parathion: *J. Econ. Entomol.* **57**, 618–627.

Melnikov, N. N. (1971) Chemistry of Pesticides: Springer-Verlag, New York.

Mingelgrin, U., Gerstl, Z. and Yaron, B. (1975) Pirimiphos-ethyl-clay surface interactions. *Soil Sci. Soc. Amer. Proc.* **39**, 834–837.

Mingelgrin, U., Saltzman, S. and Yaron, B. (1977) A possible model for the surface-induced hydrolysis of organo-phosphorus pesticides on kaolinite clays. *Soil Sci. Soc. Amer. J.* **41**, 519–523.

Mingelgrin, U., Yaziv, S. and S. Saltzman (1978) The use of differential infrared spectroscopy in the study of the surface degradation of parathion on bentonite: (submitted for publication to *Soil Sci. Soc. Amer. J.*).

Mortland, M. M. (1970) Clay-organic complexes and interactions: *Adv. Agron:* **22**, 75–117.

Mortland, M. M. and Raman, K. U. (1967) Catalytic hydrolysis of some organic phosphate pesticides by copper (II): *J. Agric. Food Chem.* **15**, 163–167.

Polon, J. A. and Sawyer, E. W. (1962) The use of stabilizing agents to decrease decomposition of malathion on high-sorptive carriers: *J. Agric. Food Chem.* **10**, 244–248.

Rosenfield, C. and Van Valkenburg, W. (1965) Decomposition of (O,O-dimethyl-0-2,4,5-trichlorophenyl) phosphorothioate (Ronnel) adsorbed on bentonite and other clays: *J. Agric. Food Chem.* **13**, 68–72.

Sahay, B. K. and Low, M. J. D. (1974) Interactions between surface hydroxyl groups and adsorbed molecules. V. Fluorobenzene adsorption on Germania: *J. Colloid. Interface Sci.* **48**, 20–31.

Saltzman, S., Yaron, B. and Mingelgrin, U. (1974) The surface-catalyzed hydrolysis of parathion on kaolinite: *Soil Sci. Soc. Amer. Proc.* **38**, 231–234.

Saltzman, S., Mingelgrin, U. and Yaron, B. (1976) The role of water in the hydrolysis of parathion and methyl-parathion on kaolinite: *J. Agric. Food Chem.* **24**, 739–743.

Shainberg, I. and Otoh, H. (1968) Size and shape of montmorillonite particles saturated with Na/Ca ions (inferred from viscosity and optical measurements): *Isr. J. Chem.* **6**, 251–259.

Theng, B. K. G. (1974) *The Chemistry of Clay-organic Reactions:* A. Holger, London.

Резюме- Адсорбционно-каталитическая деградация паратиона на поверхностях глины является гидролизным процессом, протекающим непосредственно или через стадию перегруппировки. Скорость и механизм деградации зависят от природы глины, ее состояния гидратации и насыщающего катиона. Предлагается механизм деградации паратиона в местах адсорбции на поверхностях глины, в отсутствии жидкой фазы.

Kurzreferat- Die durch Adsorption katalysierte Degradation von Paration, welches sich auf Tonoberflächen befindet, ist ein Hydrolysenprozeß, der entweder direkt oder durch eine Umlagerungsstufe vor sich geht. Die Geschwindigkeit und der Mechanismus der Degradation hängt von der Natur des Tones und der Kationen und des Tones Hydrationszustand ab. Ein Mechanismus für die Parathiondegradation an den Adsorptionsplätzen auf Tonoberflächen, in Abwesenheit einer flüssigen Phase, wird vorgeschlagen.

Résumé-La dégradation catalysée par adsorption de parathion sur des surfaces argileuses est un processus d'hydrolyse, découlant soit directement, soit d'une étape de réarrangement. La vitesse et le méchanisme de dégradation dépendent de la nature de l'argile, de son statut d'hydratation, et de son cation de saturation. Un mécanisme est proposé pour la dégradation de parathion à des sites d'adsorption sur des surfaces argileuses, sans phase liquide.

Editors' Comments
on Papers 32 Through 35

32 VOERMAN and BESEMER
Residues of Dieldrin, Lindane, DDT, and Parathion in a Light Sandy Soil after Repeated Application throughout a Period of 15 Years

33 LICHTENSTEIN, KATAN and ANDEREGG
Binding of "Persistent" and "Nonpersistent" ^{14}C-Labeled Insecticides in an Agricultural Soil

34 KAUFMAN and EDWARDS
Pesticide/Microbe Interaction Effects on Persistence of Pesticides in Soil

35 WALKER
A Simulation Model for Prediction of Herbicide Persistence

PESTICIDE PERSISTENCE

The term *pesticide persistence* has an ambiguous character; it is used to express both a positive side—the duration of efficacy—and a negative side—undesirable persistence (Frehse and Anderson, 1982). Persistence for less than the desired time may result in poor efficiency, while persistence beyond this time may lead to residue problems. The considerable amount of information about the persistence of pesticides in soils shows a very wide variation of this property among the classes of pesticides. Although the variation in the persistence of the individual compounds belonging to the same chemical group is very large, generally, the persistence of the principal groups of pesticides increases in the following order: organophosphates ≤ carbamates and aliphatic acids herbicides < phenoxy, toluidine and nitrile herbicides < benzoic acids and amides < ureas and triazines << chlorinated hydrocarbons (Hiltbold, 1974).

One of the first reports of accumulation of DDT residues in soil was from Chisholm and co-workers (1950). This work showed that annual applications of DDT to orchards, as spray or dust, resulted in the increasing accumulation of DDT in soil. Paper 32 is a representative, long-term study of the persistence of some organochlorinated and an organophosphate insecticide (parathion) in soil. The insecticides were applied to field plots by spraying the crops or by soil application. After 15 years of repeated applications the total amount of residue decreased in the order DDT > dieldrin > lindane. Parathion practically disappeared about half a year after application. Considerable amount of DDT (18-32%) and dieldrin (15-19%) of the total amounts applied were recovered. A comparison of the concentration of these compounds in soil after 15 years of application, with annual application doses, indicates that a significant buildup of residues occurred. However, the authors pointed out that in the last years of the experiment the residue level did not significantly increase, and they assumed that a stationary state had been reached.

Lichtenstein, Fuhremann, and Schulz (1971) found that 15 years after a single soil application, 10.6% of the applied DDT remained in the soil. More recently, Cooke and Stringer (1982) investigated the change of DDT residue values in the soil during an 8-year period after the last spray. It was concluded that during this post-spray period the degradation of DDT to DDE was a significant feature of DDT persistence; the total concentration of these compounds remained constant with time. Similar to other studies, the authors found that no significant transfer from the top 10 cm of the soil occurred.

Although the high persistence of chlorinated hydrocarbon insecticides in soil has been repeatedly proved, its significance is not clear. The most probable risk is connected to the biomagnification effect and the possibility of transfer to other compartments of the environment. Concerning the less persistent pesticides, the pervading opinion is that, provided a judicious use, their persistence is not likely to produce economic or ecologic damage (Hance, 1982).

The persistence, or disappearance rate of pesticides, has been usually deduced from the amount of soil residues extractable by a routine solvent extraction. The use of radio-labeled pesticides has shown that often a fraction of the residues remained in soil after extraction. This fraction, released only by combustion or high temperature distillation, is usually termed *bound residues*. The formation of bound residues of some nonpersistent (methyl parathion and fonofos) and persistent (DDT and dieldrin) insecticides was investigated by Lichtenstein, Katan, and Anderegg (Paper 33). In this work, bound residues were defined as the ^{14}C-residues remaining in soil that has been extracted three times with a mixture of benzene, methanol, and acetone (1:1:1). The bound residues were determined by oxidizing the extracted soil to $^{14}CO_2$. The results showed that the rapid disappearance of extractable residues of the nonpersistent methyl-parathion was associated with a rapid formation of bound residues. For example, 7 days after soil treatment, 41% of the applied insecticide was bound to the soil. As shown by the extractable residues, fonofos was more persistent than methyl-parathion; 28 days after soil treatment 47% of the applied fonofos was extractable. However, a considerable proportion, 35%, was recovered as bound residue. Contrary to the organophosphate insecticides, the persistent chlorinated hydrocarbon insecticides were characterized by much higher extractability and lower binding properties. With dieldrin, only 6.5% was bound after 28 days, while more than 90% remained extractable. Except for an initial relatively high binding, DDT behaved in a similar way. By taking into consideration the differences between the persistent and nonpersistent pesticides, the authors suggested that the binding of pesticides to soil must be related to their reactivity and rate of metabolism to bindable compounds. However, Khan (1980) showed that, contrary to the widely accepted opinion that bound residues are mainly degradation products bound to the soil matrix, the parent compounds can also be found in a bound form. Additional questions related to bound residues, such as their persistence, biological effects, and ultimate fate, are now being investigated (e.g., Fuhr and Mittelstaedt, 1980: Khan and Ivarson, 1981; Khan, 1982).

A problem of permanent concern, related to pesticide persistence, is the possible effect of pesticides on the nontarget organisms. The voluminous pertinent literature provides contradictory information as far as the nature and intensity of the effects of pesticides are concerned. However, according to a generally accepted opinion, most insecticides and herbicides applied at the recommended rates do not have lasting effects on soil microflora. Still, some soil fungicides and fumigants could change the microbiological equilibrium (Parr, 1974).

A new aspect of pesticide-soil microorganisms interaction, which is of considerable practical significance, is presented in Paper 34. It has been observed that the biological activity of the herbicide EPTC and of the insecticide carbofuran was reduced in some soils in which these compounds have been applied for several years. Laboratory incubation experiments showed that each pesticide was far more rapidly decomposed in such "problem" soils than in respective "nonproblem" soils. The degradation was biologically induced; in the carbofuran problem soils the number of actinomycetes was several fold greater than in the nonproblem soils, while for EPTC bacterial populations were higher in problem that in nonproblem soils. Thus, the accelerated degradation is due to the repeated applications of a specific pesticide, which affected the development of microbial populations capable of rapidly metabolizing subsequent applications of the compound.

An additional important observation is that microbial populations of the problem soils were able to adapt to other similarly structured pesticides. However, the rates of degradation of the alternative molecule were usually slower than the rates of degradation by populations directly adapted to that molecule. The authors suggested that generally the rapid loss of biodegradable pesticides could be controlled by using specific inhibitors, inducers, or activators.

Prediction of Pesticide Persistence

Over the last 10 years, mathematical models have been developed to predict the persistence of pesticides from simple laboratory experiments. The simulation model proposed by Walker (Paper 35) was intended to predict the persistence of pesticides in the field from a limited amount of laboratory data by taking into consideration the effect of temperature and soil moisture content on the degradation rate. The main assumption of this model is that degradation fits first-order kinetics. Temperature effects were characterized using the Arrhenius equation, and moisture effects by the empirical equation $H = a\ MC^{-b}$, where H is the half life, MC the moisture content, and a and b are constants. In the computer program developed, meteorological data were used to simulate soil temperature and moisture content. The model was tested for the prediction of the disappearance rate of the herbicide napropamide applied to field plots. The pattern of loss of surface-applied napropamide was very different from that of the soil-incorporated herbicide, and the model gave good approximation only for the last degradation pattern. The model failed to predict the rapid rate of disappearance in the case of surface application because it takes into consideration only chemical and biological degradation and not other factors such as losses during application, volatilization, and photodegradation.

In further tests of this model (e.g., Walker and Zimdahl, 1981) the persistence of a wide range of herbicides under assorted environmental conditions was simulated. The data obtained showed that the agreement between the predicted and observed disappearance rates was variable, and a general tendency to underestimate the rates was observed. The limitations of the model with some compounds and under some meteorological conditions are due mainly to its inaccuracy in predicting the behavior of a pesticide immediately after application. The simulation of the complex and dynamic action of several factors affecting pesticide disappearance in this initial phase is very difficult and has not yet been attempted in simulation models.

References

Chisholm, R. D., L. Koblitsky, J. E. Fahey, and W. E. Westlake, 1950, DDT Residues in *Soil, Jour. Econ. Entomol.* **43**:941-942.
Cooke, B. K., and A. Stringer, 1982, Distribution and Breakdown of DDT in Orchard Soil, *Pestic. Sci.* **13**:545-551.

Fuhr, F., and W. Mittelstaedt, 1980. Plant Experiments on the Bioavailability of Unextracted (carbonyl-^{14}C) Metabenzthiazuron Residues from Soil. *Jour. Agric. Food Chem.* **28:**122-125.

Frehse, H., and J. P. E. Anderson, 1982, Pesticide Residues in Soil—Problems between Concept and Concern, in *Pesticide Chemistry, Human Welfare and the Environment,* vol. 4, J. Miyamoto, and P. C. Kearney, eds., 5th Int. Congr. Pesticide Chemistry Proc. (Kyoto, Japan), pp. 23-32 Pergamon Press, Oxford.

Hance, R. J., 1982. Herbicide Persistence—Is it a Problem? in *Pesticide Chemistry, Human Welfare and the Environment,* vol. 4, J. Miyamoto and P. C. Kearney, eds., 5th Int. Congr. Pesticide Chemistry Proc. (Kyoto, Japan, pp. 195-200.

Hiltbold, A. E., 1974, Persistence of Pesticides in Soil, in *Pesticides in Soil and Water,* W. D. Guenzi, ed., Soil Science Society of America Inc., Madison, Wis., pp. 203-222.

Khan, S. U., 1980, *Pesticides in the Soil Environment. Fundamental Aspects of Pollution Control and Environmental Science,* Ser. 5, R. J. Wakeman, ed., 5 Elsevier, Amsterdam 240p.

Khan, S. U., 1982, Distribution and Characteristics of Bound Residues of Prometryn in an Organic Soil, *Jour. Agric. Food Chem.* **30:**175-179.

Khan, S. U., and K. C. Ivarson, 1981, Microbiological Release of Unextracted (Bound) Residues from an Organic Soil Treated with Prometryn, *Jour. Agric. Food Chem.* **19:**718-721.

Lichtenstein, E. P., T. W. Fuhremann, and K. R. Schulz, 1971, Persistence and Vertical Distribution of DDT, Lindane, and Aldrin Residues, 10 and 15 Years after a Single Soil Application, *Jour. Agric. Food Chem.* **19:**718-721.

Parr, J. F., 1974, Effects of Pesticides on Microorganisms in Soil and Water, in *Pesticides in Soil and Water,* W. D. Guenzi, ed., Soil Science Society of America, Inc., Madison, Wis., pp. 315-340.

Walker, A., and R. L. Zimdahl, 1981, Simulation of the Persistence of Atrazine, Linuron and Metolachlor in Soil at Different Sites in the U.S.A., *Weed Res.* **21:**255-265.

Residues of Dieldrin, Lindane, DDT, and Parathion in a Light Sandy Soil after Repeated Application throughout a Period of 15 Years

Simon Voerman and A. F. H. Besemer

From 1953 on, dieldrin, lindane, DDT, and parathion, in two concentrations, were sprayed on crops several times a year. In addition, soil treatment with these insecticides took place once a year. Soil and crop samples were taken regularly throughout the whole period. The results of residue analyses of soil samples taken after 15 yr in layers of 10 cm (3.9 in.) to a depth of 60 cm are reported here. In this light sandy soil, DDT and dieldrin were much more persistent than lindane. Parathion disappeared rather soon. Below 20 cm, traces of only dieldrin and DDT were found.

Literature on insecticides in soils has been excellently reviewed (Edwards, 1966). The present paper concerns an experiment started in 1953 at Wageningen and continued almost without interruption for 15 yr, up to 1968. The aim, originally, was to determine the influence of pesticides on the yield and quality of crops. It was soon obvious, however, that the experimental plots were too small to provide reliable results on crop yields. From then on, the persistence of the insecticides in the cultivated soil and their uptake by crops were investigated. Changes in the soil fauna caused by the application of lindane, DDT, and parathion were also studied (Van de Bund, 1965). Soil samples to a depth of 15 cm were regularly taken and analyzed. Finally soil samples were taken to a depth of 60 cm, subdivided into six layers of 10 cm each. The results of the analyses of these samples are the subject of this paper.

METHODS AND MATERIALS

The experimental field at Wageningen (Climatic Data, Table I) was divided into plots of 12 m² each. These plots were separated by sunken concrete slabs up to about 1 m below soil level. The ground water level is always below that level. The plots were subdivided into six parts. These six parts were planted with different field crops: potatoes, beets, turnips, carrots, chicory, lettuce, and leek. Crop rotation was applied each year; each plant species was planted in the next part of the plot in a clockwise direction. The soil was light sandy and contained about 3% organic matter. Organic manure was applied every other year (Van de Bund, 1965).

Before the start of the experiment, no insecticide was used on the plots. During the experiment only the insecticides mentioned in Table II were applied. The field was protected against spray drift or run-off from neighboring fields through a nontreated zone.

Each insecticide (of technical quality) was applied on three plots. On the plots B and C, a water emulsion (1.2 l/12 m²) of the insecticide was sprayed on the crops. Both were treated three to five times a year. The pesticide residues in the soil of these plots result only from run-off of the spray during the application, and the washing-down by rain afterwards. Plot D was given a soil treatment once a year, early in the season before sowing or planting; a water emulsion (50 l/12 m²) of the insecticide was poured directly onto the soil. The insecticide was left on the soil surface after its application. Two soil treatments took place in 1961. At harvest time the crops, mainly root crops, were removed, with the residues remaining on or in these plants.

Table II gives the total quantity of active material applied per plot of 12 m² per yr. The 1955 soil application was given during the winter 1955/1956 (in Table II mentioned under 1956 together with the application in spring 1956). Plot D of dieldrin was treated with a 2% powder in 1953 and 1954. The technical DDT contained about 25% o,p'-DDT.

The soil samples for this investigation were taken on 1/21/1969 with a special soil borer. This consists of a stainless steel tube about 75 cm long, which inside contained a tube with a capacity of 100 ml (diameter 3.6 cm, inner length 10 cm). The small tube had a loose bottom that could be fastened by a simple turning of a rod with which it and the outer tube were driven into the soil. In each plot six drilling operations were performed to a depth of 60 cm. The samples of each layer of 10 cm (0.9 ± 0.1 kg) taken from the same plot were collected in plastic bags and stored at ice-box temperature until they were analyzed.

After thorough mixing by hand, 30.0 g of soil were taken out of each bag, placed in a 100 ml flask, and shaken with a mixture of 30 ml of benzene and 15 ml of 2-propanol for 18 hr on a shaking-machine. After standing overnight, 3.0 ml of supernatant was thoroughly mixed with 7 ml of a 3% solution of sodium sulfate in water in an 11-ml test tube with glass stopper on a Vortex mixer. The layers were separated by centrifuging and the upper layer was analyzed by glc with an electron capture detector after appropriate dilution if necessary. In the case of parathion, 20 ml of extract was washed with 75 ml of a 3% sodium sulfate solution in a separation

Laboratory for Research on Insecticides, and Plant Protection Service, Wageningen, The Netherlands.

Table I. Some Climatological Data of Wageningen[a]

Season	Mean Daily Max. Temp. in °C	Mean Daily Min. Temp. in °C	Mean Precipitation in mm
Winter Dec.–Feb.	5.2	0.2	196
Spring March–May	13.7	4.6	140
Summer June–Aug.	21.5	12.3	260
Autumn Sept.–Nov.	14.2	7.1	190

[a] Climatological Data of Netherlands Stations.

Table III. Quantities of Dieldrin, Lindane, DDT, DDE, and Parathion (in ppm) after Regular Application for 15 Years of the Insecticides

Compound	Plot	Soil Layer in Cm					
		0–10	10–20	20–30	30–40	40–50	50–60
Dieldrin	B	1.25	0.23	0.02	0.01	...	...
	C	2.29	0.86	0.02	0.01	...	...
	D	7.33	2.50	0.05	0.03	0.02	0.01
Lindane	B	0.09	0.01	...	...	...	...
	C	0.34	0.07	...	...	...	...
	D	1.30	0.23	...	...	...	...
p,p'-DDT	B	3.53	0.56	0.01	0.01	...	...
	C	8.94	1.05	0.02	0.01	0.01	...
	D	59.6	9.22	0.15	0.10	0.11	0.13
o,p'-DDT	B	0.78	0.07	...	...	...	...
	C	1.77	0.22	...	...	...	...
	D	12.9	2.02	0.04	0.02	0.02	0.02
p,p'-DDE	B	0.39	0.07	...	...	...	...
	C	0.77	0.11	0.01	0.01	...	...
	D	3.33	0.51	0.02	0.02	0.02	0.02
Parathion	B	0.01	...	...	...	...	...
	C	0.02	...	...	...	...	...
	D	0.06	0.02	...	...	...	...

funnel, and the benzene layer dried with sodium sulfate. Of the dried extract, 10 ml was evaporated till nearly dry, taken up in 1.0 ml of hexane, and analyzed by glc using a phosphorus detector.

The benzene had been purified by treatment with concentrated sulfuric acid and fractionation. The 3% sodium sulfate solution in water was extracted with pure benzene prior to use. All glasswork had been rinsed with a little pure benzene.

The analyses were performed on a Varian Gas Chromatograph, Model 204-1B using colums of borosilicate glass packed with 5% Dow 11 on Chromosorb-W 70/80 ($^1/_8''$ × 129 cm, for dieldrin and lindane), 10% FS 1265 on Gaschrom Z 80/100 mixed with a same quantity support coated with 10% Dow 200 (12,500 cS) ($^1/_8''$ × 120 cm for DDT, or 5% Apiezon L on Aeropak 30 70/80 ($^1/_8''$ × 132 cm for parathion). The last mentioned specially conditioned with Silyl-8 (Pierce Chemical Co.).

The calculations were based on the over-dry weight of the soil (105° C until constant weight was attained).

Precoated tlc plates Silica Gel F 254 (E. Merck AG, Darmstadt), layer thickness 0.25 mm, were used for the separation of p,p'-DDE, o,p'-DDT, p,p'-DDT, and p,p'-DDD. They were developed with hexane in a closed tank saturated with vapor of hexane and diethyl ether. The spots showed up under UV light. Extracts representing about 2 g of soil were spotted on the plate after a cleanup with Nuchar Attaclay and a subsequent concentration to 0.2 ml.

RESULTS AND DISCUSSION

The results of the analyses are recorded in Table III. Most figures are the mean of the results of two independent determinations. The total amount of insecticide per plot of 12 m² can be calculated from the figures given in Table III. (The specific gravity of the oven-dry soil is set at 1.3, one layer is 1560 kg.) Table IV gives the results of these calculations. In plot D of lindane very small amounts of a compound with the same retention distance as γ-PCCH (γ-pentachlorocyclohexene) were found: 0.01 ppm at 0 to 10 and 10 to 20 cm (Yule et al., 1967). No paraoxon was found in the parathion plots.

The presence of p,p'-DDE in the DDT plots could be confirmed by tlc. The quantities are recorded in Tables III and IV. No o,p'-DDE was found. The column used for the estimation of DDT did not separate p,p'-DDD from p,p'-DDT completely. So it was necessary to perform this separation by tlc at first. The DDD spots (invisible under UV) were extracted with benzene and the extracts subjected to glc. The upper layers only were investigated in this way. In the plots B and C, 0 to 10 cm, was found 0.1 and 0.2 ppm, respectively, in plot D 0 to 10 cm, 1.2 ppm, and D 10 to 20 0.1 ppm. The DDT used in these experiments was of technical quality like all the other insecticides and a sample of it contained some DDD and traces of DDE. Therefore these results are not conclusive about the metabolism of DDT in the soil.

About half a year after the last application practically all the

Table II. Total Quantity Active Material Applied per 12 m² per Year in Grams
(1 g/12 m² = 0.74 lb/acre = 0.83 kg/ha)

Insecticide	Plot	Year									
		1953	'54	'56	'57, '58, '59	'60	'61	'62, '63, '64, '65	'66	'67, '68	Total
Dieldrin	B	0.4	...	0.7	0.7	1.0	1.0	1.4	2.1	1.7	16
	C	0.7	...	1.4	1.4	2.1	2.1	2.8	4.2	3.5	33
	D	3.4	6.8	9.8	4.9	4.9	9.8	4.9	4.9	4.9	83
Lindane	B	0.3	...	0.3	0.3	0.5	0.5	0.7	1.0	0.8	8
	C	0.6	...	0.7	0.7	1.0	1.0	1.3	2.0	1.7	16
	D	2.4	2.4	3.4	1.7	1.7	3.4	1.7	1.7	1.7	30
DDT (incl. 25% o,p'-DDT)	B	1.8	...	1.8	1.8	2.7	2.7	3.6	5.4	4.5	43
	C	3.6	...	3.6	3.6	5.4	5.4	7.2	10.8	9.0	86
	D	23.5	23.5	48.0	24.0	24.0	48.0	24.0	24.0	24.0	407
Parathion	B	0.5	...	0.6	0.6	0.9	0.9	1.2	1.8	1.5	14
	C	1.0	...	1.2	1.2	1.8	1.8	2.4	3.6	3.0	29
	D	3.4	3.4	19.2	9.6	9.6	19.2	9.6	9.6	9.6	151

Table IV. Total Amount Residue in Grams per Plot of 12 m² after 15 Years of Repeated Application

Compound	Plot	Applied	Recovered	% Recovered
Dieldrin	B	16	2.4	15
	C	33	5.0	15
	D	83	15.5	19
Lindane	B	8	0.2	3
	C	16	0.6	4
	D	30	2.3	8
p,p'-DDT	B	32	6.4	20
	C	64	15.6	24
	D	305	108	35
o,p'-DDT	B	11	1.3	12
	C	22	3.1	14
	D	102	23	23
p,p'-DDE	B	...	0.71	...
	C	...	1.4	...
	D	...	6.1	...
Parathion	B	14	0.02	0.1
	C	29	0.03	0.1
	D	151	0.13	0.1

parathion disappears. By comparing the figures of Table IV with reports of earlier analyses (Pesticides Reports, 1963, 1965) it can be assumed that a stationary state has been reached. In the last years there has not been a significant increase in the residue level.

In plots B and C, a quantity of insecticide probably did not reach the soil because it was applied by spraying. This can also be concluded from the figures in Table IV. Only the more persistent insecticides are found in the deeper layers, although it is a negligible percentage.

For the chlorinated hydrocarbons, the yearly dose is smaller than the quantity still present in the soil after a stationary state has been reached, especially for dieldrin and DDT (Tables II and IV).

ACKNOWLEDGMENT

The authors are indebted to N. W. H. Houx for some practical suggestions concerning the analyses and to B. H. Kwant for carefully performing them.

LITERATURE CITED

Bund, C. F. Van de, *Boll. Zool. Agr. Bachicolt. S. II*, **7**, 185 (1965).
Climatological Data of Netherlands Stations No. 2, Averages for the Period 1951–1960, 52–53, De Bilt 1969.
Edwards, C. A., *Residue Rev.* **13**, 83 (1966).
Pesticides Reports Nos. 69 and 111, Food Inspection Laboratory, Amsterdam.
Yule, W. N., Chiba, M., Morley, H. V., J. AGR. FOOD CHEM. **15**, 1000 (1967).

33

Copyright © 1977 by the American Chemical Society
Reprinted from *Jour. Agric. Food Chem.* **25**:43-47 (1977)

Binding of "Persistent" and "Nonpersistent" ^{14}C-Labeled Insecticides in an Agricultural Soil

E. Paul Lichtenstein,* J. Katan,[1] and B. N. Anderegg

The extractability and formation of bound ^{14}C-labeled residues in an agricultural loam soil were investigated with the "nonpersistent" insecticides [^{14}C]methylparathion and [^{14}C]fonofos (Dyfonate) and with the "persistent" insecticides [^{14}C]dieldrin and p,p'-[^{14}C]DDT. With [^{14}C]methylparathion only 7% of the applied radiocarbon was extractable 28 days after soil treatment, while ^{14}C-bound residues amounted to 43% of the applied dose. With [^{14}C]fonofos, however, still 47% of the applied dose was extractable and 35% of the applied radiocarbon was bound. Only a fraction of the radiocarbon extracted from [^{14}C]methylparathion treated soil was associated with the parent compound, while extractable ^{14}C-labeled residues from the other insecticide-treated soils were primarily due to the presence of the parent compounds. Smaller amounts of soil-bound residues had been formed with the "persistent" insecticides, amounting after 28 days to only 6.5% of the applied [^{14}C]dieldrin and to 25% of the applied p,p'-[^{14}C]DDT, while 95 and 72%, respectively, were still recovered by organic solvent extraction. They differed from the organophosphorus compounds in their relatively low binding properties and their high extractability from soils. Contrary to results with [^{14}C]parathion, the mechanism of binding of [^{14}C]fonofos was not dependent on the presence of soil microorganisms. At higher application rates of the insecticides, relatively less radiocarbon was bound, possibly due to saturation of binding sites. Bound residues were found to be either nontoxic to fruit flies or of drastically reduced insecticidal activity. The significance of the formation of insecticide bound residues in soils in reassessing persistence of pesticides is discussed.

During the last three decades of extensive use, insecticides were referred to as either "persistent" or "nonpersistent". In most cases, those insecticide residues which could be extracted from soils or plant material by conventional methods long after they had been applied were the "visible" ones and therefore considered to be "persistent" insecticides. The use of ^{14}C-labeled insecticides, however, has made it possible to detect unextractable ^{14}C-labeled residues by combusting the insecticide contaminated material, after exhaustive extraction, to $^{14}CO_2$. The presence of these bound ^{14}C-labeled residues changed our thinking about "persistent" or "nonpersistent" insecticides, as indicated in a recent publication from our laboratory by Katan et al. (1976). They found that the total radiocarbon (extractable and bound) recovered 28 days after treatment of an agricultural loam soil with [^{14}C]parathion still amounted to 80% of the applied dose. Of this, 35% was extractable and associated with parathion and 45% was bound. It was also found that these bound residues were a product of soil microorganism activity and were primarily amino derivatives of parathion. The production of bound residues in soil was also shown with propanil (Bartha, 1971), fonofos (Flashinski and Lichtenstein, 1974b), and others.

In view of these findings, further studies were conducted in our laboratory with silt loam soil which has been used for both field and laboratory studies with a variety of insecticides during the last 18 years. To obtain additional insight into the phenomenon of extractable and bound residues, investigations were conducted with the "nonpersistent" organophosphorus insecticides, [^{14}C]methylparathion, [^{14}C]parathion, and [^{14}C]fonofos, and the "persistent" chlorinated hydrocarbon compounds, [^{14}C]dieldrin and p,p'-[^{14}C]DDT.

EXPERIMENTAL SECTION

Materials. [*ring*-^{14}C]Methylparathion (sp act. 2.83 μCi/mg), [*ring*-^{14}C]parathion (sp act. 2 μCi/mg), and p,p'-[*ring*-^{14}C]DDT (sp act. 2.09 μCi/mg) were purchased from Amersham-Searle, [*ring*-^{14}C]fonofos (Dyfonate) (sp act. 1.78 μCi/mg) and [*ethoxy*-^{14}C]fonofos (sp act. 1.74 μCi/mg) were obtained from the Stauffer Chemical Company, and [^{14}C]dieldrin (labeled in all positions adjacent to chlorines) (sp act. 2.95 μCi/mg) was obtained from the Shell Development Company. The radiopurity of these insecticides was at least 99% after purification by thin-layer chromatography. A small amount of [^{14}C]DDE was isolated from the originally supplied

---

Department of Entomology, University of Wisconsin, Madison, Wisconsin 53706.

[1]Presently on sabbatical leave from the Hebrew University of Jerusalem, Israel.

[^{14}C]DDT. Solvents used were anhydrous methanol and redistilled acetone and benzene.

The soil used was a Plano silt loam (organic matter, 4.2%; sand, 4.8%; silt, 68%; clay, 23%; pH 6.0) and was obtained from the Experimental Farm of the University of Wisconsin near Madison.

Soil Treatment. For each test, 10-g aliquots of field moist loam soil were placed into each of 15 82 × 20 mm glass vials. Measured amounts of an acetone solution of a particular insecticide were then added to each soil at 1 or 10 ppm. Soil moistures were then adjusted to 20% and were maintained at that level throughout the test. The vials were plugged with cotton and incubated in the dark at 27 ± 1 °C for 0, 7, 14, 21, and 28 days. Additional samples containing [^{14}C]DDT were also analyzed after 1 day and of [^{14}C]methylparathion-treated soils after 1, 2, and 4 days. Each triplicated experiment was repeated once.

Since previous studies in our laboratory (Katan et al., 1976) indicated that the binding of [^{14}C]parathion was a function of soil microorganisms which reduced the insecticide to bindable amino compounds, the potential involvement of soil microorganisms in the binding of [^{14}C]fonofos was investigated in this study. Experiments were, therefore, conducted with [^{14}C]fonofos-treated soils previously sterilized by autoclaving (1 h at 120 °C and 1 atm) on two successive days or by γ irradiation (45 000 rads for 70 h), which causes less alteration of the soil structure and composition. When sterile soils were used, all procedures up to the extraction were carried out aseptically; the sterility of these soils was confirmed by incubating samples in yeast extract–dextrose medium. To test the potential effects of anaerobic microorganisms, loam soil was also flooded by pipetting distilled water onto the treated sample until it was 15 mm above the soil surface. All experiments were carried out with three replicates.

EXTRACTION AND ANALYSES

At the end of the incubation periods, the soil from each vial was quantitatively transferred into a 125-ml Erlenmeyer flask and extracted three times with 75 ml of a mixture of benzene, methanol, and acetone (1:1:1) by shaking each time for 1 min. After separation of the soil and the solvent in a Büchner funnel, the combined extracts from each soil sample were concentrated at 35 °C in a rotary evaporator and adjusted to 100 ml with the same solvent mixture. Aliquots were used for liquid scintillation counting (lsc) as described (Lichtenstein et al., 1972). Preliminary studies conducted in our laboratory with [^{14}C]parathion showed that six additional extractions with various solvent mixtures, ranging in polarity from benzene to water, yielded only a total of 2.4% of the applied radiocarbon, while 56.5% still remained in the soil as bound residues. Unextractable or bound residues were, therefore, defined as ^{14}C-labeled residues remaining in soils which had been extracted three times, as is routinely done in our laboratory. To determine bound residues, three 1.5-g aliquots of the extracted soil from each vial were oxidized to ^{14}CO$_2$ in a Packard Model 305 Tri-Carb sample oxidizer and subsequent liquid scintillation analyses were performed as described by Flashinski and Lichtenstein (1974a).

For analyses by gas–liquid chromatography (GLC), concentrated extracts of soils treated with [^{14}C]methylparathion, [^{14}C]parathion, or [^{14}C]fonofos were analyzed as described by Lichtenstein et al. (1973) except that a 1.8 m × 4 mm column, containing 10% DC-200 on 80–100 Chromosorb W, was used and the detector gas flow rates were 60 ml/min of hydrogen and 105 ml/min of air.

Extracts containing [^{14}C]DDT or [^{14}C]dieldrin were concentrated at 35 °C on a rotary evaporator to near dryness, redissolved in benzene, adjusted to volume, and analyzed by electron-capture GLC as described (Lichtenstein and Schulz, 1970).

Bioassay Procedures. Since the bound residues could not be extracted from the soils, the problem of their potential biological activity, if any, is of interest. For this purpose, fruit flies (*Drosophila melanogaster*, Meigen) were exposed to aliquots of extracted soils that contained ^{14}C-bound residues. Soil samples, previously treated at 10 ppm with [^{14}C]methylparathion or [^{14}C]fonofos, were incubated at 27 ± 1 °C for 1 or 3 weeks, respectively. The soils were then extracted as described and the amounts of bound radiocarbon were determined. Fifty flies were then introduced into each of three bioassay jars containing 4 g of extracted soil (Edwards et al., 1957). Mortality counts were performed at intervals over a 72-h exposure period.

For control purposes, fruit flies were also exposed to soils immediately after treatment with one of the insecticides at a concentration equivalent to the radiocarbon determined as "bound" in the 1- or 3-week incubated soils. To check for potential solvent toxicities, insecticide-free soils were also extracted three times as described, and then exposed to fruit flies. These latter soils, however, did not show any insect mortality over a 72-h exposure period.

RESULTS AND DISCUSSION

Methylparathion is usually referred to as considerably less persistent than parathion. Thus, Lichtenstein and Schulz (1964) determined that only 3.5% of methylparathion applied to the same loam soil could be recovered under field conditions 1 month after application, while 25% of applied parathion could still be accounted for after that time. Under laboratory conditions at 30 °C and after 12 days of incubation, methylparathion and parathion were recovered from the soil to an extent of 7 and 16%, respectively, of the applied dose. This higher "degradability" of methylparathion was later also reported by Getzin and Rosefield (1968) and by Kishk et al. (1976). If nonpersistence is, to some extent, related to nonextractability, then methylparathion could be expected to be bound to a larger degree than parathion. Figure 1A shows that this had indeed happened. (For comparison purposes, the amounts of bound [^{14}C]parathion residues as determined by Katan et al. (1976) are also shown.) In soils treated with [^{14}C]methylparathion and analyzed after 0, 1, 2, and 4 days of incubation, a relatively rapid binding of radiocarbon compounds occurred which, in turn, is reflected in the rapid decline of extractable [^{14}C]methylparathion residues. Thus, 7 days after soil treatment, only 29.2% of the applied radiocarbon was extractable while 41.2% was bound to the soil, resulting in a total recovery of 70.4% of the applied radiocarbon. At that time, however, previously determined extractable [^{14}C]parathion residues amounted to 65% of the applied dose, and only 27% was nonextractable (Katan et al., 1976). While with [^{14}C]parathion-treated soils the extractable radiocarbon was primarily associated with parathion, this was not the case with [^{14}C]methylparathion. Utilizing GLC of the soil extracts, the amounts of [^{14}C]-methylparathion were 85, 67, 44, 32, 19, and 7% and traces of the applied dose, after 0, 1, 2, 4, 7, 14, 21, and 28 days of incubation, respectively. The amounts of recovered radiocarbon, however, were considerably higher and amounted to 96, 76, 65, 45, 29, 14, 8, and 7%, respectively. This indicated that, unlike parathion, methylparathion had been metabolized to unidentified organic-soluble products. Since the total recovery of extractable and bound [^{14}C]-methylparathion derived radiocarbon amounted to only

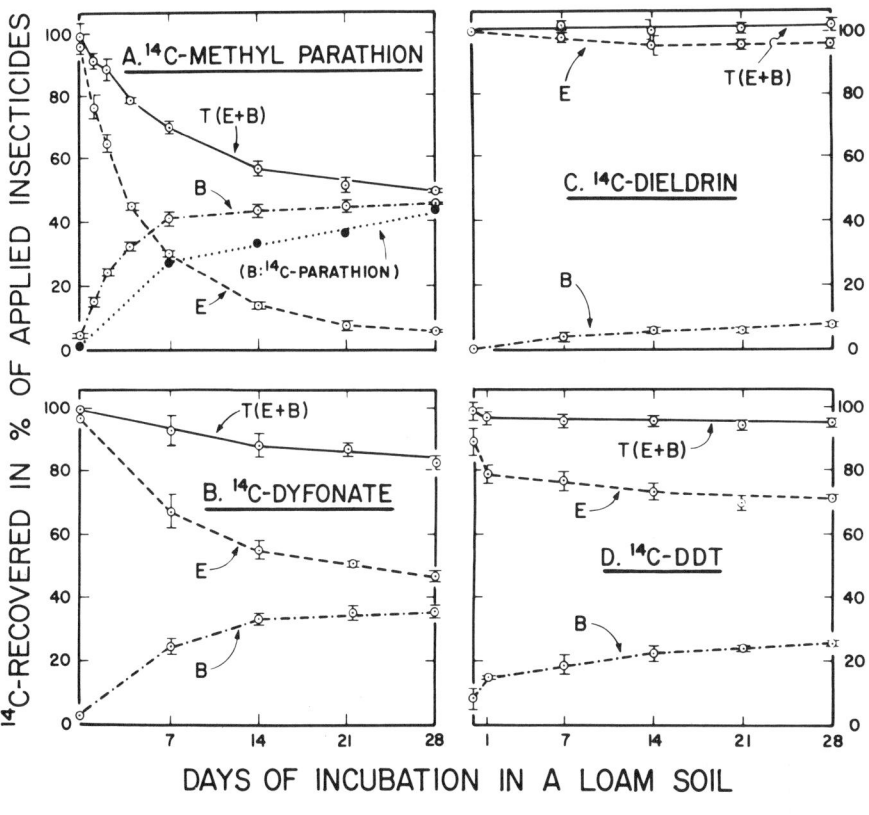

Figure 1. Binding and extractability of ¹⁴C-labeled insecticides in a silt loam soil during a 28-day incubation period, after soil treatment at 1 ppm. With the exception of [¹⁴C]methylparathion the amounts of extractable [¹⁴C]Dyfonate (fonofos), [¹⁴C]dieldrin, and [¹⁴C]DDT, as determined by gas-liquid chromatography, were similar to the amounts of extractable radiocarbon. For comparison purposes, data are inserted in A for the bound residues of [¹⁴C]parathion in soil (Katan et al., 1976).

50% of the applied dose after 28 days of incubation (this figure was 78% with [¹⁴C]parathion), volatile materials had apparently been produced which were no longer detectable. This greater reactivity or metabolism of methylparathion could be a factor in its relatively higher soil binding properties.

The organophosphorus insecticide fonofos, often used for soil insect control, had been tested with our Plano silt loam. Under field conditions 50% of the applied insecticide was extracted from the soil 28 days after soil treatment (Schulz and Lichtenstein, 1971). Under laboratory conditions [ethoxy-¹⁴C]fonofos or [ring-¹⁴C]fonofos-treated soils, through which water was percolated 1, 15, and 29 days after the insecticide application, contained, after 29 days, 59.5 and 62.0%, respectively, of the applied dose as extractable fonofos residues, while 11.4 and 16.1%, respectively, were bound (Lichtenstein et al., 1972). In the present study (Figure 1B) it is also shown that 46.8% of the applied [ring-¹⁴C]fonofos was extractable after 28 days of incubation. Results obtained for extractable radiocarbon were similar to those determined by GLC for fonofos. As with [¹⁴C]methylparathion, however, increasing amounts of bound radiocarbon were noticed with time, amounting to 35.3% of the applied insecticide at the end of the incubation period. The total recovery of extractable and bound residues 28 days after soil treatment amounted to 82.1% of the applied fonofos. Additional soil samples were also treated with [ethoxy-¹⁴C]fonofos, incubated for 14 days and extracted and analyzed as described. Results (30.0% of the applied radiocarbon bound and 54.3% extractable) were very similar to those obtained with [ring-¹⁴C]fonofos (Figure 1B), indicating that bound residues probably contain both the ethoxy and ring moieties of the fonofos molecule.

Contrary to results obtained previously with [¹⁴C]-parathion (Katan et al., 1976), the binding of [¹⁴C]fonofos in soil was not related to the presence of microorganisms. Thus, 2 weeks after soil incubation, $37.0 \pm 2.9\%$ of the radiocarbon derived from [ring-¹⁴C]fonofos was bound in control soils, $32.5 \pm 1.3\%$ in irradiated soils, $23.4 \pm 2.3\%$ in autoclaved soils, and $28.1 \pm 1.8\%$ in flooded soils. With [ring-¹⁴C]parathion, these figures were 34.8 ± 1.2, 14.0 ± 1.4, 14.7 ± 1.1, and $66.7 \pm 2.1\%$, respectively. Since fonofos does not contain a reducible nitro group, the mechanism of binding, as observed with parathion, could not be expected. Differences observed in the binding of [¹⁴C]fonofos between irradiated and nonirradiated soils were not significant, while with autoclaved soils, probably due to the change of soil structure, significantly less ¹⁴C was bound than in irradiated soils. Moreover, flooding of the soil

Table I. Insecticidal Activity of Soil Bound and Freshly Deposited Insecticide Residues in Soil[a]

Insecticide (concn, ppm)	Mode of soil contam.	% insect mortality after h of exposure to soil					
		2	3	18	24	48	72
Dyfonate (3)	Bound[b]	0	0	0	0	12 ± 2	17 ± 1
Dyfonate (3)	Fresh[c]	37 ± 5	100	100	100	100	100
Methylparathion (2.9)	Bound	0	0	0	0	4 ± 2	5 ± 3
Methylparathion (2.9)	Fresh	0	0	43 ± 7	69 ± 12	96 ± 2	96 ± 2
None[d]		0	0	0	0	0	0

[a] Test insect: *Drosophila melanogaster*, Meig. Results are means ± standard deviation of triplicated tests. [b] ^{14}C-bound residues remaining in soils after incubation with an insecticide and extraction. Concentration (ppm) is based on the amount of ^{14}C recovered by combustion. [c] Soil was treated with the insecticide at concentrations as calculated in footnote b, followed by immediate exposure of fruit flies to the soil. [d] Insecticide free soil after its extraction with a mixture of benzene, methanol, and acetone (1:1:1).

resulted in reduced binding of [^{14}C]fonofos, while with [^{14}C]parathion, a dramatic increase had occurred.

It appears that the binding of some insecticides to soil is related to their reactivity and rate of metabolism to bindable compounds. Therefore, more stable compounds, such as dieldrin and DDT, would be expected to be bound to soil to a lesser degree than organophosphorus compounds. Results obtained with ^{14}C-labeled dieldrin and DDT prove this point (Figure 1, C and D). With [^{14}C]dieldrin only 6.5% of the applied radiocarbon was bound after 28 days, while 95-97% remained extractable throughout the incubation period. As shown by GLC analyses this extractable radiocarbon was associated with dieldrin.

With p,p'-[^{14}C]DDT, however, more binding was observed, in particular during the first day after soil treatment. After that, the increase in binding was slow. Indeed, binding at zero time (actually after the first 1-2 h) was unusually high, amounting to 9.7% of the applied ^{14}C as compared to 4.2, 2.7, and 0.7% for methylparathion, fonofos, and dieldrin, respectively. After 1 day, the amount of [^{14}C]DDT derived bound radiocarbon was 17.2% of the applied dose and increased by only 7.9% during the remaining 27 days. Thus, except for the initial relatively high binding of [^{14}C]DDT, which apparently was not dependent on the activity of microorganisms, the insecticide resembled dieldrin in the slow increase of binding with time, in the high recovery of total radiocarbon, and in the high persistence of the insecticide as shown by GLC. This latter analysis also showed that the extracted radiocarbon from p,p'-[^{14}C]DDT treated soil was primarily associated with p,p'-DDT. Additional soil samples, treated with [^{14}C]DDT as described, were incubated for 56 days and extracted and analyzed. However, no further increase in binding had occurred. To test whether the high initial binding of [^{14}C]DDT was related to the type of extraction solvents used, separate [^{14}C]DDT treated soil samples were extracted after 14 days of incubation with either a 1:1 mixture of hexane-acetone or with benzene-acetone (1:1). However, the amount of bound residues was similar to that obtained with the solvents mixture used throughout this study.

Although only traces of DDE were found in DDT-treated soil, the problem of its capacity to be bound to soil was investigated. For that purpose, soil was also treated with [^{14}C]DDE at 1 ppm. After 1 day of incubation, however, only 1.3% of the applied dose was bound to the soil, indicating that this DDT metabolite was not responsible for the initial relatively high binding of [^{14}C]DDT. Porter and Beard (1968) had shown that part of the applied [^{14}C]DDT was no longer extractable after a short incubation period. Guenzi and Beard (1968) found that after 24 weeks of incubation of [^{14}C]DDT treated soil, 11% of the applied radiocarbon was detected by combustion of the previously extracted soil.

The "persistence" or "disappearance" of several chlorinated hydrocarbon insecticides in soil was also shown to be dependent on their rate of application (Lichtenstein and Schulz, 1959). Recoveries of DDT, aldrin, and lindane, expressed in percent of the applied dose, were smaller with lower application rates. To translate these 1959 findings into our knowledge today, one would assume that at lower application rates a larger proportion of the applied insecticide would be available for metabolism and binding. To study this possibility, soils were also treated at 10 ppm with p,p'-[^{14}C]DDT, [ring-^{14}C]fonofos, and [ring-^{14}C]methylparathion. After an incubation of 7 ([^{14}C]methylparathion) or 21 days, soils were extracted and analyzed as described. It was found that the amount, expressed in percent of the applied dose, of these insecticides bound to soil was indeed higher after 1-ppm application than the comparable figures obtained with soil treated at 10 ppm. These amounts for the applied dose of 1 ppm were 22.5, 33.0, and 43.2% for [^{14}C]DDT, [^{14}C]fonofos, and [^{14}C]methylparathion, respectively, while at the application of 10 ppm these figures amounted to only 15.6, 30.8, and 29.0%, respectively. Although an increase in pesticide concentrations resulted in relatively less binding, the absolute amount of bound residues was greater at the higher concentration. Since at higher application rates of insecticides to soil more binding sites might be saturated, the greater binding at lower concentrations, expressed in percent of the applied dose, could be explained.

The question of the potential biological activity of bound insecticide residues was investigated as described above by testing the insecticidal activity of bound residues from [^{14}C]fonofos and [^{14}C]methylparathion treated soils with fruit flies. Results obtained are summarized in Table I. With soils containing unextractable radiocarbon at the insecticide equivalent of 3 ppm, no mortalities were observed during a 24-h exposure period to the soil and only slight mortalities during an additional 48-h exposure period. However, with soils to which the insects were exposed immediately following the insecticide application, 50% of the flies had died within 2-3 h after fonofos application and within 18-20 h after soil treatment with methylparathion. It appears, therefore, that bound residues are not only unextractable, but they are also less active biologically.

The chlorinated hydrocarbon insecticides differed from the organophosphorus compounds in their relatively low binding properties, their high extractability from soils, and the resulting high recoveries (94-100% of applied) of the total radiocarbon, previously applied to the soil. Guenzi and Beard (1968) found that the unextractable portion of [^{14}C]DDT applied to soil was similar in both sterile and nonsterile soils. This seems to indicate that DDT is bound either as an intact molecule (Guenzi and Beard, 1968) or in the form of one of its metabolites, other than DDE, which could be rapidly formed from DDT in the soil. Hsu and Bartha (1976) reported that bound residues of 3,4-dichloroaniline were composed of both hydrolyzable and nonhydrolyzable forms. Therefore, the nature and mechanism of the formation of soil-bound residues of different pesticides may be different.

In the future, it will be important to obtain information about the mechanism of binding of pesticides, thus possibly shedding some light on the mechanism of their potential release and the conditions at which this release might occur. Since not much information is available pertaining to the nature and the potential biological activity of the compounds that are bound, extensive research in this field is highly desirable. In view of the above findings, the expression "disappearance" and "persistence" of pesticides, so widely used during the last two decades, should be reassessed to consider the bound products.

ACKNOWLEDGMENT

Special thanks are expressed to T. W. Fuhremann, T. T. Liang, and K. R. Schulz for their assistance in performing this research.

LITERATURE CITED

Bartha, R., *J. Agric. Food Chem.* 19, 385 (1971).
Edwards, C. A., Beck, S. D., Lichtenstein, E. P., *J. Econ. Entomol.* 50, 622 (1957).
Flashinski, S. J., Lichtenstein, E. P., *Can. J. Microbiol.* 20, 399 (1974a).
Flashinski, S. J., Lichtenstein, E. P., *Can. J. Microbiol.* 20, 871 (1974b).
Getzin, L. W., Rosefield, I., *J. Agric. Food Chem.* 16, 598 (1968).
Guenzi, W. D., Beard, W. E., *Soil Sci. Soc. Am. Proc.* 32, 522 (1968).
Hsu, T-S., Bartha, R., *J. Agric. Food Chem.* 24, 118 (1976).
Katan, J., Fuhremann, T. W., Lichtenstein, E. P., *Science* 193, 892 (1976).
Kishk, F. M., El-Assawi, T., Abdel-Ghafar, T., Abou-Donia, M. B., *J. Agric. Food Chem.* 24, 305 (1976).
Lichtenstein, E. P., Fuhremann, T. W., Schulz, K. R., Liang, T. T., *J. Econ. Entomol.* 66, 863 (1973).
Lichtenstein, E. P., Schulz, K. R., *J. Econ. Entomol.* 52, 124 (1959).
Lichtenstein, E. P., Schulz, K. R., *J. Econ. Entomol.* 57, 618 (1964).
Lichtenstein, E. P., Schulz, K. R., *J. Agric. Food Chem.* 18, 814 (1970).
Lichtenstein, E. P., Schulz, K. R., Fuhremann, T. W., *J. Agric. Food Chem.* 20, 831 (1972).
Porter, L. K., Beard, W. E., *J. Agric. Food Chem.* 16, 345 (1968).
Schulz, K. R., Lichtenstein, E. P., *J. Econ. Entomol.* 64, 283 (1971).

Received for review July 6, 1976. Accepted October 14, 1976. Research supported by the College of Agricultural and Life Sciences, University of Wisconsin, Madison, by the University of Wisconsin Graduate School, and by a grant from the National Science Foundation (GB-3502). Contribution by Project 1387 from the Wisconsin Agricultural Experiment Station as a collaborator under North Central Regional Cooperative Research Project 96, entitled "Environmental Implications of Pesticide Usage".

PESTICIDE/MICROBE INTERACTION EFFECTS ON PERSISTENCE OF PESTICIDES IN SOIL

Donald D. Kaufman and Debra F. Edwards

INTRODUCTION

Microbial participation in the degradation of pesticides in soil generally involves the utilization of either adaptive or constitutive enzyme systems which biochemically alter the pesticide into a utilizable nutrient and energy source. The utilization of pesticides as sole or supplemental sources of carbon, nitrogen, or other nutrients has been the subject of numerous research reports and reviews (1-3). A few pesticides are known to be "cometabolically" degraded by mechanisms which, at least initially, do not yield energy or nutrients to the microbe involved, and may in fact require expenditure of energy by the cell, and the utilization of other exogenous substrates. Microbial involvement in pesticide degradation may also occur indirectly through the creation of specific environments conducive to chemical degradation of certain pesticides. The free radical degradation of the herbicide amitrole (3-amino-s-triazole) (4,5) is affected by, and simulates microbial degradation. The chemical degradation of trifluralin (α,α,α-trifluoro-2,6-dinitro-N,N-dipropyl-p-toluidine) (6) occurs more readily under anaerobic conditions which are induced by microbial activity.

Numerous chemical, physical, and environmental factors are known to affect microbial degradation of pesticides in soil. The interactions of these various factors ultimately determines the overall persistence of each chemical under any given set of conditions. The recognition and identification of these factors are important in establishing the recommended usage patterns for each chemical. While considerable attention is generally given to the behavior of individual chemicals in the soil microbial environment, only scant attention has been given to the effects of multiple applications of either individual or combinations of pesticides on their persistence in soil. The objective of this presentation is to examine and briefly review several pesticide/microbe interactions known to affect pesticide persistence in soil, and to present new information which illustrates additional interactions.

ENHANCED OR ACCELERATED PESTICIDE DEGRADATION IN SOIL

The dissipation of certain biodegradable pesticides from soil generally follows a sigmoidal pattern (2). A lag phase occurs after the initial application in which relatively little

pesticide is lost. The lag phase is followed by a period of rapid disappearance of the
pesticide as a result of microbial metabolism. Subsequent applications of the pesticide to
soil are generally degraded more rapidly, and without the initial lag phase. The development
of microbial populations which are capable of rapidly degrading sequential applications of
pesticides has been demonstrated for various pesticides: phenoxyalkanoates (7-10); endothall
(7-oxabicyclo[2.2.1]heptane-2,3-dicarboxylic acid) (11); dalapon (2,2-dichloropropionic acid)
(12,13); chlorpropham (isopropyl m-chlorocarbanilate) (14,15); and several phenylamides
(16-19). Many of these observations, however, were made in soil perfusion or other
enrichment type systems. The occurrence of this phenomenon under actual field conditions,
however, has been recognized recently. Hurle and Rademacher (20) compared the dissipation of
DNOC (4-6-dinitro-o-cresol) and 2,4-D [(2,4-dichlorophenoxy)acetic acid] in soil treated for
the first time and soil from field plots treated annually over a period of 12 years. 2,4-D
dissipation was more rapid in previously treated soil than in soil treated for the first
time, whereas pretreatment had no effect on the rate of DNOC dissipation from soil. Similar
promotions have been obtained with 2,4-D (9,21-25), MCPA ([(4-chloro-o-tolyl)oxy]acetic acid)
(9,10,25,26), endothall (11,27), and dalapon (12,13), but not with simazine
[2-chloro-4,6-bis(ethylamino)-s-triazine] or linuron [3-(3,4-dichlorophenyl)-1-methoxy-1-
methylurea] (25). The time for MCPA applications to reach the limit of detection was reduced
from three weeks after three previous applications to four days after 10 previous
applications (25).

The actual significance of this phenomenon in soil would be of minor consequence to
pesticides which are primarily active as foliar or aerial contact chemicals. Pesticides
which are primarily active in soil or through root absorption from soil, however, could
expect to have limited effectiveness. Recent reports have indicated that several
soil-applied chemicals may be having problems of this nature in U.S. Midwestern corn-cropped
soils and elsewhere. Rahman et al., (28) reported that the herbicidal efficacy of Eradicane
[EPTC (S-ethyl dipropylthiocarbamate) + antidote(N,N-diallyl-2,2-dichloro- acetamide)] was
reduced in certain soils which had received successive annual applications of the herbicide.
They noted, however, that not all soils receiving this treatment had this "problem".
Preliminary investigations suggested that the reduced activity of Eradicane was due to a more
rapid microbial breakdown. Similar results were obtained by others (29-35). The insecticide
carbofuran (2,3-dihydro-2,2-dimethyl-7-benzofuranyl methylcarbamate) has also experienced
similar problems. Felsot, Maddox, and Bruce (36) assessed the persistence of carbofuran in
several soils with and without histories of carbofuran use. The particular history soils
selected for their study were noted as "problem" soils with a poor performance of carbofuran.
Their results suggested that an enhanced microbial degradation of carbofuran occurred in the
problem soils. Two other studies, however, were unable to define relationships between
history of insecticide use and carbofuran persistence (37,38). Thus, the relationship
between problem soils and history of application is not completely understood. While it is
true that some problem soils have a history of repeated (yearly) applications, there are
other problem soils in which there is no previous history of application of the pesticide
under concern. Research efforts of our own, therefore, were directed at first examining the
degradation of Eradicane and carbofuran in their respective problem and nonproblem soils.
"Problem" and "nonproblem" soils in our investigations (35,39) are defined as: (a) problem
soils, soil in which the chemical applied failed to control the target pest; and (b)
nonproblem soil, an identical soil type with an identical cropping history, but without any
known use of any chemical, or an identical soil type from an untreated border area adjacent
to a problem field.

We examined the degradation of ^{14}C-carbonyl carbofuran in five pairs of carbofuran
problem-nonproblem soils, and ^{14}C-ethyl- and ^{14}C-propyl-EPTC in three pairs of
Eradicane problem and nonproblem soils, all obtained from the midwestern corn producing areas
where these problems have predominantly appeared to date. All soils were received in a fresh
moist condition from field locations, and were immediately sieved through a 2mm No. 10 U.S.
Standard sieve prior to storage in polyethylene bags at 5°C. The persistence and microbial
degradation of ^{14}C-carbonyl and ^{14}C-EPTC, and several structurally related
^{14}C-herbicides, insecticides, and fungicides were then examined in both carbofuran and
Eradicane problem and nonproblem soils treated at recommended application rates. All soil
metabolism studies were performed in soil biometer flasks (40).

The results of investigations with ^{14}C-carbonyl carbofuran in carbofuran problem-
nonproblem soils are shown in Figure 1. Degradation of carbofuran with evolution of
$^{14}CO_2$ from the carbonyl position occurred far more rapidly in carbofuran problem soils
than in nonproblem soils. Similar results were obtained in two of the Eradicane problem and
nonproblem soil pairs, i.e., a more rapid evolution of $^{14}CO_2$ occurred from
^{14}C-ethyl- or ^{14}C-propyl-EPTC treated Eradicane problem soils than from nonproblem
soils. Very little difference was observed in ^{14}C-EPTC degradation between problem and
nonproblem soils of the third pair, i.e., EPTC degradation occurred rapidly in both soils.
Oat seedling bioassays of Eradicane treated soils confirmed a more rapid dissipation of EPTC

from Eradicane problem than nonproblem soils. The inhibition of degradation of these chemicals by antibiotics added to the soil, or by soil sterilization (autoclaving or gamma irradiation) confirmed the importance of an active microbial population in the degradation of these chemicals.

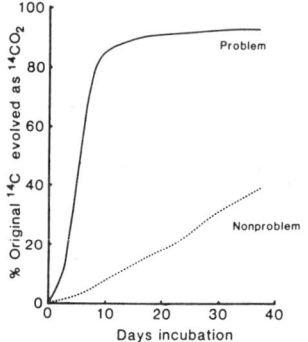

Figure 1. Evolution of $^{14}CO_2$ during degradation of ^{14}C-carbonyl-carbofuran in carbofuran problem and nonproblem soils (Mean of 5 soil pairs).

Soil bacteria, fungi, and actinomycete populations were enumerated with both standard selective dilution plating media and media which contained either carbofuran or EPTC as a sole carbon source. The total number of actinomycetes were several fold greater in all carbofuran problem soils than in the corresponding nonproblem soils, but did not correlate with either soil organic matter content or pH. Although total numbers of soil fungi could be correlated with soil pH, neither total number of fungi or bacteria could be correlated with soil organic matter content or history of carbofuran treatment when conventional selective isolation media were used (39). In all cases with Eradicane treated soil, bacterial populations were higher in problem soils than in nonproblem soils (35). When problem soils were treated with Eradicane and the soil microbial population analyzed again at the time of maximal $^{14}CO_2$ evolution, bacterial population differences were increased.

Investigations were also conducted with a pair of both carbofuran and Eradicane problem and nonproblem soils to determine what effect the microbial populations in these soils may have on the degradation of other structurally related chemicals. The soil microbial populations present in these problem soils were also capable of more rapidly degrading mobam (benzo[b]thien-4-yl methylcarbamate), CDAA (N,N-diallyl-2-chloroacetamide), and propachlor (2-chloro-N-isopropylacetanilide) (Table 1). While linuron was degraded more rapidly in carbofuran problem soil than nonproblem soil, there was no difference in the degradation rate in Eradicane problem and nonproblem soils.

Table 1. % of original ^{14}C evolved as $^{14}CO_2$ during degradation of ^{14}C-carbonyl labeled pesticides incubated for 32 days in Eradicane or carbofuran problem and nonproblem soils.

Pesticide	Soil type	% original ^{14}C evolved as $^{14}CO_2$ from soil	
		problem	nonproblem
mobam	carbofuran	39.6	8.0
	eradicane	65.9	53.5
CDAA	carbofuran	53.2	21.9
	eradicane	31.7	17.9
propachlor	carbofuran	15.2	6.7
	eradicane	26.4	23.8
linuron	carbofuran	18.4	13.1
	eradicane	14.7	14.9

The ability of microorganisms to degrade other structurally related molecules after adaptation to an original structure is a well known microbiological principle. Enzyme induction in a microbial culture is a response to an appropriate signal in the environment. Adaptation to a given substrate generally involves simultaneous adaptation to all intermediates in the breakdown chain, but it may also result in an increased ability to degrade homologous molecules as well. In the realm of pesticide metabolism this phenomenon was first demonstrated by Audus (7), who observed that a microbial population adapted to degradation of 2,4-D could also rapidly degrade MCPA and vice versa. Similar observations have been subsequently reported by others (21,26). We recently reported the isolation of several microorganisms which were each able to rapidly degrade a wide range of acetamide, acylanilide, carbamate, toluidine and urea-based pesticides (16). From these results it is obvious that there can be a cross adaptation among similarly structured molecules. The rates of degradation of the alternate molecule in most cases, however, are slower than the rates of degradation by microorganisms directly adapted to that molecule. Two alternative explanations have been suggested (41). Either each molecule induces, its own enrichment flora, each with the capacity of degrading other molecules, but with different efficiencies, or they each encourage the growth of the same organisms by respectively inducing their own specific enzymes which incidentally possess the power to degrade other similar molecules, though less efficiently. This phenomenon can be further complicated by the fact that some agricultural chemicals can induce microbial enzymes which are active in degrading other pesticides even though the inducer itself is not a substrate. Blake and Kaufman (42) observed that p-chlorophenyl methylcarbamate (PPG-124, or PCMC) induced acylamidase type enzymes in the soil fungus Fusarium oxysporum that were capable of hydrolyzing a wide variety of acetanilides although PCMC itself was not a substrate of the enzymes induced. PCMC also acted as a competitive inhibitor of the enzyme activity which it induced. Similarly, Engelhardt and associates (43,44) observed that the phenylurea herbicide monuron [3-(p-chlorophenyl)-1,1-dimethylurea] could induce an acylamidase in Bacillus sphaericus which was capable of hydrolyzing linuron although monuron itself was not a substrate. The phenylurea herbicide chlorbromuron [3-(4-bromo-3-chlorophenyl)-1-methoxy-1-methylurea], the acylanilide herbicides monalide (4-chloro-2,2-dimethyl-valeranilide) and propanil (3',4'-dichloropropionanilide) and fungicides 2-chlorobenzanilide and 2,5-dimethylfuran-3-carboxanilide, and the phenylcarbamate herbicide propham (isopropyl carbanilate) were also capable of inducing an acylamidase in B. sphaericus which hydrolyzed a wide variety of phenylamide herbicides and fungicides. The specific activity of the various extracts, however, was considerably lower than that obtained from linuron-induced cells.

The implication of these observations to agriculture are not completely clear at this time. It is evident from the experimental results described herein and literature reviewed that potential problems could develop in soils receiving various combinations of structurally related pesticides. Such problems would presumably be recognized through a loss of pesticide efficacy or an unexpectedly poor performance of the chemical affected. It is interesting to note that the problems currently being encountered with Eradicane and carbofuran were initially noted in midwestern corn fields where a preponderance of the types of agricultural chemicals used on corn are indeed structurally related. Whether the problems being encountered by either chemical are self-induced or cross-induced by other structurally related chemicals is not presently known, but is the subject of intensive investigation in our laboratory. The exceptionally rapid degradation of Eradicane in both of one history and nonhistory soil pair suggests that some predisposition to thiocarbamate degradation may have already occurred in the nonhistory soil. Additional research is needed to further clarify the significance of this phenomenon.

INHIBITED OR CONTROLLED BIODEGRADATION OF PESTICIDES IN SOIL

Persistent pesticides are necessary for adequate pest control in certain crops, but their toxicity to, or potential for contaminating succeeding crops may restrict their use. In contrast, the rapid loss of biodegradable pesticides frequently limits their usefulness and effectiveness in some cropping systems, or requires successive applications in a single season. Developing suitable methods for controlling the rate of pesticide biodegradation in soil may be a necessity for effectively dealing with the excessively rapid rates of degradation observed in the problem soils discussed in the preceding section. One way to control the persistence of biodegradable pesticides would be by using specific inhibitors, inducers, or activators for the biochemical processes involved in the degradation of the pesticide in question. In 1970 (45) we described the inhibition of chlorpropham metabolism in soil by several methylcarbamate insecticides. Two to four-fold increases in soil persistence of chlorpropham were observed when it was applied in combination with certain methylcarbamates. Detailed kinetic studies with purified enzymes from the soil bacterium Pseudomonas striata Chester revealed that certain methylcarbamates competitively inhibited the carbamate hydrolyzing enzyme. This phenomenon has since proven useful for a more prolonged control of dodder in alfalfa with fewer chlorpropham applications. The application of PCMC to chlorpropham formulations doubled the period of dodder control with one chlorpropham application (46). It was also demonstrated that simultaneous application of PCMC and propanil to soil retarded propanil degradation in soil and thus greatly reduced the formation of 3,3',4,4'-tetrachloroazobenzene (47). This resulted from PCMC inhibition of the acylamidase hydrolyzing enzymes produced by soil microorganisms that degrade propanil

(42). Propanil hydrolyzing acylamidases have been isolated from numerous soil microorganisms: P. striata Chester (45,48); Bacillus sphaericus ATC 12123 (43,44,49); Fusarium oxysporium Schlecht (42); F. solani (50); Penicillium sp. (51); and several species of Paecilomyces (52). A barban (4-chloro-2-butynyl m-chlorocarbanilate)-hydrolyzing amidase was observed in a Penicillium sp. (53). The role of these enzymes in the detoxication of acylanilides and other phenylamide type pesticides has been examined, and the individual enzymes have been characterized. Although similar in function these enzymes have differed in their specific activity, substrate specificity, pH optimum and sensitivity to various inhibitors.

The acylamidases produced by F. oxysporum and F. solani are particularly interesting from a comparative point of view. The acylamidase produced by F. oxysporum was strongly inhibited by carbaryl (1-naphthyl methylcarbamate), chlorpropham, diuron [3-(3,4-dichlorophenyl)-1,1-dimethylurea], parathion [O,O-diethyl O-(p-nitrophenyl) phosporothioate], and PCMC, but was not inhibited by carbofuran or propachlor (42). In contrast, the acylamidase produced by F. solani was not inhibited by either carbaryl or parathion, but was inhibited by propachlor (50). Thus, the type of interaction effects actually observed in soil could be dependent upon which species predominated in soil. The linuron hydrolyzing acylamidase from B. sphaericus (49) was strongly inhibited by the methylcarbamate insecticides metmercapturon (4-methylthio-3,5-dimethylphenyl methylcarbamate), aldicarb [2-methyl-2-(methylthio) propionaldehyde O-(methylcarbamoyl)oxime], carbaryl, and propoxur (O-isopropoxyphenyl methylcarbamate), and weakly inhibited by the organophosphate insecticides parathion and fenthion (O, O-dimethyl O-[4-(methylthio)-m-tolyl]phosphorothioate) and the phenylurea herbicide monuron. Enzyme activity was not affected by the thiocarbamate fungicide thiram [bis(dimethylthiocarbamoyl) disulfide], or thiocarbamate herbicide diallate [S-(2,3-dichloroallyl) diisopropylthiocarbamate]. Growth of the bacteria was not influenced by any of the compounds except thiram which was inhibitory at concentrations higher than 10^{-6} M. The chlorpropham hydrolyzing enzyme from P. striata was competitively inhibited by the insecticidal carbamates carbaryl, propoxur, mexacarbate [4-(dimethylamino)-3,5-xylyl methylcarbamate], carbanolate (6-chloro-3,4-xylyl methylcarbamate), and aminocarb [4-(diethylamino)-m-tolyl methylcarbamate] (45). Chloropropham metabolism by P. striata was also inhibited by PCMC and the organophosphate insecticides phorate (O, O-diethyl S-[ethylthiomethyl] phosphorodithioate) and diazinon [O, O-diethyl O-(2-isopropyl-6-methyl-4-pyrimidinyl) phosphorothioate] (54). PCMC inhibited the metabolism of propham, chlorpropham, and propanil by Achromobacter sp., P. Striata, and F. oxysporum, but did not inhibit microbial growth (54). PCMC also inhibited the hydrolysis of barban by an amidase of Penicillium jenseni (53).

A wide variety of methylcarbamate and organophosphate pesticides are known to increase the persistence of acylanilide, acetamide and pehnylcarbamate herbicides in soil (45-47,54). Certain of these combinations are in fact patented and available as commercial formulations for agricultural use. PCNB (pentachloronitrobenzene) (55,56) and heptachlor (1,4,5,6,7,8,8-heptachloro-3a,4,7,7a-tetrahydro-4,7-methanoindene) (55) increase the persistence of chlorpropham in soil when applied either alone or in combination with other pesticides (55). Combinations of DDT [1,1,1-trichloro-2,2-bis(p-chlorophenyl)ethane] and captan (N-[(trichloromethyl)thio]-4-cyclohexene-1,2-dicarboximide) increased soil persistence of chlorpropham, although neither pesticide alone affected chlorpropham persistence (55). It is not known whether the mechanism of inhibition caused by PCNB, heptachlor, or the DDT-captan combination is similar to, or different from that observed with the methylcarbamate or organophosphate combinations. Roslycky (57) reported that the addition of the fungicide Vorlex [20% (w/w) methylisothiocyanate and 80% (w/w) chlorinated C_3 hydrocarbons] increased the persistence of linuron in soil. This action was attributed to the reduction of soil fungal populations. Obrigawitch et al., (34) reported that the addition of R-33865 (O, O-diethyl-O-phenol phosphorothioate) to EPTC formulations extended EPTC persistence in soil and provided increased weed control. The mechanism of R-33865 action, however, is unknown. Kaufman (54) observed that PCMC and diazinon increased the persistence of EPTC in potted soils under nonleaching conditions in the greenhouse. Although the mechanism by which PCMC inhibits EPTC degradation is not fully known, it is presumed to occur in a manner similar to that observed with the acylanilides, phenylcarbamates, etc.

The practical significance of these observations is of considerable importance to agriculture. Combinations which ultimately prove injurious to crops grown in treated soil are undesireable and to be avoided. The deliberate combination of pesticides, or addition of microbial or enzyme inhibitors to pesticides formulations, for purpose of controlled persistence of biodegradable pesticides shows considerable promise.

REFERENCES

1. D.D. Kaufman, In: "Pesticides in Soil and Water" (W.D. Guenzi, ed.), Soil Sci. Soc. Am. p. 133-202, (1974).

2. D.D. Kaufman and P.C. Kearney, In: "Herbicides: Physiology, Biochemistry, Ecology" Vol. 2, (L.J. Audus, ed.), Academic Press, p. 29-64 (1976).
3. R. Bartha and D. Pramer, Adv. Appl. Microbiol. 13, 317 (1970).
4. D.D. Kaufman, J.R. Plimmer, P.C. Kearney, J. Blake, and F.S. Guardia, Weed Sci. 16, 266 (1968).
5. J.R. Plimmer, P.C. Kearney, D.D. Kaufman and F.S. Guardia, J. Agr. Food Chem. 15, 996 (1967).
6. G.H. Willis, R.C. Wander, and L.M. Southwick, J. Environ. Qual. 3, 262 (1974).
7. L.J. Audus, Plant Soil 3, 170 (1951).
8. N. Brownbridge, Ph.D. Thesis, London University, (1956).
9. K. Kirkland and J.D. Fryer, Proc. 8th Br. Weed Control Conf., 616 (1966)
10. K. Kirkland, Weed Res. 7, 364 (1967).
11. M. Horowitz, Weed Res. 6, 168 (1966).
12. J.K. Leasure, J. Agr. Food Chem. 12, (1964).
13. D.D. Kaufman, Can. J. Microbiol. 10, 843 (1964).
14. D.D. Kaufman and P.C. Kearney, Appl. Microbiol. 13, 443 (1965).
15. C.G. Clark and S.J.L. Wright. Soil Biol. Biochem. 2, 19 (1970).
16. D.D. Kaufman and J. Blake, Soil Biol. Biochem. 5, 297 (1973).
17. W.D. Burge, Soil Biol. Biochem. 4, 379 (1972).
18. S.J.L. Wright and A. Forey. Soil Biol. Biochem. 4, 207 (1972).
19. M.A. El-Dib and O.A. Aly, Water Res. 10, 1055 (1976).
20. K. Hurle and B. Rademacher, Weed Res. 10, 159 (1970).
21. K. Kirkland and J.D. Fryer, Weed Res. 12, 90 (1972).
22. A.S. Newman and J.R. Thomas, Proc. Soil Sci. Soc. Am. 14, 160 (1949).
23. A.S. Newman, J.R. Thomas and R.L. Waker, Proc. Soil Sci. Soc. Am. 16, 21 (1952).
24. O.M. Aly and S.D. Faust, J. Agr. Food Chem. 12, 541 (1964).
25. J.D. Fryer and K. Kirkland, Weed Res. 10, 133 (1970).
26. T.L. Torstensson, J. Stark and H. Goransson, Weed Res. 15, 159 (1975).
27. H.L. Jensen, Tidskr. PlAvl. 37, 553 (1964).
28. A. Rahman, G.C. Atkinson, J.A. Douglas, and D.P. Sinclair, N.Z. Jour. Agr. 139(3), 47 (1979).
29. T. Obrigawitch, A.R. Martin and F.W. Roeth, Weed Sci. Soc. Am. Abstr. No. 199, (1982).
30. R.G. Wilson, A.R. Martin, F.W. Roeth, and T. Obrigawitch, Weed Sci. Soc. Am. Abstr. No. 198, (1982).
31. J.L. Gunsolus and R.S. Fawcett, Weed Sci. Soc. Am. Abstr. No. 200, (1982).
32. D.B. Schuman and R.G. Harvey, Weed Sci. Soc. Am. Abstr. No. 201, (1982).
33. T. Obrigawitch, R.G. Wilson, A.R. Martin, and F.W. Roeth, Weed Sci. 30, 175 (1982).
34. T. Obrigawitch, F.W. Roeth, A.R. Martin and R.G. Wilson, Weed Sci. 30, 417 (1982).
35. D. F. Edwards and D.D. Kaufman, Abstr. 184th Mtg. Am. Chem. Soc., Pestic. Chem. Div., No. 49, (1982).
36. A. Felsot, J.V. Maddox, and W. Bruce, Bull. Environ. Contam. Toxicol. 26, 781 (1981).
37. G.W. Gorder, J.J. Tollefson, P.A. Dahm, Iowa State J. Res. 55, 25 (1980).
38. N. Ahmad, D.D. Walgenbach, and G.R. Sutter, Bull. Environ. Contam. Toxicol. 23, 572 (1979).
39. D.D. Kaufman, A.J. Kayser, E.H. Doyle, and T.I. Munitz, Abstr. 181st Am. Chem. Soc. Mtg., Pestic. Chem. Div., No. 28 (1981).
40. R. Bartha and D. Pramer, Soil Sci. 100, 68 (1965).
41. L.J. Audus, In: Herbicides and the Soil (E.K. Woodford α G.R. Sagar, eds.) Blackwell Scientific Publications, Oxford, pp 1-19, (1960).
42. J. Blake and D.D. Kaufman, Pestic. Biochem. Physiol. 5, 305 (1975).
43. G. Engelhardt, P.R. Wallnofer, and R. Plapp, Appl. Microbiol. 22, 284 (1971).
44. G. Engelhardt, P.R. Wallnofer, and R Plapp, Appl. Microbiol. 26, 709 (1973).
45. D.D. Kaufman, P.C. Kearney, D.W. VonEndt, and D.E. Miller. J. Agr. Food Chem. 18, 513 (1970).
46. J.H. Dawson, Weed Sci. 17, 295 (1969).
47. D.D. Kaufman, J. Blake, and D.E. Miller, J. Agr. Food Chem. 19, 204 (1971).
48. P.C. Kearney, J. Agr. Food Chem. 13, 561 (1965).
49. G. Engelhardt and P.R. Wallnofer, Appl. Microbiol. 29, 717 (1975).
50. R.P. Lanzilotta and D. Pramer, Appl. Microbiol. 19, 307 (1970).
51. N.E. Sharabi and L.M. Bordeleau, Appl. Microbiol. 18, 369 (1969).
52. T. Akatsuka, K. Suzuki, and S. Kuwazulsa, Sci. Rep. Fac. Agr. Ibarak; Univ. 17, 45 (1970).
53. S.J.L. Wright and A. Forey, Soil Biol. Biochem. 4, 207 (1972).
54. D.D. Kaufman, Soil Biol. Biochem. 9, 49 (1977).
55. D.D. Kaufman, Pestic. Chem. 6, 175 (1972).
56. A. Walker, Horticultural Res. 10, 45 (1970).
57. E.B. Roslycky, Can. J. Soil Sci. 60, 651 (1980).

A Simulation Model for Prediction of Herbicide Persistence[1]

Allan Walker[2]

ABSTRACT

A simulation model for prediction of herbicide persistence in the field is described. The model combines the effects of soil temperature and soil moisture content on the rates of herbicide loss, determined experimentally under controlled conditions, with the fluctuations in surface soil temperature and moisture content in the field. The computer program includes methods of simulating surface soil temperatures and moisture contents from standard meteorological data. In order to test the model, the degradation of napropamide (2-(α-naphthoxy)-N,N-diethylpropionamide) was examined. Under controlled conditions, loss of activity followed first-order kinetics with half-lives of 54, 63, and 90 days at 28C with soil moisture contents of 10.0, 7.5, and 3.5%, respectively. At 14C, half-lives at 10.0 and 7.5% soil moisture were 102 and 112 days, respectively. When these data were used in the simulation model in conjunction with the relevant meteorological information, the patterns of loss of napropamide incorporated 2 to 3 cm in the field could be predicted. Napropamide was lost rapidly when applied to the soil surface, and since the model only takes into account losses through microbial or chemical metabolism, this could not be simulated. Some of the limitations and the potential benefit of the simulation technique for prediction of pesticide persistence are discussed.

Additional Index Words: napropamide persistence, simulation of surface soil moisture content from meteorological data, simulation of surface soil temperatures.

An important aspect of the behavior of any pesticide in the soil is the length of its residual life or persistence.

[1] Contribution from the National Vegetable Research Station, Wellesbourne, Warwick, England. Received 17 Jan. 1974.
[2] Senior Scientific Officer, Weeds Section, National Vegetable Research Station.

This has particular relevance to herbicides in short-term high-value crops such as vegetables. A herbicide must not persist so long that limitations are placed on the nature of subsequent crops which can be grown, nor must persistence be so short that the level or duration of weed control becomes unacceptable. Before a herbicide is marketed, it is important, from both the performance and environmental viewpoints, to determine the extent of persistence under the range of conditions in which it is likely to be used. Field evaluation of persistence under a range of climatic conditions or after application at different times of the year is a labor consuming and costly operation and the possibility of using a simulation technique has therefore been investigated.

In a preliminary report, the use of a simulation model to predict the persistence of propyzamide (N-(1,1-dimethylpropynyl)-3,5-dichlorobenzamide) was described (8). From a limited amount of laboratory data, it was shown that the patterns of loss of propyzamide activity could be predicted with sufficient accuracy for practical purposes. In the present paper, full details of the model are given, and the results from experiments made to determine the persistence of napropamide (2-(α-naphthoxy)-N,N-diethylpropionamide) in soil under controlled and field conditions are described.

MATERIALS AND METHODS

The soil, a sandy loam containing 2% organic matter and 18% clay, and with pH 6.1, was taken from the surface 5 cm of Little Cherry field at the National Vegetable Research Station. The same field was used for assessment of persistence under natural conditions. Throughout the work, the herbicide used was a wettable

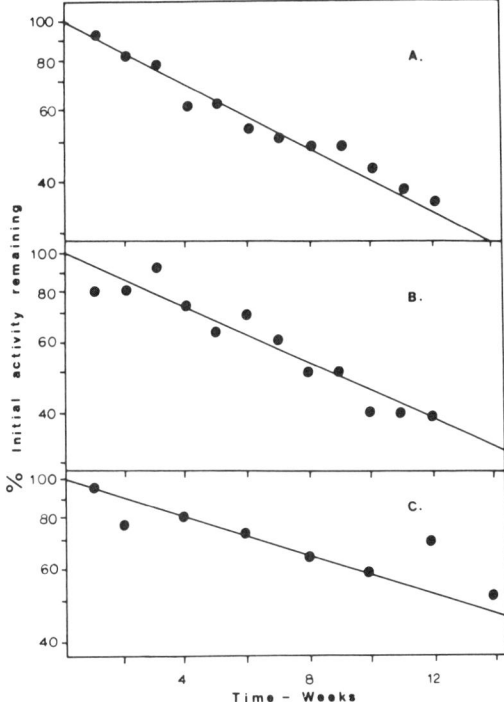

Fig. 1—Napropamide degradation under controlled conditions. A, 28C, 10% soil moisture, and 4.5 kg/ha; B, 28C, 7.5% soil moisture, and 4.5 kg/ha; and C, 28C, 3.5% soil moisture, and 4.5 kg/ha.

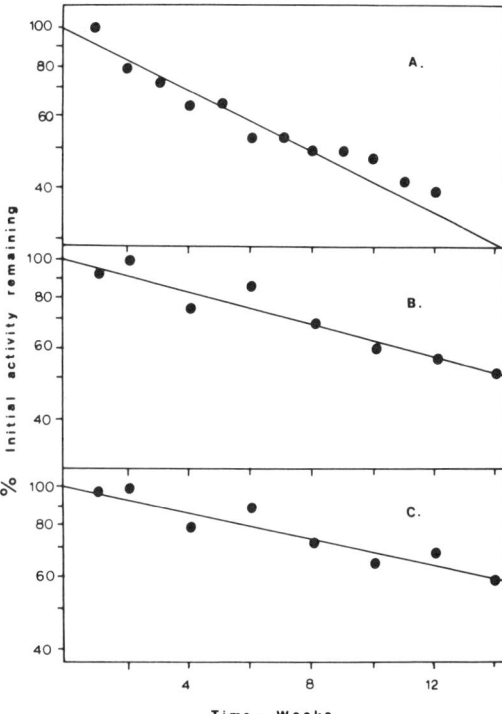

Fig. 2—Napropamide degradation under controlled conditions. A, 28C, 10% soil moisture, and 2.25 kg/ha; B, 14C, 10% soil moisture, and 4.5 kg/ha; C, 14C, 7.5% soil moisture, and 4.5 kg/ha.

powder formulation of napropamide containing 50% active ingredient.

Persistence of Napropamide Under Controlled Conditions

Separate 7-kg quantities of sieved (2 mm) air-dried soil were treated with napropamide to give a concentration of 6.4 µg/g dry soil (equivalent to 4.5 kg active ingredient/ha incorporated to a depth of 5 cm) by adding the required amount of herbicide in water to the soil. Mixing was achieved by passing the soil several times through a 2-mm sieve. The soils were stored in polyethylene bags in incubators at the required temperatures. The treatments examined were 10, 7.5, and 3.5% (w/w) soil moisture at 28C, and 10 and 7.5% soil moisture at 14C. A further treatment at 3.2 µg/g (2.25 kg/ha·5 cm) was examined at 10% moisture and 28C. Samples of soil were taken from the bags initially and at intervals during the subsequent 14 weeks and the herbicidal activity remaining was determined by a bioassay based on the shoot growth of ryegrass (*Lolium perenne* L.) which has been described previously (7). At each sampling time, the moisture content of the soil remaining was returned to that required by adding sufficient water, and it was assumed that sufficient aeration of the soil would be obtained by the periodic opening of the bags and by diffusion of oxygen through the polythene.

Persistence of Napropamide in the Field

Field plots were prepared on 26 April and 7 June 1973 and sprayed with napropamide at 4.5 kg active ingredient in 1,100 liters/ha. Two plots were prepared on each occasion and on one of these plots the herbicide was incorporated to a depth of 2 to 3 cm with a rotary power harrow. On the other plot, the herbicide remained on the surface. Further surface-sprayed plots were prepared on 14 June and 28 June. Immediately after preparation and thereafter at weekly intervals, 10 cores were taken from each plot at random positions to a depth of 5 cm. The cores for each treatment were bulked, thoroughly mixed by passing several times through a 2-mm sieve, and the herbicidal activity remaining was determined by the ryegrass bioassay.

Samples of soil were taken from the surface 2.5 cm of bare plots adjacent to the experimental persistence plots each Monday, Wednesday, and Friday for the duration of the experiments, and these samples were dried at 110C for 48 to 72 hours in order to give a continuous record of surface soil moisture content.

Measurements of soil temperature were made with thermocouples buried 2 cm deep in further bare plots close to the experimental plots. The output from these thermocouples was plotted on a chart recorder at 10-minute intervals.

RESULTS

Persistence of Napropamide Under Controlled Conditions

The results from the laboratory persistence experiments are shown in Fig. 1 and 2, in which the logarithms of the herbicidal activity remaining (% initial activity) have been plotted against time. The straight lines obtained indicate that degradation follows the first-order

398 A. Walker

Table 1—Half-lives for napropamide

Temperature	Herbicide concentration	Half-life at soil moisture content		
°C	kg/ha-5cm	10.0	7.5	3.5
		days		
28	4.50	54	63	90
14	4.50	102	112	–
28	2.25	56	–	–

rate law, a conclusion which is supported by the similarity of the slopes of the lines for the 4.5 and 2.25 kg/ha treatments at 28C. From the slopes of the lines, half-lives for the different treatments were calculated and these are shown in Table 1. It has previously been shown that the relationship between the half-life of propyzamide and soil moisture content is of the form:

$$H = a\,MC^{-b} \quad [1]$$

where H is the half-life at moisture content MC, and a and b are constants (8). The equation of best fit for the limited data for napropamide degradation at 28C was:

$$H = 189.3\,MC^{-0.550}. \quad [2]$$

Since the results in Fig. 1 and 2 demonstrate that napropamide degradation follows first-order kinetics, the relationship between the half-lives and temperature can be expressed in terms of the Arrhenius equation:

$$\log \frac{H_1}{H_2} = \frac{\Delta E}{2.303\,R}\left(\frac{1}{T_1} - \frac{1}{T_2}\right) \quad [3]$$

in which H_1 and H_2 are the half-lives at temperatures T_1 and T_2, ΔE is the activation energy and R is the gas constant. The calculated activation energy for napropamide degradation is 7.85 kcal/mole at 10% soil moisture and 7.80 kcal/mole at 7.5% soil moisture.

Persistence of Napropamide in the Field

The results from the field investigations are shown in Fig. 3 and 4. The residual activity as a percentage of that present initially has been plotted against time. The data show a marked effect of incorporation on persistence. When surface-applied on 26 April, there was a 50% loss during the following 3 weeks compared with incorporated napropamide which changed little in activity during this time. In fact, when incorporated, 50% of the initial activity remained 14 weeks after application. When applied on 7 June, almost 60% of the surface-applied herbicide was lost in the first week compared with only 9% loss from the incorporated application. Further surface-applied treatments prepared on 14 and 28 June also showed this initial rapid loss. There is evidence that napropamide is subject to loss through photodecomposition at the soil surface (Personal communication, Stauffer Chemical Co.), and the data presented here confirm a rapid loss unless incorporated. The vapor pressure of napropamide (2×10^{-6} mm mercury), would suggest that volatilization cannot be ignored as a mechanism of loss, but is unlikely to account for the magnitude of the losses recorded.

The Simulation Model

Since the degradation of napropamide was shown to follow first-order kinetics, the simulation model was based on the first-order rate equation:

$$\partial C/\partial t = -KC \quad [4]$$

where C is the concentration, t is time and K the rate constant. From the relationship between the rate constant and the half-life, this equation can be modified to:

$$\partial C/\partial t = -\frac{0.6932\,C}{H}. \quad [5]$$

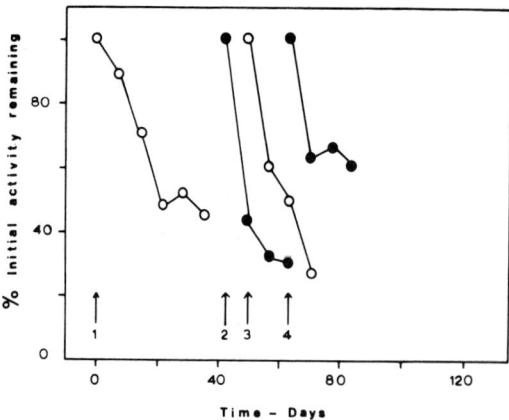

Fig. 3—Persistence of surface-applied napropamide in the field. 1, applied 26 April; 2, applied 7 June; 3, applied 14 June; and 4, applied 28 June.

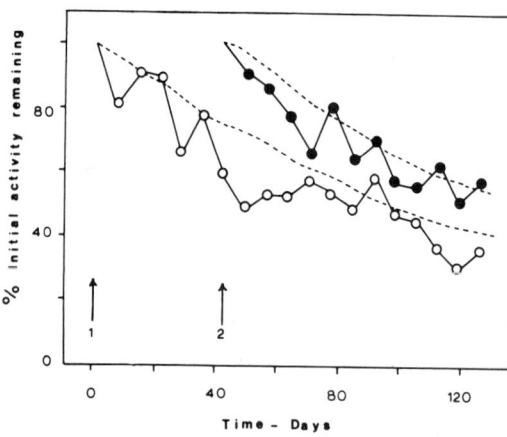

Fig. 4—Persistence of incorporated napropamide in the field. 1, applied 26 April; 2, applied 7 June. (O) determined values; (---) simulated curves.

The results in Table 1 show how the half-life, H, is dependent upon environmental parameters, and the object of the simulation model is to combine the moisture effect (Eq. 1), the temperature effect (Eq. 2), and the concentration effect (Eq. 5) with the soil moisture and temperature regimes found in the field. The program was written in the simulation language CSMP 360[3], a language designed specifically for the simulation of dynamic systems. Of the functions available in the simulation language, the arbitrary function generator with linear interpolation (AFGEN) has been used extensively in the present model. This function allows a list of data to be presented (for example, daily values of soil temperature) and when the program is run, interpolations are made linearly between the successive input values in order that calculations at time intervals shorter than one day can be made.

SIMULATION OF SOIL MOISTURE CONTENT

For the simulation of soil moisture content, the 2.5-cm layer of soil examined was assumed to be independent of any soil beneath, and calculations were made on a unit area basis. Income of water to the system (rainfall) was added to that present in the soil initially, until a limiting soil moisture content was reached—the field capacity of the soil. Any further incoming water was ignored. Loss of water from the system was calculated by multiplication of the evaporation of water from an open water surface by a weighting factor which decreased linearly from 1.0 at zero bar suction, to 0.1 at 30 bar suction and then to 0.01 at 100 bar suction. The equations for simulation are:

$$MC = (WW * 100)/(25. *BD) \qquad [6]$$

$$WW = INTGRL(IWW, LOSS) \qquad [7]$$

$$LOSS = \\ LIMIT(-WW, FC - WW, INPUT - OUTPUT) \qquad [8]$$

$$INPUT = AFGEN (RAIN, TIME) \qquad [9]$$

$$OUTPUT = AFGEN (EO, TIME) * WF \qquad [10]$$

$$WF = AFGEN (WFT, SWS) \qquad [11]$$

$$SWS = EXP(2.303 * (K3 - MC)/K4) \qquad [12]$$

where MC is the soil moisture content (% w/w), WW is the amount of water in the soil per unit surface area to a depth of 2.5 cm (mm), IWW is the initial water content (mm), FC is the field capacity of the soil per unit surface area to a depth of 2.5 cm (mm), $RAIN$ is the rainfall (mm/day), EO is the evaporation from an open water surface (mm/day), WF is the weighting factor, SWS is the soil water stress (bar), BD is the bulk density of the soil (g/cm^3), and $K3$ and $K4$ are constants which characterize the moisture release curve of the soil. In equation [8], the change in soil water content ($INPUT - OUTPUT$) is set within limits. The maximum amount of water which can be lost is limited to the amount of water present in the soil at the particular instant in time ($-WW$), and the amount of water which can be gained is limited to the difference between field capacity and the amount of water present in the soil ($FC - WW$).

The effects of soil moisture content on the half-life of napropamide at 28C are described by equation [1] with the appropriate values for the constants, and this equation completes the moisture content part of the program:

$$H = A * MC ** B. \qquad [13]$$

The results from the simulation of soil moisture content are shown in Fig. 5. The fluctuations in surface soil moisture content are predicted with sufficient accuracy over most of the experimental period of 18 weeks by this empirical technique, and the fit between predicted and observed values was particularly good during the later stages of the experiment.

SIMULATION OF SOIL TEMPERATURE

A brief description of the method used to simulate soil temperatures has been given previously (8). This method was based on the observation that the mean soil temperature at a depth of 2 cm (measured by the buried thermocouples) was similar to the standard meteorological soil temperature at 10 cm under grass during early April and

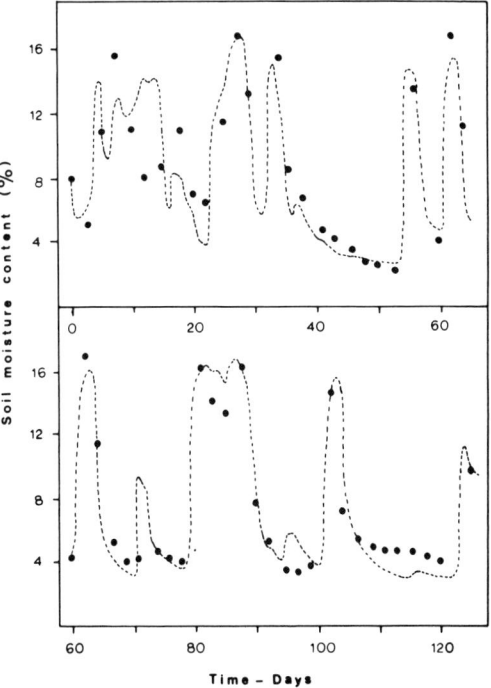

Fig. 5—Simulation of surface soil moisture content. (●) determined values; (– – –) simulated curve; Time 0 = 26 April 1973.

---

[3] IBM Users Manual H20-0367-3. Continuous System Modeling Program.

late September, but during late June and July was from 3 to 7C higher. In order to use the meteorological measurements in the computer program, an addition term was employed. This addition increased from 0C on 1 April to 5C on 1 July and then decreased to 0C on 1 October. The equations for simulation are:

$$ST = MT + ADDN + 273. \quad [14]$$

$$MT = AFGEN\ (TT,\ TIME) \quad [15]$$

$$ADDN = AFGEN\ (AT,\ TIME) \quad [16]$$

where ST is the surface soil temperature, MT is the soil temperature measured at 10 cm, and $ADDN$ is the addition. A further addition of 273K is required to convert the temperatures to degrees Kelvin, in order to use the value in the Arrhenius equation.

The equations given in the previous section allow calculation of the half-life of the herbicide under the prevailing soil moisture conditions at standard temperature T. (In the case of napropamide, $T = 28C$ i.e. 301K). In order to calculate the half-life (HL) at the corresponding soil temperature (ST), the Arrhenius equation is used. From equation [3], it can be shown that:

$$\log HL = \log H + \frac{E(T - ST)}{1.575\ T\ ST}. \quad [17]$$

For use in the simulation program, the equations are:

$$K1 = E/(4.575 * T) \quad [18]$$

$$K2 = K1 * (T - ST)/ST \quad [19]$$

$$HL = H * (EXP\ (2.303 * K2)) \quad [20]$$

To complete the dynamic section of the program, the half-life, HL, is substituted in equation [5]:

$$DCDT = -(0.6932 * CONC)/HL \quad [21]$$

$$CONC = INTGRL\ (IC,\ DCDT) \quad [22]$$

where $CONC$ is the concentration of herbicide remaining in the soil and IC is the initial concentration.

To complete the program, the values of the constants (BD, FC, $K3$, $K4$, E, A, B, and T) and the initial conditions (IWW and IC) are specified and the tables of data are fed in. The various requirements for printed output and timing of the calculations are also specified.

The results from use of the simulation program are shown for the incorporated applications of napropamide in Fig. 4. The predicted lines give good fits to the observed patterns of loss of activity for each time of application.

DISCUSSION

The results in Fig. 3 and 4 show marked differences in the patterns of loss of napropamide activity between surface and incorporated applications. When applied to the soil surface as much as 60% of the initial activity disappeared within 1 week. When incorporated, napropamide was relatively persistent under the climatic conditions of these experiments. When applied and incorporated on 26 April, more than 30% of the initial activity remained on 30 August (18 weeks later) and when applied and incorporated on 7 June, over 50% remained in the soil after 12 weeks. The data in Table 1 also reflect this relatively long persistence. Even at 28C, the half-life in moist soil was over 50 days, and in dry soil was over 90 days. The mean soil temperature at a depth of 2 cm at Wellesbourne is probably less than 20C for the 9-month period from late August until the end of May, and during the summer of 1973, the maximum mean temperature was 25.4C, reached during the first week of July. It is clear therefore that this herbicide will persist for long periods under these conditions.

Use of the simulation model gave good approximations to the observed patterns of degradation of the incorporated applications (Fig. 4) and these results illustrate the usefulness of the simulation technique when functional relationships can be established between the rates of herbicide breakdown and soil temperature, soil moisture content and herbicide concentration in the soil. One of the limitations in the present model, however, is that losses through mechanisms other than chemical or microbiological metabolism of the herbicide are not provided for. Particular care must be taken when using the model to predict persistence of surface applications when losses through photodecomposition or volatilization might occur. This is illustrated by the data for surface-applied napropamide (Fig. 3) where the rapid loss of activity will clearly not be simulated by the present technique.

The agreement between predicted and observed degradation patterns of incorporated napropamide are good for the particular soil examined and under the conditions of these experiments, but it must be stressed that the model is highly specific to the soil type for which the laboratory data are obtained. It remains to be seen whether this model, or one similar to it, can be applied more generally to other soil types and other pesticides. The methods of simulating surface soil temperatures and soil moisture contents from meteorological data are empirical, as are the relationships between half-lives and soil moisture content and soil temperature. The model requires examination with data from different soil types, and the validity of these empirical relationships requires further study. Some of these points have been discussed in more detail previously (8).

The close fit of the rate of loss of napropamide activity to first-order kinetics has simplified the construction of the simulation program described above. Propyzamide degradation also conforms to first-order kinetics (7), and the persistence of this herbicide can be simulated by a similar program (8). Although the first-order rate law has been used to interpret many studies of pesticide breakdown in soil (1, 2, 6, 7, 9, 10) there is evidence that for several herbicides and other pesticides, it does not apply (4, 5). Hamaker (4) has discussed the need for accurate determination of the kinetics of pesticide degradation in the soil, and this is certainly necessary if simulation models are to be used. Fit to the first-order rate law is not a prerequisite for successfully applying a simulation technique to predict persistence, but a knowledge of the kinetics of breakdown is required. The first stage in degradation of some pesticides in soils is conversion to a second biologically active compound, for example converting aldrin (not less than 95% of 1,2,3,4,10,10-hexachloro-1,4,4a,5,8,8a-hexahydro-1,4-*endo-exo*-5,8-dimethanonaphthalene) to dieldrin (not less than 85% of 1,2,3,4,10,10-hexachloro-6,7-epoxy-1,4,4a,5,6,7,8,8a-octahydro-1,4-*endo-exo*-5,8-dimethanonaphthalene) (3). An accurate prediction of residues remaining after use of such compounds would require knowledge of the kinetics of degradation of both compounds. In the present work, the loss of biological activity of napropamide has been examined, but from the environmental viewpoint, information on the nature and fate of its metabolites in soil is also important. If the kinetics of degradation of the metabolites were known, it might be possible to simulate a complete balance sheet for the parent compound and its breakdown products in soil.

The present program has been shown to simulate the patterns of loss of incorporated napropamide (Fig. 4) and propyzamide (8) in the field, and although with other compounds there may be further factors to be considered when simulation models are used, this technique for prediction of persistence has two uses which are of great potential value. First, the ranges of persistence of a particular compound likely after application at different times of the year, in different years, and at different sites could be obtained without resorting to detailed field investigations in the first instances. Second, the probability of a given residue level in the soil could be predicted by comparing the average with extreme weather conditions, and this information could be available at an early stage in the development of a new compound. Only limited confirmatory field experiments would then be required.

ACKNOWLEDGMENTS

Thanks are given to D. A. E. Fildes for help with the computing and to Mrs. E. J. Theodorson and Miss L. Pluck for technical assistance. Advice given by H. A. Roberts, Dr. D. J. Greenwood, and M. A. Scaife is also gratefully acknowledged.

LITERATURE CITED

1. Armstrong, D. E., G. Chesters, and R. F. Harris. 1967. Atrazine hydrolysis in soil. Soil Sci. Soc. Amer. Proc. 31:61-66.
2. Bro-Rasmussen, F., E. Noddegaard, and K. Voldum-Claussen. 1970. Comparison of the disappearance of eight organo-phosphorus insecticides from soil in laboratory and outdoor experiments. Pestic. Sci. 1:179-182.
3. Decker, G. C., W. N. Bruce, and J. H. Bigger. 1965. The accumulation and dissipation of residues resulting from the use of aldrin in soils. J. Econ. Entomol. 58:266-271.
4. Hamaker, J. W. 1972. Decomposition: Quantitative aspects. p. 253-340. *In* C. A. I. Goring, and J. W. Hamaker (ed.) Organic chemicals in the soil environment. Marcel Dekker, Inc., New York.
5. Rahn, P. R., and R. L. Zimdahl. 1973. Soil degradation of two phenyl pyridazinone herbicides. Weed Sci. 21:314-317.
6. Smith, A. E. 1969. Factors affecting the loss of tri-allate from soils. Weed Res. 9:306-313.
7. Walker, A. 1970. Persistence of pronamide in soil. Pestic. Sci. 1:237-239.
8. Walker, A. 1973. Use of a simulation model to predict herbicide persistence in the field. Eur. Weed Res. Coun. Symp. Herbicides-Soil. Proc., Paris, Dec. 1943, p. 240-250.
9. Wheatley, G. A., D. L. Suett, and J. A. Hardman. 1972. Some effects influencing the persistence of chlorfenvinphos in soil. p. 77-85. *In* A. S. Tahori (ed.) Fate of pesticides in the environment. Gordon and Breach, London.
10. Zimdahl, R. L., V. H. Freed, M. L. Montgomery, and W. R. Furtick. 1970. The degradation of triazine and uracil herbicides in soil. Weed Res. 10:18-26.

Part III

TRANSPORT PROCESSES

Editors' Comments
on Papers 36 Through 45

36 LICHTENSTEIN
Movement of Insecticides in Soils under Leaching and Non-Leaching Conditions

37 GRAHAM-BRYCE
Diffusion of Organophosphorus Insecticides in Soils

38 EHLERS et al.
Lindane Diffusion in Soils: II. Water Content, Bulk Density, and Temperature Effects

39 SCOTT, PHILLIPS, and PAETZOLD
Diffusion of Herbicides in the Adsorbed Phase

40 GERSTL, NYE, and YARON
Diffusion of a Biodegradable Pesticide: II. As Affected by Microbial Decomposition

41 HELLING
Pesticide Mobility in Soils: III. Influence of Soil Properties

42 DAVIDSON, RIECK, and SANTELMANN
Influence of Water Flux and Porous Material on the Movement of Selected Herbicides

43 VAN GENUCHTEN, WIERENGA, and O'CONNOR
Mass Transfer Studies in Sorbing Porous Media: III. Experimental Evaluation with 2,4,5-T

44 RAO et al.
Evaluation of Conceptual Models for Describing Nonequilibrium Adsorption-Desorption of Pesticides During Steady-flow in Soils

45 LEISTRA
Computed Redistribution of Pesticides in the Root Zone of an Arable Crop

DIFFUSION AND MASS FLOW

Early Research

Early studies of the redistribution of pesticides into the soil as a result of their transport following land application were mainly descriptive and did not attempt to explain the transport mechanisms. The analytical techniques used were bioassays, and only later was chemical analysis of the soil extract commonly performed. One of the first studies dealing with the effect of leaching on pesticide redistribution was that of Hanks (1947) on 2,4-D and its calcium salts. Several 2,4-D treated soils were leached with various amounts of water and the toxicity of the effluent was determined by bioassay. Differences were found between the toxicity of the leachates.

Paper 36 is an early, complex, and long-term study reporting the distribution of four insecticides in soil as affected by the moisture regime and soil characteristics. The study was carried out both in the field and under laboratory conditions. In the field experiment (for which, unfortunately, the moisture regime is not mentioned), approximately 62 to 96% of the applied DDT, aldrin, and lindane remained in the upper 3-inch layer 17 months after application. Surprisingly, more downward movement of all the insecticides was noted in the muck, compared with the loam soil. Lateral movement on sloping plots was observed 3 years after treatment. In the laboratory experiments it was found that under leaching conditions lindane was leached into the soil to an extent related to the type of soil: sand>silt loam>muck, which is inversely related to the organic matter content of these soils. Under nonleaching conditions lindane did not move significantly from the treated layers. However, under the same conditions, parathion was transported both vertically and laterally and a diffusion phenomenon was suggested. The fastest movement occurred in the sandy soil, and losses by volatilization from this soil were assumed. The results presented in this paper show clearly the complex interrelationship between many relevant factors and indicate that the transport processes may be simultaneous. The relative contribution of each process was dependent on the type of pesticide and the moisture regime.

Diffusion

In general terms, diffusion is the process whereby a material moves from a higher to a lower concentration as a result of random molecular motion; the rate of transfer is proportional to the concentration gradient normal to the direction of movement. The basic mathematics of the process was largely described by Crank (1975), and its application to pesticide behavior has been discussed by Letey and Farmer (1974), Hartley and Graham-Bryce (1980), Calvet (1984), and Jury, Spencer, and Farmer (1984).

Several factors influence the diffusion of pesticides in soils: the diffusion coefficients of pesticides in water and in air, which are related to their vapor pressure and solubility, adsorption, soil water content and temperature, and some soil physical properties such as porosity and bulk density. Letey and Farmer (1974) described some difficulties encountered in the studies of pesticide diffusion in soil as follows: (1) The diffusion coefficient cannot be assumed to be independent of concentration; (2) diffusion is confined to certain segments of the system (a molecule will not diffuse through a solid soil particle); (3) sorption of the diffusing substance by soil particles occurs often; and (4) the diffusion coefficient is dependent upon temperature and several soil properties such as mineral composition, bulk density, and water content.

Solute diffusion in soils could be described by the relationship proposed by Olsen and Kemper (1968):

$$D = \frac{D_0 f \theta}{(\rho K + \theta)} \qquad (1)$$

where D is the diffusion coefficient. The variables K (adsorption from aqueous solution), ρ (dry soil bulk density), f (impedance factor that takes into account the tortuous pathway followed by the solute through the soil pores), and D_0 (diffusion coefficient of the chemical in water) are determined empirically. This relationship assumed that herbicide movement by volatilization is negligible, that no degradation occurs during the diffusion process, and that the adsorption is linear and single-valued.

The application of diffusion theory to the movement of 2,4-D in saturated soils was done by Lindstrom, Boersma, and Gardiner (1968) on a broad range of soils with various mineralogies and organic matter contents. The diffusion coefficients for the experimental soils varied between 0.3 and 4.3 × 10^{-5} cm^2/sec. Graham-Bryce (Paper 37), studying disulfoton and dimethoate, showed that diffusion coefficients of both compounds varied little with their concentration but were affected differently by the moisture content. The moisture content had little effect on the diffusion of the less soluble and more volatile disulfoton, while with the relatively soluble dimethoate the effect was considerable, indicating that solute diffusion was the main transport pathway. Taking into consideration the relative contributions of the liquid and vapor diffusion, of adsorption, and of the geometric soil factors, the author described a possible relationship concerning pesticide diffusion in soil. The diffusion of dimethoate was successfully described; however, in the case of disulfoton, the exact contribution of liquid and vapor diffusion was difficult to establish and only approximate values could be obtained. The data indicated that practically, movement of pesticides by diffusion is slow and could not account for transport far from the application point, but that it might be important for redistribution over short distances.

The effect of soil water content on diffusion was studied in several works (e.g., Lavy, 1970; Scott and Phillips, 1972), and the results were similar to those reported in Paper 37.

In addition to the soil moisture content, the effect of the bulk density and temperature on the diffusion of lindane in soils was investigated by Ehlers and co-workers (Paper 38). Similar to disulfoton, lindane is a relatively volatile insecticide (v.p. = 9.4×10^{-6} mmHg at 20°C), and its diffusion in soils occurs in both vapor and nonvapor phases (Ehlers et al., 1969). Lindane diffusion was extremely slow at low moisture contents, increased as water content rose to 2.5 to 3%, but upon further increase in moisture resulted in a sharp decrease of the diffusion coefficient. At moisture contents higher than 4 to 5%, the diffusion coefficient remained practically unchanged. This behavior was explained by the effect of water on the adsorption-desorption process. The diffusion coefficient of lindane decreased as the bulk density was increased, a general trend for the diffusion of pesticides in soil. This effect is stronger for pesticides moving significantly in the vapor phase because the bulk density affects soil porosity.

An exponential increase in the apparent diffusion coefficient of lindane with increasing temperature was also reported. A direct relationship between temperature and diffusion in soil has been observed for other pesticides also (Lavy, 1970). This relationship is a consequence of the effect of temperature on such factors as the vapor pressure and the diffusion coefficients of pesticides in water and gaseous phases, which are determinant factors in the diffusion of pesticides in soils. Equations describing the diffusion in vapor and nonvapor phases were applied successfully in Paper 38 to predict lindane diffusion at differing temperatures. Good predictions from the relative values of the activation energy of the diffusion process were also demonstrated.

A further development in the understanding of pesticide diffusion in soil was the consideration of an additional pathway—diffusion in the adsorbed phase. Adsorbed molecules may be divided into an immobile fraction and also a mobile fraction that contributes to diffusion and has the same apparent mobility as molecules in solution. If a fraction of the adsorbed molecules is mobile, then equation (1) becomes

$$D = \frac{D_0 f(\theta + \alpha\rho K)}{(\rho K + \theta)} \qquad (2)$$

In Paper 39, Scott, Phillips, and Paetzold presented equations describing the diffusion of pesticides in the adsorbed phase (exchange diffusion along charged sites) and applied them for the calculation of metribuzin diffusivity in soils at various water contents. Metribuzin is an s-triazine herbicide relatively soluble in water (1,200 ppm), so that its movement in the vapor phase was considered negligible. The parameters needed for such calculations are soil bulk density, the distribution coefficient (determined from a linear adsorption isotherm), the apparent self-diffusion of ^{14}C-metribuzin in soil, the self-diffusion coefficient in water, and the transmission factor. The last parameter was obtained from the ratio between the apparent diffusion coefficient in soil and the self-diffusion coefficient in water of ^{36}Cl. As the authors emphasize, the assumption that the solute diffusivities of metribuzin and chloride in soils are similar could lead to erroneous results.

The data obtained show that diffusivity of metribuzin in soil was dependent on the amount in the adsorbed phase and the moisture content. The calculated diffusion coefficient in the solution phase reflected a combination of the diffusion coefficient in adsorbed phase and the apparent diffusion coefficient. Both the solution and adsorbed phase diffusivities of metribuzin decreased as the water content decreased. At the lowest water contents tested (0.34 and 0.32 cm^3/cm^3), negative diffusion coefficients were obtained in the adsorbed phase. In an attempt to find the source of this anomaly, the authors discussed the validity of the assumptions made in their calculations.

The relative contribution of the adsorbed phase diffusion to the overall diffusivity of pesticides in soil is not clear, but the authors suggested it must be related to the nature of the adsorbed molecule and soil-water-pesticide interactions, reflected in the dynamic adsorption-desorption equilibrium. The data obtained with metribuzin indicate that diffusion in the adsorbed phase made an important contribution to the soil diffusivity of this compound. However, Gerstl, Yaron, and Nye (1979) showed that in the case of parathion only a small fraction of the adsorbed chemical was undergoing diffusion, the mobile fraction being approximately 10% of the total adsorbed parathion.

Because diffusion of organic molecules in soil is slow even over a large range of moisture contents, pesticide degradation may occur during the diffusion process. In Paper 40, Gerstl, Nye, and Yaron described a model to predict the diffusion of parathion and one of its metabolites in soil by taking into consideration the microbial activity in the diffusion system. The predictions were in satisfactory agreement with the measurements carried out at different moisture contents and time intervals. Although the proposed model requires further refinement, the paper represents a new and more realistic approach. In previous investigations, pesticide diffusion was considered mainly from the physical and physico-chemical points of view, and the possibility of pesticide transformations during diffusion was neglected.

Mass Flow

As water moves through the soil, dissolved pesticides are carried along in the convective stream. The mass transport of water and solute in soils occurs through a network of pores of various forms, sizes, and shapes. During their flow through soil, pesticides are subject to a series of interactions such as adsorption, degradation, volatilization, and plant uptake. Under conditions of unsteady or transient water flow, the contact time between the solute and the soil solid phase often is insufficient to achieve an equilibrium. Rao and Jessup (1982) described the convective-diffusive, dispersive transport of pesticides during transient water flow by the equation:

$$\frac{\delta}{\delta t}(\theta C + \rho S) = \frac{\delta}{\delta x}\left[D_h \theta \frac{\delta C}{\delta x}\right] - \frac{\delta}{\delta x}(\nu \theta C) - \sum_{i=1}^{n} Q_i, \qquad (3)$$

where ρ is soil bulk density, D_h is hydrodynamic dispersion coefficient, ν is average pore-water velocity. Q_i includes various sink terms to account for pesticide losses (degradation and plant uptake), S and C, respectively, adsorbed phase and solution phase concentrations, θ is volumetric soil water content, and x is soil depth.

Reviews published by Bailey and White (1970), Letey and Farmer (1974), Leistra (1980), Jury, Spencer, and Farmer (1983), Rao and Jessup (1982) and Yaron, Gerstl, and Spencer (1985) give a comprehensive picture of convective transport of biocides through soils.

The two main factors affecting the movement of pesticides through soils by mass flow are the rate of water flow and the adsorption of pesticides by soils. This transport pathway is actually similar to the movement of a solute in solid-liquid chromatography. The theory of chromatographic movement has been taken into consideration in the development of the relationships describing the transport of pesticides through soil columns. Helling and Turner (1968) were the first to propose thin-layer chromatography as a technique for the study of pesticide mobility in soil. This simple method designates soil as the adsorbent phase and water as the eluent, and the movement of pesticides relative to the frontal movement of water (R_F) is measured. Paper 41 is one of a series of three papers by Helling describing the soil thin-layer chromatography technique and its applications to the study of pesticide mobility. In this paper the parameters influencing pesticide mobility were investigated, and the results were interpreted by correlation and regression analyses. The data obtained showed that pesticide mobility was directly related to the water flux. Movement of non-ionic compounds was inversely related to adsorption and to soil properties such as organic matter and clay content, cation exchange capacity, and field moisture capacity. Mobility of acidic compounds was directly correlated with soil pH. The use of regression equations enabled accurate prediction of the R_F of all the pesticides for all the soils investigated. The advantage of this technique for the study of the relative mobility of pesticides in soil is that a great number of pesticides and soils can be analyzed quickly and reproducibly.

Paper 42 is an example of an early investigation emphasizing the importance of water flow rate and adsorption on pesticide movement through soil. The movement of two substituted urea herbicides with water through glass beads and a soil column was dependent on the water flux or the average pore velocity. The experimental and calculated effluent concentration distributions of fluometuron in soil did not agree. Prediction was done by using a differential equation describing the process of longitudinal mixing and linear retention in porous material for one-dimensional flow. The authors suggested that their prediction could be improved by better knowledge of the distribution of the retained phase with soil depth.

Various aspects of the convective movement of pesticides into soils were described in numerous studies published during the period 1970-1980. Data of Clay and Scott (1973), Leistra, Smelt, and Zandwoort (1975), and Marriage, Khan, and Saidak (1977) showed the effect of adsorption in retarding vertical transport of pesticides under various soil and environmental conditions. Using miscible displacement techniques, Davidson and Chang (1972) measured the effect of water flow rate, bulk density, and aggregate size on the transport of picloram and reported that a decrease in water velocity caused a decrease in picloram mobility. The transport at the lower water flux velocity was predicted better than at the higher velocity. When high liquid fluxes were used, the peak of herbicide distribution moved faster than expected from equilibrium adsorption. This finding was observed with prometryne and fluometuron (Abernathy and Davidson, 1971; Davidson and McDougal, 1973). Davidson and Chang (1972), studying the movement of herbicides in soil columns consisting of either small or large aggregates, found that at high liquid fluxes the nonequilibrium effects were more marked with larger than with smaller aggregates. This difference was explained by the longer diffusion time required for the solute in the liquid phase to reach equilibrium with the internal portion of the larger aggregates. In soils with large aggregates, pesticides moving with a high liquid flux are probably adsorbed only at the surfaces of these aggregates.

Paper 43 illustrates a further development in the studies of mass transfer of pesticides. The authors used their previously developed theoretical model to describe the movement of the chlorophenoxyacetic acid herbicide 2,4,5-T through soil. In this model the soil water phase is considered to be divided into both a mobile and a stagnant region. The convective solute transport occurs in the mobile soil water region, and solute transfer between the two regions is diffusion-controlled. The model includes the effects of intra-aggregate diffusion and adsorption on the mass transfer. The treatment of this case was complicated by the fact that adsorption-desorption were hysteretic and adsorption was nonlinear. The data obtained showed that intra-aggregate exchange of 2,4,5-T and adsorption-desorption are the main mechanisms responsible for the tailing of effluent curves. The inclusion of the intra-aggregate diffusion into the model showed that hysteresis in the adsorption-desorption process is less important than previously believed in determining pesticide concentration distribution.

This diffusion-controlled adsorption model and another model (kinetics-controlled adsorption) were evaluated by Rao and co-workes (Paper 44) for describing the nonequilibrium adsorption-desorption of pesticides in soils during steady-state water flow. Similar to the model used in Paper 43, the kinetics-controlled adsorption model implies the existence of two groups of adsorption sites. In the first group adsorption is instantaneous; however, unlike the previous model, this model assumes that in the second group adsorption follows a nonlinear reversible kinetics. Both models were used to predict the effluent breakthrough curves for the amine salt of 2,4-D and atrazine in three soils. The data obtained showed that both models gave excellent agreement between the calculated and measured curves for a nonadsorbed solute. Although both models provided good descriptions of the asymmetrical breakthrough curves for pesticide displacement, they have some limitations. The diffusion-controlled adsorption model may not be applicable to the three soils investigated. The authors suggested that the conceptual representation of the adsorption-desorption process in the kinetics-controlled adsorption model must be inadequate because it failed to give a very accurate prediction of the breakthrough curves of pesticides.

In general, the studies leading to quantitative description of convection of pesticides through soil were done under controlled conditions in laboratories; however, results are often inconsistent. Differing types of soil, flow rates, and experimental techniques lead to inconsistency. Measured transport of pesticides in the field often deviates from the expected pattern. Redistribution of the applied water resulting from evapotranspiration, presence of cracks and voids, or heterogeneity of the soil constituents, may explain this deviation (Schiavon and Jacquin, 1973, Bovey et al, 1975). In addition, the presence of a growing biomass may affect the transport as a function of crop and stage of development. Leistra (Paper 45) proposed a computer model to simulate the redistribution of pesticides in the root zone of potato cultivated in a sandy loam soil. The hydraulic characteristics of the soil, the climatic data, the evapotranspiration losses, the dynamic uptake of water by plants from various depths, and soil moisture availability were taken into account. Differing pesticide adsorption for various soil layers was considered, as well as pesticide transformations affected by time and temperature changes. The data obtained from the computer program indicated that considerable convective transport could occur only in the first month after application of weakly adsorbed pesticides. The later increase in the water uptake by the growing vegetation reduced the water flux into the soil. Because of the decisive effect of adsorption, this parameter must be measured. The use of computer simulation techniques seems the best solution, enabling description of the movement of pesticides under field conditions because the computer can account for multiple changing parameters in practically infinite combinations.

The need to predict the simultaneous transport of water and herbicides in soil for better crop protection and management and for the prevention of unsaturated zone and surface water pollution has stimulated the development of a series of mathematical models. In general, the existing models for the transport of noninteracting solutes were adapted to herbicide properties, and specific models were produced that usually provide a partial agreement with experimental data. A critical review of the development and verification of existing simulation models for describing pesticide dynamics in soil was presented by Rao and Jessup (1982). Some of the existing mathematical models being developed for equilibrium conditions are inadequate for field predictions. Even though batch experiments suggest a rapid equilibrium of adsorption, in the reality of transient conditions the reactions are only partially completed. Moreover, the rate of pesticide adsorption-desorption in sites residing within immobile water is controlled by diffusive mass transfer across the liquid layer and results in an incomplete equilibrium during solute flow through soils. This situation is compounded by the heterogeneity of the field scale, where substantial lateral variations in vertical water velocity can be found. Jury (1983), summarizing the current approaches in chemical transport modeling, shows that convective models such as the transfer function model which take velocity variations into account are able to describe the spatial variability of chemical movement near surface but it is unknown to what depth the correlation inherent in the assumption of the transfer function model will still hold.

References

Abernathy, J. R., and J. M. Davidson, 1971, Effect of Calcium Chloride on Prometryne and Fluometuron Adsorption in Soil, *Weed Sci.* **19:**517-521.

Bailey, G. W., and J. L. White, 1970, Factors Influencing the Adsorption, Desorption, and Movement of Pesticides in Soil, *Residues Rev.* **32:**29-92.

Bovey, R. W., W. Burnett, C. Richardson, J. R. Baur, M. G. Merkle, and D. E. Kissel, 1975. Occurrence of 2,4,5-T and Picloram in Subsurface Water in the Blacklands of Texas, *Jour. Environ. Qual.* **4:**103-106.

Calvet, R., 1984, Behavior of Pesticides in the Unsaturated Zone. Adsorption and Transport Phenomena, in *Pollutants in Porous Media*, B. Yaron, G. Dagan, and J. Goldschmid, eds., Springer-Verlag, Heidelberg, pp. 143-151.

Clay, D. V., and K. G. Scott, 1973, The Persistence and Penetration of Large Doses of Simazine in Uncropped Soil, *Weed Res.* **13:**42-50.

Crank, J., 1975, *Mathematics of Diffusion,* 2nd ed., Clarendon Press, Oxford. 484p.

Davidson, J. M., and R. K. Chang, 1972, Transport of Picloram in Relation to Soil Physical Conditions and Pore-Water Velocity, *Soil Sci. Soc. America Proc.* **36:**257-261.

Davidson, J. M., and J. R. McDougal, 1973, Experimental and Predicted Movement of Three Herbicides in a Water-Saturated Soil, *Jour. Environ. Qual.* **2:**428-433.

Ehlers, W., J. Letey, W. F. Spencer, and W. J. Farmer, 1969, Lindane Diffusion in Soils: I. Theoretical Considerations and Mechanism of Movement, *Soil Sci. Soc. America Proc.* **33:**501-504.

Gerstl, Z., B. Yaron, and P. H. Nye, 1979, Diffusion of a Biodegradable Pesticide: I. In a Biologically Inactive Soil, *Soil Sci. Soc. American Proc.* **43:**839-842.

Hanks, R. W., 1947, Removal of 2,4-Dichlorophenoxyacetic Acid and its Calcium Salt from Six Different Soils by Leaching, *Bot. Gaz.* 187-191.

Hartley, G. S., and I. J. Graham-Bryce, 1980, *Physical Principles of Pesticide Behavior,* vol. 1, Academic Press, London, 518p.

Helling, C. S., and B. C. Turner, 1968, Pesticide Mobility: Determination by Soil Thin-Layer Chromatography, *Science* **162:**562-563.

Jury, W. A., 1983, Chemical Transport Modeling: Current Approaches and Unresolved Problems, in *Chemical Mobility and Reactivity in Soil Systems,* Soil Science Society of America Spec. Publ. No. 11, pp. 49-64.

Jury, W. A., W. F. Spencer, and W. J. Farmer, 1983, Behavior Assessment Model for Trace Organics in Soil. I. Model Description, *Jour. Environ. Qual.* **12:**558-564.

Jury, W. A., W. F. Spencer, and W. J. Farmer, 1984, Behavior Assessment Model for Trace Organics in Soil. III. Application of Screening Model, *Jour. Environ. Qual.* **13:**573-579.

Lavy, T. L., 1970, Diffusion of Three Chloro s-Triazines in Soil, *Weed Sci.* **18:**53-56.

Leistra, M., 1980, Transport in Solution, in *Interactions between Herbicides and the Soil,* R. J. Hance, ed., Academic Press, New York, pp. 31-58.

Leistra, M., J. H. Smelt, and R. Zandwoort, 1975, Persistence and Mobility of Bromacil in Orchard Soils, *Weed Res.* **15:**243-247.

Letey, J., and W. F. Farmer, 1974, Movement of Pesticides in Soil, in *Pesticides in Soil and Water,* W. D. Guenzi, ed., Soil Science Society of America, Madison, Wis. pp. 67-87.

Lindstrom, F. T., L. Boersma, and H. Gardiner, 1968, 2,4-D Diffusion in Saturated Soils: A Mathematical Model, *Soil Sci.* **106:**107-113.

Marriage, P. B., S. U. Khan, and W. J. Saidak, 1977, Persistence and Movement of Terbacil in Peach Orchard Soil after Repeated Annual Applications, *Weed Res.* **17:**219-225.

Olsen, S. R., and W. D. Kemper, 1968, Movement of Nutrients to Plant Roots, *Adv. Agron.* **20:**91-151.

Rao, P. S. C., and R. E. Jessup, 1982, Development and Verification of Simulation Models for Describing Pesticide Dynamics in Soils, *Ecol. Model.* **16:**67-76.

Schiavon, M., and F. Jacquin, 1973, Contribution a L'Etude de la Migration de Deux Triazines sous l'Influence des Precipitations, European Weed Research Council Symp. on Herbicides in Soil, Proc. Versailles, France. pp. 80-90.

Scott, H. D., and R. E. Phillips, 1972, Diffusion of Selected Herbicides in Soil, *Soil Sci. Soc. America Proc.* **36:**714-719.

Yaron, B., Z. Gerstl, and W. F. Spencer, 1985, Behavior of Herbicides in Irrigated Soils, *Adv. in Soil Sci.* **3:**121-211.

36

Copyright © 1958 by the Entomological Society of America
Reprinted from *Jour. Econ. Entomol.* **51**:380–383 (1958)

Movement of Insecticides in Soils Under Leaching and Non-Leaching Conditions[1]

E. P. LICHTENSTEIN,[2] *Department of Entomology, University of Wisconsin, Madison*

ABSTRACT

In 1954 a Miami silt loam and a muck soil were treated with aldrin at 20 lb. and 200 lb./acre, lindane at 10 lb. and 100 lb./acre and DDT at 10 lb. and 100 lb./acre. The insecticides were sprayed on the soil as an emulsion with a sprinkling can and then rototilled into the soil to a depth of 4 to 5″.

Seventeen months later 84 to 96% of the insecticides were found in the upper 3-inch level of the loam soil, 4 to 12% in the 3 to 6″ layer, and 0 to 5% in the 6 to 9″ layer. In the muck soil 62 to 74% was found in the top layer, 19 to 29% in the 3 to 6″ layer and 7 to 8% in the 6 to 9″ layer. No differences were noticed between individual insecticides. At the same time lindane was found to be unequally distributed in a horizontal direction.

Three years after treatment, slightly (5 to 15°) sloping plots contained from 1.3 to 2.2 times more insecticide in the lower half as compared with the upper half of the plot.

Experiments, conducted under laboratory conditions, showed that lindane was leached to some extent from a treated soil into an untreated one. The leaching was most noticeable in Plainfield sand and least noticeable in muck soil. Under non-leaching conditions, lindane also moved into the untreated layer, but more was retained in a muck soil than in a Plainfield sand. When radioactive parathion (P^{32}) was used, it was found that during a period of 6 days, parathion moved upwards, downwards and sidewards as well. The results obtained seem to indicate that the movement of parathion is more rapid in a Plainfield sand than in a muck soil, as the latter retains the insecticide to a greater extent. In a Plainfield sand, 6.6% of all the parathion recovered was in the untreated layers adjacent to the treated one and 3.5% in the untreated layers most distant from the treated one. The respective figures for a muck soil were 10.8% and 1.8%.

Preliminary experiments conducted with aldrin under non-leaching conditions indicate movement of this insecticide to a considerable extent.

Insecticides from sprayed crops and from direct soil application have been found to persist or even accumulate within soils (Chisholm *et al.* 1951, Fleming *et al.* 1951, Ginsburg & Reed 1954, Ginsburg 1955, Lichtenstein 1957). Such accumulation may be of value in the control of certain soil insect pests but of detrimental value to subsequent crops. Some insecticides may affect the root systems of certain plants and also be translocated into other plant tissues (Simkover & Shenefelt 1951, 1952, Stone *et al.* 1953, Fleming & Maines 1953, Terriere & Ingalsbe 1953, Allen *et al.* 1954, Boswell *et al.* 1955). If insecticides persist mainly in the upper 3 inches of soils, crops with short root systems would be most exposed to those chemicals; if below 3 inches, crops with more extended root systems could be affected. In view of these side effects and for a better understanding of soil insect control, it is of primary importance to define the movement of insecticides within soils. Field and laboratory studies were conducted to measure the movement of some typical insecticides in soils under leaching and non-leaching conditions.

METHODS.—*Field studies.*—In 1954, a Miami silt loam and a muck soil were treated with aldrin at 20 and 200 lb./acre, lindane at 10 and 100 lb./acre, and DDT at 10 and 100 lb./acre. The insecticidal applications involved thorough mixing of an emulsifiable concentrate with water. Two-gallon quantities of diluted emulsion were spread as equally as possible with a sprinkling can over successive 10×29 foot areas of each experimental plot (100′×29′). Immediately after application, each plot was rototilled to a depth of 4 to 5 inches.

Seventeen months after treatment, 40 soil cores (¾ inch dia., 9 inches long) were collected from each plot. Each core was divided into three segments (0–3″, 3–6″ and 6–9″), comparable subcores from each plot were mixed and subsequently extracted and analyzed (Schechter & Haller 1945, Schechter & Hornstein 1952, O'Donnell *et al.* 1954, Lichtenstein *et al.* 1956), to determine the vertical distribution of the insecticides.

Certain plots in the loam experiment were used to determine the movement of insecticides down gentle slopes (5 to 15°). Three years after treatment, a 40-core sample was collected in the upper half and a like number in the lower half of four plots (DDT 10 lb./acre, lindane 10 and 100 lb./acre, and aldrin 200 lb./acre). Previously mentioned analytical methods were employed to ascertain the quantity of insecticide in the thoroughly mixed samples.

Laboratory experiments.—Experiments were designed to study the movement of insecticides in different soil types under leaching and non-leaching conditions. One-quart ice cream cartons with perforated bottoms were filled with five to seven cheesecloth separated layers of soil. Columns of Plainfield sand, Miami silt loam and muck soils were used. When lindane was applied, the two upper layers of soil (150 gm. each) were treated at approximately 10 p.p.m. and placed over three untreated layers (150 gm. each). Three variables were employed: 1) No water was poured through the column during a 10-day holding period, 2) 100 ml. of water was poured through each column daily for 5 days, 3) 100 ml. of water was poured through each column daily for 10 days. The watered columns were placed on a Buchner funnel and the water, which passed through, was collected in an Erlenmeyer flask for analysis. After 5 or 10 days the columns were cut open and the different layers carefully separated and then analyzed for lindane content. Results were based on four replicates of each treatment-soil type combination.

To study the movement of an insecticide in soil under

---

[1] Approved for publication by the Director of the Wisconsin Agricultural Experiment Station. Accepted for publication December 23, 1957.

Research supported in part by funds from North Central Regional Project 19 and the Shell Chemical Company. Contribution from Wisconsin Agricultural Experiment Station as a collaborator under North Central Region cooperative research project entitled "Reduction of Hazards in the Use of Pesticides."

Reported at the Fifth Annual Meeting of the Entomological Society of America held at Memphis, Tennessee, December 2–5, 1957.

[2] Acknowledgment is given to Mrs. Ann LaPidas and Mr. Kenneth R. Schulz for assisting in the chemical analyses.

Table 1.—Vertical distribution of DDT, lindane and aldrin in a loam and a muck soil, 17 months after treatment.

INSECTI-CIDE	POUNDS APPLIED PER ACRE	SOIL TYPE	PER CENT DISTRIBUTION OF RECOVERED INSECTICIDE		
			0–3″ Layer	3–6″ Layer	6–9″ Layer
DDT	10	Loam	83.6	11.8	4.6
	100	Loam	84.0	11.9	4.1
	10	Muck	66.1	27.1	6.8
Lindane	10	Loam	96.0	4.0	0.0
	100	Loam	87.7	10.0	2.3
	10	Muck	62.3	29.4	8.3
	100	Muck	69.8	24.3	5.9
Aldrin	20	Loam	89.0	8.3	2.7
	200	Loam	88.3	8.8	2.9
	20	Muck	74.0	19.2	6.8
	200	Muck	65.7	26.7	7.6

Table 2.—Recoveries of DDT, aldrin and lindane on a sloping loam plot, 3 years after treatment.

TREATMENT	DEGREES SLOPE (APPROXI-MATE)	P.P.M. FROM		RATIO LOWER/UPPER
		Upper Half	Lower Half	
DDT, 10 lb./A.	15	1.64 ± 0.02[a]	2.12 ± 0.12	1.29
Lindane, 10 lb./A.	15	0.56 ± 0.03	1.19 ± 0.02	2.13
Aldrin, 200 lb./A.	10	7.92 ± 0.22	17.30 ± 0.20	2.18
Lindane, 100 lb./A.	5	11.2 ± 0.10	16.40 ± 0.50	1.47

[a] Standard deviation.

non-leaching conditions in both vertical and horizontal directions, radioactive parathion (P^{32}) was used and the samples were analyzed by counting with a Geiger-Mueller counter. Preliminary experiments were conducted to see whether similar results could be obtained by direct counting of the soil against counting of a soil extract. When the distribution of radioactive parathion in the various soil layers was compared by the two methods, the results were almost identical. Thus direct soil counting seemed to be the easiest practical method for such studies.

Seven 50-gram layers of soil, separated by cheesecloth, were placed into a 1-quart ice cream carton. The middle or fourth layer which had been previously treated with approximately 200 p.p.m. of radioactive parathion, was placed over three untreated layers and then covered with three additional layers of untreated soil. Four identical columns of each soil type (Plainfield sand, Miami silt loam and muck) were prepared, two of which were placed in a vertical and two in a horizontal position. No water was applied and after 6 days, the columns were opened and the layers were carefully separated and individually mixed. One-gram aliquots which proved to be "infinitely thick" were counted against a standard of the same soil. This standard had been treated with parathion at 200 p.p.m. at the beginning of the experiment.

A partition into chloroform and water was done from an acetone extract, to determine whether the P^{32} recovered within the various soil layers represented para-thion (chloroform soluble) or nontoxic hydrolyzed products of parathion (water soluble). Thirty grams of each soil layer from a loam column were extracted with warm acetone. After filtering, 5 ml. of water was added to the filtrate, the acetone was evaporated and the amount of water adjusted to 5 ml. in a 15 ml. calibrated centrifuge tube. Five ml. of chloroform were then added and the water-chloroform mixture was shaken thoroughly. After centrifuging, 1-ml. aliquots of the water and chloroform fraction were counted with a Geiger-Mueller counter.

RESULTS.—*Vertical distribution.*—Seventeen months after treatment 84 to 96% of the insecticides (DDT, lindane and aldrin) was found in the upper 3″ layer of a Miami silt loam (table 1). Four to 12% was found in the 3 to 6″ layer and 0 to 5% in the 6 to 9″ layer. In muck 62 to 74% was found in the top layer, 19 to 29% in the 3 to 6″ layer and 7 to 8% in the 6 to 9″ layer. No individual differences between insecticides were observed.

Recoveries of DDT, aldrin and lindane on a sloping plot.—Table 2 summarizes the analytical results of samples collected on the upper and lower halves of four insecticide-treated loam plots located on a gentle slope. From 1.29 to 2.18 more insecticide was found in the lower halves of these plots with DDT showing the smallest difference.

The plot treated with lindane at 10 lb./acre and having a slope of 15° contained 2.13 times more insecticide in the lower half, whereas the plot treated with lindane at 100 lb./acre but having a slope of 5°, contained only 1.47 times more in the lower half than in the upper part of the plot.

Initially after treatment only the upper half had been sampled and analyzed.

Recoveries of lindane from soil columns under leaching and non-leaching conditions.—The results of these experiments (Table 3) seem to incidate that lindane is leached

Table 3.—Vertical distribution of lindane in three soil types under non-leaching and leaching conditions.

SOIL TYPE	DAYS OF WATERING (100 ML./DAY)	P.P.M. IN VERTICAL LAYERS INDICATED					
		Treated		Untreated			
		1	2	3	4	5	TOTAL
Plainfield sand	0	9.65 ± 0.30	9.15 ± 0.40	0.00	0.00	0.00	18.80
	5	7.45 ± 0.77	7.61 ± 0.57	1.01 ± 0.43	0.21 ± 0.08	0.18 ± 0.14	16.26
	10	6.62 ± 0.44	7.98 ± 0.47	1.31 ± 0.12	0.14 ± 0.01	0.06 ± 0.06	16.11
Miami silt loam	0	10.10 ± 0.50	9.52 ± 0.43	0.27 ± 0.00	0.00	0.00	19.89
	5	9.47 ± 0.86	9.55 ± 0.32	0.44 ± 0.10	0.00	0.00	19.46
	10	7.48 ± 0.30	7.44 ± 0.56	0.57 ± 0.09	0.08 ± 0.05	0.08 ± 0.05	15.65
Muck	0	10.55 ± 0.45	10.50 ± 0.20	0.80 ± 0.23	0.16 ± 0.16	0.14 ± 0.14	22.15
	5	9.18 ± 1.44	9.59 ± 1.36	0.09 ± 0.09	0.00	0.00	18.86
	10	9.17 ± 0.07	8.88 ± 0.35	0.15 ± 0.04	0.05 ± 0.03	0.07 ± 0.07	18.32

Table 4.—Distribution (in p.p.m.) of parathion in layers of soil columns under non-leaching conditions, 6 days after treatment.

Soil Layers	Vertical Position			Horizontal Position			P^{32} in Layer Indicated with Vertical Loam Column	
	Plainfield Sand	Miami Silt Loam	Muck	Plainfield Sand	Miami Silt Loam	Muck	Direct Soil Counting	Extract Counting
1. Untreated	6.8± 0.40	4.9±0.65	4.1±0.10	5.0± 0.65	2.3±0.03	4.2±0.43	2.20	2.12
2. Untreated	8.6± 2.30	5.2±0.25	6.5±0.71	6.7± 1.47	2.6±0.62	5.7±0.10	2.15	1.75
3. Untreated	10.7± 0.45	19.9±0.85	25.7±0.85	9.4± 0.20	14.4±0.65	19.1±2.15	9.00	8.50
4. *Treated*	140.0±15.0	164.0±7.00	155.0±3.00	144.5±26.00	161.5±3.50	152.0±1.00	74.00	79.00
5. Untreated	15.4± 2.2	20.4±3.10	23.1±1.10	16.4± 1.00	14.1±1.00	18.0±0.70	9.35	5.60
6. Untreated	9.8± 0.2	4.3±0.65	5.4±0.90	6.3± 0.60	2.5±0.51	5.9±0.15	1.92	1.93
7. Untreated	6.9± 0.6	3.1±0.15	4.1±0.60	6.0± 0.35	4.0±0.27	4.8±0.12	1.38	1.10
Total	198.2	221.8	224.4	194.3	201.4	209.7	100.00	100.00

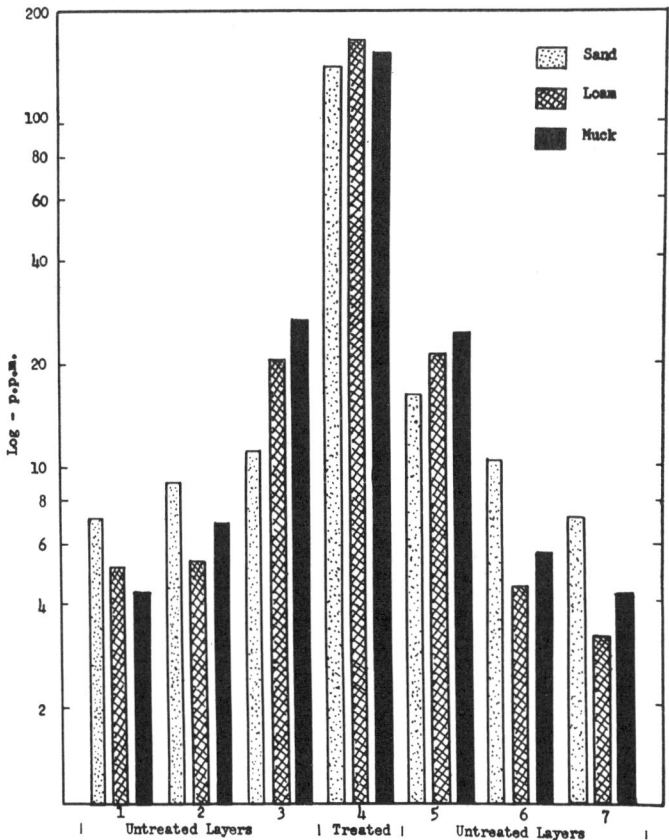

Fig. 1.—Distribution of parathion in layers of soil columns under non-leaching conditions, 6 days after treatment.

into the lower untreated layers (3, 4, 5). The third layer, which is the first untreated one adjacent to the treated layer, contained 1.31 p.p.m. of lindane in Plainfield sand, 0.57 p.p.m. in Miami silt loam and 0.15 p.p.m. in muck after 10 days leaching with 100 ml. water daily. The differences probably are due to a larger retention of the insecticides in soils containing a higher percentage of organic matter. No lindane was found in the water, which passed through the columns and was collected in an Erlenmeyer flask. Under non-leaching conditions, no lindane was found in the third layer of Plainfield sand, 0.27 p.p.m. in the third layer of loam, and 0.80 p.p.m. in the third layer of muck. Moreover the two bottom layers of sand and loam did not contain any lindane, but traces of lindane were found in these layers of the muck soil.

Movement of parathion (P^{32}) in three soil types under non-leaching conditions.—Table 4 summarizes the results obtained on the movement of radioactive parathion (P^{32}) in soils under non-leaching conditions. It was found that parathion moves within the soils in all directions, which seems to indicate a diffusion phenomenon. The layers adjacent to the treated ones contained more parathion in muck (10.8% of all the parathion recovered) than in Plainfield sand (6.6%). The layers farthest from the treated ones contained more in the sand (3.5%) than in either the loam or the muck soils (1.8%). Parathion apparently moves more rapidly in a sandy soil and possibly is lost at the soil-air interface more rapidly than from the other two soil types. The opposite seems to be true in a muck soil (fig. 1).

When water and chloroform fraction of soil extracts were counted, no counts were registered in the water fractions of the two outer layers and only traces in those of the other layers (3, 4 and 5). Therefore it might be assumed that the recoveries given in table 4 and figure 1 actually represent parathion.

References Cited

Allen, N., R. L. Walker and L. C. Fife. 1954. Persistence of BHC, DDT and toxaphene in soil and the tolerances of certain crops to their residues. U. S. Dept. Agric. Tech. Bull. 1090.

Boswell, V. R., W. J. Clore, B. B. Pepper, C. B. Taylor, P. M. Gilmer and R. L. Carter. 1955. Effects of certain insecticides in soil on crop plants. U. S. Dept. Agric. Tech. Bull. 1121.

Chisholm, R. D., L. K. Koblitzky, J. E. Fahey and W. E. Westlake. 1951. DDT residues in soil. Jour. Econ. Ent. 43(6): 941.

Fleming, W. E., W. W. Maines and L. W. Coles. 1951. Persistence of chlorinated hydrocarbon insecticides in turf treated to control the Japanese beetle. U. S. Dept. Agric. E-829.

Fleming, W. E., and W. W. Maines. 1953. Effect of chlorinated organic compounds on plants grown in treated soil. U. S. Dept. Agric. Bur. Ent. and Plant Quar. E-872.

Ginsburg, J. M. 1955. Accumulation of DDT in soils from spray practices. Jour. Agric. and Food Chem. 3(4): 322.

Ginsburg, J. M., and John P. Reed. 1954. A survey of DDT accumulation in soils in relation to different crops. Jour. Econ. Ent. 47(3): 467–74.

Lichtenstein, E. P. 1957. A survey of DDT accumulation in midwestern orchard and crop soils treated since 1945. Jour. Econ. Ent. 50(5): 545–7.

Lichtenstein, E. P., S. D. Beck and K. R. Schulz. 1956. Colorimetric determination of lindane in soils and crops. Jour. Agric. and Food Chem. 4(11): 936.

O'Donnell, A. E., M. M. Neal, F. T. Weiss, J. M. Bann, J. D. DeCino and S. C. Lau. 1954. Chemical determination of aldrin in crop materials. Jour. Agric. and Food Chem. 2(11): 573–80.

Schechter, M. S., and H. L. Haller. 1945. Colorimetric determination of DDT. Indust. and Engineer. Chem., Analyt. Ed., 17: 704.

Schechter, M. S., and I. Hornstein. 1952. Colorimetric determination of benzene hexachloride. Analyt. Chem. 24: 544.

Simkover, H. G., and R. D. Shenefelt. 1951. Effect of benzene hexachloride and chlordane on certain soil organisms. Jour. Econ. Ent. 44(3): 426.

Simkover, H. G., and R. D. Shenefelt. 1952. Phytotoxicity of some insecticides to coniferous seedlings with particular reference to benzene hexachloride. Jour. Econ. Ent. 45(1): 11–15.

Stone, M. W., F. B. Foley and D. H. Bixby. 1953. Effect of soil applications of insecticides on the growth and yield of vegetable crops. U. S. Dept. Agric. Circ. 926.

Terriere, L. C., and D. W. Ingalsbe. 1953. Translocation and residual action of soil insecticides. Jour. Econ. Ent. 46(5): 751–3.

DIFFUSION OF ORGANOPHOSPHORUS INSECTICIDES IN SOILS

By I. J. GRAHAM-BRYCE

Diffusion of disulfoton and dimethoate in a silt loam soil was studied over a range of concentrations and moisture contents. Apparent diffusion coefficients were calculated from the distribution of insecticide in a column of soil after diffusion from one half of the column to the other for a known time. The distribution was determined by slicing the column into narrow sections using a specially constructed diffusion cylinder. Diffusion coefficients varied little with concentration for both insecticides, but increased rapidly with increasing moisture content for dimethoate from $3 \cdot 31 \times 10^{-8}$ cm²/sec at 10% volumetric moisture content to $1 \cdot 41 \times 10^{-6}$ cm²/sec at 43% moisture content. In contrast, for disulfoton which is more volatile, less soluble and more strongly sorbed than dimethoate, diffusion coefficients were smaller ($2 \cdot 83 \times 10^{-8}$ cm²/sec at 41% moisture content) but did not change much as the soil became drier ($2 \cdot 74 \times 10^{-8}$ cm²/sec at 8% moisture content). The influence of partition between solid, solution and vapour phases in the soil and of the geometry of the pathway through the soil pores on the apparent diffusion coefficient is discussed. The likely behaviour of other pesticides is considered in the light of these results.

Introduction

A knowledge of the speed at which pesticides diffuse through soil is necessary to calculate distributions after they are applied to soil and to estimate amounts that can move to plant roots. Measurements of diffusion rates can also give indirect information about interactions between pesticides and soil solids. Diffusion of fumigants such as ethylene dibromide has been studied in detail[1] and measurements for 2,4-D in saturated soils have been reported recently,[2] but other pesticides do not seem to have been studied quantitatively.

Insecticides diffuse more slowly in soil than in air or free solution because the pathway through the pores is restricted and tortuous and because part of the chemical may be retarded by sorption on the solids. The exact geometry of the pathway depends on the nature of the pore space formed by the particles in a given soil and on the moisture content. The way these geometric factors influence diffusion of a pesticide depends on its solubility and volatility, which govern the relative amounts moving in the soil solution or as vapour in the soil air space. According to Fick's law, the amount of insecticide diffusing/second through unit area of soil at any point is the product of the gradient of concentration, and the apparent diffusion coefficient whose value is determined by the geometric and partition factors just discussed. Measuring diffusion coefficients has the advantages over less precise estimates of the apparent distances which pesticides diffuse in that the values can be used to predict distributions in systems different from those used in measurement; and also that the contribution of different factors to the rate of diffusion can be calculated in detail. This paper reports determinations of diffusion coefficients of the systemic insecticides dimethoate (dimethyl S-(N-methylcarbamoylmethyl)phosphorothiolothionate) and disulfoton (diethyl S-[2-(ethylthio)ethyl]phosphorothiolothionate) in soil of various moisture contents. These insecticides were chosen because they have contrasting physical properties and affinities for soil. Dimethoate is soluble in water to about 3% by wt. and has a vapour pressure of $8 \cdot 5 \times 10^{-6}$ mm Hg at 20°, whereas disulfoton is much less soluble (about 15 ppm at 20°) and is more volatile (v.p. $1 \cdot 8 \times 10^{-4}$ mm Hg at 20°; vapour pressures and solubilities from technical data given by manufacturers, and from Lord & Burt[3]).

To help analyse the contribution of different factors to the values of diffusion coefficients, sorption isotherms for these insecticides on soil were also determined.

Theoretical

The determination of diffusion coefficients for insecticides has been discussed briefly by the author;[4] fuller details are now given.

A satisfactory method for determining diffusion coefficients must allow for the possible dependence on concentration. The appropriate forms of Fick's law of diffusion are:

$$F = -D \frac{\partial C}{\partial x}$$

where
F = flux of diffusing insecticide in the x direction (g/cm²/sec)
C = concentration of insecticide in whole soil (g/ml)
D = apparent diffusion coefficient (cm²/sec)
and

$$\frac{\partial C}{\partial t} = \frac{\partial}{\partial x}\left(D \frac{\partial C}{\partial x}\right) \quad \ldots \ldots \ldots \ldots \ldots \ldots \ldots (1)$$

For the boundary conditions at $t=0$, $C=C_0$ for $x > 0$ and $C=0$ for $x > 0$ and for all values of t at $x = \pm \infty \, \partial C/\partial x = 0$. D can be obtained from Equation (1) using the method of Matano.[6] Applying the Boltzmann transformation $\lambda = x/(t)^{\frac{1}{2}}$, gives the following solutions for these boundary conditions:

$$\text{for } x < 0, \; D = -\frac{1}{2t} \frac{dx}{d(C/C_0)} \int_0^{C/C_0} x \, d(C/C_0) \ldots \ldots \ldots (2)$$

$$\text{for } x > 0, \; D = \frac{1}{2t} \frac{dx}{d(C/C_0)} \int_{C/C_0}^{1} x \, d(C/C_0) \ldots \ldots \ldots (3)$$

Hence, when the distribution of insecticide resulting from diffusion for a known time t in a system corresponding to the boundary conditions is determined, D can be evaluated at different values of C/C_0 by measuring $dx/d(C/C_0)$ and $\int x \, d(C/C_0)$ graphically on the x–C/C_0 curve. A mean value, $\bar{D}$, for the concentration range studied can be obtained from a plot of D against C/C_0. This method was recently applied by Phillips & Brown[6] for measuring ionic counter-diffusion in soil.

Insecticide may move through the gaseous and liquid phases in soil and sorbed insecticide may move along surfaces or in the solid phase. However, the pesticide associated with the solid almost certainly diffuses much more slowly than in the pores and therefore makes a negligible contribution to the total flux, even when most of the pesticide occurs on the solid because of strong sorption. Sorption isotherms for pesticides are usually approximately linear (for examples see literature cited by Graham-Bryce[7]) and if Henry's law holds for the insecticide vapour, partition between solid, solution and vapour should be independent of concentration. D should then also be independent of concentration, assuming that equilibration between the phases is rapid compared with the change in concentration resulting from diffusion and neglecting the minor influence of concentration on diffusion coefficients in free air and free solution.

The relative contributions of the liquid and vapour pathways to the value of the diffusion coefficient will depend on the volatility and solubility of the pesticide. Volatile but sparingly soluble compounds may diffuse almost entirely through the vapour phase whereas this pathway would be insignificant for soluble, involatile materials. With such extreme compounds that diffuse effectively through only one pathway:

$$D\frac{dC}{dx} = D_Z V_Z f_Z \frac{dC_Z}{dx}$$

where the subscript Z refers to either the liquid (L) or gaseous (G) phase, V is the fraction of the soil occupied by the phase referred to, f is a geometric factor which allows for the tortuosity of the diffusion path through the soil pores, D_Z is the diffusion coefficient in free solution or free air and C_Z is the concentration/unit volume of the phase. Hence:

$$D = D_Z V_Z f_Z \frac{dC_Z}{dC}$$

If Henry's law holds and the adsorption isotherm is linear, dC_Z/dC is a constant so that:

$$D = D_Z V_Z f_Z \frac{C_Z}{C} \quad \cdots\cdots\cdots\cdots\cdots (4)$$

Nye[8] developed similar equations for ionic diffusion in soils.

Experimental

Soil

Soil from the top 15 cm of the plot on Broadbalk field at Rothamsted Experimental Station which has been receiving farmyard manure annually since 1843 was air-dried and sieved (<2 mm) before use. This soil is a silt loam and the properties of the sample used were clay 18%; cation exchange capacity 19·8 mequiv./100 g, pH (1:2·5 suspension in water) 7·8; organic carbon 2·7%.

Determination of diffusion coefficients

The experimental system consists of a cylinder of soil mixed uniformly with insecticide at a concentration C_0 joined to a similar cylinder initially free from the insecticide. After time t, the insecticide has diffused so that it is distributed as shown in Fig. 1. This system corresponds to the boundary conditions of the Matano method provided that t is short enough for the assumption that x extends infinitely in both directions to be valid. The concentration along the cylinder is found by slicing it into narrow sections and measuring the insecticide in each. The values of $dx/d(C/C_0)$ and $\int x\, d(C/C_0)$ in Equations (2) and (3) correspond to the tangents and areas indicated in Fig. 1 where examples are given for $C/C_0 = 0·2$ and 0·8.

Fig. 2 shows the apparatus used to obtain the experimental system. The two halves of a brass tube (5·5 cm long × 3 cm dia.) are joined by a brass coupling ring, which runs on a thread machined on the outside of the tube. A small rabbet

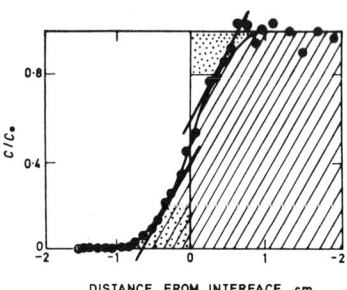

FIG. 1. *Experimental system for measuring diffusion coefficients*

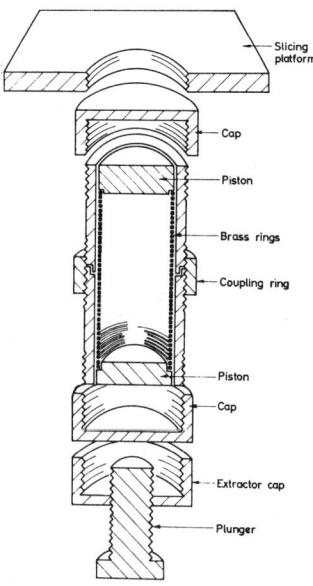

FIG. 2. *Diffusion cylinder*

ensures that the two halves fit tightly together. A cap with washer closes each end of the tube. Fitting exactly inside this outer tube is a second cylinder composed of 40 brass rings (axial thickness 1 mm, annular thickness 1 mm) machined flat to fit flush with each other. At each end of the inner cylinder is a brass piston, which also fits exactly inside the outer cylinder. A 1 mm rabbet in the pistons fits into the inner cylinder when the caps are tightened and presses the soil towards the centre.

Soil to be used in diffusion experiments is first equilibrated with the appropriate solutions by shaking on an end-over-end shaker. For disulfoton, which is appreciably sorbed, about 25 g portions of soil were shaken eight times with 100 ml of nearly saturated disulfoton solution in 0·01 M-CaCl$_2$. The suspensions were centrifuged and the supernatant fluid was discarded between shakings. For dimethoate, shaking twice with 0·1% solutions was sufficient and gave soil concentrations of the same order as those obtained with disulfoton. With each insecticide a similar portion of soil was shaken under identical conditions with 0·01 M-CaCl$_2$ only, to provide soil for the half of the cylinder free from insecticide. After the final equilibration, the suspensions were filtered using a Buchner funnel and suction was applied for a further 30 min after all supernatant fluid had been removed. This gave a soil with a reproducible moisture content, which was passed damp through a 2 mm sieve to produce small aggregates. When drier soil was required, these aggregates were dried for an appropriate time with constant stirring under an infra-red lamp using standardised conditions. For measurements in soil wetter than the aggregates, the required additional portions of the final equilibrating solutions were added dropwise with a pipette to layers of the prepared soil during packing into the diffusion cylinders.

For packing, the cylinders were separated into the two halves, and the pistons reversed so that the rabbet would not compress the soil. Successive weighed portions of prepared soil, sufficient to occupy 4 rings of the inner cylinder, were carefully tamped down into the appropriate volume so that there was a uniform bulk density of 1·25 g dry soil/ml through the column. When full, the half-cylinders were sealed and left for at least 24 h to allow moisture conditions to equilibrate in the aggregates. The treated and untreated halves of each cylinder were then joined tightly using the coupling ring, the pistons reversed and the caps screwed tight so that the slight compression produced by the rabbets ensured contact between the two halves of the soil column. The cylinders were kept at 20° in constant temperature rooms for a suitable period while diffusion took place. One end cap was then replaced by an extractor cap fitted with a threaded plunger (1 mm pitch) and the other end cap replaced by a flat slicing platform which was screwed down flush with the top of the diffusion cylinder. The cylinder was held vertically in a clamp and the inner cylinder extruded slowly upwards from the outer tube by turning the plunger. As each ring emerged, it was removed and the soil sliced off using a single hollow-ground razor. The soil sections were dried with sodium sulphate and the insecticide was extracted for analysis by gas–liquid chromatography. Disulfoton was extracted as described for measuring adsorption isotherms (see below) and dimethoate was extracted using 10 ml acetone/g soil. From a plot of the distribution of insecticide along the soil column, diffusion coefficients were calculated using the appropriate equations. Each experiment was replicated three times. Moisture contents were checked gravimetrically for sample sections along each cylinder.

Adsorption isotherms

The determination of sorption isotherms for disulfoton has been described previously.[7] 2 g portions of soil were equilibrated for 16 h with 100 ml disulfoton solutions having a range of initial concentrations 2–14 ppm in 0·01 M-CaCl$_2$. After centrifuging, the equilibrium concentration of disulfoton in solution was found by extracting into hexane and analysis by gas–liquid chromatography. Disulfoton sorbed was determined by extracting the soil, after decanting the supernatant solution and drying with anhydrous sodium sulphate, using 15–30 ml of solvent containing 3 parts acetone to 2 parts hexane. Acetone was removed from the mixed extract by shaking with 2% sodium sulphate solution and disulfoton in the remaining hexane was determined by gas–liquid chromatography.

Much less dimethoate was sorbed than disulfoton, so that 10 g portions of soil were equilibrated with 20 ml dimethoate solutions (10 ppm–2%) in 0·01 M-CaCl$_2$. Even with this large soil/solution ratio, gas–liquid chromatography had too large an experimental error to measure reliably the decrease in solution concentration during equilibration. Also, it was difficult to determine amounts sorbed by extracting the soil, because the solution retained by the soil after centrifuging contained very much more dimethoate than was sorbed. Sorption was therefore studied using ^{32}P-labelled dimethoate and amounts on the solid calculated from the decrease in solution concentration only, assuming that all material lost from solution was taken up by the solid. Radioactivity before and after equilibration was measured by counting 10 ml portions of the aqueous solution using a liquid Geiger-Müller counter and a Panax GX9 autoscaler. For adsorption isotherms, amounts taken up by the soil were plotted against equilibrium solution concentrations.

Gas–liquid chromatography

An Aerograph 1520 gas chromatograph with a thermionic phosphorus detector was used. Operating temperatures were injector 210°, column 195° and detector 215°. The stainless-steel columns were 3 mm × 75 cm for dimethoate and 3 mm × 150 cm for disulfoton and were packed with 5% SE30 on 60/80 Chromosorb W. The instrument was calibrated each day with standard solutions, and the calibration was checked at intervals during a run. Each extract was injected at least twice.

Results and Discussion

The equations for calculating diffusion coefficients were derived assuming that C/C_0 was a function of $x/(t)^{\frac{1}{2}}$, i.e. that changes in the distribution of insecticide in the cylinders with time were caused only by normal diffusion so that the distance moved by any value of the relative concentration (C/C_0) was proportional to $(t)^{\frac{1}{2}}$. The validity of this assumption must be tested, especially because organophosphorus insecticides are transformed chemically and microbially in soil. Evidence of decomposition was obtained by extracting test samples of soil. The amounts extracted decreased to about 70% of the initial value after one week with disulfoton and to about 60% with dimethoate. Although diffusion experiments were usually completed in 3–5 days therefore, the distribution of insecticide could be significantly affected by processes other than diffusion. If the assumption that C/C_0 is a function of $x/(t)^{\frac{1}{2}}$ is valid, a plot of x, the distance moved by a given value of C/C_0 against $(t)^{\frac{1}{2}}$ should be a straight line passing through the origin. Fig. 3 shows results of experiments to test this. For this purpose C_0 was taken as the concentration determined in sections of the treated soil sufficiently far from the boundary to be unaffected by diffusion.

With both disulfoton and dimethoate, agreement with theory is reasonable, showing that any decomposition of the parent molecules which occurred did not affect the shape of the distribution curve much. This would be the case if the transformations were first order with respect to insecticide concentration.

Table I gives results of diffusion measurements at different soil moisture contents. Because of the limited movement and the few points available, and because of the experimental errors associated with gas–liquid chromatographic analysis of the soil extracts, the exact shape of the disulfoton distribution curve was often difficult to draw at large or small values of C/C_0 where the curvature is steep. Also, with both insecticides, the slopes represented by $dx/d(C/C_0)$ are more difficult to measure accurately where the curvature is steep. The method is therefore least reliable at extreme values of C/C_0 and results in Table I are confined to values of C/C_0 from 0·2 to 0·8.

Concentration had little effect on the diffusion coefficient except that values tended to be somewhat smaller at intermediate concentrations. However, this effect was not consistent and seemed independent of the value of C_0, so it may be attributed to experimental error. Fig. 4 shows that the adsorption isotherm for disulfoton is linear and although the dimethoate isotherm is better described by the Freundlich equation, the curvature is small so that a straight line is a good approximation over the concentration range studied. Unless C_G/C varied markedly with concentration, therefore, only a small effect of concentration on D would be expected.

From the results in Table I, average diffusion coefficients, $\bar{D}$, can be obtained for the concentration range studied. Fig. 5 shows how soil moisture content influences these average values. Moisture content has little effect on the diffusion of disulfoton, but with dimethoate the effect is considerable, indicating the very much greater importance of the solution pathway in this case.

Consideration of the relative importance of liquid and vapour diffusion for these two compounds allows some further interpretation of these results. As already discussed, vapour diffusion can be neglected for pesticides of small volatility

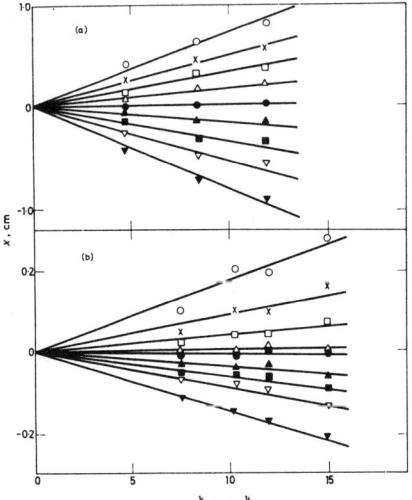

FIG. 3. *Test of validity of diffusion equations*
(a) Dimethoate; (b) disulfoton
C/C_0: = 0·9 (○); 0·8 (×); 0·7 (□); 0·6 (△); 0·5 (●); 0·4 (▲); 0·3 (■); 0·2 (▽); 0·1 (▼)

TABLE I
Apparent diffusion coefficients ($D \times 10^7$ cm^2/sec) for dimethoate and disulfoton at different concentrations in Broadbalk soil over a range of moisture contents

Insecticide	Volumetric moisture content, %	C_0, mg/ml whole soil	C/C_0						
			0·2	0·3	0·4	0·5	0·6	0·7	0·8
Dimethoate	10·4	0·425	0·32	0·27	0·29	0·31	0·34	0·37	0·50
	17·4	0·396	1·31	1·20	1·02	0·96	1·13	1·22	1·32
	23·4	0·350	2·74	2·45	2·26	1·94	1·82	1·91	2·15
	31·6	0·430	5·94	4·63	4·34	4·10	4·29	4·68	4·84
	32·8	0·370	6·11	6·16	5·75	5·68	5·47	5·54	4·80
	35·6	0·421	9·42	9·65	9·70	9·44	9·00	8·56	8·82
	42·9	0·535	14·05	13·27	12·59	13·08	14·03	16·00	15·91
Disulfoton	7·7	0·298	0·35	0·24	0·23	0·23	0·27	0·31	0·37
	16·4	0·246	0·23	0·22	0·23	0·23	0·26	0·29	0·30
	21·6	0·096*	0·20	0·18	0·17	0·19	0·18	0·24	0·27
	23·0	0·256	0·20	0·20	0·18	0·18	0·18	0·22	0·28
	30·1	0·271	0·19	0·15	0·14	0·16	0·18	0·34	0·47
	31·9	0·226	0·20	0·16	0·13	0·13	0·13	0·22	0·29
	39·3	0·243	0·36	0·29	0·25	0·24	0·26	0·31	0·35
	40·8	0·267	0·32	0·27	0·25	0·24	0·25	0·30	0·46

* Preliminary experiment at smaller concentration

For disulfoton and dimethoate, however, comparison with published values for molecules of similar size (e.g. in the International Critical Tables) suggests that D_L would be approximately 5×10^{-6} cm^2/sec whereas D_G would be approximately 10^{-1} cm^2/sec. The vapour concentration must therefore be very much less than the solution concentration before vapour movement can be neglected. Provided the vapour obeys Henry's law and with liquid pesticides, provided the liquid is not readily miscible with water, the ratio of the concentration in saturated aqueous solution to that in saturated air may be used to calculate the partition of pesticide between soil solution and soil air as suggested by Hartley.[9] Values of these partition coefficients would be approximately $2 \cdot 5 \times 10^8$ for dimethoate and $5 \cdot 5 \times 10^3$ for disulfoton. Vapour diffusion should therefore be insignificant for dimethoate. If it is assumed that the isotherm is linear, with slope b and the quantity in the vapour is negligible, then $C_L/C = 1/(bB+V_L)$, where B is the bulk density. Equation (4) may then be written:

$$D = D_L V_L f_L/(bB+V_L) \quad \ldots \ldots \ldots \ldots \ldots \ldots (5)$$

Similar equations were derived by Call[1] for the diffusion of ethylene dibromide and by Olsen et al.[10] for diffusion of phosphate ions in soil, assuming diffusion was effectively through only one phase in the soil. The equation shows quantitatively how the different factors influence the value of the apparent diffusion coefficient; for example, the contribution of adsorption is determined by the value of b whereas f_L allows for the tortuosity of the pathway. Values of f_L found for dimethoate in a given soil should also apply for other pesticides that diffuse predominantly in solution. Values for B and V_L are known and b is obtained from the slope of the best fitting linear isotherm, so that if D_L is assumed to be 5×10^{-6}, f_L may be calculated at different moisture contents from the measured D values. Fig. 6 shows this for dimethoate. The values agree well with those for the equivalent 'transmission factor' found by Porter et al.[11] for diffusion of chloride in the Ca form of Pierre Clay soil over a smaller range of moisture content (Fig. 6). The shape of the curve resembles that found by Rowell et al.[12] for chloride diffusion in an Upper Greensand soil, although the values of f_L obtained for dimethoate are smaller as might be expected using the heavier Broadbalk soil. Because D_L was assumed for dimethoate, f_L values are only approximate, but comparison with these previous studies shows that the behaviour is consistent with diffusion predominantly in the liquid phase.

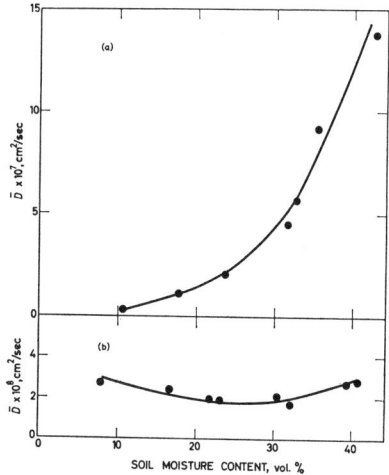

FIG. 5. *Effect of soil moisture content on average diffusion coefficients*
(a) Dimethoate; (b) disulfoton

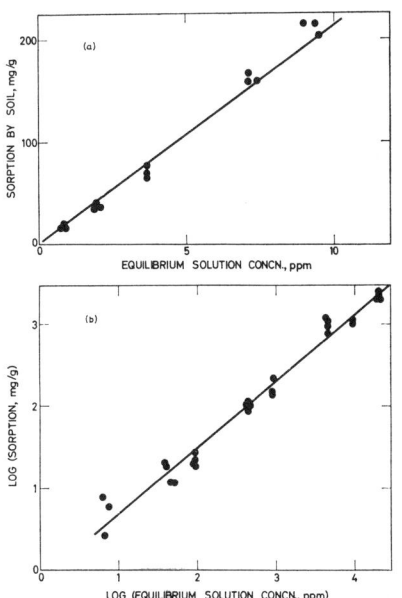

FIG. 4. *Sorption isotherms for* (a) *disulfoton and* (b) *dimethoate*

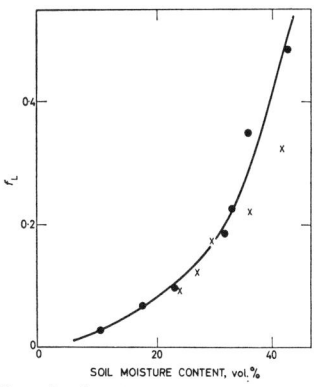

FIG. 6. *Effect of soil moisture content on tortuosity factors for dimethoate*
● Experimental points; × points from Porter et al.[11]

In contrast, moisture content has little effect on diffusion of the more volatile disulfoton and the contribution of vapour diffusion seems to increase with decreasing water content in approximately the same proportion as diffusion in solution decreases. The exact contribution of each phase is difficult to calculate. Rowell et al.[12] analysed the contributions of different pathways to the diffusion of ions in soil by equating the total flux with the sum of the fluxes in the different phases and estimating the liquid contribution by applying f_L values obtained for chloride which was assumed to move entirely in solution. Simple addition of fluxes in this way is questionable for pesticides if the fluxes in the two phases are comparable. It would be permissible if the soil consisted of independent air and water channels parallel to the direction of diffusion, analogous to a set of electrical conductors in parallel. Equally, in a hypothetical soil consisting of air and water laminae normal to the direction of diffusion, analogous to electrical conductors in series, the resultant flux would be zero if either component were non-conducting. Real soil is a complex mixture of these two extremes. The tortuosity factors f are empirical functions of this complexity and must be functions of both diffusivities, as well as of the geometry, unless one diffusivity is negligible. Values of f_L for dimethoate cannot therefore be used to estimate the solution phase contribution to the diffusion of disulfoton. However, the extent to which vapour movement for disulfoton modifies the movement of an involatile pesticide with the same adsorption characteristics may be seen by comparing the measured effect of moisture content on D with that on $D_L f_L V_L/(bB+V_L)$ calculated from the disulfoton sorption isotherm and using f_L values for dimethoate (Fig. 7). The calculated solution diffusion coefficient is a very small fraction of the observed value for total diffusion at small moisture contents, showing the pronounced effect of vapour movement. With increasing moisture, the estimated value increases rapidly and, in accordance with theory, agrees well with the measured value at the large moisture contents when vapour movement would be negligible. For the wettest soil, the calculated value in fact slightly exceeds the measured figure. In view of the assumptions made, this over-estimate is not large and may be attributed to errors in the assumed quantities. Alternatively, it may indicate that distribution coefficients (b values) obtained from equilibrium adsorption experiments with continuous shaking at large solution/soil ratios cannot be applied safely for diffusion calculations. Slow desorption could cause the effective distribution coefficients during diffusion to be larger than measured b values. The likely behaviour of other pesticides may be considered in the light of these results. The results obtained with dimethoate suggest that, provided f_L values are known, diffusion coefficients of relatively involatile materials may be calculated from independent measurements of sorption, moisture content etc., by using Equation (5). Although f_L values shown in Fig. 6 are only approximate, they are accurate enough to allow reasonable estimates of tortuosity effects for similar pesticides in this soil. Further, curves relating f_L to V_L for different soils seem to lie fairly close,[11] so that, in the absence of measurements for a given soil, f_L values reported here or elsewhere[11,12] could probably be used to indicate behaviour.

Estimating the additional vapour contribution to diffusion of more volatile pesticides is more difficult. However, the results for disulfoton suggest that for pesticides of similar volatility, values of D could be calculated by Equation (5) for wet soils where vapour movement was small and then assumed to change little over a wide range of moisture contents. Finally at the other extreme, diffusion in solution could be neglected for very volatile pesticides and diffusion coefficients estimated using the form of Equation (4) appropriate for vapour movement, such as that given by Call[1] for fumigants.

The measurements reported here emphasise that diffusion of both these insecticides is a relatively slow transport process whatever the soil moisture content. The root mean square displacement of the diffusing molecules (given by $(2Dt)^{\frac{1}{2}}$) would be at most about 2·5 cm in a month for dimethoate in the wettest soil studied, whereas the corresponding figure for disulfoton is about 0·3 cm. Diffusion also becomes less effective with time and cannot therefore transport pesticides of this type for long distances from where they are applied in soil. However, the differences in behaviour found could have important practical consequences for movement over short times and short distances, for example in the micro-regions around plant roots or granules in soil.

Acknowledgments

The author wishes to thank Dr. G. S. Hartley for helpful discussions about diffusion in porous media. Technical disulfoton and dimethoate were kindly supplied by Messrs. Baywood Chemicals Ltd. and Fisons Pest Control Ltd.

Rothamsted Experimental Station,
Harpenden,
Herts.

Received 2 January, 1969

References

1. Call, F., *J. Sci. Fd Agric.*, 1957, **8**, 143
2. Lindstrom, F. T., Boersma, L., & Gardiner, H., *Soil Sci.*, 1968, **106**, 107
3. Lord, K. A., & Burt, P. E., *Chemy Ind.*, 1964, p. 1262
4. Graham-Bryce, I. J., *S.C.I. Monogr. No. 29*, 1968, p. 251 (London: Society of Chemical Industry)
5. Matano, C., *Jap. J. Phys.*, 1933, **8**, 109
6. Phillips, R. E., & Brown, D. A., *J. Soil Sci.*, 1966, **17**, 200
7. Graham-Bryce, I. J., *J. Sci. Fd Agric.*, 1967, **18**, 72
8. Nye, P. H., *J. Soil Sci.*, 1966, **17**, 16
9. Hartley, G. S., in 'The Physiology and Biochemistry of Herbicides', (Ed. Audus, L. J.) 1964, p. 115 (London: Academic Press)
10. Olsen, S. R., Kemper, W. D., & van Schaik, J. C., *Proc. Soil Sci. Soc. Am.*, 1965, **29**, 155
11. Porter, L. K., Kemper, W. D., Jackson, R. D., & Stewart, B. A., *Proc. Soil Sci. Soc. Am.*, 1960, **24**, 460
12. Rowell, D. L., Martin, M. W., & Nye, P. H., *J. Soil Sci.*, 1967, **18**, 204

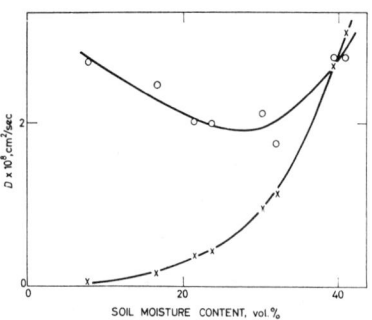

FIG. 7. *Comparison of measured diffusion coefficients for disulfoton with calculated solution diffusion coefficients*
○ Measured points; × calculated solution diffusion coefficients

Lindane Diffusion in Soils: II. Water Content, Bulk Density, and Temperature Effects[1]

WILFRIED EHLERS, W. J. FARMER, W. F. SPENCER, AND J. LETEY[2]

ABSTRACT

The diffusion of lindane in Gila silt loam is strongly influenced by soil water content, bulk density, and temperature. The diffusion coefficient is nearly zero in soil of 1% water content. With an increase to 3% water content, which is equivalent to two layers of water between the montmorillonite clay plates, water is able to displace the lindane from the adsorbing surface so that the diffusion coefficient becomes maximal. A small additional increase in water content reduces the diffusion coefficient to about one-half of the maximal value. This value then remains constant with increasing water content up to saturation. Decreasing bulk density or increasing temperature raises the diffusion coefficient. The influence of bulk density and temperature on diffusion is in good agreement with theoretical calculations.

Additional Key Words for Indexing: pesticide residues, soil water and lindane adsorption, vapor phase movement, activation energy.

IN A PREVIOUS paper (1) we outlined the theory of the diffusion of volatile insecticides in soils, and described the influence of concentration, time and pressure on diffusion of lindane in Gila soil. In the present article we shall give information on how lindane diffusion is influenced by soil water content, soil bulk density and temperature.

MATERIALS AND METHODS

The details of the procedures followed to measure lindane diffusion coefficients in soils were presented in a previous paper (1). Briefly, a half-cell containing soil treated with ^{14}C-labeled lindane was brought in contact with a half-cell containing untreated soil. The diffusion coefficient was calculated from the fraction of lindane transferred form one half-cell to the other in a given time.

Soil water contents less than 4% (weight basis) were established by equilibrating the soil at various relative humidities using saturated salt solutions in desiccators. Small changes in water content occurred during packing of the soil into the cell; therefore, the actual water content was determined in parallel samples. Soil water contents of 10 and 20% were established by mixing air-dried soil in a vial with the appropriate water quantity. A water content of 35% and water saturation were achieved by adding water to the soil while filling the cell with air-dry soil. Experiments with different soil water contents were run at 20 and 30C. The bulk density was 1.26.

Different bulk densities of the soil with 10% water content were achieved by taking different soil quantities for filling the

---

[1] Contribution of the Department of Soils & Plant Nutrition, University of California, Riverside and the Soil & Water Conservation Research Division, ARS, USDA. This work has been supported in part by USDA Cooperative Agreement no. 12-14-100-9016 (41). Received Oct. 30, 1968. Approved. Feb. 27, 1969.

[2] Former Postgraduate Research Soil Scientist and Assistant Soil Chemist, University of California, Riverside; Soil Scientist, USDA; and Professor of Soil Physics, University of California, Riverside. Present address of senior author: Institute für Pflanzenbau und Pflanzenzüchtung 34 Göttingen, von Siebold str. 8, Germany.

volume of the diffusion half-cells (0.795 cm³). The air- and water-filled porosities for the chosen densities are given in Table 1. The temperature was maintained at 30C.

The influence of 20, 30, and 40C temperatures on diffusion of lindane was checked, with the bulk density of the soil 1.26 and the water content 10%.

The ^{14}C-lindane activity was determined in the presence of soil after desorbing the lindane in the scintillation solution (3). Labeled lindane was desorbed from the soils with 1 to 4% water content by shaking the soil samples in a scintillation solution consisting of 5 g of PPO in 1 liter of toluene. Soil samples having 10% water or more were shaken in a solution consisting of 10.4 g of PPO and 166 g of naphthalene dissolved in 800 ml of xylene, 800 ml of dioxane and 473 ml of absolute ethanol.

RESULTS AND DISCUSSION

Water Content of the Soil

Diffusion coefficients for lindane at 20 and 30C are plotted against water content of Gila soil in Fig. 1. The curves for each temperature have a similar shape. The increase in temperature causes an increase in diffusion over the whole range of water contents. However, at 1% water content [corresponding to 34% relative humidity (RH)], so little lindane diffused (2% of total lindane) at either temperature that the calculated diffusion coefficients approach zero (0.14 mm²week⁻¹). An increase in water content above 1% results in a sharp increase in lindane

Table 1—Volumetric distribution of air, water and solid particles in Gila silt loam at different bulk densities

Bulk density	Soil per diffusion half-cell	Air-filled porosity	Water-filled porosity	Total porosity
g/cm³	g	%	%	%
1.00	0.795	51.5	10.0	61.5
1.26	1.000	38.9	12.6	51.5
1.52	1.210	26.3	15.2	41.5
2.06*		0.0	20.6	20.6

* Theoretical value.

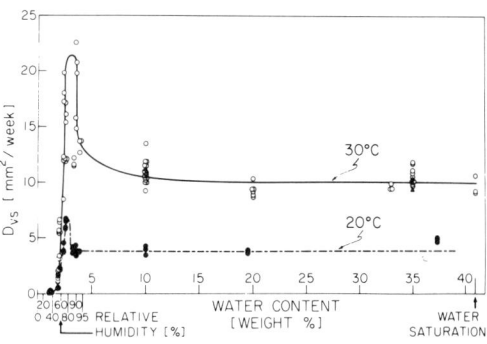

Fig. 1—Effect of soil water content on the diffusion of lindane in Gila silt loam. The equivalent relative humidities were determined at 20C.

275

diffusion, reaching a maximal value at about 2.5 to 3% water content, corresponding to 72–82% RH. Between 3 and 4% water content, corresponding to 82–94% RH, the diffusion coefficient decreases again very sharply. Diffusion coefficients remain relatively constant at water contents higher than 4 to 5%. Farmer and Jensen (3) found a sharp increase in dieldrin diffusion with increasing water content for different soils that had been equilibrated with water vapor corresponding to 75% to 92% RH.

Certainly the low diffusion coefficient found for lindane at 1% soil water content is due to the high adsorption affinity of the relatively dry soil for the insecticide. With an increase in water content up to 3% the lindane molecules are apparently displaced from the adsorbing surface by water molecules, thus allowing a higher diffusion rate.

The effect of water on the desorption of lindane can be illustrated by calculating the number of molecular water layers necessary to cause maximal lindane diffusion. The ethylene glycol surface area of Gila silt loam is 90 m²/g. Approximately 10% of the total surface is external surface (4, 8). If the density of the adsorbed water is assumed to be one, a water content of 1.55% (corresponding to 44% RH) is sufficient to produce one molecular layer of water between the montmorillonite plates and one layer on the external surfaces. At 2.81% water content (about 80% RH) sufficient water is present to produce two layers of water between the montmorillonite plates and one layer on the external surfaces. (A water content of 3.11% is equivalent to two water layers on both the internal and external surfaces.) This value of 2.8 to 3.2% water content corresponds to the water content at which we obtained the highest diffusion coefficient.

Two possible explanations may be given for the decrease of the diffusion coefficient at water contents higher than about 3%. Both explanations will assume that lindane is first adsorbed only at the external surfaces. Lindane was added in a hexane solution to the air-dry soil. Hexane cannot enter the interlattice space of montmorillonite (5), and as the soil is in equilibrium with approximately 50% RH in the laboratory, it is not likely that the lindane can penetrate into the 3A wide interlattice space. (The average diameter of a lindane molecule as calculated from density and molecular weight is 7.89A.)

The first explanation of the decrease in the lindane diffusion coefficient will assume that after completion of the second layer of water between the montmorillonite plates a third layer of water is immediately built up, causing the lattice to open from about 6A to about 9A. Now the lindane is able to penetrate between the montmorillonite plates, and diffusion of lindane will decrease. However, Emerson (2) and Mering (6) found that Ca montmorillonite showed a base distance of approximately 15A up to a water content corresponding to 92% RH. At this RH Gila soil has a water content of 3.8% and the diffusion coefficient is already greatly reduced.

The second explanation for the decrease in the diffusion coefficient is based on the assumption that between approximately 3 and 3.8% water content the additional water is adsorbed only at the external surface while the internal lattice space is maintained constant at two layers of water. This would result in a rapid increase in the thickness of the water film on the external surface since this surface area is small compared with the total surface. Orchiston (7) studied the uptake of water at the external and internal surfaces of a montmorillonite with increasing relative humidity. He found that after 1.5 layers of water on the average were adsorbed at the external surfaces and between the montmorillonite layers the thickness of the water layer on the external surface increased more rapidly than the thickness of the water layer between the clay plates. From this we may deduce that the decrease in lindane diffusion above 3% water content is caused by the rapid increase in water on the external surface area. The thickness of three or more molecular layers of water would exceed the diameter of the lindane molecule. The lindane would thus have the opportunity to be "immersed" in water and would probably orient at the liquid-gas or liquid-solid interfaces. This "immersion" could conceivably decrease the diffusion rate.

With 10% water content the diffusion coefficient is nearly the same as that with water saturation. We have already shown (1) that at 10% water content and 30C, 50% of the diffusion occurs in the vapor phase and the other 50% in "nonvapor" phase. Increasing water to saturation causes vapor phase movement to decrease to zero. This decrease is fully compensated by an increase in "nonvapor" phase movement.

That the movement of lindane at higher water contents is not caused by water movement from the upper treated half-cell to the lower untreated half-cell is established by parallel samples where the lindane had to diffuse from the lower treated half-cell to the upper untreated half-cell (see Fig. 1). The diffusion coefficients are independent of the direction of the diffusion.

Bulk Density

The relationship between bulk density (β), the diffusion coefficient and the fraction of diffused lindane after 2 days are plotted in Fig. 2. The relation between β and the diffused fraction is linear. Extension of the line to the point of no transfer crosses the β-coordinate at a value of 2.6.

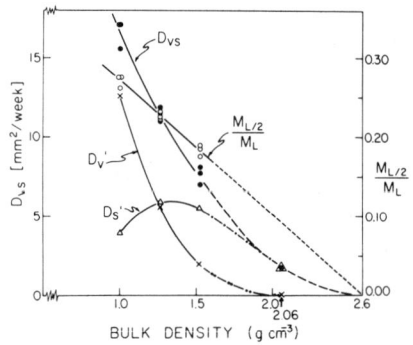

Fig. 2—Effect of bulk density on the diffusion of lindane in Gila silt loam at 10% water content and 30C.

This is the value for the mineral density of the soil as determined with a pycnometer.

In a previous paper (1) the combined diffusion in vapor and "nonvapor" phase was described by the equation:

$$\partial c/\partial t = \partial/\partial x\, [(D_v S^{10/3}/\beta S_T^2)\, \partial \rho/\partial c$$
$$+ \theta\, (L/L_e)^2 (1-\gamma) D_s] \partial c/\partial x \quad [1]$$

where

D_v = vapor diffusion coefficient in air

β = bulk density

S = air-filled porosity

S_T = total porosity

ρ = vapor density

θ = volumetric water content

D_s = "nonvapor" diffusion coefficient

γ = term for interaction between lindane and soil

$(L/L_e)^2$ = turtuosity factor

$(D_v S^{10/3}/\beta S_T^2)\partial \rho/\partial c = D_v{'}$ (apparent vapor diffusion coefficient) [2]

$\theta (L/L_e)^2 (1-\gamma) D_s = D_s{'}$ (apparent "nonvapor" diffusion coefficient) [3]

$D_v{'} + D_s{'} = D_{vs}$ (apparent total diffusion coefficient). [4]

In the first paper (1) we determined that with 10% water, β of 1.26, and temperature of 30C, $D_v{'}$ was 5.6 mm²week⁻¹ and $D_s{'}$ was 5.8 mm²week⁻¹. With known $D_v{'}$, β, S, and S_T, we can calculate $D_v(\partial \rho/\partial c)$. This term should be independent of β. Thus $D_v{'}$ can be calculated for different β. Since D_{vs} is measured and $D_v{'}$ can be calculated, we can compute $(L/L_e)^2(1-\gamma)D_s$ for each β. Results of this calculation are given in Table 2. $D_v{'}$ decreases with increasing β. $D_s{'}$ initially increases with increased β from 1.00 to 1.26 and then decreases as β is increased to 1.52.

In Fig. 2 the curve for the total diffusion coefficient, D_{vs}, was extended to β of 2.6 by using the straight line that describes the relation between β and the fraction of diffused lindane. (For construction of this curve we employed equation [14] given in Part I.) For β of 2.06 the total porosity of 20.6% is filled with water (Table 1) and diffusion of lindane could occur only in the "nonvapor" phase. Both theoretical values for the diffusion coefficient

Table 2—Apparent diffusion coefficient of lindane in vapor and "nonvapor" phase as influenced by bulk density

Bulk density	D_{vs}	$D_v{'}$	$D_{vs} - D_v{'} = D_s{'}$	$(L/L_e)^2 (1-\gamma)(D_s)$
g/cm³		mm²/week		
1.00	16.5*	12.6	3.9	39
1.26	11.5*	5.6*	5.9	59
1.52	7.5*	2.0	5.5	55

* Experimentally determined.

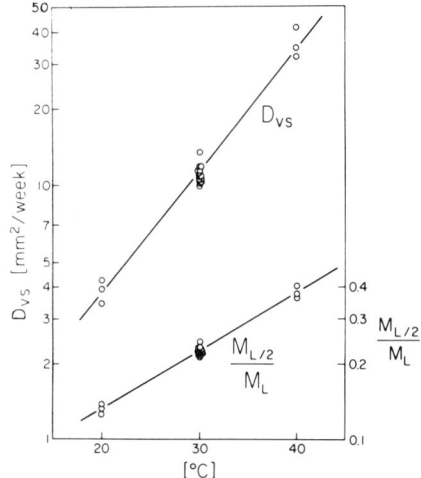

Fig. 3—Effect of temperature on the diffusion of lindane in Gila silt loam at 10% water content.

at β of 2.06 are indicated in Fig. 2. They seem to conform to the calculated values of $D_v{'}$ and $D_s{'}$.

Temperature

The effect of temperature on fraction of lindane transferred and D_{vs} is illustrated in Fig. 3. In the temperature range chosen there is an exponential relationship between temperature and both D_{vs} and $M_{L/2}/M_L$.

The temperature-dependent components of $D_v{'}$ are D_v and $\partial\rho/\partial c$ (equation [2]). The general relationship between the vapor diffusion coefficient and temperature is $D_v/D_{vo} = (T/T_o)^{1.5+n}$ where D_{vo} and T_o are reference diffusion coefficient and temperature. The term n usually has the value from 0 to 0.5 depending upon the specific vapor. We will assume that n is 0.25 for lindane. (An error in this assumption will have only minor effects on the results.) Spencer and Cliath (unpublished data, 1968), using procedures previously described for dieldrin (9), found the saturated vapor density of lindane to be 0.49, 1.48 and 5.89 μg/liter at 20, 30, and 40C, respectively. We will assume that the relative change in $\partial\rho/\partial c$ with temperature is the same as the relative change in saturated vapor density with temperature. Since $D_v{'}$ was found to be 5.6 at 30C, it is possible to calculate $D_v{'}$ at the other temperatures from the above-listed considerations.

The factors in $D_s{'}$ which are temperature dependent are probably γ and D_s (Equation [3]). It is reasonable that D_s would be dependent upon the water fluidity. There is probably another effect of temperature on γ which we have no basis to estimate. However, in Fig. 1 it can be noted that D_{vs} at water saturation is about 2.5 times higher at the 30C than at the 20C temperature. (At water saturation $D_{vs} = D_s{'}$). The relative increase in fluidity over this temperature range is 1.26, which indicates than $(1-\gamma)$ approximately doubles in going from 20 to 30C. We will assume that this factor again doubles as the temperature increases from 30 to 40C.

The validity of this analysis and thus our ability to predict diffusion rates using the theoretical equations can be tested by calculating D_v' and D_s' for 20 and 40C. From these data we can calculate D_{vs} at 20 and 40C and compare the calculated values with the experimental values. These calculations and comparisons are made in Table 3.

The agreement between calculated and observed D_{vs} is very good. These results provide evidence that the developed equations are basically sound and that they are valuable for predictive purposes. For example the effect of temperature on D_{vs} could have been predicted very closely.

The proportion of lindane diffused in the vapor and "nonvapor" phase at various temperatures can also be calculated from the relative values of the activation energy, E_{vs}, for the diffusion process and the heat of vaporization, ΔH_v, for vaporization of lindane. The activation energy, E_{vs} can be calculated from the data in Fig. 3, using the Arrhenius equation.

$$\log D_2/D_1 = (E/2.303R)(1/T_1 - 1/T_2) \quad [5]$$

where D_2 and D_1 are D_{vs} for lindane at 313K and 293K, respectively. The heat of vaporization of lindane can be calculated from the change in vapor pressure with temperature using the Clausius-Clapeyron equation.

$$\log P_2/P_1 = (\Delta H_v/2.303R)(1/T_1 - 1/T_2) \quad [6]$$

where P_2 and P_1 are vapor pressures calculated from vapor densities measured at 313K and 293K, respectively.

The activation energy, E_{vs}, calculated with equation [5] is 20.0 kcal/mole. The heat of vaporization of lindane calculated with equation [6] is 24.1 kcal/mole. Assuming the activation energy for vapor phase diffusion, E_v to be equal to ΔH_v for lindane, the activation energy, E_s, for "nonvapor" phase diffusion can be calculated with equation [7] using the experimentally determined values of D_{vs}, D_v', and D_s' at 30C given in Table 3.

$$E_{vs} = (D_s'/D_{vs})E_s + (D_v'/D_{vs})E_v \quad [7]$$

From this relationship, E_s was calculated to be 16.1 kcal/mole. The relative values of E_v and E_s indicate that as the temperature increases a greater proportion of the diffusion will be in the vapor phase. With E_v and E_s known, equation [5] can be used to predict vapor and "nonvapor" diffusion coefficients for lindane at various temperatures from the experimentally determined values of D_v' and D_s' at 30C. The predicted values of D_v', D_s', and D_{vs} at 20 and 40C are shown in Table 4 for comparison with those in Table 3.

Table 3—Calculation of the influence of temperature on the apparent diffusion coefficients in vapor and "nonvapor" phase

T [°K]	ρ [μg/liter]	Relative ρ	$T^{1.75}$	Relative $T^{1.75}$	Relative $\rho T^{1.75}$	D_v'
293	0.52	0.294	2.13×10^4	0.965	0.284	1.6
303	1.77	1	2.21×10^4	1	1	5.6*
313	5.42	3.06	2.29×10^4	1.04	3.18	17.8

T	Fluidity	Relative fluidity	Relative γ	Relative (fluidity ×γ)	D_s'	$D_v' + D_s' =$ D_{vs} calc.	D_{vs} obs.
293	99.5	0.80	0.5	.40	2.3	3.9	3.8*
303	125	1	1.0	1.00	5.8*	11.4	11.5*
313	152	1.22	2.0	2.44	14.1	31.9	34.0*

* Experimentally determined.

Table 4—Influence of temperature on apparent diffusion coefficients calculated from relative energies of activation

Temperature	D_v'	D_s'	D_{vs}
°K	mm²/week		
293	1.6	2.1	3.7
303	5.6*	5.8*	11.4*
313	18.1	15.1	33.2

* Experimentally determined values.

LITERATURE CITED

1. Ehlers, Wilfried, J. Letey, W. F. Spencer, and W. J. Farmer. 1969. Diffusion of lindane in soil: I. Theoretical considerations and mechanism of movement. Soil Sci. Soc. Amer. Proc. 33:501–504. (this issue)
2. Emerson, W. W. 1962. The swelling of Ca-montmorillonite due to water adsorption. I. Water uptake in vapor phase. J. Soil Sci. 13:31–39.
3. Farmer, W. J., and C. R. Jensen. 1969. Diffusion and analysis of carbon-14 labeled dieldrin in soils. Soil Sci. Soc. Amer. Proc. In Press (Manuscript submitted).
4. Grim, R. E. 1968. Clay mineralogy. McGraw Hill, New York.
5. MacEwan, D. M. C. 1948. Complexes of clays with organic compounds. I. Trans. Faraday Soc. 44:349–367.
6. Mering, J. 1946. On the hydration of montmorillonite. Trans. Faraday Soc. 42B:205–219.
7. Orchiston, H. D. 1954. Adsorption of water vapor. II. Clays at 25C. Soil Sci. 78:463–480.
8. Scheffer, F., and P. Schachtschabel. 1966. Lehrbuch der Bodenkunde, Ferdinand Enke, Stuttgart.
9. Spencer, W. F., and M. M. Cliath. 1968. Vapor density of dieldrin. Environmental Sci. Technol. In Press. Vol. 3.

Diffusion of Herbicides in the Adsorbed Phase[1]

H. D. Scott, R. E. Phillips, and R. F. Paetzold[2]

ABSTRACT

Equations were presented which describe the diffusion of the soil-adsorbed herbicide phase. Assuming (i) no precipitation and volatilization, (ii) a linear adsorption isotherm, and (iii) no net change in herbicide concentration in the adsorbed phase, the data needed to make such a calculation can be determined experimentally and are as follows: soil bulk density, distribution coefficient, apparent self-diffusion coefficient of the herbicide in soil, self-diffusion coefficient of the herbicide in water and the transmission factor. The results indicated that the magnitude of the diffusion coefficient of metribuzin [4-amino-6-*tert*-butyl-3-(methylthio)-*as*-triazin-5(4H) one] in Dubbs soil depends primarily upon the amount in the adsorbed phase and on the moisture content. The magnitude of the adsorbed phase diffusion coefficients of ^{14}C-metribuzin was less than the apparent self-diffusion coefficients and decreased with decreasing soil water content. Since negative adsorbed phase diffusion coefficients were observed at the lower soil water contents, the data suggested that the amount of metribuzin in the adsorbed phase increased as soil water content decreased.

Additional Index Words: pesticides, transport model, soil phases, adsorption, metribuzin.

RECENTLY, several measurements of diffusion coefficients of pesticides in soil have been reported (3, 6, 7, 13, 15). Soil physical factors such as bulk density, moisture content, and temperature, and soil chemical factors such as pH and extent of adsorption have been shown to affect the diffusion rate.

For a detailed understanding of the transport mechanism involved, consideration should be given to the possible diffusion pathways. In general, pesticides may diffuse through soil in the following phases: *solution* (diffusion as a molecule or ion-pair in the soil solution), *vapor*, and *adsorbed* (exchange-diffusion along charged sites). The concentration and mobility of the pesticide in each of these soil phases will vary depending upon the relative partitioning between these phases and the interaction between the physical and chemical properties of the pesticide and the soil. The apparent diffusion coefficient of a pesticide in the soil will then depend upon the sum of the rates of diffusion in each soil phase.

The objectives of this paper are to present apparent diffusion coefficients for the solution and adsorbed phases of ^{14}C-metribuzin at various soil water contents. In presenting these solutions it was assumed that (i) the pesticide was nonvolatile, (ii) did not precipitate, (iii) that a linear relation can be used to describe the concentration of pesticide in the adsorbed and solution phases, and (iv) the net pesticide concentration in each soil phase was dependent only on concentration of the herbicide and was independent of space, time, and soil water content.

THEORETICAL CONSIDERATIONS

Let the total flux of herbicide in soil, J_t, be the sum of the flux in the solution phase, J_s, and the adsorbed phase, J_a. Then

$$J_t = J_a + J_s \quad [1]$$

where

$$J_s = -D_s^m \frac{\partial C_s^m}{\partial x} \quad [2]$$

and

$$J_a = -D_a^m \frac{\partial C_a^m}{\partial x} \quad [3]$$

In the above equations D_a^m is the diffusion coefficient in the adsorbed phase, D_s^m is the diffusion coefficient in the solution phase, C_s^m and C_a^m are the concentration of herbicide in the solution phase and adsorbed phase, respectively, and x is the space coordinate. The superscript m indicates that the measurements are based on volume of the entire soil mass. Equations [2] and [3] assume that the volume fraction of each phase remains constant for any given concentration, C_{sa}, of herbicide. However, a molecule of one phase may exchange with a molecule of the other phase but it is assumed that they do so instantaneously.

We define the variable of concentration, C_{sa}^m, to be

$$C_{sa}^m = C_s^m + C_a^m. \quad [4]$$

These may be expressed with respect to the solution phase as

$$\theta C_{sa} = C_{sa}^m$$
$$\theta C_s = C_s^m$$
$$\theta C_a = C_a^m \quad [5]$$

where θ is the volumetric water content. C_{sa}, C_s, and C_a then have units of µg per cm³ of total soil water present.

Substituting the values of J_s and J_a of Eq. [2] and [3] into Eq. [1] and expressing the concentration with respect to the liquid phase, we obtain

$$J_t = -D_s^m \frac{\theta \partial C_s}{\partial x} - D_a^m \frac{\theta \partial C_a}{\partial x}. \quad [6]$$

Since we are assuming that C_a and C_s are independent of x and t and are single valued functions of C_{sa} and can be determined in experiments independent of the diffusion experiments (i.e., adsorption isotherms), the chain-rule of differentiation can be applied to Eq. [6] to give

---

[1] A joint contribution from the Agr. Exp. Sta. of the University of Arkansas and the University of Kentucky. The research was supported in part by the Office of Water Resources Research, USDI Project A-021-Ark. Published with the approval of the director of Arkansas Agr. Exp. Sta. Received 9 July 1973. Approved 4 March 1974.

[2] Assistant Professor, Department of Agronomy, University of Arkansas, Fayetteville, Ark. 72701; Professor of Soil Science, University of Kentucky, Lexington, Ky. 40506; and Graduate Assistant, University of Arkansas, respectively.

$$J_t = -\left[D_s^m \frac{dC_s}{dC_{sa}} + D_a^m \frac{dC_a}{dC_{sa}} \right] \frac{\partial C_{su}}{\partial x}. \quad [7]$$

Combining Eq. [7] with the equation of continuity, $\partial C_{su}^m/\partial t = \partial J_t/\partial x$, gives

$$\frac{\partial C_{su}}{\partial_t} = \frac{\partial}{\partial x} \left\{ \left[\left(D_s^m \frac{dC_s}{dC_{sa}} + D_a^m \frac{dC_a}{dC_{sa}} \right) \middle/ \left(\frac{dC_s}{dC_{sa}} + \frac{dC_a}{dC_{sa}} \right) \right] \frac{\partial C_{su}}{\partial x} \right\} \quad [8]$$

The terms within the brackets of Eq. [8] are defined to be the apparent diffusion coefficient, D_e^m.

The development of Eq. [8] is similar to that given by Ellis et al. (4) except they included additional terms for the precipitated phase; in this development we are assuming no precipitate to be present. Equation [8] is the same as Eq. [5] of Ellis et al. (4) when terms referring to precipitate are equated to zero. They have shown that the denominator of the quantity within the brackets of Eq. [8] is equal to unity; i.e.,

$$\frac{dC_s}{dC_{sa}} + \frac{dC_a}{dC_{sa}} = 1. \quad [9]$$

The apparent diffusion coefficient, D_e^m, for this case is

$$D_e^m = D_s^m \frac{dC_s}{dC_{sa}} + D_a^m \frac{dC_a}{dC_{sa}}. \quad [10]$$

Since the objective of this study is to evaluate D_a^m, the adsorbed diffusion coefficient (exchange diffusion coefficient), all quantities other than D_a^m occurring in Eq. [10] must be known. D_e^m can be measured in diffusion experiments. A procedure of obtaining estimates of D_s^m will be outlined below. The quantities, dC_a/dC_{sa} and dC_s/dC_{sa} can be determined from standard adsorption experiments. In this study the adsorption of herbicide molecules by soil is considered to be rapid as compared with the diffusion process, thus the concentration of herbicide in the solution and adsorbed phases can be assumed to be in equilibrium at each point in the soil. If the adsorption isotherm is nonlinear, the quantities dC_a/dC_{sa} and dC_s/dC_{sa} can be graphically measured from plots of C_a versus C_{sa} and C_s versus C_{sa}, respectively. This procedure for cations in soil is illustrated in Fig. 3 and 4 of Phillips et al. (10). As pointed out previously in Eq. [9], the sum of dC_s/dC_{sa} and dC_a/dC_{sa} is equal to unity and serves as a check on errors involved in measuring the quantities C_s, C_a, and C_{sa}.

If a linear adsorption isotherm of the herbicide exists over the range of concentration of interest (see Fig. 1), then

$$\frac{dC_a}{dC_{sa}} = \frac{C_a}{C_{sa}} \text{ and } \frac{dC_s}{dC_{sa}} = \frac{C_s}{C_{sa}} \quad [11]$$

and

$$D_e^m = D_a^m \frac{C_a}{C_{sa}} + D_s^m \frac{C_s}{C_{sa}}. \quad [12]$$

The quantity D_e^m can be evaluated as outlined above. If on the other hand, the Freundlich constant or distribution coefficient, Kd, is available from a plot as shown in Fig. 1, the following procedure may be used to calculate D_a^m. For a linear adsorption isotherm we define

$$R = C_a/C_s = Kd\rho_b/\theta \quad [13]$$

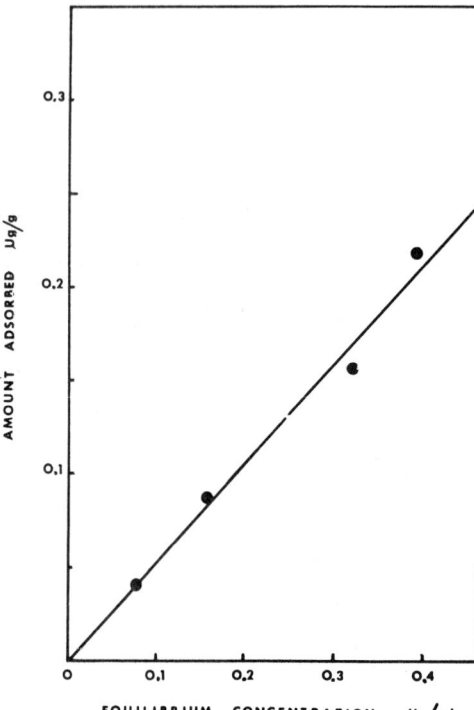

Fig. 1—Adsorption of ^{14}C-metribuzin by Dubbs silt loam. To obtain C_a^m values multiply the values on the ordinate by the bulk density, ρ_b, C_a can be obtained by dividing C_a^m by the water content, θ.

where ρ_b is the soil bulk density, θ is the volumetric water content and other terms are as defined above. In Fig. 1 the quantity plotted on the ordinate when multiplied by ρ_b and divided by θ is C_a; the quantity plotted on the abscissa is C_s. From Eq. [4], [5], and [13] it follows that

$$\frac{C_a}{C_{sa}} = \frac{R}{1+R} \text{ and } \frac{C_s}{C_{sa}} = \frac{1}{1+R} \quad [14]$$

Substituting the relations of Eq. [14] into Eq. [12] and rearranging, D_a^m is

$$D_a^m = \frac{1}{R} [(1+R) D_e^m - D_s^m]. \quad [15]$$

Again, D_a^m can be calculated if all quantities on the right side of Eq. [15] are known. In this study, D_e^m, θ, and ρ_b was obtained from diffusion experiments; D_s^m was estimated from Cl diffusion coefficients as outlined in the "Results and Discussion" section; and Kd was determined from a linear adsorption isotherm.

PROCEDURES

To calculate the diffusion coefficient of a herbicide in the adsorbed phase, apparent diffusion coefficients and adsorption isotherms of a ^{14}C-herbicide were determined at several soil water contents. Similarly, diffusion and adsorption experiments

Table 1—Physical and chemical properties of the Dubbs silt loam

Particle size distribution (%)	
sand	32.2
silt	58.5
clay	9.2
pH	6.8
Total carbon (%)	1.03
CEC (meq/100 g)	10.1
Ethylene glycol retention (mg/g)	21.7
Base saturation (%)	71.0
Percent moisture (weight basis)	
1/3 bar	16.5
1 bar	8.7
4 bar	5.7
15 bar	4.7

were conducted with ^{36}Cl to determine transmission factors. In this way the influence of the soil on the relative rates of diffusion of metribuzin was compared with the amount held in the adsorbed phase.

The herbicide chosen in the experiment was 4-amino-6-*tert*-butyl-3-(methylthio)-*as*-triazin-5(4H)one (metribuzin) a pre-emergence and early post-emergence herbicide used for control of several broadleaf and grass weed species. This compound has a molecular weight of 214.3 and solubility in water of 1,200 ppm (Chemagro Chemical Co., 1972. Sencor technical data sheet. Chemagro Chemical Co., Kansas City, Mo.). Preliminary observations both in the laboratory and field have indicated that metribuzin is one of the more mobile herbicides presently being studied for soybean weed control (personal communication with Dr. Ron Talbert, 1973, University of Arkansas).

The soil used was the A horizon of a Dubbs silt loam (Typic Hapludalf) from Crittenden County, Arkansas. The soil was air dried, ground to pass a 10-mesh screen, and mixed thoroughly. Some of the physical and chemical properties of the disturbed soil samples as described by Segraves are presented in Table 1 (D. J. Segraves, 1973. M.S. Thesis University of Arkansas, Fayetteville).

The adsorption of ^{14}C-metribuzin by the Dubbs soil was determined using the batch technique and a soil/water ratio of 1:2. This technique involves the equilibration of an aliquot of soil (5 g) with a given volume (10 ml) of aqueous solution containing various concentrations of ^{14}C-metribuzin for a given amount of time, centrifuging, and determining the amount of herbicide in the supernatant by liquid scintillation techniques. The amount of metribuzin adsorbed by the soil was assumed to be the difference between the amount added and the amount recovered in solution. Preliminary data indicated that an equilibrium between the amount of metribuzin adsorbed by the soil and the amount in solution was established in less than 2 hours on a rotary shaker. For convenience a 4-hour equilibration period was selected. Similar procedures were followed in determining the amount of ^{36}Cl adsorbed by the soil.

Apparent diffusion coefficients, D_e^m, of ^{14}C-metribuzin and ^{36}Cl in the Dubbs soil were determined by the method of Scott and Phillips (13). Preliminary measurements of the apparent diffusion coefficient of ^{14}C-metribuzin indicated that the rate of diffusion was not concentration dependent at low rates of addition. The selected rate of metribuzin added to the soil was 0.67 μg/g. The diffusion time ranged from 5 hours to 1 day depending upon the fraction of radioactive molecules which moved across the interface. Seven replications were made at each soil water content.

The self-diffusion coefficient of ^{14}C-metribuzin in water, D_o, was determined using the method of Phillips and Ellis (11) as modified by Brown (1). Twelve replications were made. The self-diffusion rate in water was needed to determine the diffusion coefficient in the soil-solution phase.

RESULTS AND DISCUSSION

The adsorption isotherm for ^{14}C-metribuzin in the Dubbs silt loam (Fig. 1) indicates that a linear relationship existed between the amount adsorbed and the amount in solution

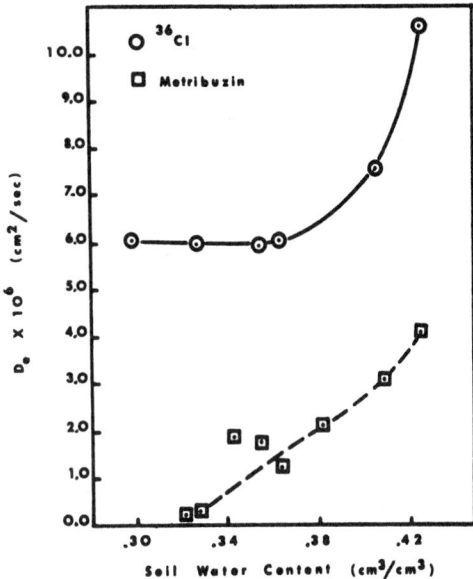

Fig. 2—Self-diffusion of ^{14}C-metribuzin and ^{36}Cl in Dubbs silt loam.

over the concentration range studied. Each data point is the average of three determinations. The Kd value obtained by taking the slope of the line was 0.54. A Kd of 0.003 (not shown) was obtained with ^{36}Cl and the Dubbs soil. These data indicate that little positive adsorption of the chloride occurred and of that amount added to the soil approximately 80% of the metribuzin and 99.7% of the chloride was contained in the solution phase. The actual mechanism of adsorption of metribuzin by soil is not presently known.

Apparent diffusion coefficients, D_e^m, for ^{14}C-metribuzin and ^{36}Cl as a function of soil water content are shown in Fig. 2. Each data point is the mean of seven determinations. The increase in diffusion rate as the water content increases can be attributed to a number of factors inherent in the soil-water system. With an increase in moisture content these factors, as given by Olsen and Kemper (8), include (i) a decrease in the length of the diffusion path (L/Le), (ii) a greater mobility of the soil water (α), and (iii) a greater homogeneity of pore size (γ). As given by Eq. [16] and [17], the selfdiffusion coefficient of an ion or molecule is the product of each of these factors and the self-diffusion coefficient in bulk water (D_o).

$$D_p^m = D_o (L/Le)^2 \gamma \alpha \theta \quad [16]$$

and

$$D_e^m = D_p^m / B\theta \quad [17]$$

where B is the capacity factor and D_p^m is the porous-diffusion coefficient. As shown by Scott and Phillips (14), B is equal to $(R + 1)$.

At a given water content the diffusion coefficient of chloride was found to be greater than that of metribuzin. This difference can be attributed to the greater amount of metribuzin adsorbed by the soil. Thus, for a compound that exists only in the solution and adsorbed phases, the data seem to suggest that the greater the interaction with the surfaces of the Dubbs soil the slower the diffusion coefficient. This has been previously shown by Walker and Crawford (15) and Scott and Phillips (13) using other herbicides and other soil types.

The data also indicate that a change in diffusion coefficient with a corresponding change in water content is greater for the herbicide metribuzin than with the chloride ion. As the water content decreases from 42 to 32%, the diffusion coefficient decreased by a factor of 2 and 20 for chloride and metribuzin, respectively. It is thought that the greater decrease of the herbicide diffusion coefficient can also be attributed to differences in the interactions with the soil surfaces as the soil moisture content changes. At least two explanations of this phenomenon can be given. First, a decrease in soil water content would tend to decrease the thickness of water films around the soil particles which would result in a greater interaction of the herbicide and soil surfaces and greater exclusion from the small pores by the chloride anion. Since the extent of adsorption of the herbicide on these surfaces is greater than the chloride anion, one would expect its diffusion rate to be reduced by a greater extent. Therefore, reaction of the herbicide with the adsorptive surfaces of the soil is thought to have a greater effect on decreasing the herbicide diffusion coefficient than anion exclusion has on decreasing the chloride diffusion coefficient. Second, in transient-state diffusion systems the concentration of herbicide in the solution phase changes with total concentration of the herbicide due to continuing adsorption and desorption reactions with the soil surfaces. Data presented by Hornsby and Davidson (5) and Davidson and McDougal (2) indicate that the adsorption and desorption isotherms of 1,1-dimethyl-3-(α,α,α-trifluoro-m-tolyl) urea (fluometuron) and 4-amino-3,5,6-trichloropicolinic acid (picloram) were not the same. Therefore, since the kinetics of the adsorption reactions may be different from the kinetics of the desorption reactions, these processes cause a corresponding change in the concentrations of herbicide in the adsorbed phase. If the adsorption-desorption relations between metribuzin and the Dubbs soil are not single valued or reversible, then the curve probably consists of many curves in a similar manner as exist with water in coarse-textured soils. Thus, with a decrease in water content a greater amount of herbicide would occur in the adsorbed phase for longer periods of time and since adsorbed molecules diffuse at lower rates, a lower apparent diffusion coefficient results. Also, since the amount of chloride adsorbed by soil was essentially zero, it would not be affected by this phenomenon.

Before the apparent diffusion coefficient of metribuzin in the adsorbed phase can be calculated by Eq. [15], values for D_s^m must be known. We calculated the values of D_s^m shown in Table 2 by using the data obtained from the ^{36}Cl diffusion experiments to determine transmission factors (TF). Values for TF at a given water content were calculated by dividing the chloride apparent-diffusion coefficient in soil by the chloride self-diffusion coefficient in aqueous solution of 1.96×10^{-5} cm^2/sec at 25C. This is the same procedure as used by Porter et al. (12) to calculate TF values for chloride in Ca- and Na- saturated soil. Olsen et al. (9) used these same TF values to estimate phosphorus diffusion in soil. Our calculations (Table 2) gave similar TF values at soil water contents of 0.36, 0.34, and 0.32 which can be attributed to the similar diffusivities in soil. At a given water content, D_s^m for metribuzin was calculated by multiplying TF by the self-diffusion coefficient of metribuzin determined in aqueous solution (0.95×10^{-5} cm^2/sec at 25C). In using this approach to calculate D_s^m the assumption is made that the effects of tortuosity, electrostatic restrictions due to irregular pore size, and mobility differences of the water in the solution phase would be the same with metribuzin as with chloride.

As the soil water content decreased, the solution and adsorbed phase diffusivities of metribuzin also decreased with the overall decrease being greater in the adsorbed phase (Table 2). The calculated values of D_s^m were greater in magnitude than D_e^m and D_a^m. This relation shows that D_e^m is a combination of the diffusion rate in the solution as well as adsorbed phases. Two of the values of D_a^m for metribuzin calculated according to Eq. [15] were negative. These negative values occurred at the lowest soil water contents and are not physically possible. Evaluation of Eq. [15] suggests that the explanation for these negative values for D_a^m can be attributed to the calculated values of R and/or D_s^m. Since the values of D_a^m were calculated assuming no net changes in the amount of metribuzin in the adsorbed and solution phases, a constant Kd value of 0.54 was used in the calculating R at all soil water contents. If one assumed that ρ_b remained constant and that R increased with the decrease in θ, then this would imply that the amount of herbicide in the adsorbed phase also increased. Larger values of R would give smaller negative (i.e., more positive) values for D_a^m. For example, to give a D_a^m value of 0.05×10^{-6} cm^2/sec at θ of 0.34 and 0.32, Kd would have to increase to 0.87 and 2.92, respectively. Since the adjustment of Kd to larger values is in agreement with the necessity of obtaining positive values for D_a^m, the amount of metribuzin in the adsorbed phase is thought to have increased as the soil water content decreased. There are at least two ways this increase in the adsorbed phase concentration might occur. First, the soil/water ratio used in the

Table 2—Bulk density, ρ_b, apparent self-diffusion coefficient, D_e^m, transmission factor, TF, diffusion coefficient in the solution phase, D_s^m, and diffusion coefficient in the adsorbed phase, D_a^m, of metribuzin at several soil water contents of Dubbs soil.

θ	ρ_b	$D_e^m \times 10^6$	TF*	$D_s^m \times 10^6$	$D_a^m \times 10^6$†
cm^3/cm^3	g/cm^3	cm^2/sec		—cm^2/sec—	
0.42	1.36	3.75	0.49	4.66	3.23
0.40	1.43	2.70	0.37	3.52	2.28
0.38	1.48	2.05	0.33	3.14	1.53
0.36	1.52	1.45	0.31	2.95	0.82
0.34	1.50	0.65	0.31	2.95	−0.32
0.32	1.48	0.25	0.31	2.95	−0.80

* TF values were determined by dividing the chloride self-diffusion coefficient in soil by the chloride self-diffusion coefficient in aqueous solution.
† D_a^m values were calculated according to Eq. [15] in the text.

adsorption experiment was 1:2 whereas the soil/water ratio used in the diffusion experiment varied from 1:0.30 to 1:0.15 on a weight basis. For a given amount of herbicide added to the soil, more herbicide may be adsorbed at the lower soil/water ratio. Obviously, we would not observe any differences with the equilibrium adsorption technique used in this experiment. Second, of the information needed to calculate $D_a{}^m$, only ρ_b and $D_e{}^m$ were previously thought to be dependent on soil water content. It can be shown that ρ_b does not significantly influence the observed results (Table 2) and also, that the value of $D_e{}^m$ is inversely related to Kd (14). Since transient state diffusion systems involve both adsorption and desorption reactions, it may be that the amount of metribuzin in the adsorbed phase is greater during the desorption reactions. Davidson and Hornsby (5) have shown that for a given amount of fluometuron in the solution phase, the amount in the adsorbed phase is greater during the desorption reaction. The Kd value used in Eq. [15] was obtained from an adsorption experiment only, and of course is not valid for the desorption process. However, we recognize that in using our technique to measure $D_e{}^m$, one side of the diffusion cell is undergoing desorption reactions primarily whereas the other side is undergoing adsorption reactions primarily. Using microtome techniques we observed a symmetrical distribution of metribuzin around the interface of the two half cells which indicates that $D_e{}^m$ was relatively constant on both sides. Therefore, the values of $D_e{}^m$ given in Table 2 represent a combination of both adsorption and desorption reactions and the contribution that each makes to the observed diffusivity of metribuzin obviously depends on the soil water content.

Another factor which may have caused the negative $D_a{}^m$ values was the use of chloride diffusion data to calculate the transmission factor. Any differences between metribuzin and chloride with respect to tortuosity, electrostatic restrictions, and soil water mobility would cause differences in the calculated values of TF. The transmission factor was used to calculate $D_s{}^m$.

In conclusion, the magnitude of the diffusion coefficient of ^{14}C-metribuzin in soil depends on the amount in the adsorbed phase and moisture content. In transient state systems other properties such as compaction are of secondary importance compared with the above. The relative contribution of the adsorbed molecules to the total diffusion depends not only upon the relative proportion of herbicide molecules in the adsorbed and solution phases, but also on the nature of the adsorbed molecule and its "ease of release" from the adsorption sites. At the two lowest soil water contents negative adsorbed phase diffusion coefficients were observed. Possible explanations of this result included soil/water ratio effects, changes in the amount of herbicide in the adsorbed phase during adsorption-desorption process, and differences in the values of TF between metribuzin and chloride.

ACKNOWLEDGMENT

We would like to thank Chemagro Chemical Company, Kansas City, Missouri, for supplying the ^{14}C-metribuzin used in this study.

LITERATURE CITED

1. Brown, D. A. 1974. A capillary tube diffusion cell for measuring ion diffusion in aqueous solutions. Soil Sci. Soc. Amer. Proc. 38:533–535.
2. Davidson, J. M., and J. R. McDougal. 1973. Experimental and predicted movement of three herbicides in a water-saturated soil. J. Environ. Qual. 2:428–433.
3. Ehlers, W., W. J. Farmer, W. F. Spencer, and J. Letey. 1969. Lindane diffusion in soils: II. Water content, bulk density and temperature effects. Soil Sci. Soc. Amer. Proc. 33:505–508.
4. Ellis, J. H., R. I. Barnhisel, and R. E. Phillips. 1970. The diffusion of copper, manganese, and zinc as affected by concentration, clay mineralogy and associated anions. Soil Sci. Soc. Amer. Proc. 34:866–870.
5. Hornsby, A. G., and J. M. Davidson. 1973. Solution and adsorbed fluometuron concentration distribution in a water saturated soil: I. Experimental and predicted evaluation. Soil Sci. Soc. Amer. Proc. 37:823–828.
6. Lavy, T. L. 1970. Diffusion of three chloro s-triazines in soil. Weed Sci. 18:53–56.
7. Lindstrom, F. T., L. Boersma, and H. Gardiner. 1968. 2,4-D diffusion in saturated soil: a mathematical theory. Soil Sci. 106:107–113.
8. Olsen, S. R., and W. D. Kemper. 1968. Movement of nutrients to plant roots. Advan. Agron. 20:91–151.
9. Olsen, S. R., W. D. Kemper, and R. D. Jackson. 1962. Phosphate diffusion to plant roots. Soil Sci. Soc. Amer. Proc. 26:222–227.
10. Phillips, R. E., R. I. Barnhisel, and J. H. Ellis. 1972. Percent of copper diffusing in the adsorbed and electrolyte-solution phases in kaolinite and montmorillonite. Soil Sci. Soc. Amer. Proc. 36:35–39.
11. Phillips, R. E., and J. H. Ellis. 1970. A rapid method of measurement of diffusion coefficients in aqueous solutions. Soil Sci. 110:421–425.
12. Porter, L. K., W. D. Kemper, R. D. Jackson, and B. A. Stewart. 1960. Chloride diffusion in soils as influenced by moisture content. Soil Sci. Soc. Amer. Proc. 24:460–463.
13. Scott, H. D., and R. E. Phillips. 1972. Diffusion of selective herbicides in soil. Soil Sci. Soc. Amer. Proc. 36:714–719.
14. Scott, H. D., and R. E. Phillips. 1973. Self diffusion coefficients of selected herbicides in water and estimates of their transmission factors in soil. Soil Sci. Soc. Amer. Proc. 37:965–967.
15. Walker, A., and D. V. Crawford. 1970. Diffusion coefficient of two triazine herbicides in six soils. Weed Res. 10:126–132.

Diffusion of a Biodegradable Pesticide: II. As Affected by Microbial Decomposition[1]

Z. Gerstl, P. H. Nye, and B. Yaron[2]

ABSTRACT

The diffusion of parathion (O,O-diethyl O-p-nitrophenyl phosphorothioate) was investigated and a model which involves treating microbial activity in the diffusion system was developed. The kinetics of parathion degradation during 7 days of incubation at three moisture content (θ = 0.14, 0.24, and 0.34) was determined and compared with the calculated values. It was found that the rate of parathion decomposition rises with increasing water content. The amount of parathion decomposed at any time increased with the increase in the initial concentration. The calculated course of decomposition of parathion fits the experimental results obtained over 7 days, which was the longest period in the diffusion. The experimental results and predicted distribution of parathion and of the metabolite formed (diethylthiophosphate) are presented as affected by time and soil moisture content. The proposed model—which considers the rate of decomposition at any distance and time, and depends on the local concentration of parathion and on the microbial activity—satisfactorily fits the experimental results.

Additional Index Words: parathion, diethyethiophosphate, adsorption, microbial activity.

Gerstl, Z., P. H. Nye, and B. Yaron. 1979. Diffusion of a biodegradable pesticide: II. As affected by microbial decomposition. Soil Sci. Soc. Am. J. 43:843–848.

D IFFUSION of potentially biodegradable pesticides in soils will be affected by the rate of degradation. Until now, plant uptake and volatalization have been the only processes responsible for irreversible loss considered in the diffusion of pesticides in soils (e.g. Ehlers et al., 1969; Graham-Bryce, 1969; Schearer et al., 1973; Letey and Farmer, 1974). Biological decomposition will also lead to an irreversible removal of the diffusing substance and often to the formation of a metabolite which itself undergoes diffusion through the soil.

Parathion (O,O-diethyl O-p-nitrophenyl phosphorothioate) was selected as an example of a biodegradable pesticide, whose diffusion in a sterile soil has been described previously (Gerstl et al., 1979). In natural soil parathion is metabolized by soil microorganisms. Lichtenstein and Schultz (1964) reported that 97% of the parathion applied to a split loam field disappeared within 3 months, confirming the previous results of Chisholm et al., (1955) for a sandy loam soil. Iwata et al. (1973) showed that the concentration of parathion declined sharply within a few days after its application to a wide range of soils. In a majority of cases studied unextractable residue of parathion was found. This was termed "bonded" fraction and was considered to be more tightly bound to the soil than the parent insecticide (Katan and Lichtenstein, 1977).

In general it was observed that in natural soils parathion decomposed at the beginning at an increasing rate and subsequently at a declining rate (Sethunathan, 1977), and that biologically induced degradation is the main process affecting parathion persistence (Yaron, 1975).

This paper describes a mechanistic model able to predict the diffusion of parathion and one of its metabolites in a nonsterile soil. The model involves consideration of microbial activity in the diffusion system. The predictions are compared with experimental results.

THEORETICAL CONSIDERATIONS

Biologically Induced Decomposition of Parathion

Preliminary experiments on the kinetics of biologically induced degradation of parathion in a nonsterile soil (Gilat — Israel) showed two distinct decomposition stages. In the first stage parathion degraded at an increasing rate. The initial rate was roughly proportional to the initial concentration except at a very low concentration. The second stage was characterized by a declining rate. We thought that the initial stage might correspond to an increase in microbial activity and the second stage to a diminishing rate controlled by the decreasing concentration of parathion.

Let us postulate that in a system in which microbial activity is not diffusable, but solute may be diffusible, the rate of increase in microbial activity (Verstraete and Vanloocke, 1974) is given by the equation

$$(\delta m/\delta t)_x = kCm \quad [1]$$

and that in a closed system (i.e. one receiving no additions of solute from outside, as in an incubation experiment) by the equation

$$\frac{dm}{dt} = -\frac{dC}{dt}. \quad [2]$$

The symbols used in the paper are presented in Table 1.

Equation [1] can account for the initial increasing rate of degradation because dm/dt is proportional to m. It can account for the declining rate in a closed system because dm/dt is proportional to C which declines. It can also describe the development of microbial activity in a soil zone which initially contains no parathion, but into which parathion diffuses, because the activity is governed by the product of m and C.

Equation [2] is equivalent to saying that substrate is consumed proportionately as microbial activity increases, (Dean and Hinshelwood, 1966, p. 77). m is expressed in units of substrate concentration.

Table 1—List of symbols.

C	= conc. parathion (μg cm^{-3} whole soil)
C_o	= initial conc. parathion (μg cm^{-3} whole soil)
$C_o{}^* $	= $C_o + m_o$
C'	= conc. metabolite (μg cm^{-3} whole soil)
D	= diffusion coefficient of parathion in whole soil (cm^2 sec^{-1} or cm^2 day^{-1})
D'	= diffusion coefficient of metabolite in whole soil (cm^2 sec^{-1} or cm^2 day^{-1})
K	= adsorption coefficient (ml g^{-1})
k	= rate constant (μg^{-1} cm^3 day^{-1})
m	= microbial activity (by Eq. [2] μg cm^{-3} parathion)
m_o	= initial microbial activity (by Eq. [2] μg cm^{-3} parathion)
ρ	= dry soil density (g cm^{-3})
t	= time (sec or day)
x	= distance from junction of half-cells (cm)

---

[1] Contribution from Department of Agricultural Science, University of Oxford, Oxford, U.K. Received 21 Dec. 1978. Approved 18 Apr. 1979.

[2] Research Associate, Reader in Soil Science, and Visiting Fellow respectively. The permanent address of Z. Gerstl and B. Yaron is: Institute of Soils and Water, Volcani Center, Bet-Dagan, Israel.

Integrating Eq. [2] yields

$$m - m_o = C_o - C$$

where m_o is the initial microbial activity, when $C = C_o$. m_o may include an element of adaptation but considering the range of C_o the value of m_o is not great.

$$\text{Hence } m = (C_o^* - C) \quad [3]$$

where $C_o^* = C_o + m_o$. [4]

The unknown m may now be eliminated between Eq. [1] and [3]. From Eq. [1],

$$\frac{d \ln m}{C} = k dt,$$

and substituting [3] for m results in

$$\frac{d \ln (C_o^* - C)}{C} = k dt,$$

which upon integrating becomes

$$\frac{1}{C_o^*} \ln \frac{C}{C_o^* - C} = -kt + B.$$

The integration constant B may be determined since $C = C_o$ when $t = 0$. Hence

$$\ln \frac{C_o^* - C_o}{C_o} + \ln \frac{C}{C_o^* - C} = -C_o^* kt. \quad [5]$$

Now

$$\frac{C}{C_o^* - C} \simeq \frac{C}{C_o - C}$$

except when C is near C_o. By regressing $\ln [C/(C_o - C)]$ on t an approximate value of C_o^* may be obtained from the intercept and of k from the slope. The values of C used are sufficiently different from C_o to obtain the linear plot. The first estimate of $\ln [C/(C_o^* - C)]$ is now regressed on t, and revised values of C_o^* and k obtained. The process is repeated until successive values of C_o^* and k are sufficiently close.

Diffusion

If a block of soil uniformly mixed with parathion is placed in contact with a parathion-free block, the problem is to predict the distribution of parathion in this system, when the concentration of parathion is affected both by microbial decomposition and diffusion.

The continuity equation for parathion is

$$\frac{\delta C}{\delta t} = D \frac{\delta^2 C}{\delta x^2} - R_{zt} \quad [6]$$

where R_{zt} is the rate of microbial decomposition.

Since parathion is converted quantitatively mole for mole into its decomposition product, the continuity equation for the product is

$$\frac{\delta C'}{\delta t} = D' \frac{\delta^2 C'}{\delta x^2} + R_{zt}. \quad [7]$$

Clearly the rate of decomposition at any distance and time will depend on the local concentration of parathion and on the microbial activity. These factors will reflect earlier changes not only in concentration but also in microbial activity at the distance in question. The model developed in the last section allows us to express the rate of microbial activity by Eq. [1]. By Eq. [2] this is also the value of R_{zt}. The boundary conditions are

$$t = 0 \quad x < 0 \quad C = C_o \quad m = m_o \quad C' = 0$$
$$\quad\quad\quad x > 0 \quad C = 0 \quad m = m_o \quad C' = 0$$
$$t > 0 \quad x = \pm \infty \quad dC/dx = 0.$$

Equation [6] was solved numerically using the Crank and Nicholson finite difference method (Crank, 1975, p. 144). The computer program, written in FORTRAN, incorporates the algorithm described by Richtmyer (1957, Ch. 6 Sect. 5). At each time step the updated concentration profile was calculated first, and then the profile of microbial activity using Eq. [1]. Finally, the concentration profile of the decomposition product was calculated from the parallel finite difference form of Eq. [7]. Distance steps of .02 cm and time steps of 1/10 day were used.

MATERIALS AND METHODS

The materials and methods have been described in the previous paper (Gerstl et al., 1979) and only those particular to the present research will be described below.

Materials—The decomposition product ^{14}C-labelled diethylthiophosphate (MW: 170) was prepared by heating an aqueous solution of parathion adjusted to pH 10.5 at 60°C for 1 week. After cooling and readjusting the pH to 7.0, a 5-ml sample was extracted with hexane to determine the presence of unhydrolyzed parathion. No trace of parathion was observed, with all the labelled material remaining in the aqueous phase, thus showing 100% hydrolysis.

Methods—Microbial decomposition of parathion on natural Gilat soil was studied by incubating samples of different initial concentrations at 3 moisture contents ($\theta = 0.14, 0.24, 0.34$). ^{14}C-labelled parathion was added to the air-dry soil in a hexane solution which was slowly evaporated, followed by addition of $0.01N$ $CaCl_2$ aqueous solution to obtain the desired moisture contents. The flasks were sealed and kept in an incubator at 25 ± 1°C. At various times, duplicate samples of each treatment were extracted with 5:2 water-hexane mixture, and the radioactivity determined in both the aqueous and organic phases.

Adsorption of ^{14}C-diethylthiophosphate was determined in a batch study at a soil/solution ratio of 2:5. The isotherm obtained was linear and the K value determined as 0.215 ml/g.

The diffusion coefficient of diethylthiophosphate was calculated from the equation $D' = D_o f \theta / (\rho K + \theta)$ where D_o was as-

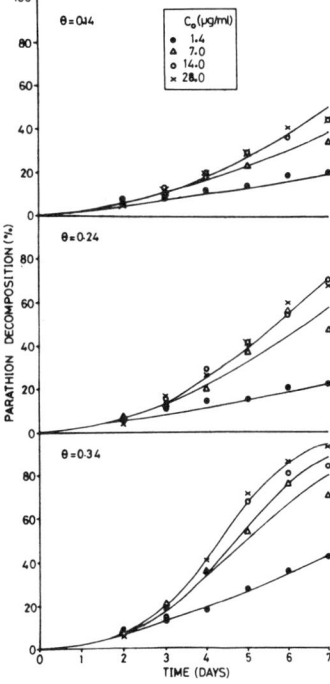

Fig. 1—Parathion degradation in a biologically active Gilat soil as affected by initial concentration and soil moisture content. Solid line calculated from Eq. [5]; points represents experimental measurements.

Table 2—Values of initial microbial activity (m_o) and decomposition constant (k) at different initial concentration of parathion (C_o) and moisture level (θ).†

C_o	$\theta = 0.34$		$\theta = 0.24$		$\theta = 0.14$	
	m_o	k	m_o	k	m_o	k
µg cm⁻³						
28	0.325	0.0363	0.724	0.0229	1.066	0.0164
14	0.214	0.0629	0.386	0.0450	0.910	0.0264
7	0.290	0.090	0.372	0.0643	0.577	0.0421
1.4	0.150	0.203	0.426	0.0621	0.482	0.0471

† Calculated from Eq. [5].

signed the value of 0.685×10^{-6} cm² sec⁻¹ by comparison with compounds of similar molecular weight. The diffusion coefficient of parathion was provided by the experiments with sterile soil (Gerstl et al., 1979).

Diffusion studies were similar to those previously described (Gerstl et al., 1979) with the exception that unsterilized soil was used. Each section was analyzed for parathion and its labelled degradation product.

RESULTS AND DISCUSSION

Biologically Induced Decomposition of Parathion

Figure 1 shows the kinetics of parathion decomposition in the nonsterile Gilat soil over 7 days of incubation, which was the longest period in the diffusion runs. For all initial concentrations and soil moisture contents no detectable lag period was observed. The rate of decomposition rises with increasing water content. The percentage of parathion decomposed at any time is greater the greater the initial concentration. At a $\theta = 0.34$, for example, the remaining parathion in soil after 7 days of incubation was 2.4 µg/ml when the initial concentration was 28 µg/ml and 0.8 µg/ml when the initial concentration was 1.4 µg/ml. The amount of water soluble parathion degradation product formed in soil after 7 days of incubation—measured by ¹⁴C counting—was in a 1:1 ratio with the parathion biodegraded. The degradation of parathion was never complete. The percentage degraded after 11 days ranged from 96.0 at high concentration and moisture level ($C_o = 28$ µg/ml, $\theta = 0.34$), to 20.0 at low concentration and moisture level. Further degradation was very slow.

Figure 1 shows also the curves obtained by plotting Eq. [5] using the values of C_o^* and k derived from the interaction procedure previously described. The values of k and m_o as a function of C_o are presented in Table 2.

If Eq. [1] and [2] are correct for all conditions, k and m_o should be independent of C_o. It is evident that this is not the case, and this limitation must be born in mind examining the diffusion data. Using values of k and m_o appropriate for $C_o = 28$ µg/ml, the measured degradation of parathion at lower initial concentrations is underestimated at all moisture levels.

Diffusion of Parathion and of the Metabolite Formed

EFFECT OF TIME

In all diffusion experiments the initial concentration in the treated half-cell was 28 µg/ml. Figure 2 shows the parathion distribution in the nonsterilized Gilat soil after 2, 4, and 7 days when the soil was packed at a bulk density (ρ) of 1.4 g/cm³ and at a moisture content $\theta = 0.34$. We observed that in the very beginning (after 2 days of incubation) when the percentage decomposed is still very low the distribution of parathion along the diffusion cell is similar to that occurring in the sterile soil. After 4 days of incubation, with the buildup of microbial activity and the increase of decomposition the distribution pattern is drastically changed. At this time the parathion distribution in the nonsterile soil differs from that in

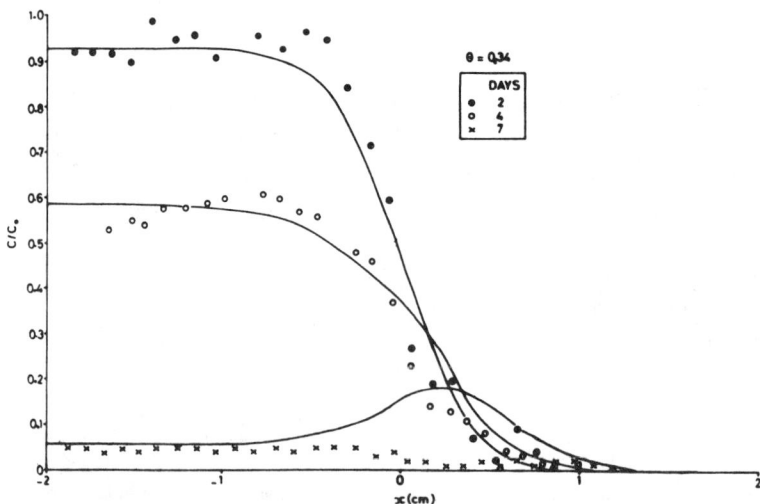

Fig. 2—Daily calculated (solid line) and measured points distribution of parathion in a biologically active Gilat soil during the process of diffusion. Volumetric soil moisture content $\theta = 0.34$.

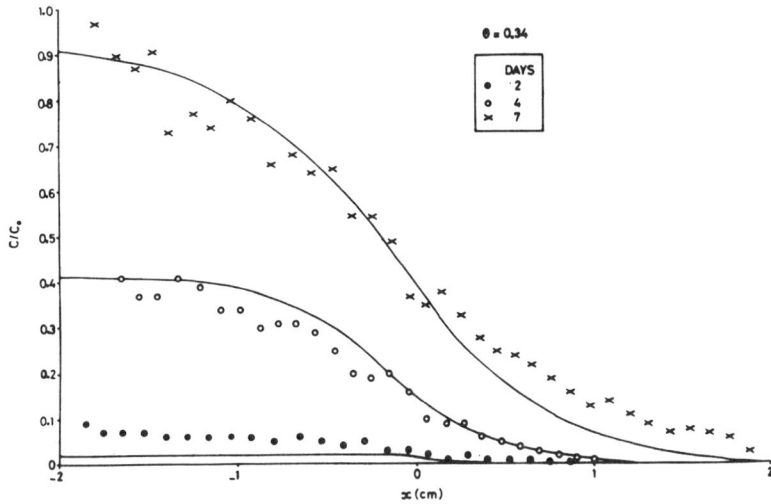

Fig. 3—Daily calculated (solid line) and measured (points) distribution of parathion degradation product (diethylthiophosphate) in a biologically active Gilat soil during the process of diffusion. Volumetric soil moisture content $\theta = 0.34$.

sterile soil mainly in the half-cell where the pesticide was initially added; it differs less in the second half-cell which was initially parathion free. After 7 days of incubation the parathion is almost completely decomposed and its distribution is affected by the decomposition process in both half-cells. This may be explained by the development of microbial activity in the half-cell which initially was free of parathion.

The predicted distribution agrees well with the experimental results for 2 days and fairly well for 4 days (Fig. 2 — solid lines). A 7 days the decomposition of parathion in the treated half-cell has been satisfactorily predicted. However, in the untreated half-cell it is predicted that there should be up to 18% parathion close to the boundary, but in practice there is little. In the untreated half-cell microbial activity is predicted to increase only slowly, since it increases only in response to parathion diffusing into it. In practice it seems that the microbial activity is greater.

In predicting the distribution of parathion, values of m_o and k for an initial concentration of 28 μg/ml were used. While these are satisfactory for the treated half-cell, their use in the untreated half-cell into which

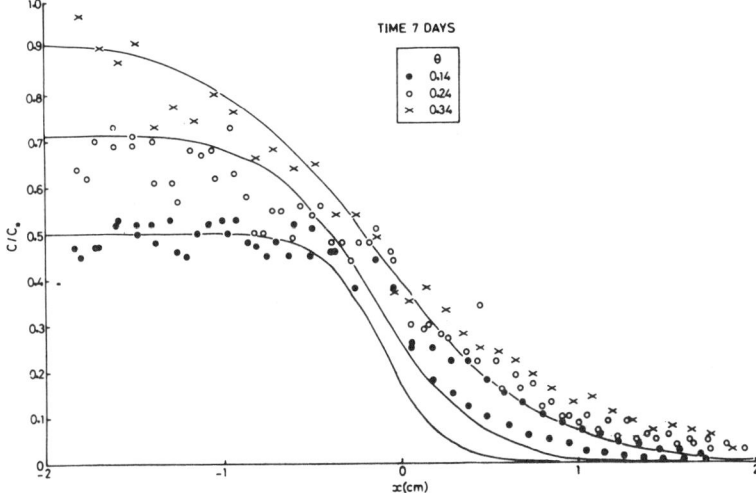

Fig. 4—Effect of the volumetric content (θ) on the parathion distribution in biologically active Gilat soil, after 7 days of diffusion. Solid line calculated from Eq. [6]; points represents experimental measurements.

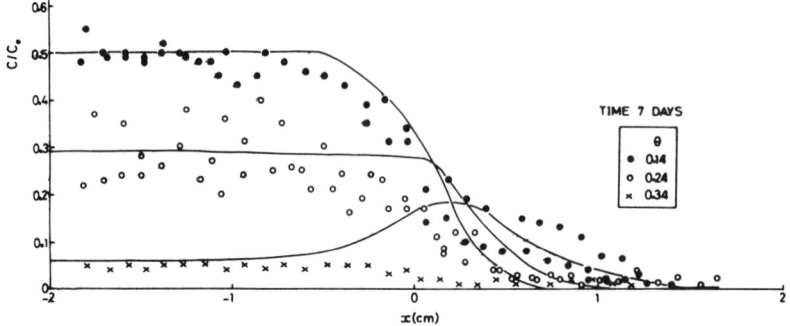

Fig. 5—Effect of the volumetric moisture content (θ) on the distribution of parathion degradation product (diethylthiophosphate) after 7 days of diffusion. Solid line calculated from the Eq. [7]; points represents experimental measurements.

parathion is diffusing is arbitrary, and may well underestimate the rate of degradation.

Figure 3 shows the distribution of the metabolite (^{14}C-diethylthiophosphate) after 2, 4, and 7 days in the wettest soil as measured in the aqueous phase of the soil extract. The numerical solution of Eq. [7] describes the course of the metabolite distribution quite satisfactorily. The decomposition product, which is not degraded, is more mobile than the parathion, since its K value is lower. After 7 days much of it has diffused into the untreated half-cell. The predicted distribution of the product agrees well with the measured distribution after 2 days and satisfactorily after 4 days. The predicted amount of product in the untreated half-cell after 7 days of incubation is somewhat less than the amount found, possibly because microbial activity there was greater than expected, as mentioned previously.

Moisture Level

Figure 4 and 5 show the distribution of parathion and its decomposition product after 7 days in soils at three moisture contents. In the treated half-cell, the decomposition decreases as the soil moisture decreases, and the computer simulation has satisfactorily predicted the amount remaining. In the untreated half-cell the distribution of parathion has also been satisfactorily predicted when $\theta = 0.24$. When $\theta = 0.14$ the parathion has diffused somewhat further than expected. The discrepancy when $\theta = 0.34$ has already been discussed.

The distribution of the metabolite in the treated half-cell has been satisfactorily predicted at all moisture contents. In the untreated half-cell the amount of metabolite and its spread are somewhat greater than expected.

The model has assumed that microbial activity does not migrate, though there is evidence that parathion diffusing into the untreated half-cell is rapidly decomposed. It seems unlikely that bacteria would be sufficiently mobile, though fungal hyphae could develop from the treated into the untreated half-cell. Exoenzymes produced by microorganisms in the treated half-cell could also diffuse into the untreated half-cell.

CONCLUSION

This work suggests a way of predicting the distribution of a diffusing solute undergoing microbial decomposition, and also the distribution of its metabolites. The predictions agree fairly well with measurements over a range of moisture contents and time. Further refinement of the model requires a more detailed examination of microbial activity in an open system when the concentration of solute is changing. The possibility that microbial activity can migrate into adjacent soils needs further study. Identification of the source of microbial activity will clearly assist in devising a more realistic kinetic model.

ACKNOWLEDGEMENT

This work was supported in part by a Research Fellowship Award granted (to B. Yaron) by the International Development Research Center, Ottawa, Canada.

LITERATURE CITED

1. Chisolm, D., A. W. MacPhee, and C. R. MacEachern. 1955. Effects of repeated applications of pesticides to soil. Canada J. Agric. Sci. 35:433–438.
2. Crank, J. 1975. Mathematics of diffusion. 2nd ed. Clarendon Press, Oxford.
3. Dean, A. R. C., and C. N. Hinshelwood. 1966. Growth function and regulation in bacterial cells. Clarendon Press, Oxford.
4. Ehlers, W., J. Letey, W. F. Spencer, and W. J. Farmer. 1969. Lindane diffusion in soils: I. Theoretical considerations and mechanism of movement. Soil Sci. Soc. Am. Proc. 33:501–504.
5. Gerstl, Z., B. Yaron, and P. H. Nye. 1979. Diffusion of biodegradable pesticide: I. In a biologically inactive soil. Soil Sci. Soc. Am. J. 43:839–842 (this issue).
6. Graham-Bryce, I. J. 1969. Diffusion of organo-phosphorus insecticides in soil. J. Sci. Food Agric. 20:489–492.
7. Iwata, Y., W. E. Westlake, and F. A. Gunter. 1973. Persistence of parathion in six Californian soils under laboratory conditions. Arch. Environ. Contam. and Toxicol. 1:84–96.
8. Katan, J., and E. P. Lichtenstein. 1977. Mechanisms of production of soil-bound residues of ^{14}C-parathion by microorganisms. J. Agric. Food Chem. 25:1404–1408.
9. Letey, J., and W. J. Farmer. 1974. Movement of pesticides in soil. In W. D. Guenzi (ed.). Pesticides in soil and water. Soil Sci. Soc. Am., Madison, Wis. p. 67–87.
10. Lichtenstein, E. P., and K. R. Schultz. 1964. The effects of moisture and microorganisms on the persistence and metabolism of some organophosphorous insecticides in soils, with special emphasis on parathion. J. Econ. Entomol. 5:618–622.

11. Richtmyer, R. D. 1957. Difference method for initial value problems. Interscience, New York.
12. Schearer, R. C., J. Letey, W. J. Farmer, and A. Klute. 1973. Lindane diffusion in soil. Soil Soc. Am. Proc. 37:189-193.
13. Sethunathan, N., R. Siddaramappa, K. P. Rajaram, S. Barik, and P. A. Wahid. 1977. Parathion: Residues in soil and water. Residues Rev. 68:91-122.
14. Verstraete, W., and R. Vanloocke. 1974. Mathematical modelling of biodegradation processes. p. 603-616. Proc. Seminar on Ground Water Quality Control and Management, Brussels.
15. Yaron, B. 1975. Chemical conversion of parathion on soil surfaces. Proc. Soil Sci. Soc. Am. 39:639-642.

Pesticide Mobility in Soils III. Influence of Soil Properties[1]

CHARLES S. HELLING[2]

ABSTRACT

Soil parameters influencing pesticide movement were isolated using simple correlation and multiple linear regression analyses. Mobilities of 12 pesticides on 14 soils were first characterized by soil thin-layer chromatography. Mobility of nonionic compounds was inversely related to adsorption of similar compounds, field moisture capacity, organic matter and clay contents, and cation-exchange capacity. Mobility of acidic compounds (dicamba, picloram, fenac, and 2,4-D) was directly correlated with soil pH and inversely with picloram adsorption. Pesticide mobility tended to be directly related to increased water flux.

When soils were grouped according to their clay mineralogy, there was a tendency for movement of acidic pesticides to be directly related to montmorillonitic clay content and inversely related to nonmontmorillonitic clay content.

Regression equations usually contained field moisture capacity, water flux, and often simazine or chlorpropham adsorption terms for predicting movement. These parameters are highly correlated with soil organic-matter content, which does not itself appear in the regression equations. The average deviation of predicted from observed mobility, across all soils and pesticides, was 0.04 R_F units.

Additional Key Words for Indexing: herbicide, insecticide, leaching, movement of pesticides, clays, organic matter, field moisture capacity.

THE observation that pesticides applied to coarse-textured, sandy soils are subject to greater leaching than those found in soils of higher clay and organic content is now virtually a truism. Numerous references supporting this are found in two recent reviews (1, 8). Adsorption of pesticides to various soils usually follows the inverse generalization, supporting the contention that adsorption governs movement. Thus, those soil factors influencing pesticide adsorption—especially soil organic matter, clay, and (sometimes) soil pH—have often been related to movement. Movement of organophosphorus insecticides (15) and s-triazine herbicides (5) was inversely related to their adsorption to four soils.

Statistical analyses are sometimes used to confirm and measure the significance of soil parameters on properties such as pesticide adsorption (14) and bioactivity (12). Atrazine[3] retention against leaching in a miscible displacement experiment was highly correlated with organic matter content, surface area, and cation exchange capacity (CEC), according to Snelling, Hobbs, and Powers (16). Adsorption itself was negatively correlated with movement of 29 nonionic herbicides, although leaching was performed with ethanol/water in a partition thin-layer chromatographic (TLC) system (5).

The objective of the present study was to examine soil from the standpoint of parameters influencing pesticide mobility. The use of soil TLC (9, 10) permitted examination of pesticide movement in many soils, facilitating the subsequent correlation and regression analyses.

MATERIALS AND METHODS

Soils—The properties of soils used in this and other (9, 10, 11) studies are indicated in Table 1. Characteristics and the methods used to determine them include: organic matter, by the Walkley-Black procedure (13); clay content, by the hydrometer method; cation exchange capacity (CEC), by the Ca saturation and titration method (13); field moisture capacity (FMC), after drainage of excess water in soil columns; and pH, electrometrically in a 1:1 soil/water paste. Mineralogy of the clay fraction was determined using X-ray diffraction after standard preparatory procedures. All soils used, except Hagerstown, Lakeland, and Duffield, were previously characterized by Harris and Sheets (7).

Adsorption Experiments—The procedure of Harris and Sheets (7) was used. Duplicate 1-g (oven-dry basis) samples

---

[1] Presented in part before Div. S-1 and S-2, Soil Science Society of America, Nov. 13, 1968 at New Orleans, La. Received Feb. 10, 1971. Approved May 14, 1971.
[2] Research Soil Scientist (Chemist), Plant Science Research Division, ARS, USDA, Beltsville, Md. 20705.
[3] Pesticide chemicals mentioned in this text are given in Table 1 of a preceeding publication (9).

Influence of Soil Properties

Table 1—Soil properties

Soil Name	Number	Origin (State)	Organic matter	Clay	Moisture at Field capacity	Moisture at Air-dry	pH	CEC	Water flux on soil TLC*	Adsorption Picloram	Adsorption Simazine	Adsorption Diuron	Adsorption Chlorpropham	Dominant clay mineral†
			%	%	%	%		meq/100 g	cm/hr	%	%	%	%	
Norfolk sl	26	?	0.14	11.3	6.5	0.17	5.1	0.2	11.8	8	5	2	10	Int Vm, Kn
Lakeland sl	40	Maryland	0.90	12.0	8.5	0.14	6.4	3.0	30.3	4	8	10	18	Kn, Qz
Christiana l	10	Maryland	0.99	24.4	19.7	0.47	4.4	5.6	1.8	9	20	10	30	Kn, Mica
Ascalon scl	11	Colorado	1.48	26.6	18.0	1.50	7.3	12.7	4.2	6	13	19	36	Mica, Int Mt
Sterling cl	13	Colorado	1.64	30.7	23.8	2.82	7.7	22.5	14.3	6	15	19	38	Mt
Dundee sil	4	Mississippi	1.67	29.0	24.3	0.85	5.0	18.1	7.5	6	50	17	51	Mt
Wehadkee sil	37	Maryland	1.93	25.2	23.7	0.82	5.6	10.2	5.3	13	25	16	42	Kn, Vm
Duffield cl	42	Maryland	2.20	33.7	21.5	1.38	6.3	10.9	5.1	4	23	14	52	Mica, Vm
Beltsville sl	7	Maryland	2.42	22.4	22.5	0.58	4.3	4.2	4.5	11	30	15	60	Vm, Kn
Hagerstown sicl	41	Maryland	2.50	39.5	25.8	0.82	6.8	14.7	4.6	3	28	35	56	Vm, Kn
Chillum sil	27	Maryland	4.40	22.1	24.6	0.86	4.6	7.6	5.3	9	35	29	69	Vm, Kn
Iredell sil	29	?	5.27	23.2	31.3	1.61	5.4	17.0	6.9	13	35	47	75	Mt
Barnes cl	16	Minnesota	6.90	34.4	28.5	2.99	7.4	33.8	6.9	11	43	63	76	Mt
Berkley sic	33	West Virginia	8.02	50.5	30.8	3.23	7.1	33.7	5.2	7	38	66	79	Vm, Kn

* Derivation is discussed in the text.
† Abbreviations: Int = Interstratified, Vm = Vermiculite, Kn = Kaolinite, Qz = Quartz, Mt = Montmorillonite.

of soil (sieved at 250 or 500 μ) were shaken 2 hours with 5.00 ml herbicide solution in 13 by 100 mm test tubes having teflon-lined screw caps. The herbicides and their concentrations (ppm) were: picloram (400), diuron (25), chlorpropham (80), and simazine (4), all in 0.01M $CaCl_2$. Suspensions were then centrifuged 30 min at 3,300 $\times$ g and the decantates, again for 20 min at 9,000 $\times$ g. The ^{14}C-simazine decantate was analyzed by liquid scintillation counting. The others, nonradioactive, were analyzed using ultraviolet spectroscopy at the λ_{max} of each: picloram (222 mμ), diuron (252 mμ), and chlorpropham (238 mμ). Before analysis they were diluted 66 $\times$, 4 $\times$, or 6 $\times$, respectively. Adsorption is therefore presented (Table 1) as the percentage of herbicide removed from the near-saturated solutions. Length of shaking, from 15 min to 2 hours, did not affect picloram adsorption onto Hagerstown soil.

Soil Thin-layer Chromatography—The technique has been described in detail (9, 11). For these experiments 13 different radioactive pesticides were applied in a statistically random order across two 20 by 20-cm soil plates. Monuron (an internal standard) occurred once on each plate. Four such repetitions were made per soil. The development time was recorded for each plate; water flux data (Table 1) is based directly on this time, an average from eight plates.

RESULTS AND DISCUSSION

The 14 surface soils used in this study of mobility versus soil properties were selected to include a broad range in both clay (11 to 51%) and organic matter (0.1 to 8.0%) contents, and in pH (4.3 to 7.7). Other properties in Table 1, such as moisture capacity, CEC, and adsorption of various pesticides, are usually correlated with clay and/or organic matter. As expected, they also vary widely in magnitude.

The soils can be grouped broadly into five montmorillonitic soils (numbers 4, 11, 13, 16, 29) and nine non-montmorillonitic soils. The latter characteristically contain kaolinite and vermiculite and are from the eastern USA.

A summary of pesticide mobility appears in Table 2. The soils are ranked according to increasing organic-matter content. For pesticides from 2,4-D to azinphosmethyl there is a general trend toward reduced mobility with increased organic matter. Relative order of pesticide mobility is usually the same among soils. The pesticides, which are ranked (with the exception of diphenamid) by decreasing R_F on Hagerstown, have average R_F values for the 14 soils that correspond to the order of Table 2, except that diphenamid (0.44) is slightly less mobile than simazine (.46). Thus, both the "relative mobility classification" concept (11), and the use of Hagerstown soil to define this classification, appear justified.

Simple Correlation

Correlation coefficients of pesticide mobility and soil properties are presented in Table 3. Trifluralin is omitted, as it was immobile in all soils. Fine clay content was also omitted because correlations were nonsignificant.

Mobilities of 10 of the 12 pesticides were directly correlated with water flux. Flux was negatively correlated (−.577*) with field capacity but was not as closely related to other soil parameters. Field capacity itself was highly negatively correlated with movement of nonionic pesticides. As expected, field capacity was also highly correlated

Table 2—Mobility of 13 pesticides on 14 soils, using soil thin-layer chromatography

Soil Name	Number	R_F of Dicamba	Picloram	Fenac	2,4-D	Monuron	Atrazine	Diphenamid	Simazine	Diuron	Chlorpropham	Azinphosmethyl	Diquat	Trifluralin
Norfolk sl	26	1.00	0.96	0.84	0.88	0.84	0.74	0.68	0.80	0.61	0.52	0.46	0.04	0.00
Lakeland sl	40	1.00	0.96	1.00	1.00	0.89	0.89	0.94	0.96	0.60	0.59	0.28	0.19	0.00
Christiana l	10	0.83	0.67	0.44	0.50	0.56	0.49	0.45	0.45	0.36	0.23	0.19	0.00	0.00
Ascalon scl	11	0.99	0.92	0.90	0.87	0.57	0.58	0.34	0.51	0.33	0.24	0.13	0.01	0.00
Sterling cl	13	1.00	0.96	0.97	0.94	0.59	0.59	0.42	0.52	0.36	0.25	0.16	0.00	0.00
Dundee sil	4	0.94	0.89	0.82	0.82	0.41	0.35	0.16	0.16	0.23	0.22	0.07	0.00	0.00
Wehadkee sil	37	0.95	0.85	0.72	0.71	0.61	0.56	0.46	0.51	0.36	0.22	0.18	0.00	0.00
Duffield cl	42	0.90	0.82	0.74	0.66	0.48	0.45	0.45	0.45	0.23	0.15	0.12	0.04	0.00
Beltsville sl	7	0.92	0.78	0.38	0.41	0.59	0.51	0.50	0.38	0.32	0.16	0.17	0.00	0.00
Hagerstown sicl	41	0.96	0.84	0.84	0.69	0.48	0.47	0.49	0.45	0.24	0.18	0.15	0.06	0.00
Chillum sil	27	0.96	0.85	0.62	0.50	0.44	0.35	0.39	0.31	0.23	0.13	0.11	0.04	0.00
Iredell sil	29	0.92	0.78	0.66	0.64	0.37	0.34	0.34	0.30	0.18	0.11	0.09	0.03	0.00
Barnes cl	16	1.00	0.85	0.82	0.71	0.34	0.37	0.22	0.32	0.14	0.13	0.08	0.00	0.00
Berkley sic	33	0.98	0.84	0.80	0.75	0.39	0.38	0.36	0.36	0.20	0.15	0.12	0.03	0.00
Average R_F		0.95	0.86	0.75	0.72	0.54	0.51	0.44	0.46	0.31	0.23	0.17	0.03	0.00

Table 3—Simple correlation coefficients (r) among pesticide R_F values and soil properties, for 14 soils

Pesticide	Organic matter	Clay	Moisture at		pH	CEC	Water flux on soil TLC	Adsorption			
			Field capacity	Air-dry				Picloram	Simazine	Diuron	Chlorpropham
Dicamba	.151	-.005	-.192	.328	.616*	.298	.504	-.207	-.174	-.225	-.078
Picloram	-.251	-.227	-.467	.067	.484	.015	.645*	-.413	-.382	-.189	-.412
Fenac	-.077	.083	-.252	.283	.761**	.300	.595*	-.557*	-.279	.069	-.286
2,4-D	-.262	-.109	-.435	.150	.618*	.146	.683**	-.481	-.424	-.131	-.504
Monuron	-.728**	-.709**	-.924**	.595**	-.115	-.704**	.672**	-.235	-.856**	-.729**	-.891**
Atrazine	-.656**	-.609**	-.878**	-.447	.113	-.557*	.749**	-.314	-.870**	-.616**	-.858**
Diphenamid	-.488	-.569**	-.766**	-.549**	-.084	-.644**	.717**	-.287	-.764**	-.502	-.658**
Simazine	-.537*	-.553*	-.850**	-.399	.154	-.521	.737**	-.324	-.885**	-.505	-.793**
Diuron	-.718**	-.722**	-.930**	-.583*	-.149	-.680**	.661*	-.208	-.837**	-.716**	-.911**
Chlorpropham	-.606*	-.649**	-.909**	-.496	-.007	-.520	.797**	-.340	-.723**	-.570*	-.851**
Azinphosmethyl	-.561*	-.619**	-.851**	-.518	-.196	-.610**	.456	-.107	-.762**	-.565*	-.786**
Diquat	-.174	-.344	-.513	-.360	.118	-.341	.809**	-.477	-.408	-.155	-.339

* Significant at 5% level.   ** Significant at 1% level.

with more fundamental parameters—organic matter, clay, and CEC. These parameters were generally negatively correlated with movement of nonionic pesticides.

Soil pH was important only for movement of the acidic compounds, though picloram was significantly correlated only at the 10% level. That is, the higher the pH, the greater the movement of these compounds. This corroborates data obtained for dicamba and fenac by direct modification of soil pH (10).

Pesticide movement is often thought to be governed largely by its adsorption to soil. Adsorption (Table 3) was an accurate single-factor predictor of the movement of most nonionic compounds. Prediction of movement was often best for chemically similar pesticides, e.g., simazine (—.885**) or atrazine (—.870**) mobility with simazine adsorption. Chlorpropham's adsorption was less well correlated with its own mobility than with the mobility of three other herbicides. Perhaps this reflects the extensive diffusion chlorpropham undergoes, a process more likely subject to variability in soil TLC than the mass transfer movement that characterizes other compounds. Picloram adsorption is significantly related only to fenac mobility, although movement of 2,4-D and picloram tend also to be inversely related. Picloram adsorption data (Table 1) seemed erratic, perhaps because it was always rather low. It is clear from correlations of movement with adsorption and other soil parameters that acidic pesticides behave in a strikingly different manner than do nonionic compounds.

Adsorption itself was highly correlated with soil organic-matter content: simazine (.671**), diuron (.961**), chlorpropham (.884**). Picloram adsorption was nonsignificantly correlated, however. Adsorption of diuron (.695**) and chlorpropham (.650*) was related to total clay content; simazine was less closely related (.534). All three compounds were correlated with CEC. These trends substantially agree with Harris and Sheets (7), who correlated adsorption and phytotoxicity with properties of 32 soils, many identical to those used in this study.

To summarize the simple correlation results of Table 3, for 14 soils, mobility of nonionic compounds was directly related to water flux and inversely related to adsorption of similar compounds, field moisture capacity, organic-matter and clay contents, and CEC. Mobility was generally not related to adsorption of a dissimilar compound (picloram), fine clay content, pH, and moisture content of air-dry soil. Mobility of acidic compounds was directly related to water flux and pH, and inversely related to picloram adsorption.

The relationship of mobility and water flux was unexpected and therefore prompted the direct experimentation reported earlier (9). It was concluded from the latter that there may be some direct relationship between mobility and flux, or penetrability. Because of continued uncernearly correlated (5% level) with retardation of diquat movement, in contrast to the effect of nonmontmorillonitic clays. This trend is expected, since diquat is more strongly adsorbed to montmorillonite than to kaolinite or vermiculite (3, 17). Clay content had the opposite effect on R_F values for acids, especially dicamba: movement tended to be directly related to montmorillonitic clay, suggesting tainty, two soils giving extreme values of flux, Christiana l and Lakeland sl, were consecutively omitted, and correlations were rerun on the 13 and 12 remaining soils. Omission of the loam had little effect, but when Lakeland was also removed, water flux was then significantly related (5% level) only to picloram, 2,4-D and chlorpropham movement. Flux was correlated (inversely) with a single soil parameter, field moisture capacity.

The principal changes noted in eliminating Christiana l were nearly always increased correlation of picloram mobility with factors such as organic matter, field capacity, and adsorption of pesticides. This acidic (pH 4.4) soil caused unusual retardation of picloram movement, probably depressing the 14-soil correlation coefficients. Christiana soil appeared to contain a relatively high iron content. Hydrated iron oxides and low pH were both shown to enhance picloram adsorption (4).

The original 14 soils were subdivided, as previously described, into montmorillonitic and nonmontmorillonitic soils. The rationale behind these distinctions is that montmorillonite is an expandable layer silicate clay with high cation-exchange capacity, conditions favorable for positive adsorption of neutral or cationic pesticides and for negative adsorption of anionic species. The nonmontmorillonitic soils contain kaolinite and/or vermiculite clays that generally adsorb nonionic pesticides *less,* but anionic species *more* than montmorillonite. Simple correlation coefficients for each group appear in Table 4.

The number of significant correlations in the montmorillonite group is small, probably because the degrees of freedom have been greatly reduced.[4] Soil organic matter is negatively correlated with chlorpropham R_F and is likely important in movement of diuron, 2,4-D, monuron, and perhaps picloram and atrazine. Montmorillonitic clay was

---
[4] For significance at the 10% level, $r \geq 0.805$ and 0.582 for the groups of five and nine soils, respectively.

Table 4—Simple correlation coefficients (r) among pesticide R_F values and soil properties, for soils grouped by clay mineralogy

Pesticide	Organic matter	Clay	Moisture at		pH	CEC	Water flux on soil TLC	Adsorption			
			Field capacity	Air-dry				Picloram	Simazine	Diuron	Chlorpropham
5 Montmorillonitic Soils											
Dicamba	-.074	.721	-.481	.716	.951*	.457	.272	-.398	-.504	-.006	-.392
Picloram	-.757	.394	-.800	.172	.565	-.101	.475	-.917*	-.562	-.719	-.895*
Fenac	-.613	.516	-.751	.350	.714	.057	.474	-.828	-.586	-.565	-.811
2,4-D	-.843	.249	-.812	.066	.477	-.243	.486	-.936*	-.603	-.812	-.945*
Monuron	-.803	-.089	-.797	.058	.533	-.469	.377	-.757	-.883*	-.765	-.934*
Atrazine	-.629	.059	-.766	.273	.737	-.284	.340	-.655	-.932*	-.575	-.842
Diphenamid	-.238	-.318	-.166	.375	.517	-.278	.457	-.051	-.881*	-.211	-.385
Simazine	-.308	.018	-.524	.483	.835	-.141	.286	-.314	-.965**	-.249	-.575
Diuron	-.866	-.122	-.783	-.026	.429	-.515	.420	-.798	-.824	-.837	-.955*
Chlorpropham	-.931*	.111	-.876	-.161	.308	-.410	.335	-.981**	-.907*	-.978**	
Azinphosmethyl	-.523	-.021	-.521	.390	.695	-.232	.552	-.476	-.939*	-.481	-.714
Diquat	.283	-.860	.423	-.311	.418	-.466	-.357	.650	-.079	.257	.397
9 Non-montmorillonitic Soils											
Dicamba	.185	-.118	-.265	.082	.453	-.128	.573	-.205	-.175	.234	-.066
Picloram	-.117	-.356	-.545	-.148	.389	-.137	.719*	-.287	-.474	-.089	-.351
Fenac	-.007	.010	-.350	.083	.775*	.176	.666*	-.615	-.407	.136	-.291
2,4-D	-.183	-.197	-.575	-.047	.620	.019	.773*	-.457	-.634	-.064	-.542
Monuron	-.728*	-.825**	-.941**	-.693*	-.139	-.684*	.783*	-.022	-.915**	-.700*	-.896**
Atrazine	-.664	-.710*	-.891**	-.602	.055	-.563	.861**	-.163	-.894**	-.599	-.857**
Diphenamid	-.594	-.681*	-.861**	-.588	.110	-.544	.938**	-.334	-.829**	-.527	-.762*
Simazine	-.612	-.647	-.899**	-.523	.173	-.487	.880**	-.299	-.913**	-.536	-.856**
Diuron	-.695*	-.815**	-.948**	-.656	-.168	-.651	.742**	.007	-.923**	-.667*	-.916**
Chlorpropham	-.579	-.696*	-.923**	-.534	.063	-.507	.856**	-.222	-.905**	-.518	-.860**
Azinphosmethyl	-.618	-.684*	-.884**	-.541	-.169	-.553	.500	.010	-.866**	-.574	-.847**
Diquat	-.192	-.314	-.526	-.243	.453	-.170	.928**	-.647	-.487	-.107	-.376

* Significant at 5% level.   ** Significant at 1% level.

probable negative adsorption. Dicamba was not adsorbed by montmorillonite, vermiculite, and several soils, but was adsorbed by kaolinite in one study (2). Soil pH was correlated (.951*) only with dicamba movement. Adsorption of picloram, simazine, and chlorpropham was negatively correlated with their observed mobilities and those of related compounds. The relationship between picloram adsorption and chlorpropham movement, and vice versa, is not understood.

Soil organic matter, in a group of nine nonmontmorillonitic soils, was significantly correlated with reduction of monuron and diuron movement. Atrazine was just below significance (—.664 vs. the required .666*). The remaining nonionic pesticides had r values much higher than those of ionic compounds. Clay was highly negatively correlated with monuron and diuron movement, and less well, with movement of other nonionic compounds. Although picloram is not significantly correlated with clay, the trend (—.356) is opposite that in the previous soil group (.394) suggesting that the kaolinite/vermiculite clay group is less negatively charged, as expected.

Soil pH was correlated with fenac R_F and nearly correlated (5% level) with 2,4-D R_F. This continues the general observation that pH is primarily related to movement of acidic compounds.

In contrast to montmorillonitic soils, water flux was correlated with movement of nearly all pesticides in the second soil group. From the previous discussion, this may reflect artificially strong influence of soils 10 and 40, especially as they are now in a group of 9 rather than 14 soils.

Adsorption of simazine, diuron, and chlorpropham was well correlated with reduced movement of chemically similar compounds. Picloram adsorption was related only to fenac mobility (10% level) in this soil group.

Multiple Linear Regression

Multiple linear regression equations were also developed (Table 5) for prediction of mobility and to determine the relative importance of the soil parameters, when considered together. Independent variables were added only so long as significance at the 10% level or better was indicated by the calculated F. The first parameter added is always that giving the largest r value.

It is significant to note that soil pH appears only among the organic acids as an important parameter affecting pesticide mobility. Although pH is absent from picloram's equation in Table 5, the next variable added to the regression equation would have been pH. This, however, is strongly influenced by Christiana loam. By omitting this soil the following regression equation was obtained for picloram R_F:

$$\hat{Y} = 0.99 - 0.0034 X^{***}_{CIPC\ ads} + 0.0035 X^{**}_{CEC} \quad (R^2 = 0.828)$$

The multiple coefficient of determination, R^2, indicates that 83% of the variation in picloram mobility can be predicted from data on chlorpropham (CIPC) adsorption and and CEC. The improvement is marked over the use of water flux ($R^2 = 0.416$) for 14 soils. Omission of Christiana soil had much less effect on regression equations for other compounds. Water flux remained a significant term in six equations.

The independent variables that appear in Table 5 are nearly always derived parameters; i.e., they are correlated with the fundamental soil components, clay and organic matter, and/or with soil pH. Since measurements of chlorpropham or simazine adsorption reflect interaction with several soil parameters, and since adsorption appears to be

Table 5—Multiple regression data relating pesticide mobility ($\hat{Y}$) with soil parameters (X), for 14 soils

Pesticide	Regression equation	R^2
Dicamba	$\hat{Y} = 0.80 + 0.035 X^*_{pH} - 0.0021 X_{clay}$	0.525
Picloram	$\hat{Y} = 0.80 + 0.0073 X^*_{flux}$	0.416
Fenac	$\hat{Y} = 0.07 + 0.100 X^{***}_{pH} + 0.0113 X^*_{flux}$	0.768
2,4-D	$\hat{Y} = 0.21 + 0.0087 X^{**}_{flux} + 0.099 X^*_{pH}$ $- 0.0066 X^*_{CIPC\,ads} + 0.0067 X_{simaz\,ads}$	0.864
Monuron	$\hat{Y} = 0.88 - 0.0123 X^{***}_{FMC} - 0.0040 X_{simaz\,ads}$ $+ 0.0048 X_{flux}$	0.920
Atrazine	$\hat{Y} = 0.73 - 0.0066 X^{***}_{FMC} - 0.0056 X_{simaz\,ads}$ $+ 0.0048 X_{flux}$	0.935
Diphenamid	$\hat{Y} = 0.10 - 0.0094 X^{***}_{FMC} + 0.0211 X^*_{picl\,ads}$ $- 0.0038 X^*_{simaz\,ads} + 0.0043 X^*_{diuron\,ads}$ $+ 0.0225 X^*_{flux} + 0.0186 X_{clay} - 0.0227 X_{CEC}$	0.988
Simazine	$\hat{Y} = 0.65 - 0.0107 X^{***}_{simaz\,ads} + 0.0120 X^{***}_{flux}$	0.918
Diuron	$\hat{Y} = 0.72 - 0.0183 X^{***}_{FMC}$	0.865
Chlorpropham	$\hat{Y} = 0.46 - 0.0134 X^{***}_{FMC} + 0.0082 X^{***}_{flux}$	0.938
Azinphosmethyl	$\hat{Y} = 0.43 - 0.0118 X^{***}_{FMC}$	0.724
Diquat	$\hat{Y} = -0.02 + 0.0057 X^{***}_{flux}$	0.654

*Significant at 5% level. **Significant at 1% level. ***Significant at 0.5% level.
Remaining independent variables are significant at 10% level.

a key factor affecting pesticide movement, it is not surprising that these data are useful predictors of mobility. Of the more common measurements of soil properties, field moisture capacity (FMC) is perhaps the most useful predictor of mobility.

When the regression equations of Table 5 are actually used to predict pesticide R_F, the average absolute deviation from the observed mobility was only 0.04. For all pesticides except fenac (deviation was 0.07), estimated R_F deviated from 0.02–0.05, averaged across 14 soils. If every independent variable in Table 3 is included, the average deviation is 0.02, indicating the improved accuracy of this prediction.

The average relative contribution of each variable to the R_F of a pesticide is expressed by comparing their standardized partial regression coefficients, b'. For dicamba movement in 14 soils, b' is 0.85 and —0.45 for pH and clay, respectively; pH is thus ca. twice as important as clay content. With 2,4-D, the order is chlorpropham adsorption > pH > simazine adsorption > water flux. These first three terms account for ca. 75% of soil-to-soil variability in 2,4-D mobility. For monuron and atrazine, the relative contributions were FMC > simazine adsorption > water flux. With simazine movement, b' values were —0.69 (simazine adsorption) and 0.42 (water flux).

ACKNOWLEDGEMENT

I thank E. J. Koch, Biometrical Services, ARS, USDA, Beltsville, Md. for his assistance with experimental design and statistical analysis.

LITERATURE CITED

1. Bailey, G. W., and J. L. White. 1970. Factors influencing the adsorption, desorption, and movement of pesticides in soil. Residue Rev. 32:29–92.
2. Burnside, O. C., and T. L. Lavy. 1966. Dissipation of dicamba. Weeds 14:211–213.
3. Coats, G. E., H. H. Funderburk, J. M. Lawrence, and D. E. Davis. 1966. Factors affecting persistence and inactivation of diquat and paraquat. Weed Res. 6:58–66.
4. Hamaker, J. W., C. A. I. Goring, and C. R. Youngson. 1966. Sorption and leaching of 4-amino-3,5,6-trichloropicolinic acid in soils. Advan. Chem. Ser. 60:23–37.
5. Hance, R. J. 1967. Relationship between partition data and the adsorption of some herbicides by soils. Nature (London) 214:630–631.
6. Harris, C. I. 1966. Adsorption, movement, and phytotoxicity of monuron and s-triazine herbicides in soil. Weeds 14:6–10.
7. Harris, C. I., and T. J. Sheets. 1965. Influence of soil properties on adsorption and phytotoxicity of CIPC, diuron, and simazine. Weeds 13:215–219.
8. Helling, C. S. 1970. Movement of s-triazine herbicides in soils. Residue Rev. 32:175–210.
9. Helling, C. S. 1971. Pesticide mobility in soils I. Parameters of soil thin-layer chromatography. Soil Sci. Soc. Amer. Proc. 35:732–737 (this issue).
10. Helling, C. S. 1971. Pesticide mobility in soils II. Applications of soil thin-layer chromatography. Soil Sci. Soc. Amer. Proc. 35:737–743 (this issue).
11. Helling, C. S., and B. C. Turner. 1968. Pesticide mobility: Determination by soil thin-layer chromatography. Science 162:562–563.
12. Hermanson, H. P., and C. Forbes. 1966. Soil properties affecting dieldrin toxicity to *Drosophila melanogaster*. Soil Sci. Soc. Amer. Proc. 30:748–752.
13. Jackson, M. L. 1958. Soil chemical analysis. Prentice-Hall Inc., Englewood Cliffs, N.J. 498 p.
14. Koren, E., C. L. Foy, and F. M. Ashton. 1969. Adsorption, volatility, and migration of thiocarbamate herbicides in soil. Weed Sci. 17:148–153.
15. McCarty, P. L., and P. H. King. 1966. The movement of pesticides in soils. Purdue Univ. Eng. Bull., Ext. Ser. no. 121, 156–171.
16. Snelling, K. E., J. A. Hobbs, and W. L. Powers. 1969. Effects of surface area, exchange capacity, and organic matter content on miscible displacement of atrazine in soils. Agron. J. 61:875–878.
17. Weber, J. B., P. W. Perry, and R. P. Upchurch. 1965. The influence of temperature and time on the adsorption of paraquat, diquat, 2,4-D and prometone by clays, charcoal, and an anion-exchange resin. Soil Sci. Soc. Amer. Proc. 29:678–688.

INFLUENCE OF WATER FLUX AND POROUS MATERIAL ON THE MOVEMENT OF SELECTED HERBICIDES[1]

J. M. Davidson, D. E. Rieck, and P. W. Santelmann[2]

ABSTRACT

The rate at which two substituted urea herbicides move through a water-saturated glass bead and uniformly packed soil column depends upon the water flux or average pore velocity. A 200-ml slug of 0.01N CaSO$_4$ solution containing either fluometuron (3-(m-trifluoromethylphenyl)-1,1-dimethylurea) or diuron (3-(3,4-dichlorophenyl)-1,1-dimethylurea) was introduced at one end of a 30-cm long column at a specific flux and the herbicide displaced through a column with 0.01N CaSO$_4$ at the same water flux. Effluent samples collected every 5-ml show fluometuron and the chloride ion to mix in a similar manner at the two fluxes studied. Experimental and calculated effluent concentration distributions did not agree when fluometuron was retained by the porous material.

Additional Key Words for Indexing: molecular diffusion, herbicide transport, fluometuron.

An understanding of the movement and retention of herbicides in the soil is of significant importance to future crop production. Because of potential chemical residue problems, considerable qualitative work has been conducted on the fate of specific herbicides in the soil (1, 2, 6, 10, 13, 14, 21). These studies suggest that the main parameters involved in herbicide movement through soils are: (i) water flux, (ii) molecular diffusion of the herbicide, (iii) retention properties of the medium, (iv) solubility of the herbicide, (v) soil-water content, and (vi) biological degradation.

The behavior of a herbicide in a soil cannot be easily described quantitatively if all of the above parameters are operative simultaneously. However, if the only effects are those of dispersion in the fluid flow phenomenon with equilibrium linear retention superimposed, the problem is simplified considerably. Models which most closely resemble the conditions of herbicide movement through porous materials are found in the theory of chromatography (15, 16, 19). Recent attempts to use chromatography models to describe the mixing and retention of solutes in soil have met with varying degrees of success. (4, 9, 18).

The differential equation frequently assumed to describe the process of longitudinal mixing (simultaneous mixing by molecular diffusion and convective flow) and linear retention in porous materials for one dimensional flow is:

$$D_o \frac{\partial^2 C}{\partial x^2} - v_o \frac{\partial C}{\partial x} - \left[1 + \frac{\rho K}{\theta}\right] \frac{\partial C}{\partial t} = 0 \quad [1]$$

where D_o is the apparent molecular diffusion coefficient (cm^2/hr), C the concentration of solute (g/cm^3), x the distance from the input (cm), v_o the average pore velocity of flow in the x direction (cm/hr) [$v_o = V/\theta$ where V is the water flux (cm/hr) and θ the volumetric water

---

[1] Contribution from the Department of Agronomy, Oklahoma Agr. Exp. Sta. as part of Projects S-1310 and S-1324. Journal Manuscript no. 1567. Received Dec. 14, 1967. Approved June 3, 1968.

[2] Associate Professor, Instructor, and Professor, respectively, Agronomy Department, Oklahoma State Univ., Stillwater.

content [cm^3/cm^3)], ρ the bulk density (g/cm^3), K the distribution coefficient (cm^3/g), and t the time (hr). The distribution coefficient is obtained by assuming the following linear relation between the quantity of solute in solution and that retained:

$$S = KC \qquad [2]$$

where S is the quantity of solute retained per unit weight of adsorbent and C the concentration of solute per unit volume of soil water. The distribution coefficient is assumed concentration independent over the herbicide concentration range in question. Also, the retention phenomenon described by equation [2] is applicable to all processes where a linear relation exists between the concentration of the chemical in the liquid phase and that in the reacted phase.

A problem of interest to agriculture is the uniform application of a herbicide solution to the surface of a semi-infinite body of soil. At a predesignated time, T, a herbicide-free solution is applied and the herbicide slug of $v_o T$ width displaced through the soil profile. The initial and boundary condition describing the herbicide application and subsequent displacement are:

(a) $C_{(x,o)} = 0$

(b) $v_o C_o = v_o C - D_o \frac{\partial C}{\partial x}$ (for $x = 0$, $0 < t < T$)

(c) $v_o C_{(o,t)} = 0$ (for $x = 0$, $t > T$)

$$[3]$$

where C_o represents the concentration of the incoming solution. The above boundary conditions at $x = 0$ are an improvement over those used by Lapadus and Amundson (15) in their solution of equation [1]. For example, the boundary condition at the inlet or soil surface requires only a constant flux of concentration C_o at $x = 0$ and not $C = C_o$ at $x = 0$ for times greater than zero but less than T. The latter constraint may be difficult to achieve for small water fluxes and solutes and adsorbents with large distribution coefficients.

The solution to equation [1] subject to the initial and boundary conditions [3] and for times greater than T is:

$$\frac{C}{C_o} = \frac{1}{2} \left\{ \mathrm{erfc}\left[\frac{x - vt}{\sqrt{4Dt}}\right] - \mathrm{erfc}\left[\frac{x - v(t-T)}{\sqrt{4Dt}}\right] \right.$$

$$+ 2v\sqrt{\frac{t}{D\pi}} \exp - \left[\frac{x - vt}{\sqrt{4Dt}}\right]^2$$

$$- 2v\sqrt{\frac{(t-T)}{D\pi}} \exp - \left[\frac{x - v(t-T)}{\sqrt{4D(t-T)}}\right]^2$$

$$- \frac{V}{D}\left[\frac{D}{V} + vt + x\right] \exp\left[\frac{xv}{D}\right] \mathrm{erfc}\left[\frac{x + vt}{\sqrt{4Dt}}\right]$$

$$+ \frac{V}{D}\left[\frac{D}{V} + v(t-T) + x\right] \exp\left[\frac{xv}{D}\right]$$

$$\left. \mathrm{erfc}\left[\frac{x + v(t-T)}{\sqrt{4D(t-T)}}\right] \right\} \qquad [4]$$

where

$$D = \frac{D_o}{1 + (\rho K/\theta)} \quad \text{and} \quad v = \frac{v_o}{1 + (\rho K/\theta)} \qquad [5]$$

The quantity $1 + (\rho K/\theta)$ in equation [5] is called the retardation factor by Hashimoto et al. (12). The value $\rho K/\theta$ represents the apparent increase in pore volume owing to the linear retention process. It should be noted that D and v are the two parameters that determine the shape and position of the distribution curve obtained from equation [4]. Equation [4] is a modification of Brenner's (5) solution of equation [1] subject to equations [3a] and [3b]. Brenner did not treat the problem of solute retention.

The objective of this investigation was to illustrate the importance of water flow rate and retention on herbicide movement, and using an existing mathematical model, attempt a quantitative description of herbicide movement through a porous medium.

EXPERIMENTAL

An apparatus similar to that described by Nielsen and Biggar (17) was used to confine the porous material in a glass cylinder between two fritted glass bead plates. The cross sectional area and length of the column were 43.2 cm^2 and 30 cm, respectively. The flow rate was controlled with a constant volume pump and 5-ml effluent samples were collected sequentially at atmospheric pressure with an automatic fraction collector. Volume of solution passing through the column with time was read from a burette connected to the column through the pump.

A calcium-saturated Norge loam soil and 250-μ glass beads were the porous materials used in this investigation. A sample of the air-dry loam, previously screened through a 2-mm sieve, was uniformly packed in the container and then saturated with a 0.01N CaSO$_4$ solution. The glass beads were packed in 0.01N CaSO$_4$ solution to insure complete water saturation. Each material was maintained water saturated for the duration of the experiment.

The solution in the reservoir behind the inflow fritted plate and all glass tubes between the burette and plate were flushed thoroughly when a change from herbicide free to herbicide solution was made or vice versa. When the solution change-over was conducted, the outlet tube was closed to prevent a displacement of the herbicide during the change-over to the new solution.

The herbicides used in this study were fluometuron (3-(m-trifluoromethylphenyl))-1,1-dimethylurea) and diuron (3-(3,4-dichlorophenyl)-1,1-dimethylurea), both substituted ureas. Solutions of 29 ppm of herbicide in 0.01N CaSO$_4$ or CaCl$_2$ were used. Each herbicide was in the commercial 80% wettable powder form. A 200-ml slug of herbicide solution was introduced into each column, previously saturated with 0.01N CaSO$_4$, at a preset flux. After 200 ml of herbicide solution had been added, the system from the inflow plate reservoir to a second burette was flushed with 0.01N CaSO$_4$ and the flow rate was resumed until the end of the study. The analysis of the chloride ion in the effluent was by titration with AgNO$_3$. The herbicide concentration was measured using a UV spectrophotometer at a wavelength setting of 242 mμ for fluometuron and 247.5 mμ for diuron. Effluent samples containing fluometuron from the soil were mixed on an equal volume basis with *n*-pentane and the *n*-pentane was analyzed for the herbicide (7). The latter procedure eliminated interference problems caused by dissolved substances in the soil-water effluent samples.

Retention isotherms were determined from 1:1 extracts using 20 g of the loam and glass beads and 20 ml of various concentrations of fluometuron in 0.01N $CaSO_4$. Each sample was shaken for 2 hours and centrifuged, and the supernatant liquid was analyzed. Previous studies of fluometuron retention by the Norge loam with time showed 2 hours was more than adequate for equilibrium. Blanks consisting of 0.01N $CaSO_4$ plus herbicide solution were included in each determination; the difference in herbicide concentration between the blanks and the supernatant was attributed to retention.

Pore volume was measured by oven drying the saturated material at 105C for 48 hours at the completion of each study. The volume of solution in the input fritted plate and effluent plate and reservoir was measured and subtracted from the total effluent volume for a measurement of the number of pore volumes displaced through the column.

RESULTS AND DISCUSSION

The 200-ml slug of solution is equivalent to applying 2.206 kg/ha and simulates a field situation in which the material is dissolved and entering the porous medium at a constant rate. The procedure also provides a measurement of the amount of chemical retained between the input and effluent end of the column. If the concentration for kill or reduction in plant growth is known, then ideally, a calculation of the amount of water necessary for this concentration to be at a specific depth can be made from previously measured parameters.

The fluometuron and chloride concentration distribution from 250-μ glass beads for a displacement velocity (average pore water velocity) of 4.81 cm/hour is given in Fig. 1. The pore volume of the 250-μ glass bead system was 480 cc. The two materials appear to mix at the same rate at the high velocity where the influence of molecular diffusion is minimal. The solid line was calculated from equation [4] for a D_0 value of 0.123 cm^2/hour and a distribution coefficient of zero. The shape of the curve is good, but the data are displaced to the left of the calculated line. This same displacement has been observed by Biggar and Nielsen (3) and Nielsen and Biggar (18), and represents possibly an excluded volume to the transport process. The displacement of the data to the left of one pore volume is a relative measure of the volume of water not displaced within the sample. This area has been described as "holdback" by Danckwerts (8).

Reducing the average pore water velocity 20-fold (0.252 cm/hr) in the 250-μ glass beads significantly changes the chloride and fluometuron concentration distribution (Fig. 2). Again, the chloride and fluometuron appear to mix at the same rate as previously illustrated in Fig. 1. The solid line was calculated from equation [4] for a D_0 value of 0.0426 cm^2/hour. Note the relative concentration peak for fluometuron occurs at a larger pore volume than the chloride and also note the slight skewness of the peak. The skewness and transposition of the fluometuron concentration distribution peak is characteristic of retention but an integration under the fluometuron data reveals that

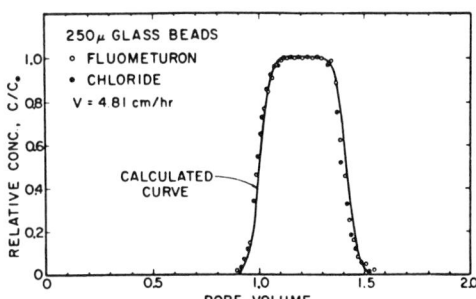

Fig. 1—Experimental and calculated relative fluometuron and chloride concentration distribution for saturated 250-μ glass beads with an average pore water velocity of 4.81 cm/hour.

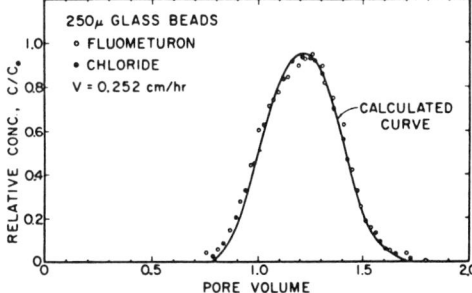

Fig. 2—Experimental and calculated relative fluometuron and chloride concentration distribution for saturated 250-μ glass beads with an average pore water velocity of 0.252 cm/hour.

2.0% more fluometuron was recovered in the effluent than added. The shape of the fluometuron concentration distribution at the higher fluometuron concentrations was reproducible and is taken as real. A continuous input of fluometuron, contrasted to the slug input, revealed the skewed line was extremely asymptotic in its approach of $C/C_0 = 1$ which is also characteristic of retention by the porous material.

A similar apparent diffusion coefficient for fluometuron and chloride was not anticipated, since the molecular weight of fluometuron is 232.2, over six and one-half times larger than the chloride ion. Because of this, chloride and fluometuron displacements were made through 250-μ glass beads in the absence of one another; identical results were obtained indicating no measurable chemical or physical interaction in the displacements shown in Fig. 1 and 2. The data in Fig. 2 again leads the calculated curve as illustrated in Fig. 1 but not by the same magnitude, indicating a better mixing of the total pore volume at the lower flux.

The retention isotherms for two porous materials and fluometuron are given in Fig. 3. The solid lines were obtained using regression analyses. From the retention isotherms, the distribution coefficient was calculated (equation 2) and an indirect measure of the amount of herbicide retained during the 200-ml displacement was calculated using equation [4]. It is interesting to note that the 250-μ

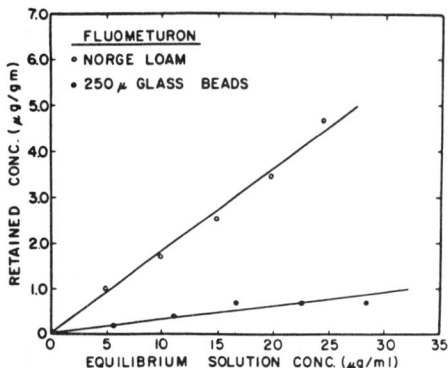

Fig. 3—Retention of fluometuron on calcium-saturated Norge loam and 250-μ glass beads.

glass bead and fluometuron system shows retention (Fig. 3); this is not obvious from Fig. 2 except perhaps in the shape and location of the maximum fluometuron concentration. For an equilibrium system, the results of Fig. 3 justify the earlier assumption that the distribution coefficient is concentration independent and retention is linear over the concentration range in question. Linear retention isotherms for other herbicides have been illustrated by Harris and Warren (11) and Talbert and Fletchall (20).

Comparison of the differences in the amount of retention of diuron and fluometuron, both substituted ureas, in a saturated 250-μ glass bead system is shown in Fig. 4. The diuron concentration in the effluent is displaced to the right of fluometuron owing to a greater retention. Note the clockwise rotation of the left leg of the diuron distribution curve and the flatter peak with increased retention. The position of the diuron data with respect to fluometuron indicates a larger distribution coefficient for diuron than fluometuron. The area under the two curves is representative of the amount of herbicide retained in the porous material. The distribution of the retained material along the column, however, was not measured. Both solid lines were eye-fit to the data points.

Equation [4] does not, when applied to effluent data, predict the amount of herbicide retained by a reduction in area under the curve as does the experimental data. Therefore, for all calculated curves presented in this manuscript where retention is operative, the shape and position of the left leg of the distribution curve is the only portion that should be used to describe herbicide retention. The area under all calculated curves is equal to the 200-ml slug added.

Figure 5 shows the real and calculated relative fluometuron concentration distributions using the distribution coefficient for fluometuron and 250-μ glass beads (Fig. 3). The D_o value used was the same as that used for the chloride and fluometuron in Fig. 2. The measured retardation factor for this system is 1.124. This represents an increase in pore volume of 0.124 from the measured value (Fig. 5). Recent work (9, 12) where theory was "fitted" to the experimental data, has shown a satisfactory description of portions of the data. However, independent measurements of the distribution coefficient and retardation factor were not used by these researchers and are necessary before an acceptance of the model is justified. Results of Biggar and Nielsen (3) for exchange of Mg^{2+} for Ca^{2+} using an independent measurement of the exchange constant in the model proposed by Lapadus and Amundson (15) over estimated adsorption in a similar way to that shown in Fig. 5.

The mixing and retention of fluometuron in a Norge loam is illustrated in Fig. 6. The pore volume of the Norge loam system was 450 cc. Calculated curve 1 was "fitted" to the data disregarding the retention isotherm (Fig. 3). A D_o value of 1.016 cm²/hour and retardation factor of 1.62 were found to describe the left-hand portion of the data satisfactorily. However, when the measured retardation factor (1.93) for fluometuron with Norge loam was used (calculated curve 2), a greater retention was predicted (Fig. 6). This is in agreement with the results obtained for glass beads (Fig. 5). This problem has been treated by Biggar et al. (4) using a model employing equilibrium conditions in various size soil plates and "holdback." The procedure requires an estimate of the "holdback" which at present would appear difficult. Previous work with the Norge loam using $0.01N$ $CaCl_2$ and $CaSO_4$ suggest this soil possesses a large "holdback" volume.

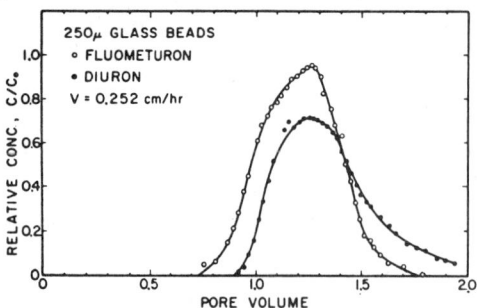

Fig. 4—Relative fluometuron and diuron concentration distributions for saturated 250-μ glass beads with an average pore water velocity of 0.252 cm/hour.

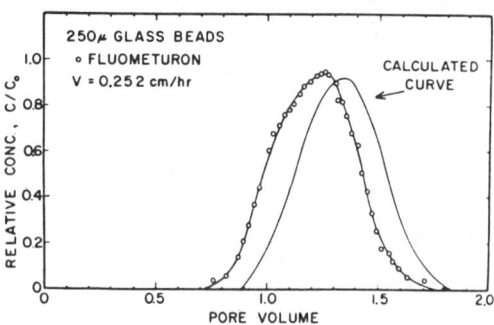

Fig. 5—Experimental and calculated relative fluometuron concentration distribution for saturated 250-μ glass beads with an average pore water velocity of 0.252 cm/hour. Calculated curve is based on retention isotherm data.

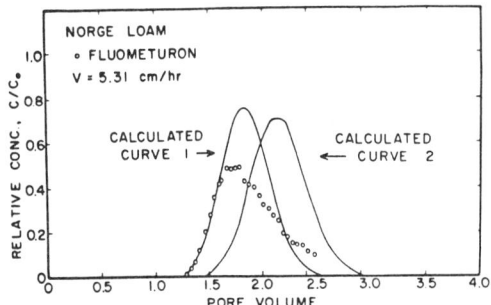

Fig. 6—Experimental and calculated relative fluometuron concentration distributions for saturated Norge loam with an average pore water velocity of 5.31 cm/hour. Calculated curve 1 is theory "fitted" to the data and curve 2 is based on retention isotherm (Fig. 3).

It is well to recall that linear retention theory requires a fixed amount of retention per gram of porous material for a specific herbicide concentration. It further requires that the total soil mass be in equilibrium with the solution in the porous material. These conditions may not be met, however, for the soil mass in the above miscible displacement study. When the herbicide first enters the soil column at a concentration C_o, it is retained; the solution containing the remaining herbicide moves forward at a lower concentration to be retained and so on, until it traverses the complete length of the soil mass. This procedure allows a large quantity of herbicide to be retained at the input, but a smaller amount at the effluent end. This lack of equilibrium or failure of the retained herbicide concentration to be the same throughout the total soil mass offers a possible explanation for the early arrival observed for fluometuron in glass beads and Norge loam to that calculated using the retention isotherm results. This may be further complicated by the mixing processes. For example, the flow process may cause the concentration in the liquid phase to change more rapidly than the porous material can adjust and establish an equilibrium condition. A knowledge of the distribution of the retained phase with soil depth would assist in evaluating this problem.

CONCLUSIONS

Fluometuron, a substituted urea, was shown to have an apparent molecular diffusion coefficient in 250-μ glass beads and Norge loam similar to that of the chloride ion. This is of particular interest since this material is commonly used as a pre-emergence herbicide and must permeate the soil mass quickly for effective weed control. Also, owing to its high mobility and low retention within the soil mass, a reduction in weed control could be anticipated following a heavy rain or irrigation. This information is of significant importance to frequency and application rate of herbicides to irrigated soils.

Retention isotherms were linear for fluometuron in 250-μ glass beads and Norge loam with the soil retaining more material per weight than the glass beads. However, the experimental results show the herbicide, fluometuron, appearing in the effluent earlier than predicted by the selected model. Further work on the distribution of retained material in the soil with depth following an exchange process is necessary to adequately describe the movement of fluometuron through a soil.

It can be concluded that fluometuron was sufficiently influenced by water flux and pore geometry to necessitate a knowledge of these parameters before estimating the amount of herbicide movement through a soil. Also, two herbicides from the same chemical family were found to have distinctly different soil retention properties.

LITERATURE CITED

1. Ashton, F. M. 1961. Movement of herbicides in soil with simulated furrow irrigation. Weeds 9:612-619.
2. Bayer, D. E. 1967. Effect of surfactants on leaching of substituted urea herbicides in soil. Weeds 15:249-252.
3. Biggar, J. W., and D. E. Nielsen. 1963. Miscible displacement: V. Exchange processes. Soil Sci. Soc. Amer. Proc. 27:623-627.
4. Biggar, J. W., D. R. Nielsen, and K. K. Tanji. 1966. Comparison of computed and experimentally measured ion concentrations in soil column effluents. Trans. Amer. Soc. Agr. Eng. 9:784-787.
5. Brenner, H. 1962. The diffusion model of longitudinal mixing in beds of finite length. Numerical values. Chem. Eng. Sci. 17:229-243.
6. Burnside, O. C., C. R. Fenster, and G. A. Wicks. 1963. Dissipation and leaching of monuron, simazine, and atrazine in Nebraska soils. Weeds 11:209-213.
7. Davidson, J. M., C. E. Rieck, and P. W. Santelmann. 1968. Quantitative extraction of fluometuron from water samples by n-Pentane. J. Weed Sci. 16:356-358.
8. Danckwerts, P. V. 1953. Continuous flow systems: distribution of residence times. Chem. Eng. Sci. 2:1-13.
9. Elrick, D. E., K. T. Erh, and H. K. Krupp. 1966. Application of miscible displacement techniques to soils. Water Resources Res. 2:717-727.
10. Harris, C. I., and G. F. Warren. 1964. Adsorption and desorption of herbicides in soil. Weeds 12:120-126.
11. Harris, C. I. 1967. Movement of herbicides in soils. Weeds 15:214-216.
12. Hashimoto, I., K. B. Deshpande, and H. C. Thomas. 1964. Peclet numbers and retardation factors for ion exchange columns. Ind. and Eng. Chem. Fundamentals. 3:213-218.
13. Hill, G. D., J. W. McGahen, H. M. Baker, D. W. Finnerty, and C. W. Bingeman. 1955. The fate of substituted urea herbicides in agricultural soils. Agron. J. 47:93-104.
14. Lambert, S. W., P. E. Porter, and R. H. Schieferstein. 1965. Movement and sorption of chemicals applied to the soil. Weeds 13:185-190.
15. Lapadus, L., and N. R. Amundson. 1952. Mathematics of adsorption in beds. VI. The effect of longitudinal diffusion in ion exchange and chromatographic columns. J. Phys. Chem. 56:984-988.
16. Littlewood, A. B. 1962. Gas chromatography. Academic Press, New York.
17. Nielsen, D. R., and J. W. Biggar. 1961. Miscible displacement in soils: I. Experiment information. Soil Sci. Soc. Amer. Proc. 25:1-5.
18. Nielsen, D. R. and J. W. Biggar. 1962. Miscible displacement: III. Theoretical considerations. Soil Sci. Soc. Amer. Proc. 26:216-221.
19. Purnell, H. 1962. Gas chromatography. Academic Press, New York.
20. Talbert, R. E., and G. H. Fletchall. 1965. The adsorption of some s-triazines in soils. Weeds 13:46-52.
21. Wiese, A. F., and R. G. Davis. 1964. Herbicide movement in soils with various amounts of water. Weeds 12:101-103.

Mass Transfer Studies in Sorbing Porous Media: III. Experimental Evaluation with 2,4,5-T

M. TH. VAN GENUCHTEN,[2] P. J. WIERENGA,[3] AND G. A. O'CONNOR[3]

ABSTRACT

Comparisons are made between observed and calculated effluent concentration distributions for the movement of 2,4,5-T (2,4,5-trichlorophenoxyacetic acid) through 30-cm long unsaturated soil columns. The comparisons are made using both analytical and numerical solutions of a previously published model, which included the effects of intra-aggregate diffusion and adsorption. The results in this study indicate that intra-aggregate diffusion and adsorption/desorption are the main mechanisms responsible for effluent tailing. An estimated 60% of the adsorption was found to occur in the stagnant region of the soil. When intra-aggregate diffusion was included in the model, the observed adsorption/desorption hysteresis phenomenon found to be significant in several earlier studies, was shown to be much less important in describing the observed concentration distributions.

Additional Index Words: intra-aggregate diffusion, adsorption-desorption, hysteresis, miscible displacement, 2,4,5-T.

IN PART I OF THIS STUDY (van Genuchten and Wierenga, 1976a), a theoretical model was developed to describe the movement of chemicals through unsaturated, aggregated, sorbing porous media. Analytical solutions were derived for both the mobile and immobile liquid regions of the medium, assuming the presence of a linear equilibrium isotherm. These solutions were subsequently used in part II (van Genuchten and Wierenga, 1977; this issue) to describe tritium movement through unsaturated, aggregated Glendale clay loam. In this paper, part III of the study, the theoretical model will be used to describe experimental data on the movement of the herbicide 2,4,5-T (2,4,5-trichlorophenoxyacetic acid) through the same Glendale soil. As was mentioned in Part I, no analytical solutions exist when either the adsorption isotherm is nonlinear, or when the adsorption-desorption process is not single-valued (hysteretic). For these conditions numerical techniques must be used. The numerical solutions used in this analysis were programmed in IBM S/360 CSMP as described by van Genuchten and Wierenga (1974; 1976b).

[1]Journal article no. 568, Agric. Exp. Stn., New Mexico State Univ., Las Cruces, NM 88003. The work upon which this report was based was supported in part by funds obtained from the U.S. Dep. of the Interior, Office of Water Resources Research, as authorized under the Water Resources Act of 1964. Received 8 March 1976. Approved 9 Nov. 1976.
[2]Formerly Graduate Student, Dep. of Agronomy, New Mexico State Univ.; presently Research Associate, Dep. of Civil Engineering, Princeton Univ., Princeton, NJ 08540.
[3]Associate Professor, Dep. of Agronomy, New Mexico State Univ., Las Cruces, NM 88003.

Table 1—Soil-physical data for data for various displacements through Glendale clay loam.

Exp. no.	Tracer	Bulk density, ρ	Water content, θ	Flux, q	Pulse period T_1	t_1	Largest aggregate size
		g/cm³	cm³/cm³	cm/day		days	mm
			Experiment 1				
1-1	³H	1.360	0.460	5.09	2.334	6.331	2.0
1-2	³H	1.360	0.460	5.09	0.512	1.389	2.0
1-4	2,4,5-T	1.360	0.473	5.11	2.761	7.672	2.0
1-5	³H	1.360	0.460	2.55	1.990	10.76	2.0
			Experiment 2				
2-1	³H	1.361	0.464	4.55	2.361	7.215	2.0
2-2	³H	1.361	0.468	4.57	0.452	1.389	2.0
2-4	2,4,5-T	1.361	0.479	4.59	2.460	7.708	2.0
2-5	³H	1.361	0.467	2.29	1.762	10.77	2.0
			Experiment 3				
3-2	³H	1.222	0.454	17.0	3.102	2.475	6.3
3-3	2,4,5-T	1.222	0.475	17.0	3.994	3.212	6.3
3-4	³H	1.222	0.445	4.20	1.909	6.076	6.3
3-6	³H	1.222	0.434	1.32	1.788	9.653	6.3
			Experiment 4				
4-1	2,4,5-T	1.309	0.456	16.8	4.948	4.028	6.3
4-2	³H	1.309	0.433	4.04	1.871	6.021	6.3

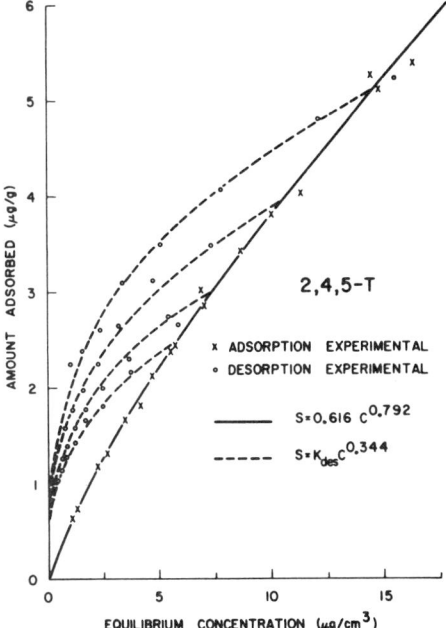

Fig. 1—Equilibrium adsorption and desorption data for 2,4,5-T sorption on Glendale clay loam.

MATERIALS AND METHODS

The same experimental setup as discussed in part II (van Genuchten and Wierenga, 1977) was used to study the movement of 2,4,5-T through 30-cm long soil columns uniformly packed with the same aggregated (Glendale clay loam soil as was used for the tritium experiments. Concentrations of 10 ppm of 2,4,5-T were used for the leaching studies. Radioactive 2,4,5-T with a specific activity of 4.93 µCi/mg, and labeled at the carboxyl position was added to unlabeled herbicide solution. Activities of the 2,4,5-T¹⁴C were determined on a Packard TriCarb Model 3310 liquid scintillation counter, using Aquasol (New England Nuclear, Boston, Mass.) as the counting medium. Relative effluent concentrations (c_m) were obtained by dividing the herbicide concentrations in the effluent (C_m) by the concentration of the herbicide pulse (C_o) applied to the columns.

A total of five runs were made with four columns, using Glendale clay loam of two different aggregate sizes and employing soil water fluxes (q), ranging from 4.0 to 17.0 cm/day. The experimental data for the various 2,4,5-T displacements are given in Table 1. Included in this table are the soil physical data of several tritium displacements at similar water contents and flow velocities through the same columns. The experiments in Table 1 are numbered by column and flux (q).

The equilibrium adsorption isotherm for 2,4,5-T and the Glendale clay loam was determined using duplicate 1:1 mixtures of 25 g of soil and 25 ml of various herbicide concentrations in 0.01N CaCl₂. Each sample was shaken for 24 hours and then centrifuged at 1,500 rpm for 15 min. The 24-hour shaking period was experimentally determined to be more than sufficient time to obtain equilibrium between sorbed and solution concentrations (G. A. O'Connor, 1972)[4]. Ten milliliters of the supernatant was removed and analyzed for ¹⁴C activity and the count rates were converted to herbicide concentrations. The difference between the original herbicide concentration and that in the supernatant was assumed to be the result of adsorption.

Equilibrium desorption isotherms were obtained using the same duplicate 1:1 soil-to-herbicide solution mixtures used for the adsorption study. The 10-ml supernatant sample removed from the centrifuged samples to determine the initial adsorption was replaced by 10 ml of 2,4,5-T-free 0.01N CaCl₂. The soil was loosened from the tubes by vigorous hand shaking. The samples were then shaken mechanically for 24 hours to establish a new equilibrium, centrifuged, and another 10-ml extraction and analysis of the supernatant was made. This procedure was followed for each successive desorption step. In total, seven data points were determined for each of four desorption curves.

The adsorption studies for 2,4,5-T were repeated using two different soil solution ratios: 20 g of soil mixed with 30 ml of herbicide solution, and 30 g of soil with 20 ml of herbicide solution. No differences in adsorption were found. The adsorption points obtained with these two additional adsorption studies were used, together with the earlier results to construct the equilibrium adsorption isotherm.

RESULTS AND DISCUSSION

The equilibrium adsorption and desorption data for 2,4,5-T and Glendale clay loam are given in Fig. 1. The data show that adsorption and desorption cannot be described by the same relation, and hence, are nonsingle valued (hysteretic). Both the adsorption and desorption isotherms could be described by Freundlich-type equations of the form:

$$S = KC^N. \qquad [1]$$

The coefficients K and N in Eq. [1] are different for adsorption and desorption. Values for K and N are listed in Table 2; they were obtained following the procedures outlined by Hornsby and Davidson (1973) and van Genuchten et al. (1974). In this study a value of 2.3 for the ratio N_{ads}/N_{des} was obtained, as shown by the data in Table 2. This value was found to be independent of the maximum concentration (C_{max}) before desorption is initiated. Similar values were reported in several earlier studies on the adsorption and desorption of several pesticides from different soils (Swan-

[4] G. A. O'Connor. 1972. Soil adsorption and desorption of 2,4,5,-T. Agron. Abstr. p. 92.

Table 2—Parameters used in Eq. [1] to describe the adsorption and desorption of 2,4,5-T on Glendale clay loam. C_{max} is the solution concentration before desorption is initiated. The subscripts "ads" and "des" indicate adsorption and desorption, respectively.

Process	Chemical	K_{ads}	N_{ads}	K_{des}	N_{des}
Adsorption	2,4,5-T	0.616	0.792	--	--
Desorption	2,4,5-T				
C_{max} = 5.87 µg/cm³		--	--	1.36	0.344
C_{max} = 7.41 µg/cm³		--	--	1.51	0.344
C_{max} = 10.30 µg/cm³		--	--	1.75	0.344
C_{max} = 14.35 µg/cm³		--	--	2.03	0.344

son and Dutt, 1973; Hornsby and Davidson, 1973; van Genuchten et al., 1974). Somewhat smaller values (closer to one) however, were obtained by Farmer and Aochi (1974) and Wood and Davidson (1975).

Observed and calculated 2,4,5-T effluent curves for column experiment number 2-4 are shown in Fig. 2. The solid line represents the analytical solution for the concentration in the mobile liquid (c_m). Since this solution requires a linear adsorption isotherm (see van Genuchten and Wierenga, 1976a), it was necessary to approximate the nonlinear adsorption isotherm in Fig. 1 by a straight line. The following procedure was used to linearize the equilibrium isotherm. We denote the linearized isotherm by $S = K^1_{ads} C$ and require that the areas under the isotherms over the range 0-10 ppm (all columns were leached with 10-ppm solutions) for both the linearized and the nonlinear Freundlich equations be the same (Fig. 1). Thus

$$\int_0^{10} K^1_{ads} C \, dC = \int_0^{10} 0.616 C^{0.792} \, dC \quad [2]$$

where K^1_{ads} (= 0.426) is referred to as the linearized adsorption constant. From the value of 0.426 a retardation factor R (for notation see the appendix) of 2.210 was calculated. A three-parameter curve fitting was then carried out with the previously described GAUSHAUS nonlinear curve fitting program (van Genuchten and Wierenga, 1977) to obtain estimates for the dimensionless parameters β, P, and $\bar{\alpha}$ which appear in the analytical solution. Results obtained for four 2,4,5-T displacements are given in Table 3. Included in the table are data obtained from several tritium displacements through the same columns. The fraction of the adsorption sites located in the dynamic region of the soil (f), was subsequently calculated for each 2,4,5-T displacement using Eq. [17] of van Genuchten and Wierenga (1977) i.e.

$$\beta R = \phi + f(R - 1) \quad [3]$$

or

$$f = \frac{\beta R - \phi}{R - 1}. \quad [4]$$

For the data of experiment no. 2-4, one obtains

$$f = \frac{(0.596)(2.21) - (0.862)}{(1.21)} = 0.376. \quad [5]$$

The solid line in Fig. 2 was calculated with the analytical solution and shows a fairly good agreement with the experimental data. The dashed-dotted line was obtained with the numerical solution of van Genuchten and Wierenga (1976b), using the value of f as calculated above and including the nonlinearity of the adsorption isotherm. The differences between the nonlinear (numerical) model and the linear (analytical) model are rather small. The breakthrough part of the curve becomes somewhat steeper and the peak concentration decreases when the nonlinearity is taken into account. These two effects of the nonlinear isotherm on the calculated curves are easily explained by considering the retardation factor R_m of the dynamic region (van Genuchten and Wierenga, 1976a)

$$R_m = 1 + f \rho K N C_m^{N-1} / \theta_m. \quad [6]$$

Since $N < 1$, R_m will decrease with increasing concentration C_m. Retardation of the chemical will be more extensive at the lower and less extensive at the higher concentrations as compared to the linear case. The net result will be a steeper

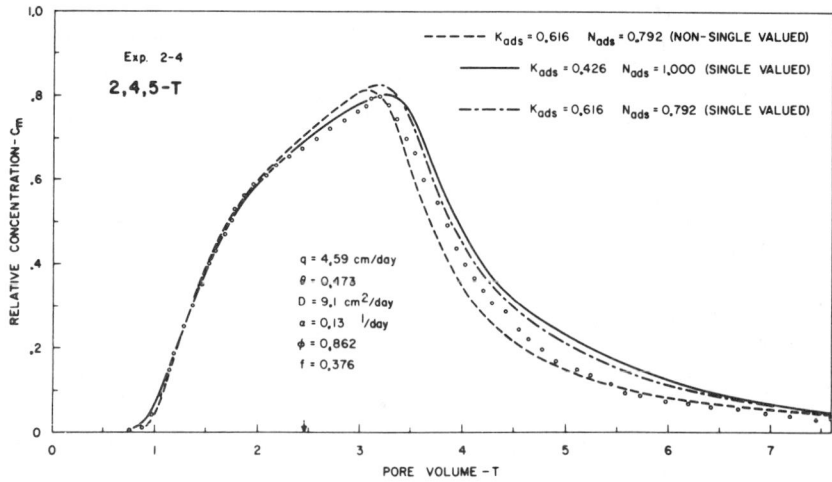

Fig. 2—Observed and calculated 2,4,5-T effluent curves for experiment 2-4. The open circles represent observed data points.

Table 3—Summary of various 2,4,5-T and tritium displacements through Glendale clay loam. The tritium data were obtained from van Genuchten and Wierenga (1977). The superscripts l and f indicate linearization and fitted, respectively.

Exp. no.	Tracer	R	β	P	$\bar{\alpha}$	f	ϕ	D cm^2/day	α 1/day	R^f	K^f_{ads}	K^f_{ads}/K^l_{ads}
1-1	^{3}H	1.027	0.926	95	1.47	0.400†	0.940	3.7	0.25	..	0.009	..
1-2	^{3}H	1.027	0.926	95	1.47	0.400†	0.940	3.7	0.25	..	0.009	..
1-4	2,4,5-T	2.225	0.661	41	0.55	0.433	0.940	8.4	0.09	2.301	0.453	1.06
1-5	^{3}H	1.027	0.928	95	0.56	0.400†	0.942	1.9	0.048	..	0.009	..
2-1	^{3}H	1.026	0.850	45	1.50	0.400†	0.862	7.6	0.23	..	0.009	..
2-2	^{3}H	1.026	0.887	45	1.31	0.400†	0.900	7.2	0.20	..	0.009	..
2-4	2,4,5-T	2.210	0.596	37	0.88	0.376	0.862	9.1	0.13	2.279	0.450	1.06
2-5	^{3}H	1.026	0.830	45	1.05	0.400†	0.841	3.9	0.080	..	0.009	..
3-2	^{3}H	1.024	0.841	56	0.39	0.400†	0.852	24.0	0.22	..	0.009	..
3-3	2,4,5-T	2.139	0.616	19	0.21	0.409	0.852	68.0	0.12	1.855	0.338	0.79
3-4	^{3}H	1.025	0.872	56	0.72	0.400†	0.884	5.7	0.10	..	0.009	..
3-6	^{3}H	1.025	0.717	56	1.96	0.400†	0.725	2.2	0.086	..	0.009	..
4-1	2,4,5-T	2.223	0.602	20	0.39	0.399	0.850	64.0	0.22	2.476	0.489	1.15
4-2	^{3}H	1.027	0.788	57	0.93	0.400†	0.798	6.2	0.12	..	0.009	..

† Assumed to be 0.400.

front and a higher peak concentration for the curve based on the nonlinear model.

Figure 2 also shows the effect of the observed hysteresis on the numerical results. The break-through side of the effluent curve is not affected by the hysteresis, but the relative concentration during elution decreases much faster and more prolonged tailing occurs at the higher pore volumes if hysteresis is taken into account. From Fig. 2 one may conclude that both the nonlinearity of the adsorption isotherm, as well as the observed hysteresis do not affect the shape and position of the effluent curve to any great extent. All three calculated curves describe the experimental data reasonably well. Note that an f-value of 0.376 was obtained from the data presented in Fig. 2. This indicates that more than 60% of the adsorption occurs in the stagnant region of the soil.

The dispersion coefficient D used to construct the three curves in Fig. 2 is higher than the dispersion coefficient used to describe tritium displacement through the same column at the same velocity. Table 3 shows that the Peclet numbers based on the 2,4,5-T displacements are smaller than those determined from the tritium data. Theoretically, only one value of the Peclet number should hold for all displacements through the same column, independent of the tracer used, provided the flow velocities are sufficiently high that longitudinal diffusion is negligible compared to mechanical dispersion (Perkins and Johnson, 1963). The discrepancies between the Peclet numbers for 2,4,5-T and tritium are in our opinion due to the existence of kinetic phenomena during adsorption of the herbicide. This study assumes equilibrium adsorption, a condition which may not have been met during the leaching experiments. In a study of the equilibrium times for 2,4,5-T and Glendale clay loam, G. A. O'Connor (1972)[4] showed that although the initial adsorption rate was very fast, complete adsorption could be obtained only after herbicide and soil were shaken for several hours. When a kinetic adsorption mechanism is present, the front portion of the effluent curve will be less steep, as shown in previous studies (van Genuchten et al., 1974; van Genuchten and Wierenga, 1974). By neglecting the kinetic effects and using an equilibrium instead of a kinetic model, a higher dispersion coefficient will hence be necessary in order to obtain the same description of the front portion of the curve. It should be mentioned that in addition to the occurrence of kinetic phenomena, the curve-fitted value of the Peclet number will be influenced also by the fact that a linearized adsorption relation is used in the model, instead of the nonlinear one observed. However, this effect is small as shown by the differences between the solid and dashed lines in Fig. 2. It should be further noted that a slightly different value of D will somewhat affect the value of the mass transfer coefficient α that is obtained; however, it will not influence the value of f obtained with the curve-fitting procedure.

The solid line in Fig. 2 was obtained with the linear adsorption model, using a value of 0.426 for the (linearized) adsorption constant, obtained from the batch equilibrium studies. This coefficient may also be estimated directly from the effluent curve, using a four-parameter rather than a three-parameter curve fit. Results for experiment No. 1-4 are shown in Fig. 3, where the solid line was calculated with a fitted adsorption constant of 0.453, while the dashed line was obtained with the linearized adsorption constant of 0.426. Both calculated curves are essentially the same. Note that the values of the parameters D, α, and f obtained with the four-parameter curve fitting are also very close to those determined with the three-parameter fitting. Values of the fitted adsorption constant (K^f_{ads}) for the four 2,4,5-T displacements are listed in Table 3. The values of K^f_{ads} agree reasonably well with the linearized adsorption constant K^l_{ads}, the average K^f_{ads} being nearly identical to K^l_{ads}. This suggests that, when assuming a linear adsorption isotherm, the adsorption constant can be estimated with reasonable accuracy from the effluent data. However, Green et al. (1972) in a study on the movement of picloram through two aggregated soils obtained much higher values for K^f_{ads} than could be accounted for by means of equilibrium measurements. It appears that additional experimental verificiation, using different pesticides and soils is necessary before definite conclusions can be reached.

Figure 4 compares observed and calculated 2,4,5-T effluent curves for experiment no. 4-1. The largest aggregate size used for this experiment was 6.3 mm, compared to 2.0 mm for the data shown in Fig. 2 and 3. The solid line in Fig. 4 again represents the analytical solution using the linearized adsorption constant. A reasonable description of the

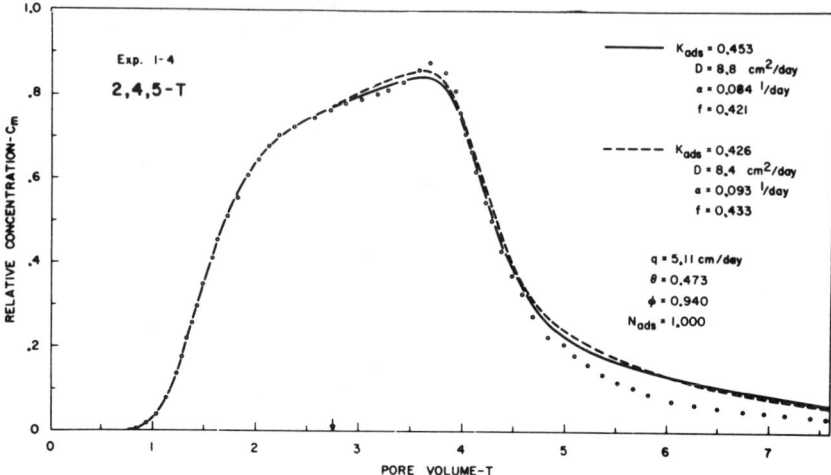

Fig. 3—Observed and calculated 2,4,5-T effluent curves for experiment 1-4. Calculations were based on the linear adsorption model, with and without fitting the adsorption constant K_{ads} to the effluent data. The open circles represent observed data points.

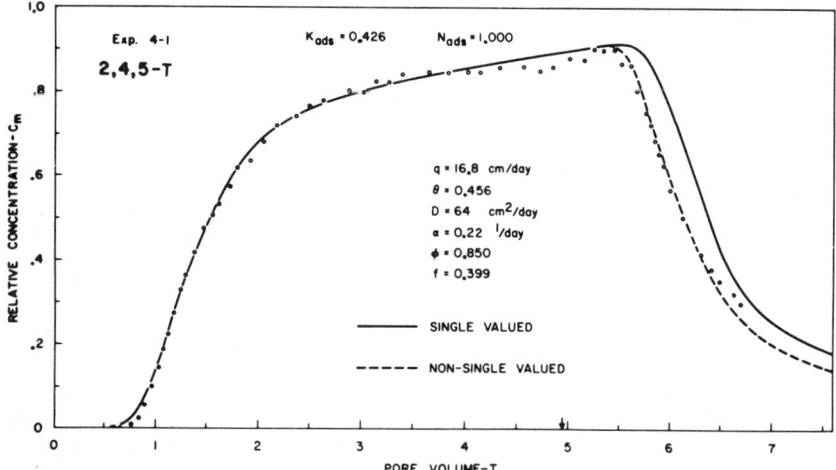

Fig. 4—Observed and calculated 2,4,5-T effluent curves for experiment 4-1. The adsorption constant K^1 was obtained by linearization of the equilibrium adsorption isotherm. The open circles represent observed data points.

experimental data is obtained, even though in this case the fitted adsorption constant was found to be 15% higher than the linearized value (Table 3). Note that an f-value of approximately 0.40 was obtained from the data in Fig. 4. Hence 40% of the adsorption occurred in the dynamic and 60% in the stagnant region of the soil. Similar values for f were obtained for the other displacements (Table 3), suggesting that the relative location of the sorption sites is not influenced significantly by either aggregate size or bulk density.

Table 3 further shows that the values of the mass transfer coefficient, α, obtained from the 2,4,5-T displacements are lower than those obtained from the tritium displacements at similar flow velocities. This may be expected since the mass transfer coefficient is proportional to the diffusion coefficient (Coats and Smith, 1964), and the diffusion coefficient of tritium is very large compared to those of most organic chemicals (Wang et al., 1953; International Critical Tables, 1929). The data in Table 3, except for the displacements in column 4, show that the ratio of the mass transfer coefficients of 2,4,5-T and tritium is approximately 0.5 (± 0.15) at similar flow rates. A comparison of experiments 4-1 and 4-2 is meaningless, since the displacements of 2,4,5-T and tritium in this column were carried out at different velocities (α is a function of the average pore-water velocity).

The inclusion of hysteresis in the model does not appreciably affect the calculated curves (Fig. 2 and 4). Thus, for the conditions of this study, intra-aggregate exchange of

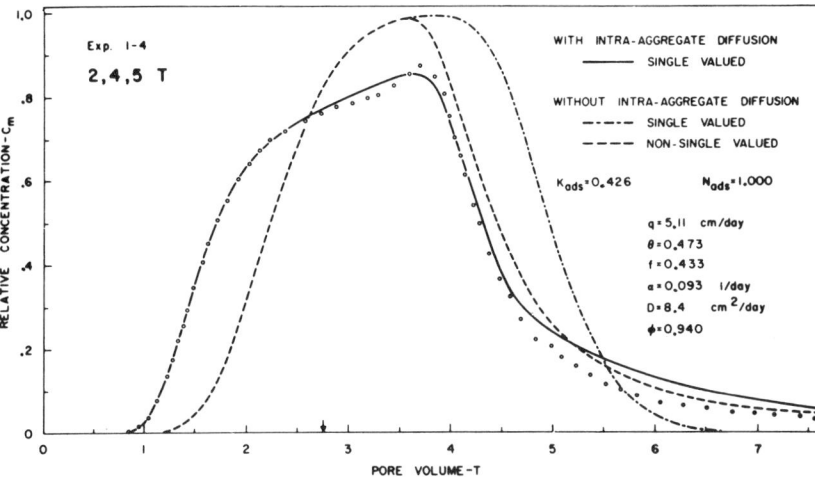

Fig. 5—Observed and calculated 2,4,5-T effluent curves for experiment 1-4, with and without including intra-aggregate diffusion. The open circles represent observed data points.

2,4,5-T between mobile and immobile regions was far more important than the hysteresis in the adsorption-desorption process. The effects of hysteresis, however, become more pronounced when intra-aggregate diffusion is neglected, as was suggested in several earlier studies (e.g., Swanson and Dutt, 1973; van Genuchten et al., 1974). This is clearly demonstrated in Fig. 5, where the experimental data of experiment no. 1-4 are compared with calculated curves, which include (solid line) or neglect intra-aggregate diffusion (dashed and dashed-dotted lines). The dashed line includes the effect of hysteresis and was obtained with the numerical program of van Genuchten and Wierenga (1974, example 1). The dashed-dotted line assumes no hysteresis. The linear adsorption model without intra-aggregate diffusion ($\alpha=0$, $\beta=1$) reduces then to the solution of Lindstrom et al. (1967). The more or less symmetrical curve obtained with this solution fails to describe the data in Fig. 5. When hysteresis is included in the calculations, the observed tailing during elution of the chemical is fairly well described. However, hysteresis has no effect on the break-through side of the curve, and hence, cannot explain the early appearance of the chemical in the effluent nor the tailing at the higher concentrations. Tailing during break-through will remain largely unnoticed when relatively short pulses are used. This may have influenced earlier conclusions about the importance of hysteresis on solute movement in soils (Swanson and Dutt, 1973; van Genuchten et al., 1974).

The break in the effluent concentration distributions during break-through of the chemical, and its subsequent tailing off, has been observed also by Kay and Elrick (1967) for the movement of lindane through Honeywood silt loam, and by J. M. Davidson (unpublished data, 1973) for the movement of picloram through Norge loam. Kay and Elrick (1967) attributed the tailing to intra-aggregate diffusion and adsorption, a phenomena which is substantiated by both experimental and theoretical descriptions of 2,4,5-T effluent curves in this study.

In Fig. 4 a comparison was made between the linear adsorption model and the observed data from experiment No. 4-1. An accurate description of the data was obtained, especially during break-through of the chemical. The parameter values presented in Fig. 4 were also used to calculate concentration profiles inside the column for both the mobile (c_m) and the immobile liquids (c_{im}). The results are given in Fig. 6. The differences in concentration between the two liquid regions are significant. For example, the relative concentration in the mobile region at a depth of 20 cm reaches a value of 0.5 after only 0.8 days, while it takes about 2.3 days for the concentration in the immobile region to reach this same value. Previously, (van Genuchten and Wierenga, 1977) similar distributions for tritium were shown. Although still considerable, the differences between c_m and c_{im} for tritium were less than those shown in Fig. 6. This may be expected, since tritium adsorption is very small ($K = 0.009$), and material which diffuses into the stagnant region of the soil can be stored only in the (immobile) liquid.

The fraction of the chemical diffusing into the stagnant region which will remain in solution may be calculated by considering the transport equation which describes the diffusional exchange between dynamic and stagnant regions. Assuming linear adsorption, this equation is given by (van Genuchten and Wierenga, 1976a, Eq. [16]

$$[\theta_{im} + (1 - f)\rho K] (\partial C_{im}/\partial t) = \alpha (C_m - C_{im}). \quad [7]$$

The first term on the left represents the material stored in the immobile liquid, while the second term gives the material adsorbed by the stagnant region. The fraction remaining in solution hence is given by

$$\theta_{im}/[\theta_{im} + (1 - f)\rho K].$$

Using the data from experiment no. 4-1, one may calculate that only 17% of 2,4,5-T will remain in solution, while the remaining 83% will be adsorbed (for tritium nearly 100%

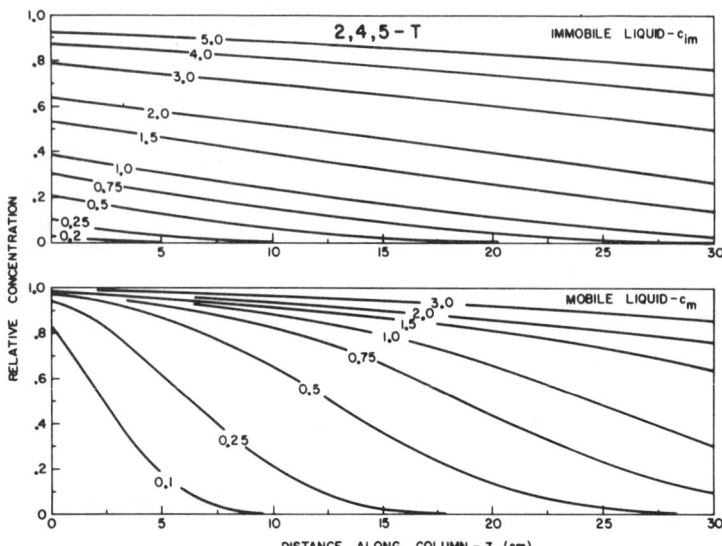

Fig. 6—Concentration profiles inside column 4, at different times during leaching with 2,4,5-T. Values of the parameters used to construct the curves are given in Fig. 4. Numbers on the curves indicate time (days) after leaching was initiated.

will remain in solution). Thus, when sorption increases, relatively more material can diffuse into the stagnant region of the soil. Since this diffusion is very slow compared to the rate at which the bulk of the chemical moves through the soil, the differences between the concentrations c_m and c_{im} are expected to increase progressively as sorption increases.

APPENDIX-NOTATION

C_0 = influent concentration ($\mu g/cm^3$)
C_1 = concentration break-through side effluent curve
C_m, C_{im} = concentration in mobile and immobile liquid, respectively ($\mu g/cm^3$)
c_m, c_{im} = relative concentrations (C_m/C_0; C_{im}/C_0, respectively).
D = dispersion coefficient (cm^2/day)
f = fraction adsorption sites in dynamic region
I_0, I_1 = modified Bessel functions
K, N = constants in Freundlich sorption isotherm: $S = KC^N$
L = length of column (cm)
P = peclet number of column: $P = v_m L/D$
q = flux
R = total retardation factor: $R = 1 + \rho K/\theta$
R_m = retardation factor of dynamic region: $R_m = 1 + f\rho K/\theta_m$
R_{im} = retardation factor of stagnant region: $R_{im} = 1 + (1-f)\rho K/\theta_{im}$
S_m, S_{im} = adsorption in dynamic and stagnant regions, respectively ($\mu g/g$)
t = time (days)
t_1 = pulse period (days)
T = pore volume $T = v_0 t/L = v_m t \phi/L$
T_1 = dimensionless pulse period: $T_1 = v_0 t_1/L$
v_0 = average pore-water velocity (cm/day): $v_0 = q/\theta$
v_m = average pore-water velocity in dynamic region: $v_m = q/\theta_m$ (cm/day)
x = dimensionless distance: $x = z/L$
z = distance (cm)
α = mass transfer coefficient (1/day)
$\bar{\alpha}$ = dimensionless mass transfer coefficient: $\bar{\alpha} = \alpha L/q$
β = $\phi R_m/R$
θ = water content: $\theta = \theta_m + \theta_{im}$ (cm^3/cm^3)
θ_m = mobile water content (cm^3/cm^3)
θ_{im} = immobile water content (cm^3/cm^3)
ϕ = fraction mobile water: $\phi = \theta_m/\theta$
ρ = bulk density (g/cm^3)

LITERATURE CITED

1. Coats, K. R., and B. D. Smith. 1964. Dead-end pore volume and dispersion in porous media. Soc. Pet. Eng. J. 4:73–84.
2. Farmer, W. J., and Y. Aochi. 1974. Picloram sorption by soils. Soil Sci. Soc. Am. Proc. 38:418–423.
3. Green, R. E., P. S. C. Rao, and J. C. Corey. 1972. Solute transport in aggregated soil: tracer zone shape in relation to pore-velocity distribution and adsorption. In Proc. 2nd Symp. Fundamentals of transport phenomena in porous media. IAHR-ISSS. (Guelph, Canada) August 7–11, 1972. Vol. 2:732–752.
4. Hornsby, A. G., and J. M. Davidson. 1973. Solution and adsorbed fluometuron concentration distribution in a water-saturated soil: experimental and predicted evaluation. Soil Sci. Soc. Am. Proc. 37:823–828.
5. International Critical Tables. 1929. McGraw-Hill Book Co., Inc., New York. Vol. V.
6. Kay, B. D., and D. E. Elrick. 1967. Adsorption and movement of lindane in soils. Soil Sci. 104:314–322.
7. Lindstrom, F. T., R. Hague, V. H. Freed, and L. Boersma. 1967. Theory on the movement of some herbicides in soils. Environ. Sci. Techn. 1:561–565.
8. Perkins, T. K., and O. C. Johnson. 1963. A review of diffusion in porous media. Soc. Pet. Eng. J. 19:70–84.
9. Swanson, R. A., and G. R. Dutt. 1973. Chemical and physical processes that affect atrazine movement and distribution in soil systems. Soil Sci. Soc. Am. Proc. 37:872–876.
10. van Genuchten, M. Th., and P. J. Wierenga. 1974. Simulation of one-dimensional solute transfer in porous media. N. Mex. Agric. Exp. Stn. Bull. 628. 40 p.
11. van Genuchten, M. Th., J. M. Davidson, and P. J. Wierenga. 1974. An evaluation of kinetic and equilibrium equations for the prediction of pesticide movement through porous media. Soil Sci. Soc. Am. Proc. 38:29–35.
12. van Genuchten, M. Th., and P. J. Wierenga. 1976a. Mass transfer studies in sorbing porous media: I. Analytical solutions. Soil Sci. Soc. Am. Proc. 40:473–480.
13. van Genuchten, M. Th., and P. J. Wierenga. 1976b. Numerical solution for convective dispersion with intra-aggregate diffusion and non-linear adsorption. p. 275–292. In G. C. van Steenkiste (ed) System Simulation in Water Resources North Holland Publ. Co., Amsterdam.
14. van Genuchten, M. Th., and P. J. Wierenga. 1977. Mass transfer studies in sorbing porous media: II. Experimental evaluation with tritium (^3H_2O). Soil Sci. Soc. Am. J. 41:272–278 (this issue).
15. Wang, J. H., C. V. Robinson, and I. S. Edelman. 1953. Self-diffusion and structure of liquid water: III. Measurement of the self-diffusion of liquid water with H^2, H^3 and O^{18} as tracers. J. Am. Chem. Soc. 75:466–470.
16. Wood, A. L., and J. M. Davidson. 1975. Fluometuron and water content distributions during infiltration: measured and calculated. Soil Sci. Soc. Am. Proc. 39:820–825.

EVALUATION OF CONCEPTUAL MODELS FOR DESCRIBING NONEQUILIBRIUM ADSORPTION-DESORPTION OF PESTICIDES DURING STEADY-FLOW IN SOILS[1]

P. S. C. Rao, J. M. Davidson, R. E. Jessup, and H. M. Selim[2]

ABSTRACT

Breakthrough curves (BTC) from miscible displacement of two pesticides through three soils were measured for two input concentrations of each pesticide. These BTC data were used to evaluate two conceptual models for describing the nonequilibrium adsorption-desorption of pesticides in soils under steady-state water flow conditions. In both models, adsorption on one group of sites was assumed to be instantaneous, while the rate of adsorption on the second group of sites followed either nonlinear reversible kinetics (Model I) or was a diffusion-controlled process (Model II). Parameters in both models were estimated by curve-fitting model predictions to one set of measured BTC data using a nonlinear least-squares optimization procedure. These parameter values were then used to verify the conceptual models by comparing simulated and measured BTC for a different input concentration. A different set of model parameters were required to describe the BTC data for each input concentration for the same soil-pesticide combination. The measured ^3H_2O BTC for all three soils were symmetrical in shape with no apparent tailing. Evaluation of Model I and II using these BTC suggested that the mobile-immobile water concept may not be applicable to the three soils used in this study.

Additional Index Words: two-site models, diffusion-controlled adsorption, parameter estimation, model verification, least-squares optimization, pesticide transport, and kinetics.

Rao, P. S. C., J. M. Davidson, R. E. Jessup, and H. M. Selim. 1979. Evaluation of conceptual models for describing nonequilibrium adsorption-desorption of pesticides during steady-flow in soils. Soil Sci. Soc. Am. J. 43:22–28.

THE IMPORTANCE of solute transport through soils as it relates to hazardous chemicals and soil-applied agro-chemicals has been well documented in the literature. For steady-state water flow conditions, the following partial differential equation for describing the transport of adsorbed solutes through soils has been proposed (Lapidus and Amundson, 1952)

$$\frac{\partial C}{\partial t} + \frac{\rho}{\theta}\frac{\partial S}{\partial t} = D\frac{\partial^2 C}{\partial x^2} - v\frac{\partial C}{\partial x} \quad [1]$$

where C and S are solution-phase ($\mu g/cm^3$) and adsorbed-phase ($\mu g/g$) concentrations, D is dispersion coefficient (cm^2/day), v is average pore-water velocity (cm/day), ρ is soil bulk density (g/cm^3), θ is volumetric soil-water content (cm^3/cm^3), x is distance (cm), and t is time (days).

Several conceptual models have been proposed and evaluated for describing the solute adsorption-desorption term ($\partial S/\partial t$) in Eq. [1]. Many of these models predict symmetrical and sigmoidal breakthrough curves (BTCs), while experimental BTCs are frequently asymmetrical or skewed in shape. As a result, earlier models based on the assumption of instantaneous adsorption and linear isotherms failed to describe experimental data (e.g., Davidson et al., 1968). Models based on first-order or other reversible kinetic adsorption-desorption processes were found to simulate measured BTC reasonably well at low pore-water velocities but generally failed at high pore-water velocities (e.g., Davidson and Mc Dougal, 1973).

More recent attempts to model the asymmetry or "tailing" in experimental BTCs may be classified into two groups. In the first group, physical processes were assumed responsible for the observed tailing. In these models, the soil-water regime was divided into mobile and immobile regions. Although the solute adsorption was instantaneous, the rate at which adsorbate molecules approached a fraction of adsorption sites was governed by diffusion through the stagnant soil-water region (Skopp and Warrick, 1974; van Genuchten and Wierenga, 1976 a, b). The assumption that a wide range in pore-water velocities would result in the observed tailing has been evaluated using a capillary bundle model (Rao et al., 1976). In a variation of the latter approach, Skopp et al. (1977)[3] considered convective-dispersive solute transport in a system consisting of two soil-water phases where the mass transfer of the solute between these phases obeyed a first-order kinetic process.

In the second group of conceptual models, the observed asymmetry in the BTCs was attributed to chemical processes. Adsorption-desorption isotherms for

---

[1] Contribution from Soil Sci. Dept., Univ. of Florida. Florida Agric. Exp. Stn. J. Series No. 1115. Received 27 Mar. 1978. Approved 28 Aug. 1979.
[2] Assistant Research Scientist, Professor, and Scientific Programmer, Soil Sci. Dept., Univ. of Florida, Gainesville, FL 32611; and Assistant Professor, Agronomy Dept., Louisiana State Univ., Baton Rouge, LA 70803, respectively.

[3] J. Skopp, E. J. Tyler, and W. R. Gardner. 1977. An interacting two-pore model for solute dispersion in aggregated soils. p. 137. Agron. Abst. Am. Soc. Agron.

some of these models were assumed to be nonsingular. Although good agreement could be achieved between predicted and measured BTCs for low average pore-water velocities, these models failed to describe the experimental data for high pore-water velocities (Hornsby and Davidson, 1973; van Genuchten et al., 1974). Furthermore, the physical and/or chemical explanation for the nonsingularity in pesticide adsorption-desorption isotherms has recently been questioned.[4] Selim et al. (1976) and Cameron and Klute (1977) have proposed a two-site adsorption-desorption model for describing asymmetrical breakthrough data. Justification for this conceptualization may be found in the chromatography literature (cf. Giddings, 1965); thus, a thorough evaluation of the two-site adsorption-resorption model appears warranted.

Agreement between model simulations and experimental data is generally used as a criterion for verification of conceptual processes included in the model. Davidson et al. (1976) discussed the limitations of such an approach when model parameters are estimated by "best-fit" to experimental data and not by independent measurements. However, due to the present inadequacy of experimental techniques, independent determination of model parameters may not always be feasible. For curve-fitting procedures to be a valid procedure for model verification, the same set of parameters estimated from a given experiment should be used to predict experimental results obtained under different conditions (e.g., column length, input concentrations, etc.).

The major objective of the present study was to critically evaluate two conceptual models for describing nonequilibrium adsorption-desorption of pesticides in soils during steady-state water flow. In both of these models, adsorption on one group of sites was assumed to be instantaneous, while adsorption on the other group of sites followed either nonlinear reversible kinetics (Selim et al., 1976) or was a diffusion-controlled process (van Genuchten and Wierenga, 1976 a, b). Laboratory data obtained from miscible displacement experiments reported by Rao and Davidson (1979) were used to evaluate the two conceptual models.

CONCEPTUAL MODELS
Model I

An equilibrium plus a kinetic model to describe the adsorption term, $\partial S/\partial t$, in Eq. [1] was proposed by Selim et al. (1976). In their model, two groups of adsorption sites—one group achieved instantaneous equilibrium (type-1) while the other group was time-dependent (type-2)—were assumed responsible for solute adsorption by soils. Using this approach, the time rate of change in the adsorbed phase concentration ($\partial S/\partial t$ in Eq. [1]) may be expressed as

$$\frac{\rho}{\theta}\frac{\partial S}{\partial t} = [\frac{\rho K_1 N C^{N-1}}{\theta}]\frac{\partial C}{\partial t} + [k_1 C^N - \frac{k_2 \rho}{\theta} S_2] \quad [2]$$

where K_1 and N are constants associated with instantaneous adsorption on type-1 sites, k_1 and k_2 are, respectively, forward and backward rate coefficients (days^{-1}) for kinetic adsorption on type-2 sites, S_2 is adsorbed phase concentration on the kinetic sites,

[4] P. S. C. Rao, J. M. Davidson, and D. P. Kilcrease. 1978. Examination of nonsingularity of adsorption-desorption isotherms for soil-pesticide systems. p. 34. Agron. Abst., Am. Soc. Agron.

and other terms are as defined previously. Note that at equilibrium, adsorption on both types of sites is described by the Freundlich equation

$$S = S_1 + S_2 = K_1 C^N + K_2 C^N = K C^N \quad [3]$$

where $K = (K_1 + K_2)$, K_1, K_2, and N are Freundlich constants, S_1 and S_2 (μg/g) are adsorbed phase concentrations on type-1 and type-2 sites, respectively, and K and S are defined on the basis of total amount of solute adsorbed. Note that the exponent N in Eq. [3] was assumed identical for both sites. Assuming that type-1 sites represent some fraction F of the total available adsorption sites, the following relationships may be stated:

$$K_1 = FK \quad [4a]$$
$$K_2 = (1-F)K. \quad [4b]$$

By assuming a linear equilibrium adsorption isotherm (i.e., $N = 1$ in Eq. [2] and [3]), an analytical solution to Model I (i.e., Eq. [2] coupled with Eq. [1]) has been obtained by Cameron and Klute (1977). Numerical solutions were presented by Selim et al. (1976) for the case of nonlinear adsorption isotherms ($N \neq 1$) as well as the more general case when adsorption on both sites is kinetics-controlled.

Model II

Equation [1] implies that all the soil-water freely participates in the convective-dispersive solute transport process and that all adsorption sites are readily accessible to the solute. In the model developed by van Genuchten and Wierenga (1976 a, b), the soil-water phase was partitioned into mobile and immobile regions. Convective-dispersive solute transport was limited to the mobile soil-water region. Solute transfer between the immobile and mobile soil-water regions was assumed to be diffusion-controlled. Solute adsorption-desorption in both regions was considered instantaneous; at equilibrium the relationship between adsorbed and solution concentration was described by a Freundlich equation.

Based on the above conceptualization of the system, the solute transport model may be restated as follows:

$$\phi R_1 \frac{\partial C_1}{\partial t} + (1 - \phi) R_2 \frac{\partial C_2}{\partial t} = \phi D \frac{\partial^2 C_1}{\partial x^2} - \phi v_1 \frac{\partial C_1}{\partial x} \quad [5]$$

$$\theta_2 R_2 \frac{\partial C_2}{\partial t} = \alpha [C_1 - C_2] \quad [6]$$

$$R_1 = [1 + (\rho F K N C_1^{N-1}/\theta_1)] \quad [7]$$

$$R_2 = [1 + \frac{\rho (1-F) K N C_2^{N-1}}{\theta_2}] \quad [8]$$

$$\phi = \theta_1/\theta; \theta = \theta_1 + \theta_2 \quad [9]$$

$$v_1 = v/\phi = v\theta/\theta_1 \quad [10]$$

In the above set of equations, the variables C, D, v, ρ, θ, K, N, x, and t are as defined for Eq. [1] and [3] with the subscripts 1 and 2 denoting, respectively, the mobile and immobile regions. Also, α is the mass transfer coefficient (days^{-1}) for solute exchange between the mobile and immobile regions, ϕ is fraction of soil-water content associated with the mobile region, F is the fraction of the total adsorption sites residing in the mobile region, and R is the retardation factor (Davidson et al., 1968).

Model II described above may also be viewed as a two-site adsorption model; although solute adsorption-desorption on all sites is considered instantaneous, the rate of adsorption-desorption within the immobile phase is diffusion-controlled. Numerical solutions to Eq. [5] through [10] for the case of nonlinear adsorption isotherms ($N \neq 1$) were given by van Genuchten and Wierenga (1976a), while analytical solutions for the linear adsorption isotherm case ($N = 1$) were presented by van Genuchten and Wierenga (1976b).

MATERIALS AND METHODS
Experimental Data

Effluent breakthrough curves (BTCs) for two pesticides were measured using water-saturated columns of Webster (Typic Haplaquolls), Cecil (Typic Hapludults), and Eustis (Typic

Table 1—Physical data for saturated soil columns used in the miscible displacement studies.

Soil	Pesticide	Input concentration, C_o	Column length L	Bulk density, ρ	Water content, θ	Average pore-water velocity, v	Dispersion coefficient D	Pore-volumes of pesticide solution applied
		µg/ml	cm	g/cm³	cm³/cm³	cm/day	cm²/day	
Cecil	2,4-D Amine	50	15.90	1.505	0.420	10.200	4.054	3.928
		5,000	15.90	1.505	0.420	13.056	5.190	2.646
Webster	2,4-D Amine	50	15.79	1.318	0.509	11.952	2.550	6.881
		5,000	15.79	1.318	0.509	11.448	2.443	2.670
Eustis	Atrazine	5	15.79	1.622	0.349	15.984	2.404	6.346
		50	14.90	1.633	0.337	17.880	2.537	6.001

Quartzipsamments) soils. These BTC data and the experimental methods employed were discussed earlier by Rao and Davidson (1979). The two pesticides used by these authors were: dimethylamine salt of 2,4-D [2,4-dichlorophenoxyacetic acid] and atrazine [2-chloro-4-ethylamino-6-isopropylamino-s-triazine]. BTCs were measured for 2,4-D amine displacement at two input concentrations (C_o = 50 and 5000 µg/ml) using Webster and Cecil soils, and two concentrations (C_o = 5 and 50 µg/ml) of atrazine using Eustis soil. Pertinent physical data for these displacements are listed in Table 1.

Equilibrium adsorption isotherms were measured using the batch shaking procedure (Rao and Davidson, 1979). Values of the Freundlich constants, K and N, derived from these isotherms for the soil-pesticide combinations of interest in the present study are listed in Table 2. Note that the adsorption isotherms in all three cases were nonlinear ($N < 1$).

Model Parameter Estimation

In both conceptual models described above, the values of the experimental variables K, N, θ ρ, and v are known. The value of the dispersion coefficient, D, for each soil was estimated from the tritiated water BTC by the method proposed by Rose and Passioura (1971). Experimental methods are presently unavailable to independently measure the parameters F, k_1, and k_2 in Model I and the parameters ϕ, F, and α in Model II. It may be noted from Eq. [2], [3], and [4] that

$$K_2 = (1-F)K = (\theta k_1 / \rho k_2) \quad [11]$$

and

$$k_2 = \frac{\theta k_1}{(1-F)\rho K}. \quad [12]$$

Since the values of k_2 can be calculated given K, F, and k_1, the problem now reduces to that of estimating the two unknown parameters F and k_1 in Model I.

The model parameters were estimated using a nonlinear least-squares (NLLS) optimization procedure (Meeter and Wolfe, 1968) by fitting the model prediction to measured BTC for the low input concentration. This iterative technique is based on minimizing the differences between simulated and measured BTC data by successive refinement of the initial parameter values; the estimates at each iteration are obtained by a combination of Gaussian method and steepest descent method described by Marquardt (1963). These estimates of model parameters were then used to verify the conceptual models by comparing the predicted and the measured BTCs obtained at a higher input concentration. Conversely, the model parameters obtained by fitting to high input concentration BTC data were also used to predict low input concentration BTC data. Model verification procedures similar to this have been employed by O'Connor et. al. (1976), van Genuchten et al. (1977) and Gaudet et al. (1977).

A finite-difference scheme was used to numerically solve Model I (Eq. [1] and [2]) and Model II (Eq. [5] through [10]), both subject to the following initial and boundary conditions:

$$C = 0, S = 0, 0 \leqslant x \leqslant L, t = 0 \quad [13a]$$

$$vC - D(\partial C/\partial x) = \begin{cases} vC_o, & x = 0, t \leqslant t_1 \\ 0, & x = 0, t > t_1 \end{cases} \quad [13b]$$

$$\partial C/\partial x = 0, x = L, t > 0. \quad [13c]$$

These conditions are applicable for a soil column of L cm length, initially void of the pesticide, to which a pesticide solu-

Table 2—Freundlich adsorption constants for the three soil-pesticide combinations studied.

Soil	Pesticide	K	N
Webster	2,4-D†	4.62	0.70
Cecil	2,4-D†	0.66	0.83
Eustis	Atrazine	0.62	0.79

† Dimethylamine salt of 2,4-D.

tion of C_o (µg/ml) concentration is applied at an average pore-water velocity of v cm/day for a period of t_1 days, followed by pesticide-free solution. Note that for Model II, v and C in Eq. [13b] and [13c] were replaced, respectively, by v_1 and C_1. Also, the initial condition given by Eq. [13a] is modified as $C_1 = 0$, $C_2 = 0$, $S = 0$ for $t = 0$. All computations were performed on an AMDAHL 470 V/6 − II digital computer with the aid of computer programs written in FORTRAN IV language.

RESULTS AND DISCUSSION

Uniqueness of Estimated Parameter Values

The convergence of the nonlinear least-squares (NLLS) procedure may depend on the initial estimates of the model parameters allowed to vary in the curve-fitting procedure. Although a unique set of parameter values may exist for a given set of data, curve-fitting procedures may converge to different parameter values depending upon their initial estimates (cf. Bekey, 1970). A calculated BTC, simulated using Model I, was used as "measured" data to test the uniqueness of the model parameters estimated by the NLLS procedure. Curve-fitting of Model I to the synthetic "measured" data was considered by varying K, N, F, and k_1. Reasonable initial estimates for N and F were possible because the value of N ranges between 0.7 and 1.0 for most soil-pesticide systems (Hamaker and Thompson, 1972) while the value of F is bounded between 0 and 1. Equilibrium adsorption isotherms provide excellent starting values for K and N. Methods to estimate the values of k_1 (and k_2) are generally unavailable; thus, it is possible that the initial estimates of k_1 used in the NLLS procedure may be in error by as much as 100-fold.

The model parameters selected to calculate the synthetic "measured" data used in the NLLS procedure were: $K = 1.604$, $N = 0.7305$, $F = 0.2659$, and $k_1 = 0.515$ day^{-1}. Several curve-fitting runs were made where the initial values of K and k_1 were varied by 10-fold and 100-fold, respectively, while in all cases the initial values of F and N were 0.5 and 0.8, respectively. For all test cases, the maximum deviation between the parameter values estimated by the NLLS procedure and the "true" values was less than 0.004%. Therefore, it was concluded that for the present case, the NLLS procedure yields a unique set of parameters even when the initial estimates of K and k_1 may be considerably different from the true values.

Kinetics-Controlled Adsorption (Model I)

Measured and simulated BTCs for 2,4-D amine displacement through a water-saturated Cecil soil column are presented in Fig. 1. Pesticide mobility was greater for the high input concentration as indicated by the lefthand shift of the BTC for $C_o = 5,000$ μg/ml compared to that for $C_o = 50$ μg/ml. This increased pesticide mobility at the higher concentration was due to the nonlinear adsorption isotherm (Rao and Davidson, 1979). The BTCs calculated using Model I with the parameters F and k_1 estimated by curve-fitting to $C_o = 50$ μg/ml BTC data are shown as solid lines in Fig. 1. Independently measured values of K and N (Table 1) along with the estimated values of F, k_1, and k_2 were used to predict the BTC for $C_o = 5,000$ μg/ml. The agreement between the simulations (solid lines in Fig. 1) and the measured BTC is reasonable, but not perfect. Therefore, the model parameters K, N, F, k_1, and k_2 were re-estimated from a four-parameter fit using $C_o = 50$ μg/ml BTC data. The revised parameter estimates improved the prediction (dashed lines in Fig. 1) of the measured BTCs.

The values of the model parameters K, N, F, k_1, and k_2 estimated from 2,4-D amine data for $C_o = 50$ μg/ml by varying two parameters (F and k_1) or four parameters (K, N, F and k_1) in the NLLS procedure are presented in Table 3. More than a twofold increase in the K value and a small decrease in the N value over that obtained from an equilibrium adsorption isotherm was necessary to describe the BTC data for both input concentrations. Furthermore, the two estimates of the k_1 and k_2 values were also different. Note that about 50 to 70% of the total asorption sites were required to be kinetic (i.e., type-2 sites) in order to describe the extensive tailing observed in the measured BTC as C/C_o approached 0.0 or 1.0.

The four-parameter fit technique was used to estimate the model parameters for Model I which best described the 2,4-D amine BTC data from Webster soil (Fig. 2) and atrazine BTC data from Eustis soil (Fig. 3). Although reasonable predictions were achieved at low input concentrations, parameters estimated from these data and used in Model I failed to describe the BTC for high input concentration for either soil. The estimated value of the Freundlich

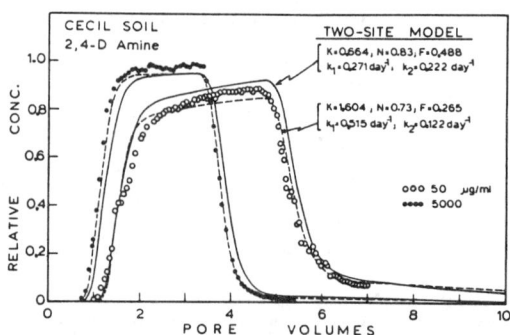

Fig. 1—Measured and simulated breakthrough curves for 2, 4-D amine displacement through Cecil soil column. Parameter values used to calculate the solid lines were obtained from a 2-parameter fit, and those for dashed lines were estimated from a 4-parameter fit to $C_o = 50$ μg/ml data.

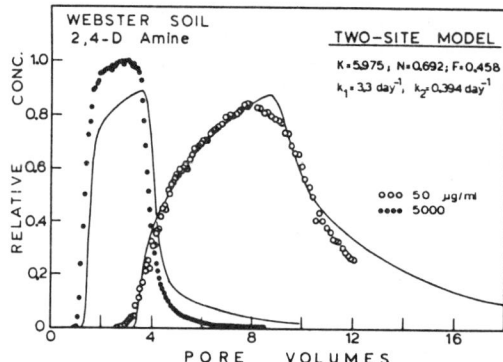

Fig. 2—Measured and simulated breakthrough curves for 2, 4-D amine displacement through Webster soil column. Simulated curves (solid lines) were calculated using Eq. [1] and [2] where the model parameters were estimated from $C_o = 50$ μg/ml data.

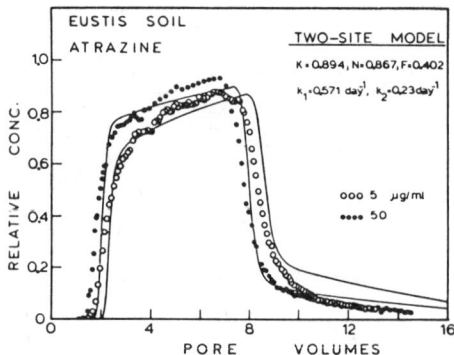

Fig. 3—Measured and simulated breakthrough curves for atrazine displacement through Eustis soil column. Simulated curves (solid lines) were calculated using Eq. [1] and [2] where the parameters were estimated from $C_o = 5$ μg/ml data.

Table 3—Comparison of Model I parameter values estimated from the 2,4-D amine breakthrough data ($C_o = 50$ μg/ml) for Cecil soil by varying either two or four parameters in the nonlinear least-squares curve-fitting procedure.

Parameter	Cecil-2,4-D amine	
	2-parameter fit	4-parameter fit
N	0.83†	0.73
K	0.664†	1.604
K_1	0.324	0.425
F	0.488	0.265
k_1 (day^{-1})	0.271	0.515
k_2 (day^{-1})	0.222	0.122
k_1/k_2	1.220	4.221

† These values of the Freundlich constants were obtained independently from the equilibrium adsorption isotherm.

Table 4—Comparison of Model I parameter values obtained by nonlinear least-squares estimation from pesticide BTC data measured at two input concentrations (C_o in μg/ml) in two soils.

Parameter	Cecil—2,4-D amine		Eustis—Atrazine	
	$C_o = 50$	$C_o = 5,000$	$C_o = 5$	$C_o = 50$
N	0.7305	0.7214	0.8675	0.9555
K	1.604	0.4858	0.8941	0.4975
F	0.2649	0.6585	0.4018	0.5042
K_1	0.4249	0.3199	0.3592	0.2508
k_1 (day^{-1})	0.515	0.7296	0.5705	0.60
k_2 (day^{-1})	0.122	1.2276	0.2299	0.5242
(k_1/k_2)	4.2213	0.5943	2.4815	1.1446

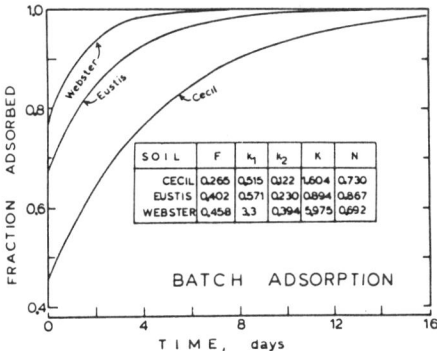

Fig. 4—Rate of approach to equilibrium for three soils in batch adsorption experiments, simulated using Eq. [2]. The parameter values shown were obtained by curve-fitting to BTC data in Fig. 1, 2, and 3.

adsorption constant (K) was larger than the measured value in both the Webster and Eustis soil, as was the case for the Cecil soil (Table 3). Also, it was necessary that about half of the adsorption sites be of the kinetic type.

Further attempts to verify Model I involved prediction of the low concentration data using K, N, F, and k_1 values estimated from a four-parameter curve-fit to the high concentration BTC data. Note that this case is converse to that described above. Model I failed to describe the measured low input concentration BTC data when the model parameters were estimated from the fit to the BTC from the high input concentration. The values of the model parameters estimated by curve-fitting to either low or high input concentration BTC data are summarized in Table 4. A completely different set of model parameters were necessary to describe the BTC data for the two input concentrations. Estimates from the high concentration data base were characterized by a larger K value with a high fraction of the total adsorption capacity associated with the kinetic sites and larger kinetic rate coefficients (Table 4).

Numerical solutions to Eq. [2] were used to predict the time required for equilibrium to be achieved in batch adsorption experiments for the three soil-pesticide combinations. The parameters K, N, F, k_1, and k_2 were those from the BTC data for the low input concentration. In each simulation, the initial solution concentration was chosen such that the equilibrium solution concentration was 50 μg/ml for 2,4-D amine and 5 μg/ml for atrazine. A plot of the quantity of pesticide adsorbed vs. time is shown in Fig. 4. The simulated curves suggest that adsorption equilibrium was achieved after 5, 12, and 24 days, respectively, for Webster, Eustis, and Cecil soils. However, preliminary experiments (data not shown) to measure adsorption kinetics indicated that for each soil-pesticide combination, equilibrium was apparently reached in less than 2 to 4 hours. Thus, the values of k_1 and k_2 estimated by curve-fitting to the BTC data are much smaller than those suggested by independent experiments.

Haque et al. (1968) have reported that the values of the kinetic rate coefficients for 2,4-D adsorption on kaolinite and montmorillonite were 0.19 and 0.05 day^{-1}. Because these values were of the same order of magnitude as diffusion coefficients, these authors concluded that in their experiments the rate of adsorption may have been limited by diffusion to the sites. The estimated values of k_1 and k_2 in the present study range from 0.12 to 3.3 day^{-1} and are of the same magnitude as those reported by Haque et al. (1968). It is likely that in the column experiments, the adsorption-desorption processes were diffusion-controlled and not limited by the reaction kinetics at the soil-solution interface. The Cecil soil appears to be the best case upon which to test this assumption using Model II because (i) more than 70% of the adsorption sites were estimated to be time-dependent for adsorption, and (ii) the estimated values of k_1 and k_2 are the lowest among the three soils considered here.

Diffusion-Controlled Adsorption (Model II)

Rao and Davidson (1979) presented the BTCs for tritiated water (^3H_2O) displacement through water-saturated columns of Cecil, Webster, and Eustis soils. These data and the 2,4-D amine BTC data for Cecil soil (Fig. 1) were used to evaluate Model II. The measured BTCs for ^3H_2O in all three soils, shown in Fig. 5, were symmetrical and sigmoidal in shape with no apparent tailing. Therefore, excellent agreement was obtained between the measured BTCs and those predicted (solid lines in Fig. 5) using a numerical solution to Eq. [1] subject to Eq. [13] for a nonadsorbed solute. The Peclect numbers used to calculate the BTCs were 74, 40, and 105, respectively, for Webster, Cecil, and Eustis soils. Note that Eq. [1] assumes all soil-water to be mobile. The ^3H_2O BTC in Fig. 5 were also fitted to Model II for a nonadsorbed solute (i.e., $K = 0$); a value of $\phi \geq 0.99$ was obtained for all three soils. Model II parameters estimated by curve-fitting to Cecil ^3H_2O BTC, allowing for adsorption, were: $\phi = 0.986$, $F = 0.7055$, $K = 0.0047$, and $\alpha = 2.3 \times 10^{-4}$ day^{-1}.

Model II parameters estimated by curve-fitting to 2,4-D amine BTC ($C_o = 50$ μg/ml) for Cecil soil were: $\phi = 0.817$, $F = 0.529$, and $\alpha = 0.093$ day^{-1}. These values along with independently measured Freundlich adsorption coefficients ($K = 0.664$ and $N = 0.83$ from Table 2) were then used to simulate the BTC for $C_o = 5,000$ μg/ml. The predicted (solid lines) and measured curves are compared in Fig. 6. Although the position of the measured BTC matched well with that

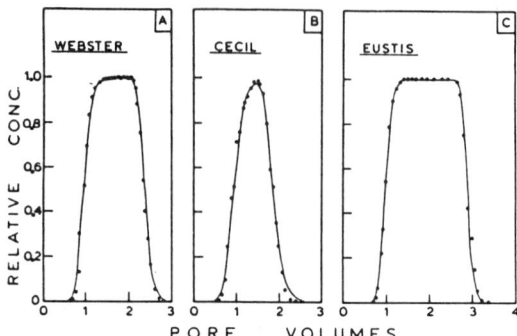

Fig. 5—Measured and simulated breakthrough curves for ^3H_2O displacement through three soils. Simulated curves shown as solid lines in A, B, and C were calculated using Model I.

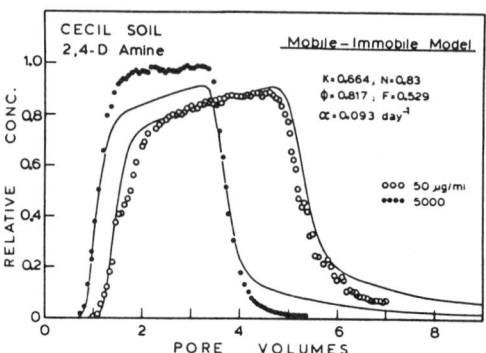

Fig. 6—Measured and simulated breakthrough curves for 2,4-D amine displacement through Cecil soil column. Measured data are the same as in Fig. 1, while the simulated curves (solid lines) were calculated using Eq. [5] — [10], where the model parameters were estimated by curve-fitting to $C_o = 50$ µg/ml data.

predicted, Model II over-predicted the tailing in the BTC for $C_o = 5,000$ µg/ml. Hence, the parameter values were re-estimated using the high input concentration data. These values were: $\phi = 0.820$, $F = 0.549$, and $\alpha = 0.00082$ day^{-1}. The BTCs calculated using these revised parameter values correctly predicted the arrival of 2,4-D amine in the Cecil soil column effluent for $C_o = 50$ µg/ml, but under-estimated the tailing in the BTC. The values of ϕ and F estimated using the BTC data for two input concentrations (C_o) were nearly identical, while the value of the mass transfer coefficient (α) varied by an order of magnitude for a 100-fold change in C_o. The estimated quantity of mobile water, ϕ, in the Cecil soil, as determined by Model II, was not the same for ^3H_2O and 2,4-D amine. Using the 2,4-D BTC data, ϕ was estimated to be about 0.820 compared to about 0.99 for the ^3H_2O BTC data.

SUMMARY

In the present study, two conceptual models describing time-dependent adsorption-desorption of pesticides during steady-state water flow in soils were evaluated. In both models, adsorption-desorption on a group of sites was assumed to be instantaneous, while the rate of adsorption-desorption on the other group of sites followed either nonlinear reversible kinetics (Model I) or was a diffusion-controlled process (Model II). All soil-water was considered to be mobile in Model I and the nonequilibrium adsorption-desorption processes were attributed to the reaction kinetics at the soil-solution interface. In Model II, the soil-water was partitioned into mobile and immobile regions and convective-dispersive solute transport was limited to the mobile region. Also, the rate of adsorption-desorption on the sites residing within the immobile region was assumed to be governed by diffusive mass transfer across the mobile-immobile soil-water interface.

Independent measurements of all model parameters were not feasible due to inadequate experimental methods. Hence, some model parameters were estimated by curve-fitting to measured BTC data. Both conceptual models (Models I and II) contain a sufficient number of parameters to provide a good description (by curve-fitting) of the asymmetrical BTCs for pesticide displacement through three soils. However, two different sets of parameter values in both models were required to predict the BTCs at two different input concentrations. Unlike the pesticide data, the ^3H_2O BTCs for all three soils were symmetrical and sigmoidal in shape. Excellent agreement was found between these measured BTCs and those predicted by Model I for a nonadsorbed solute. A value of $\phi \geqslant 0.99$ was obtained by curve-fitting Model II to the ^3H_2O BTC assuming no adsorption of ^3H_2O. The model parameters were also estimated by curve-fitting to the Cecil ^3H_2O data, where allowance was made for adsorption of tritiated water. These values ($\phi = 0.986$, $F = 0.705$, $K = 0.0047$, and $\alpha = 2.3 \times 10^{-4}$ day^{-1}) indicated that the concept of mobile-immobile soil-water (Model II) may not be applicable to the three soils used in this study. Although Model I described the ^3H_2O data well, it provided only a fair prediction of the BTCs for adsorbed solutes. Thus, the conceptual representation of the adsorption-desorption processes in Model I (Eq. [2]) is apparently inadequate. Further studies to investigate the causes for nonequilibrium pesticide adsorption-desorption during flow through soils are currently being carried out in our laboratory.

ACKNOWLEDGEMENTS

This research was supported in part by the Grant No. R-803849 from the Solid and Hazardous Waste Research Division, Municipal Research Lab., USEPA and in part by special funds from the Center for Environ. Programs, Institute of Food and Agric. Sci., Univ. of Florida. The computer program to perform the nonlinear least-squares optimization procedure was generously provided by Mr. Joe Skopp, Univ. of Wisconsin, Madison, Wis.

LITERATURE CITED

1. Bekey, G. A. 1970. System identification: An introduction and survey. Simulation. 15:151–166.
2. Cameron, D. A., and A. Klute. 1977. Convective-dispersive solute transport with a combined equilibrium and kinetic adsorption model. Water Resour. Res. 13:183–188.
3. Davidson, J. M., C. E. Rieck, and P. W. Santelman. 1968. Influence of water flux and porous material on the move-

ment of selected herbicides. Soil Sci. Soc. Am. Proc. 32:629–633.
4. Davidson, J. M., and J. R. McDougal. 1973. Experimental and predicted movement of three herbicides in a water-saturated soil. J. Environ. Qual. 2:428–433.
5. Davidson, J. M., P. S. C. Rao, R. E. Green, and H. M. Selim. 1976. Evaluation of conceptual models for solute behavior in soil-water systems. Proc. Int. Congr. for Agro-Chemicals in Soils. Int. Soil Sci. Soc. 14–18 June 1976. Jerusalem, Israel (in press).
6. Gaudet, J. P., H. Jegat, G. Vachaud, and P. J. Wierenga. 1977. Solute transfer, with exchange between mobile and stagnant water, through unsaturated sand. Soil Sci. Soc. Am. J. 41:665–671.
7. Giddings, J. C. 1965. Dynamics of chromatography. Part I. Principles and theory. Marcel Dekker, Inc., New York.
8. Hamaker, J. W., and J. M. Thompson. 1972. Adsorption. p. 49–144. In C. A. I. Goring and J. W. Hamaker (ed.) Organic chemicals in the soil environment, Vol. 1, Marcel Dekker, Inc.
9. Haque, R., F. T. Lindstrom, V. H. Freed, and R. Sexton. 1968. Kinetic study of the sorption of 2, 4-D on some clays. Environ. Sci. Technol. 2:207–211.
10. Hornsby, A. G., and J. M. Davidson. 1973. Solution and adsorbed Fluometuron concentration distributions in a water-saturated soil: Experimental and predicted evaluation. Soil Sci. Soc. Am. Proc. 37:823–828.
11. Lapidus, L., and N. R. Amundson. 1952. Mathematics of adsorption in beds: 6. The effect of longitudinal diffusion in ion-exchange and chromatographic columns. J. Phys. Chem. 56:984–988.
12. Marquardt, D. L. 1963. An algorithm for least-squares estimation of non-linear parameters. J. Soc. Ind. Appl. Math. 2:431–441.
13. Meeter, D. A., and P. J. Wolfe. 1968. UWHAUS-nonlinear least-squares fitting and function minimization. Supplementary Program Series no. 50, 25 Oct. 1968. Univ. of Wisconsin Computing Center, Madison, Wis.
14. O'Connor, G. A., M. Th. van Genuchten, and P. J. Wierenga. 1976. Predicting 2, 4, 5-T movement in soil columns. J. Environ. Qual. 5:375–378.
15. Rao, P. S. C., R. E. Green, L. R. Ahuja, and J. M. Davidson. 1976. Evaluation of a capillary bundle model for describing solute dispersion in aggregated soils. Soil Sci. Soc. Am. J. 40:815–820.
16. Rao, P. S. C., and J. M. Davidson. 1979. Adsorption and movement of selected pesticides at high concentrations in soils. Water Res. Vol. 13 (in press).
17. Rose, D. A., and J. B. Passioura. 1971. The analysis of experiments on hydrodynamic dispersion. Soil Sci. 111:252–257.
18. Selim, H. M., J. M. Davidson, and R. S. Mansell. 1976. Evaluation of a two-site adsorption-desorption model for describing solute transport in soils. Proc. Summer Computer Simulation Conf., 12–14 July 1976. Washington, D. C. p. 444–448.
19. Skopp, J., and A. W. Warrick. 1974. A two-phase model for the miscible displacement of reactive solutes through soils. Soil Sci. Soc. Am. Proc. 38:545–550.
20. van Genuchten, M. Th., J. M. Davidson, and P. J. Wierenga. 1974. An evaluation of kinetic and equilibrium equations for the prediction of pesticide movement in porous media. Soil Sci. Soc. Am. Proc. 38:29–35.
21. van Genuchten, M. Th., and P. J. Wierenga. 1976a. Numerical solutions for convective dispersion with intra-aggregate diffusion and nonlinear adsorption. p. 275–291. In G. C. van Steenkiste (ed.) System simulation in water resources. North Holland Publishing Co., Amsterdam.
22. van Genuchten, M. Th. 1976b. Mass transfer studies in sorbing porous media: I. Analytical solutions. Soil Sci. Soc. Am. J. 40:473–480.
23. van Genuchten, M. Th., P. J. Wierenga, and G. A. O'Connor. 1977. Mass transfer studies in sorbing porous media: III. Experimental evaluation with 2, 4, 5-T. Soil Sci. Soc. Am. J. 41:278–285.

Copyright © 1978 by Martinus Nijhoff Publishers
Reprinted from *Plant and Soil* **49**:569–580 (1978)

COMPUTED REDISTRIBUTION OF PESTICIDES IN THE ROOT ZONE OF AN ARABLE CROP

by M. LEISTRA

Laboratory for Research on Insecticides, Marijkeweg 22, Wageningen, The Netherlands

SUMMARY

A series of computations was made to simulate the redistribution in the soil of four model pesticides that had been incorporated to a depth of 8 cm in spring. Water flow in the soil after low, average and high rainfall, respectively, was described in some detail. Data for potato crops were used for the schematized description of root uptake activity as a function of depth in soil and time. It was found that redistribution of the pesticides in the soil occurred mainly in the first month after application, when water uptake by the root system was comparatively small. Under average rainfall conditions only the weakly adsorbed compounds were redistributed markedly. With the moderately to strongly adsorbed pesticides, and with any pesticide under low rainfall conditions, redistribution was limited to a few centimetres. The risk of weakly adsorbed pesticides being leached too deeply as a result of high rainfall was clearly demonstrated.

INTRODUCTION

Pesticides to control soil-inhabiting nematodes or insects are often applied just before or during the planting or seeding of the crop. For good protection the active compound must be present in the root zone at sufficient concentrations during a period of at least a few months to allow adequate extension of the root system. If methods of applying pesticides could be improved, it might be possible to attain the desired effectiveness with comparatively low dosages. It is often necessary to minimize the dosage of pesticides to make their use economical and to keep the residue-level in the crops as low as possible. However, sound quantitative recommendations can only be made if the behaviour of pesticides in soil under field conditions is well understood.

Homogeneous incorporation into a top layer of soil several centimetres thick has often been found to give the best results; rototillage seems to be most suitable for attaining this[18]. To be able to recommend the optimum depth of incorporation, we must know to what extent the compound is likely to be redistributed through the root zone during the first months after application. Shallow incorporation might result in inadequate protection of the roots in the lower part of the plough layer. Very deep incorporation is expensive, and its effectiveness, especially in the top of the plough layer, might become too low. Several factors including the properties of the pesticide, soil characteristics, climatic conditions, and nature of the crop are likely to affect the redistribution of pesticides in soil.

A computer model was used to ascertain the extent of the redistribution to be expected and the relative importance of the various factors. Special attention was paid to the effect of the adsorption strength and of different amounts of rainfall in the first months after application.

PROCEDURES*

The behaviour of pesticides in a sandy loam soil under field conditions was simulated. The crop data used were taken from potato. The dosage of the model pesticides was 3 kg ha^{-1}, corresponding to 150 nmol per cm^2 soil surface for a relative molar mass of 200. This dosage was homogeneously distributed in the top 8 cm layer just before planting the hypothetical crop at the beginning of April.

The upper 80 cm of the hypothetical soil was assumed to be a sandy loam with hydraulic characteristics as given by Stroosnijder[19] (Fig. 26i) concerning the relationship between tensiometer pressure p_t and volume fraction of liquid ε_l, and that between hydraulic conductivity K and ε_l. The layer from 80 to 150 cm depth was assumed to consist of loamy fine sand with corresponding hydraulic characteristics (Ref. [19], Fig. 26h). A constant groundwater level was assumed at 150 cm depth, and this depth also served as a reference level for the gravity pressure. At the start of the computations, the soil water profile was almost in equilibrium with the groundwater table: the hydraulic pressure p_h initially approached zero throughout the soil profile.

The flux of the liquid phase in soil is described by Darcy's law:

$$J_l = -K\, \partial p_h/\partial z.$$

where J_l is the volumetric liquid flux, K is the hydraulic conductivity, and z is the depth in soil. The hydraulic pressure p_h, equals the sum of the gravitational pressure p_g and tensiometer pressure p_t all in mbar.,

* See List of Symbols and Units at the end of this paper.

The climatic data used were based on observations given in the Klimaatatlas van Nederland[11]. Four rainfall patterns were included; pattern R1 represents average rainfall amounting to 4.6 in the first, 5.0 in the second and 5.6 cm^3 (water) cm^{-2} (soil surface) in the third month. This rain fell in three periods (four days each) per month, with maximum intensities ranging from 0.4 to 1.2 cm^3 cm^{-2} day^{-1}. The other rainfall regimes included dry or wet monthly periods at different times after planting, with the monthly rainfall during the remainder of the computation period identical to that for R1. In pattern R2, the amount of rainfall during the first month was half that for the same period with R1. This low value corresponds with a monthly amount that is exceeded with a probability of about 90%[11]. In rainfall pattern R3, the amount of rain during the first month was 1.75 times the amount for the first month in R1, while for pattern R4 the rainfall in the second month was multiplied by the factor 1.75. These high monthly amounts are exceeded with a probability of about 10%[11].

The rate at which water was taken up by the plant roots for transpiration and directly for growth increased slowly from zero at day 5 to 0.03 cm^3 cm^{-2} day^{-1} at day 20. There was then a more rapid linear increase to 0.3 cm^3 cm^{-2} day^{-1} at day 61, after which the rate of water uptake remained constant. This transpiration pattern was derived from the data of Endrödi and Rijtema[3]. The cumulative amount of water taken up at day 90 was 15.7 cm^3 cm^{-2}. The rate of water evaporation from the soil surface amounted to 0.08 cm^3 cm^{-2} day^{-1} during the first 61 days, after which it decreased to 0.03 at 90 days. The tendency for evaporation to increase with time was counteracted by the increasing crop cover[1]. According to this pattern, the cumulative amount of water evaporated from the soil surface at day 90 was 6.4 cm^3 cm^{-2}.

The uptake of water from various depths in the soil was described in terms of the distribution of root activity and of soil moisture availability. Starting from the description of root system development[12] and from observations[2] for a potato crop, root activity in the computation compartments was introduced as a function of time. The root system gradually penetrated into deeper soil layers at a maximum rate of 0.8 cm day^{-1}. During the first two months after planting, the increase in root activity was comparatively fast. Root activity increased at a lower rate during the third month, with some extension mainly in the deeper soil layers. The final pattern of root activity was attained at day 90. Root activity then increased from zero at the soil surface to maximum values in the 4 to 12 cm layer, below which there was a gradual decrease in activity down to about 24 cm depth. It then decreased comparatively steeply down to a depth of about 45 cm, and continued at rather low values to a maximum depth of 80 cm. Soil moisture was assumed to be available without restriction at tensiometer pressures, p_t, of -1000 mbar and higher. In the drier soil layers, water availability decreased linearly with decreasing p_t down to zero at -16000 mbar. For each of the computation compartments in the root zone, the product of relative root activity and water availability was calculated, and divided by the sum of the products for the complete root zone. This quotient determined the fraction of the transpired water withdrawn from the compartment in a particular time interval Δt.

The conservation equation for the liquid phase in soil states that the rate of change in volume fraction of liquid, ε_l, equals the change in the liquid flux, J_l, plus $R_{u,l}$, the rate of liquid uptake by plant roots:

$$\partial \varepsilon_l / \partial t = - \partial J_l / \partial z - R_{u,l}$$

Transport of the pesticide was confined to the liquid phase, assuming that the volatility was negligible. The total flux of substance, J_s is composed of contributions by convection, hydrodynamic dispersion and diffusion:

$$J_s = J_l c_l - D_d \, \partial c_l / \partial z - D_{lp} \, \partial c_l / \partial z$$

Convective transport is described by the product of J_l and c_l, the concentration in the liquid phase. The coefficient of hydrodynamic dispersion, D_d, was calculated[4] with $D_d = L_d |J_l| I$. The proportionality constant L_d, also called the dispersion distance, was taken to be 2.0 cm (medium). The coefficient for diffusion of the substance in the liquid phase is computed from:

$$D_{lp} = f_l \, \varepsilon_l \, D_l$$

For the coefficient of diffusion in bulk liquid, D_l, a value of 0.35 cm² day⁻¹ was taken according to Reid and Sherwood[15]. The impedance factor f_l for diffusion in the liquid phase was introduced as a function of ε_l. The values of f_l were 0.03, 0.10, 0.20, 0.34 and 0.50 cm² cm⁻² for ε_l-values of 0.10, 0.20, 0.30, 0.40 and 0.50 cm³ cm⁻³, respectively [7 14 16 17].

For the adsorption of pesticides, the soil system was divided from top to bottom into layers 30, 38, and 82 cm thick, respectively. The adsorption coefficients, $K_{s/l}$, for the second and third layer were 0.7 and 0.2 times the coefficient for the top layer, respectively. For the bulk density ρ_b of the layers, values of 1.25, 1.35 and 1.40 g cm⁻³, respectively were taken.

The pesticide was converted in the soil according to a first-order rate equation:

$$(\partial c_m / \partial t)_c = - k_c \, c_m$$

The concentration of the pesticide in the medium, c_m, was related to the concentration in the liquid phase via $c_m = (\varepsilon_l + \rho_b \, K_{s/l}) \, c_l$.

The rate constant k_c was set according to the temperature. The average soil temperature was assumed to range from 6°C at the beginning of April to 16°C at the end of June. During this period, the value of k_c increased from 0.007 day⁻¹ to 0.019 day⁻¹. The values of the rate constant k_c were reduced when the soil dried: at ε_l-values of 0.05, 0.10, 0.20, and 0.30 cm³ cm⁻³, the values of k_c were 0.47, 0.60, 0.85, and 1.0 times the maximum value at the prevailing temperature, respectively. The pesticides were taken up by the plant roots with the transpiration flow of water:

$$R_{u,s} = R_{u,l} c_l$$

The conservation equation for the pesticide in the soil medium then reads:

$$\partial c_m / \partial t = - \partial J_s / \partial z - k_c \, c_m - R_{u,s}$$

The differential equations, boundary conditions and basic relationships were programmed in computer language CSMP[9] and the program was run on the DECsystem-10 computer of the Agricultural University, Wageningen. (A copy of the computer program can be obtained from the author on request). For these computations, the soil system was divided into 27 computation compartments: 15 compartments (2 cm thick) in the upper part of the soil profile, while below 30 cm depth the compartments became gradually thicker (multiplication factor 1.25). The computations were carried out with Euler's integration method using a time step Δt of 0.04 day, over a total period of 90 days. At high rainfall intensities, the time step had to be reduced to 0.005 day. At regular intervals during the computations, the material balances of the liquid and of the pesticide in the soil system were checked.

To compute the liquid fluxes, the arithmetic averages of the hydraulic conductivities for adjacent compartments were taken: this has been found to give the best description of liquid flow[10]. For the description of the convective flux of substance, $J_l\ c_l$, the concentrations in adjacent compartments were averaged, which minimizes numerical dispersion[6]. In the first weeks of the simulated period, when there was a steep concentration gradient at the lower end of the initial distribution, there was a risk of artificial downward flow by hydrodynamic dispersion, although liquid flow was upward because of evaporation at the soil surface. These artificial fluxes were suppressed by making the total substance flux smaller than or equal to the diffusion flux, when J_l was negative at this depth.

RESULTS AND DISCUSSION

Soil moisture contents, computed for different times in a period of three months with average amounts of rainfall, are represented in Fig. 1.

The times 25, 55 and 85 days fell after periods of four to six days without rainfall, whereas at 30, 60 and 90 days, there had been a rainy period of four days just before. The initial volume fraction of liquid, ε_l, in the top 30 cm was around 0.24 cm^3 cm^{-3}. Rainfall in the first month caused ε_l to increase somewhat, especially just after rain. In the dry spells there was a marked redistribution of soil moisture, with drying by evaporation at the soil surface. At the end of the second month, substantial amounts of water had been withdrawn by the root system: this was most marked at depths of about 15 cm. Great fluctuations in soil moisture content occurred near the soil surface as a result of rainfall, redistribution, withdrawal, and evaporation. At the end of the third month, the root system had withdrawn the available water from the 10 to 30 cm layer of the soil. The depth from which water was taken up gradually increased. Re-

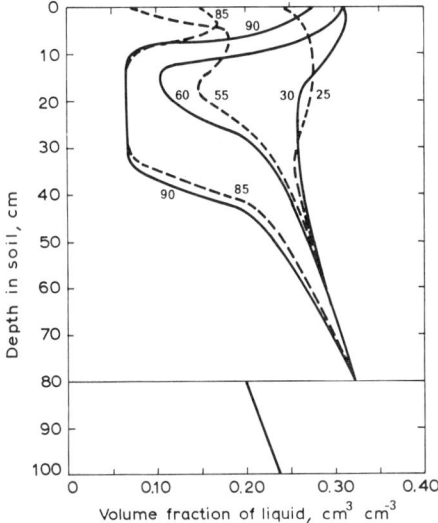

Fig. 1. Volume fractions of the liquid phase computed for the root zone of an arable crop on sandy loam soil. Times in days from the planting date in the beginning of April. Average amounts of rainfall.

distribution and uptake of recent rainwater occurred in the top region of the root zone. There was little upward movement of water from the phreatic level in this 90-day period: only 0.3 cm³ cm⁻².

The computed pattern of water withdrawal from soil under field conditions corresponds in general lines with measured patterns for various annual crops, as reviewed by Gardner[5] and Hillel[8]. Durrant et al.[2] measured water withdrawal by a potato crop using neutron scattering equipment. They found that most water was withdrawn from the shallow soil layers that had the highest rooting density, until the water supply was greatly depleted. Meanwhile the root systems penetrated into deeper soil layers and water-uptake activity gradually extended to these greater depths.

The computed relationships between the concentration of the pesticides and soil depth for average amounts of rainfall are given in Fig. 2.

During the first month, a considerable fraction of the pesticide that was most weakly adsorbed ($K_{s/l} = 0.01$ for the top layer), moved downwards. At day 25, which fell after a dry period with water evaporating from the soil surface, there was a tendency for some of

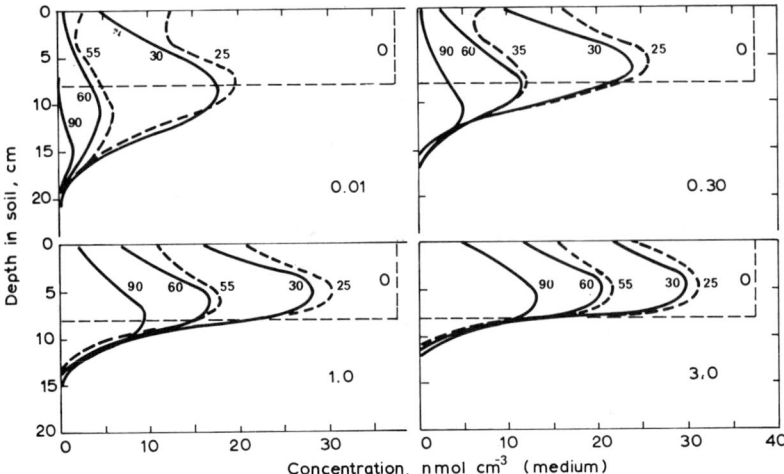

Fig. 2. Distributions of pesticides in soil computed for a field situation with an arable crop. Rainfall pattern R1: average amounts of rainfall. Times 25, 30, 55, 60, and 90 days after pesticide incorporation in soil and planting. Adsorption coefficient for the substances in the top layer 0.01, 0.30, 1.0, and 3.0 cm³ g⁻¹, respectively.

the substance to accumulate near the soil surface. In the rainy period preceding day 30, there was a distinct downward movement of the pesticide in the upper 10 cm, but below that depth, movement was only slight. At day 55 the amount of pesticide had greatly decreased as a result of uptake and decomposition. At the end of the period considered (day 90), only very low concentrations remained. It is remarkable that, after the first month, there was no distinct further penetration of the pesticide into the lower part of the plough layer.

The results of the computations for model pesticides with higher adsorption coefficients are also shown in Fig. 2. They show how downward movement decreases as adsorption onto the solid phase becomes stronger. At moderate values of the adsorption coefficient ($K_{s/l}$ = 1.0 and 3.0), downward redistribution was limited to only a few centimetres. Again, by far the major downward movement occurred in the first month after incorporation in the soil. The temporary slight accumulation of pesticide near the soil surface in dry periods became less marked as the adsorption coefficient increased.

Fig. 2 also shows that the decrease in the amount of pesticide in soil was most rapid for substances that were only weakly adsorbed. For $K_{s/l}$ -values of 0.01, 0.30, 1.0, and 3.0, the amounts remaining after 90 days were 3, 14, 26, and 34% of the dosage, respectively. The amounts taken up in this period by the plant roots were lower as adsorption was stronger; they were computed to be 63, 41, 22, and 9% of the dosage, respectively. These figures illustrate the significant part that uptake by plant roots may play in the decline of pesticide concentrations in soil, especially for weakly-adsorbed compounds.

The results of the computations for rainfall pattern R2, including a comparatively dry first month, are given in Fig. 3. During the first month, downward movement from the initial distribution was limited to only a few centimetres, even for the most weakly adsorbed compounds with $K_{s/l} = 0.01$ and 0.30. The presence of rather steep concentration gradients caused some further penetration during the second month, although limited to 1 to 2 cm at most. During the third month, hardly any further penetration occurred.

About 3% more dosage equivalents of the most weakly adsorbed compound ($K_{s/l} = 0.01$) had been taken up at the end of the first month than under average rainfall conditions, while for the other model compounds there was hardly any difference. At day 30, the cumulative amounts decomposed were about 2% dosage equivalents lower than under average rainfall conditions for all the compounds.

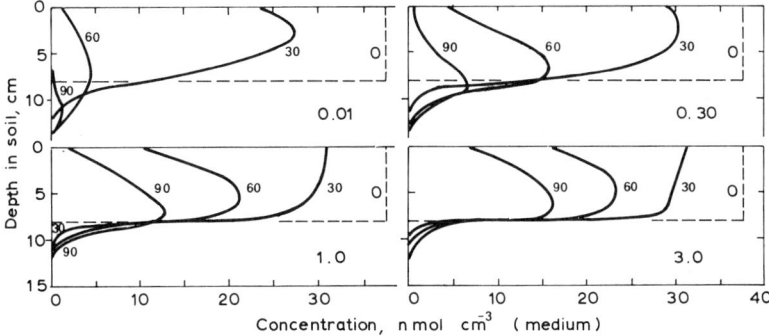

Fig. 3. Concentrations of the model pesticides in soil, computed for rainfall pattern R2, including a comparatively dry first month. Times 30, 60, and 90 days after application. Adsorption coefficients 0.01, 0.30, 1.0, and 3.0 cm^3 g^{-1}, respectively

COMPUTED REDISTRIBUTION OF PESTICIDES 577

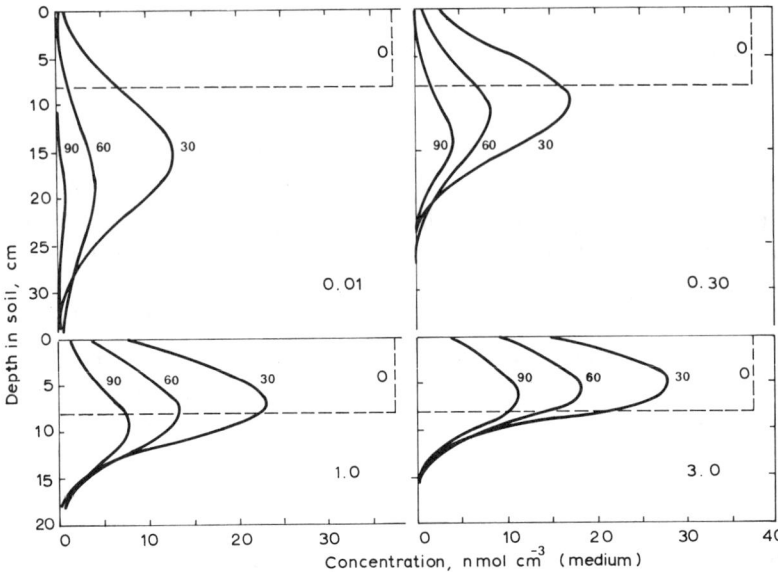

Fig. 4. Computed concentration-depth relationships for model pesticides in the soil of an arable field at 30, 60, and 90 days after incorporation. Comparatively high rainfall in the first month: rainfall pattern R3.

Fig. 4 gives the computed distribution of pesticides in the soil after a comparatively high rainfall during the first month after incorporation. The downward movement of the weakly-adsorbed compounds increased considerably, and their concentrations near the soil surface dropped to rather low values. During the second and third month with average amounts of rainfall, there was hardly any further redistribution of the compounds. The total amount decomposed after 30 days amounted to around 0.21 dosage equivalents, which was only about 0.006 dosage equivalents higher than for average rainfall conditions.

The amount of the most weakly adsorbed compound taken up by the root systems was 0.10 dosage equivalents as compared with 0.13 under average rainfall conditions. Evidently, this compound moved out of the soil region that had the highest uptake by the root systems. For the compounds with $K_{s/l} = 1.0$ and 3.0, high rainfall had no effect on the amount taken up.

With rainfall pattern R4, the effect of high rainfall in the second month could be checked. After 60 days, the peak of the distribu-

322

tion for the compound with $K_{s/l} = 0.01$ was at about 17 cm, compared with that at about 10 cm depth with pattern R1. For the substance with $K_{s/l} = 0.30$, the peak was at about 12 cm compared with 9 cm with R1. The concentrations in the upper part of the plough layer were comparatively low and approached those for the corresponding time with rainfall pattern R3 (first month with high rainfall). At 60 days with R4, penetration to depths below 20 cm was somewhat less than with rainfall pattern R3.

The possible influence of a reduced uptake by plant roots was checked by introducing an uptake efficiency factor of 0.5. This implies that only half the amount of pesticide supplied by the transpiration stream was taken up. For the case with $K_{s/l} = 0.01$ and an average rainfall pattern (R1), downward penetration was only increased by about 1 cm. Although concentrations at the various depths were comparatively high, there was again hardly any further penetration after the first month.

Because the values of the dispersion distance L_d may differ considerably for various soils, the effect of doubling the value of L_d to 4.0 cm was checked. The spread of the distribution was somewhat increased, especially in the upper part of the concentration pattern. After 30 days, the concentration level of 5 nmol cm^{-3} was only about 1 cm deeper than in the standard cases with $L_d = 2.0$ cm. Although the effective dispersion coefficient is lower at higher adsorption coefficients, the concentration gradients were comparatively steep, so a similar effect was obtained.

GENERAL DISCUSSION

The redistribution of pesticides in soil after incorporation in a top layer in spring was computed to be very limited. Under average conditions, only in the first month after application was there considerable convective transport of weakly-adsorbed pesticides by water flow. After that, uptake of water from the soil by the developing root system progressively slowed down the water fluxes over greater depths. The convective transport of pesticide below the top layer then became very slow and there was hardly any spreading by hydrodynamic dispersion. Spreading by diffusion through the liquid phase continued, but the diffusion rates were low, especially in drier soil regions, because the effective diffusion coefficients were very low.

The values for the adsorption coefficient $K_{s/l}$ were related to combinations of pesticides and soils. Besides the predisposition of the compound to adsorption, the adsorption characteristics of the soil are decisive. Therefore, the adsorption of pesticides by different soils must be measured. In practice, the low extent of redistribution to be expected, especially with pesticide-soil combinations showing strong adsorption, should be considered, because this dictates the depth to which pesticides have to be incorporated. In a series of field trials [13,20], the effects of shallow (7.5 cm) and deep (15 cm) incorporation of aldicarb and oxamyl granules on nematode population at potato harvest have been compared. In spite of the weak adsorption of these compounds in soils, deep rotavation of 5.6 kg active ingredient per ha was found to be more effective than shallow rotavation. The results of the present study indicate that incorporation to a depth of 7.5 cm might have been too shallow for a sufficient redistribution in the top 20 cm of the soil. At comparatively low dosages, the difference in efficacy between the two depths of incorporation was only small [13,20], possibly because the dilution effect became more predominant. To interpret correctly the results of trials with different depths of pesticide incorporation, it is essential to measure and take account of the rainfall pattern in the first months after application, especially with weakly adsorbed substances.

Received 6 April 1977

REFERENCES

1 Arkin, G, F., Ritchie, J. T. and Adams, J. E., A method for measuring first-stage soil water evaporation in the field. Soil Sci. Soc. Am. Proc. **38**, 951–954 (1974).
2 Durrant, M. J., Love, B. J. G., Messem, A. B. and Draycott, A. P., Growth of crop roots in relation to soil moisture extraction. Ann. Appl. Biol. **74**, 387–394 (1973).
3 Endrödi, G. and Rijtema, P. E., Calculation of evaporation from potatoes. Neth. J. Agric. Sci. **17**, 283–299 (1969).
4 Frissel, M. J., Poelstra, P. and Reiniger, P., A simulation model for the evaluation of the apparent diffusion coefficient in undisturbed soils with tritiated water. Plant and Soil **33**, 161–176 (1970).
5 Gardner, W. R., Dynamic aspects of soil-water availability to plants. Annu. Rev. Plant Physiol. **16**, 323–342 (1965).
6 Goudriaan, J., Dispersion in simulation models of population growth and salt movement in the soil. Neth. J. Agric. Sci. **21**, 269–282 (1973).
7 Graham-Bryce, I. J., Diffusion of organophosphorus insecticides in soils. J. Sci. Food. Agric. **20**, 489–494 (1969).
8 Hillel, D., Soil and water: physical principles and processes. Academic Press, New York (1971).

9. IBM, System/360 Continuous System Modeling Program, User's Manual. Program Number 360A-CX-16X. Technical Publications Department, White Plains, New York (1972).
10. Keulen, H. van and Beek, C. G. E. M. van, Water movement in layared soils. A simulation model. Neth. J. Agric. Sci. **19**, 138–153 (1971).
11. KNMI, Klimaatatlas van Nederland. Royal Netherlands Meteorological Institute. Staatsuitgeverij, The Hague, The Netherlands (1972).
12. Kutschera, L., Wurzelatlas mitteleuropäischer Ackerunkräuter und Kulturpflanzen. DLG-Verlags-GmbH, Frankfurt am Main (1960).
13. Moss, S. R., Crump, D. and Whitehead, A. G., Control of potato cyst-nematodes, *Globodera rostochiensis* and *G. pallida*, in different soils by small amounts of oxamyl or aldicarb. Ann. Appl. Biol. **84**, 355–359 (1976).
14. Porter, L. K., Kemper, W. D., Jackson, R. D. and Stewart, B. A., Chloride diffusion in soils as influenced by moisture content. Soil Sci. Soc. Am. Proc. **24**, 460–463 (1960).
15. Reid, R. C. and Sherwood, T. K., The properties of liquids and gases. McGraw-Hill, New York (1966).
16. Rowell, D. L., Martin, M. W. and Nye, P. H., The effect of moisture content and soil-solution concentration on the self-diffusion of ions in soils. J. Soil Sci. **18**, 204–222 (1967).
17. Scott, H. D. and Phillips, R. E., Diffusion of selected herbicides in soils. Soil Sci. Soc. Am. Proc. **36**, 714–719 (1972).
18. Smelt, J. H., Voerman, S. and Leistra, M., Mechanical incorporation in soil of surface-applied pesticide granules. Neth. J. Plant Pathol. **82**, 89–94 (1976).
19. Stroosnijder, L., Infiltratie en herverdeling van water in grond (Infiltration and redistribution of water in soils). Agric. Res. Rep. **847**, Pudoc. Wageningen, The Netherlands (1976).
20. Whitehead, A. G., Bromilow, R. H., Lord, K. A. and Moss, S. R., Incorporating granular nematicides in soil. Proc. 8th British Insectic. Fungic. Conf. 133–144 (1975).

List of Symbols and Units

ρ_b	= bulk density, g(soil) cm^{-3} (medium)
ε_l	= volume fraction of liquid, cm^3 (liquid) cm^{-3} (medium)
K	= hydraulic conductivity, cm^3 (liquid) cm^{-1} (medium) mbar^{-1} day^{-1}
p_t	= tensiometer pressure, mbar
p_g	= gravity pressure, mbar
p_h	= hydraulic pressure, mbar
J_l	= volumetric liquid flux, cm^3 (liquid) cm^{-2} (medium) day^{-1}
J_s	= substance flux, nmol cm^{-2} (medium) day^{-1}
L_d	= dispersion distance, cm (medium)
D_d	= coefficient of hydrodynamic dispersion, cm^3 (liquid) cm^{-1} (medium) day^{-1}
D_l	= coefficient for diffusion in liquid, cm^2 (liquid) day^{-1}
D_{lp}	= coefficient for diffusion in liquid phase, cm^3 (liquid) cm^{-1} (medium) day^{-1}
f_l	= impedance factor for diffusion through liquid phase, cm^2 (medium) cm^{-2} (liquid)
c_l	= concentration in liquid phase, nmol cm^{-3} (liquid)
c_m	= concentration in the medium, nmol cm^{-3} (medium)
$K_{s/l}$	= adsorption coefficient, {nmol g^{-1} (soil)}/{nmol cm^{-3} (liquid)}
z	= depth in soil, cm (medium)
t	= time, day
$R_{u,l}$	= rate of liquid uptake by roots, cm^3 (liquid) cm^{-3} (medium) day^{-1}
$R_{u,s}$	= rate of substance uptake by roots, nmol cm^{-3} (medium) day^{-1}
k_c	= rate constant for conversion, day^{-1}

Editors' Comments
on Papers 46 Through 48

46 HALL, PAWLUS, and HIGGINS
Losses of Atrazine in Runoff Water and Soil Sediment

47 LORBER and MULKEY
An Evaluation of Three Pesticide Runoff Loading Models

48 VINTEN, YARON, and NYE
Vertical Transport of Pesticides into Soil when Adsorbed on Suspended Particles

TRANSPORT IN ADSORBED PHASE

Strongly adsorbed pesticides may be transported laterally in sediments removed by erosion and runoff water. In addition, downward transport into the soil may occur, mainly in aggregated soils characterized by macropores and cracks.

Movement in Runoff Water

Runoff, induced either by rains or irrigation, is an important means of pesticide transport from one soil location to another and from the soil to surface water bodies. Characteristic of pesticide transport in runoff water is the fact that concomitant movement, both as a solute and in the adsorbed state, may occur. In this way, highly insoluble and adsorbed pesticides that do not move significantly as solutes can be transported.

The ratio of the amount of pesticides transported as solute to the amount adsorbed on suspended particles is dependent on their partitioning between the solid and liquid phases. In addition to pesticide and soil properties, the amount of pesticides found in runoff water is affected by the intensity of falling water and by the slope of the land. The formulation of the pesticide and its method of application may also affect its transport by runoff.

Paper 46 presents a study of the loss of the herbicide atrazine in runoff water and soil sediments. The results showed that the amounts of atrazine lost in eroded sediments and runoff water were affected significantly by the rate of application. The concentration of atrazine in the eroded sediment was higher than in runoff water because this compound is adsorbed by soil colloids. However, a greater total amount of atrazine was lost in runoff water. The pattern of the loss in runoff water during the growing season of corn was similar to the pattern of water loss. The data showed that atrazine was significantly removed from sloping cropland both in eroded sediments and in runoff water, indicating potential contamination of the adjacent area. However, the authors point out that these results were obtained in an experimental setting intentionally designed to maximize the losses (e.g., slope intensity of 14%, corn rows oriented parallel to the slope). The results are difficult to extrapolate to field conditions. The authors assumed that under good management and conservation practices the transport of atrazine in sediments and runoff water would be minor and not constitute an environmental hazard. The data of Leonard, Longdale, and Fleming (1979) confirmed that the strongly adsorbed herbicide paraquat was transported only in adsorbed sediments and that diphenamid and atrazine were transported mainly as solutes in the runoff water.

Pesticide formulation as a factor affecting the transport of herbicides in runoff was studied by Wauchope (1978). In both seasonal and long-term studies wettable powders consistently produced the highest runoff losses, up to 5%. The runoff loss of water-insoluble pesticides, usually applied as emulsions, was 1% or less; that of water-soluble pesticides, usually applied as aqueous solutions, and of soil-incorporated pesticides was only 0.5% or less.

Pesticide runoff loading models are required in order to assess environmental risks related to pesticide use. Models for runoff and erosion from agricultural lands were adapted to pesticide transport (Agricultural Runoff Management Model—ARM: Donigian et al., 1977; Chemicals, Runoff and Erosion from Agricultural Management Systems—CREAMS: Knisel, 1980; Continuous Pesticide Simulation—CPS: Steenhuis, 1979). Paper 47 presents a critical evaluation of the three models. In studies carried out in two watersheds, the models were tested for ability to simulate periodic losses of runoff water, sediment, and pesticides, for response to main assumption and input parameters, and for options and predictions in a long-term simulation of toxaphene and atrazine runoff. It was found that all models were able to reproduce the field data. For instance, the models differed by no more than 10% in predicting toxaphene loss. However, in the long-term experiment, the models differed in predicting the peak events, and they also behaved differently in the sensitivity test.

As a result of a study of pesticides in surface irrigation water, Spencer and co-workers (1985) concluded that the development of an operational model for estimating pesticide runoff from irrigated fields should include, besides the accepted parameters for rainfall areas, the hydraulics parameters derived from the use of various irrigation techniques.

Vertical Transport

Little attention has been paid to the vertical transport of pesticides adsorbed on suspended particles. The presence of macropores and cracks in soils enables the downward movement of water containing suspended colloids and adsorbed pesticides. Paper 48 proves experimentally that strongly adsorbed pesticides such as paraquat and DDT, which are considered virtually immobile in soil, may be transported vertically, adsorbed on suspended particles. The experiments were carried out in the laboratory, using soil columns leached either with paraquat adsorbed on Li-montmorillonite or with DDT adsorbed on the suspended organics of a sewage effluent. The extent of the vertical transport was directly related to the flow rate and soil physical properties. The authors suggested that this transport pathway of strongly adsorbed pesticides may occur in the field under conditions favorable to soil dispersion and release of mineral and organic colloids that can move through the soil as suspensions.

References

Donigian, A. S., D. C. Beyerlein, H. H. David, Jr., and N. H. Crawford, 1977, *Agricultural Runoff Management (ARM) Model Version II: Refinement and Testing,* EPA-600/3-77-098, USEPA, Athens, Ga.

Knisel, W. G., ed., 1980, CREAMS: A Field Scale Model for Chemicals Runoff and Erosion from Agricultural Management Systems, *USDA Soil Conservation Res. Rep. 26.*

Leonard, R. A., G. W. Longdale, and W. G. Fleming, 1979, Herbicide Runoff from Upland Piedmont Watersheds: Data and Implications for Modeling Pesticide Transport, *Jour. Environ. Qual.* **8:**223-229.

Spencer, W. F., M. M. Cliath, J. Blair, and R. A. Le Mert, 1985, Transport of Pesticides from Irrigated Fields in Surface Runoff and Tile Drainage Water, *U.S. Dept. Agric. Res. Rep. No. 31.*

Steenhuis, T. S., 1979, Simulation of the Action of Soil and Water Conservation Practices in Controlling Pesticides, in *Effectiveness of Soil and Water Conservation Practices for Pollution Control,* A. D. Haith and R. C. Loehr, eds., EPA-600/3-79-106, USEPA, Athens, Ga.

Wauchope, R. D., 1978, The Pesticide Content of Surface Water Draining from Agricultural Fields—A Review, *Jour. Environ. Qual.* **7:**459-472.

Losses of Atrazine in Runoff Water and Soil Sediment[1]

J. K. Hall, M. Pawlus, and E. R. Higgins[2]

ABSTRACT

Atrazine losses in runoff water and soil sediment were determined in 1967 and 1968 after seven rates (0, 0.6, 1.1, 2.2, 4.5, 6.7, and 9.0 kg/ha) of atrazine were applied pre-emergent to corn (Zea mays L.) seeded on field plots of Hagerstown silty clay loam (14% slope). Average losses for all rates in 1967 in runoff water and soil sediment equaled 2.4% and 0.16% of the total applied, respectively. In 1967, at the recommended rate (2.2 kg/ha) for pre-emergence applications to Pennsylvania soils, composite losses were 2.5% of the applied or approximately 0.05 kg/ha. In 1968, one year after atrazine application, the average loss over all rates for the combined substrates was 0.01%.

Analyses of soil core samples taken from all plots in 1967 revealed that 1 month after atrazine application an average of 67.9% remained in the soil, and 3 months later recoveries had decreased to 21.4% of that applied. The following year atrazine remaining in the soil had decreased to 15.9% in April and to 5.4% in September. At the recommended rate of application, recoveries decreased from 39% of that applied to 9% for the same time period in 1967. In 1969, typical atrazine toxicity symptoms were found in oats growing on plots which had received 6.7 and 9.0 kg/ha of atrazine in 1967. Damage was confined to the uppermost parts of the slope on these treatments.

Additional Index Words: pesticide, herbicide, field runoff plots.

The concern for maintaining high quality water resources has resulted in a continued focus on sources of water contamination. Herbicides applied to agricultural lands have been suggested as sources of pollution, yet little data are available on quantitative losses of herbicides in runoff water and eroded soils from agricultural watersheds. Considerable information has been gained from studies on simulated rainfall induced losses of 2,4-D (2) and atrazine (19) from fallow plots and 2,4,5-T, dicamba and picloram losses from both sod and fallow plots (17). In addition, losses of 2,4-D, 2,4,5-T, amitrole, and picloram applied to forested lands have been assayed under natural environmental conditions (9, 10). Additional information is needed on the runoff losses of herbicides from agronomic areas under these same conditions.

Atrazine is a widely used selective herbicide that will control most annual broadleaf and grassy weeds in corn (Zea mays, L.) and certain other crops. Its wide usage, moderate persistence in soils, and mode of application in crops such as corn, suggested that significant losses of this herbicide in runoff from sloping cropland may occur after intense rainfall. A field experiment planted in corn was established in the spring of 1967 to evaluate losses of atrazine in runoff water and eroded soil sediment. The specific objectives of this study were to determine (i) the concentration and amounts of atrazine lost in runoff water and sediment, (ii) the rate of dissipation of atrazine in the soil over two cropping seasons, and (iii) whether any atrazine residues remained in the soil during the third cropping season by using a qualitative oat bioassay.

METHODS AND MATERIALS

This investigation was conducted on a Hagerstown silty clay loam using field runoff plots of 14% slope. The soil had an organic carbon content of 1.38%, an exchange capacity of 15.4 meq/100 g soil, and a pH of 6.6. The 14 plots were individually tiered and divided by steel barriers extending approximately 30 cm above and below the soil surface throughout the plot length. Each plot (1.8 by 22.3 m) was fitted with a trough and chute at the base of the slope which facilitated the transport of runoff water to a collection point within an enclosed facility. Individual plots were prepared for planting each year by applying fertilizer and liming amendments to satisfy soil test recommendations and rototilling to a depth of approximately 15 cm. Each plot was smoothed by hand such that each surface had a uniform slope and erodibility. The plots were seeded to corn 'Pa 70RF X Pa 887P' in 1967 and 1968 and thinned to a common stand of 48,165 plants/ha (19,500 plants/acre) each year. Corn rows were oriented parallel to the slope. The plots were hand-cultivated only in 1968. The plots which had received the first five treatments in 1967 were weeded on June 21 and July 2, 32 and 44 days after planting.

Pre-emergent applications of atrazine (2-chloro-4-ethylamino-6-isopropylamino-s-triazine, 80% wettable powder supplied by Geigy Agricultural Chemicals) were made on May 19, 1967 at 7, twice-replicated rates (0, 0.6, 1.1, 2.2, 4.5, 6.7, and 9.0 kg active ingredient/ha) using a pressure-regulated hand sprayer. No further applications of atrazine were made. No atrazine residues were present in this soil prior to treatment. This was confirmed by an oat bioassay experiment conducted in a greenhouse, where the growth of oats (Avena sativa L.) on this untreated soil was compared with growth on the same soil to which various atrazine levels had been added. Runoff was collected from date of planting to harvest during each year of the study, which extended from May 18 to November 7 in 1967 and from May 20 to October 25 in 1968. An on-site weather station provided rainfall data throughout the course of the study. Runoff suspensions were collected and weighed. In most instances, a 3-liter sample was collected from each runoff plot after thoroughly suspending the sedimented soil in each container. The runoff samples were separated into soil and water fractions based on the calculated percent of suspended material. This was determined by mechanically shaking each sample, withdrawing specific volume aliquots, and weighing before and after drying. The bulk of the soil in each sample was separated from suspension by centrifuging at 2,000 rpm for 10 min. The separated soil was dried at 40C, sieved (<2 mm), and saved for atrazine residue analysis. The decanted centrifugate was further centrifuged at 50,000 rpm for 10 min to remove most of the finely suspended soil material. The centrifugate was stored in tightly sealed plastic containers at room temperature (27C). One milliliter of chloroform was added to each sample immediately after collection and again after centrifuging. Periodic analysis of selected samples indicated that no atrazine degradation occurred during storage.

Soil core samples were collected at periodic intervals from each plot to obtain a measure of the rate of dissipation of atrazine over two growing seasons. Soil samples were collected at depths of 0 to

---

[1] Authorized for publication on March 25, 1971, as paper no. 3943 in the Journal Series of the Pennsylvania Agr. Exp. Sta., University Park, Pa. Presented at NEBASA Meetings, June 14–17, 1970. University of Maryland, College Park, Md. Received April 21, 1971.

[2] Assistant Professor of Soil Chemistry, Visiting Agronomist and former Graduate Assistant, The Pennsylvania State Univ., University Park, Pa. 16802. The present address of the second author is Dept. of Particular Plant Cultivation, College of Agriculture, Szczecin, Poland. The third author is presently Research Representative, Geigy Agricultural Chemicals, Ardsley, New York.

10 and 10 to 20 cm at approximately 3-m intervals within each plot. Six samples from each depth within a plot were composited separately, air-dried, ground, sieved (<2 mm), and retained for atrazine residue analysis.

The plots were seeded to 'Pennfield' oats (*Avena sativa* L.) on April 17, 1969 to provide a visual bioassay of any residual atrazine activity in the soil.

Water samples were analyzed for atrazine by the pyridine-alkali-ethyl cyanoacetate (PAE) procedure as described by Radke et al. (11). Two modifications of this procedure were employed. The atrazine-pyridine coupling reaction was allowed to proceed in a water bath at 85C for 2 hours instead of 100C for 30 min. Using a lower reaction temperature prevented excessive volatile losses of reactants without an apparent loss of reaction efficiency. A comparison of standard curves indicated that reaction times greater than 2 hours yielded an incomplete reaction when concentrations of atrazine were greater than 1.0 ppm. A reaction time of 2 hours at 100C was reported to be optimum (8), however, it appeared that a 2-hour exposure at 85C was satisfactory at low concentrations of atrazine. Water samples having a concentration greater than 1.0 ppm were diluted with distilled water prior to analysis and later corrected for proper concentration. The intensity of the red chromophore developed by this procedure was read at 550 mμ using a Bausch and Lomb Spectronic 20. The concentration of atrazine in each water sample was determined from a standard curve prepared from technical grade atrazine (98% purity) and carried through the same procedure. Atrazine recoveries from water samples using this procedure were 98%.

In order to establish the optimum percent recovery of atrazine from this soil with minimum contamination from co-extracted organic constituents, untreated soil was collected from the experimental site, air-dried, ground, sieved (<2 mm), and treated with various rates of atrazine. Atrazine concentrations were formulated by diluting appropriate aliquots of a stock atrazine solution (50 ppm in 95% ethyl alcohol) with distilled water. Sufficient volume was used to saturate the soil in each container. After 2 hours, the soils were placed in an oven and dried at 40C. This wetting and drying cycle was repeated using a comparable volume of distilled water. Triplicate samples (25 g) of dried soil were extracted with with 100 ml of various concentrations of ethyl alcohol for 16 hours using Soxhlet extraction. Soil extracts were treated with 1 ml of 10% $(NH_4)_2SO_4$ to precipitate co-extracted organic constituents. After 24 hours, the extracts were gravity filtered (Whatman no. 42 paper) which yielded a nonturbid, yellow extract. Samples were analyzed by the PAE method and concentrations were determined from an appropriate standard curve prepared from 98% atrazine. Percent recoveries were determined for each extractant concentration. For the selected extractant (50% ethyl alcohol), recoveries ranged from 90 to 150% when the soil concentration varied from 5.0 to 0.25 ppm. Recovery was greater than 150% at soil concentrations less than 0.25 ppm. The atrazine concentrations of all soil samples (eroded sediment and plot core samples) were corrected to 100% recovery. Since "zero concentrations" were not obtained for extracts of soil from check plots due to the yellow color present, values were subtracted from the atrazine concentrations of treated soils for a respective runoff or sampling date. Because the "concentrations" for runoff check samples were comparable throughout the study, an average value was subtracted in instances where no sample was collected from the check plots.

RESULTS

The amount of precipitation from May through September in 1967 was above normal for this region. In terms of total rainfall for this period, the deviation from the normal 83-year (1887 through 1970) average was +4.45 cm. The majority of this increase came during the months of July, August, and September where the total rainfall deviated from average monthly levels by +9.39, +.02, and +2.66 cm, respectively. This is best reflected by the number of runoff collections obtained during this period (Fig. 1). Runoff was heaviest during July where five collections were obtained following periods of precipitation ranging from 1.02 to 5.72 cm. In general, increasing amounts of water and soil were lost in each runoff with increasing rate of herbicide applied. As the season progressed more total rainfall was necessary to induce runoff losses of water and soil. In August where precipitation was recorded on eight different occasions, some of which were equal to or greater in total amount than several runoff-inducing rains recorded in previous months, only one runoff collection was obtained on August 28. The results for the total amounts of water lost with each runoff in 1967 indicated that amount and frequency of precipitation were important parameters in regulating these losses, particularly early in the growing season. Crop density and evapotranspirational losses of water appeared to have an important influence on regulating the number of runoff collections and amounts of water and soil lost from mid- to late-season.

Increasing levels and amounts of atrazine were found in runoff water as the rate of application increased from 0.6 to 9.0 kg/ha (Fig. 2). In the first collection on June 12, 23 days after application, concentrations ranged from 0.39 to 4.68 ppm from the lowest to the highest rate of application. Approximately 1 month after application (June 23), the concentration of atrazine in runoff water was 4 to 5 times less than that previously detected. These lower concentrations, which continued to decrease with each runoff collection throughout the growing season, were reflective of the rate of atrazine dissipation in this soil type which is discussed hereafter. The total amounts of atrazine lost in runoff water in 1967 (Fig. 2) varied with the amounts of water lost. With the exception of the first collection, where water loss was low but concentrations were high, the pattern of total losses throughout most of the growing season is coincident with the amount of water lost. The fact that rainfall frequency is important in regulating runoff losses can be seen from a comparison of the amounts lost in the third and fourth runoff collections. More total atrazine was lost on July 5 than July 3 despite the fact that the amount of precipita-

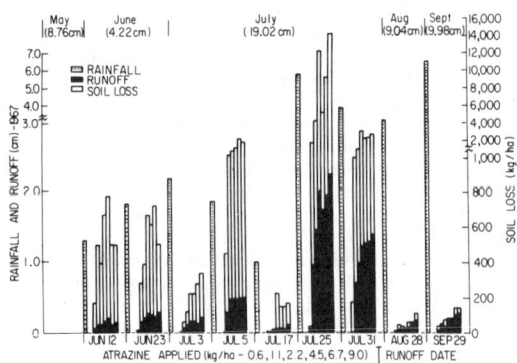

Fig. 1—Runoff and soil sediment losses from nine runoff-inducing rains in 1967. The rainfall totals for each month are presented at the top of the graph.

tion and the atrazine concentration in the soil were less. More water was lost on July 5 which could have resulted from a higher residual soil moisture level and/or storm intensity.

The pattern of loss of atrazine in eroded soil sediment was similar to that found for runoff water throughout the growing season. Increasing levels of herbicide were detected in soil sediment with increasing rates of application (Fig. 3). Likewise, the levels of atrazine in soil sediment showed a steady decline with each succeeding runoff. The concentrations of atrazine in soil sediment in the first collection were found to be 0.33, 0.96, 1.39, 2.87, 5.37, and 6.23 ppm for the respective rates of application. These levels were higher than those detected in runoff water for the same collection, and this relationship held in general for each succeeding runoff where soil sediment was collected. Higher concentrations of atrazine in soil sediment than in runoff water would be expected since results of previous investigations have shown that atrazine molecules can be adsorbed on soil colloidal material (16) and be transported in soil sediment (19). Despite the higher concentrations in soil sediment, greater amounts of atrazine were lost in runoff water. As a comparison, the amounts of atrazine lost in runoff water on June 12 ranged from 5.0 to 61.0 g/ha with increasing herbicide rate. The amounts lost in soil sediment on the same date ranged from 0.2 to 3.0 g/ha (Fig. 2 and 3).

The total amounts and the percentage of atrazine lost in each substrate during 1967 are shown in Table 1. The amounts of atrazine lost were significantly affected by the different rates of application. This was not true with the percent of total atrazine lost, which was variable with application rate. Annual losses in runoff water and soil sediment during 1967 equaled 2.4 and 0.16%, respectively. Losses in water were fairly constant, ranging from a 1.7% loss at the 0.6-kg/ha rate to a 3.6% loss at the 1.1-kg/ha rate. Percent losses in eroded sediment increased from 0.03 to 0.28% as the rates increased from 0.6 to 9.0 kg/ha. At the recommended rate for surface application to Pennsylvania soils, 2.2 kg/ha, composite losses (water and soil) amounted to 2.5% of that applied or 55 g/ha. This percentage loss was comparable to the average composite loss over all rates of application. In 1968, atrazine was detected in the first three of nine collections of runoff water and in very small amounts (Table 2). Concentrations of atrazine in eroded sediment were again higher than in runoff water, but less than 1967 values. Average losses of atrazine in the combined substrates in 1968 amounted to 0.53 g/ha or 0.01% of that applied in 1967.

A previous study by White et al. (19) showed that a simulated rainfall of 3.1 cm, which was comparable to a 1-year frequency storm, induced total losses of 12.0 and 5.3% after 3.4 kg/ha of atrazine was applied 1 and 96 hours before rainfall application. In comparison, 9 runoff-producing rains totaling 26.7 cm of rainfall in 1967 induced total atrazine losses of 2.5% and 2.2%, respective-

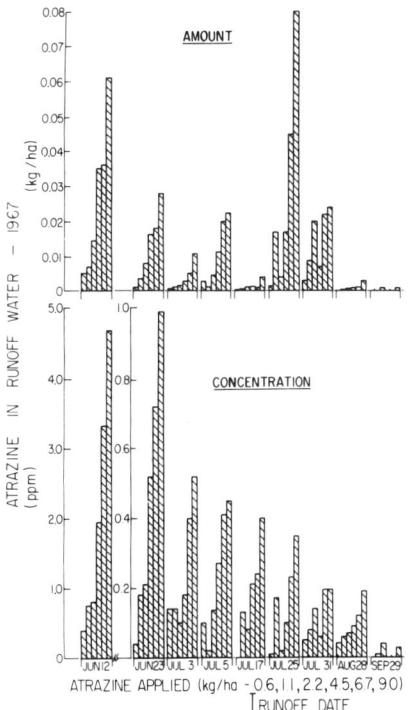

Fig. 2—The concentrations and amounts of atrazine in runoff water in 1967.

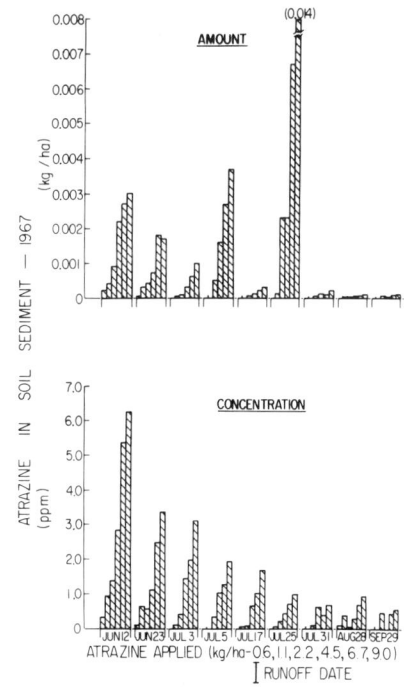

Fig. 3—The concentrations and amounts of atrazine in soil sediment in 1967.

Table 1—Total amounts and the percentage of atrazine lost in runoff water and soil sediment in 1967

Rates applied kg/ha	Amounts, g/ha			Percent, %		
	Water	Soil	Water + Soil	Water	Soil	Water + Soil
0.6	10.0	0.2	10.2	1.7	0.03	1.73
1.1	40.0	0.8	40.8	3.6	0.07	3.67
2.2	50.0	4.3	54.3	2.3	0.20	2.50
4.5	90.0	7.5	97.5	2.0	0.17	2.17
6.7	140.0	14.9	154.9	2.1	0.22	2.32
9.0	240.0	24.9	264.9	2.7	0.28	2.98
Means	95.0	8.8	103.8	2.4	0.16	2.56

	Significance level	
	Amounts	Percent
Rates applied (R)	**	NS
Substrate (S)	NS	NS
R × S	NS	NS

NS = Nonsignificant; ** = Significant at the 1% level of probability.

Table 2—Average concentrations of atrazine in runoff water and soil sediment and the total amounts and percentage of atrazine lost in the combined substrates in 1968

Rates applied† kg/ha	Average concentrations, ppm‡		Amounts, g/ha	Percent, %
	Water	Soil	Water + Soil	Water + Soil
0.6	0	<0.01	<0.01	<0.01
1.1	<0.01	0.10	0.07	<0.01
2.2	0.01	0.12	0.47	0.02
4.5	0.01	0.15	0.27	<0.01
6.7	0.02	0.37	1.56	0.02
9.0	0.02	0.41	1.16	0.01
Means	0.01	0.19	0.53	0.01

	Significance level		
	Average concentration	Amounts	Percent
	Water	Soil		
Rates applied	NS	**	**	NS

NS = Nonsignificant; ** = Significant at the 1% level of probability.
† Atrazine was applied only once in the spring of 1967.
‡ Atrazine was only detected in the first three of nine runoff water collections in 1968. Average concentrations reported for each substrate at each application rate were calculated over nine collections.

ly, at the 2.2- and 4.5-kg/ha rates of application. The relatively small losses of atrazine in the combined substrates is a function of the rate of attenuation and degradation of this compound in soil.

In addition to the precipitation-induced losses discussed here, atrazine may degrade as a result of photodecomposition (4), volatilization (4, 6), microbial transformation (5, 15), catalytic processes (1, 15), and decomposition by plant metabolism (14). As a result of these processes, atrazine degrades at a moderate rate, which is evident from the analysis of soil cores collected during 1967 and 1968 (Table 3). One month after application, an average of 68% of the total atrazine applied was recovered, which corresponded to an average dissipation of 32%. By September 5, 79% of the total atrazine had dissipated on the average. At the prescribed rate, approximately 60% of the atrazine was lost in 1 month and 91% by the last sampling. The highly significant interactions of rates × sampling depth and sampling depth × sampling date indicated that movement of atrazine downward in the soil profile was not uniform over treatment levels or time. The data show that the atrazine concentrations within the 10 to 20-cm soil depth fluctuated throughout the growing season. However, in most cases atrazine levels were higher within the upper 10 cm of soil, indicating that the majority of the atrazine was retained in that layer. At the first sampling in 1968 (Table 3) an average of 15.9% of the applied atrazine was recovered. Some experimental error, either in sampling or analytical technique, did occur, as indicated by the higher levels detected at some of the rates compared with values recorded the previous year. By September 1968, 95% of the applied atrazine had dissipated in this soil and most of the residual activity was concentrated within the two highest rates. The recovery levels on September 3, 1968, in the upper 20 cm of soil represented concentrations of 0.04, 0.02, 0.05, 0.04, 0.08, and 0.09 ppm for the six rates of application, respectively.

In the spring of 1969, oats were seeded on all plots to obtain a qualitative measure of whether sufficient atrazine activity remained in the soil to cause injury in oats. Typical atrazine toxicity symptoms were found in oats growing on plots which had received 6.7 and 9.0 kg/ha of atrazine in 1967. Damage was confined primarily to the tops of the slopes where erosion was less. No damage was detected on any other treatments.

DISCUSSION

The magnitude of losses of any herbicide compound from sloping cropland is governed by the volume and intensity of rainfall and proximity to the application date, the quantity of chemical applied, formulation, mode of application, water-solubility, persistence in the soil, type of crop and plant density, topography, and soil characteristics. Some of these parameters have been demonstrated by previous investigators (2, 9, 10, 17, 19) to be important in regulating runoff losses of a herbicide. The consequences of runoff losses of a herbicide should be considered in two ways. One is its potential effect on adjacent areas contacted by runoff flow and planted to crops sensitive to herbicides, such as atrazine. The other consideration is the amount that might reach and contaminate surface waters. The results of this study indicated that the potential for contamination of adjacent areas exists. However, by the very nature of the experimental

Table 3—Atrazine levels and the mean percentage of applied atrazine recovered at two soil depths in 1967 and 1968

Rates applied, kg/ha	ppm, 1967								ppm, 1968			
	June 22		July 13		August 9		September 5		April 10		September 3	
	0-10cm	10-20cm	0-10cm	10-20cm	0-10cm	10-20cm	0-10cm	10-20cm	0-10cm	10-20cm	0-10cm	10-20cm
0.1	0.09	0.13	0.08	0.11	0.06	0.03	0.06	0.07	0.08	0.01	0.00	0.04
1.1	0.17	0.09	0.11	0.14	0.09	0.03	0.04	0.04	0.08	0.01	0.01	0.01
2.2	0.33	0.06	0.16	0.13	0.06	0.03	0.08	0.01	0.13	0.01	0.04	0.01
4.5	1.57	0.11	0.79	0.16	0.48	0.02	0.28	0.04	0.15	0.00	0.02	0.02
6.7	1.68	0.18	1.12	0.10	0.59	0.19	0.46	0.03	0.23	0.02	0.06	0.02
9.0	3.12	0.15	1.06	0.08	1.05	0.09	0.72	0.05	0.40	0.05	0.06	0.03
Means, ppm	1.16	0.12	0.55	0.12	0.39	0.07	0.27	0.04	0.18	0.02	0.03	0.02
Percent, %	52.7	15.2	28.9	16.4	19.7	5.1	14.6	6.8	14.4	1.5	1.8	3.6

	Significance level	
	1967	1968
Rates applied (R)	**	**
Sampling depth (Dp)	**	**
Sampling date (Dt)	**	**
R × Dp	**	NS
R × Dt	**	**

** = Significant at the 1% level of probability.   NS = Nonsignificant.

design and plot construction used in this study, conditions for runoff losses were maximized. Aside from the slope intensity (14%), corn rows were oriented parallel to the slope, thereby creating little resistance to downslope movement of water. The stand density of 48,165 plants/ha (19,500 plants/acre) can be considered a medium plant population by today's standards. A greater population with consequent higher leaf area index would further limit raindrop impact at the soil surface, which would reduce surface puddling and excessive soil and water loss. Furthermore, the losses reported in this study resulted from direct movement of atrazine in runoff water and/or eroded soil from a field site to a collection facility. No physical impedance to runoff flow existed from the uppermost portion of each tier to its base, except for later insurgence of weedy growth after residue levels declined to low levels. Under field conditions, good soil and water conservation practices on sloping cropland would limit the total amounts of water and soil which could be transported from a given area by runoff to nontreated areas. Consequently, even though the rainfall pattern, land-slope and plant density within an area might be similar to that used here, it would be expected that runoff losses, if any, would be markedly reduced. Differences in amount of loss from variable rainfall patterns would likely be insignificant where the recommended rate is applied pre-emergent. In a present study, where runoff losses of different s-triazine herbicides are being evaluated on this site, losses of atrazine at this rate were 2.0% and 0.4% higher in runoff water and eroded soil than in 1967. These higher losses resulted from a different rainfall spectrum which yielded five runoff collections within the first month after application. The first runoff came 6 days after application, compared to 23 days in 1967. The more intense rainfall distribution early in the growing season of 1970 did not produce drastically higher losses at this rate.

Where runoff does occur within a watershed, a natural dilution effect would result from mutual contact between runoff flows from treated and nontreated areas. Paradoxically, a small concentration of atrazine lost in runoff water and contacting adjacent areas planted to atrazine-sensitive crops may have a stimulating effect on growth and crop quality. Protein levels and in some instances yields have increased (3, 12, 13) in some s-triazine-sensitive crops after exposure to low concentrations of atrazine and a related chlorotriazine, simazine (2-chloro-4,6 bis-ethylamino-s-triazine).

Although it is difficult to extrapolate the conditions prevailing at this experimental site to field conditions, which vary widely, it seems reasonable to assume that runoff losses (where they occur) from recommended levels of atrazine application would probably be minor in terms of seriously damaging sensitive crop species or contaminating surface waters adjacent to a treated area. However, the results speak strongly for the maintenance of sound soil and crop management systems. The interaction effect of atrazine with other chemicals in the environment, and the impact of long-term exposure of small levels of atrazine and most of the pesticides in use today on other biological systems, is not known. On the other hand, studies have indicated that the toxicity of atrazine to warm-blooded animals and fish is low. Daily incorporations of 200 ppm of atrazine in the diet of rats over a 2-year period (7), and exposure of minnows to 0.5 ppm of atrazine for 48 hours (18), yielded no significant effects over the controls in either case. Concentrations of 10 and 5 ppm of atrazine were lethal to minnows in 6 hours and 48 hours, respectively.

Considering the nature of the experimental site, along with the retention and moderate rate of dissipation of atrazine in the soil and the small amounts lost in runoff from standard erosion plots at the prescribed rate for Pennsylvania soils (2.2 kg/ha), it seems unlikely that atrazine would contribute to contamination of any non-treated area enough to cause severe crop damage, particularly where sound management and conservation practices are used.

LITERATURE CITED

1. Armstrong, D. E., and G. Chesters. 1968. Adsorption catalyzed chemical hydrolysis of atrazine. Environ. Sci. Tech. 9:683-689.
2. Barnett, A. P., E. W. Hauser, A. W. White, and J. H. Holladay. 1967. Loss of 2,4-D in washoff from cultivated fallow land. Weeds 15:133-137.
3. Eastin, E. F., and Davis, D. E. 1967. Effects of atrazine and hydroxyatrazine on nitrogen metabolism of selected species. Weeds 15:306-309.
4. Jordan, L. S., W. J. Farmer, J. R. Goodin, and B. E. Day. 1970. Nonbiological detoxication of the s-triazine herbicides Residue Rev. 32:267-286.
5. Kaufman, D. E., and P. C. Kearney. 1970. Microbial degradation of triazine herbicides. Residue Rev. 32:235-266.
6. Kearney, P. C., T. J. Sheets, and J. W. Smith. 1964. Volatility of seven s-triazines. Weeds 12:83-87.
7. Knusli, E. 1964. Herbicides. In G. Zweig (ed.) Analytical methods for pesticides, plant growth regulators and feed additives. IV:33-36. Academic Press, New York.
8. McGlamery, M. D., F. W. Slife, and H. Butler. 1967. Extraction and determination of atrazine from soil. Weeds 15:35-38.
9. Norris, L. A. 1967. Chemical brush control and herbicide residues in the forest environment. p. 103-123. In Symposium proceedings: Herbicide and vegetation management in forest, ranges and noncrop lands. School of Forestry, Oregon State Univ., Corvallis, Oreg.
10. Norris, L. A. 1969. Herbicide runoff from forest lands sprayed in summer. Res. Prog. Rep., Western Soc. Weed Sci. p. 24-26.
11. Radke, R. O., D. E. Armstrong, and G. Chesters. 1966. Evaluation of the pyridine-alkali-colorimetric method for determination of atrazine. J. Agr. Food Chem. 14:70-73.
12. Ries, S. K. 1968. Spray-on protein boosters. Crops Soils Mag. 20(8):15-17.
13. Schweizer, C. J., and Ries, S. K. 1969. Protein content of seed: Increase improves growth and yield. Science 165: 73-75.
14. Shimabukuro, R. H., and H. R. Swanson. 1969. Atrazine metabolism, selectivity, and mode of action. J. Agr. Food Chem. 14:199-205.
15. Skipper, H. D., C. M. Gilmour, and W. R. Furtick. 1967. Microbial versus chemical degradation of atrazine in soils. Soil Sci. Soc. Amer. Proc. 31:653-656.
16. Talbert, R. E., and O. H. Fletchall. 1965. The adsorption of some s-triazines in soils. Weeds 13:46-52.
17. Trichell, D. W., H. L. Morton, and M. G. Merkle. 1968. Loss of herbicides in runoff water. Weed Sci. 16:447-449.
18. Vivier, P., and M. Nisbet. 1965. Toxicity of some herbicides, insecticides, and industrial wastes. US Pub. Health Serv. Pub. no. 999-WP-25, 167-169.
19. White, A. W., A. P. Barnett, B. G. Wright, and J. H. Holladay. 1967. Atrazine losses from fallow land caused by runoff and erosion. Environ. Sci. Tech. 1:740-744.

An Evaluation of Three Pesticide Runoff Loading Models[1]

MATTHEW N. LORBER AND LEE A. MULKEY[2]

ABSTRACT

Three nonpoint source runoff models were tested and compared for their abilities to predict the movement of the pesticides toxaphene and atrazine (2-chloro-4-(ethylamino)-6-(isopropylamino)-1,3,5-triazine) from a 15.6-ha watershed in the Mississippi Delta region and a smaller watershed in the Southern Piedmont. The three models are the Agricultural Runoff Management (ARM), Continuous Pesticide Simulation (CPS), and the Chemical, Runoff, and Erosion from Agricultural Management Systems (CREAMS). Published data on runoff, erosion, toxaphene, and atrazine runoff were used to test the models. Testing exercises indicated that all models accurately reproduced field data. For the total period of study, model predictions of total runoff differed from field observations by 15% or less. For the CPS and ARM models, predictions of total erosion differed from observations by 6%, whereas CREAMS underpredicted erosion by 25%. All models are within 10% of observations in overland toxaphene loss predictions. Five-year simulations indicated that the models can differ in their predictions of peak events. Sensitivity analysis indicated that ARM can predict higher losses of soluble chemicals than CPS or CREAMS, due to an interflow component unique to the ARM model. Similarly, estimation of a sediment enrichment in the CREAMS model resulted in higher toxaphene loss predictions than the other two models.

Additional Index Words: mathematical simulation, runoff, toxaphene fate and transport.

Lorber, M. N., and L. E. Mulkey. 1982. An evaluation of three pesticide runoff loading models. J. Environ. Qual. 11:519-529.

The use of mathematical models in evaluating water quality issues has increased in recent years. Hydrologic transport models have contributed significantly to such efforts especially when the water quality impact from nonpoint sources was of concern. The early models attempted to simulate runoff-related water quality processes including the transport and fate of pesticides (1, 2, 8). Since the initial models were published, a number of studies have illustrated their value in evaluating the potential effectiveness of Best Management Practices (4, 16) and their use as a component in more comprehensive modeling of water quality impacts in river systems draining agricultural areas (7).[3] More recently the requirement to perform environmental exposure assessments for toxic substances, including pesticides, has led to increased use of pesticide runoff loading models to enhance estimates of environmental risk arising from specific chemical usage.[4,5]

Recognition that pesticide runoff models offer conceptually valid means to aid in the evaluation of environmental risks has always been tempered by the questions of model verification, validation, and performance. The most appropriate procedures to validate models are not clearly defined and in most cases are somewhat dependent upon the anticipated use of the model. In all cases, some measure of "model performance" is desirable if modeling results are to become part of a planning, design, or regulatory decision-making effort.

This study evaluated the performance of three pesticide-runoff models when applied to a typical environmental risk-assessment problem. The models evaluated were the Agricultural Runoff Management (ARM), version II; the Continuous Pesticide Simulation (CPS), based upon an earlier model developed by Steenhuis (1979); and the Chemical, Runoff, and Erosion from Agricultural Management Systems (CREAMS). The problem to which all models were applied is an estimate of the field-scale runoff losses from the pesticides toxaphene applied to cotton (*Gossypium hirsutum* L.) grown in the Mississippi River Delta and atrazine (2-chloro-4-(ethylamino)-6-(isopropylamino)-1,3,5-triazine)[6] applied to a watershed in the Southern Piedmont. Because some field data were available, model evaluations included comparison of simulated and observed data. Comparisons were also made of each model's performance in sensitivity testing. Because a potential use of models for exposure assessment is long-term simulations in which patterns of pesticide runoff can be examined quantitatively, all models were used for a 5-year simulation under identical conditions.

DESCRIPTION OF MODELS EVALUATED

The general conceptual basis for all the models evaluated is similar. All are based on the premise that processes influencing pesticide concentration or mass in the soil-plant profile must be combined with the runoff and erosion transport mechanisms that move the pesticide from its point of application to the boundaries of the area being analyzed. These fate and transport mechanisms result from time series inputs to the watershed of interest and are modeled as continuous (in time) processes. Because pesticide loading is of interest in this study, the geographical scale of interest is very small watersheds typified by field units.

Complete and detailed descriptions of each model are found in several references and user manuals (5, 11, 16); summaries of the key differences among the models are given in Tables 1-3. The most notable of the differences are found in the hydrologic and erosion components. Approaches for the hydrologic components vary from the classical water balance model first published as the Stanford Watershed Model (3) to the more theoretically based Green-Ampt in-

---

[1] Contribution of the Environmental Research Laboratory, U.S. Environmental Protection Agency, College Station Rd., Athens, GA 30613. Received 21 July 1981.
[2] Agricultural Engineers, EPA Environmental Research Lab.
[3] A. D. Nicks, G. A. Gander, M. H. Frere, and R. G. Menzel, 1979. Evaluation of chemical transport models on range and cropland watersheds. Presented at 1979 summer meetings of Am. Soc. Agric. Eng. and Can. Soc. Agric. Eng., University of Manitoba, Winnipeg. 24-27 June 1979.
[4] L. A. Mulkey, and K. F. Hedden. 1980. Assessment of toxaphene exposure levels in the Yazoo River resulting from basin-wide application of toxaphene to cotton and soybeans. EPA Environ. Res. Lab., Athens, GA 30613 (unpublished report).
[5] J. W. Falco, L. A. Mulkey, K. F. Hedden, C. N. Smith, T. O. Barnwell, J. D. Dean, R. E. Lipcsei, and M. C. Smith. 1978. Estimated degradation and transport of dimilin in selected rivers of the southern U.S. EPA Environ. Res. Lab., Athens, GA 30613 (unpublished report).
[6] This paper reports the results of research only. Mention of a pesticide does not constitute a recommendation for use by the U.S. EPA nor does it imply registration under FIFRA, as amended. Mention of trade names or commercial products is for information purposes only and does not constitute endorsement or preferential treatment by U.S. EPA.

Table 1—Hydrology components of ARM, CPS, and CREAMS models.

Model	Zones modeled	Runoff	Infiltration	Evapotranspiration	Snowmelt
CPS	Three zones: 1) plow layer, 2) root zone layer, and 3) below root zone.	SCS curve number equation	Infiltration of water above field capacity occurs in 1 d. No infiltration < field capacity.	Two options: 1) estimated using pan evaporation, and 2) seasonal sinusoidal function.	Degree-day formulation for snow accumulation and melt
CREAMS Option 1	Seven layers extending to bottom of root zone.	SCS curve number equation	Infiltration calculation for water above field capacity.	Potential ET is a function of monthly average temperature and average solar radiation, divided into soil evaporation and plant transpiration.	Degree-day formulation for snow accumulation and melt
CREAMS Option 2	Two zones: 1) surface and 2) root.	$q_i = r_i - f_i$†	Green-Ampt infiltration formulation. Divided into soil evaporation and plant transpiration.	Potential ET is a function of monthly average temperature and average solar radiation, and plant transpiration.	Degree-day formulation for snow accumulation and melt
ARM	Three zones: 1) surface, 2) upper, and 3) lower.	Kinematic wave	Phillip's Equation	Areal ET as a function of soil moisture, vegetal cover, and pan-input data	Energy balance

† q_i = runoff, r_i = rainfall, and f_i = infiltration, respectively, period i.

filtration equations (9). The descriptions of erosion mechanics vary from modifications of the Universal Soil Loss Equation (USLE) for total erosion to bed-load transport equations for rill erosion (13, 22). Pesticide-related processes include degradation, transport between vertical spatial zones in the soil column, and partitioning between water and particulate phases. In all models, degradation processes are assumed to obey first-order kinetics. A series of first-order processes that vary in time and space are permissible only in the ARM model. Transport between vertical zones represented in the ARM and CREAMS models is simulated by simple mass-balance equations for each zone. The CPS model maintains a mass balance for pesticides in an active surface zone (approximately 0.5-3 cm in depth). Downward movement is by plug flow. A major difference between the ARM model and the others is the extent to which storms are represented. Storm totals only (mean storm concentration, total storm mass, etc.) are simulated by the CPS and CREAMS models, whereas the ARM model calculates such information for each time step (e.g., 15 min, hourly). Conceptually, the ARM structure enables simulation of pesticide concentrations within storm events rather than on a storm-averaged basis. Another key difference among the models (described in more detail in the sensitivity analysis) is the treatment of interflow. The ARM model explicitly simulates interflow and associated pesticide transport—the other models do not.

PROCEDURE

Each model was subjected to three "tests": (i) the ability to simulate observed monthly and annual losses of runoff water, sediment, and pesticide from an experimental watershed in the Mississippi Delta; (ii) the sensitivity (model response) of each model to key assumptions, input parameters, and model options; and (iii) the comparison of model predictions in a long-term (5-year) simulation of toxaphene runoff using rainfall records from the Mississippi Delta. In addition to these "tests," previous evaluations of all three models for their ability to predict atrazine (2-chloro-4-ethylamino-6-isopropylamino-1,2,5-triazine) runoff were examined and included.

The data base used originated from a cooperative study between the USDA-Agricultural Research Service Soil and Water Pollution Unit in Baton Rouge, La., and the USDA Sedimentation Laboratory in Oxford, Miss. Initiated in 1972, this study investigated the relationship between runoff, sediment yield, and chemical yield from flat agricultural land in the Mississippi Delta (20). The site selected was a 15.6-ha (38.5-acre) watershed on the G. L. McWilliams farm near Clarksdale, Miss., in the Yazoo River Basin. The site had been formed for drainage with mean slopes of 0.2%. The soil was a Sharkey Silty Clay with 1% sand, 52% silt, 47% clay, and 2.5% organic matter. Runoff was routed off the field via turn-rows (shallow V-ditches) and directed into a 1.6-ha (4-acre) pond located in the watershed. Instrumentation was installed where runoff entered the pond. Collection of runoff samples began in July 1972 and pesticide sampling began in March 1973. The calibration period for the three models was 1 Mar. 1974–28 Feb. 1975 (hereafter referred to as 1974 for simplicity).

Cultural practices on the McWilliams farm were typical of those in the Mississippi Delta Region. The land was planted to continuous cotton (*Gossypium hirsutum* L.) several years before the study began. In general, the cotton was planted in 1-m (40-in) rows upslope and downslope. Following harvest, cotton stalks were shredded and the field was disked in winter or early spring. Additional spring operations included pulverizing the soil with DO-ALL® or similar equipment, bedding with disk hipper, application of preplant herbicides and N fertilizer, and planting cotton. Several cultivations and application of postemergence pesticides occurred during the growing season.

In an earlier study, the sediment and hydrology portions of the ARM model were calibrated for the McWilliams farm for 1974.[7] De-

[7] Hydrocomp Inc., 1979. Estimation of toxaphene concentrations in the mouth of the Yazoo River, Mississippi, resulting form basin-wide application of toxaphene to cotton and soybeans. Prepared for the U.S. EPA Environ. Res. Lab., Athens, GA 30613 (unpublished report).

Table 2—Sediment components of ARM, CPS, and CREAMS models.

Model	Model structure	Detachment	Transport	Deposition	Sediment vs. in situ soil
CPS	USLE assumes field site is a uniform plane.	NA	NA	NA	Sediment characteristics are not different from in situ soil.
CREAMS	Sequence of detachment and transport of sediment is characterized by either of six options.	Modified USLE	Yalin's transport equation	Deposition occurs when sediment load (detached + upslope load) exceeds potential transport load.	Sediment-enrichment calculated by user-specified sediment characteristics or else model claculates sediment characteristics given in situ soil characteristics.
ARM	Calcualtes erosion for each time step within storms.	Power function of precipitation with output to storage reservoir	Removal from storage reservoir as power function of flow	None	Sediment characteristics are not different from in situ soil.

Table 3—Pesticide components of ARM, CPS, and CREAMS models.

Model	Pesticide reservoir structure	Degradation	Adsorption/desorption	Foliar washoff
CPS	Reservoir maintained for active surface zone (depth of zone is an input parameter); pesticides applied as bands that move downward at constant width.	First order, lumped decay rate	Equilibrium single-value linear isotherm	No algorithm
CREAMS	Mass balance maintained in each vertical zone of model, including a surface active zone of 1 cm.	First order, lumped decay rate	Equilibrium single-value linear isotherm	Empirical algorithm including separate reservoir, plant reservoir half-life, etc.
ARM	Reservoir maintained for active surface zone (depth of active zone is an input parameter), surface upper zone, and ground-water zone.	Time-phased series of first order, lumped decay rates	Equilibrium single and non-single-valued, Freundlich isotherm	No algorithm

tailed rainfall records from the McWilliams farm were not available, so hourly records from nearby Clarksdale were obtained from the National Weather Service. Daily (and hourly) rainfall totals from Clarksdale were adjusted so that monthly totals were equal to those at the McWilliams farm as reported by Willis et al. (20). Similarly, monthly predictions of runoff and sediment loss were compared with reported values in the calibration process.

The same procedure for developing the meterological data input stream was used in CPS and CREAMS model testing. Initial parameter estimates for the sediment and hydrology portions of CREAMS were made using the CREAMS user manual (11). Evaluation of resulting simulations led to the reassignment of a key model parameter, which influenced runoff totals. Parameters for the CPS model were developed from available sources pertaining to the SCS runoff equation (18) and the USLE equation (12), upon which the hydrology and sediment algorithms of CPS are based.

The model evaluations were completed by simulating toxaphene losses using the calibrated models. Information for this exercise was found in Willis et al. (20). Willis and coworkers reported soil residue levels of toxaphene prior to March 1974, as well as the rates and timings of applications. Other, more detailed information about toxaphene behavior on the plant and in the soil were not available. This lack of detail raised several questions: (i) Because toxaphene was applied to the plant at full canopy cover, what fraction of the application reached the ground and was hence immediately available for runoff? (ii) How should the washoff of toxaphene from the plant be dealt with following application? (iii) At defoliation and leaf fall, what contribution to the soil reservoir of toxaphene should be made from residual toxaphene on the leaves; and (iv) What is the critical soil depth in which all toxaphene present is available for runoff? Assumptions made to address these issues were: (i) 15% of applied toxaphene reached the ground and is immediately available for runoff, (ii) washoff from the plant surface can be neglected, (iii) volatilization from and degradation of toxaphene on the plant surfaces accounted for all toxaphene assumed to be intercepted by the plants and hence none was left to be added to the soil reservoir at defoliation, and (iv) a critical soil depth of 1 cm is appropriate. These assumptions do not represent a unique calibration solution, nor do they imply that reality is most accurately portrayed with them. Rather, they represent reasonable assumptions that appear not to conflict with the observed data.

The pesticide-dissipation rate and the partition coefficient for toxaphene were derived from McWilliams farm field data.' Final pesticide-related parameters for all models are given in Table 4.

Several tests of model sensitivity were performed using McWilliams farm data from 1974. Three methods of estimating storm erosivity in the event-based Universal Soil Loss Equation of the CPS model were examined. Both hydrology options and the foliar-washoff algorithm of the CREAMS model were also evaluated. In a sensitivity test for toxaphene involving all three models, erosion estimates were normalized so that monthly totals of erosion were equal in all models. Because toxaphene is a chlorinated hydrocarbon, much of the pesticide sorbs to soil. Hence, this exercise provided a test of how the models would predict toxaphene losses given equal sediment transport predictions.

Model performance was also evaluated by comparing results of sensitivity tests that simulate conditions not evlauated in the comparison with observed data. For these evaluations, the transport predictions from all models were left unaltered and the sensitivity of pesticide-loss predictions to varying pesticide parameters was determined. Specifically, the decay coefficient, adsorption partition coefficient, and water solubility values were examined. Increasing the decay coefficient (decreasing the pesticide half-life) will decrease the magnitude of total loss without influencing the relative partitioning of pesticide loss between the runoff and sediment fractions. Decreasing the adsorption partition coefficient from the toxaphene-assigned value of 6,000 will shift the equilibrium concentrations of pesticides from the sediment to the water portions. One would expect this to increase total predicted losses of pesticide because the total volume of runoff water is greater than that of eroded sediment. For model testing, the partition coefficient (K_D) was varied between 5,000 and 1, including the values, 5,000, 500, 50, 25, 10, 5, 2, and 1. Water solubilities assigned to these partition coefficients varied from 2 ppm (K_D = 5,000) to 10,000 ppm (for K_D = 1) (10). The first-order decay coefficient was held constant at 0.028, which corresponds to a half-life of 25 d. These choices of K_D and the choice of a constant decay rate do not represent a particular compound or a particular situation. Rather, the test is performed keeping all model parameters equal and

Table 4—Toxaphene parameters for testing and production runs.

Parameter	Value
Adsorption partition coefficient	6,000
First-order decay rate	0.0014 (half-life = 500 d)
Solubility, ppm	3
Initial soil storage, kg/ha	4.5 for vertification; 64.6 for cotton production
Active surface zone depth, cm	1; 0.5 for CPS and ARM production
Application rates and dates	
Testing	6 applications at 0.255 kg (15% of total application of 1.7 kg/ha); 2 Aug.–16 Sept. 1974
Production	6 applications at 2.7 kg/ha (30% of total application of 9 kg/ha); mid-July–mid-August
	6 applications at 1.3 kg/ha (simulating leaf fall); 10 Oct.–20 Oct.

Table 5—Testing results—comparison of predicted and observed runoff at the McWilliams Farm.

Month	Observed	ARM	CPS	CREAMS
		cm		
		1974		
Mar.	0.23	0.46	1.36	1.10
Apr.	6.07	5.94	8.84	6.50
May	26.64	21.11	31.36	21.32
June	15.88	15.07	10.95	12.51
July	3.43	4.11	5.08	7.35
Aug.	3.53	4.15	1.54	7.34
Sept.	6.45	3.93	3.13	5.48
Oct.	0.58	1.24	0.41	1.22
Nov.	4.93	2.91	4.55	1.17
Dec.	6.25	5.95	3.72	1.53
		1975		
Jan.	6.10	4.37	4.16	3.87
Feb.	7.91	6.68	5.64	5.67
Total	88.01	75.92	80.74	75.06

Table 6—Testing results—comparison of predicted and observed sediment loss at the McWilliams Farm.

Month	Observed	ARM	CPS	CREAMS
		t/ha		
		1974		
Mar.	0.04	0.05	1.20	0.63
Apr.	2.29	2.09	3.95	2.93
May	10.09	11.43	10.21	6.52
June	7.71	4.59	5.67	4.05
July	1.77	1.46	1.47	2.41
Aug.	0.40	1.35	0.72	1.25
Sept.	1.03	0.65	0.64	0.58
Oct.	0.02	0.03	0.11	0.16
Nov.	0.13	0.31	0.45	0.11
Dec.	0.58	0.60	0.72	0.09
		1975		
Jan.	0.96	1.55	1.33	0.47
Feb.	1.23	1.62	1.46	0.43
Total	26.25	25.73	27.93	19.63

Table 7—Testing results—comparison of predicted and observed total toxaphene losses at the McWilliams farm.

Month	Observed	ARM	CPS	CREAMS
		g/ha		
		1974		
Mar.	0.14	0.12	2.50	2.24
Apr.	6.07	4.57	7.75	9.92
May	26.64	23.41	18.14	19.40
June	17.74	9.05	9.20	11.61
July	4.12	2.79	2.20	6.10
Aug.	6.71	14.85	6.97	20.18
Sept.	18.06	8.15	7.68	10.25
Oct.	0.58	0.41	1.29	3.01
Nov.	2.47	3.70	5.53	2.03
Dec.	5.00	6.64	7.83	2.18
		1975		
Jan.	6.10	16.19	13.43	10.18
Feb.	3.17	16.10	14.19	8.76
Total	96.53	105.98	96.67	105.85

varying only the partition coefficient and the closely related parameter, water solubility.

An earlier study using the ARM model was completed by performing 5-year simulations of toxaphene application to continuous cotton and soybeans (*Glycine max* (L.) Merr.).[7] The assumptions relative to distribution of application between foliage and soil surface, defoliation, and soil depths used in the initial ARM study were different from those selected for use in the evaluations of all three models that used McWilliams farm data. To enable utilization of the previous results, earlier assumptions made for ARM were used for the CPS and CREAMS production runs. Thus, the comparative results were valid. Cotton production runs were then duplicated with CPS and CREAMS models. The period of simulation was 1 Mar. 1971, to 31 Dec. 1975, and hourly rainfall records were taken from Clarksdale, Miss. Cultural practices for cotton production were typical for the Mississippi Delta.

RESULTS AND DISCUSSION

Overall Results for Toxaphene

Final results for runoff, sediment loss, and total toxaphene loss are given in Tables 5-7. Runoff predictions with the CREAMS model were obtained using hydrology option 2, which employs breakpoint rainfall data and is based on the Green-Ampt infiltration formulation. In this application, hourly data are represented as breakpoint data and the necessary adjustments in the data are made within the CREAMS model. The CREAMS sedment model also allows one to choose from among six different watershed representations. These choices specify the sequence of detachment and transport of sediment particles. Possible sequences include overland, overland-pond, overland-channel, overland-channel-channel, overland-channel-pond, and overland-channel-channel-pond. The overland-channel-channel option was chosen for this application. The cotton rows represent the overland portion, the row middles are simulated as rectangular-shaped channels, and the turn-row V-ditches at the field edge represent the second channel simulated as a triangular-shaped channel.

As can be seen from Table 5, all models simulate monthly total runoff quite well. The peak runoff months of May and June are portrayed well in model predictions. The CPS and ARM models simulated sediment loss equally well, as shown in Table 6. The CREAMS model underpredicted sediment loss, however, primarily in the peak months of May and June. Apparently, excessive deposition in the channels is the cause of this underprediction.

All models appear to predict annual totals of toxaphene loss equally well. The CREAMS model results, however, must be carefully considered. Toxaphene is a chlorinated hydrocarbon and hence adsorbs to soil particles. Therefore, sorbed losses of toxaphene comprise the bulk of total losses, and the CREAMS model predicted approximately 25% less erosion than either the CPS or the ARM model and 25% less erosion than was observed at the McWilliams farm. The CREAMS sediment module calculates a soil-enrichment ratio for each storm and sends this value to the pesticide module for estimation of sorbed losses of chemical on sediment. The average enrichment ratio calculated for 1974 was 1.60. Willis and coworkers found enrichment ratios for organic matter and clay to be 2.5 and 1.5, respectively, on the McWilliams farm (20).

"Enrichment" is a commonly observed phenomenon. However, enrichment is not accounted for in the CPS and ARM models. For the CPS model partitioning of chemical between the runoff sediment and water is the same as the partitioning in situ. As such, sorbed losses of pesticide are estimated as: concentration of chemical in in situ soil (based on K_D) × mass of eroded sediment. The ARM model partitions in situ and removes pesticide in each phase from the soil with no recalculation of distribution within runoff. The CREAMS model does account for enrichment, and sorbed losses of chemical are estimated as the product of: concentration of chemical in the in situ soil (based on the same K_D), mass of eroded sediment, and enrichment of ratio. Therefore, the K_D parameter required for all models represents the same process. If the enrichment ratios determined for CREAMS are uniformly set to 1.00, then the output of CREAMS is analogous to CPS and ARM. In this case, the annual predicted loss of toxaphene drops from 106 to 65 g.

In this application, the partition coefficient was determined from toxaphene measurements made from sediment samples and not from in situ soil samples. Hence, an "enrichment" was built into its estimation.

Table 8—The standard error (SE), coefficient of variation (CV), and simple correlation coefficient (CC) for observed vs. predicted runoff, sediment, and toxaphene loss for 1974 at the McWilliams farm.

Loss	ARM			CPS			CREAMS		
	SE	CV	CC	SE	CV	CC	SE	CV	CC
Runoff, cm	2.08	0.28	0.98	2.72	0.37	0.95	3.14	0.43	0.91
Sediment, t/ha	1.04	0.48	0.94	0.86	0.39	0.96	1.56	0.71	0.94
Pesticide, g/ha	6.58	0.82	0.61	6.18	0.77	0.62	5.76	0.71	0.69

Simple statistical information including standard errors, coefficients of variation, and correlation coefficients for the pairs of predicted and observed monthly totals of runoff, erosion, and toxaphene losses are given in Table 8. Although this information is somewhat limited because monthly and not event totals are compared and only a small sample is being tested, certain observations can be made. The coefficients of variation for all models are on the order of 80%. The correlation coefficients are higher for runoff and sediment predictions than for pesticide predictions. This is not surprising considering the several assumptions that were made and the uncertainty in toxaphene parameters. Nevertheless, observed losses following application equal 0.4% of application and the model's predicted 0.5% loss. Indeed, one can conclude that all models reproduce field data quite well in this application.

Calibration and Boundary Conditions for Toxaphene

The claim has been made that both the CPS (actually its predecessor upon which CPS is based) and CREAMS models are calibration-independent (11, 16). Reported "best" results for the CPS and CREAMS models were reached, however, only after some deliberation and reassignment of initial parameter estimates. For the CPS model, the user is given a choice of three models to estimate the erosivity of individual storms for use in the event-based USLE. The first model was developed by Wischmeier and was used in the development of annual estimates of the erosivity factor, R (21). The second model was proposed by Onstad and Foster (14) and incorporates Wischmeier's erosivity factor plus a second term based on runoff volume and peak rate of runoff. The third model was developed by Williams (19) and is a function of field area, runoff volume, and peak rate of runoff. Predicted sediment losses with all three models are given in Table 9. The Williams model overpredicted by a factor of 2, and the Wischmeier model significantly underpredicted observed loss. The major discrepancy among the three models occurred for the peak months of April, May, and June. In previous testing of the CPS model using data from a Georgia Piedmont field site in Watkinsville, Ga., all three models were found to overpredict sediment loss for peak events (16). In this study, the Williams model was also found to simulate the highest totals, followed by the Onstad-Foster and Wischmeier models, respectively.

The user is also given several options for various calculational procedures within the CREAMS model. In the absence of hourly or breakpoint precipitation data, one may use daily rainfall totals in hydrology option 1, which is based on the SCS curve number equation. After simulation of 1974 with daily rainfall totals, the sediment model was run using the hydrology pass file created by option 1. Sediment and runoff predictions for these runs are given in Table 10. Runoff predictions were higher using option 1, and annual totals of observed and predicted runoff compare more favorably than with hydrology option 2. Sediment predictions, however, compare less favorably because of an underestimation of peak runoff rates and storm erosivity when using option 1.

Testing of the CREAMS model also led to reassignment of a major model parameter: saturated hydraulic conductivity (RC). The CREAMS manual (11) suggests a value for RC of 0.76 mm/h (0.03 in/h) for the watershed description: row crops, straight row, poor hydrologic condition, and hydrologic soil group D (the relevant description for the McWilliams farm). Runoff predictions are underestimated with this value of RC, however, yielding a total for 1974 of 44 cm. Assigning an RC value of 0.025 cm/h (0.01 in/h) adjusts model predictions to more closely agree with observations. This value of RC is at the bottom of the range of RC values suggested for a D soil in the CREAMS manual (0.01-0.20), and is also closer in value to the parameter

Table 9—Calibration of CPS—comparison of sediment predictions with three different erosivity models.

Month	Observed	Wischmeier	Onstad-Foster	Williams
		t/ha		
		1974		
Mar.	0.04	1.46	1.20	1.85
Apr.	2.29	2.34	3.95	5.23
May	10.09	3.99	10.21	20.20
June	7.71	3.83	5.67	11.22
July	1.77	1.67	1.47	2.65
Aug.	0.40	0.96	0.72	1.28
Sept.	1.03	0.47	0.64	0.94
Oct.	0.02	0.13	0.11	0.27
Nov.	0.13	0.26	0.45	0.88
Dec.	0.58	0.57	0.72	1.27
		1975		
Jan.	0.96	1.05	1.33	2.36
Feb.	1.23	0.97	1.46	2.68
Total	26.25	17.70	27.93	50.83

Table 10—CREAMS model testing—hydrology and sediment using hydrology option 1, SCS curve number methodology.

Month	Runoff		Soil loss	
	Observed	Predicted	Observed	Predicted
	cm		t/ha	
	1974			
Mar.	0.23	1.05	0.04	0.56
Apr.	6.07	8.49	2.29	2.31
May	26.64	29.11	10.09	5.06
June	15.88	16.82	7.71	2.75
July	3.43	5.42	1.77	1.55
Aug.	3.53	3.95	0.40	0.74
Sept.	6.45	3.97	1.03	0.78
Oct.	0.58	0.73	0.02	0.20
Nov.	4.93	1.88	0.13	0.38
Dec.	6.25	4.88	0.58	0.27
	1975			
Jan.	6.10	6.17	0.96	0.58
Feb.	7.92	8.91	1.23	0.90
Total	88.01	91.37	26.25	16.08

INFIL (infiltration rate) determined for the ARM model 0.038 cm/h (0.015 in/h).

The model calibration exercise was completed with the development of toxaphene assumptions. As noted earlier, four assumptions derived earlier for the ARM model[7] were systematically tested and revised using data from 1974 at the McWilliams farm. Results of these tests as applied to the CPS model are shown in Table 11.

For the earlier study, the assumption was made that toxaphene on the plant degrades at a rate of 0.01/d. At leaf fall, the amount remaining on the plant from all applications was summed and divided into six pseudo applications over a 10-d period encompassing the defoliation date. Total overland loss of toxaphene decreased by 44% when it was assumed that defoliated leaves contained no residual toxaphene.

The second assumption examined was the depth assigned to the surface-zone layer. Conceptually, this zone is one in which rain, shallow flow, and pesticide intermix. Estimating its vlaue poses problems in that it has never been determined experimentally, and literature values can be found ranging from 0.3 to 3.0 cm (2). Steenhuis and Walter (17) provide an indirect way of obtaining this parameter by regressing losses of pesticide in overland flow with cumulative rainfall. For practical purposes, determining its value often requires calibration. In the CREAMS model, however, this zone is set at 1 cm and is not an input parameter. The initial estimate for the ARM model was 0.5 cm. Increasing its value to 1 cm in CPS model testing resulted in slight increases up until the time of application, and in 50% reductions thereafter. These reductions are the result of what can. be termed the "dilution" effect; that is, surface applications are now adsorbed to twice the amount of surface soil and concentrations are havled in runoff water and sediment.

The final revised assumption was the estimate of the percentage of applied pesticide reaching the ground and available for runoff. Initially, the estimate was 30%. Reducing this to 15% reduced the predicted losses following application by one-half.

The issue of calibration vs. noncalibration is not a trivial one and has several implications concerning model usage. If a model requires calibration, then it can only be applied when field data exists to calibrate the model. As experience with the model is gained, ranges of parameter values can be determined and summarized. The ease of model usage is often related to the calibration vs. noncalibration issue. If the model requires calibration, then the model user needs to be familiar with the sensitivity of model output to changes in model parameters. As well, he must be familiar with the model theory in order to properly locate the parameters that require calibration. The CPS model is the simplest of the three models to use (estimate parameters), because the hydrology and erosion algorithms are based on the empirical SCS runoff and USLE equations and a wide body of literature exists to assist in the development of the relevant parameters (18, 21). Hydrology option 2 and the sediment algorithms of CREAMS are more theoretically based than the algorithms of CPS and subsequently more parameters are required. The CREAMS model developers have prepared a comprehensive user guide that assists in developing parameter sets based on commonly available information such as soil characteristics or agronomic practices. For these reasons, input parameters for both CPS and CREAMS can be estimated with relative ease in the absence of field data. The ARM model can also be applied in the absence of field data, but input parameters are less well-defined and, hence, ARM is perhaps the most difficult of the three to use in a noncalibrated mode.

Field data can assist in the estimation of model parameters. In situ pesticide-decay rates and soil-water partitioning can be assessed with core data taken at regular intervals. Adjustment of model parameters and choice of optional calculational procedures is warranted if field data exist to justify the decisions. Presenting the erosion results of CPS based on the erosivity term of Onstad and Foster was a "calibration" decision simply because the predicted losses matched the observed losses more closely than in the other options in CPS. The reassignment of RC in CREAMS was also a calibration decision and is justified for the following reasons: (i) sensitivity analysis presented in the CREAMS manual indicated that runoff totals are particularly sensitive to RC; (i) the manual suggests a value of 0.03, but also gives an "expected range of values," (11) since the saturated hydraulic conductivity is a site-specific soil characteristic; (iii) calibration of ARM resulted in a value of 0.038 cm/h (0.015 in/h) for infiltration rate; and (iv) the change of RC from 0.03 to 0.01 resulted in a significant improvement in model predictions of runoff. This discussion is not meant to imply that CPS and CREAMS require calibration, rather, that model performance can and should be enhanced by calibration. The assertion that a model "can be aplied in the absence of field data" (11, 16) was not violated in the calibration techniques presented for CPS and CREAMS. Field data increases one's knowledge of reality and the information gained should be used, particularly when model results are to be part of a decision-making framework.

Foliar Washoff of Toxaphene

Of the three models, only CREAMS has a foliar washoff algorithm for pesticides, and it was tested in

Table 11—CPS model testing—derivation of toxaphene assumptions using CPS.

Month	Observed	No. 1†	No. 2‡	No. 3§	No. 4¶
1974					
Mar.	0.14	2.44	2.44	2.46	2.46
Apr.	6.07	7.57	7.57	7.75	7.75
May	26.64	16.79	16.79	18.14	18.14
June	17.47	8.02	8.02	9.20	9.20
July	4.12	1.86	1.86	2.20	2.20
Aug.	6.71	24.84	24.84	13.01	6.97
Sept.	18.06	27.79	27.79	14.50	7.68
Oct.	0.58	7.90	4.64	2.43	1.29
Nov.	2.47	44.85	19.99	10.45	5.53
Dec.	5.00	64.69	28.07	14.82	7.83
1975					
Jan.	6.10	109.73	47.66	24.40	13.43
Feb.	3.17	114.65	49.78	26.83	14.19
Total	96.57	431.13	239.45	147.19	96.67

† Leaf fall contributions, surface zone = 0.5 cm, 30% reaching ground.
‡ No leaf fall contributions, surface zone = 0.5 cm, 30% reaching ground.
§ No leaf fall contributions, surface zone = 1.0 cm, 30% reaching ground.
¶ No leaf fall contributions, surface zone = 1.0 cm, 15% reaching ground.

Table 12—CREAMS model testing—testing of the foliar washoff algorithms.

Month	Observed	Predicted, without	Predicted, with
		g/ha	
		1974	
Mar.	0.14	2.24	2.24
Apr.	6.07	9.92	9.92
May	26.64	19.40	19.40
June	17.47	11.61	11.61
July	4.12	6.10	6.10
Aug.	6.71	20.18	24.51
Sept.	18.06	10.25	12.40
Oct.	0.58	3.01	3.64
Nov.	2.47	2.03	2.40
Dec.	5.00	2.18	2.58
		1975	
Jan.	6.10	10.18	12.07
Feb.	3.17	8.76	10.38
Total	96.53	105.85	117.24

this application. This part of the model requires an input foliar residue half-life, fraction on the foliage available for rainfall washoff, and rainfall threshold for washoff. Following the first rainfall after application that exceeds the threshold, the total available residue present is removed and added to the active surface-zone pesticide reservoir. Assuming a rainfall threshold of 0.30 cm and toxaphene foliar parameters suggested by the CREAMS manual, the resultant change in total toxaphene loss is shown in Table 12. Total losses were increased 1-4 g/h for all months following application. The large amount of observed toxaphene loss in September credited to foliar washoff by Willis et al. (20) was not simulated well. Predictions of monthly losses were also higher with foliar applications through February, despite leaf defoliation 5 months earlier. This is due to a consistently higher available reservoir of toxaphene resulting from additions made in the foliar washoff event several months earlier.

Overall Results for Atrazine

In previous studies, all three models were evaluated in simulations of atrazine (5, 11, 16). The study site is located in the Southern Piedmont region of Georgia and is designated P-2. Detailed information on soil characteristics, agronomic practices, and other relevant data for P-2 are contained in Smith et al. (15). The annual summaries of observed and predicted total losses of atrazine are provided in Table 13. All models predicted atrazine loss well except CPS in 1975, and this was due to an overprediction of runoff in that year. In this application, sufficient information was available to allow for calibration of the interflow component of ARM. For

Table 13—Model testing for total runoff (sorbed and soluble) of the compound atrazine on the P-2 watershed in Watkinsville, Ga.

Year	Observed	ARM	CPS	CREAMS
		g/ha		
1973	82	56	82	54
1974	8	8	5	2
1975	13	10	64	13
Total	103	74	151	69

the 3 years of simulation, 5% of runoff occurring during the intense summer storms was from interflow, and 21% of total atrazine loss was via interflow (5).

Unlike the toxaphene study described in this paper, parameters for the three models were determined independently and, as a result, the pesticide parameters were somewhat different for each model. The decay rate and partition coefficient, K_D, for all models are listed in Table 14. The decay rates assigned for each model were calibrated from observed behavior. Atrazine was found to decay rapidly prior to the first rainfall event following application, and then to decay more slowly after this event. This was credited to a combination of high temperatures at the soil surface leading to volatilization loss, and to the leaching of atrazine below the surface following the first event (5). The ARM and CPS models were calibrated by assigning different decay rates before and after the first storm. CREAMS applied one decay rate derived from Watkinsville data and subsequently underpredicted atrazine runoff for later events. Recorded data at Watkinsville also showed the tendency for atrazine adsorption on sediment to increase over time. The ARM model is the only one of the three models that has an option for a non-single-valued adsorption/desorption isotherm, and the predictions in Table 13 are made using this option. Testing with the single-valued option led to underprediction of atrazine loss in 1973 and a 3-year total loss of 54 g. The CPS model applied to the same site made an empirical adjustment to the single-valued adsorption/desorption isotherm and made the overall distribution coefficient, K_D, a function of the reservoir present in the soil, as shown in Table 14. The CREAMS manual includes a section on determining K_D values, which includes a table summarizing literature values of K_D for several compounds. The assigned K_D for atrazine was determined using this information, and resulted in a mean value of 4.3.

Sensitivity Testing for Toxaphene

The data from 1974 were used in two sensitivity tests of the pesticide algorithms of the three models. In one, sediment predictions in the CPS and CREAMS models were normalized to monthly ARM model predictions. (This approach was selected only for convenience and is not meant to imply that ARM predictions are more nearly accurate.) Results are shown in Table 15. The

Table 14—Atrazine parameters for ARM, CPS, and CREAMS.

Parameter	ARM	CPS	CREAMS
Decay rate (d⁻¹)	0.10†	0.12†	0.14
	0.04‡	0.04†	
K_d	1.0§	1.5 400§	4.0
		5 150-400	
		20 91-150	
		30 78-91	
		100 50-78	
		1000 20-50	
		10000 10-20	
		100000 10	

† Prior to 1st rainfall event after application.
‡ After 1st rainfall event after application.
§ Freundlich non-single-value adsorption/desorption coefficient.
¶ Atrazine quantity in soil surface reservoir, g/ha.

CPS model consistently predicted slightly less total toxaphene loss than did the ARM model, and the CREAMS model consistently predicted higher losses than the ARM or CPS models. The CREAMS model predictions with soil-enrichment ratios equal to 1.00 are also listed in Table 15. In this case, predictions match those of the CPS model.

A separate run with the CREAMS model was performed using the same normalized erosion totals, and the K_D value of 4,000 suggested for toxaphene in the CREAMS manual. This test provides an answer to the question: How would the CREAMS model have predicted toxaphene loss, given more accurate erosion predictions than were simulated in the validation test, and a K_D value suggested for toxaphene for in situ conditions rather than an "enriched" K_D calibrated for the ARM model? Total predicted loss of toxaphene increased by 1%; sorbed loss (which comprise 97% of total loss) decreased by 0.3%, and soluble loss increased by 49%. This indicates that predictions of total loss are relatively insensitive to the change in K_D from 6,000 to 4,000. Changing the enrichment ratios to 1.00 in the previous test indicated that the enrichment factor calculated by CREAMS is solely responsible for the discrepancy between CREAMS toxaphene predictions and the predictions of CPS and ARM.

Results from the model sensitivity tests related to pesticide parameters, K_D, and water-solubility values, were examined in terms of total annual loss and were fit to the smooth curves shown in Fig. 1. The CPS and CREAMS models behaved similarly over the range of partition coefficients, although CREAMS consistently predicted approximately twice the amount of total loss than did CPS. Both showed a peak loss at $K_D = 5$, with the CREAMS model predicting 358 g/ha and the CPS model predicting 149 g/ha loss. The subsequent drop in predicted losses at $K_D < 5$ can be explained in both models by reduction of surface concentrations of pesticide due to water infiltration and chemical leaching prior to runoff. In the ARM model, surface concentrations are also reduced prior to runoff. Leached pesti-

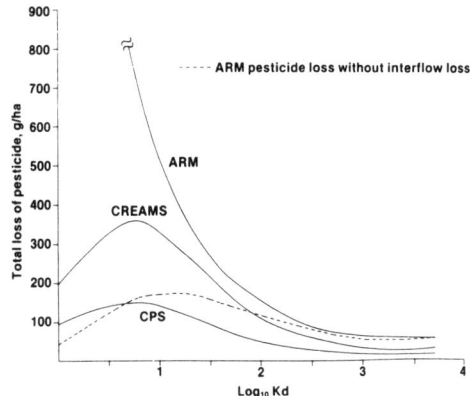

Fig. 1—Model-sensitivity analysis. Total loss of pesticide with different partition coefficients.

cide can be picked up by interflow, however, and resurface at the edge of field, contributing to total predicted losses. If the interflow component is neglected, then total loss would follow the pattern shown by the dotted line in Fig. 1. The magnitude and pattern of loss in this case is comparable to that of the CPS model, although the peak is reached at $K_D = 10$. With the interflow component total losses continue to rise as K_D decreases, and at $K_D = 1$ the ARM model predictions exceed CREAMS predictions by a factor of 10 and CPS predictions by a factor of 20. One must be cautious in oversimplifying the interflow result, however, because no calibration of interflow was attempted. Interflow is largely a hydrograph-shape factor in the ARM model, and no hydrograph data were available.

The extent to which chemicals move with interflow is obviously important. A study by Edwards et al. (6) sheds some light on this issue. Runoff samples from several plots at the North Appalachian Experimental Watershed near Coshocton, Ohio, were analyzed for the presence of glyphosate (N-(phosphonomythyl)glycine), which had been applied as a preceding herbicide in the no-tillage establishment of fescue (*Festuca arundinacea*). Glyphosate is a soluble compound and was found to persist in runoff about as long after application as does atrazine. In general, concentrations were found to be highest for runoff events immediately following application and to appear in runoff up to 4 months later. In one instance, however, concentrations of glyphosate were higher for the second storm following application, which occurred 10 d after the first storm. The first storm was a low-intensity, long-duration storm, and runoff continued 3.5 h after the rainfall had stopped. This post-storm runoff was caused by interflow. The second storm was short and intense, and, unlike the first storm, essentially all the runoff was from overland flow. Edwards and coworkers speculated that the topsoil acted as a filtering system in the long storm and that infiltrating water was purged of glyphosate picked up from the surface soil. If this result can be generalized, then one can conclude that the ARM model may assign a disproportionally high amount of infiltrat-

Table 15—Model sensitivity analysis—comparison of total losses of toxaphene when soil loss predictions are equal for all models.

Month	Soil loss: ARM, CPS, CREAMS	Toxaphene				
		Observed	ARM	CPS	CREAMS	CREAMS†
	t/ha	g/ha				
		1974				
Mar.	0.05	0.14	0.1	0.1	0.2	0.1
Apr.	2.09	6.07	4.6	4.3	7.0	4.3
May	11.43	26.64	23.4	21.6	33.4	21.3
June	4.59	17.47	9.1	6.7	12.3	8.1
July	1.46	4.12	2.8	2.2	3.6	2.4
Aug.	1.35	6.71	14.9	12.8	23.6	13.7
Sept.	0.65	18.06	8.2	7.8	11.3	8.1
Oct.	0.03	0.58	0.4	0.4	1.0	0.4
Nov.	0.31	2.47	3.7	4.0	5.6	3.0
Dec.	0.60	5.0	6.6	6.7	10.7	6.0
		1975				
Jan.	1.55	6.10	16.2	15.6	26.9	15.7
Feb.	1.62	3.17	16.1	15.5	26.7	14.7
Total	25.73	96.53	106.1	97.7	162.3	97.3

† CREAMS soil-enrichment ratios = 1.00.

Table 16—Summary of 5-year cotton production runs.

Model	Mar.	Apr.	May	June	July	Aug.	Sept.	Oct.	Nov.	Dec.	1971	1972	1973	1974	1975	Total
3704								Runoff, cm								
ARM	4.2	2.7	6.0	4.8	3.3	3.2	0.6	0	0	10.1	34.9	67.5	82.2	75.2	75.1	334.9
CPS	3.2	6.3	6.0	3.7	4.6	1.2	0.9	0	1.5	9.7	37.1	63.9	84.1	75.1	75.1	335.3
CREAMS	4.3	4.8	3.0	7.8	5.9	4.4	3.1	0	0.1	7.8	41.2	73.6	85.6	68.7	72.0	340.1
								Soil loss, t/ha								
ARM	0.9	0.6	1.1	3.2	0.3	1.6	0	0	0	2.0	9.7	19.2	22.7	18.0	22.8	92.4
CPS	1.3	2.4	1.5	3.0	0.8	0.4	0.3	0	0.1	1.1	10.9	19.1	24.0	21.7	20.1	96.5
CREAMS	0.9	1.6	0.4	2.9	0.8	0.6	0.5	0	0	0.5	8.2	13.2	15.5	13.2	12.3	62.4
								Toxaphene, g/ha								
ARM	26	16	27	71	21	193	7	0	0	62	423	1959	975	493	833	4783
CPS	37	65	40	72	48	47	66	0	5	41	421	1063	900	682	903	3969
CREAMS	42	80	15	120	46	80	87	0	2	37	509	898	971	719	896	3993

ing chemical to interflow removal. Donigian[a] also notes the general problem of measuring and modeling interflow solute transport. Morel-Seytoux (12) states that the interflow component of the Stanford Watershed model has never been physically based.

The interflow component of ARM is sensitive to parameters required in the ARM hydrology module. The partition of total edge-of-field runoff between overland and interflow losses can be calibrated if detailed within-storm runoff information is available. For the McWilliams farm, this information was unavailable. Subsequently, the calibration to total runoff resulted in 39% contribution from interflow for 1974. This amount of interflow carried significant amounts of chemical off the field for the lower K_D values tested in Fig. 1. For the K_D values of 1, 2, 5, 10, 25, and 50, the percentage of total loss (soluble plus sorbed) carried by interflow was 98, 94, 81, 67, 46, and 33%, respectively. Only 1% of total loss was attributed to interflow removal for the toxaphene calibration K_D of 6,000.

Model testing was completed with 5-year cotton production runs with all three models. Toxaphene assumptions pertaining to toxaphene availability for runoff were those determined in an earlier study[7] and not those revised in the other model evaluations. Results from these runs are summarized in Table 16. Many of the trends noted earlier reappear in these production runs. Annual totals of runoff are indistinguishable in the three models, and only as little as 5 cm separates the 5-year totals of runoff predictions. Monthly variations in runoff predictions can be noted, such as in June of 1971, although peak months such as December 1971 are portrayed in all three models. The CPS model predicted slightly more sediment loss than did the ARM model, as was also the case in the earlier evaluation. The CREAMS model sediment predictions were again lower than either the CPS or ARM models.

Simulated toxaphene loss was similar for all three models. In the CREAMS model, soil-enrichment ratios again increased toxaphene-loss predictions. Monthly variations in simulated loss of toxaphene explain an apparent discrepancy when comparing 1972 totals of ARM and CPS models. Despite equivalent annual sediment and runoff predictions, the ARM model predicted twice as much toxaphene loss as did the CPS model. In examining monthly totals of sediment loss, however, the ARM model predicted three times as much sediment loss as did the CPS model in the last 3 months of 1972. This corresponds to the time of cotton defoliation and to the addition of residual toxaphene on the cotton leaves to the soil reservoir. Therefore, the ARM model's toxaphene-loss predictions for these months were significantly higher than the CPS model predictions.

Quite often environmentally significant events result from intense storms and subsequent runoff. Listed in Table 17 are the model responses to three individual events. As can be seen, model responses can differ significantly, a fact that is particularly evident in the 13 Nov. 1972 comparison. The CPS model predicted less total loss than did the ARM model, because of lower sediment predictions. The CREAMS model predicted less than either of the other two models as a result of low sediment predictions. A second reason for this is the dilution effect, which was described earlier in the calibration of toxaphene assumptions. The CREAMS model has a constant active zone layer 1 cm deep, and the CPS and ARM models have variable depths that were set to 0.5 cm for production testing. Toxaphene-loss predictions for the other events listed were more consistent among the various models. Despite more rainfall and higher predictions of runoff and erosion, simulated losses for 26 Nov. 1975, were lower than for 7 Nov. 1972. This is due to a lower initial reservoir of toxaphene prior to the storm on 26 November as compared with the reservoir on 7 November, which had recently been replenished with residual toxaphene from defoliated leaves.

Production runs were further examined in the frequency-distribution diagram shown in Fig. 2. Monthly totals of toxaphene loss were normalized such that 100 on the x-axis corresponded to the peak monthly total found by all models (in this case, by the ARM model). The most notable feature of these results is the difference between ARM model predictions and the CPS and CREAMS model predictions. One conclusion that can be drawn is that the ARM model predicts more months of low toxaphene loss totals than do the other two models. If one examines the sediment results in Table 6, it is apparent that the ARM model also predicted the low months of March and October better than the other two models. Over the course of 5 years, several months

[a] A. S. Donigian, Jr., 1981. Water quality modeling in relation to watershed hydrology. Presented at the Int. Symp. on Rainfall-Runoff Modeling, Mississippi State Univ., Mississippi State. 18–21 May 1981.

Table 17—Single-event sensitivity—comparison of predictions of three events in cotton-production testing.

Date	Precipitation	Model	Runoff	Soil loss	Toxaphene
	cm		cm	t/ha	g/ha
7 Nov. 1972	1.90	ARM	0.47	0.05	32
		CPS	0.06	0.04	41
		CREAMS	0.66	0.07	13
13 Nov. 1972	4.57	ARM	2.54	1.30	560
		CPS	1.13	0.30	305
		CREAMS	3.15	0.40	73
26 Nov. 1975	2.49	ARM	0.74	0.06	3
		CPS	1.24	0.10	5
		CREAMS	0.41	0.02	2

of low-intensity rainfall result in more predictions of low toxaphene loss with the ARM model. ARM also predicted peak months of a magnitude not matched by either of the other models. Indeed, the peak month of CREAMS model simulation is less than half of ARM model's peak month, and the peak month as predicted by the CPS model is <40% of that found by the ARM model.

CONCLUSIONS

The three models evaluated required calibration in order to closely reproduce observed data. Calibration procedures for each model were quite different; in general, the CPS model required the least adjustment, followed by the CREAMS and ARM models. All calibrated models predicted observed monthly and total toxaphene losses equally well, although the CREAMS model predictions were tempered by the fact that the calculation of soil-enrichment ratios resulted in equivalent toxaphene losses despite underprediction of erosion.

Sensitivity testing showed that simulated pesticide losses for the three models diverged for partition coefficients <10, with the greatest difference being a much higher loss simulated by the ARM model. With partition coefficients >400, all three models quickly converged. When interflow losses predicted by the ARM model were neglected, the ARM and CPS models converged; the CREAMS model followed the same trend but at a higher value. Previous testing with all three models on the soluble compound atrazine did not result in significant differences between ARM predictions and those of CPS and CREAMS, as might be inferred from the toxaphene test results. This was due to a combination of fators: (i) the calibration of ARM hydrology parameters resulted in 95% of total runoff from overland flow during the short intense summer storms; and (ii) atrazine was applied during this time and the rapid decay of atrazine resulted in depletion of the reservoir before the winter storms, which had a greater contribution of total runoff from interflow.

Long-term simulated toxaphene losses followed the same general frequency trends for each model. The ARM model, however, yielded different monthly extremes for both low and high monthly totals. Such results have important implications for exposure assessment because quite often peak events are of most concern. For this study, insufficient data precluded an evaluation of the accuracy of any of the models to represent extreme events. In a complete assessment, overland losses would be related to instream concentrations of chemical and further to resulting impact on the aquatic environment. Obviously, the instream concentrations that result from the three models also need to be evaluated.

ACKNOWLEDGMENTS

The authors would like to thank G. R. Foster (Hydraulic Engineer, USDA-ARS, Purdue University, West Lafayette, Ind.) for his assistance in parameter development in the erosion submodule of the CREAMS model; and S. Hill (Computer Sciences Corp., Athens, Ga.) for his assistance in running the CPS model.

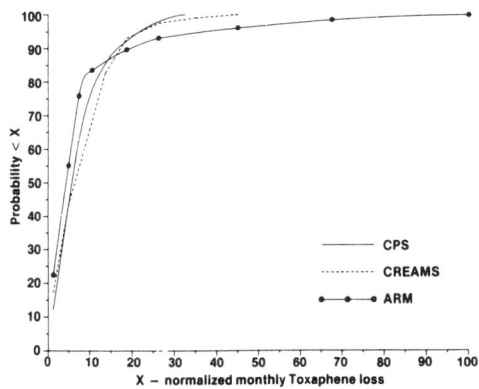

Fig. 2—Cumulative frequency-distribution diagram for monthly totals in 5-year production runs.

LITERATURE CITED

1. Adams, R. T., and F. M. Kurisu. 1976. Simulation of pesticide movement on small agricultural watersheds. U.S. Environ. Prot. Agency Rep. no. EPA-600/3-76-066. U.S. Government Printing Office, Washington, D.C.
2. Crawford, N. H., and A. S. Donigian, Jr. 1973. Pesticide transport and runoff model for agricultural lands. U.S. Environ. Prot. Agency Rep. no. EPA-600/2-74-013. U.S. Government Printing Office, Washington, D.C.
3. Crawford, N. H., and R. K. Linsley. 1966. Digital simulation in hydrology: Stanford Watershed Model IV. Tech. Rep. no. 39. Stanford Univ., Stanford, Calif.
4. Dean, J. D., and L. A. Mulkey. 1979. Interactive effects of pesticide properties and selected conservation practices on runoff losses: a simulation study. p. 715–734. In R. C. Loehr et al. (ed.) Best management practices for agriculture and silviculture. Ann Arbor Science Publishers Inc., Ann Arbor, Mich.
5. Donigian, A. S., Jr., D. C. Bryerlein, H. H. Davis, and N. H. Crawford. 1977. Agricultural Runoff Management Model Version II: refinement and testing. U.S. Environ. Prot. Agency Rep. no. EPA-600/3-77-098. U.S. Government Printing Office, Washington, D.C.
6. Edwards, W. M., G. B. Triplett, Jr., and R. M. Kramer. 1980. A watershed study of glyphosate transport in runoff. J. Environ. Qual. 9:661–665.
7. Falco, J. W., and L. A. Mulkey. 1976. Modeling the effect of pesticide loading on riverine ecosystems. In W. R. Ott (ed.) Proc. Conf. on Environmental Modeling and Simulation, Cincinnati, Ohio. 19–22 Apr. 1976. U.S. Environ. Prot. Agency Rep. no. EPA-600/9-76-016. U.S. Government Printing Office, Washington, D.C.
8. Frere, M. H., C. A. Orstad, and H. N. Holtan. 1975. ACTMO: An Agricultural Chemical Transport Model. Rep. no. ARS-H-3. USDA-ARS, Hyattsville, Md.

9. Green, W. H., and G. A. Ampt. 1911. Studies on soil physics: I. The flow of air and water through soils. J. Agric. Sci. 4:1-24.
10. Karickhoff, S. W. 1981. Semi-empirical estimation of sorption of hydrophobic pollutants on natural sediments and soils. Chemosphere 10:833-846.
11. Knisel, W. G. (ed.). 1980. CREAMS—a field scale model for Chemicals, Runoff, and Erosion from Agricultural Management Systems. Conserv. Res. Rep. no. 26. USDA, Washington, D.C.
12. Morel-Seytoux, H. T. 1979. Flow forecasting based on preseason conditions. p. 41-46. *In* Hsieh Wen Shen (ed.) Modeling of rivers. John Wiley & Sons, New York.
13. Negev, M. A. 1967. Sediment model of a digital computer. Tech. Rep. no. 76. Dep. of Civil Engineering, Stanford University, Stanford, Calif.
14. Onstad, C. A., and G. R. Foster. 1975. Erosion modelling on a watershed. Trans. ASAE 18:288-292.
15. Smith, C. N., R. A. Leonard, G. W. Langdale, and G. W. Bailey. 1978. Transport of agricultural chemicals from small upland piedmont watersheds. U.S. Environ. Prot. Agency Rep. no. EPA/600-3-78-056. U.S. Government Printing Office, Washington, D.C.
16. Steenhuis, T. S. 1979. Simulation of the action of soil and water conservation practices in controlling pesticides. Sect. 7. *In* D. A. Haith and R. C. Loehr (ed.) Effectiveness of soil and water conservation practices for pollution control. U.S. Environ. Prot. Agency Rep. no. EPA-600/3-79-106. Athens Environmental Research Lab., Athens, Ga.
17. Steenhuis, T. S., and M. R. Walter. 1980. Closed form solution for pesticide loss in runoff water. Trans. ASAE 23:615-620, 628.
18. Stewart, B. A., D. A. Woolhiser, W. H. Wischmeier, J. H. Caro, and M. H. Frere. 1976. Control of water pollution from cropland. Vol. D. An overview. U.S. Environ. Prot. Agency Rep. no. EPA-600/2-75-0266. U.S. Government Printing Office, Washington, D.C.
19. Williams, J. R. 1975. Sediment yield prediction with universal equation using runoff energy factor. Present and prospective technology for predicting sediment yields and sources. Rep. no. ARS-S-40. USDA-ARS, Washington, D.C. p. 244-252.
20. Willis, G. H., L. L. McDowell, J. F. Parr, and C. E. Murphree. 1976. Pesticide concentrations and yields in runoff and sediment from a Mississippi Delta watershed. p. 353-364. *In* Proc. 3rd Federal Interagency Sedimentation Conf., Denver, Colo. 22-25 Mar. 1976. Water Resources Council, Denver.
21. Wischmeier, W. H., and D. D. Smith. 1978. Predicting rainfall erosion losses—a guide to conservation planning. Agric. Handb. no. 537, USDA (superscedes Handb. no. 282.). U.S. Government Printing Office, Washington, D.C.
22. Yalin, Y. S. 1963. An expression for bedload transportation. J. Hydraulics Div. Am. Soc. Civ. Eng. 89(HY3):221-250.

VERTICAL TRANSPORT OF PESTICIDES INTO SOIL WHEN ADSORBED ON SUSPENDED PARTICLES

A. J. A. Vinten and B. Yaron
Agricultural Research Organization, Bet Dagan, Israel

P. H. Nye
Department of Agricultural Sciences, University of Oxford, Oxford

---

It is generally assumed that pesticides with very high K_D values are virtually immobile in the soil, though they may be transported laterally by erosion. Data presented in this paper show that vertical transport of [^{14}C]DDT and [^{14}C]paraquat adsorbed on suspended material can occur. Under favorable conditions for transport, 18% of the applied [^{14}C]DDT was transported on solids in sewage effluent to a depth greater than 9 cm in a sandy loam soil. Dispersed Li-montmorillonite suspension transported over 50% of applied [^{14}C]paraquat to a depth of 12 cm. The conditions required for such transport to occur are described.

It is observed in the literature that pesticides with high K_D values such as DDT [1,1,1-trichloro-2,2-bis(4-chlorophenyl)ethane] and paraquat (1,1'-dimethyl-4,4'-dipyridylium chloride) are so strongly adsorbed by soils as to be virtually immobile (Guenzi and Beard, 1967). Such pesticides may, however, be transported laterally in sediments removed by erosion [see Wauchope (1978) for a review] and in runoff water. Rao and Davidson (1982) point out that unless the K_D (partition coefficient) is greater than about 100 or the sediment load is high (>0.1%) most of the pesticide transported during lateral flow is in the aqueous phase because of the low concentration of solids in suspension. K_D is defined by the equation $X_e/n = K_D S_e$ where K_D = partition coefficient (mL/g), n = suspension concentration (g/mL), X_e = concentration of adsorbate on the adsorbed phase (g/mL), and S_e = concentration of adsorbate in the solution phase (g/mL). By contrast, during the flow of water through soil the contribution of pesticides in solution to vertical transport is always low if K_D is large and if the pesticide is rapidly transferred to adsorbing surfaces. Instantaneous equilibration is a good approximation of pesticide adsorption on the soil solid phase unless the flow rate is high (Davidson and Chang, 1972). Thus, only if the pesticide is complexed or adsorbed on mobile colloids will any significant vertical transport occur. Ballard (1971) had demonstrated the role of humic substances, dispersed by addition of urea to a forest soil, in complexing and transport of DDT. Guenzi and Beard (1967) postulated transport of DDT adsorbed on soil colloids as a mechanism of downward transport. The experiments reported here were part of two studies on the transport of suspended solids through soil. [^{14}C]Paraquat adsorbed on Li-montmorillonite in suspension was used in one study (Vinten, 1981) and [^{14}C]DDT adsorbed on the organic suspended solids in sewage effluent was used in the other (Vinten et al., 1983). As a side shoot of these investigations these data show that under favorable conditions transport of pesticides with high K_D values on mobile colloids is feasible.

Table I. Characteristics of Soils

soils	% clay	% organic matter	% carbonates	pH	packing density, g/cm^3
[^{14}C]DDT Experiments					
Bet Dagan, sandy loam	13.7	0.68	2.3	7.9	1.30
Gilat, silty loam	23.1	0.95	12.9	7.8	1.59
Bene Darom, coarse sand	1.2	0.21	2.8	7.8	1.30
[^{14}C]Paraquat Experiments					
Begbroke, sandy loam	18	2.5	n.a.[a]	6.1	1.4

[a] Not available.

EXPERIMENTAL SECTION

Materials. [^{14}C]Paraquat and [^{14}C]DDT were obtained from Amersham International, Ltd., United Kingdom. They had specific activities of 425 and 80.8 mCi/g, respectively. Sewage effluent was obtained from Kibbutz Givat Brenner, Israel. It had a pH of 7.6, EC of 2.0 mmho/cm, COD of 316 mg/L, and a suspended solids concentration of 98 mg/L. Unaltered effluent (A) and effluent after filtration through a Whatman No. 92 filter (B) were used. Details of soils used in the experiments are given in Table I. Counting of samples was done on

Table II. Adsorption Data for Pesticides Studied

pesticide	adsorbent	K_D, mL/g	concn of suspended solids, ppm (w/v)	$\frac{X_e}{S_e}$[a]
paraquat	Li-montmorillonite, <0.15 μm esd	7.3×10^4	30	2.2
DDT	sewage effluent solids (fraction A)[b]	7.8×10^4	98 (84% organic)[b]	7.7
DDT	sewage effluent solids (fraction B)[b]	2.1×10^5	38 (100% organic)[b]	8.0

[a] X_e = adsorbed label concentration at equilibrium. S_e = solution label concentration at equilibrium. [b] Determined by ignition at 600 °C.

Beckman ([^{14}C]paraquat) or Packard Prias ([^{14}C]DDT) liquid scintillation counters.

Procedure. (a) [^{14}C]Paraquat was added to 30 ppm of Li-montmorillonite suspension at a rate of 15 nCi/mL (0.27 μequiv/L) and shaken for 2 h. Supernatant activity was measured giving a K_D value (see Table II). Kinetics studies showed that desorption of pesticide from solids was of minor importance in the time course of experiments.

Columns of Begbroke sandy loam, varying from 2.1 to 12 cm in length, were packed to a bulk density of about 1.3 g/cm^3, saturated and leached with [^{14}C]paraquat-labeled Li-montmorillonite suspended in distilled water or 1 mM CaCl$_2$. Leachate activity and volume were monitored. A constant activity was achieved after 2-5 pore volumes, depending on column length.

(b) [^{14}C]DDT was adsorbed on suspended solids in sewage effluent by adding 300 nCi to 300 mL of effluent and gently stirring for 1 h. Supernatant activity gave the K_D value (Table II). Desorption kinetics were measured by monitoring the rate of partitioning of [^{14}C]DDT into a hexane phase in the presence of effluent and from distilled water. These again showed that release of DDT from the adsorbed phase was not important in the time course of experiments, though a correction needs to be made for DDT present in the equilibrium solution. Further details of experimental methods are given in Vinten et al. (1982) for DDT and Vinten (1981) for paraquat.

Columns of Gilat (silt loam), Bet Dagan (sandy loam), and Bene Darom (coarse sand) soils were leached with [^{14}C]DDT-labeled effluent. Following leaching of the soils, columns were sectioned and the [^{14}C]DDT was extracted by shaking with a 1:1 hexane-acetone mixture for 1 h. The distribution of solids deposited in the soil with depth was thus determined. Activity in the leachate was also measured. By assuming that the rate of deposition of particles is a function of depth through the column only (and not of time) for a given run, it is possible from the distribution in the column to calculate the concentration of pesticide remaining in the mobile phase as a function of depth. Thus, if half the pesticide used in the experiment is located in the soil in the top 5 cm, then it is assumed that the concentration of pesticide in suspension at that depth is half that of the input suspension and that this remains constant.

RESULTS AND DISCUSSION

Figure 1A shows the vertical transport of [^{14}C]paraquat adsorbed on Li-montmorillonite suspension through columns 2.1-12-cm in length. Data on experimental conditions in columns are presented in Table III. When distilled water is the suspension medium, 50% of the pesticide penetrates beyond 12 cm. However, in 1 mM CaCl$_2$ only 5% of the pesticide penetrates deeper than 1 cm. The high [Ca^{2+}] results is rapid immobilization of the clay in the soil through flocculation and straining or through adsorption

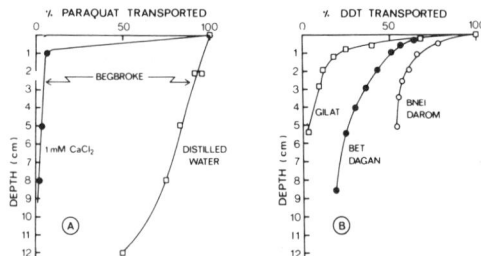

Figure 1. Vertical transport of DDT and paraquat adsorbed on suspended particles (in percent of total applied). (A) [^{14}C]Paraquat adsorbed on Li-montmorillonite. (B) [^{14}C]DDT adsorbed on suspended solids in sewage effluent.

Table III. Data on Column Experiments

[^{14}C]Paraquat Experiments			
column length, cm	suspension medium	initial flow rate, cm/s	leachate vol, cm
2	distilled water	3.5×10^{-4}	30
2	distilled water	2.6×10^{-3}	23
5	distilled water	1.1×10^{-3}	24
8	distilled water	5.4×10^{-3}	35
12	distilled water	1.2×10^{-2}	45
1	1 mM CaCl$_2$	4.3×10^{-3}	32
2	1 mM CaCl$_2$	1.1×10^{-2}	35
8	1 mM CaCl$_2$	3.1×10^{-3}	38

[^{14}C]DDT Experiments			
soil	effluent	initial flow rate, cm/s	leachate vol, cm
Bene Darom	A	9.4×10^{-2}	20
Bet Dagan	A	1.2×10^{-2}	54
Gilat	B	8.9×10^{-4}	36

on pore surfaces, and consequently little pesticide transport occurs. Under conditions when the clay is dispersed there is no flocculation and little interaction with the soil solid phase, so the pesticide is readily transported through the soil.

In the second case a range of behavior of [^{14}C]DDT is observed in three soils (Figure 1B). In the Gilat soil—a silt loam—the flow rate is slow, the pore structure is fine, and little transport of pesticide occurs (only 3% reaches 5.4 cm). This is a result of efficient removal of suspended solids by the soil, despite the use of the finer effluent fraction B. Fraction B, the effluent filtered through a Whatman No. 92 paper, was used to prevent sedimentation of coarse suspended solids at the soil surface. This occurred because of the low flow rates occurring in the Gilat soil. In the case of the Bet Dagan soil—a sandy loam—there is considerable pesticide transport as the flow rate is higher and the organic colloids are more mobile. Eighteen percent of the pesticide is transported to a depth greatr than 9 cm. In the case of the Bene Darom coarse sand, still more transport of pesticide occurs, with 54% penetrating deeper than 5 cm. Again these differences are due to the differences in mobility of the solids which is directly related to soil type and flow rate.

Some qualification of the assumption of the steady rate deposition of suspended particles (see Experimental Section) is needed here. As deposited solids accumulate in the soil, clogging of soil pores may occur, reducing flow rate. This may result in increased particle deposition and, hence, pesticide removal near the soil surface. Data presented in Figure 1B thus give the time-averaged concentration of pesticide in the mobile phase at any depth. This simplification was valid for the experimental conditions used here as the leachate activity was nearly constant with time, despite the reduction in the flow rate that occurred.

Having demonstrated the feasibility of transport of strongly adsorbed pesticides on mobile colloids during leaching, it is important to indicate under what field conditions such transport might occur and contribute to edge-of-field loss. DDT or paraquat may reach the soil following foliar application and will be strongly adsorbed close to the soil surface. Other soil-applied herbicides such as the s-triazine group and other organochlorine pesticides have high K_D values and will not be transported much in solution. Hartley and Graham-Bryce (1980) point out the potential hazard in lateral runoff for such pesticides but consider vertical transport unimportant. However, if soils are leached with rainwater, or with sodic water and subsequently with low EC water, dispersion and release of clay can occur [e.g., Shainberg et al. (1981)]. Under such conditions, when release of soil colloids occurs, pesticides adsorbed in the surface soil may be transported to drainage water. Pesticides applied as wettable powders may also be transported in suspension form if leaching conditions occur soon after application.

The extent to which this type of transport occurs depends on the amount of clay or organic matter released by the surface soil on dispersion, the mobility of these colloids in the soil profile, the rate at which soil clogging occurs, the K_D value, and the kinetics of desorption of pesticide from the mobile colloids.

Registry No. Paraquat, 4685-14-7; Li-montmorillonite, 67034-72-4; DDT, 50-29-3.

LITERATURE CITED

Ballard, T. M. *Soil Sci. Soc. Am. Proc* **1971**, *35*, 145–147.

Davidson, J. M.; Chang, R. K. *Soil Sci. Soc. Am. Proc.* **1972**, *36*, 257–261.

Guenzi, W. D.; Beard, W. E. *Soil Sci. Soc. Am. Proc.* **1967**, *31*, 644–647.

Hartley, G. S.; Graham-Bryce, I. J. "Physical Principles of Pesticide Behavior"; Academic Press: London, 1980; Vol. I, Chapter 5.

Rao, P. S. C.; Davidson, J. M. In "Environmental Impact of Nonpoint Source Pollution"; Overcash, M. R.; Davidson, J. M., Eds.; Ann Arbor Science Publishers: Ann Arbor, MI, 1982; Chapter 1, pp 23–68.

Shainberg, I.; Rhoades, J. D.; Prather, R. J. *Soil Sci. Soc. Am. J.* **1981**, *45*, 273–277.

Vinten, A. J. A. Doctor of Philosophy Thesis, University of Oxford, 1981.

Vinten, A. J. A.; Mingelgrin, U.; Yaron, B. *Soil Sci. Soc. Am. J.* **1983**, in press.

Wauchope, R. D. *J. Environ. Qual.* **1978**, *7*, 459–472.

Received for review June 15, 1982. Accepted December 27, 1982. Contribution No. 683-E, 1983 series, from the Agricultural Research Organization.

Editors' Comments
on Papers 49 Through 51

49 SPENCER, CLIATH, and FARMER
Vapor Density of Soil-Applied Dieldrin as Related to Soil-Water Content, Temperature, and Dieldrin Concentration

50 MAYER, LETEY, and FARMER
Models for Predicting Volatilization of Soil-Incorporated Pesticides

51 JURY, FARMER, and SPENCER
Behavior Assessment Model for Trace Organics in Soil: II. Chemical Classification and Parameter Sensitivity

VOLATILIZATION

The movement of pesticides in the vapor phase from soil to the atmosphere is an important pathway for their loss from treated agricultural lands. Potential volatility of a pesticide is related to its inherent vapor pressure, but actual vaporization rates depend on the environmental conditions and on all the other factors that control behavior of the chemical at the solid-air-water interface. Early studies of pesticide gaseous transport through soils investigated mainly behavior of fumigants. Call (1957), studying the transport of ethylene dibromide through soils, showed that this process is governed by diffusion and can be adequately explained by the diffusion theory. The author pointed out the great influence of sorption on the unsteady-state diffusion coefficient.

The volatilization of pesticides from soils and the factors governing this phenomenon were studied initially by bioassay procedures. In an early study Harris and Lichtenstein (1961) showed that the rate of aldrin volatilization from the soil rose with increasing insecticide concentration, soil moisture, and temperature and rate of air movement over the soil surface. Lower volatilization rates were noted in dry soils rich in clay and organic matter, and in wet soils with high organic matter contents, suggesting that less aldrin volatilization occurred under such conditions, favoring its adsorption instead.

Later studies using improved analytical techniques enabled a better understanding of the volatilization process. Soil-incorporated pesticides volatilize at a rate dependent not only on their equilibrium distribution between the air, water, and soil matrix as related to vapor pressure, solubility, and adsorption coefficients, but also on their rate of movement to the soil surface. This involves desorption and transport in the vapor phase to the soil surface.

The soil water content effect is especially important in the volatilization of relatively nonpolar pesticides from soil. Measurement of vapor density of pesticides in soil at various water contents demonstrated conclusively that the greater extent of vaporization from wet than from dry soils is due mainly to increased vapor density, resulting from displacement of the chemical from the soil adsorbing surfaces by water. The effect of soil water on the volatility of pesticides was described by Bowman, Schechter, and Carter (1965) as a co-distillation phenomenon, but this concept was proven inadequate by Hartley (1968) and Igue and co-workers (1972).

Paper 49 presents a study of the factors affecting the movement and volatilization of a persistent chlorinated hydrocarbon insecticide, dieldrin, in soil, with emphasis on the influence of water content and temperature on its solid phase-vapor distribution. An interesting effect of the moisture content on the vapor density of dieldrin was observed: the water content had no effect on vapor density until the soil was dried to a water content corresponding to about a monolayer of water. A decrease in the water content below this point caused a steep drop in the vapor density. The drying effect was reversible. This relationship between soil moisture and vapor density is apparently due to the competition between water molecules and dieldrin for adsorption sites. When sufficient water was present to shield the adsorbing surfaces, the vapor density approached that of pure dieldrin. Based on the results obtained, it was assumed that surface application of dieldrin will result in a volatilization rate similar to that of pure material until the concentration at the surface decreases approximately to 25 ppm. Because the data in this paper were obtained with a desert soil, having a very low organic matter content, this conclusion may not be valid for soils with higher adsorptive capacities.

Volatilization of pesticides from soils can be estimated from a consideration of the physical and chemical factors controlling their concentration at the soil surface. With soil-incorporated pesticides, the initial volatilization rate will be a function of their vapor density at the surface as affected by adsorptive interactions with the soil. The fraction of the exposed material that remains on the soil surface after mixing is readily lost. Volatilization then becomes dependent upon the rate of the movement of the pesticide to the soil surface by diffusion and convection in evaporating water. Usually the two mechanisms, diffusion and convection, are simultaneous. Movement to the surface by bulk flow or convection in the soil water is the dominant factor in controlling volatilization of pesticides incorporated in moist soil. Most models developed for estimating volatilization rates are based upon equations describing the rate of movement of the chemicals to the surface by diffusion and/or by convection and from the surface through the air boundary layer by diffusion.

In Paper 50, Mayer, Letey, and Farmer proposed a series of five models to describe the volatilization of soil-incorporated pesticides under various environmental conditions. The basic assumption of these models was that movement occurred only by diffusion, so that volatilization can be predicted by solving the diffusion equations for various boundary conditions. Data calculated by these equations were compared with published results of the volatilization of lindane and dieldrin. The models assuming that pesticide concentration at the soil surface or in the air above the soil is maintained at zero by air movement give a good assessment of the volatilization losses under laboratory conditions. Calculations done according to the model assuming a nonmoving air layer above the surface showed that under these conditions volatilization was reduced appreciably. As mentioned by the authors, the proposed models can be applied to predict the volatilization of moderately and strongly adsorbed pesticides moving mainly by diffusion. Jury and co-workers (1980) include in the modeling of vapor losses if soil-incorporated triallate such factors as desorption and subsequent transport of the herbicide to the surface by diffusion and mass flow in both the presence and absence of evaporative water.

Jury, Spencer, and Farmer (1983) assume that the soil-surface boundary consists of a stagnant boundary layer connecting the soil and air, through which pesticides and water vapor must move to reach the atmosphere. They assume further that the gas and liquid concentrations are related by Henry's law, that the adsorption isotherms are linear, and that degradation occurs by a first-order rate process. Their model is intended to classify and screen organic chemicals for environmental behavior based on physical and chemical properties such as vapor pressure, solubility, Henry's constant, organic carbon partition coefficient, and degradation rate.

A simplification of the screening model by dividing chemicals into volatilization and mobility categories is presented in Paper 51. In this paper the volatilization classification is based on whether or not the predominant resistance to volatilization loss lies in the soil or in the boundary layer above the soil surface. The extent to which the boundary layer limits the volatilization flux is used as a criterion for classifying volatilization of pesticides. In this model a distinction is made between volatilization with and without evaporation. In general (independent of chemical properties of the pesticide) the volatilization rate is enhanced by evaporation. A complementary paper published by the same research group (Jury et al., 1984) presents the results of volatilization tests for 30 pesticides both with and without evaporation. For the majority of the compounds tested (except for parathion and methylparathion, characterized by short half lives), the dependence of volatilization on evaporation water was proved.

References

Bowman, M. C., M. S. Schechter, and R. L. Carter, 1965, Behavior of Chlorinated Insecticides in a Broad Spectrum of Soil Types, *Jour. Agric. Food Chem.* **13:**360-365.

Call, F., 1957, Soil Fumigation. V. Diffusion of Ethylene Dibromide through Soils, *Jour. Sci. Food Agric.* **8:**143-150.

Harris, C. R., and E. P. Lichtenstein, 1961, Factors Affecting the Volatilization of Insecticidal Residues from Soils, *Jour. Econ. Entomol.* **54:**1038-1045.

Hartley, G. S., 1968, Evaporation of Pesticides, in Pesticidal Formulation Research, *Adv. Chem. Ser.* **86:**115-134.

Igue, K., W. J. Farmer, W. F. Spencer, and J. P. Martin, 1972, Volatility of Organochlorine Insecticides from Soil: II. Effect of Relative Humidity and Soil Water Content on Dieldrin Volatility, *Soil Sci. Soc. America Proc.* **36:**447-450.

Jury, W. A., W. F. Spencer, and W. J. Farmer, 1983, Behavior Assessment Model for Trace Organics in Soil. I. Model Description, *Jour. Environ. Qual.* **12:**558-564.

Jury, W. A., R. Grover, W. F. Spencer, and W. J. Farmer, 1980, Modeling Vapor Losses of Soil Incorporated Triallate, *Soil Sci. Soc. America Jour.* **44:**445-450.

Jury, W. A., W. F. Spencer, and W. J. Farmer, 1984. Behavior Assessment Model for Trace Organics in Soil. III. Application of Screening Model, *Jour. Environ. Qual.* **13:**573-579.

Vapor Density of Soil-Applied Dieldrin as Related to Soil-Water Content, Temperature, and Dieldrin Concentration[1]

W. F. SPENCER, M. M. CLIATH, AND W. J. FARMER[2]

ABSTRACT

Vapor densities of dieldrin (principal constituent HEOD-hexachloro-epoxyoctahydro-endo, exo-dimethanonaphthalene) in dieldrin-soil mixtures increased with temperature and dieldrin concentration but were not affected by soil-water content until the water content decreased below that equivalent of one molecular layer of water. Vapor densities dropped to very low values when the water content fell below this level, but increased again as water was added to the dry soil, indicating that the drying effect is reversible. When more than a monomolecular layer of water was present in the soil, vapor density increased with increasing soil dieldrin (HEOD) concentration until a saturation vapor density equal to that of HEOD without soil [54, 202, and 676 ng HEOD/liter at 20, 30, and 40C, respectively (9)] was reached at approximately 25 ppm HEOD. This implies that surface applications of dieldrin and probably other similar chlorinated hydrocarbon insecticides will volatilize as rapidly from mineral soils as from the pure materials until the concentration at the surface falls to relatively low levels.

The data indicate that loss of water, per se, is not required for significant rates of volatilization to occur from soils or other surfaces on which water can successfully compete for adsorption sites.

Additional Key Words for Indexing: insecticide residues, volatilization, pesticides.

DIELDRIN is one of the more persistent chlorinated hydrocarbon insecticides and often becomes a soil residue problem. It is important to understand the soil and environmental factors that affect this persistence in order to develop better means of enhancing its dissipation from soils. Considerable evidence indicates that volatilization from the soil surface may be an important pathway for loss of dieldrin and other such persistent insecticides (4, 7). Acree, Bowman, and coworkers (1, 2, 3) reported that loss of water contributed to insecticide volatilization by an apparent "codistillation" process. Spencer and Cliath (9) reported that the vapor densities associated with solid-phase dieldrin (HEOD) and dieldrin-soil mixtures measured by a gas saturation technique were three to 12 times greater than predicted from published vapor pressure values (8). Their measured vapor densities were 54, 202, and 676 ng HEOD/liter equivalent to an apparent vapor pressure of 2.6×10^{-6}, 1.0×10^{-5}, and 3.47×10^{-5} mmHg at 20, 30, and 40C, respectively. The vapor density of dry HEOD was the same as that of HEOD plus water and the vapor density of HEOD in moist soil at 100 ppm was the same as that of HEOD without soil; however, at 10 ppm the vapor density was reduced approximately 80%. The heat of vaporization of HEOD, either with or without soil, was 23.6 K cal/mole.

The objective of the present work was to pinpoint the factors affecting volatilization and movement of chlorinated hydrocarbon compounds, such as dieldrin, in soils. This paper presents data relating solid-phase dieldrin concentration in soil to vapor density as affected by the soil-water content and temperature, and discusses the implication of the results to volatilization from soils and to diffusion through soils.

METHODS AND MATERIALS

Vapor density of soil-applied dieldrin as affected by dieldrin concentration, soil-water content, and temperature was measured by a gas saturation method. The details of the apparatus and procedures have been previously described (9). Briefly, in the gas saturation method a current of inert gas is passed through or over the material at a sufficiently slow rate to insure equilibrium vapor saturation. In our studies the vapor density of recrystallized dieldrin, 99% HEOD (1,2,3,4,10,10-hexachloro-6,7-epoxy-1,4,4a,5,6,7,8,8a-octahydro-1,4-endo,exo-5,8-dimethanonaphthalene), was determined by measuring the amount of HEOD in a stream of nitrogen gas slowly moving through a column of soil containing various concentrations of HEOD. The HEOD was removed from the nitrogen gas stream in gas-washing bottles containing hexane. The HEOD content of the hexane was determined with a gas-liquid chromatograph equipped with an electron-capture detector.

The Gila silt loam used in these studies is a desert soil containing 18.4% clay—predominately montmorillonite—0.6% organic matter, with a surface area of approximately 90 m²/g and an exchange capacity of 18 meq/100 g. Dieldrin in acetone was added to moist, autoclaved soil with an atomizer. The soil was aerated with moist air to remove the acetone, then adjusted to the desired water content by adding a predetermined amount of water with an atomizer, or as a weighed amount of finely ground ice at temperatures below freezing. After packing the columns with approximately 900 g of treated soil, they were incubated at 30C for 30 days before measurements were initiated. Nitrogen gas flow rates of from 3 to 6 ml/min were generally used to provide a total flow through the saturator of from 10 to 80 liters. In all measurements with soils at different water contents, the humidity of the nitrogen carrier gas was controlled in equilibrium with the moisture content of the soils. This resulted in no net loss, or gain, of water from the soil columns during the periods of measurement. Vapor density at 20, 30, and 40C was determined on each column made up at a particular HEOD concentration and/or water content. From three to nine measurements were made on each set of columns at each temperature, which resulted in an average coefficient of variation in vapor density of 6%.

RESULTS AND DISCUSSION

Desorption isotherms relating vapor density and HEOD concentration in Gila silt loam at water contents of 10% or greater are shown in Fig. 1. At all temperatures, vapor density increased rapidly as concentration increased until

---

[1] Contribution from the Southwest Branch, Soil & Water Conservation Research Division, ARS, USDA, and the California Agr. Exp. Sta., Riverside, Calif. Presented before Div. S-1 and S-2, Soil Science Society of America, Nov. 13, 1968, New Orleans, Louisiana. Received Jan. 27, 1969. Approved March 19, 1969.

[2] Soil Scientist, Chemist, USDA, and Asst. Chemist, UCR, Riverside, Calif.

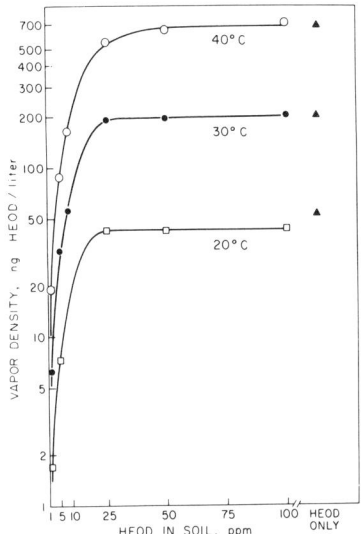

Fig. 1—Vapor density of HEOD (dieldrin) in Gila silt loam at water contents of 10% or greater as affected by temperature and concentration of HEOD.

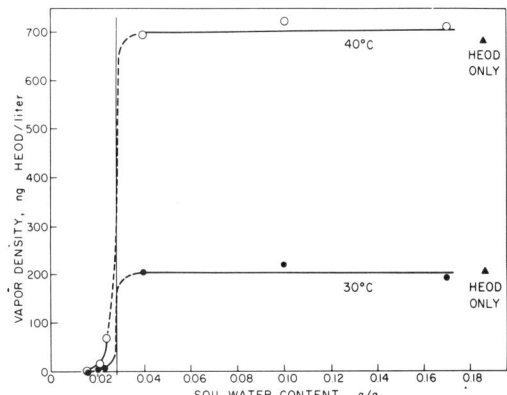

Fig. 2—Effect of soil-water content on vapor density of HEOD (dieldrin) in Gila silt loam at 100 ppm HEOD.

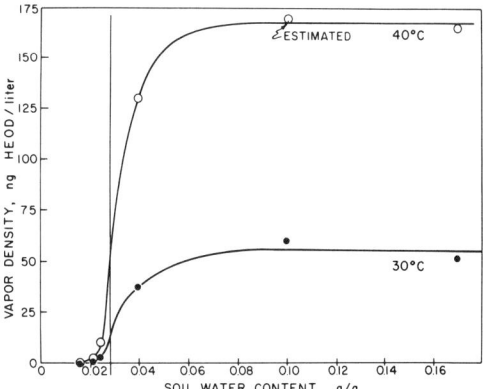

Fig. 3—Effect of soil-water content on vapor density of HEOD (dieldrin) in Gila silt loam at 10 ppm HEOD. Vapor density at 40C and 10% water content estimated from other HEOD concentrations at 10% water.

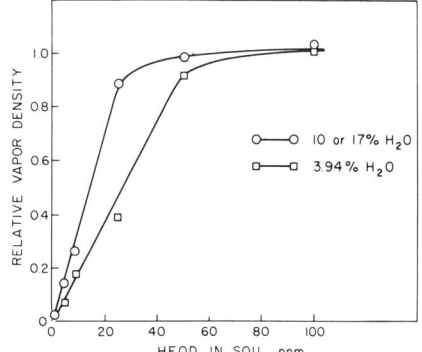

Fig. 4—Relative vapor density of HEOD (dieldrin) vs concentration of HEOD in Gila silt loam as affected by soil-water content. Vapor density of HEOD without soil equals 1.0.

a saturation vapor density, equal to that of HEOD without soil, was reached at concentrations near 25 ppm HEOD. Vapor density of soil-applied HEOD increased with temperature in the same manner as HEOD without soil. This corroborates the earlier finding by Spencer and Cliath (9) that heats of vaporization for HEOD in soil and HEOD only are similar.

Figures 2 and 3 show the effect of soil-water content on vapor density of HEOD in Gila silt loam at 100 ppm and 10 ppm HEOD, respectively. At 100 ppm HEOD the soil-water content had essentially no effect on vapor density, until the soil was dried to a water content approaching 1 molecular layer of water. When the water content dropped only slightly below this to 2.1%, the vapor density decreased markedly. The vertical line in Fig. 2 and 3 at 0.028 g water/g soil indicates the calculated amount of water equivalent to a monomolecular layer, assuming a surface area of approximately 90 m²/g. In Gila silt loam 17% water is equivalent to field capacity, 10% water is equivalent to approximately 2 atm matrix suction, and 3.94% water to 94% relative humidity or 90 atm suction. Thus the soil is extremely dry before the vapor density decreases appreciably. The vapor density approached a concentration near the lower limit of measurement when the soil was air dried to 1.6% water. However, this drying effect is reversible. Upon rewetting the air-dry soil vapor density increased to its original maximum value. For example, at 100 ppm HEOD and 40C vapor density of HEOD in the air-dry soil was 1.8 ng/liter. When water was added, to 17%, the vapor density immediately increased to 710 ng HEOD/liter.

At 10 ppm HEOD the vapor density begins to decrease at a slightly higher moisture content than at 100 ppm HEOD, as illustrated by lower vapor density at 3.9 than at 10% water in Fig. 3. This is probably due to the interaction between moisture and HEOD in competition for adsorption sites on the soil.

The great difference in vapor density between wet and dry soils could conceivably result in significant aerial transfer of chlorinated hydrocarbon insecticides from relatively wet treated areas to dry untreated fields.

Figure 4 combines all variables studied—HEOD concentration, water content, and temperature into a generalized picture of the relationship between concentration in the soil and vapor density. Relative vapor densities or the ratio of the vapor density of soil-applied HEOD to vapor density of HEOD without soil was calculated using all data obtained at temperatures of 30 and 40C. These relationships should hold at any temperature for this particular soil. Vapor density increases linearly with concentration until essentially a saturated vapor density is reached. The curve for 10 or 17% water would apply to most field situations, particularly during the growth of a crop. The right-hand curve in Fig. 4 was obtained at 3.94% water, or 90 atm matrix suction. As the soil continues to dry below this level to air dryness, the curves will approach the X axis, or negligible vapor density. At 2.1% water, equivalent to 50% relative humidity in this soil, the relative vapor density was just above the horizontal line even at 100 ppm HEOD.

The very marked effect of small amounts of water on vapor density of dieldrin in the dry soil range would explain greater volatilization of chlorinated hydrocarbons from wet than from dry soils. Apparently, water increases the vapor density in this very dry range due to competition for adsorption sites on the soil. When sufficient water is present to cover the surface, the dieldrin volatility approaches that of the pure material. There is no evidence from our data to indicate that the evaporation of water, per se, increases vapor density by a "codistillation" process. The actual amount of dieldrin volatilized during a given period, a week for example, would be related to the time it takes to dry the soil sufficiently to reduce the vapor density to an insignificantly low value. Competition between water and the chlorinated hydrocarbons for adsorption sites would also occur on other surfaces and the effect of humidity on their volatility from surfaces such as glass, metal, or skin could be explained on the basis of the higher humidity furnishing water molecules to displace the hydrocarbons from the solid surfaces. The fact that water also had no effect on vapor density of HEOD without soil (9) is additional evidence to indicate that water loss does not increase vapor density.

The fact that the vapor density and heat of vaporization of HEOD applied to a slightly moist soil at 25 ppm, or greater, is the same as HEOD without soil indicates that the adsorption forces between HEOD and soil are quite weak and probably the HEOD is present as globules or adsorbed at the air-water interface. Since HEOD is a relatively nonpolar molecule it would not be expected to be strongly adsorbed by soil mineral surfaces. Therefore, surface applications of dieldrin and probably other similar chlorinated hydrocarbon insecticides will volatilize as rapidly from mineral soils as from the pure materials until the concentration at the surface falls below approximately 25 ppm. The high vapor density, until relatively low concentration levels are reached, indicates that dissipation curves for dieldrin from moist mineral soils would be similar to those postulated by Gunther and Blinn (6) for the disappearance of insecticides on and in plant tissues. In their models, relatively high rates of loss by volatilization are predicted soon after application of the materials to the plant surfaces. The loss from soils would probably be similar, but on a somewhat different time scale.

Diffusion of a volatile compound through soil can be either in the vapor or nonvapor phase. The apparent vapor diffusion coefficient is proportional to the slope of the vapor density-concentration curve. Thus, the data presented in Fig. 1 and 4 aid in the calculation of the apparent vapor diffusion coefficient. The linear relationship between vapor density and concentration of HEOD in soil means that the ratio of vapor density to soil concentration is a constant, therefore the apparent vapor diffusion coefficient is constant over this concentration range (5). Different soils and different insecticides will result in different slopes of lines relating vapor density to concentration in soils, and it would appear to be worthwhile to determine these relationships for other soils, particularly those higher in organic matter, and for other insecticides.

LITERATURE CITED

1. Acree, F., Jr., M. Beroza, and M. C. Bowman. 1963. Codistillation of DDT with water. J. Agr. Food Chem. 11:278–280.
2. Bowman, M. C., F. Acree, Jr., L. S. Lofgren, and M. Beroza. 1964. Chlorinated insecticides: Fate in aqueous suspension containing mosquito larvae. Science 146: 1480–1481.
3. Bowman, M. C., M. S. Schecter, and R. L. Carter. 1965. Behavior of chlorinated insecticides in a broad spectrum of soil types. J. Agr. Food Chem. 13:360–365.
4. Edwards, C. A. 1966. Insecticides in soils. Residue Rev. 13:82–132.
5. Ehlers, W., J. Letey, W. F. Spencer, and W. J. Farmer. 1969. Lindane diffusion in soils. I. Theoretical calculations and mechanism of movement. Soil. Sci. Soc. Amer. Proc. 33:501–504. (this issue)
6. Gunther, F. A., and R. C. Blinn. 1955. Analysis of insecticides and acaracides, p. 141–147. Interscience Publishers Inc., New York.
7. Harris, C. R., and E. P. Lichtenstein. 1961. Factors affecting the volatilization of insecticides from soils. J. Econ. Entmol. 54:1038–1045.
8. Porter, P. E. 1964. Dieldrin. p. 143–153. In G. Zweig (ed.) Analytical methods for pesticides, plant growth regulators and food additives. Vol. II. Insecticides. Academic Press, New York.
9. Spencer, W. F., and M. M. Cliath. 1969. Vapor density of dieldrin. Environ. Sci. Tech. (In press). Vol. 3.

Models for Predicting Volatilization of Soil-Incorporated Pesticides[1]

R. Mayer, J. Letey, and W. J. Farmer[2]

ABSTRACT

In the absence of appreciable mass transfer due to water movement, diffusion processes in the soil account for the movement of pesticides to the soil surface to replace that lost by volatilization. Published solutions for heat flow equations have been applied to the volatilization of lindane and dieldrin from Gila silt loam for a number of different initial and boundary conditions. Predicted fluxes agreed well with experimental values. Five models have been proposed to describe various environmental conditions found in the field. Models I, II, and III assume pesticide concentration at the soil surface is maintained at zero concentration by air movement. Model IV assumes surface pesticide concentrations greater than zero with air turbulence sufficient to maintain zero pesticide concentration gradient in the air above the soil. Model V assumes a nonmoving air layer of various depths above the soil surface so that the pesticide concentration gradient in the air controls the rate of volatilization.

Additional Index Words: dieldrin, lindane, organochlorine insecticides, diffusion, transport processes.

SOME ORGANIC compounds, especially pesticides, applied to soils are frequently transported to nontarget areas. In order to predict losses from soil, it is desirable to know something of the mechanisms involved in the transport of pesticides through the soil. This paper will concentrate on loss from soil by volatilization. Several investigators have demonstrated that the volatility rate for soil-incorporated pesticides is initially high and decreases with time at a rate dependent on the pesticide and on the experimental conditions (Farmer et al., 1972; Igue et al., 1972; Spencer and Cliath, 1973).

As a pesticide volatilizes at the soil surface, it may be replaced by pesticides within the soil profile. Pesticides move to the soil surface by mass transport in water flow to the surface, or by diffusion of the pesticide. Farmer et al. (1973) has shown the importance of diffusion processes in controlling dieldrin volatility.

The objective of this study is to develop mathematical equations for predicting volatilization of soil-incorporated pesticides as a diffusion-controlled process and to compare calculated with measured results.

THEORETICAL CONSIDERATIONS

The basic assumption in the mathematical treatment of the movement of pesticides in soils under a concentration gradient is the applicability of the diffusion laws. The changes in pesticide concentration within the soil as well as the loss of pesticides at the soil surface by volatilization can then be predicted by solving the diffusion equation for different boundary conditions. Recognizing the analogy between the heat transfer equation (Fourier's law) and the transfer of matter under a concentration gradient (Fick's law), solutions of the heat transfer equation given by the mathematical theory of conduction of heat may be used. The mathematical model for predicting volatilization of pesticides is then given as a set of boundary conditions sufficient to solve the diffusion equation.

Consider a system where a pesticide is uniformly mixed with a layer of soil and volatilizing at the soil surface. If diffusion is the only mechanism supplying pesticide to the surface, and if we assume isotropic conditions in the soil as well as constancy of the diffusion coefficient D, then the general diffusion equation is

$$\frac{\partial^2 c}{\partial x^2} - \frac{1}{D}\frac{\partial c}{\partial t} = 0 \qquad [1]$$

where c is the pesticide concentration in the soil (g/cm^{-3} total volume), x is the distance measured normal to the soil surface (cm), D is the diffusion coefficient (cm^2 sec^{-1}), and t is the time (sec).

Model 1

As a first approximation we may assume that pesticide volatilizes and is removed rapidly maintaining a zero concentration at the soil surface. The depth of the soil layer treated uniformly with pesticide is L and the initial concentration C_o (g/cm^3). The initial and boundary conditions for which Eq. [1] is to be solved are then

$$c = C_o \text{ at } t = 0, \ 0 \le x \le L$$
$$c = 0 \text{ at } x = 0 \text{ and } t > 0$$
$$\partial c/\partial x = 0 \text{ at } x = L.$$

Carslaw and Jaeger [1959, p. 97, Eq. (8)] give a solution for these boundary conditions that can be used by accepting the following analogies:

v (temperature) $= c$ (concentration)
V_o (initial temperature) $= C_o$ (initial concentration)
κ (thermal diffusivity)
K (thermal conductivity) $\Big\} = D$ (apparent diffusion coefficient).

The concentration units for c and C_o must be expressed on a total volume basis in order for D to be analogous to K and κ.

The solution for Eq. [1] is then

$$c = \frac{4C_o}{\pi} \sum_{n=0}^{\infty} \frac{(-1)^n}{(2n+1)} \{\exp[-D(2n+1)^2 \pi^2 t/4L^2]\}$$
$$\cos\frac{(2n+1)\pi(L-x)}{2L}. \qquad [2]$$

(Carslaw and Jaeger used the boundary condition $x = L$ and we have used $x = 0$ at the soil surface. Identical results are achieved by our using $(L - x)$ in the equations where Carslaw and Jaeger used x). The pesticide flux, f (g/cm^2/sec), through the surface is given as the concentration gradient at $x = 0$ times the diffusion coefficient D

---

[1] Contribution from the Department of Soil Science & Agricultural Engineering, University of California, Riverside 92502. Supported by Environmental Protection Agency Grant no. 13020GRQ. Received 16 March 1973. Approved 1 March 1974.
[2] Former Post-doctoral Research Soil Scientist, Professor of Soil Physics, and Associate Professor of Soil Science, respectively, Univ. of California, Riverside. Present address of senior author: Institut für Bodenkunde und Waldernährung, 34 Göttingen-Weende, Büsgenweg 2, Germany.

$$f = D[\partial c/\partial x]_{x=0}$$

$$= \frac{DC_o}{(\pi Dt)^{1/2}} [1 + 2 \sum_{n=1}^{\infty} (-1)^n \exp(-n^2L^2/Dt)]. \quad [3]$$

Model II

The summation term in Eq. [3] decreases with increasing L and decreasing D and t. If this term is small enough to be negligible, Eq. [3] reduces to

$$f = DC_o/(\pi Dt)^{1/2}. \quad [4]$$

This equation is identical with the solution given by Carslaw and Jaeger [1959, p. 59, Eq. (3)] for heat flow in an infinite solid. The concentration for the semi-infinite case is given by

$$c = C_o \operatorname{erf} [x/2(Dt)^{1/2}]. \quad [5]$$

Equations [4] and [5] are applicable also on a finite system (in the region $0 < x < L$) as long as the concentration at the lower boundary of the soil layer, $x = L$, is not decreased by pesticide moving in the upward or downward direction. To estimate the maximum time at a given set of parameters for which Model II is adequate, let us assume the critical value to be a drop of 1% in the initial concentration, C_o, at the lower boundary of the soil layer. With this assumption the boundary conditions used in deriving Eq. [5] are violated if

$$\operatorname{erf} [L/2(Dt)^{1/2}] \leq .99$$

or

$$t > L^2/14.4\, D.$$

Model III

Model I assumed that no diffusion occurs across the lower boundary $x = L$. This boundary condition is appropriate for laboratory studies which will be reported later, but not realistic for field conditions where diffusion can occur downward across the boundary $x = L$. The boundary conditions for the latter case are

$$c = C_o \text{ at } t = 0, 0 \leq x \leq L$$

$$c = 0 \text{ at } t = 0, x > L$$

$$c = 0 \text{ at } t > 0, x = 0$$

Solution of Eq. [1] with these initial and boundary conditions is given by Carslaw and Jaeger (1959, p. 62, Eq. [14]) as

$$c = (C_o/2) \{2 \operatorname{erf} [x/2(Dt)^{1/2}] - \operatorname{erf} [(x-L)/2(Dt)^{1/2}]$$
$$- \operatorname{erf} [(x+L)/2(Dt)^{1/2}]\}. \quad [6]$$

The flux is obtained by differentiating Eq. [6] with respect to x, determining $\partial c/\partial x$ at $x = 0$, and multiplying by D. The result is

$$f = [DC_o/(\pi Dt)^{1/2}] [1 - \exp(-L^2/4Dt)]. \quad [7]$$

Note that Eq. [7] reduces to Eq. [4] for large values of $L^2/4Dt$. Less than 1% error will result from using Eq. [4] if

$$\exp(-L^2/4Dt) < .01$$

or

$$t < L^2/18.4D.$$

Model IV

The solutions for the diffusion equation become more difficult if the concentration at the soil surface is variable instead of being maintained at zero. The rate of removal of pesticide in the air layer above the soil may then become a limiting factor in the rate of volatilization.

A model which accounts for a possible incomplete depletion at the soil surface may be described as follows. A soil layer ($0 \leq x < L$) has a uniform initial distribution of pesticide C_o with no flux allowed through the lower boundary at $x = L$. The soil surface, $x = 0$, is in contact with a volume, V, of well stirred air (uniform concentration within the air) from which a volume per unit time and unit area, v, is withdrawn and replaced by the same volume of air at zero concentration. The airflow thus removes an amount of pesticide which is equal to the flow velocity, v, times the concentration within the air, c_a. The initial and boundary conditions are

$$c = C_o \text{ at } t = 0, 0 \leq x \leq L$$

$$\partial c/\partial x = 0 \text{ at } x = L$$

$$f = v\, c_a \text{ at } t > 0, x = 0.$$

This model is analogous to that of Carslaw and Jaeger (1959, p. 129, Eq. [11]) in which a slab of uniform initial temperature is in contact with a mass per unit area of well-stirred fluid which loses heat by radiation at a rate H times its temperature. The initial temperature of the fluid is zero C. Their solution is

$$c = 2C_o \sum_{n=1}^{\infty} \frac{\exp(-D\alpha_n^2 t)(h - k\alpha_n^2) \cos[\alpha_n(L-x)]}{[L(h - k\alpha_n^2)^2 + \alpha_n^2(L+k) + h] \cos \alpha_n L} \quad [8]$$

where α_n are the roots of

$$\alpha \tan(\alpha L) = h - k\alpha^2. \quad [9]$$

Carslaw and Jaeger define $h = H/K$ where K is the thermal conductivity. For our case K is replaced by the diffusion coefficient, D, and H by the air flow velocity, v.

The concentration in the air, c_a, must be expressed in terms of the concentration at the soil surface, c_s. This is done by the use of the adsorption isotherm. Adsorption isotherms for lindane and dieldrin given in literature (Spencer and Cliath, 1970; Spencer, Cliath and Farmer, 1969) suggests that we may, for a small concentration range of c_s consider the isotherm as

$$c_a = R\, c_s. \quad [10]$$

Thus, h in Eq. [8] appears to be

$$h = Rv/D. \quad [11]$$

(Note that both concentrations c_a and c_s in Eq. [10] must be expressed on a total volume basis.)

The term k, appearing in Eq. [8] is defined by Carslaw and Jaeger as the ratio of the heat content of the fluid to the heat content of the slab. The ratio in the heat flow model is equal to R in Eq. [11] so that

$$k = R. \quad [12]$$

It can be shown that for all cases treated here

$k \ll h$ or $k\alpha_n^2 \ll h$.

Equations [8] and [9] can therefore be simplified to

$$c = 2C_o \sum_{n=1}^{\infty} \frac{\exp(-D\alpha_n^2 t) h \cos[\alpha_n(L-x)]}{(Lh^2 + \alpha_n^2 L + h)\cos\alpha_n L} \quad [13]$$

and

$$\alpha \tan(\alpha L) = h. \quad [14]$$

The pesticide flux through the soil surface at $x = 0$ is then given by

$$f = 2DC_o \sum_{n=1}^{\infty} \frac{\exp(-D\alpha_n^2 t) h^2}{Lh^2 + \alpha_n^2 L + h}. \quad [15]$$

Carslaw and Jaeger (1959, p. 491) give a table of the first six roots of the equation

$$\alpha \tan \alpha = M \quad [16]$$

which can be used instead of Eq. [14] when $L = 1$ or by scaling all variables containing units of length in terms of L.

Model V

Model IV assumes the air above the soil to be sufficiently stirred so that its concentration may be taken to be constant throughout. This may be due to turbulence as well as to a large diffusion coefficient within the air. However, Model IV does not take into account any diffusive transfer of pesticide. When air velocity becomes zero, the predicted flux through the soil surface becomes zero as well. This model thus can only be appropriate if the convective flow due to air movement is considerably greater than the flow due to a diffusion gradient.

If there is a non-moving air layer in contact with the soil surface, the following model may be adequate. The diffusion coefficient of a pesticide in air is D' and the thickness of the air layer is d. As a first approximation we may neglect the capacity of the air layer. For example, Spencer and Cliath (1970) report that at 30C a soil with 10^4 ng/cm^3 lindane is in equilibrium with a lindane vapor density in air of 0.22 ng/cm^3. Then, if the concentration in soil at the soil surface is c and the concentration in air at the soil surface is Rc, we have the flux through the air layer given by

$$f = D'Rc/d$$

when the concentration at the upper edge of the air layer is zero. The initial and boundary conditions are

$$c = C_o \text{ at } t = 0, 0 \leq x \leq L$$

$$\partial c/\partial x = 0 \text{ at } x = L$$

$$f = (D'Rc_s)/d \text{ at } t > 0, x = 0$$

$$c = c_s \text{ at } t > 0, x = 0.$$

These conditions are equivalent to the "radiation" boundary conditions described by Carslaw and Jaeger (1959). We may then adopt the solution given by these authors [p. 316, Eq. (24)] after substituting h in their equations by

$$h = D'R/Dd. \quad [17]$$

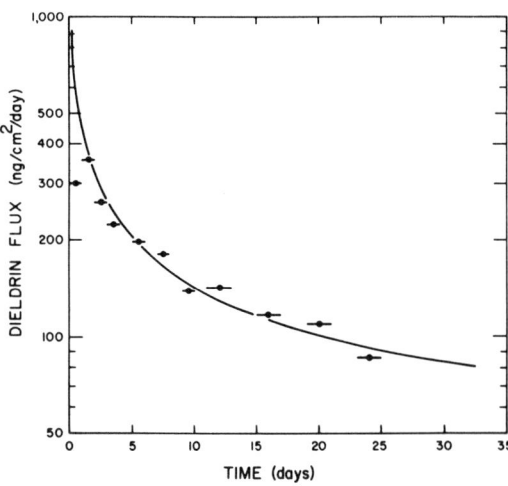

Fig. 1—Comparison between calculated values (*solid curve*) from Model II and experimental values (*horizontal lines*) for dieldrin flux. The experimental values were taken from Spencer and Cliath (1973). The length of the horizontal lines indicate the time over which the experimental values were taken.

The concentration of pesticide in soil is

$$c = \frac{2D'RC_o}{Dd}$$

$$\sum_{n=1}^{\infty} \frac{\exp(-D\alpha_n^2 t)\cos[\alpha_n(L-x)]}{[L(D'R/Dd)^2 + L\alpha_n^2 + D'R/Dd]\cos\alpha_n L} \quad [18]$$

with

$$\alpha \tan(\alpha L) = D'R/Dd. \quad [19]$$

The flux through the soil surface at $x = 0$ is

$$f = 2DC_o \sum_{n=1}^{\infty} \frac{\exp(-D\alpha_n^2 t)(D'R/Dd)^2}{L(D'R/Dd)^2 + L\alpha_n^2 + D'R/Dd}. \quad [20]$$

Note that Eq. [18], [19], and [20] are identical with Eq. [13], [14] and [15], but instead of

$$h = Rv/D$$

we have defined h by Eq. [17].

RESULTS

A comparison will now be made between calculated results from our equations and results of published volatilization experiments. All these experiments, using lindane and dieldrin, were designed to measure the volatilization of pesticides from the surface of a soil column by leading an air stream over the surface. To prevent net loss of soil water and thus prevent transport of pesticide by mass flow to the soil surface, the air was brought to 100% relative

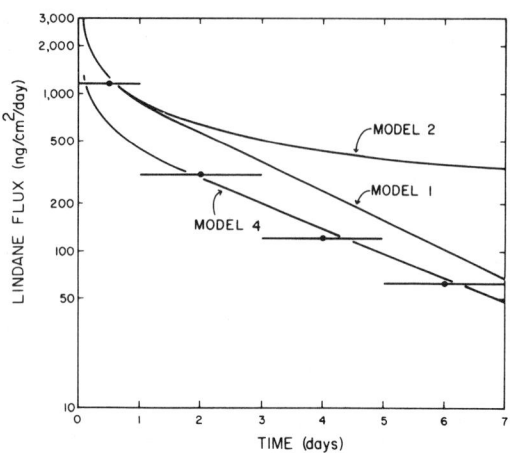

Fig. 2—Comparison between calculated and experimental values for lindane flux from treated Gila silt loam. Experimental data are from Farmer et al. (1972).

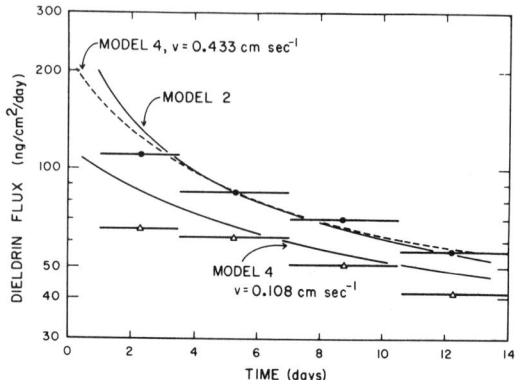

Fig. 3—Comparison of calculated and experimental values for dieldrin volatilization flux from uniformally treated Gila silt loam for air moving over the soil surface at two velocities. Experimental data from Farmer et al. (1972).

humidity. The amount of pesticide carried by the air stream was measured for distinct periods of time.

The rate of volatilization of dieldrin measured by Spencer and Cliath (1973) is presented in Fig. 1. The soil was Gila silt loam at a bulk density of 1.4 g/cm^3. The initial concentration, C_o, in the soil was 1.4×10^4 ng/cm^3 (10 ppm). Soil water content was maintained at 23% (w/w). The temperature was kept at 30C. Igue (1969) gives an apparent diffusion coefficient of $D = 2.3$ mm^2/week for dieldrin under these conditions. Air flow velocity was maintained at 2.15 cm sec^{-1}. The depth of the column was 11 cm. According to the evaluations given for Model II, we may assume the column to be infinite in length for at least 2×10^3 days after the start of the experiment. The solid curve in Fig. 1 was calculated according to Eq. [4]. The good agreement between measured and calculated values indicates that zero concentration at the soil surface is, under these conditions, a reasonable assumption. For very short times after the start of volatilization, the boundary conditions are violated in that the pesticide concentration at the soil surface is greater than zero. This may lead to some over estimation of the initial volatility rates. The low initial measured flux as compared to the calculated flux may be due to the experimental technique (e.g., losses by adsorption on walls).

Figure 2 represents data taken from a volatilization experiment carried out with lindane by Farmer et al. (1972). Parameters for this experiment were: Gila silt loam, soil water content 10% (w/w), bulk density 0.75 g/cm^3, temperature 30C, initial concentration 7.5×10^2 ng/cm^3 (10 ppm), and depth of the column 0.5 cm. With the diffusion coefficient for lindane under these conditions equal to $D = 30$ mm^2/week (taken from Ehlers et al., 1969), we calculate that after 22 hours from the start of the experiment the concentration at the bottom of the column decreases by more than 1%. This accounts for the considerable difference between the two solid curves in Fig. 2, calculated from Model I and Model II.

Figure 2 also shows a curve calculated from Model IV. The air velocity across the surface of the column was $v = 0.433$ cm sec^{-1}. R, as calculated from the adsorption isotherms given by Spencer and Cliath (1970), was $R = 3 \times 10^{-5}$. Therefore $h = R\, v/D$ becomes equal to 26.1 cm^{-1} compared to $k = R = 3 \times 10^{-5}$, the Eq. [13], [14], and [15] being applicable. The agreement between values calculated from Model IV and the measured values is quite good at the longer times and indicates that the flow velocity was not sufficient to create a zero concentration at the surface.

Volatilization experiments were conducted with dieldrin by Farmer et al. (1972) in which different air velocities were used. The soil was a Gila silt loam, soil water content 10% (w/w), bulk density 0.75 g/cm^3, initial dieldrin concentration was 7.5×10^3 ng/cm^3 (10 ppm), depth of the column was 0.5 cm, and temperature was kept at 20C. The diffusion coefficient of 1.5 mm^2/week was derived from Igue (1969), assuming a reduction of 50% compared to the value given by this author to account for the lower temperature (20C instead of 30C). A similar reduction with decreasing temperature was found by Ehlers et al. (1969) for lindane. R was taken from Spencer, Cliath, and Farmer (1969) to be 2.4×10^{-6}. With air velocities of 0.433 and 0.108 cm sec^{-1}, h became 39.46 and 9.86 cm^{-1}, respectively. Again, Eq. [13], [14], and [15] apply to this situation to allow calculation of pesticide flux.

Figure 3 shows the measured rates of dieldrin volatilization at two different air velocities together with the curves calculated from Model IV as well as a curve calculated for Model II. Under the conditions given in the experiment, it can be seen at the higher air flow velocity the boundary condition "zero concentration at the soil surface" is appropriate. This is not true for the lower air flow velocity where Model IV better accounts for the lower volatilization rates.

Note that the calculated curves for Model II and Model IV at the higher air velocity are very similar. Model IV predicts a lower flux during the initial time period than Model II. This is reasonable because initially there is pesticide at the soil surface and a higher air velocity would be necessary to maintain a concentration equal to zero at the surface which is required by Model II. As time progresses and volatilization has occurred, the surface is depleted of pesticide and the surface concentration can be kept at zero with a lower air velocity. The low fluxes measured, as compared to the calculated, in the beginning of the experiment are probably due in part to losses by adsorption on walls.

DISCUSSION

From the equations presented for Model I and II, it can be seen that for a given initial concentration and depth of the soil layer, the flux at any time depends only on the apparent diffusion coefficient. Since the apparent diffusion coefficient depends on various parameters including bulk density, water content, concentration, temperature, and adsorption (Ehlers et al., 1969; Farmer and Jensen, 1970), all equations presented can only be used if during the time of observation, D remains constant. Ehlers et al., (1969) showed for lindane that this assumption holds true up to concentrations of about 20 ppm. This agreement between measured and calculated pesticide fluxes, if an appropriate model is chosen, indicates that constancy for the conditions met in the experiments may also be assumed for the dieldrin diffusion coefficient.

Models I, II, and IV prove to be valid for describing volatilization losses from experiments carried out in the laboratory with appropriate boundary conditions. Prediction of volatilization rates of pesticides under field conditions with these or similar models becomes more difficult, mainly because the boundary conditions are not as well defined as in a laboratory experiment. Yet, the models are valuable at least in making reasonable estimates of pesticide losses by diffusion.

Let us take, for example, a pesticide incorporated in a soil surface layer. If the soil surface is in contact with a moving layer of air, and if diffusion to the surface is the only transport mechanism for pesticide in the soil, then Model IV should apply. The wind velocity can be measured. One difficulty is that the model assumes that the air increment passing over each unit area of surface is initially devoid of pesticide. When a large land area is treated with pesticide, the wind will become enriched with pesticide as it passes over the field.

The results presented in Fig. 1 and 3 indicate that only very low wind velocities (less than about 0.016 km/hr^{-1} or 0.04 mi/hr^{-1}) are required to keep the pesticide concentration at zero at the soil surface. Thus it would appear Models I, II, or III could be used with considerable confidence if there is much air movement.

The situation is different if we have a non-moving air layer above the surface, for example where the air movement is restricted by a standing crop. Under these conditions Model V applies.

Figure 4 gives some lindane flux curves calculated for different values of h. The diffusion coefficient D' for lindane in air is not known, but may be derived [Ehlers et al. (1969), p. 502, Eq. (9)] as being 10^3 cm^2/day. The initial concentration was assumed to be 10^4 ng/cm^3, the diffusion coefficient for lindane in the soil 20 mm^2/week, R as used above was 3×10^{-5}. The depth of the soil layer treated with pesticide was assumed to be 10 cm. Equation [17] then gives the values of h which correspond with different values of d, the thickness of the air layer.

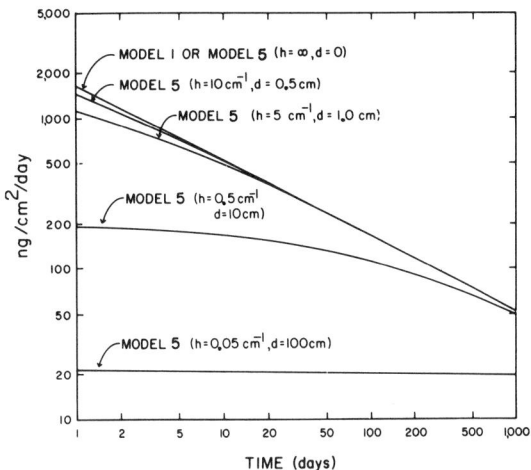

Fig. 4—Calculated values for the influence of depth of non-moving air layer on lindane volatility flux from uniformally treated Gila silt loam.

From Figure 4 it can be seen that a non-moving air layer may restrict the rate of volatilization considerably. Finally it may be mentioned that since we have under most conditions water movement in the soil profile, the models presented above predict volatilization rates from a soil surface better for moderately to strongly adsorbed pesticides of low mobility with moving water.

LITERATURE CITED

1. Carslaw, H. S., and J. C. Jaeger. 1959. Conduction of heat in solids. 2nd Ed. Oxford University Press, Oxford.
2. Ehlers, W., W. J. Farmer, W. F. Spencer, and J. Letey. 1969. Lindane diffusion in soils: II. Water content, bulk density and temperature effects. Soil Sci. Soc. Amer. Proc. 33:505–508.
3. Farmer, W. J., and C. R. Jenson. 1970. Diffusion and analysis of carbon-14 labeled dieldrin in soils. Soil Sci. Soc. Amer. Proc. 34:28–31.
4. Farmer, W. J., K. Igue, W. F. Spencer, and J. P. Martin. 1972. Volatility of organochlorine insecticides from soil: I. Effect of concentration, temperature, air flow rate and vapor pressure. Soil Sci. Soc. Amer. Proc. 36:443–447.
5. Farmer, W. J., K. Igue, and W. F. Spencer. 1973. Effect of bulk density on the diffusion and volatilization of dieldrin from soil. J. Environ. Qual. 2:107–109.
6. Igue, K. 1969. Volatility of organochlorine insecticides from soil. Ph.D. Diss., University of California, Riverside Univ. Microfilms, Ann Arbor, Mich. (Diss. Abstr. 70-19,242).
7. Igue, K., W. J. Farmer, W. F. Spencer, and J. P. Martin. 1972. Volatility of organochlorine insecticides from soil. II. Effect of relative humidity and soil water content on dieldrin volatility. Soil Sci. Soc. Amer. Proc. 36:447–450.
8. Spencer, W. F., and M. M. Cliath. 1970. Vapor density and apparent vapor pressure of lindane. J. Agr. Food Chem. 18:529–530.
9. Spencer, W. F., and M. M. Cliath. 1970. Desorption of lindane from soil as related to vapor pressure. Soil Sci. Soc. Amer. Proc. 34:574–578.
10. Spencer, W. F., and M. M. Cliath. 1973. Pesticide volatilization as related to water loss from soil. J. Environ. Qual. 2:284–289.
11. Spencer, W. F., M. M. Cliath, and W. J. Farmer. 1969. Vapor density of soil-applied dieldrin as related to soil-water content, temperature, and dieldrin concentration. Soil Sci. Soc. Amer. Proc. 33:509–511.

Behavior Assessment Model for Trace Organics in Soil: II. Chemical Classification and Parameter Sensitivity[1]

W. A. JURY, W. J. FARMER, AND W. F. SPENCER[2]

ABSTRACT

In this paper, the organic chemical transport screening model developed in Jury et al. (1983) is simplified by dividing chemicals into volatilization and mobility categories. The volatilization classification is based on whether or not the predominant resistance to volatilization loss lies in the soil or in the boundary layer above the soil surface. This categorization reduces to a condition on the Henry's constant (K_H) and organic C partition coefficient (K_{oc}) when standard values are used to represent soil and chemical parameters. The mobility categories are based on the calculated time to convect or diffuse a given distance through the soil.

Simulations are conducted for chemicals falling into one or another of these volatilization or mobility categories to examine the sensitivity of these processes to variations in water evaporation, water content, organic C fraction, and boundary layer thickness. The dependence of both volatilization flux and leaching flux on these parameters is summarized.

Additional Index Words: chemical movement, diffusion, volatilization, leaching.

Jury, W. A., W. J. Farmer, and W. F. Spencer. 1984. Behavior assessment model for trace organics in soil: II. Chemical classification and parameter sensitivity. J. Environ. Qual. 13:567-572.

In a previous paper (Jury et al., 1983), we introduced a screening model for describing pesticide volatilization, leaching, and degradation in soil. The soil surface boundary consisted of a stagnant boundary layer connecting the soil and air through which pesticide and water vapor must move to reach the atmosphere. Assuming constant water flow and uniform soil properties, we derived an analytical solution, which describes pesticide concentration and flux as a function of chemical, environmental, and soil properties.

In its present form, the theory is too complex to allow a simple analysis to be made of the influence of soil and management properties on chemical behavior. Furthermore, it is not clear to what extent uncertainties in the values of the measured chemical properties will influence the predictions made by the model. Since our proposed use of the model will be as a screening tool to classify pesticides and other trace organics, such knowledge of input parameter sensitivity is essential. In this paper we examine the three major loss pathways: degradation, mobility, and volatilization, and simplify the general theory in such a way as to allow general pesticide classification into specific behavioral groups. Within these groups, we will conduct a sensitivity analysis that will examine the influence of various soil and chemical properties on the loss pathways.

THEORY

Degradation

The processes contributing to biological or chemical degradation of an organic compound in soil are complex, and their functional dependence on such soil and environmental parameters as water content, temperature, organic C, and soil pH are not well understood (Hamaker, 1972). In the absence of such quantitative information, the degradation potential of a given chemical is described with an effective first-order rate constant, μ, or half-life, $T_{1/2}$ (Nash, 1980; Rao & Davidson, 1980). This parameter represents the combined influence of degradation in all phases, and is usually measured by determining the fraction $M(T)/M(0)$ of a given initial quantity of applied chemical $M(0)$ remaining after a time t according to Eq. [1]

$$M(t) = M(o) \exp(-\mu t). \qquad [1]$$

Published measurements of μ differ widely (Hamaker, 1972; Nash, 1980; Rao & Davidson, 1980), not only because of different conditions, but because the degradation process may not best be described as first-order, or because unmeasured volatilization losses and soil measurement errors may interfere with the measurement of degradation losses by Eq. [1]. Nevertheless, the first-order rate coefficient is useful as a relative index of persistence.

In the model of Jury et al. (1983), the soil concentrations and surface volatilization fluxes are proportional to the factor $\exp(-\mu t)$. The uncertainty in μ is likely to be as high as 100% or more (Nash, 1980), which could create a large error in the estimates made for compounds with short half-lives (large μ).

Mobility

CONVECTIVE MOBILITY

In the model of Jury et al. (1983), it was shown that a chemical with a linear, equilibrium partitioning between its vapor, liquid, and adsorbed phases will move with convective velocity

$$V_E = J_W/R_L = J_W/(\varrho_b K_D + \Theta + aK_H). \qquad [2]$$

The ratio of the total concentration to the liquid concentration is R_L, where J_W is water flux, K_D is adsorbed-liquid distribution coefficient, ϱ_b is soil bulk density, K_H is Henry's constant, a is volumetric air content, and Θ is volumetric water content. When the model is used to conduct leaching screening tests, the convective mobility may be classified in a variety of ways. One useful index, in analogy with chromatography, is to define a convection time t_c to move a distance l when a water flux J_W is present (Eq. [3]).

$$t_c = l/V_E = (\varrho_b K_D + \Theta + aK_H) l/J_W. \qquad [3]$$

---

[1] Contribution of Dep. of Soil and Environ. Sci., Univ. of California-Riverside, and USDA, Riverside, CA 92521. Received 10 Sept. 1982.
[2] Professor of soil physics, professor of soil science, Univ. of California-Riverside, and soil scientist, USDA.

When adsorption is relatively high [i.e., $K_D > 4 \times 10^{-1}$ (m^3/kg)], the water content Θ and aK_H may be neglected and t_c will be proportional to K_D and ϱ_b. For chemicals such as nonionic pesticides, which primarily adsorb to organic matter, the distribution coefficient K_D may be written as $f_{oc}K_{oc}$, where f_{oc} is organic C fraction and K_{oc} is organic C partition coefficient. In this case,

$$t_c \approx \varrho_b f_{oc} K_{oc} l / J_w. \quad [4]$$

This convection time is a useful index of relative mobility and also will approximately describe the movement of a front or of the peak of a narrow pulse of chemical.

DIFFUSIVE MOBILITY

When mass flow by convection is small or negligible, the chemical is able to move through the soil only by liquid or vapor diffusion. In analogy with the convection time, we may define a characteristic diffusion time t_D to move a distance l, which may be written as (Carslaw & Jaeger, 1959).

$$t_D = l^2/D_E, \quad [5]$$

where D_E is the effective soil diffusion coefficient (m^2/d), given in Jury et al. (1983) as

$$D_E = \frac{D_G^{air} K_H a^{10/3}/\phi^2 + D_L^{water} \Theta^{10/3}/\phi^2}{\varrho_b K_D + \Theta + aK_H}, \quad [6]$$

where K_H is Henry's constant, D_G^{air} is gaseous diffusion coefficient in air, D_L^{water} is liquid diffusion coefficient in water, a is air content, and ϕ is porosity. Only those chemicals that move predominantly in the vapor phase will have a relatively small t_D. For these chemicals, the first term in Eq. [6] dominates the second, and Eq. [5] may be written as

$$t_D \approx (\varrho_b f_{oc} K_{oc} + \Theta + aK_H) \phi^2 l^2 / D_G^{air} K_H a^{10/3}. \quad [7]$$

Unlike the convection time t_c, this index will strongly depend on water or air content.

Volatilization

As described in our previous paper (Jury et al., 1983), the soil and atmosphere are connected by a stagnant air boundary layer through which water vapor and chemical vapor are assumed to move by diffusion. The extent to which this boundary layer limits the volatilization flux may be used as a criterion for classifying pesticides and other volatile organics into general categories, similar to the volatilization groups used to classify chemical losses from water bodies (Smith et al., 1980, 1981). To achieve this, it is convenient to distinguish between processes where no water flow (E) is occurring ($E = 0$) and processes where both volatilization and evaporation are occurring.

CASE 1. NO WATER EVAPORATION $E = 0$

When a chemical is initially uniformly incorporated in the soil at a total concentration C_0 (g/m^3), the maximum volatilization flux rate J_1 through the soil surface to the atmosphere that could occur is given by Jury et al. (1980)

$$J_V = C_0 (D_E/\pi t)^{1/2}, \quad [8]$$

where D_E is given in Eq. [6] and where t is time (days). This flux rate is that which would occur with no boundary layer resistance in the air or equivalently when the surface concentration $C_T(0,t)$ is held at zero for all $t > 0$.

When a boundary layer of thickness d is present, the maximum flux J_{V_2} that can move through the boundary layer occurs when no soil resistance is present and the gas concentration C_G at the soil surface is held at its initial value $C_G(0) = C_0/R_G$

$$J_{V_2} = D_G^{air} C_0/R_G d, \quad [9]$$

where $R_G = R_L/K_H = (\varrho_b K_D + \Theta + aK_H)/K_H$ is the ratio of the total chemical concentration to the concentration in the vapor phase. Equation [9] assumes that the concentration of the chemical in the free air above the boundary layer is zero.

When a boundary layer is present, it will act to restrict volatilization fluxes only if the maximum flux through the boundary layer J_{V_2} is small compared with the rate at which chemical moves to the soil surface, which we may represent approximately as J_{V_1}. Thus, if $J_{V_2} \ll J_{V_1}$, then

$$D_G^{air} C_0/R_G d \ll C_0 (D_E/\pi t)^{1/2}. \quad [10]$$

By plugging the definitions for R_G and D_E into Eq. [10], we may rewrite the condition expressed there in terms of the soil and chemical parameters in the various terms. To simplify the interpretation, we will use standard values for many of the soil and chemical properties other than the properties that differ greatly for different chemicals. These are summarized in Table 1.

When the soil water content is reasonably high (e.g., $\Theta > 0.2$), then the second term in the numerator of Eq. [6] will dominate the first term under the same circumstances (small K_H) when the inequality in Eq. [10] is valid. Thus, using $D_E \approx D_L/R_L$, we may rewrite Eq. [10] as

$$K_H^2/K_{oc} \ll \frac{D_L^{water} d^2 \Theta^{10/3}}{(D_G^{air})^2 \pi t \phi^2} \varrho_b f_{oc}, \quad [11]$$

Table 1—Standard values of soil and chemical properties used in simulations.

Parameter	Symbol	Units	Standard value
Porosity	ϕ	m^3/m^3	0.5
Bulk density	ϱ_b	kg/m^3	1350
Organic C fraction	f_{oc}		0.0125
Liquid diffusion coefficient	D_L^{water}	m^2/d	4.3×10^{-5}
Air diffusion coefficient	D_G^{air}	m^2/d	4.3×10^{-1}
Water content	Θ	m^3/m^3	0.3

where it has been assumed in going from Eq. [10] to Eq. [11] that $R_L = \varrho_b K_D = \varrho_b f_{oc} K_{oc}$.

If we plug in the standard values from Table 1, along with $t = 2$ d and $d = 5$ mm from Jury et al. (1983), we obtain a benchmark criterion for a boundary layer influence when volatilization occurs without water evaporation

$$K_H^2/K_D \ll 9 \times 10^{-8} \, (\text{kg/m}^3) \qquad [12]$$

with $f_{oc} = 0.0125$.

CASE 2. WATER EVAPORATION $E \neq 0$

If upward water flow carries an insignificant amount of pesticide compared with upward diffusion, then the analysis is identical to case 1. However, if upward convection is dominant, as it will be if the solution concentration is high, or if evaporation and volatilization both occur for a long time period, then the upward flux of chemical J_{V_1} toward the boundary layer is approximately equal to

$$J_{V_1} = C_L E \approx C_o E/R_L, \qquad [13]$$

where C_L is solution concentration. The criterion for a boundary layer restriction on volatilization in this case occurs when $J_{V_2} \ll J_{V_1}$, or

$$D_G^{\text{air}} C_o/R_G d \ll C_o E/R_L. \qquad [14]$$

Further, if we assume, as in Jury et al. (1983), that water evaporation is also regulated by the boundary layer, we may write a water vapor diffusion equation across the boundary layer as

$$E = [D_{wv}^{\text{air}} \varrho_{wvs} (1 - \text{RH})/2\varrho_{wL} d \qquad [15]$$

where ϱ_{wL} is liquid water density, ϱ_{wvs} is saturated water vapor density and RH is relative humidity. The factor of 2 is inserted, as explained in Jury et al. (1983), because our model uses a steady-state evaporation flux, whereas normal field evaporation rates are small during the evening hours. When Eq. [15] is plugged into Eq. [14] we obtain

$$K_H \ll [D_{wv}^{\text{air}} \varrho_{wvs}(1 - \text{RH})]/2D_G^{\text{air}} \varrho_{wL}$$

$$\approx 2.5 \times 10^{-5}. \qquad [16]$$

Note that Eq. [12] and [16] are identical for $K_D \approx 7 \times 10^{-3}$ (m³/kg), which is a value representing moderate adsorption.

Relationship to Chemical Volatilization from Water Bodies

Volatilization of dissolved chemicals from water bodies has been modeled using a linear two-resistance film model (Liss & Slater, 1974), and by a two film model using penetration theory to represent transport from the liquid to the air water interface (Smith et al., 1980, 1981). Irrespective of the model use, however, one concludes that there is substantially less resistance to volatilization from the water body than has been found here (see Eq. [2]) for volatilization from soil. As a result, in water systems the air boundary layer forms a barrier to chemical loss at a much higher value of K_H than that predicted by Eq. [16] for soil systems. For example, the criterion equivalent to Eq. [16] obtained using the approach of Smith et al. (1981) for chemical loss from rivers is

$$K_H \ll 3.8 \times 10^{-3}. \qquad [17]$$

The reason that one obtains such a different answer in water bodies than in soil is that in a soil system upward chemical movement is restricted both by adsorption and by tortuosity effects (increased path length, decreased cross-sectional area) on diffusion compared with water transport. Since resistance to transport to the atmosphere through the stagnant air boundary layer is similar in both cases, the transition point where volatilization loss is regulated by the vapor phase shifts upward by over two orders of magnitude when water is analyzed instead of soil.

Pesticide Volatilization Categories

To simplify subsequent discussion, pesticides whose properties obey the inequalities Eq. [10] or Eq. [16] will be called category III, those whose properties obey the opposite inequality ($\gg$) will be called category I, and those for which Eq. [10] or Eq. [16] represent an equality will be called category II. Table 2 presents hypothetical but representative benchmark properties for pesticides in each of the three categories. For simplicity, each pesticide is given the same K_{oc}, water solubility (C_L^*), and μ, and Henry's constant K_H variations are achieved by varying saturated vapor density, C_G^*. With the above choices, the three pesticides fall unambiguously into different categories by both Eq. [10] and [16].

RESULTS

Mobility Classification

Equation [3] or [4] defines the dependence of the convective mobility time t_c on soil and chemical properties. To illustrate its use we may calculate the time required to move the chemical $l = 10$ cm when water is applied at $J_W = 1.0$ cm/d, for the standard conditions given in Table 1, with the result that $t_c \sim 170 K_{oc} + 10 \Theta$ (days). Thus, the chemicals 2,4-D [(2,4-dichlorophenoxy) acetic acid] and lindane (γ-1,2,3,4,5,6-hexachlorocyclo-

Table 2—Hypothetical pesticide benchmark properties and their categorical designation.

Property	Category I	Category II	Category III
Vapor density C_G^* (g/m²)	10^{-1}	10^{-3}	10^{-5}
Solubility C_L^* (g/m³)	40	40	40
K_{oc} (m³/kg)	0.5	0.5	0.5
K_D (m³/kg)	6.3×10^{-3}	6.3×10^{-3}	6.3×10^{-3}
μ (d⁻¹)	0	0	0
K_H	2.5×10^{-3}	2.5×10^{-5}	2.5×10^{-7}
K_H^2/K_{oc}	1.25×10^{-5}	1.25×10^{-9}	1.25×10^{-11}

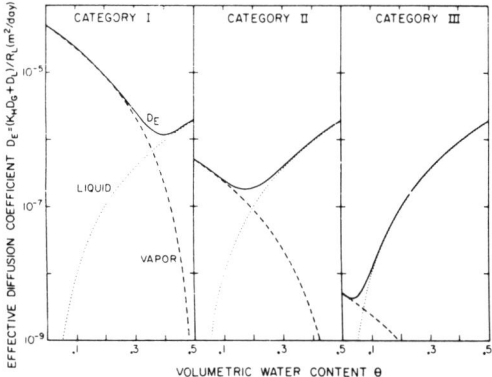

Fig. 1—Effective diffusion coefficients as a function of water content for prototype chemicals representing the three volatilization categories.

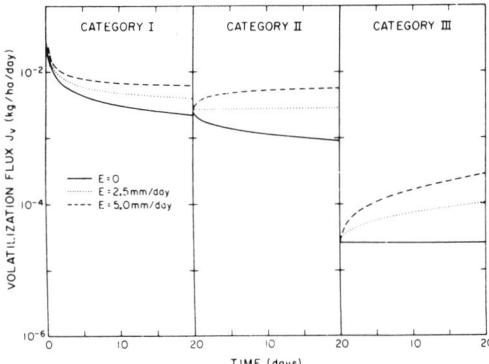

Fig. 2—Volatilization flux rates for the three chemical prototypes for three rates of water evaporation.

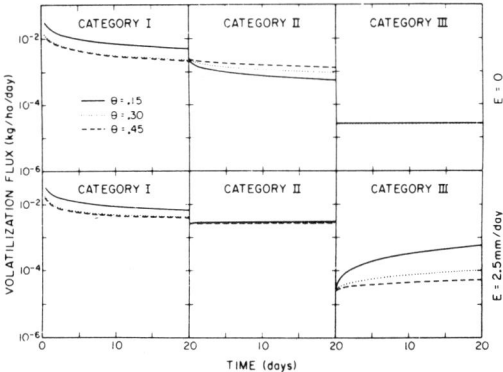

Fig. 3—Effect of changes in water content on volatilization flux rates for the three chemical prototypes. Top curves are for zero evaporation and bottom curves for evaporation of 2.5 mm/d.

hexane), which have $K_{oc} \doteq 0.02$ and 1.3 m³/kg, respectively (Jury et al., 1983) have convective times of 6.4 and 224 d, respectively. These chemicals would represent highly mobile and relatively immobile compounds in a leaching classification scheme such as that of Helling (1971).

The diffusive mobility as defined by Eq. [7] will be important only for vapor dominated compounds with large K_H and small K_{oc}. To see this, if we require that in dry soil ($a \sim \phi$) the diffusive time to move 10 cm be < 20 d for a soil with properties given in Table 1; Eq. [7] reduces to the condition $K_{oc}/K_H < 20$. This condition is met only for fumigants and other vapor-dominated compounds of low adsorption. The compounds 2,4-D and lindane, for example, have $K_{oc}/K_H = 3.5 \times 10^6$ and 1×10^4, respectively (Jury et al., 1983).

Volatilization Classification

Figure 1 shows a plot of effective diffusion coefficient D_E (Eq. [6]) as a function of volumetric water content for the prototype chemicals (Table 2) chosen to represent the three categories. From this figure it is clear that a category I chemical is dominated by vapor diffusion and a category III chemical is dominated by liquid diffusion over most of the water content range. Category II chemicals are vapor-dominated at low water content and liquid-dominated at high water content.

Figure 2 shows volatilization flux rates vs. time for the three prototype chemicals under three cases of (i) no evaporation, (ii) steady evaporation at 2.5 mm/d, and (iii) steady evaporation at 5.0 mm/d. Again, a clear distinction is apparent between the behavior of category I and category III chemicals. For category I, the volatilization flux shows a characteristic decrease with time in all three cases, whereas the flux rate of the category III chemical tends to increase with time when upward water flow is occurring and to decrease slowly with time when evaporation is not present. The category II volatilization flux decreases with time when no evaporation occurs and increases with time when high evaporation occurs.

Figure 3 shows the influence of changes in water content on volatilization flux rates for the three chemicals for both volatilization without evaporation and volatilization with a water evaporation rate of 2.5 mm/d. The results suggest a very complicated dependence on water content for both cases. For example, category III chemicals show no water content dependence when water is not evaporating, but are strongly water content dependent when evaporation is occurring.

Figure 4 shows the effect of changing organic C fraction f_{oc} on volatilization flux rates. Since decreasing adsorption increases both convective and diffusive transport to the surface, in all cases volatilization increases with decreasing organic C fraction. However, the extent of the dependence seems somewhat stronger in category III than category I.

Figure 5 shows the influence on volatilization of arbitrarily changing the thickness d of the stagnant boundary layer while forcing evaporation to be either 0 or 2.5 mm/d. This arbitrary action has the effect of decoupling water evaporation and boundary layer thickness (Eq. [10]), but could be accomplished in prin-

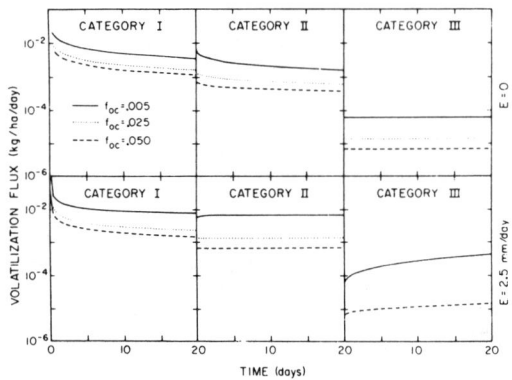

Fig. 4—Effect of changes in organic C fraction on volatilization flux rates for the three chemical prototypes. Top curves are for zero evaporation and bottom curves are for evaporation of 2.5 mm/d.

ciple by adjusting the relative humidity of the air above the boundary layer so as to maintain E = constant. Here it is obvious that the category I chemical has a volatilization rate that is independent of boundary layer thickness in the range $d < 5$ cm and that the category III chemical has a volatilization rate that is inversely proportional to boundary layer thickness over the range $0.05 < d < 5$ cm.

DISCUSSION

Volatilization without Evaporation

Since Eq. [8] and [9] are respectively the maximum rate of chemical movement through the soil to the surface and from the surface to the air, it is worthwhile to compare these fluxes with the actual predicted volatilization rates for the three chemical categories. This is shown in Fig. 6, where the dashed curve gives the boundary layer flux J_{V_z} (Eq. [9]) where $C_G(o)$ is held at its initial value, the dotted curve gives the maximum soil loss rate J_{V_z} (Eq. [8]) and the solid curve gives the actual flux. From this figure it is clear that the category I

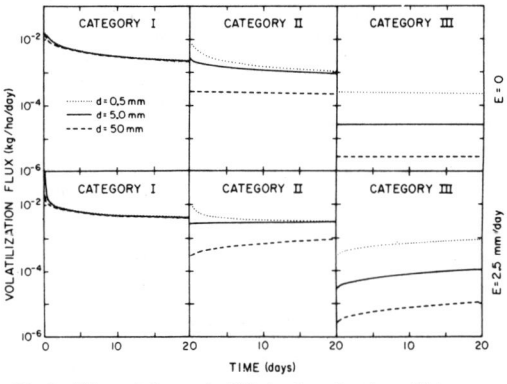

Fig. 5—Effect of changes in diffusion boundary layer thickness on volatilization flux rate for the three chemical prototypes. Top curves are for zero evaporation rate and bottom curves are for evaporation of 2.5 mm/d.

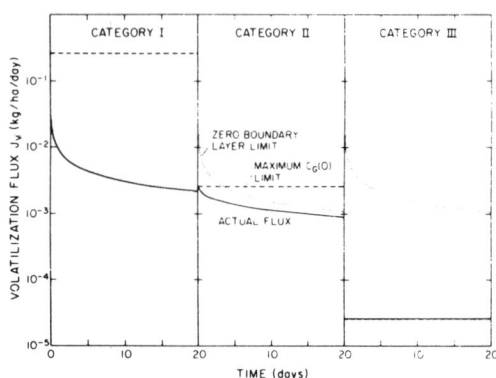

Fig. 6—Calculated volatilization flux rates for the three chemical prototypes when water evaporation is not occurring (solid lines), along with maximum flux through boundary layer given by Eq. [3] (dashed lines) and maximum flux possible when no boundary layer is present, calculated from Eq. [2] (dotted lines).

chemical behaves as though there is no boundary layer resistance [$C_G(o) = 0$] and the category III chemical behaves as though there is no soil resistance [$C_G(o)$ remains at its initial value]. The category II chemical has properties that create soil and boundary layer resistance of approximately similar size.

Thus, Eq. [8] and [9] may be used to represent the volatilization rates of category I and III chemicals, respectively, when no water evaporation is present. In particular, the functional dependence of the volatilization rate on various soil and chemical parameters may be obtained by plugging in the defining equations for D_E and R_G into Eq. [8] and [9], respectively.

Category I Chemicals

$$J_V \propto C_o K_H^{1/2} a^{5/3} K_{oc}^{-1/2} f_{oc}^{-1/2} t^{-1/2} \quad [18]$$

Category III Chemicals

$$J_V \propto C_o K_H K_{oc}^{-1} f_{oc}^{-1} d^{-1} \quad [19]$$

Equations [18] and [19] explain the relevant functional dependences of the $E = 0$ volatilization curves shown in Fig. 1–5. For example, Eq. [19] approximately predicts that a category III chemical will have a volatilization rate that has no Θ dependence (Fig. 3), no time dependence (Fig. 3), will be inversely proportional to f_{oc} (Fig. 4), and inversely proportional to d (Fig. 5).

Volatilization with Evaporation

The limiting behavior for volatilization with evaporation is more complex than the evaporation-free case for several reasons. First, both diffusion and convection may contribute to the movement of chemical toward the soil surface. Therefore, category I chemicals will not have volatilization rates that are equal to the rate at which the chemical is moved with water in accordance with Eq. [13], except at large times if the system approaches a steady-state rate of loss. Second, for a boundary-layer limited chemical (category III), upward

Table 3—Theoretical influence of various soil and environmental parameters on pesticide leaching and volatilization rate summarized by chemical category.

Parameter	Category I	Category III
	Convective mobility time, t_c	
Θ	Small, unless $K_{oc} \ll 0.1$ (m³/kg)	Small, unless $K_{oc} \ll 0.1$
f_{oc} (or K_{oc})	f_{oc}	f_{oc}
K_H	None	None
	Volatilization without evaporation	
Θ	$D_G(\Theta)^{1/2}$	None
f_{oc} (or K_{oc})	$f_{oc}^{-1/2}$	f_{oc}^{-1}
K_H	$K_H^{1/2}$	K_H
t	$t^{-1/2}$	None
d	None	d^{-1}
$C_T(o)$	$C_T(o)$	$C_T(o)$
	Volatilization with evaporation	
Θ	$D_G(\Theta)^{1/2}$ for small t; none for large t	Increases as $D_L(\Theta)$ decreases
f_{oc} (or K_{oc})	$f_{oc}^{-1/2}$ for small t; f_{oc} for large t	Increases as f_{oc} decreases
d	None	d^{-1}
t	Decreases as $t^{-1/2}$ to to constant value	Increases as t increases
$C_T(o)$	$C_T(o)$	$C_T(o)$
e	Moderately sensitive for small K_{oc}; insensitive for large K_{oc}	Very sensitive for small K_{oc}

convection can cause concentrations to build up at the surface above the initial value just as salt accumulates at an evaporation surface. In this case, the vapor pressure and vapor density of the chemical at the soil-air boundary can increase with time, limited only by its saturation value. As the vapor density increases, so does the volatilization flux (Eq. [9]). This explains the time dependence shown in Fig. 2 for the category III chemical. As the concentration builds up at the surface, diffusion tends to move chemical back into the soil. Thus, for a given convective flux, the surface concentration at a given time will be highest when the soil diffusion coefficient D_E is lowest. This explains the somewhat strange effect of water content changes on volatilization flux (Fig. 3) predicted to occur for a category III chemical when water evaporation is present. The volatilization flux in Fig. 3 decreases as water content increases. From Fig. 1 we see that D_E increases as water content increases and hence that the surface concentration decreases as water content increases.

Influence of Evaporation on Volatilization

For all three chemical categories, the volatilization rate is enhanced by evaporation (Fig. 2). By comparing the top and bottom curves in Fig. 4, however, it is seen that the extent of the enhancement for a given E decreases as f_{oc} increases. For category I chemicals this is explained by the fact that upward diffusion varies as $f_{oc}^{-1/2}$ (Eq. [18]), whereas upward convection is proportional to C_L, which varies as f_{oc}^{-1}. For large f_{oc}, therefore, convection has a relatively small influence on J_V. For category III chemicals the flux rate is proportional to $C_G(o)$, which remains near its initial value for evaporation-aided volatilization due to upward convection. As f_{oc} increases, the rate of decrease of $C_G(o)$ with time lowers because upward convection decreases. Thus, we may conclude that in all cases evaporation most strongly influences volatilization for weakly adsorbed chemicals with non-negligible vapor density.

SUMMARY AND CONCLUSIONS

Table 3 summarizes the significant functional dependencies for leaching mobility and volatilization of category I and category III chemicals on various soil and environmental parameters found in the simulation studies conducted in this paper. The dependencies shown are for chemicals that clearly fall into the appropriate categories by the criteria given in Eq. [6] or [11]. For category II chemicals, the behavior will in all cases be intermediate between category I and III. Since the model is intended as a screening tool, the functional dependencies given in Table 3 and shown in Fig. 1-5 will help to characterize the susceptibility of a candidate chemical to various loss pathways.

In a future paper we will illustrate the use of the screening model on a number of organic chemicals for which we have obtained benchmark chemical data (Jury et al., 1984a) and will show how these chemicals fall into the groups discussed in this paper. In the final paper in this series (Jury et al., 1984b) we will review the experimental literature to show support for many of the assumptions and predictions of our model.

REFERENCES

1. Carslaw, H. S., and J. C. Jaeger. 1959. Conduction of heat in solids. Oxford University Press, London.
2. Hamaker, J. S. 1972. Decomposition: quantitative aspects. p. 253-340. *In* C. A. I. Goring and J. W. Hamaker (ed.) Organic chemicals in the soil environment. Marcel Dekker, New York.
3. Helling, C. S. 1971. Pesticide mobility in soils: III. Influence of soil properties. Soil Sci. Soc. Am. Proc. 35:743-748.
4. Jury, W. A., R. Grover, W. F. Spencer, and W. J. Farmer. 1980. Modeling vapor losses of soil-incorporated triallate. Soil Sci. Soc. Am. J. 44:445-450.
5. Jury, W. A., W. F. Spencer, and W. J. Farmer. 1983. Behavior assessment model for trace organics in soil: I. Description of model. J. Environ. Qual. 12:558-564.
6. Jury, W. A., W. F. Spencer, and W. J. Farmer. 1984a. Behavior assessment model for trace organics in soil: III. Application of screening model. J. Environ. Qual. 13:573-579 (this issue).
7. Jury, W. A., W. F. Spencer, and W. J. Farmer. 1984b. Behavior assessment model for trace organics in soil: IV. Review of experimental evidence. J. Environ. Qual. 13:580-586 (this issue).
8. Liss, P. S., and P. G. Slater. 1974. Flux of gases across the air-sea interface. Nature 247:181-184.
9. Nash, R. G. 1980. Dissipation rate of pesticides from soils. p. 560-594. *In* W. G. Knisel (ed.) CREAMS. Vol. 3. U.S. Department of Agriculture, Washington, DC.
10. Rao, P. S. C., and J. M. Davidson. 1980. Estimation of pesticide retention and transformation parameters required in nonpoint source pollution models. p. 23-67. *In* M. R. Overcash and J. M. Davidson (ed.) Environmental impact of nonpoint source pollution. Ann Arbor Science Publishers, Ann Arbor, MI.
11. Smith, H. H., D. C. Bomberger, and D. L. Haynes. 1980. Prediction of the volatilization rates of high volatility chemicals from natural water bodies. Environ. Sci. Technol. 14:1332-1337.
12. Smith, J. H., D. C. Bomberger, and D. L. Haynes. 1981. Volatilization rates of intermediate and low volatility chemicals from water. Chemosphere 10:281-289.

AUTHOR CITATION INDEX

Abdel-Ghafar, T., 239
Abernathy, J. R., 263
Abou-Donia, M. B., 239
Acree, F. Jr., 354
Adams, J. E., 324
Adams, R. S. Jr., 37, 158
Adams, R. T., 343
Adamson, A. W., 119, 158
Aebi, H., 47
Ahlrichs, J. L., 24, 87
Ahmad, N., 245
Ahuja, L. R., 313
Akashi, K., 158
Akatsuka, T., 245
Alberty, R. A., 216
Aleem, M. I. H., 199
Alexander, M., 5, 199, 203, 216, 220
Allaway, W. H., 87
Allen, N., 268
Allen, R. J. L., 183
Allison, F. E., 220
Aly, O. M., 245
Amiel, A., 158
Ampt, G. A., 344
Amundson, N. R., 299, 313
Anderson, J. P. E., 231
Arkin, G. F., 324
Armstrong, D. E., 216, 251, 333
Arnold, J. S., 66
Arnon, D. I., 183
Arthur, B. W., 183
Ashton, F. M., 14, 86, 294, 299
Atkinson, G. C., 245
Audus, L. J., 171, 178, 245
Augustinsson, K. B., 203
Austin, N. M., 159

Baetcke, K. P., 226
Bailey, G. W., 14, 32, 37, 86, 123, 133, 148, 158, 263, 294, 344
Baker, H. M., 299
Baldwin, B. C., 220
Ballard, T. M., 347
Banin, A., 158
Bann, J. M., 21, 268
Banwart, W. L., 158
Barik, S., 289
Barlin, G. B., 48

Barlow, F., 5, 50, 133
Barnes, S., 21
Barnett, A. P., 333
Barnhisel, R. I., 283
Bartell, F. E., 158
Bartha, R., 112, 195, 239, 245
Bartholomew, W. V., 220
Bates, D. H., 133
Baughman, G. L., 113
Baur, J. R., 263
Baxter, R. F., 183
Bayer, D. E., 299
Beard, W. E., 172, 207, 239, 347
Beck, S. D., 21, 239, 268
Beck, W., 87
Beek, C. G. E. M, 325
Behrens, R., 216
Bekey, G. A., 312
Belasco, I. J., 159
Bellamy, L. J., 112
Bender, R., 158
Benesi, H. A., 86, 172, 207
Bergmann, K., 86
Berkheimer, H. E., 148
Bernstein, F., 86
Beroza, M., 354
Beyerlein, D. C., 328
Beynon, K. I., 141
Biggar, J. W., 66, 158, 159, 299
Bigger, J. H., 251
Bijl, D., 112
Bingeman, C. W., 299
Birrell, K. S., 207
Bisque, R. E., 32
Bixby, D. H., 268
Black, C. A., 24
Black, C. C. Jr., 112
Blair, J., 328
Blake, J., 245
Blinn, R. C., 354
Bodenheimer, W., 128
Boersma, L., 264, 274, 283, 306
Bolt, G. H., 47, 95
Bomberger, D. C., 365
Bordas, E., 50, 207
Bordeleau, L. M., 245
Bose, S., 74
Boswell, V. R., 268
Bovey, R. W., 263

Bower, C. A., 220
Bowman, B. T., 128, 158
Bowman, J. S., 183
Bowman, M. C., 351, 354
Boyce, C. B. C., 143
Bozarth, G. A., 220
Bradford, G. R., 74
Bradley, W. F., 86
Bray, G. A., 220
Bray, M. F., 220
Breese, K., 143
Bremer, J. M., 37
Brenner, H., 299
Briggs, G. G., 137, 141, 158
Brindley, G. W., 86, 87, 158
Bristol Roach, B. M., 183
Bro-Rasmussen, F., 251
Broadbent, F. E., 74
Bromilow, R. H., 325
Brown, C. B., 171
Brown, D. A., 274, 283
Brown, D. S., 137, 158
Brown, J. W., 171, 178
Brownbridge, N., 245
Bruce, W., 245
Bruce, W. N., 251
Brunton, G., 87
Bryerlein, D. C., 343
Büchel, K. H., 112
Buehring, N., 159
Bull, H. B., 143
Bund, C. F. V. D, 234
Burchill, S., 15, 67
Burge, W. D., 245
Burger, K., 199
Burnett, G., 50
Burnett, W., 263
Burns, R. G., 172
Burnside, O. C., 216, 294, 299
Burschel, P., 148, 216
Burt, P. E., 274
Busvine, G. R., 21
Butler, H., 333
Butler, J. H. A., 74

Cain, R. B., 199
Call, F., 274, 351
Calvet, R., 32, 37, 263
Calvin, M., 112

367

Author Citation Index

Cameron, D. A., 312
Camper, N. D., 203
Carlson, E. C., 21
Caro, J. H., 344
Carringer, R. D, 67
Carslaw, H. S., 359, 365
Carter, N., 183
Carter, R. L., 268, 351, 354
Cartwright, N. J., 199
Casida, J. E., 183, 203
Castellan, G. W., 119
Chambers, C. W., 199
Chambers, H. W., 226
Chang, R. K., 263, 347
Chanutin, A., 183
Chapman, S. L., 133
Chatterjee, B., 74
Chaussidon, J., 32, 87
Chesters, G., 113, 216, 251, 333
Cheung, M. W., 66, 158
Chiba, M., 234
Chiou, C. T., 137, 158
Chisaka, H., 195
Chisholm, R. D., 21, 230, 268
Chisolm, D., 288
Chodan, J. J., 159
Christenson, I., 203
Clark, C. G., 195, 245
Clark, L. B., 183
Classtone, S., 158
Clay, D. V., 263
Cliath, M. M., 278, 328, 354, 359
Clore, W. J., 268
Cluett, M. L., 203
Coats, G. E., 220, 294
Coats, K. R., 306
Coerman, S., 325
Coggins, C. W., 47
Cohen, J. M., 57
Coleman, N. T., 32, 119
Coles, L. W., 21, 268
Collander, R., 141
Cook, H. W., 128
Cooke, B. K., 230
Corey, J. C., 306
Couch, R. W., 216
Cowan, C. T., 86
Crafts, A. S., 47
Craig, R. P., 158
Crank, J., 119, 263, 288
Cranwell, P. A., 112
Crawford, D. V., 283
Crawford, N. H., 328, 343
Crosby, D. G., 112
Crump, D., 325
Cruz, M., 32, 37, 67
Cuthbert, F. J., 87

D'Hondt, C., 158
Dahm, P. A., 67, 158, 245
Danckwerts, P. V., 299
Daniels, F., 216

Danish, A. A., 21
Dao, T. H., 37
David, H. H. Jr., 328
Davidson, J. M., 159, 263, 283, 299, 306, 312, 313, 347, 365
Davis, D. E., 216, 220, 294, 333
Davis, H. H., 343
Davis, R. G., 299
Dawson, J. H., 245
Day, B. E., 333
Day, P. R., 37
De Bock, J., 159
De Rose, H. R., 178
DeCino, J. D., 268
DeCino, T. J., 21
DeTella, R., 57
Dean, A. R. C., 288
Dean, J. A., 74
Dean, J. D., 343
Decker, G. C., 251
Detling, K. D., 172, 207
Deno, N. C., 148
Deshpande, K. B., 299
Diamond, S., 87
Dieter, C. T., 141
Dishburger, H. J., 159
Dittert, L. W., 203
Donigian, A. S. Jr., 328, 343
Dostel, K. A., 57
Douglas, J. A., 245
Downs, W. G., 207
Doyle, E. H., 245
Draycott, A. P., 324
Du Pont, M., 86
Ducros, P., 86
Dumford, S. W., 112
Dunigan, E. P., 112, 119
Durrant, M. J., 324
Dutt, G. R., 306

Earle, N. W., 21
Eastin, E. F., 333
Ebert, E., 112
Edelman, I. S., 306
Edwards, C. A., 5, 57, 234, 239, 354
Edwards, D. F., 245
Edwards, J. A., 172
Edwards, W. M., 343
Ehlers, W., 263, 278, 283, 288, 354, 359
El-Assawi, T., 239
El-Dib, M. A., 245
Elkins, D., 141
Ellgehausen, H., 158
Ellis, J. H., 283
Elrick, D. E., 159, 299, 306
Emerson, W. W., 278
Emodi, B. S., 95
Endrödi, G., 324
Engelhardt, G., 195, 245
Eno, C. F., 210

Ensminger, L. E., 220
Erh, K. T., 299
Estermann, E. F., 87, 220
Eyring, H., 119

Fahey, J. E., 21, 230, 268
Falco, J. W., 343
Farmer, V. C., 86, 123, 128
Farmer, W. J., 66, 137, 263, 264, 278, 283, 288, 289, 351, 354, 359, 365
Faust, S. D., 57, 245
Fava, A., 119
Fawcett, R. S., 245
Felbeck, G. T. Jr., 32, 67, 113, 119
Felsot, A., 67, 158, 245
Fenster, C. R., 299
Fest, C., 226
Fife, L. C., 268
Finalyson, C. M., 87
Finnerty, D. W., 299
Flashinski, S. J., 239
Fleck, E. E., 207
Fleming, I., 113
Fleming, W. E., 21, 268
Fleming, W. G., 328
Fletchall, O. H., 216, 299, 333
Folckemer, F. B., 172, 207
Foley, F. B., 268
Forbes, C., 294
Forey, A., 245
Foster, G. R., 344
Foster, J. W., 199
Foster, R., 112
Fowkes, F. M., 172, 207
Foy, C. L., 294
Freed, V. H., 47, 137, 148, 158, 251, 306, 313
Frehse, H., 231
Frere, M. H., 343, 344
Freundlich, H., 158
Friedel, R. A., 113
Friend, R. B., 5, 57
Fripiat, J. J., 86, 87
Frissel, M. J., 47, 86, 95, 324
Fryer, J. D., 245
Fu, Y., 158
Fuerer, R., 158
Fuhr, F., 231
Fuhremann, T. W., 231, 239
Fujita, T., 48, 141
Fukuto, R. R., 183
Funderburk, H. H. Jr., 216, 220, 294
Furmidge, C. G. L., 48
Furtick, W. R., 251, 333

Gaillardon, P., 32
Gardiner, H., 264, 274, 283
Gardner, W. R., 324
Gargantivi, H., 57
Gaudet, J. P., 313
Geissbühler, H., 47, 195

Genrich, D. A., 37
Geoghegan, M. J., 220
Gerloff, G. C., 183
Gerstl, Z., 158, 226, 263, 264, 288
Getzen, F. W., 133
Getzin, L. W., 210, 239
Ghosh, S., 143
Gianotti, O., 57
Gibbs, J. W., 158
Giddings, J. C., 313
Gieseking, J. E., 220
Giles, C. H., 32, 67, 87, 158
Gilmer, P. M., 268
Gilmour, C. M., 333
Gilmour, J. T., 32, 119
Ginsburg, J. M., 21, 268
Ginsburg, M. J., 21
Glass, B., 172
Glasstone, S., 119
Glebova, G. I., 119
Goodin, J. R., 333
Goransson, H., 245
Gorder, G. W., 245
Gore, R. C., 128, 226
Goring, C. A. I., 67, 141, 158, 220, 294
Goudriaan, J., 324
Gould, E. S., 87
Gould, J. P., 119, 133
Graham-Bryce, I. J., 5, 158, 172, 263, 274, 288, 324, 347
Gramlich, J. V., 216
Green, R. E., 15, 67, 119, 306, 313
Green, W. H., 344
Greenland, D. J., 15, 67, 87, 158
Grim, R. E., 87, 95, 278
Gross, P. M., 57
Grover, R., 37, 67, 351, 365
Guardia, F. S., 195, 245
Guenzi, W. D., 172, 207, 239, 347
Guggenheim, E. A., 158
Gunsolus, J. L., 245
Gunter, F. A., 288
Gunther, F. A., 354
Gupta, U. C., 113
Gysin, H., 143, 216

Hadaway, A. B., 5, 50, 133
Hague, R., 306
Haider, K., 172
Hair, M. L., 128
Haller, H. L., 183, 207, 268
Hamaker, J. W., 15, 67, 141, 158, 172, 251, 294, 313, 365
Hammett, L. P., 148
Hance, R. J., 32, 48, 119, 133, 137, 141, 143, 148, 158, 172, 231, 294
Hanks, R. W., 178, 263
Hansch, C., 48, 141, 112
Hansen, R., 158
Haque, R., 119, 133, 313

Hardman, J. A., 251
Harris, C. I., 37, 47, 148, 294, 299
Harris, C. R., 351, 354
Harris, R. F., 251
Harris, W. G., 207
Harter, R. D., 87
Hartley, G. S., 87, 148, 158, 216, 263, 274, 347, 351
Hartzell, A., 21
Harvey, R. G., 245
Harward, M. E., 123
Haselbach, C., 47
Hashimoto, I., 299
Hassett, J. J., 158
Hata, U., 158
Hauck, R. D., 199
Hauser, E. W., 333
Haworth, R. D., 112
Hayes, M. H. B., 15, 67, 112, 119
Haynes, D. L., 365
Hayward, D. O., 119, 133
Healy, M. J. R., 21
Heller, L., 128
Helling, C. S., 141, 263, 294, 365
Hemwall, J. B., 87
Hendricks, S. B., 74, 220
Hendry, D., 112
Hermanson, H. P., 294
Heukelekian, H., 199
Hickey, R. J., 183
Hill, G. D., 87, 299
Hillel, D., 324
Hiltbold, A. E., 231
Hilton, H. W., 47, 87
Hilton, J. L., 112
Hinshelwood, C. N., 288
Hoagland, D. R., 183
Hobbs, J. A., 294
Hobbs, M. E., 57
Hoffmann, R. W., 86, 87
Holladay, J. H., 333
Holly, K., 143
Holmes, H. N., 158
Holmgren, G. G. S., 207
Holtan, H. N., 343
Holton, W. F., 220
Hook, B. J., 87
Hornsby, A. G., 283, 306, 313
Hornstein, I., 268
Horowitz, M., 245
Horrobin, S., 216
Hsu, T. S., 112, 239
Huang, P. M., 37
Huber, W., 210
Hughes, D. E., 199
Huitson, A., 158
Hunter, J., 87
Hurle, K., 245
Huyskens, P. L., 159

Ibaraki, K., 113
Igue, K., 351, 359

Iler, R. K., 87, 95
Ingalsbe, D. W., 21, 268
Isozaki, Y., 158
Ivarson, K. C., 231
Iwan, J., 195
Iwasa, J., 48, 141
Iwasaki, I., 195
Iwata, Y., 288

Jackson, M. L., 37, 133, 216, 220, 294
Jackson, R. D., 274, 283, 325
Jacquin, F., 264
Jaeger, J. C., 359, 365
James, D. W., 123
Jamet, P., 158
Janssen, M. J., 87
Jansson, G., 203
Jegat, H., 313
Jensen, C. R., 278, 359
Jensen, H. L., 245
Johnson, M. R., 172, 207
Johnson, O. C., 306
Johnson, R. S., 87
Johnson, S. L., 87
Joiner, R. L., 226
Jordan, L. S., 333
Jordon, J. W., 87
Jury, W. A., 137, 264, 351, 365

Kabler, P. W., 199
Kahanovich, Y., 159
Kainer, H., 112
Kameda, Y., 199
Kamprath, E. J., 74
Kanazawa, J., 158
Karickhoff, S. W., 137, 158, 344
Katan, J., 239, 288
Kaufman, D. D., 141, 172, 195, 203, 216, 244, 245, 333
Kay, B. D., 159, 306
Kayser, A. J., 245
Kearney, P. C., 113, 172, 195, 203, 216, 245, 333
Kearns, D. R., 112
Kemper, W. D., 264, 274, 283, 325
Kempson-Jones, G. F., 172
Kercher, R. B., 32
Kerndorff, H., 113
Khan, A., 158
Khan, S. U., 32, 67, 112, 119, 159, 231, 264
Kimura, Y., 199
King, P. H., 159, 294
Kinter, E. B., 87
Kipling, J. J., 57, 119, 159
Kirkland, K., 245
Kirson, B., 128
Kishk, F. M., 239
Kissel, D. E., 263
Kliger, L., 159
Klingbiel, U. I., 195

Author Citation Index

Klute, A., 289, 312
Knisel, W. G., 328, 344
KNMI, Klimaatatlas van Nederland, 325
Knüsli, E., 143, 216, 333
Koblitsky, L., 21, 230, 268
Kodama, H., 113, 119
Kohl, R. A., 87
Kohnert, R. L., 137, 158
Kononova, M. M., 74
Koren, E., 294
Kozak, J., 67
Kramer, R. M., 343
Kreissal, J. F., 57
Kries, O. H., 178
Krupp, H. K., 299
Kurisu, F. M., 343
Kutschera, L., 325
Kuwazulsa, S., 245
Kwong, N. K. K. F, 37

Laby, R. H., 158
Lagerkrantz, C., 112
Lambert, S. M., 48, 137, 141, 148, 159, 299
Langdale, G. W., 344
Langmuir, I., 148, 159
Lanzilotta, R. P., 245
Lapidus, L., 299, 313
Laskowski, D. A., 159
Lau, S. C., 268
Laug, E. P., 21
Lauge, W. H., 21
Lavy, T. L., 37, 264, 283, 294
Lawrence, J. M., 220, 294
Leasure, J. K., 245
Leenheer, J. A., 24, 133
Lees, H., 178, 216
Leistra, M., 264, 325
Lemcoe, M. M., 87
Le Mert, R. A., 328
Leo, A., 141
Leonard, A., 86
Leonard, R. A., 328, 344
Leopold, A. C., 47, 133, 148
Letey, J., 263, 264, 278, 283, 288, 289, 354, 359
Lewis, C. C., 207
Lewis, D., 119
Lewis, G. N., 57
Li, G., 67, 119
Liaw, W. K., 37
Lichtenstein, E. P., 21, 57, 133, 172, 226, 231, 239, 268, 288, 351, 354
Lidov, R. E., 21
Lilly, J. H., 21
Lind, E. L., 57
Lindstrom, F. T., 264, 274, 283, 306, 313
Ling Ong, H., 32
Linke, H. A. B., 195

Linsley, R. K., 343
Liss, P. S., 365
Litchfield, J. R., 21
Littlewood, A. B., 299
Liu, C. L., 37
Loeffler, E. S., 172, 207
Lofgren, L. S., 354
Longdale, G. W., 328
Loos, M. A., 199
Lopez-Gonzalez, J. D., 207
Lord, K. A., 207, 274, 325
Love, B. J. G., 324
Low, M. J. D., 226

McCall, P. J., 159
McCarty, P. L., 159, 294
McCaulley, D. F., 128
McCollum, J. P., 216
McDougal, J. R., 263, 283, 313
McDowell, L. L., 344
MacEachern, C. R., 288
MacEwan, D. M. C., 87, 220, 278
MacEwan, T. H., 32, 67, 87
McGahen, J. W., 299
McGlamery, M. D., 37, 112, 333
McGowan, J. C., 148
McIntosh, T. H., 119, 112
McKelvey, J. B., 158
McLaren, A. D., 87, 210, 220
McLean, E. O., 207
MacPhee, A. W., 288
MacRae, I. C., 199, 210, 216
Mackenzie, R. C., 87
Maddox, J. V., 245
Maines, W. W., 21, 268
Manes, M., 137
Mansell, R. S., 313
March, R. B., 183
Markham, R., 87
Marquardt, D. L., 313
Marriage, P. B., 264
Marth, P. C., 178
Martin, A. R., 245
Martin, J. P., 141, 148, 351, 359
Martin, M. W., 274, 325
Matano, C., 274
Mathews, L. J., 37
Matsumara, F., 172
Matsuo, M., 159
Means, J. C., 158
Meeter, D. A., 313
Meggitt, W. F., 24, 87
Melnikov, N. N., 226
Mercado, A., 159
Mering, J., 278
Merkle, M. G., 263, 333
Merritt, L. L. Jr., 74
Messem, A. B., 324
Metcalf, R. L., 183
Mill, T., 112
Miller, D. E., 245
Mills, A. C., 159

Mingelgrin, U., 128, 158, 172, 226, 347
Mitchell, J. W., 171, 178
Mittelstaedt, W., 231
Moe, P. G., 133
Molyneux, P., 47
Monaco, T. J., 67
Montgomery, M. L., 251
Moore, G. T., 183
Moreale, A., 159
Morel-Seytoux, H. T., 344
Moreland, D. E., 112, 203
Morin, R. E., 47
Morley, H. V., 234
Morris, R. T., 159
Morrison, F. O., 21
Morse, P. M., 37
Mortland, M. M., 32, 37, 67, 86, 87, 123, 128, 159, 226
Morton, H. L., 333
Moss, S. R., 325
Mounter, L. A., 183
Moyer, J. R., 32
Mulkey, L. A., 343
Müller Wegener, U., 112
Munitz, T. I., 245
Murari, K., 47
Murphree, C. E., 344
Myer, J., 183
Myers, L., 112

Nakamoto, K., 123, 128
Nakhwa, S. N., 32, 67, 87
Nash, R. G., 207, 365
Navarro, L., 50, 207
Neal, M., 47, 133, 148
Neal, M. M., 268
Nearpass, D. C., 216
Negev, M. A., 344
Negi, N. S., 220
Nels, P. C., 37
Newman, A. S., 245
Nielsen, D. R., 299
Nisbet, M., 333
Noddegaard, E., 251
Norris, L. A., 333
Nutman, P. S., 178
Nye, P. H., 263, 274, 288, 325
Nyquist, R. A., 123

O'Connor, G. A., 313
O'Donnell, A. E., 268
O'Konski, C. T., 86
Obrigawitch, T., 245
Olsen, S. R., 264, 274, 283
Onstad, C. A., 344
Orchiston, H. D., 278
Orlov, D. S., 119
Orstad, C. A., 343
Osgerby, J. M., 48
Otoh, H., 226
Ozawa, T., 195

Author Citation Index

Parr, J. F., 207, 231, 344
Parris, G. E., 112
Passioura, J. B., 313
Payne, W. R., 37
Pease, H. L., 159, 195, 203
Pepper, B. B., 268
Perkins, T. K., 306
Perrin, D. D., 48
Perry, P. W., 67, 113, 220, 294
Peters, L. J., 137, 158
Peterson, G. H., 210, 220
Petzer, W. E., 21
Philen, O. C. Jr., 74
Phillips, F. T., 47
Phillips, R. E., 264, 274, 283, 325
Pickett, A. G., 87
Piedaller, M. A., 158
Pinck, L. A., 220
Plapp, R., 195, 245
Plimmer, J. R., 113, 172, 195, 245
Poelstra, P., 324
Polon, J. A., 226
Pommer, A. M., 87
Pope, J. D. Jr., 37
Popenoe, H., 207
Porter, L. K., 239, 274, 283, 325
Porter, P. E., 137, 299, 354
Potts, W. J., 123
Powers, W. L., 294
Pramer, D., 195, 245
Prather, R. J., 347
Puri, B. R., 47
Purnell, H., 299

Quastel, J. H., 178
Quayle, O. R., 148
Quirk, J. P., 158

Rademacher, B., 245
Radke, R. O., 216, 333
Ragab, M. T. H., 216
Raghu, K., 210
Rahman, A., 37, 245
Rahn, P. R., 251
Rajaram, K. P., 210, 289
Raman, K. V., 37, 226
Rand, M. C., 199
Randall, M., 57
Rao, P. S. C., 306, 313, 347, 365
Rastetko, L., 210
Ray, S., 158
Raymond, R., 113
Reed, J. P., 21, 268
Reid, R. C., 325
Reiniger, P., 324
Retcofski, H. L., 113
Rhoades, J. D., 347
Rhodes, R. C., 159
Richardson, C., 263
Richardson, H., 112
Richtmyer, R. D., 289
Rieck, C. E., 299, 312

Ries, S. K., 333
Rijtema, P. E., 324
Ritchie, J. T., 324
Robeck, G. D., 57
Robinson, C. V., 306
Roe, J. W., 87
Roeth, F. W., 245
Rose, D. A., 313
Rose-Innes, A. C., 112
Rosefield, I., 210, 239
Rosen, J. D., 195
Rosenfield, C., 226
Roslycky, E. B., 245
Rothberg, T., 32, 158
Rowell, D. L., 274, 325
Rowlands, J. R., 113
Rumon, K. A., 87
Russell, J. D., 32, 37, 67, 128
Ryland, L. B., 172, 207

Sabljic, A., 137
Sahay, B. K., 226
Saidak, W. J., 264
Saltzman, S., 15, 24, 128, 159, 172, 226
Sans, W. W., 158
Santelmann, P. W., 159, 299, 312
Sawyer, E. W., 226
Sawyer, W. M., 172, 207
Schachtschabel, P., 278
Schearer, R. C., 289
Schechter, M. S., 268, 351, 354
Scheffer, F., 278
Schiavon, M., 264
Schieferstein, R. H., 137, 299
Schlotzhauer, P. F., 113
Schmedding, D. W., 137, 158
Schmidt, E. L., 216
Schmidt, K. J., 226
Schnitzer, M., 113, 119, 159
Schofield, R. K., 95
Schulz, K. R., 21, 172, 226, 231, 239, 268, 288
Schuman, D. B., 245
Schweizer, C. J., 333
Scott, D. C., 220
Scott, H. D., 264, 283, 325
Scott, K. G., 263
Scott, T. A., 137, 158
Selim, H. M., 313
Senesi, N., 113
Servais, A., 86
Sethunathan, N., 172, 210, 289
Sexton, R., 119, 133, 313
Shainberg, I., 226, 347
Sharabi, N. E., 245
Shaw, W. M., 37
Sheets, T. J., 14, 47, 67, 199, 216, 294, 333
Shenefelt, R. D., 268
Shepard, H., 57
Sherburne, H. R., 47

Sherwood, T. K., 325
Shimabukuro, R. H., 333
Shin, Y., 159
Siddarampappa, R., 210, 289
Siewerski, M., 195
Simkover, H. G., 268
Sinclair, D. P., 245
Skipper, H. D., 37, 333
Skopp, J., 313
Slater, P. G., 365
Slife, F. W., 37, 112, 333
Smelt, J. H., 264, 325
Smit, N. S. H., 37
Smith, A. E., 251
Smith, B. D., 306
Smith, C. N., 344
Smith, D., 32, 67, 87, 158
Smith, D. D., 344
Smith, H. H., 365
Smith, J. W., 333
Smith, R. J., 195
Smith, S., 207
Snell, C. T., 47
Snell, F. D., 47
Snelling, K. E., 294
Sobieszczanski, J., 195
Sofer, Z., 128
Soulides, D. A., 220
Southwick, L. M., 245
Spencer, W. F., 137, 263, 264, 278, 283, 288, 328, 351, 354, 359, 365
Stacey, M., 67, 119
Stark, J., 245
Steenhuis, T. S., 328, 344
Stephenson, H. F., 199
Stevenson, F. J., 74, 113, 119
Stewart, B. A., 274, 283, 325, 344
Stojanovic, B. J., 172
Stone, M. W., 268
Storrs, E. E., 21
Stringer, A., 230
Stroosnijder, L., 325
Stroube, E. W., 216
Stuhl, M. S., 159
Suett, D. L., 251
Suffet, I. H., 57
Sullivan, J. D., 32, 113, 119
Sun, Y. P., 21, 172, 207
Sutter, G. R., 245
Suzuki, K., 245
Swann, R. L., 159
Swanson, C. L. W., 5, 57
Swanson, H. R., 333
Swanson, R A., 306
Swisher, R. D., 199
Swoboda, A. R., 24, 159
Syers, J. K., 37, 133
Synge, R. L. M., 141

Tabak, H. H., 199
Tahoun, S. A., 87

Author Citation Index

Talbert, R. E., 216, 299, 333
Tanji, K. K., 299
Taylor, C. B., 268
Taylor, D. L., 178
Taylor, S. A., 87
Teasley, J. I., 37
Tensmeyer, L. G., 87
Terce, M., 32, 37
Terriere, L. C., 21, 268
Testini, C., 113
Tettenhorst, R. C., 87
Thaysen, A. C., 183
Theng, B. K. G., 172, 226
Thomas, G. W., 24, 159
Thomas, H. C., 299
Thomas, J. R., 245
Thompson, G. P., 113
Thompson, J. M., 15, 67, 119, 158, 313
Thompson, S. O., 113
Thornton, H. G., 178
Thorp, F. C., 5, 57
Tinsley, J., 47
Tollefson, J. J., 245
Tollin, G., 112
Torstensson, T. L., 245
Toyoura, E., 199
Trapnell, B. M. W., 119, 133
Trebst, A., 113
Trichell, S. W., 333
Triplett, G. B., 343
Tu, C. M., 220
Turner, B. C., 263, 294

Umbreit, W. W., 183
Underkofler, L. A., 183
Unger, S. M., 159
Upchurch, R. P., 47, 67, 148, 220, 294
Utsumi, S., 195
Uytterhoeven, J. B., 87, 159

Vachaud, G., 313
Valenzuela-Calahorro, C., 207
Vallet, M., 32
Van Bladel, R., 159
Van Genuchten, M. Th., 159, 306, 313
Van Keulen, H., 325
Van Schaik, J. C., 274

Van Schaik, P., 47, 133, 148
Van Valkenburg, W., 226
Vanloocke, R., 289
Varner, R. W., 87
Veith, G. D., 159
Verstraete, W., 289
Vivier, P., 333
Vogler, K. G., 183
Voldum-Claussen, K., 251
Volk, V. V., 37
Von Endt, D. W., 172, 245
Vrona, S. A., 159

Wäckers, R. W., 183
Wagner, G. H., 74
Wahid, P. A., 210, 289
Waker, R. L., 245
Waksman, S. A., 183
Walcott, A. R., 159
Walgenbach, D. D., 245
Walker, A., 231, 245, 251, 283
Walker, R. L., 268
Walker, W. W., 172
Walkley, A., 216
Walling, C., 87
Wallnofer, P. R., 195, 245
Walter, M. R., 344
Wander, R. C., 245
Wang, J. H., 306
Wang, M. K., 37
Wang, T. S., 37
Ward, T., 119
Ward, T. M., 32, 74, 113, 119, 133, 143, 148, 220
Warkentin, B. P., 95
Warren, G. F., 37, 47, 299
Warrick, A. W., 313
Wauchope, R. D., 328, 347
Weaver, R. J., 5
Weaver, R. M., 37
Weber, J. B., 15, 32, 37, 67, 74, 113, 119, 159, 220, 294
Weber, W. J., 119, 133
Weed, S. B., 15, 32, 67, 74, 113, 119, 159, 220
Weiss, F. T., 268
Welch, C. D., 74
Westlake, W. E., 21, 230, 268, 288
Wheatley, G. A., 251
White, A. W., 333
White, D., 86

White, J. L., 14, 32, 37, 67, 86, 123, 133, 158, 171, 263, 294
Whitehead, A. G., 325
Wicklander, L., 216
Wicks, G. A., 299
Wierenga, P. J., 159, 306, 313
Wiese, A. F., 299
Wilcoxon, F., 21
Wilde, S. A., 21
Willard, H. H., 74
Williams, D. H., 113
Williams, J. R., 344
Willis, G. H., 245, 344
Wilson, J., 158
Wilson, R. G., 245
Winston, P. W., 133
Winten, A. J. A., 347
Wischmeier, W. H., 344
Wolcott, A. R., 24
Wolf, D. E., 87
Wolf, J. P. III, 183
Wolfe, P. J., 313
Wood, A. L., 306
Wood, S. G., 158
Woolheiser, D. A., 344
Wormald Taylor, A., 95
Wright, A. N., 141
Wright, B. G., 333
Wright, S. J. L., 195, 245
Wu, C. H., 159
Wu, M. H., 37
Wylie, W. D., 21

Yalin, Y. S., 344
Yamane, V. K., 15, 119
Yariv, S., 128, 226
Yaron, B., 15, 24, 128, 158, 159, 172, 226, 263, 264, 288, 289, 347
Yhland, M., 112
Yoshida, T., 172
Youngson, C. R., 67, 294
Yount, J. B., 195
Yuen, Q. H., 47, 87
Yule, W. N., 234

Zandwoort, R., 264
Zepp, R. G., 113
Ziechmann, W., 113
Zimdahl, R. L., 231, 251

SUBJECT INDEX

Acetamide compounds, 165, 184-195, 240, 243, 244
Acidic pesticides
  adsorption, 12, 63-64, 75, 84-86, 92-93
  mobility, 260, 290-294
Activation energy, 64, 114-119, 133
Activity, solute, 14, 53-54, 150, 156
Acylanilide compounds. See also Anilide compounds, 165, 184-195, 240, 243, 244
Adsorption
  adsorbate properties effect, 11-12, 38-47
    acidity/basicity, 11
    hydrophobic/hydrophylic balance, 135-136, 142-143
    solubility, 14, 38, 45
  adsorbent properties effect, 9-11, 18-19, 34-37
    clay characteristics, 11, 13
    clay hydration status, 13, 65, 124-128
    metal oxides, 18, 33-37
    organic matter, 22-24
    organo-mineral association, 22-24, 83, 152
    pH, 25-32, 59-61, 63-64, 73, 75-95
    surface area, 10, 11, 132, 151
    water-holding capacity, 18-19
  capacity, 9, 11, 33-47, 114-119, 129-134
  environmental conditions effect, 12-14, 47
  isotherms
    corrected, 51-57
    Freundlich-type, 34-35, 44, 56, 75-87, 114-116, 147-148
    interpretation, 25-32, 59
    Langmuir-type, 46
  kinetics
    humic acid, 114-119
    organic matter, 65-66, 129-134
    soil fractions, 34-36
  mechanisms
    association or bridging complexes, 63, 64, 82
    coordination complexes, 60, 65, 120-122, 124-128
    electron donor-acceptor reactions, 62, 96-108
    hydrogen-bonding, 60, 63, 65, 82-86
    hydrophobic bonding, 62, 64, 66, 119, 155
    ion-dipole, 60, 65
    physical forces, 60, 61, 63-65, 82-86, 133
    surface protonation, 29-32, 36-37
  models, 134-137, 142. 144-159

  negative, 63, 80-81, 85, 92
  prediction, 134-159
  sites, 10-12, 48, 60, 64, 66, 70-74
Alachlor, 185, 192-194
Aldicarb, 140, 244
Aldrin
  adsorption, 9, 16, 18-21
  degradation products, 21
  leaching, 265-266
  residues, 17-21
  volatilization, 348
Ametryne, 14, 96-113
Amiben, 63, 75-87
Aminocarb, 244
3-aminotriazole. See Amitrole
Amitrole, 61, 240, 329
Am. Cyanamid 12008, 179-183
Aminoparathion, 208-210
Anilide compounds
  adsorption, 75-77
  microbial degradation, 243
Aniline compounds
  adsorption, 75-77, 147-148
  microbial degradation, 184-195
Arrhenius equation, 117, 133
Atratone, 75-87
Atrazine
  adsorption, 11, 14, 33-37, 75-87
  hydrolysis, 163, 170, 211-216
  mass flow, 261, 291, 309
  runoff, 326-327, 340
Azinphosmethyl, 291
Azobenzene compounds, 165, 184, 243

Barban, 185, 192-194, 244
Basic pesticides, 12, 60-62, 75, 82-84, 93-95
Benefin, 185, 192-194
Benzoic acid compounds
  adsorption, 63, 75-77
  β and γ benzene-hexachloride, 9, 13, 51-57. See also Lindane microbial degradation, 165
Bioactivity, 13, 120-123
Bioassays, 17-21, 120-123, 255
Biodegradability, 200-203
Biological degradation, 163, 165
  accelerated, 229, 240-245

Subject Index

adsorption effect, 170, 176, 217-220
controlled, 243-244
effect on detoxification, 173-178
factors
    chemical structure, 196-199, 200-203
    combinations of pesticides, 244
lag-phase, 163, 174-176, 240-241
pathways, 179-183, 184-194
Bipyridilium herbicides, 60, 68-74, 217
Bound residues, 228-229, 235-239
Bromacil, 149-159

Captan, 244
Carbamate compounds. *See also* Phenylcarbamate compounds
  adsorption, 12, 48, 75, 120-123, 129-134, 142
  chemical degradation, 167
  microbial degradation, 165-167, 184-195, 243
Carbanilate compounds, 165, 184-195
Carbanolate, 244
Carbaryl, 65-66, 244
Carbofuran, 229, 241-242
Carbophenothion, 149-159
Cation, exchangeable
  effect on adsorption, 11, 59, 63-66, 75-87, 93-94, 120-128
  effect on surface reactions, 221-226
Cation exchange capacity, 10, 11, 68, 75, 78-80, 93
Cationic pesticides, 59-60
CDAA, 185, 192-194, 242
CDEC, 185, 192-194
Chemical degradation
  adsorption effect, 211-216
  factors
    aeration, 168-169, 208-210
    moisture, 204-207
    organic matter, 211-216
    pH, 211-216
    temperature, 204-207
  hydrolysis reactions, 163, 211-216
  instantaneous, 169, 208-210
  pathways, 171, 221-226
Chlorazine, 88-95
Chlorbromuron, 185, 192-194, 243
Chlorinated hydrocarbon insecticides, 2
  adsorption, 16, 17-21, 51-57
  degradation, 168
  persistence, 9, 227, 229
  residues, 19-21
  toxicity, 16, 17-21
Chloroaniline, 184-195
2-chlorobenzanilide, 243
Chloroxuron, 38-47, 147, 185, 192-194
Chlorphenamidine, 185, 192-194
Chlorpropham
  adsorption, 75-87
  degradation, 185, 192-194, 200-203, 241, 243-244
  mass flow, 291
Chlorpyrifos, 149-159
CIPC. *See* Chlorpropham
Clay-humic acid mixtures, 25-32

Clays
  adsorption of pesticides, 11-13, 25-37, 51-57, 60-61, 63-65
  microbial degradation of pesticides, 170-171
  surface-catalyzed degradation of pesticides, 171, 221-226
CMU. *See* Monuron
Collander's equation, 135
Complexes, pesticide-adsorbent 62, 64-65, 83, 120-128
Connectivity index, molecular, 136
Cycloate, 185, 192-194
Cyclodiene insecticides, 3

2,4D
  adsorption, 1, 63-64, 75-87, 88-95
  degradation, 186, 173-178, 241, 243
  diffusion, 256
  mass flow, 255, 261, 291, 309
Dalapon, 241
DBCP, 149-159
DCMU. *See* Diuron
DDE, 168, 204-207, 228
DDT
  adsorption, 16, 49-50, 149-159
  bound residues, 228-229, 235-239
  chemical degradation, 168, 204-207
  combinations of pesticides, 244
  general consideration, 1, 2
  leaching, 265-268
  mass flow, 255, 265-268
  toxicity, 16, 49-50
  vertical transport, 323, 345-347
Degradation rate
  factors
    molecular structure, 166-167, 196-199, 200-203
    soil properties, 167-168, 211-216
    repeated applications, 240-245
  order of reaction, 166, 214-216, 247-251
  use in models, 246-251
Desmetryne, 96-113
Desorption, 2
  factors, 9-14, 22-24
  hysteresis, 10, 24, 25, 29, 31
  isotherm, 25, 32
  kinetics, 16, 116
Diallate, 185, 192-194, 244
Diazinon, 221, 244
Dicamba, 291, 329
Dichlorophenoxyacetic acid. *See* 2,4D
2,6-dichlorothiobenzamide, 140
Dicryl, 75-87, 185, 192-194
Dieldrin
  adsorption, 9, 13, 16, 21, 49, 149-159
  bound residues, 228-229, 235-239
  persistence, 228
  residues, 232-234
  vapor density, 352-254
  volatilization, 349, 355-359
Diffusion
  adsorption effect, 257, 258, 279-283
  coefficient (apparent), 257, 269-274, 277, 279-283

Subject Index

effect on adsorption rate, 13, 114-119, 129-134
factors
    bulk density, 257, 276-277
    concentration, 256, 272, 284
    microbial activity, 259, 284-289
    moisture, 256, 272, 279-283
    temperature, 257, 277-278
models, 261, 279-283, 284-289
non-vapor phase, 257, 277-278
theoretical considerations, 255-256
vapor phase, 257, 277-278
Dimefox, 179, 183
Dimethoate, 256, 257, 269-274
2,5-dimethylfuran-3-carboxanilide, 243
Dinoseb, 63, 88-95
Diphenamid
    degradation, 185, 192-194
    mass flow, 291
Diquat
    adsorption, 59, 68-74
    mass flow, 291
    microbial degradation, 170, 217-220
Dissociation constant, 11, 75-87
Distribution coefficients, 23, 48, 134, 146-148, 150, 156
Disulfoton
    adsorption, 149-159
    diffusion, 257, 269-274
Diuron
    adsorption, 38-41, 75-95, 147
    degradation, 185, 192-194, 244
    mass flow, 295-299
DMU, 185, 192-194
DNBP. See Dinoseb
DNC. See DNOC
DNOC, 88-95, 241
Dow ET-57, 179, 183
Dursban, 221
Dyfonate. See Fonofos

Endothall, 241
EPTC
    adsorption, 65, 120-123
    degradation, 229, 241, 244
EDB. See Ethylene dibromide
Eradicane, 241, 242
Ethion, 149-159
Ethylene dibromide
    adsorption, 149-159
    diffusion, 348
Exchange resins, 68-74, 131-132

Fenac, 291
Fenthion, 244
Fenuron
    adsorption, 38-47, 75-87, 142-143, 147
    degradation, 185, 192-194
Fluometuron
    degradation, 185, 192-194
    mass flow, 261, 295-299
Fluorodifen, 149-159
Fonofos, 228, 235-239

Freundlich equation, 27-29, 34, 44, 51-57, 59, 64, 75-81, 114-119, 142, 146-147
Fulvic acids, 11, 62
    humic acids, 62, 96-113
    pesticides, 12, 60, 66
    soil organic matter, 11, 13

Henry's law, 350
Heptachlor, 244
Hydroxyatrazine, 211-216
Humic acids, 11, 25-32, 60, 62, 64, 96-119
Humic substances, 10, 13, 14, 62
Humin, 11, 62

Illite, 63, 93-94
IPC. See Propham

Kaolinite
    adsorption of pesticides, 92-94
    degradation of pesticides, 170, 217-220

Langmuir equation, 38, 46, 59, 75-81, 150
Leaching, 2, 255, 265-268
Lindane
    adsorption, 9, 14, 18-21, 49-51, 149-159
    diffusion, 255, 257, 275-278
    leaching, 257, 265-268
    residues, 17-21, 228, 232-237
    toxicity, 49-51
    volatilization, 355-359
Linuron
    adsorption, 38-47
    degradation, 241-244

Mass flow
    adsorption-desorption effect, 261-262, 300-306, 310-312
    diffusion effect, 261
    dispersion coefficients, 259, 309
    factors
        aggregate size, 260, 300-306
        biomass, 262
        bulk density, 260
        soil properties, 290-294
        water flow rate, 260, 295-299
    models, 261, 262, 307-313, 314-325
    spatial variability, 263
    theoretical considerations, 259
MCPA, 88-95, 241
Methoprotryne, 96-113
Methylcarbamate compounds, 240, 243, 244
Methylparathion
    adsorption, 149-159
    bound residues, 228, 235-239
Metmercapturon, 244
Metobromuron, 185, 192-194
Metribuzin, 258, 279-283
Mexacarbate, 244
MIT, 149-159
Mobam, 242
Mobility, of pesticides, 260, 290-294, 360-365

Subject Index

Moisture, soil
  effect on adsorption, 13, 49-50
  effect on degradation, 167-168, 204-207
  effect on persistence, 246-251
  effect on volatilization, 349, 352-354
Molecular parameters
  effect on adsorption, 11, 12, 48, 60-61, 135-137, 144-148, 157
  effect on degradation, 196-199, 200-203
Molecular structure, of pesticides. *See* Molecular parameters
Molinate, 149-159
Monalide, 243
Monolinuron, 38-47, 147
Montmorillonite
  adsorption of pesticides, 10, 14, 25, 32, 51-57, 63, 65, 75-87, 92-94
  degradation of pesticides, 170, 217-220
Monuron
  adsorption, 38-47, 75-87, 88-95, 147, 157
  degradation, 243-244
  mass flow, 291

Napropamide
  adsorption, 149-159
  persistence, 230, 246-251
Neburon
  adsorption, 38-47, 147
  degradation, 185, 192-194
NIA-11092, 185, 192-194
Nitralin, 185, 192-194
Non-ionic pesticides
  adsorption
    factors, 12, 48
    mechanisms, 13, 64-66, 129-134
    models, 134-135, 137-143, 149-159
  mobility, 290-294
Norea, 185, 192-194

Organic matter, soil
  adsorption of pesticides, 60, 66, 68-74, 134, 138-141, 144-159
  charge density, 59-60, 68
  effect on adsorption, 9, 12, 18-20, 22-24, 33-38, 45-46, 48-53, 152
  effect on degradation, 167, 170, 211-216
  hydration, 13
  hydrophobic properties, 14, 48, 66, 129-134, 149-159
Organic soils, 14, 18-20, 51-57, 73, 129-134
Organophosphorus insecticides
  adsorption
    isotherms, 22-24, 129-134
    kinetics, 129-134
    mechanisms, 124-134
  degradation
    microbial, 179-183
    surface-catalyzed, 171
  enzymes inhibition, 240, 244
  pesticides combinations, 244

Parachor, 136, 144-148
Paraquat
  adsorption, 59, 68-74
  runoff, 327
  vertical transport, 328, 345-347
Parathion
  adsorption-desorption, 10, 14, 22-24, 125, 132
    mechanisms, 65-66, 124-128, 129-134
    prediction, 149-158
  bound residues, 235
  degradation, 168-169
    microbial, 179, 183, 284-286
    surface reactions, 171, 208-210, 221-226
  diffusion, 258, 286-289
  enzyme inhibition, 244
  residues, 228, 232-234
Partition chromatography. *See* Adsorption, models
Partition coefficients, 12, 135-141, 149-159. *See also* Distribution coefficients
PCMC, 243-244
PCNB, 244
Persistence, pesticide
  adsorption effect, 16
  classification, 227
  pesticide/microbe interaction, 240-245
  prediction, 230, 246-251
pH
  effect on adsorption, 10, 23, 25-32, 59-61, 63-64, 73, 75-95
  effect on degradation, 169-170, 211-216
Phenoxyacetic acid herbicides, 2, 3
  adsorption, 63-64, 75-77, 83, 85, 114-119, 140
  degradation, 164
Phenoxyalkanoate compounds, 240-241. *See also* Phenoxyacetic acid herbicides
Phenylcarbamate compounds
  adsorption, 12, 28, 75, 80-84
  microbial degradation, 200-203, 240
  persistence, 244
Phenylurea compounds, 75-87
Phorate
  adsorption, 149-159
  degradation, 164, 244
Picloram
  adsorption, 63-64, 75-87, 114-119
  mass flow, 260-291
  runoff, 329
Piperophos, 149-159
Polarity, pesticide, 11, 51-57, 153, 155, 157. *See also* Molecular parameters
Polychlorinated biphenyl compounds, 140
Prometone, 75-87, 96-113
Prometryne, 261
Propachlor, 185, 192-194, 242, 244
Propazine, 75-87, 211
Propham
  adsorption, 75-87
  degradation, 165, 184-195, 200-203, 244
Propoxur, 244
Propyzamide, 246, 250-251
Pyrimiophos ethyl, 221

R-33865, 244
Redox potential, soil, 168–169, 208–210
Residues, pesticide, 17–21, 227–229
Ronnel
    adsorption, 149–159
    degradation, 221
Runoff
    factors, 326, 327, 329–333
    models, 326–327, 334–344

Schradan, 179, 183
Self-diffusion coefficient, 258, 279–283
Sesquioxides, 11, 33–37
Simazine
    adsorption, 75–87, 88–95
    degradation, 211, 241
    mass flow, 291
Simetone, 75–87
Soil microorganisms
    accelerated degradation, 229, 240, 242–243
    degradation of pesticides, 179–183, 184–195
Solan, 75–87, 165, 184–195
Solubility, pesticide, 11, 12, 38, 45, 51–57, 75–87, 135, 137, 149–159
Substituted benzene compounds, 196–199
Substituted urea herbicides, 12, 38–48, 75, 79–80, 140, 142, 147, 243
Surface acidity, clays
    effect on adsorption, 13, 29–31, 60, 75–87
    effect on degradation, 171
Surface-catalyzed reactions, 170–171
    factors, 221–226
    instantaneous degradation, 208–210
    mechanisms, 221–226
Swep, 165, 184–195

2,4,5-T
    adsorption, 63, 75–87, 88–95, 300–302
    mass flow, 300–306
Temperature
    effect on adsorption, 14, 36, 51–57, 116–119, 129–134
    effect on degradation, 167–168, 204–207
    effect on persistence, 246–251
    effect on volatilization, 349, 351–354
Terbufos, 149–159
Terbutryn, 10, 25–32, 62
Thermodynamic parameters, 64, 75–77, 114–119, 129–134, 136–137, 144–148, 150–151
Thimet, 179–183
Thin-layer chromatography, soil, 135–136, 138–143
Thiocarbamate compounds, 240. *See also* Carbamate compounds
Thiram, 244
Toluidine compounds, 165, 184–195, 243
Toxaphene, 327, 334–344
Toxicity, 18–21, 49–51
s-Triazine herbicides, 3
    adsorption, 10, 25–37, 60–62, 75, 78–79, 96–113, 142
    degradation, 171, 211
Trietazine, 75–95, 149–158
Trifluralin, 240–291

Uracil herbicides, 142

Vertical transport, 328, 345–347
Volatilization
    adsorption effect, 16, 348–349
    diffusion effect, 349–350
    factors
        air movement, 355–359
        concentration, 352–354
        evaporating water, 350–351, 364–365
        moisture, 352–254
        temperature, 352–354
        vapor density, 349, 352–354
    mass-flow effect, 349–350
    models, 350, 355–359, 360–365
Vorlex, 244

Water, competition for adsorption sites, 64–65, 116, 120–123, 129–134, 148, 151–152

Zytron, 221

About the Editors

SARINA SALTZMAN received the M.Sc. in soil science from the University of Agricultural Sciences of Bucharest, Romania, and the D.Sc. in soil science from the University of Louvain, Belgium. She worked as a soil scientist at the Institute of Agricultural Research and at the Institute of Agricultural Planning in Bucharest, Romania, until 1970. Since then she has worked at the Institute of Soils and Water of the Agricultural Research Organization, Volcani Center, Bet Dagan, Israel, on soil-pesticides interactions. She has published about forty research papers in this and other fields.

BRUNO YARON received the degrees in soil science and agricultural sciences from the Agricultural University of Bucharest, Romania and Gembloux, Belgium. He joined the Institute of Soils and Water of the Agricultural Research Organization, Volcani Center in Bet Dagan, Israel in 1960 and became the head of that institute in 1973. Dr. Yaron spent sabbatical leaves in the United States (Texas A&M University and University of California at Riverside) and in Europe (Institut National de Recherche Agronomique, Versailles, France and Department of Agricultural Sciences, and St. Cross College, University of Oxford, Oxford, United Kingdom). Dr. Yaron started his research on clay- and soil-pesticides interaction in 1967 and since then has published a series of research papers and review articles in this particular field of interest. Dr. Yaron has published about 80 research papers and edited 3 books. Presently, he is the coordinating editor of "Advanced Series in Agricultural Sciences" and serves on the editorial board of *Advances in Soil Science* and *Agronomie*.